Physics and Chemistry of Ice

Physics and Chemistry of Ice

Edited by

Werner F Kuhs
University of Göttingen, Göttingen, Germany

RSC Publishing

The proceedings of the 11th International Conference on the Physics and Chemistry of Ice held at Bremerhaven, Germany on 23-28 July 2006.

Special Publication No. 311

ISBN: 978-0-85404-350-7

A catalogue record for this book is available from the British Library

Published by The Royal Society of Chemistry,
Thomas Graham House, Science Park, Milton Road,
Cambridge CB4 0WF, UK

Registered Charity Number 207890

For further information see our web site at www.rsc.org

Printed by Henry Ling Ltd, Dorchester, Dorset. UK

PREFACE

The 11th International Conference on the Physics and Chemistry of Ice (PCI-2006) took place in Bremerhaven, Germany, 23-28 July 2006. It was jointly organized by the *University of Göttingen* and the *Alfred-Wegener-Institute (AWI)*, the main German institution for polar research. The attendance was higher than ever with 157 scientists from 20 nations highlighting the ever increasing interest in the various frozen forms of water. As the preceding conferences PCI-2006 was organized under the auspices of an International Scientific Committee. This committee was led for many years by John W. Glen and is chaired since 2002 by Stephen H. Kirby. Professor John W. Glen was honoured during PCI-2006 for his seminal contributions to the field of ice physics and his four decades of dedicated leadership of the International Conferences on the Physics and Chemistry of Ice. The members of the International Scientific Committee preparing PCI-2006 were J.Paul Devlin, John W. Glen, Takeo Hondoh, Stephen H. Kirby, Werner F. Kuhs, Norikazu Maeno, Victor F. Petrenko, Patricia L.M. Plummer, and John S. Tse; the final program was the responsibility of the editor of this volume. The oral presentations were given in the premises of the *Deutsches Schiffahrtsmuse*um *(DSM)* a few meters away from the *Alfred-Wegener-Institute*. It was probably the hottest week of summer 2006 in Germany with all ice breaking and melting between participants in the social part of the conference, yet with solid and profound exchange in the cooler, pleasantly air-conditioned lecture hall hosting excellent presentations followed by rich and lively discussions. In the spirit of promoting the exchange between scientists from various fields of ice research no parallel sessions were organized. Thus every participant got a chance to follow the entire program – seemingly a success as indicated by the excellent attendance and expressed also in various feedbacks of participants. Ample room was given to poster presentations resulting in a lively exchange and accompanied by very welcome German beer – it was hot as I mentioned above. The poster prize was shared by Sebastian Schöder und Sergei Skiba – both of them receiving a copy of the book *Physics of Ice* signed by the authors Viktor F. Petrenko and Robert W. Whitworth, members of the prize committee.

Papers were collected during the conference for inclusion in the proceedings, including a number of invited papers reviewing important fields of research. The refereeing procedure was rigorous and fast and I would like to thank all referees and authors for the excellent cooperation during the three months following the conference. It may be useful here to give a list of the preceding conferences and their proceedings:

1. **Erlenbach (Switzerland) 1962**. Proceedings were not published.
2. **Sapporo (Japan) 1966**. *Physics of Snow and Ice* (Ed. H.Oura). International Conference on Low Temperature Science, Sapporo, August 14-19, 1966, Hokkaido University, Institute of Low Temperature Science, 1967 (712 pages).
3. **München (Germany) 1968**. *Physics of ice* (Eds. N.Riehl, B.Bullemer, H.Engelhardt), Proceedings of the International Symposium on Physics of Ice, Munich, September 9-14, 1968, Plenum Press, New York, 1969 (642 pages).
4. **Ottawa (Canada) 1972**. Physics and Chemistry of Ice (Eds. E.Whalley, S.J.Jones, L.W.Gold), Papers presented at the Symposium on the Physics and Chemistry of Ice, Ottawa, August 14-18, 1972, Royal Society of Canada, Ottawa, 1973 (403 pages).

5. **Cambridge (England) 1977**. Symposium on the Physics and Chemistry of Ice, Cambridge, September 12-16, 1977. Published in *Journal of Glaciology*, 1977, Vol.21 (714 pages).

6. **Rolla (Missouri, USA) 1982**. Sixth International Symposium on the Physics and Chemistry of Ice, Rolla, August 2-6, 1982. Published in *Journal of Physical Chemistry*, 1983, Vol.87(21), pp. 4015-4340.

7. **Grenoble (France) 1986**. VIIth Symposium on the Physics and Chemistry of Ice, Grenoble, September 1-5, 1986. Published in *Journal de Physique*, 1987, Vol.48(3), Supplément Colloque N°1, (707 pages).

8. **Sapporo (Japan) 1991**. *Physics and Chemistry of Ice* (Eds. N.Maeno, T.Hondoh). Proceedings of the International Symposium on the Physics and Chemistry of Ice, Sapporo, September 1-6, 1991, Hokkaido University Press, Sapporo, Japan, 1992 (516 pages).

9. **Hanover (New Hampshire, USA) 1996**. Physics and Chemistry of Ice, Hanover, August 27-31, 1996. Published in *Journal of Physical Chemistry B*, 1997, Vol.101(32), pp. 6079-6312.

10. **St. John's (Canada) 2002**. 10th International Conference on the Physics and Chemistry of Ice, St. John's, July 14-19, 2002. Published in Canadian Journal of Physics, 2003, Vol.81(1/2), pp.1-544.

A good review of the scope of this series of conferences (prior to 1996) is found in an article by John W.Glen in Journal of Physical Chemistry B 101 (1997) 6079-6081. The richness of topics and disciplines in the wide field of physics and chemistry of ice is reflected therein. PCI-2006 certainly has confirmed this aspect and has reinforced the beneficial presence of biology and snow research. The ever larger and wider audience suggests that this series of ice conferences is as young as 40 years ago. During the meeting in Bremerhaven the International Scientific Committee has agreed on the location of the next conference – it will be organized in Sapporo in 2010 by our Japanese colleagues lead by Takeo Hondoh.

The success of this conference was largely due to the efficiency and determination of the local organization committee chaired by Frank Wilhelms from the *Alfred-Wegener-Institute* with the support of Hans-Walter Keweloh (*DSM*), Anja Benscheidt und Kerstin Dürschner (*Historisches Museum Bremerhaven*), Gerhard Bohrmann (*RCOM Bremen*), Renate Wanke, Sepp Kipfstuhl und Heinz Miller (*AWI*). Crucial financial support was obtained from the *Deutsche Forschungsgemeinschaft* (*DFG*) and the *City of Bremerhaven*, the *AWI* and the *GZG/ University of Göttingen* with further contributions from the *DSM*, the *Historical Museum of Bremerhaven, RCOM/ University of Bremen, bremenports GmbH & Co KG*, and the *Deutsche Gesellschaft für Polarforschung e.V.*

Finally, the organization of this conference and the editing of these proceedings would have been impossible without the professional skill and personal dedication of Ines Ringel (*GZG University of Göttingen*) – Ganz herzlichen Dank!

Göttingen, November 2006

Werner F. Kuhs

Contents

Contents

Contents

Corresponding Authors

Koji	Abe	kabe@gauguin.pc.uec.ac.jp
Masahiko	Arakawa	arak@lowtem.hokudai.ac.jp
András	Baranyai	bajtony@chem.elte.hu
Ivan	Brovchenko	brov@heineken.chemie.uni-dortmund.de
Jean-Bruno	Brzoska	jean-bruno.brzoska@meteo.fr
Bertrand	Chazallon	chazallon@phlam.univ-lille1.fr
Evgeny	Chuvilin	chuvilin@geol.msu.ru
Maurice	de Koning	dekoning@ifi.unicamp.br
Florent	Dominé	florent@lgge.obs.ujf-grenoble.fr
Andrzej	Falenty	afalent@gwdg.de
Frederic	Flin	flin@lowtem.hokudai.ac.jp
Thomas	Hansen	hansen@ill.fr
Kazunari	Ohgaki	ohgaki@cheng.es.osaka-u.ac.jp
Akira	Hori	horiak@mail.kitami-it.ac.jp
Hans-Werner	Jacobi	hwjacobi@awi-bremerhaven.de
Samantha	Jenkins	Samantha.Jenkins@htu.se
Yasushi	Kamata	kamata@rtri.or.jp
Volker	Kempter	Volker.Kempter@tu-clausthal.de
Stephen	Kirby	skirby@usgs.gov
Mikhail	Kirov	kirov@ikz.ru
Alice	Klapproth	aklappr@gwdg.de
Marta	Krzyzak	mkrzyza@gwdg.de
Thomas	Loerting	thomas.loerting@uibk.ac.at
Edward	Lozowski	Edward.Lozowski@ualberta.ca
Soenke	Maus	Sonke.Maus@bjerknes.uib.no
Markus Maria	Miedaner	miedaner@uni-mainz.de
Maurine	Montagnat	maurine@lgge.obs.ujf-grenoble.fr

Benjamin	Murray	B.J.Murray@leeds.ac.uk
Mohammad Mangir	Murshed	mmurshe@gwdg.de
Hiroki	Nada	hiroki.nada@aist.go.jp
Namsrai	Tsogbadrakh	Tsogbadrakh@num.edu.mn
David	Nutt	david.nutt@iwr.uni-heidelberg.de
Hiroshi	Ohno	hohno@mines.edu
Takuo	Okuchi	okuchi@eps.nagoya-u.ac.jp
Nikolaos	Papadimitriou	nikpap@ipta.demokritos.gr
Philippe	Parent	parent@ccr.jussieu.fr
Victor	Petrenko	victor.petrenko@dartmouth.edu
Patricia	Plummer	PlummerP@missouri.edu
László	Pusztai	lp@szfki.hu
Hugh	Richardson	richards@helios.phy.ohiou.edu
John	Ripmeester	john.ripmeester@nrc.ca
Vladislav	Sadtchenko	vlad@gwu.edu
Christoph	Salzmann	christoph.salzmann@chem.ox.ac.uk
Shigeo	Sasaki	ssasaki@cc.gifu-u.ac.jp
Judith	Schicks	schick@gfz-potsdam.de
Masaya	Shishido	m4410@rtri.or.jp
Mary Jane	Shultz	mary.shultz@tufts.edu
Sherwin	Singer	singer@chemistry.ohio-state.edu
Sergey	Skiba	sergey-s-s@mail.ru
Giovanni	Strazzulla	gianni@oact.inaf.it
Iwao	Takei	i-takei@hokuriku-u.ac.jp
Atsushi	Tani	atani@ess.sci.osaka-u.ac.jp
Bernhardt	Trout	trout@mit.edu
Tsutomu	Uchida	t-uchida@eng.hokudai.ac.jp
Kazuto	Ueno	spryx563@ybb.ne.jp
Lubos	Vrbka	lubos.vrbka@uochb.cas.cz
Virginia	Walker	walkervk@biology.queensu.ca
Frank	Wilhelms	fwilhelm@awi-bremerhaven.de
Salvador	Zepeda	zepeda@lowtem.hokudai.ac.jp

Invited Papers

MOLECULAR SIMULATIONS OF GAS HYDRATE NUCLEATION

B.J. Anderson[1], R. Radhakrishnan[2], Baron Peters[4], G.P. Borghi[3], J.W. Tester[4], and B.L. Trout[4]

[1]Department of Chemical Engineering, West Virginia University, Morgantown, WV 26506, USA
[2]Department of Bioengineering, University of Pennsylvania, Philadelphia, PA 19104, USA
[3]Center for Upstream Oil & Gas Technologies, EniTecnologie S.p.A., Via Maritano 26 20097, San Donato Milanese (MI), Italy
[4]Department of Chemical Engineering, Massachusetts Institute of Technology, 77 Massachusetts Avenue 66-458, Cambridge, Massachusetts 02139, USA

1 INTRODUCTION

We present an overview of cutting edge molecular simulation methods applied to understanding the nucleation of ice[1-3] and gas hydrates[4-7]. These methods allow us to probe the molecular details of these complex processes, quantify their kinetics, and engineer new ways of changing the kinetics. Specifically, we first select order parameters, mathematical functions that can describe the nucleation process. We use these order parameters within Monte Carlo and molecular dynamics simulations in order to explore the parts of the free energy surfaces related to nucleation. Along the way, we also test the order parameters for their ability to describe the nucleation process and determine the molecular mechanism for nucleation. We also describe how we have used molecular dynamics simulations to understand the inhibition of clathrate-hydrate formation[6], of tremendous interest to oil and gas companies, and use that understanding to design new kinetic inhibitors.

2 NUCLEATION OF HYDRATE CLATHRATES

Equilibrium properties of the CO_2/sea-water system have been well researched from an experimental standpoint. In particular, the clathrate hydrate forming conditions ($T < 285\,K$ and $P > 4\,MPa$) are well established. Several experiments have been performed under conditions mimicking the direct injection process and have attempted to study the dissolution rate of CO_2 in seawater. Under direct injection conditions, the injected CO_2 is in the form of a liquid droplet and a thin spherical shell of CO_2 clathrate hydrate of structure I is observed to form around the CO_2 drop, separating it from the sea water. The process of hydrate formation has many similarities with that of crystallization, i.e., it can be divided into a nucleation phase and a growth phase. For CO_2 clathrates, the nucleation phase involves the formation of a hydrate nucleus of a critical size at the liquid–liquid interface of CO_2 and water. This homogeneous nucleation process is believed to be stochastic in nature, i.e., the critical nucleus is formed because of a local thermodynamic fluctuation in the system. The formation of the critical nucleus is followed by the spontaneous growth of the hydrate phase at the interface.

2.1   Methods

We use an order-parameter formulation, in conjunction with non-Boltzmann sampling to study the nucleation of clathrate hydrates from water–CO_2 mixtures, using Monte Carlo simulations. A set of order parameters are defined: Φ_i^{gg} (i=1,2,...,n and gg for guest–guest), which characterize the spatial and orientational order of the CO_2 molecules, and Φ_i^{hh} (hh for host–host), which govern the ordering of the water molecules. Tetrahedral and Steinhardt[8] order parameters satisfactorily describe the ordering of the water (host) molecules in the clathrate. The Steinhardt order parameters are bond-orientational order parameters based on the average geometrical distribution of nearest-neighbor bonds. The tetrahedral order parameter measures the degree to which the nearest-neighbour water molecules are tetrahedrally coordinated with respect to a given water molecule.[9,10] The tetrahedral order parameter, ζ^{hh}, is defined as follows:

$$\zeta^{hh} = \frac{1}{N}\sum_N\left[1-\frac{3}{8}\sum_{i=1}^{3}\sum_{j=i+1}^{4}\left(\cos\psi_{ij}+1/3\right)^2\right]$$ (1)

where N is the number of water molecules, the indices i, j run over the four nearest neighbours of a given water molecule, and ψ_{ij} is the angle between the nearest-neighbour bond associated with molecule i and that of molecule j. The tetrahedral order parameters are not sufficient to characterize the order of the guest molecules in the clathrate; therefore, we define the order parameter ζ_i^{gg} as the ratio of the area under i-th peak of the guest-guest radial distribution ($g_{gg}(r)$) function of any given configuration to the same area for the clathrate phase. More details regarding the validation of these order parameters can be found in Radhakrishnan and Trout, 2002.[5]

The free-energy hypersurface as a function of the order parameters is calculated using the Landau–Ginzburg approach[11]. The critical cluster size that leads to the nucleation of the clathrate phase is determined accurately by analyzing the free energy surface. The free energy hypersurface of implanted clusters were mapped as a function of the cluster order parameters to analyze the thermodynamic stability of different cluster implants. The global minimum in the Landau free energy hypersurface of a stable (growing) cluster occurs at values of the cluster order parameters close to that of the clathrate phase. On the other hand, the global minimum in the Landau free energy hypersurface of an unstable cluster occurs at the values of the cluster order parameters close to that of the liquid phase. Figure 1 shows the Landau free energy hypersurface as a function of the $\zeta_1^{gg,cluster}$ order parameter for the implementation of a 14.5 Å cluster. Point 1 corresponds to a liquid-like solution ($\zeta_1^{gg,liquid} = 0.43$) while point 6 corresponds to the clathrate phase ($\zeta_1^{gg,clathrate} = 1.0$).

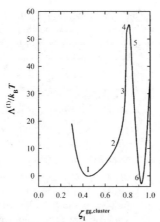

Figure 1   *The first-order distribution function of the Landau free energy hypersurface, $\Lambda^{(1)}[\zeta_1^{gg,cluster}]$ for a cluster of size 14.5 Å, showing the transformation from a liquid-like cluster to a clathrate-like cluster, eventually leading to the nucleation of the clathrate phase.*

2.2   Results

We find that the nucleation proceeds via "the local structuring mechanism,[12]" i.e., a thermal fluctuation causing the local ordering of CO_2 molecules leads to the nucleation of the clathrate, and not by the labile cluster hypothesis, one current conceptual picture. The local ordering of the guest molecules induces ordering of the host molecules at the nearest- and next-to-nearest-neighbor shells, which are captured by a three-body host–host order parameter, ζ^{hh}; these thermodynamic fluctuations lead to the formation of the critical nucleus.

Based on the Landau–Ginzburg free energy calculations, the critical cluster size for the nucleation of CO_2 clathrate hydrate at the liquid–liquid interface of CO_2 and H_2O at 220 K and 4 MPA was calculated to be between 9.6 and 14.5 Å. This is to be compared with the result of classical nucleation theory, which Larson and Garside[13] used to estimate a critical size of 32 Å. Classical nucleation theory clearly overestimates the size of the critical nucleus for CO_2 clathrates and therefore would underestimate the ability for CO_2 hydrates to spontaneously nucleate. A quantitative estimation of the free energy barrier, ΔF, to nucleation was obtained using a path integral method, which samples the four-dimensional order-parameter space. The precise free energy difference between the liquid phase and the transition state ($55\ k_BT$), and the transition state and the hydrate phase ($58\ k_BT$) was calculated. Teng et al.[14,15] and Mori et al.[16,17] have reported that the time scale for the nucleation of the CO_2 clathrate hydrate in their experiments to be of the order of 1–5 s. Using transition state theory and our simulation results we predict the time constant, $\lambda=1/k$, to be 1.2 s. This result is for the volume available for nucleation around a 1 mm CO_2 droplet with an assumed CO_2–H_2O interface thickness of 10 Å. Therefore our prediction agrees reasonably well with the experimental observations.

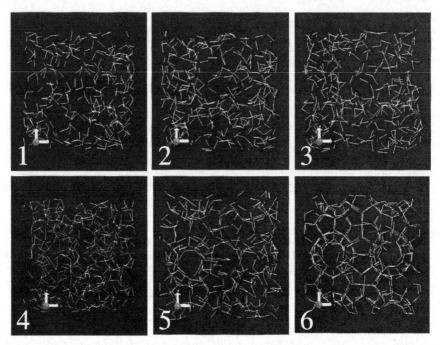

Figure 2   *Snapshots of the molecular configurations are shown in the form of distribution of hydrogen bonds in space. The cubical subvolume of the whole system encompassing the cluster is shown. Six different snapshots along the path of nucleation are depicted. The numbers correspond to the numbers marked along the free energy surface in Fig 1.*

3 INHIBITION OF HYDRATE FORMATION

Over the last decade or so, many research efforts have been focused on developing what are termed "low-dosage hydrate inhibitors", or LDHIs, that potentially can kinetically inhibit hydrate formation.[18] LDHIs operate via a much different mechanism than thermodynamic inhibitors such as methanol. They are often effective at concentrations as low as 0.5 wt%[18] and act by delaying the onset of hydrate formation, while thermodynamic inhibitors are effective only at much higher concentrations and act by changing the conditions of hydrate thermodynamic stability, thus shifting the phase diagram.

We have developed a molecular simulation approach that is based on sophisticated methods from theoretical chemistry to characterize the nucleation process of natural gas hydrates[2,12,19]. Recently, Rodger's group at Warwick[20,21] used molecular simulations and found that LDHIs (specifically tributylammonium-propylsulfonate [TBAPS], poly-vinylpyrollidone [PVP], poly-vinylcaprolactam [PVCap], and poly-dimethylaminoethyl methacrylate [PDMAEMA]) reduce the degree of structure in the surrounding water which would presumably increase the barrier to hydrate nucleation. The focus of our study is the examination of the energetics of the interaction of LDHIs on ensuing crystallites of hydrates within a reasonable framework of nucleation and crystallization.

3.1 Proposed Inhibition Mechanism

Our approach is similar to the classical theory of nucleation, in that our approach treats nucleation as an activated event, one that is more or less irreversible. Once the system surpasses the free energy barrier to nucleation, crystal growth occurs spontaneously. We therefore have proposed that hydrate inhibition occurs via a two-step mechanism. (1) Inhibitor molecules disrupt the local organization of the water and guest molecules, increasing the barrier to nucleation and nuclei propagation. (2) Once nucleation occurs, the inhibitor binds to the surface of the hydrate nanocrystal and retards further growth along the bound growth plane.

In the first step, potential disruption of newly forming nuclei occurs as proposed by Storr et al.[20] who used simulations and demonstrated that localized structure inconsistent with hydrate formation was induced by tributylammoniumpropylsulfonate (TBAPS) over several solvation shells. This element of the mechanism has not yet been experimentally verified. Our work focuses on step (2) and, as we will demonstrate, step (2) is consistent with several qualitative experimental results.

While TBAPS was shown to have an inhibition activity comparable to poly(N-vinyl-2-pyrrolidone), known as PVP, the resulting crystal morphology was quite different. PVP and PVCap have been shown to result in plate-like hydrate crystals upon crystallization[20,22-24], consistent with part (2) of the proposed mechanism while the hydrate crystals grown in the presence of TBAPS have been observed to be deformed, and particularly elongated, octahedra.

Once an inhibitor molecule such as PVCap binds to one face of the hydrate nanocrystal, growth along that growth face is slowed significantly. King et al[25] have shown that in the presence of a hydrate-crystal/liquid slurry three active inhibitors, PVP, PVCap, and N-methyl, N-vinylacetamide/N-vinyl-2-caprolactam copolymer (VIMA/PVCap), are adsorbed to the hydrate-crystal surface while a non-inhibiting polymer, poly(ethylene oxide) was not adsorbed. This experimental evidence further supports the surface binding hypothesis. Given these initial results, *we hypothesize that the stronger its binding to the hydrate surface, the more disruptive an inhibitor is to the structure of forming hydrate nuclei.*

3.2 Methods

There are differences between our approach and previous studies[20,21,26-30]: (1) the use of a liquid water phase in equilibrium with the hydrate crystal, (2) the quantitative analysis of the energetics of inhibitor binding, (3) the use of fully dynamic water molecules in the hydrate crystal, and (4) the placement of the water-soluble inhibitor in the liquid water phase as opposed to in the gas or vacuum phase. Previous computational studies focused on the morphologic effects,[20] the topology[26-29] of the hydrate–inhibitor interaction, or the structural behavior of inhibitor molecules in solution,[30] all structural studies. Our approach focuses on estimating the binding energy of the inhibitor on the (111) structure II hydrate crystal surface as determined by Carver et al.[31] to be the most active growth plane for kinetic inhibition.

We use molecular interaction parameters that we developed in previous studies[32,33] to accurately model the interaction of the guest-host interactions. 18,000 methane-water *ab initio* energies were fit to the CHARMM® potential minimizing a Boltzmann-weighted square error between the *ab initio* potential energy surface and the CHARMM® potential energy surface. The CHARMM® model with this set of intermolecular potential parameters was then verified by simulation of a 34.6 Å cubic volume consisting of 8 structure II unit cells (2×2×2) with full methane occupancy.

A molecular-scale slab model with periodic boundary conditions was used to model the hydrate–water interface in surface interaction calculations involving the sII hydrate molecules and inhibition molecules. The hydrate molecules are embedded in 4 sII unit cells that were placed in a 34.6 Å × 34.6 Å × 17.3 Å box. On top of the solid layer of crystalline hydrate is placed a layer of liquid water another 17.3 Å thick (see Figure 3). Molecular dynamics simulations of a single inhibitor monomer bound to the hydrate surface and in the bulk liquid water phase of about 3–4 ns were performed following approximately 2.5 ns of equilibration. This study focused on determining the strength of binding of the monomer building block of polymeric inhibitor molecules. Comparing the total energy from the MD simulations with inhibitor molecules inserted into the bulk liquid water and inhibitors bound to the hydrate surface, the energy of interaction can be calculated. Furthermore, the Gibbs free energy of binding is calculated as the difference in the Gibbs free energy of inserting an inhibitor molecule on the surface of the hydrate and in the liquid water phase using thermodynamic integration[34].

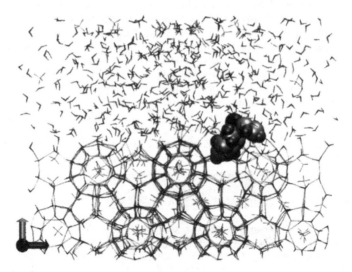

Figure 3   *Snapshot from the Simulation of PVCap in the Presence of a Hydrate Surface.*

3.3 Results

It is found through the molecular dynamic simulations that inhibitor molecules that experimentally exhibit better inhibition strength also have higher free energies of binding, providing support for our proposed mechanism. Inhibitors increasing in effectiveness, PEO<PVP<PVCap<VIMA, have increasingly negative (exothermic) binding energies of -0.2 < -20.6 < -37.5 < -45.8 kcal/mol and binding free energies of increasing favorability (+0.4 ≈ +0.5 < -9.4 < -15.1 kcal/mol). The increasingly negative binding free energies would result in increased equilibrium surface concentrations.

Furthermore, the effect of an inhibitor molecule on the local liquid water structure under hydrate-forming conditions was examined and correlated to the experimental effectiveness of the inhibitors. Two molecular characteristics that lead to strongly binding inhibitors were found: (1) a charge distribution on the edge of the inhibitor that mimics the charge separation in the water molecules on the surface of the hydrate and (2) the congruence of the size of the inhibitor with respect to the available space at the hydrate-surface binding site. These two molecular characteristics could potentially be amplified and combined to improve the binding ability of kinetic inhibitor molecules to the growth surface of natural gas clathrate hydrates.

4 KINETICS OF METHANE DISPLACEMENT BY CARBON DIOXIDE

Using advanced molecular modelling techniques can potentially assist in the development and evaluation of proposed technologies. It has been suggested that clathrate hydrates could provide a sequestration media for CO_2. Carbon dioxide and methane can form binary structure I hydrate mixtures[35]. This observation suggests the strategy of Figure 3 for sequestering CO_2 while simultaneously displacing CH_4 fuel from natural gas hydrates. The technology could potentially provide carbon neutral energy and clean burning natural gas from hydrate deposits without mining.

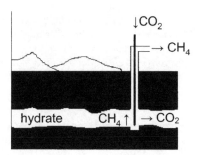

Figure 4 *Scheme for sequestering CO_2 in natural gas hydrate deposits while displacing natural gas for fuel.*

Processes like nucleation and guest diffusion in binary clathrate mixtures are difficult to study because of limitations on accessible simulation timescales and lengthscales. Peters and Trout[7] have recently developed a method that calculates reaction coordinates of complex processes with approximately two orders of magnitude less computation than any previous algorithm. The new method is widely applicable because it is based on two very general concepts, likelihood maximization and transition path sampling. We are currently applying the method to understand the kinetics of methane displacement by carbon dioxide in structure I hydrates.

To relate our simulations to experimental observables, we are performing simulations on three lengthscales as shown in Figure 4. At the molecular level, we are using the new method to determine hopping rates between cages in the hydrate structure. The hopping rates will depend on the route between the initial and final cage, the type of guest that is hopping, the presence of a guest vacancy in the final cage, and the presence of structural water vacancies. At the mesoscale, diffusion in the hydrate is represented by a lattice model where each vertex represents a small or large cage occupied by CH_4, CO_2, or a vacancy. Kinetic Monte Carlo with hopping rates from the molecular simulation will reveal the dependence of the diffusion "constants" on local vacancy concentration. Finally, the concentration dependent diffusion constants are incorporated into a macroscopic model of binary diffusion in the hydrate with boundary conditions corresponding to experiments being performed by ENI. Such concentration dependent diffusion equations have been solved by Crank, Babu, Stockie, and others. The third panel of Figure 4 shows an example from a paper by Stockie[36].

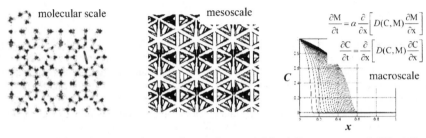

Figure 5   *Multiscale approach to understand rate of CO_2 diffusion into and CH_4 diffusion out of a structure I hydrate. (left) Molecular simulation for individual hopping rates. (middle) Mesoscale kinetic Monte Carlo simulation of hopping on the hydrate lattice to determine dependence of diffusion constants on vacancy, CO_2 and CH_4 concentrations. (right) Macroscopic coupled non-linear diffusion equations to describe rate of CO_2 infusion and methane displacement. Graph from Stockie[36].*

4 CONCLUSIONS

We have developed new methods to allow us to probe the molecular details of complex processes involving the nucleation of hydrates, quantify their kinetics, and engineer new ways of changing the kinetics. Based on Landau–Ginzburg free energy calculations, we have estimated the size of the critical nucleus for homogeneous CO_2 hydrate nucleation to be between 9.6 and 14.5 Å at 220 K and 4 MPa. We found that the activation free energy for this nucleation process to be 55 k_BT resulting in a time constant for nucleation around a 1 mm CO_2 droplet to be ~1.2 seconds. This result agrees quite well with published experimental values.

We have quantified the binding energies of four monomer molecules on a hydrate growth surface and found a positive correlation between the binding free energies and the inhibitor effectiveness. Two molecular features, a localized dipole and dynamic monomer size, have been identified as contributors to the binding capability of monomer molecules and manipulation of these two features will assist in the design of new inhibitor molecules. Finally, we are developing methods to quantify the diffusion rate of CO_2 within a methane hydrate lattice using a new reaction coordinate determination method. This new method is applicable to a number of different applications and improves on previous methods by reducing the computational time by approximately two orders of magnitude.

References

1   Radhakrishnan, R. and Trout, B. L., *Journal of the American Chemical Society*, 2003, **125**, 7743.
2   Radhakrishnan, R. and Trout, B. L., *Physical Review Letters*, 2003, **90**, 158301.
3   Radhakrishnan, R. and Trout, B. L., Order parameter approach to understanding and quantifying the physico-chemical behavior of complex systems, in *Handbook of Materials Modeling, Part B*, Dordrecht Springer, The Netherlands, 2005.
4   Radhakrishnan, R., Demurov, A., Herzog, H., and Trout, B. L., *Energy Conversion and Management*, 2003, **44**, 771.
5   Radhakrishnan, R. and Trout, B. L., *Journal of Chemical Physics*, 2002, **117**, 1786.
6   Anderson, B. J., Tester, J. W., Borghi, G. P., and Trout, B. L., *Journal of the American Chemical Society*, 2005, **127**, 17852.
7   Peters, B. and Trout, B. L., *Journal of Chemical Physics (in press)*, 2006.
8   Steinhardt, P. J., Nelson, D. R., and Ronchetti, M., *Physical Review B*, 1983, **28**, 784.
9   Chau, P. L. and Hardwick, A. J., *Molecular Physics*, 1998, **93**, 511.
10   Errington, J. R. and Debenedetti, P. G., *Nature*, 2001, **409**, 318.
11   Landau, L. D. and Lifshitz, E. M., *Statistical Physics*, 3rd ed. Pergamon, London, 1980.
12   Radhakrishnan, R. and Trout, B. L., *Journal of Chemical Physics*, 2002, **117**, 1786-1796.
13   Larson, M. A. and Garside, J., *Chemical Engineering Science*, 1986, **41**, 1285.
14   Teng, H. and Yamasaki, A., *International Journal of Heat and Mass Transfer*, 1998, **41**, 4315-4325.
15   Ogasawara, K., Yamasaki, A., and Teng, H., *Energy & Fuels*, 2001, **15**, 147-150.
16   Ohmura, R. and Mori, Y. H., *Journal of Chemical and Engineering Data*, 1999, **44**, 1432-1433.
17   Mori, Y. H. and Mochizuki, T., *Energy Conversion and Management*, 1998, **39**, 567-578.
18   Lederhos, J. P., Long, J. P., Sum, A., Christiansen, R. L., and Sloan, E. D., *Chemical Engineering Science*, 1996, **51**, 1221-1229.
19   Radhakrishnan, R. and Trout, B. L., *Journal of the American Chemical Society*, 2003, **125**, 7743-7747.
20   Storr, M. T., Taylor, P. C., Monfort, J. P., and Rodger, P. M., *Journal of the American Chemical Society*, 2004, **126**, 1569-1576.
21   Hawtin, R. W., Moon, C., and Rodger, P. M., Simulation of Hydrate Kinetic Inhibitors: The Next Level, in *The Fifth International Conference on Gas Hydrates*, Trondheim, Norway, 2005, pp. 317-321.

22   Makogon, T. Y., Larsen, R., Knight, C. A., and Sloan, E. D., *Journal of Crystal Growth*, 1997, **179**, 258-262.

23   Larsen, R., Knight, C. A., and Sloan, E. D., *Fluid Phase Equilibria*, 1998, **151**, 353-360.

24   Sakaguchi, H., Ohmura, R., and Mori, Y. H., *Journal of Crystal Growth*, 2003, **247**, 631-641.

25   King, H. E., Hutter, J. L., Lin, M. Y., and Sun, T., *Journal of Chemical Physics*, 2000, **112**, 2523-2532.

26   Carver, T. J., Drew, M. G. B., and Rodger, P. M., *Journal of the Chemical Society-Faraday Transactions*, 1995, **91**, 3449-3460.

27   Carver, T. J., Drew, M. G. B., and Rodger, P. M., *Journal of the Chemical Society-Faraday Transactions*, 1996, **92**, 5029-5033.

28   Kvamme, B., Huseby, G., and Forrisdahl, O. K., *Molecular Physics*, 1997, **90**, 979-991.

29   Carver, T. J., Drew, M. G. B., and Rodger, P. M., *Gas Hydrates: Challenges for the Future*, 2000, **912**, 658-668.

30   Carver, T. J., Drew, M. G. B., and Rodger, P. R., *Physical Chemistry Chemical Physics*, 1999, **1**, 1807-1816.

31   Carver, T. J., Drew, M. G. B., and Rodger, P. M., *Journal of the Chemical Society-Faraday Transactions*, 1996, **92**, 5029.

32   Anderson, B. J., Tester, J. W., and Trout, B. L., *Journal of Physical Chemistry B*, 2004, **108**, 18705.

33   Cao, Z. T., Tester, J. W., and Trout, B. L., *J. Chem. Phys.*, 2001, **115**, 2550.

34   Frenkel, D. and Smit, B., *Understanding molecular simulation: from algorithms to applications*, 2nd ed. Academic Press, San Diego, 2002.

35   Sloan, E. D., *Nature*, 2003, **426**, 353.

36   Stockie, J. M., in *Boundary and Internal Layer Analysis, Conf. Proc.*, 2004.

EXTRATERRESTRIAL ICE WITH EMPHASIS ON AGGREGATION/INTERACTION WITH ORGANIC MATTER: COLLISIONAL AND ACCRETIONAL PROPERTIES OF MODEL PARTICLES

M. Arakawa

Graduate School of Environmental Studies, Chikusa-ku Furo-cho, Nagoya 464-8602, Japan

1 INTRODUCTION

Water ice exists not only on Earth but also in space. Astronomical spectroscopy indicates that water ice is one of the main components of dust in interstellar molecular clouds. Infrared spectroscopy has been used to study the chemical composition of interstellar dust, and as a result, we know that the most abundant condensate of volatiles is H_2O ice, followed by CO, CO_2, NH_3, and CH_3OH ices.[1] Dust particles in molecular clouds are smaller than 1 μm in diameter and have a layered structure consisting of a silicate core and an organic mantle covered with a thick ice layer[2]. This model for interstellar dust is called the Greenberg model.[2] The formation mechanism of organic matter on interstellar dust has been studied to reveal its relationship to UV irradiation into the ice layer.[2] These dusts are the main source of planets and satellites in the solar system. In protosolar nebula, these dusts are proposed to coagulate and become dust aggregates, and this aggregation is the first step toward planetary growth. In regions far from the sun, these aggregates would accumulate to become icy planetesimals with bodies like comets. This accumulation process of aggregates is the second step toward becoming a planet. Comets are believed to gather primordial material from molecular clouds. Therefore, researchers have attempted to determine the constituent material in the interest of discovering the primordial material that formed planets.[3] The third step is the collision among icy planetesimals to form icy satellites and icy planets. Our research goal is to clarify the physical mechanisms related to these three steps. Especially, I focus on the formation process of icy bodies, which are mainly composed of water ice and organic matter.

Planetesimals are coagulated bodies of dust aggregate and the embryo of a planet. It is an intriguing mystery to know how these planetesimals are formed in a protosolar nebula. Because of the turbulence in nebula gas, dust and aggregate were estimated to collide on the order of 10 m/s.[4] Materials must stick together at an impact velocity higher than ~10 m/s in order to grow from aggregate to planetesimals by direct sticking. However, the sticking velocities have not been revealed yet for dust and aggregate. There is no experimental study on the dust aggregation process of water ice and organic matter although silicate dust aggregation has been investigated by both theories and laboratory experiments. So, I present our recent experimental studies on the aggregation of ice and organic matter in Section 2.

There are many icy bodies in addition to comets in the solar system: they are icy satellites and Kuiper belt objects. Icy satellites of the Jupiter and Saturn systems were observed by spacecraft to clarify their densities and surface compositions. As a result, it is widely accepted that the main component of icy satellites is water ice, and the existence of water ice is confirmed by the observation of near infrared reflectance spectra. Icy satellites were revealed to have various surface morphologies and geologic activities depending on their origin and the thermal evolution process. Most of the icy satellites have densities from 1 to 2 g/cm^3, which means that these bodies are a mixture of ices and silicates[4]. Icy satellites were formed by collisional accretion of small porous bodies. These bodies could be ice-silicate mixture and the porosity was corrupted according to their growth. Therefore, impact properties of an ice-silicate mixture with various porosities are necessary to be clarified in order to study the formation process of icy satellites. I review systematic experimental results on impact of ice-silicate mixture in Section 3.

2   AGGREGATION OF ICE AND ORGANIC DUST IN PROTOSOLAR NEBULA

2.1   Sticking Velocity of Icy Dust

Chokshi et al. (1993) made a theoretical study of the coagulation of interstellar dust. They formulated a theory of two colliding particles with adhesion force caused by surface energy in order to derive critical velocity (v_{cr}), which is the upper limit of sticking velocity.[5] In their theory, a Van der Waals force working on contacting dusts dynamically was included in the adhesion force because the Van der Waals force is the most common adhesion force working on any surfaces. According to their theory, the critical velocity depends on the radius (R) of the colliding particles and varies as $R^{-5/6}$. Thus, micron-sized quartz particles are able to stick together at a velocity less than 1 m/s, while centimeter-sized quartz particles require a velocity less than 10^{-4} m/s for sticking. v_{cr} also depends on the material properties, so that ice particles with a diameter of 0.1 μm are able to stick together below the velocity of 20 m/s, which is rather higher than the v_{cr} of quartz particles, 1 m/s. Therefore, we can expect that ice grains can more easily coagulate to become large aggregates compared to silicate materials.

Sticking properties of ice grains were experimentally studied related to the dynamics of Saturn's ring.[6] The restitution coefficient of ice is a key parameter of ring dynamics, so Bridges et al. (1984) and other researchers in later studies measured the velocity dependence of the restitution coefficient (ε) of ice at very slow velocity range from 10^{-3} to 10^{-2} m/s.[7] The measurement was made by using a special pendulum with an ice ball on one side to achieve a very low impact velocity of less than 10^{-2} m/s. Bridges et al. noticed that ε is strongly affected if there is a frost layer covering the ice surface. With successive studies to reveal the effect of a frost layer on ε, they found that the frost layer could not only dissipate kinetic energy but also affect the sticking of ice surfaces to each other. A detailed study of the sticking force of a frosted layer on ice in colliding bodies showed that the force increases up to 100 dyn for the frost thickness between 10 and 100 μm and indicates the maximum at 10^{-3} m/s. There was a critical velocity that corresponded to v_{cr} of 3×10^{-4} m/s, although they used a frosted ice ball with a 5-cm diameter. This v_{cr} is smaller than that of the theoretical value for ice with the same particle size of 5 cm, $\sim 2 \times 10^{-3}$ m/s. The sticking force of the frost layer was explained by the Velcro model: geometrical interaction between the frosted surface induces the sticking, a mechanism different from that proposed by Chokshi. Therefore, the result obtained by Bridges et al. could be recognized as the sticking properties of a frost layer constructed by aggregates of ice grains.

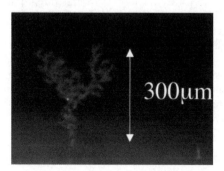

Figure 1 *Ice aggregate formed under electric field. It has a tree-like shape.*

Arakawa (2001) made direct measurements of the sticking velocity of small ice grains and aggregates.[8] Ice grains were condensed from water vapor in air cooled by liquid nitrogen. The condensed ice grains were polygonal, from 5 to 10 μm in diameter. These ice grains were set in a strong electrostatic field up to 1 kV/mm and floated in motion with thermal convection of air. As ice grains were charged, they were accelerated by electrostatic force to collide with other grains. One possible charging mechanism was the grains' polarization. Water ice has a large dielectric constant of about 100, so that under a strong electric field water molecules could be polarized and an icy dust would have a polarized charge. This polarized water ice was attracted by electrostatic force. The collision was observed by a high-speed video camera under a microscope. Arakawa observed that ice aggregates were attracted by both electrodes. This means that the electrical charge on ice aggregates should be both plus and minus. Ice grains colliding at a velocity of ~ 0.1 m/s were found to stick on the other grain and form aggregates, which were shaped like trees. The theoretical estimation of v_{cr} is about $0.4 \sim 0.8$ m/s for corresponding size, which is consistent with the obtained $v_{cr} > 0.1$ m/s in the experiment.

Ice aggregates were formed as a result of the sticking of ice grains, which had a tree-like structure with the size of a few hundred micrograms, as shown in Figure 1. The sticking property of ice aggregates was examined by using this aggregate. The aggregate was set under an electrostatic field, so that it was accelerated between electrodes by electrostatic force. The impact velocities of ice aggregates can be varied from 0.1 to 20 m/s. The collision of ice aggregate was observed by the high-speed video camera to determine the impact strength of aggregates. At a low impact velocity of less than ~ 1 m/s, the aggregate sticks on a copper electrode without disruption. Below this velocity, the tree-like structure is maintained on an accreting body. Beyond the velocity of 1 m/s, the aggregate disrupts completely as a result of the collision. Once the aggregate was disrupted, compaction on an accreting body could densify the aggregate. In this experiment, the adhesion force among ice grains included electrostatic force in addition to Van der Waals force, and thus the obtained v_{cr} should be at the upper limit without an electric field.

The v_{cr} measured for ice aggregates is smaller than 10 m/s. This means that it might be difficult for ice grains to grow into icy planetesimals in a turbulent nebula. However, it is probable that the observed compaction of aggregates may force v_{cr} higher because of the higher strength expected from the dense structure acquired.

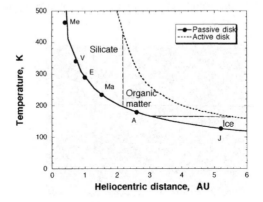

Figure 2 *Temperature distribution of protoplanetary nebula marked with the present date positions of planets to Jupiter. The dotted line is for an active disk, and the solid line is for a passive disk. Material distribution is also shown with two boundaries proposed by Kouchi et al. (2002).*

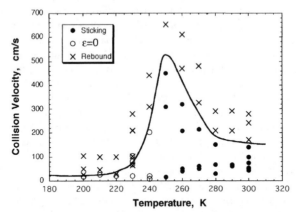

Figure 3 *Sticking condition of the IOM-layer with the thickness of 1 mm impacted by a copper ball with a diameter of 1 cm.*

2.2 Sticking Condition of Organic Matter

Interstellar dusts could have a layered structure, as proposed by Greenberg (1998), of an icy mantle covering an organic-silicate core. On the way from an interstellar molecular cloud to protosolar nebula, interstellar dust could be modified by any thermal events, e.g., nebula gas heated by shock waves on an accreting disk or the absorption of solar radiation.

The icy mantle could easily evaporate as a result of such heating, so the remnant, organic-silicate dust, could be distributed in the particular region of protosolar nebula where rising temperatures are appropriate.[9] The material distribution in the protosolar nebula expected by Kouchi et al. (2002) is shown in Figure 2. There are two typical temperature distributions in the protosolar nebula: the upper curve corresponds to that expected in an active disk heated by a shock wave of infalling gas, and the lower curve corresponds to a passive disk calculated for the equilibrium temperature by solar radiation. According to the evaporation experiment of simulated interstellar organic matter made by Kouchi et al. (2002), the distribution of organic matter together with water ice and silicate would be as shown in Figure 2; the organic matter can be found between 2 and 5 AU in the active disk and between 2 and 3 AU in the passive disk.

To study aggregation of the organic-silicate dust, Kudo et al. examined the effect of an organic layer on the coagulation by collisional experiments using simulated interstellar organic matter (IOM).[10] v_{cr} was obtained at various temperatures from 200 to 300 K (Figure 3). They found that v_{cr} changes with temperature: it has the maximum of about 5 m/s at 250 K and v_{cr} decreases above and below 250 K. As viscosity of IOM strongly depends on temperature, it becomes hard like a solid below 200 K. This is why elastic rebounding between dust particles occurs at low temperatures. On the other hand, IOM has low viscosity at temperatures higher than 300 K. The colliding body does not dissipate the kinetic energy so much when it goes through this liquid layer. v_{cr} becomes small beyond 300 K. At the middle temperature range around 250 K, the viscosity and the surface tension of IOM can work as mechanisms to cause colliding bodies to stick effectively. The maximum v_{cr} of 5 m/s is rather higher than that of other materials, so that we can expect that this organic layer works as a glue for coagulation of dusts in the protosolar nebula where it exists on the dust surface. Although the size dependence of v_{cr} has not yet been clarified for dust covered with a viscous layer, it is quite noticeable that the result was derived from a mm to cm scale experiment. That supports that IOM could be effective even for the dust size as large as mm to cm. Because of the radial dependence of adhesion force, v_{cr} of elastic bodies is quite small (e.g., 10^{-5} m/s for ice grains 1 mm in diameter).

Kouchi et al. (2002) discussed the early growth of planetesimals in the region of an asteroid belt, where conditions are almost consistent with those in regions in which an organic layer exists on dust. The derived v_{cr} of 5 m/s is the same order of collisional velocity as that caused by nebula gas turbulence. Thus, dust with an organic layer could coagulate to grow into planetesimals in the protosolar nebula. This early growth of planetesimals might explain the origin of an achondrite parent body and the deficiency of the material in an asteroid belt.

3 IMPACT CRATERING AND DISRUPTION IN THE SOLAR SYSTEM

3.1 Cratering on Snow

On its way to becoming a planet, a planetesimal can grow by accretion; the collisional velocity is between 10 and 1000 m/s.[4] Therefore, collisional experiments using simulated planetsimals are necessary to study the impact conditions needed for planetary growth. Planetesimals are supposed to have high initial porosity, so snow is a good material to simulate icy planetesimals.

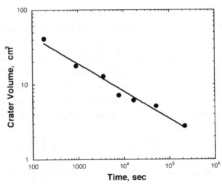

Figure 4 *Relationship between crater volume and sintering duration at –10°C. Snow projectile with 40% porosity and a size of 7 mm was impacted at 100 m/s.*

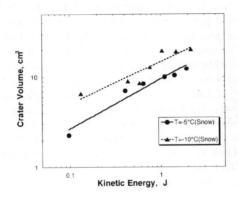

Figure 5 *Relationship between crater volume and kinetic energy. The snow target was sintered for 15 min. The snow projectile had a porosity of 40%, and a diameter of 7 mm was used.*

Impact experiments of snow on snow were made by Arakawa (2005) to examine cratering efficiency.[11] In his experiments, the cratering efficiency was studied using sintered snow. We can expect that icy planetesimals could thermally evolve by radioactive heating like other bodies, which means that icy planetesimals could be sintered according to the state of thermal evolution. Our snow samples had a porosity of ~40% and were sintered at –10°C from 3 min. to 60 hrs. These samples were impacted by using a 7-mm snow projectile at a velocity of 100 m/s. The crater volume was measured after the impact and decreased with increasing sintering duration. The empirical relationship shown by these data is V_{cr} (cm³)=236 t(s)$^{-0.37}$ (Figure 4). This was a first report of snow cratering experiments in order to study the effect of sintering on the cratering efficiency, so that other data for the comparison was not available. The mechanical strength was also measured for the sintered sample, and the relationship between the strength and the sintering duration was obtained as follows: σ (kPa)=5.75 t(s)$^{0.28}$. Therefore, we can derive the dependence of the crater volume on the mechanical strength as follows:

$$V_{cr}(\text{cm}^3) = 1852 \ \sigma(\text{kPa})^{-1.26} \tag{1}$$

The crater volume increases with the projectile kinetic energy. This relationship for the snow sintered for 15 minutes at different temperatures is shown in Figure 5. The empirical equation shows that the crater volume is almost proportional to the root square of the projectile kinetic energy, which means that it is simply proportional to the impact velocity (v_i). Thus, we can describe the cratering efficiency as follows: $V_{cr} \sim v_i^{1.0} \ \sigma^{1.26}$. This could be useful for estimating the mass depletion of the icy planetesimals when we consider the effect of sintering processes on planetary growth.

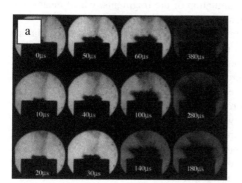 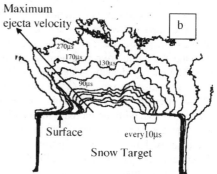

Figure 6 *(a) Successive images taken by image-converter camera. (b) Contour of cratering ejecta expanding with time.*

3.2 Impact Fragmentation of Snow and Snow-Silicate Mixture

With the growth of an icy body by accretion of icy planetesimals, the internal pressure induced by self-gravity increased to cause the pressure sintering of snow. Therefore, Arakawa et al. (2002) conducted impact experiments using snow with various porosities made by pressure sintering and studied the impact strength dependence on the porosity.[12] The icy planetesimals include not only water ice but also silicate materials. The effect of silicate materials on the impact strength is also important when applying laboratory results to the planetary accretion process in the protosolar nebula. Therefore, snow mixed with silicate powder was also examined to reveal the effect of silicate mixing.[13]

Collision of planetesimals can disaggregate the icy body in which the pressure sintering proceeds. The icy body could have large self-gravity that an escape velocity becomes important after fragmentation so that reaccumulation can occur. In the planetary accretion process, reaccumulation is an important mechanism for the growth of bodies. Reaccumulation means that the fragments of a disrupted body with the ejection velocities lower than the escape velocity of the original target body are captured by the mixture gravity between the fragments.

To evaluate the reaccumulation condition, it is necessary to know about the fragment velocities. Thus, Arakawa and Tomizuka (2004) measured the maximum ejecta velocity from an impact site by using an image-converter camera. Figure 6a shows the images obtained for targets of snow-silicate mixtures with a porosity of 38% at the impact velocity of 667 m/s. The mixtures were prepared by mixing serpentine powder with snow in equal parts by weight. The cratering process is clearly observed in these successive images, and the contour map of the ejecta shown in Figure 6b was used to analyze the maximum ejecta velocity.

Figure 7 shows the relationship between the normalized maximum ejecta velocity ($v_{e\text{-}max}$) and the target porosity for snow and snow-silicate mixture. The $v_{e\text{-}max}$ normalized by the impact velocity simply deceases with the increase of the target porosity (ϕ). The slopes of the empirical equations are slightly different in the two cases, and they are very small below 0.2 at porosities larger than 50%. This reduction of the ejection velocity with increasing porosity should enhance the reaccumulation of the fragments, which means that a porous body can more efficiently grow when hit by another body.

The impact strength was defined by a parameter that describes the impact conditions such as energy density and the degree of disruption (e.g., the normalized fragment mass showing the cumulative fragment mass corresponding to half of the original target mass, $f_{0.5}$). The energy density is defined as the projectile kinetic energy divided by the target mass. The impact strength ($Q_{0.5}*$) was obtained by the energy density when $f_{0.5}$=0.5.[12] The impact strength strongly depends on the porosity, as shown in Figure 8.[13] The $Q_{0.5}*$ for snow increases with the increase of porosity, which means that the porous target is stronger than the compacted target. On the other hand, the $Q_{0.5}*$ for a snow-silicate mixture decreases with the increase of porosity, a trend opposite to that obtained for snow. Porosity dependence on the impact strength of ice and ice-silicate mixture was firstly investigated by us, so that other data for the comparison was not available.

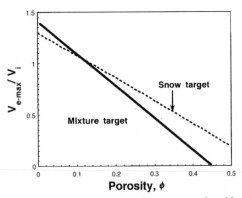

Figure 7   *Dependence of the maximum ejecta velocity normalized by impact velocity on target porosities*

This curious behavior, wherein the $Q_{0.5}*$ decreases or increases as porosity increases, might be explained by the shock pressure propagation in addition to the mechanical strength varying with the porosity and silicate contents. The mechanical strength obtained by the static compression tests is shown in Figure 9 for snow and snow-silicate mixture with various porosities. We notice that the snow is always stronger than the mixture and the slope of the mixture is steeper than that of snow. These results are empirically formulated as follows:

$$\sigma = \sigma_0 (1-\phi)^n \qquad (2)$$

where we obtain σ_0 and n are 9.8 MPa and 3.4, 9.5 MPa and 6.4 for pure ice and mixture, respectively. Following a theoretical consideration proposed by Mizutani et al. (1990)[14], Arakawa et al. (2002) formulated an impact strength of porous matter, $Q(\phi)$, related to a non-porous matter, $Q(0)$, as follows:

$$\frac{Q(\phi)}{Q(0)} = \left(\frac{P_0}{\sigma_0}\right)^{3\left(\frac{1}{m_0}-\frac{1}{m}\right)} \cdot (1-\phi)^{\frac{3(n-l)-m}{m}}, \qquad (3)$$

where P_0 is a shock pressure induced in a non-porous matter with a relationship to a shock pressure of porous matter described as $P_{0\phi}=P_0(1-\phi)^l$. A shock pressure distribution in the target at the distance r is described as follows:

$$P(r) = P_{0\phi} \cdot \left(\frac{r_0}{r}\right)^m, \qquad (4)$$

where r_0 is the radius of an isobaric core, and m is assumed to be $m=m_0(1-\phi)^k$. When we apply appropriate values for each parameter shown in Eq.(3), the porosity dependence on impact strength could be reproduced as shown in Figure 10. In this figure, only the parameter n in Figure 9 was changed; other parameters were assumed to be the same for both lines, $l=1.5$, $k=-0.5$, $m_0=2$, and $P_0/\sigma_0=100$. We notice that the relative impact strength decreases with increasing porosity when n is 6.4 of the mixture sample and increases with increasing porosity when n is 3.4 of the pure ice sample. Eq.(3) looks a little bit messy, but the physical meaning of each parameter is very clear and it is successful at reproducing the curious behavior found in Figure 8. There are two key parameters to control the impact strength: they are the pressure decay constant and the power law index of the porosity dependence on the strength.

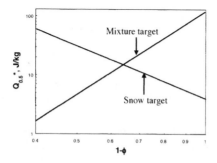

Figure 8 *Porosity dependence of impact strength for snow and snow-silicate mixture.*

A qualitative explanation for the opposite dependence of impact strength on the porosity found in Figure 8 is illustrated in Figure 11. We assume that the impact-induced pressure decreases with increasing porosity by porosity-collapse and the decay constant increases with increasing porosity. In this figure, the shock pressure at the antipodal point is compared to the mechanical strength to assess whether the target is disrupted. Because of the difference of the slopes in Figure 9, the shock pressure propagating in the more porous target is lower than the mechanical strength of snow at the antipodal point, which means that the target is not disrupted, but it is higher than that of the snow-silicate mixture, which means disruption has occurred. For a less porous target, the mechanical strength of snow is lower than the shock pressure at the antipodal point, but that of the snow-silicate mixture is higher than the shock pressure at the antipodal point, which might explain the opposite dependence of the impact strength on the porosity discovered.

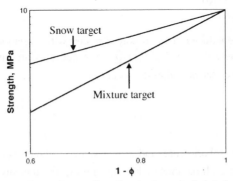

Figure 9   *Porosity dependence of mechanical strength (σ) derived by static compression tests for snow and snow-silicate mixture.*

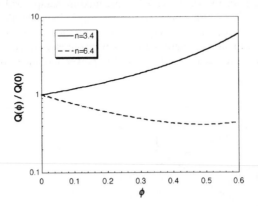

Figure 10   *Empirical relationship between the porosity and the relative impact strength according to Eq.(3).*

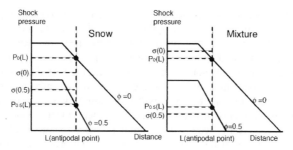

Figure 11 *Schematic illustration explaining porosity dependence of impact strength.*

3.3 Accretion Conditions of Small Icy Bodies

The reaccumulation condition of icy bodies is estimated by comparing fragment velocities with the escape velocity of the original body and by comparing energy density with the impact strength of the body. Impact disruption occurs when the energy density overcomes the impact strength defined by $Q_{0.5}$, as shown in Figure 8. The critical velocity of the disruption, $v_{i\text{-disp}}$, is derived as follows with the empirical equations for porous bodies:

$$v_{i-disp} = \left[2a(1-\phi)^{b+1} \frac{\rho_t}{\rho_p} \left(\frac{R}{r_p} \right)^3 \right]^{1/2} \tag{5}$$

where a is 20.7 for snow and 124 for the mixture, b is -1.6 for snow and 2.5 for the mixture, R is the radius of the target body, r_p is the radius of a projectile body, ϕ is the porosity of the target body, and ρ_p and ρ_t are the density of the projectile and target body, respectively. We assume that the projectile body is composed of water ice to derive Eq.(5).

We can expect that all of the fragments ejected from the target could be captured by the self-gravity of the original target body when the maximum ejecta velocity is lower than the escape velocity of the original target body. Therefore, this physical condition is adopted for the reaccumulation condition of a disrupted body. The critical velocity showing the reaccumulation is described as follows when we include the empirical equations for porous bodies:

$$v_{i-acc} = \frac{\left(\frac{8}{3}\pi G \rho_t (1-\phi) \right)^{1/2}}{-3.2\phi + 1.4} R. \tag{6}$$

Consider what happens in collisions of porous icy bodies when we apply two empirical equations, Eq.(5) and Eq.(6). Figure 12 shows a collisional state map separated by $v_{i\text{-disp}}$ and $v_{i\text{-acc}}$ described on the velocity-porosity plane. This map is calculated for $R=100$ km and $r_p=20$ km. The icy body will be disrupted above the dashed line, and the mixture body will be disrupted above the dashed-dotted line estimated from Eq.(5). Reaccumulation of the fragments will occur below the solid line for both target bodies estimated from Eq.(6). To contribute to planetary growth, the impact condition must be lower than the solid line, which corresponds to the compaction and regolith region and the rubble pile region. A large difference between the snow and snow-silicate mixture bodies can be recognized in cratering with the mass loss region. The presence of this region indicates that the crater has formed on the surface and the ejecta has escaped from the body to reduce the mass but the body was not disrupted. The cratering with a mass loss region is quite narrow for snow and rather wide for the snow-silicate mixture. It might be advantageous for planetary growth to have wider cratering with a mass loss region, because the mixture body can withstand a higher impact velocity that would cause disruption for snow. However, it is necessary to conduct a detailed numerical simulation to estimate the quantitative effect of the difference caused by the presence of the mass loss region on planetary growth.

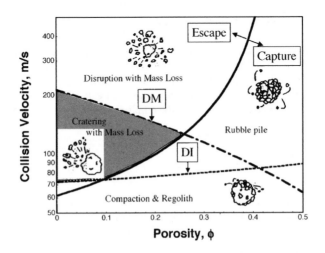

Figure 12 *Collisional state map calculated for $r_p=20km$ and $R=100km$. DI and DM are the boundaries showing the impact disruption of the ice and the mixture bodies, respectively. Solid line is the boundary showing the reaccumulation of impact fragments.*

Acknowledgements

I am grateful to the reviewer, Dr. S. Kirby for his detailed study of the manuscript and his constructive remarks. This research series was partly supported by a grant-in-aid for scientific research from the Japan Ministry of Education, Sciences, Sports and Culture.

References

1   J. Crovisier, in *Formation and Evolution of Solis in Space*, eds. J.M. Greenberg and A. Li, Kluwer, Dordrecht, 1999, p.389.
2   J.M. Greenberg, *Astron. Astrophys.*, 1998, **330**, 375.
3   M. F. A'Hearn, M. J. S. Belton, W. A. Delamere, J. Kissel et al., *Science*, 2005, **310**,254.
4   M. Arakawa and A. Kouchi, *"Impact processes of ice in the solar system"* in *High -*

Pressure Shock Compression of Solids V, Springer-Verlag, New York, 2002, p.199.

5   A. Chokshi, A.G.G.M. Tielens and D. Hollenbach, *Astrophys. J.*, 1993, **407**, 806.

6   A.P. Hatzes, F. Bridges, D.N. Lin and S. Sachtjen, *Icarus*, 1991, **89**, 113.

7   F. Bridges, A. Hatzes and D.N. Lin, *Nature*, 1984, **309**, 333.

8   M. Arakawa, 2001, Japan Earth and Planetary Science Joint Meeting, P0-003.

9   A. Kouchi, T. Kudo, H. Nakano, M. Arakawa and N. Watanabe, *Astrophys. J.*, 2002, **566**, L121.

10   T. Kudo, A. Kouchi, M. Arakawa and H. Nakano, *Meteoritics & Planetary Science*, 2002, **37**, 1975.

11   M. Arakawa, Proc. 38th ISAS Lunar and Planet. Symp., 2005, 68.

12   M. Arakawa, J. Leliwa-Kopystynski and N. Maeno, *Icarus*, 2002, **158**, 516.

13   M. Arakawa and D. Tomizuka, *Icarus*, 2004, **170**, 193.

14   H. Mizutani, Y. Takagi and S. Kawakami, *Icarus*, 1990, **87**, 307.

INTERACTIONS BETWEEN SNOW METAMORPHISM AND CLIMATE: PHYSICAL AND CHEMICAL ASPECTS

F. Domine[1]*, A.-S. Taillandier[1], S. Houdier[1], F. Parrenin[1], W.R. Simpson[2] and Th.A. Douglas[3]

[1]CNRS, Laboratoire de Glaciologie et Géophysique de l'Environnement, BP 96, 38402 Saint-Martin d'Hères Cedex, France. *E-mail : florent@lgge.obs.ujf-grenoble.fr
[2]Geophysical Institute, University of Alaska Fairbanks, Fairbanks, Alaska, USA, and Department of Chemistry, University of Alaska Fairbanks, USA
[3]Cold Regions Research and Engineering Laboratory, Fort Wainwright, Alaska, USA.

1 INTRODUCTION

The snowpack forms an interface between the atmosphere and the ground or sea ice that affects the energy balance of the Earth's surface[1,2] and the exchange of chemical species between the surface and the atmosphere.[3] Studies of snow areal extent (e.g. ref. 4) show that during the boreal winter snow covers about 14% of the Earth's surface, therefore affecting the surface energy balance at near-planetary scales, while its overall effect on tropospheric chemistry still needs to be assessed.

The physical impact of the snowpack depends on its physical properties, such as albedo and heat conductivity. Its chemical impact depends on its chemical composition and its reactivity, determined in part by the light flux inside the snowpack. All of these properties change with time, because of a set of physical and chemical processes regrouped under the term "snow metamorphism", defined below.

Snow is a porous medium formed of air, ice crystals and small amounts of chemical impurities. Because ice has a high vapor pressure (165 Pa at -15°C, 610 Pa at 0°C), the vertical temperature gradient that is almost always present within the snowpack generates sublimation and condensation of water vapor that change the size and shape of snow crystals. This results in changes in physical variables such as density, albedo, heat conductivity, permeability and hardness. These physical changes have formed the basis for the definition of snow metamorphism.[5]

Interest has recently developed in snowpack compositional changes associated with metamorphism. Motivation includes the interpretation of chemical analyses of ice cores[6] and the understanding of the impact of the snowpack on tropospheric chemistry.[3] Indeed, sublimation and condensation of water vapor entrains trace gases dissolved within ice crystals, which has been invoked to explain concentration changes of species such as HCl, HNO_3, HCHO and H_2O_2.[7-10] The light that penetrates inside the snowpack[11] also drives photochemical reactions[12,13] resulting in changes in snow chemical composition. The growing documentation of these chemical changes during metamorphism imposes their consideration as a new aspect of snow metamorphism. We therefore propose to define "physical metamorphism" as "the changes in physical properties undergone by snow after deposition" while "chemical metamorphism" can be defined as "the changes in chemical composition undergone by snow after deposition".

Next to the temperature gradient inside the snowpack, an important driving force for snow metamorphism is wind, that lifts, transports and redeposits snow crystals, changing snowpack mass and density[14,15] and deposits aerosols inside the snowpack.[16,17] Wind and temperature are climatic variables that determine metamorphism and snowpack physical properties such as albedo and heat conductivity. These properties affect the energy balance of the snow-atmosphere and of the soil-snow interfaces, which in turn affect climate.

Snow-climate feedbacks therefore exist. The most discussed is probably the snow areal extent–climate feedback[4], as a reduced snow cover caused by warming changes the surface albedo from high to low values, exerting a positive feedback. Predicted increases in soot deposition to snow are also expected to decrease snow albedo, constituting another positive feedback.[18] Changes in vegetation patterns in the Arctic may also produce complex feedbacks. For example, warming-induced shrub growth on the tundra will trap snow, limit the effect of wind and this will result in snow with a lower heat conductivity.[19] The soil temperature will therefore rise, leading to increased shrub growth and increased emissions of CO_2, a greenhouse gas, by the soil due to enhanced microbial activity. While this snow-vegetation interaction may produce a positive feedback, very complex effects are involved, so that its sign and magnitude are uncertain, as detailed in Sturm et al.[19]

Snowpack chemical emissions include oxidants and aerosol precursors, that also interact with climate. Oxidants determine the lifetime of greenhouse gases and aerosols impact the atmospheric radiation budget.[20] If snowpack chemical emissions are determined by climate, feedback loops involving snow chemistry also need to be studied to assess the extent of climate change in snow-covered regions.

The purpose of this paper is to reflect upon the existence of feedback loops between climate and snow metamorphism, both physical and chemical. Snow metamorphism has been much studied (e.g. refs. 5,21-23) and we present a brief reminder to feed our discussion. We first recall how climate affects snow physical properties by discussing the major seasonal snowpack types. We then explain the role of temperature and wind speed in the formation of these types. We then focus on four important snowpack variables : albedo, light penetration, heat conductivity, air permeability, and discuss how climate change will affect these variables. Attempts to quantify snow-climate feedbacks are made for some of these properties. Finally, we discuss how physical changes in snowpacks may affect changes in snow chemical composition. Snow-climate feedbacks are extremely complex and quantifying them, or even just determining their sign, will require much further research. However, we wish to put forward ideas to stimulate discussion and this will require making some highly speculative propositions that will need further testing.

A further limitation of this paper is that it focuses on dry metamorphism, i.e. in the absence of melting. Melting obviously affects snowpack physics and chemistry in an important manner and some limited discussion of its impact will nevertheless be included here, but detailed considerations of these additional aspects will require a separate paper.

2 BACKGROUND

Among several existing classifications of seasonal snowpacks, we use that of Sturm et al.[24] that describes the tundra, taiga, Alpine, maritime, prairie and ephemeral snowpacks. Sturm et al. also mention the mountain snowpack, that displays such large spatial variations that it defies a precise definition. The ephemeral snowpack affects only episodically temperate areas and has a limited long-term effect. It will not be discussed. The prairie snowpack covers the Asian steppe and North-American grasslands. Its impact is limited because it is often spatially and temporally discontinuous. We therefore limit our discussions to the taiga, tundra and maritime/Alpine snowpacks, these last two types presenting sufficient similarities to be discussed together.

2.1 The taiga snowpack

The Taiga snowpack covers cold forested regions in North America and Eurasia, as represented in detail in Sturm et al.[24] It is typically 50 cm thick (Figure 1) and covers the ground from October-November to April. In mid-winter, it is composed of a thick basal layer formed of centimetric depth hoar crystals, that has a very low mechanical strength, and a density near 0.2 g.cm^{-3}. It is topped by a layer of faceted crystals 1 to 2 mm in size, that eventually transform into depth hoar.[22] Layers of decomposing crystals and of fresh snow are observed after snow falls. All of these snow layers have a low density, typically < 0.2 g.cm^{-3} as shown in Figure 1.

In these forested and mostly inland areas, wind speed remains low, typically below 2 m.s^{-1}. Winter temperatures frequently drop below -30°C, while ground temperatures remain above -10°C and the temperature gradient has a typical value of 50°C.m^{-1}. In the fall, the combination of cold spells and of a thin snow cover can raise the temperature gradient to 100, even 200°C.m^{-1}.[22]

2.2 The tundra snowpack

The tundra snowpack covers cold treeless regions around the Arctic ocean and treeless parts of central Asia, south of the taiga region. It is typically 40 cm thick (Figure 1) and covers the ground from September-October to June in the Arctic. In mid-winter, it is composed of a basal layer of centimetric depth hoar crystals of density 0.25 to 0.35 g.cm^{-3}. It is topped by a hard to very hard layer of small wind-packed rounded grains 0.2 to 0.3 mm in size, of density 0.35 to 0.55 g.cm^{-3}.[25] Near the surface, recent snowfalls transform into faceted crystals and eventually into depth hoar in the absence of wind. If they are wind-blown, the crystals are then broken up and sublimate while air-borne,[15] contributing to the build-up of the windpack layer.

In these unforested and mostly coastal areas, wind speeds reach values up to 15 to 35 m.s^{-1}. Winter temperatures often drop below -35°C and the longer snow season, coupled to a higher heat conductivity of the snow than on the taiga, contributes to an efficient ground cooling that limits the value of the temperature gradient to typical values of 15 to 40°C.m^{-1}, except in the fall, when the snowpack is thin and the ground still near 0°C.

2.3 The maritime and Alpine snowpacks

The maritime and Alpine snowpacks are found in temperate regions with abundant precipitation and reach thicknesses of 1 to 2 m, sometimes much more. They cover the ground from November-December to March-May. An example of maritime snowpack stratigraphy is shown in Figure 1. The warm climate results in numerous periods when the temperature exceeds 0°C. The resulting snow melting and subsequent refreezing forms layers of ice and of melt-freeze polycrystals, resulting in an overall density of the base of the snowpack around 0.4 g.cm^{-3}. These layers, when frozen, are mechanically very strong. In the absence of melting, fresh snow transforms into small rounded grains, that form most of the snowpack. Melting events happen at any time of the year. When sufficient liquid water forms, percolation takes place. Upon refreezing, horizontal ice layers and vertical ice columns (percolation channels) form in the snowpack.

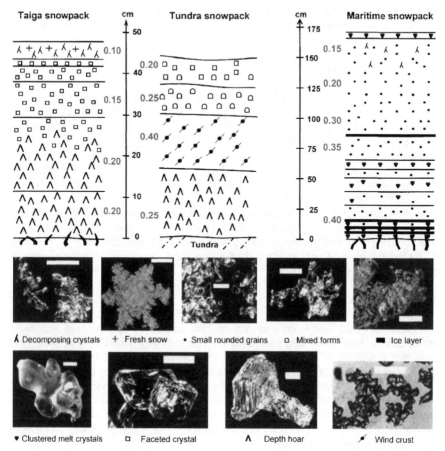

Figure 1   *Typical stratigraphies of the taiga, tundra and maritime snowpacks, with photomicrographs of typical snow crystal types. Scale bars: 1 mm. Bold numbers next to the stratigraphies are density values, in g.cm^{-3}.*

Temperate maritime areas are usually forested, resulting in low wind speeds. The mild temperatures, combined with a thick snowpack, lead to a very low temperature gradient, of the order of 2 to 5 °C.m^{-1}, except near the surface, where daily temperature variations can lead to elevated transient gradients that change sign during the day. The density of the maritime snowpack usually increases monotonously with depth.

The Alpine and maritime snowpacks have many similarities. A specific figure is not necessary here and can be found in Sturm et al.[24] Alpine snowpacks form in regions slightly colder than the maritime snowpack and signs of melting are less frequent. The temperature gradient in late fall can be sufficient to form a depth hoar layer 5 to 20 cm thick. With sufficient snowpack build-up, however, typical gradient values are 5 to 10 °C.m^{-1}. The Alpine snowpack forms in regions with little vegetation and windpacks can be found. Overall, the density increases with depth, but not as monotonously as in the maritime case. The basal depth hoar layer, when present, has a density between 0.25 to 0.30 g.cm^{-3}, and windpacks can be denser than underlying layers not affected by wind.

2.4 Temperature and snow metamorphism

Dry metamorphism operates by the transfer of water vapor between ice surfaces, caused by a gradient in water vapor pressure. The vapor pressure over an ice surface is determined by temperature following Clapeyron's law and by the curvature of the surface following Kelvin's law. A snow crystal located in a layer at -12°C within a temperature gradient of 20°C.m^{-1} is subjected to a gradient in water vapor pressure of 5 Pa.m^{-1}. In an isothermal snowpack at -12°C, the gradient in water vapor pressure will be determined solely by differences in curvature. Radii of curvature for fresh snow crystals are of the order of 10 μm for sharp edges[23,26] and 1000 μm for flat faces. For a crystal 4 mm in size, the resulting gradient between a sharp angle and a flat face will be 0.2 Pa.m^{-1}, illustrating that the greatest water vapor fluxes will be found when high temperature gradients are present.

At the beginning of the snow season, ground temperatures are usually close to 0°C, so that the temperature gradient is determined by the atmospheric temperature. In temperate areas with a maritime or Alpine snowpack, the temperature gradient will usually be less than 10°C.m^{-1}, while in subarctic or Arctic areas, it will usually be greater than 20°C.m^{-1}.

Under high temperature gradient (typically >20°C.m^{-1},[27]), high water vapor fluxes lead to rapid crystal growth producing large faceted crystals with sharp edges, becoming hollow as size increases.[21-23] These large hollow faceted crystals can reach several cm in size and are called depth hoar crystals. No preferential growth takes place at grain boundaries and layers of faceted or depth hoar crystals are highly uncohesive and their presence considerably increases the avalanche risk in mountain areas.[28] This intense metamorphism is hereafter called high gradient metamorphism (HGM).

Under low temperature gradient (typically <10°C.m^{-1}), crystal growth is too slow to lead to the formation of large faceted crystals.[23] The curvature dependence of the water vapor pressure has an effect : highly positively curved (convex) areas sublimate while condensation takes place on flat and concave areas. This process leads to rounded crystals that form cohesive layers because concavities exist at grains boundaries, that grow preferentially. Growth being slow, crystals remain small, typically 0.5 mm or less. This low-intensity metamorphism is hereafter called quasi-isothermal metamorphism (QIM).

The impact of the temperature gradient on metamorphism explains many of the features of Figure 1. The typical HGM-type metamorphism of the taiga snowpack eventually transforms most of the snowpack into depth hoar,[22] while the QIM-type metamorphism of the maritime and Alpine snowpacks forms, in the absence of melting, layers of small rounded grains 0.2 to 0.4 mm in diameter. However, considering the effects of other climate variables such as wind speed is necessary to explain features such as the presence of windpacks formed of small rounded grains in the tundra snowpack.

2.5 Wind and snow metamorphism

Wind of moderate speed can drag snow crystals on the surface, or lift them so that they undergo saltation or become suspended in air.[14] Saltation, i.e. bouncing and breaking on the snow surface, leads to grain fragmentation and sublimation[29] and crystals are usually 0.1 to 0.3 mm in diameter. Crystals accumulate in wind-sheltered areas, such as the lee of sastrugi[15,25] and form dense layers of small grains, mostly rounded. (Sastrugi are small dune-like structures 5 to 30 cm high formed by deposition/erosion processes during wind storms[25]).The density of windpacks thus formed depends on wind speed and may exceed 0.5 g.cm^{-3} for wind speeds greater than 20 m.s^{-1}.

In dense windpacks, crystals do not appear to have the space required to grow to large sizes and this has been invoked to explain why depth hoar does not grow in dense snow.[27] However, we found depth hoar of density up to 0.4 g.cm^{-3} on the tundra, so that space is probably not the only factor. As detailed below, windpacks have a high heat conductivity that hinders the establishment of a high temperature gradient across them and this may also explain why little transformation is observed in such snow layers.

In the absence of wind, snow density increases because of loading by subsequent layers and of destruction of small structures by sublimation, leading to the collapse of crystals higher up. Without wind and melting and under QIM conditions, snow density is around 0.35 g.cm^{-3} at 1 m depth. Under HGM conditions, compaction is compensated by upward vapor fluxes,[22] limiting the density of basal depth hoar layers to about 0.2 g.cm^{-3}.

3 PHYSICAL PROPERTIES OF THE SNOWPACK

3.1 Selection of variables

Variables such as density, albedo, light e-folding depth, specific surface area (SSA), crystal size and shape, heat conductivity, permeability, diffusivity and shear resistance are required for a complete physical description of the snowpack. Not all these variables have major relevance to climatic issues. Albedo, i.e. the fraction of incident light that is reflected, has obvious climatic relevance and is discussed here. It depends in part on crystal size and shape and this dependence can in fact more simply be related in part to the SSA of the snow[30] i.e. the surface area accessible to gases per unit mass.[31]

The incident light that penetrates the snowpack undergoes multiple reflections, to the point that a few cm down, the actinic flux can be up to 4 times as large as the incident flux. This amplification is strongly dependent on the solar zenith angle and is greater when the sun is higher on the horizon.[11] Below that surface region, the actinic flux decreases exponentially and the snow depth over which the flux decrease by a factor of e is called the e-folding depth. Similarly to the albedo, the e-folding depth depends on snow SSA. Hence, we discuss snowpack albedo, e-folding depth and SSA as a single section.

The second physical variable of obvious climatic relevance is the heat conductivity, that determines the heat flux between the atmosphere and the ground or sea ice. Another section is therefore devoted to this variable.

Snowpack permeability determines air flow within the snowpack, as driven by differences in surface pressure between different parts of the snow surface. Air flow through the snow leads to the exchange of heat and of chemical species between the snow and the atmosphere and is the last physical variable discussed here.

3.2 Albedo, e-folding depth and specific surface area

Snow reflects light at all visible wavelengths and pure snow has an albedo in the visible and near UV in the range 0.96 to 0.98.[32] In the IR, snow albedo is much lower, decreasing to values below 0.1 around 1500 and 2000 nm, so that the albedo of pure small-grained snow averaged over the solar spectrum is of the order of 0.8.[33]

The interaction of light with snow is described by the combination of two processes : scattering by the numerous surfaces offered by snow crystals and absorption by ice and impurities contained in the snow, such as particles of soot and of mineral dust.[32,34] These processes are quantified by the coefficients Sc (for scattering) and K (for absorption), both in units of inverse length. Numerous theories with various degrees of sophistication have been proposed to describe light-snow interactions. With sufficiently sophisticated theories,[35,36] it is possible to relate Sc and K to snow physical properties and chemical composition. Specifically, the scattering coefficient is related to snow SSA and the absorption coefficient is related to the wavelength-dependent absorption of ice and to absorption caused by impurities. Grenfell and Warren[37] and Neshyba et al.[38] recommend the use of independent spheres with the same surface area-to-volume ratio, S/V, as a non-spherical ice crystal for calculating optical parameters. This S/V ratio is closely related to the specific surface area (SSA) of snow : SSA=S/(V.ρ) where ρ is the density of pure ice and therefore the scattering coefficient of snow should be related to snow SSA.

Recent developments to measure snow SSA[31] have allowed the testing of the relationship between SSA and optical properties. In the visible and UV, snow albedo is very high, so that the sensitivity to snow SSA is not large. In the IR this sensitivity is large and Domine et al.[30] have recently established a quasi-linear relationship between SSA and snow albedo in the wavelength range 1650-2260 nm. Likewise, Simpson et al.[39] observe a proportionality between the optical scattering coefficient of snow and its SSA.

Modeling light-snow interactions will thus use as basic data snow SSA and absorption coefficients by ice and impurities. Our goal is to discuss the modifications of light-snow interactions induced by climate change. Ice absorption is an intrinsic property insensitive to climate. The amount of impurities in snow will be affected by climate, through complex effects on emissions, atmospheric chemical transformations, large scale transport and deposition processes, that are beyond our scope. We instead focus on the physical aspect, i.e. on how climate will affect light scattering by snow and therefore snow SSA.

Most snow SSA values were obtained by measuring the adsorption isotherm of methane on snow at 77 K.[31] For dry snow, values range from around 1500 $cm^2.g^{-1}$ for fresh dendritic snow to about 100 $cm^2.g^{-1}$ for depth hoar. Melting severely decreases SSA and values of 18 $cm^2.g^{-1}$ have been measured for melt-freeze crusts.[40]

The grain growth almost always observed during metamorphism results in a decrease in snow SSA. The rate of decrease greatly affects snow albedo and e-folding depth, considered over large spatial and temporal scales. The initial decrease is very fast, with a factor of 2 decrease in 1 to 2 days. Experimental and field studies have quantified the rate of decrease of snow SSA as a function of temperature and temperature gradient.[41-45] In all cases, the best empirical fit of SSA decay plots was of the form :

$$SSA = B - A \ Ln(t + \Delta t) \tag{1}$$

where t is time and A, B and Δt are adjustable parameters. Several attempts have been made recently to establish a physical basis for these equations[44,46,47] and it was shown that the rate of isothermal SSA decrease followed the laws of Ostwald ripening. However, the approximations that had to be made render the applicability of these equations highly uncertain. In particular snow crystals were approximated as spheres and Legagneux and Domine[46] have shown that this resulted in huge errors in SSA estimations. It must therefore be concluded that there is today no reliable theoretical basis to predict the evolution of snow SSA and we must rely on empirical equation (1).

Taillandier et al.[45] propose formulations of *A, B* and *Δt* that depend on temperature and on the initial SSA of the snow, SSA_0. They propose two sets of parameterizations, one for QIM conditions (temperature gradient <10°C.m⁻¹) and one for HGM conditions, (gradient >20 C.m⁻¹). Taillandier et al. observed that the value of the temperature gradient did not affect the rate of SSA decrease, as long as it was above the 20°C.m⁻¹ threshold.

Using the equations of Taillandier et al., we show (Figure 2) that a rise in temperature, at constant temperature gradient, accelerates SSA decrease, therefore decreasing the time-averaged SSA. However, climate warming will lead to a decrease in the temperature gradient in the snowpack, that may be amplified by increased snowpack thickness due to enhanced precipitation in polar regions (see Fig. 19 of Hansen et al.[48]). The possibility thus exists that in some areas, the metamorphic regime will change from HGM to QIM. Figure 2 indeed shows that, except for the first day after the snowfall, the snow SSA remains higher under the warmer QIM conditions than under the colder HGM conditions.

Climate warming will therefore have a complex effect on snow SSA and therefore albedo. Warming without a change in metamorphic regime from HGM to QIM will decrease snow SSA, while warming with a change from HGM to QIM conditions will increase snow SSA. Moreover, an increased frequency of melting events will form melt-freeze crusts on the surface that have a very low SSA, so that the physical aspect of the climate warming-snow albedo feedback is very complex and will certainly show large geographical variations, both in magnitude and sign. These feedbacks can have large magnitudes, as estimated below.

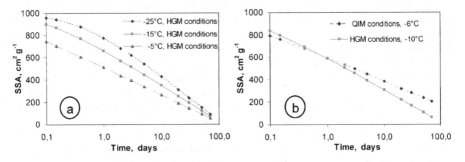

Figure 2   *(a) Calculated rate of decrease of the specific surface area (SSA) of snow of initial SSA 1000 $cm^2.g^{-1}$ evolving under a temperature gradient greater than 20°C.m⁻¹ (HGM conditions, see text) at three different temperatures.*
(b) ibid, for snow evolving under two different conditions : quasi isothermal (QIM, see text) conditions at -6°C and under a temperature gradient greater than 20°C.m⁻¹ (HGM conditions) at -10°C. This shows that after one day, a change in metamorphic regime from HGM to QIM offsets a warming of 4°C and SSA will then decrease slower under a warmer climate.

Several models have quantified the climate warming-snow albedo feedbacks caused by an increase in soot in polar snow. Hansen and Nazarenko[18] found that predicted soot increases would decrease the spectrally-averaged albedo by 1.5%. This results in a 1.5 W.m⁻² forcing at high latitude, producing a warming between 1 and 2 °C.

Following the scenario of Figure 2b, we assume that climate-induced changes in snow metamorphism will increase snow SSA from 100 to 200 cm².g⁻¹. Using the method of

Stamnes et al.[36] we calculate that this will increase the spectrally-averaged snow albedo by 4% (from 0.75 to 0.79). With an incident solar flux of 100 W.m^{-2}, and assuming that the relationship between forcing and warming of Hansen and Nazarenko[18] is linear, we predict that changes in snow physics will produce a cooling of 3 to 5°C. A slightly different parameterization[48] leads to a similar range : 2.5 to 4°C, still a significant cooling.

If fine-grained snow of SSA 200 cm^2.g^{-1} is replaced by a melt-freeze crust of SSA 20 cm^2.g^{-1}, the reverse would be true and a positive feedback of 3 to 5°C would be produced. From Figure 2a, we predict that more limited positive feedbacks will be produced by a 5°C warming without a change in metamorphic regime, of the order of 1°C.

These considerations warrant further investigation but they show that the snow-albedo physical feedback, via SSA changes, can be strong. The extreme feedback strengths of 4 to 5°C, warming or cooling, are probably reserved to exceptional conditions and the sign of the feedback will probably vary with time and space, so that the overall effect of the warming-SSA feedback cannot be determined. We wish to stress, however, that snow-climate feedbacks may not all be positive, and that the climate-SSA feedback may be important and deserves investigations.

Changes in SSA will affect light penetration in the snowpack. The light extinction coefficient depends on Sc$^{1/2}$, and therefore on SSA$^{1/2}$.[39] If the SSA increases from 100 to 200 cm^2.g^{-1}, the e-folding depth and the light intensity integrated over the whole snowpack will decrease by a factor of 1.4.

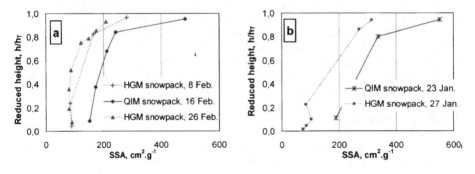

Figure 3 *Measured snow SSA profiles in central Alaska in two different snowpacks : a natural snowpack on the ground, where a strong temperature gradient led to depth hoar formation (HGM snowpack); and the same snowpack on Tables, under which the air circulation prevented the establishment of a significant temperature gradient (QIM snowpack). (a) comparison of the QIM snowpack of 16 February with the HGM snowpack sampled 8 days before and 10 days after; (b) comparison of both snowpacks sampled just 4 days apart. In both cases the SSA is much higher in the QIM snowpack. Typical snowpacks heights were 50 cm for HGM and 40 cm for QIM.*

Our suggestion that an increase in snow SSA will result from a reduction in the temperature gradient was confirmed by field studies in Alaska during the 2003-2004 winter. Figure 3 compares the SSA evolution of the natural taiga snowpack to that of a snowpack where the temperature gradient was suppressed by allowing falling snow to accumulate on tables, under which air circulation prevented the establishment of a temperature gradient. Until March, low insolation at our 65°N site limited surface heating and transient surface temperature gradients. In the natural snowpack on the ground, the temperature gradient reached 100°C.m^{-1} in the fall, decreasing to 20°C.m^{-1} in late winter. In the snowpack on the tables, the gradient mostly remained below 10°C.m^{-1}. Sampling and measurements were done alternatively on the ground and on the tables. Figure 3a clearly shows that the SSA vertical profile in the QIM snowpack on the tables displays higher values than both profiles in the HGM snowpack on the ground. Figure 3b shows samplings done just 4 days apart and the conclusion is the same. In most cases, values on the ground are a factor of 1.5 to 2 lower than on the tables, in reasonable agreement with Figure 2. This confirms that negative snow-climate feedback loops may exist.

Regions where this negative feedback may be observed include the southern edge of the taiga, where depth hoar formation may be reduced by warming and this may be enhanced by increased precipitation in the southern taiga of Canada and Eastern Siberia.[48] Other regions concerned by this effect may be the colder Alpine areas in the fall, when depth hoar formation will be hindered and conditions become more maritime-like. On the contrary, positive feedbacks are to be expected in the maritime and the warmer Alpine snowpacks, due to more frequent melting and to a faster SSA decrease under the QIM regime. Likewise, warming in the tundra and Northern taiga, without a change in metamorphic regime, is expected to enhance SSA decrease. The transformation of the southern tundra into shrub tundra and taiga[19] may transform the tundra windpack into faceted crystals or depth hoar of lower SSA, leading to a decreased albedo and longer e-folding depth. The impact of the changing vegetation on the light intensity reaching the snow surface would also have to be accounted for, for a full quantification of effects.

Other factors may further complicate the climate-SSA feedback. Albedo is mostly determined by surface snow, although underlying layers also have an effect.[49] A critical factor to determine snow SSA is the age of the surface snow, and hence the time elapsed between snow falls. Thus, the time distribution of precipitation events, as well as that of wind storms, is an extra factor to consider when modeling the warming-albedo feedback.

3.3 Heat conductivity

The heat conductivity of snow, k_T, relates the vertical heat flow q through the snowpack to the temperature gradient dT/dz, where z is the vertical coordinate[50] :

$$q = -k_T \, dT/dz \qquad (2)$$

Values of k_T for dry snow vary by a factor of 25 from 0.026 to 0.65 W m^{-1} K^{-1}.[50] The range is large because heat conduction in snow results from several processes (Figure 4) and the contribution of each process depends on the snow type, determined by metamorphism. For example, the most efficient process is conduction through the interconnected network of ice crystals. Depth hoar crystals are poorly interconnected and this snow type has the lowest k_T values, 0.026 to 0.10 W.m^{-1}.K^{-1}. Hard windpacks have strongly connected crystals and display the highest values for snows not subjected to melting, up to 0.65 W.m^{-1}.K^{-1}. Fined-grained snow that forms in temperate regions also have high k_T values, 0.3 to 0.4 W.m^{-1}.K^{-1}.

Climate change will modify k_T values in complex ways. Warming will limit depth hoar formation, increasing k_T. More frequent melting events in temperate climates will form ice layers with high k_T values. In contrast, the growth of shrubs on the tundra will limit the effect of wind,[19] transforming windpacks into depth hoar of much lower k_T values.

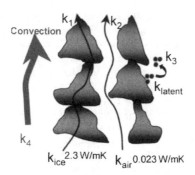

Figure 4 *The main processes contributing to the heat conductivity of snow, k_T :*
k_1=conduction through the network of interconnected ice crystals.
k_2=conduction through the air in the pore spaces.
k_3=latent heat transport across pore spaces due to ice
condensation/sublimation cycles during metamorphism.
k_4=air convection in the pore space.

The heat conductivity of snow will therefore change in many areas, affecting ground temperature, permafrost extent and sea ice growth. Resulting effects on climate include:

- Change in microbial activity and in CO_2 and CH_4 emissions by soils.[51]
- Change in the duration of the snow cover, in the duration of the growing season and of carbon sequestration by the terrestrial biosphere.[52]
- Modification of sea ice growth and areal extent and thus of the albedo of polar oceans.

To illustrate the climatic potential of changes in snow k_T values, we develop here the example of sea ice growth, based on observations and measurements on Storfjord, on the East coast of Spitzbergen during the winters of 2005 and 2006. While many effects intervene in sea ice growth,[2,53,54] our objective is to use a simplified system to isolate the effect of changes in snow k_T. Our campaign site was perfect for this purpose: it was on fast ice, in a shallow sheltered bay with homogeneous sea ice and snow cover over more than 1 km in all directions, justifying the use of a 1-D model. In that bay, heat transport by oceanic current is greatly reduced and is neglected here. We measured sea ice thickness, snow stratigraphy and heat conductivity, and perform calculations using 2 temperature scenarios differing on average by 5.7°C (Figure 5a) : a cold one and a warm one, based on the unusually warm 2006 winter. Other simplifying assumptions are :

- The initial system consists of an ice layer 10 cm thick, over an isothermal ocean at the freezing temperature of sea ice, -1.8°C, on 1st October.
- The temperature was kept constant for a given month, at the values shown in Figure 5a.
- The snow stratigraphy was also kept constant for a given month.
- Given this lack of temperature variations, steady state was assumed, i.e. heat storage in snow and ice is negligible.
- All heat exchanges at the water/ice interface are latent heat. The value of latent heat used is that of pure ice : $3,3 \ 10^8$ J.kg^{-1}.
- The temperature of the snow surface is equal to the air temperature.
- The heat conductivity of sea ice used is 2.1 W.m^{-1}.K^{-1}.[55]

Under these conditions, the heat fluxes, q, at the air-snow and sea ice-ocean interfaces are always equal and are simply related to h_i, the thickness of snow or ice layer i, and to k_{Ti}, the heat conductivity of layer i by :

$$q = (T_{ocean} - T_{air}) / \sum_i \frac{h_i}{k_{Ti}} \qquad (3)$$

Besides temperature (Figure 5a), the cold and warm scenarios differ by the structure of the snowpack. In both cases, the snow water equivalent have the same temporal evolution : 2 cm at the end of October, 11 cm at the end of January and 15.7 cm in late April. Stratigraphies and heat conductivities are very different. In the cold scenario, depth hoar layers of low densities (0.21 to 0.26 g.cm^{-3}) alternate with denser windpacks (0.38 to 0.48). Transient layers of fresh snow and of faceted crystals are also present. k_T values range from 0.06 W.m^{-1}.K^{-1} for aged depth hoar to 0.46 W.m^{-1}.K^{-1} for dense windpacks. In the warm scenario, two melt-freeze layers (densities 0.40 to 0.55) alternate with hard windpacks (0.34 to 0.41) while layers of fresh snow are sometimes included in the mean monthly stratigraphies. k_T values are 0.45 and 0.63 W.m^{-1}.K^{-1} for the melt-freeze layers and range from 0.36 to 0.48 W.m^{-1}.K^{-1} for dense windpacks. Recent snow has values around 0.2 W.m^{-1}.K^{-1}. Overall, the warm snowpack has a greater heat conductivity than the cold one.

Results are shown in Figure 5b. At the end of April, the ice thickness is 37 cm with the cold scenario and 58 cm with the warm scenario, so that changes in snow k_T more than offset atmospheric warming. This increased sea ice thickness clearly constitutes a negative snow-climate feedback loop that deserves consideration in climate models. Of course, this fast ice example cannot be taken at face value, as sea ice growth is much more complex than described here. Other effects must be taken into account such as convection in the water that depend on growth rate,[53] lateral heat fluxes in the ice and turbulent fluxes at the snow-air and sea-ice interfaces.[2,54] However, this example does illustrate the importance of one factor, the heat conductivity of snow, that will be affected by climate change.

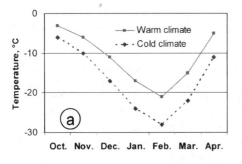 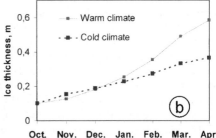

Figure 5   *(a) Mean monthly average temperatures of the warm and cold temperature scenarios used to calculate sea ice growth. In the warm scenario, spells above freezing formed melt-freeze crusts and different snow stratigraphies and heat conductivities were used for each scenario (see text).*
(b) Ice thicknesses during the winter according to the warm and cold scenarios. At the end of winter, sea ice is thicker under the warm scenario, because of the higher k_T values.

3.4 Permeability

The snowpack is a porous medium through which air flows if pressure differences exist within the snowpack. Such pressure gradients may be generated by the action of wind on sastrugi. Under most conditions relevant to snowpacks, the flow velocity v is proportional to the pressure gradient $\partial P / \partial x$ and the proportionality factor is the snow permeability K_p divided by the air viscosity η (Darcy's law) :

$$v = -\frac{K_p}{\eta} \frac{\partial P}{\partial x} \tag{4}$$

The permeability of snow ranges from 20×10^{-10} m^2 to about 500×10^{-10} m^2 and is strongly dependent on snow type : it is highest for depth hoar, which has large crystals and a low density, and lowest on windpacks, that have small crystals and a high density.[56] Shimizu[57] proposed an empirical relationship to relate K_p (in m^2) to snow density ρ_s (in kg.m^{-3}) and grain diameter D (in m):

$$K_p = 0.077 \, e^{-0.0078\rho_s} \, D^2 \tag{5}$$

During metamorphism, grain size increases faster under high temperature gradients and culminates with centimetric depth hoar crystals in the taiga snowpack.[22] A decrease in temperature gradient caused by global warming will therefore reduce grain size. Furthermore, upward vapor fluxes that compensate densification will also be reduced under a lower gradient. Slower crystal growth and a higher density caused by warming will therefore combine to reduce snow permeability. This prediction was confirmed by our field experiments in Alaska, described two sections up, where we compared the evolution of a given snowpack under a temperature gradient and under quasi-isothermal conditions.

In the natural snowpack on the ground, evolving under HGM conditions, depth hoar of density 0.20 g.cm^{-3} formed rapidly and the permeability in the lower half of the snowpack increased to beyond 500×10^{-10} m^2 in late March. In contrast, on the tables under QIM conditions, fine-grained snow of density 0.28 g.cm^{-3} formed and the permeability decreased to values between 30 and 70×10^{-10} m^2 in early March.

Warming will also increase the occurrence of melting events that produce melt/freeze crusts of low permeability. We therefore suggest that in most cases, climate change will result in a decrease in snow permeability. Of course, as mentioned above, in the southern tundra vegetation growth will reduce the effect of wind and facilitate depth hoar formation, enhancing snow permeability. But overall, the lower permeability induced by warming will reduce air flow through snow and therefore reduce the impact of one heat transfer process. This may partly counterbalance the increase in heat conductivity discussed earlier. However, the effect will be felt only in regions where highly permeable snow forms, essentially the taiga, where there is little wind. We do not attempt to quantify this thermal effect here but we suspect that it will be small. Its main impact may be on the transfer of chemical species from the snowpack to atmosphere, as discussed in the next chapter.

4 CHEMICAL PROPERTIES OF THE SNOWPACK

4.1 Chemical species in the snowpack

Species analyzed in snow include mineral and organic ions,[58,59] organic molecules such as aldehydes, peroxides, pesticides and hydrocarbons,[60-63] organic macromolecules present as particles[64] and mineral dust.[32] Motivations for these analyses include the understanding of the quality of water resources and the interpretation of ice core analyses.[6] Recently it has also become clear that snow is a complex multiphase (photo)chemical reactor producing species such as aldehydes, nitrogen oxides and halocarbons that impact atmospheric composition and reactivity.[3,12,65,66] Atmospheric chemistry interacts with climate, so that a full treatment of snow-climate feedbacks must discuss snowpack chemical reactivity and how climate-induced changes in snow metamorphism will affect it. We first briefly discuss how chemical species are incorporated in snowpacks and how this affects their reactivity. We then speculate on how each type of reactivity is affected by climate change. The mechanisms through which snow can incorporate chemical species (Figure 6) include :

- Adsorption. The concentration of adsorbed species in snow is determined by snow SSA, temperature and by the partial pressure of the species in snowpack interstitial air.[25,67,68] Thus, the reduction in SSA usually observed during metamorphism will lead to the emissions of adsorbed species. Adsorbed species are also readily available for dark and light-induced chemical reactions.[69]
- Formation of a solid solution. Small molecules such as HCl, HNO_3, HCHO and H_2O_2 dissolve within the ice lattice to form a solid solution.[10,61,70-73] The reactivity of these molecules will be limited by their trapping in the ice lattice. For example, photofragmentation will often lead to recombination through cage effects. Physical release will require solid state diffusion or sublimation during metamorphism.
- Trapping of aerosol particles. Falling snow efficiently scavenges particles suspended in air.[74] Wind blowing over snow also deposits particles within the snow, that acts as an efficient filter.[16,17,75] The fate of these particles is not well known. Hydrophilic particles such as sulphate aerosols may interact with the snow and spread on the surface, but this is not documented. Some particles remain essentially unaffected, as observed by SEM[76] and just sit on the surface of snow crystals.

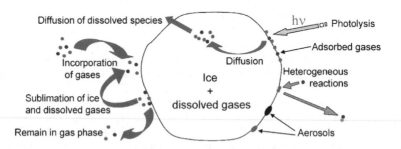

Figure 6   *Location of impurities in snow and the mechanisms through which they can react and/or be released to snowpack interstitial air.*

4.2 Adsorption/desorption of species

Modeling air-snow exchange of adsorbed species such as volatile and semi-volatile organic molecules requires the understanding of the rate of decrease of snow SSA,[68] that depends on temperature and on whether the snowpack is quasi-isothermal (QIM, temperature gradient <10°C.m[-1]) or under a high temperature gradient (HGM, gradient >20°C.m[-1]). Between these two values, data is lacking but a transition regime is likely.

Since both albedo and gas adsorption depend on SSA, the climate response of the concentration of species adsorbed within the snowpack will be similar to that of albedo : increase in regions where warming is accompanied by a change from HGM to QIM, such as the southern taiga and the warmer Alpine areas in the fall and decrease in the other regions. These effects will be modulated by the temperature increase, that will decrease the concentration of adsorbed species. For example, a temperature rise from -15 to -10 °C will desorb 40% of adsorbed acetone molecules, that have an adsorption enthalpy of 57 kJ/mol,[67], at constant SSA. The combined effect of warming and SSA change will then probably lead to a decrease in the concentration of adsorbed species in most areas.

4.3 Species forming a solid solution

In general, the concentration of soluble species in snow such as HCHO has been found to decrease with time.[9,61] A hypothesis for this is that they are incorporated out of equilibrium in clouds, possibly because snow crystals there grow fast and solute concentration is governed by kinetics rather than equilibrium.[70,77] Another possibility is that temperature and the partial pressure of the solute are very different in the cloud and at the surface, and re-equilibration takes place. Due to a lack of chemical measurements in clouds, it is difficult to test this hypothesis. In any case, since soluble species are usually emitted after precipitation, we will discuss how warming may affect emissions of these species.

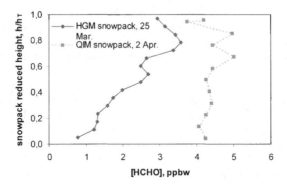

Figure 7 *Vertical profiles of formaldehyde, HCHO, in the snowpacks in central Alaska described in Figure 3 and subjected to different thermal regimes : a temperature gradient (HGM) regime, and a quasi-isothermal (QIM) regime.*

Warming will reduce the temperature gradient in the snowpack and the intensity of sublimation/condensation cycles that are the most likely to lead to the release of ice-soluble species. We then propose that emissions will decrease in a warmer climate. To test this idea, HCHO was measured during the Alaska field experiment described above, where two snowpacks were compared : one with the natural temperature gradient (HGM) and one where this gradient had been suppressed (QIM). Results (Figure 7) clearly demonstrate that emissions were suppressed in the QIM snowpack, as deeper layers retain their initial concentrations, while most of the HCHO is eventually lost from the HGM snowpack. We then suggest that in many cases (except again in the southern tundra) warming will limit the release of ice-soluble species to the atmosphere by ice sublimation. This effect may however be counterbalanced by increased diffusion rates at higher temperature, enhancing release.[9] Soluble species such as HCHO and H_2O_2 are oxidant precursors, as their photolysis releases HO_x radicals. These processes may then modify the oxidative capacity of the atmosphere over snow-covered regions. They therefore deserve attention.

4.4 Photochemical reactions

Many reactive species emitted by the snowpack such as NO, NO_2, HONO, aldehydes, alkenes and hydrocarbons[12,13,64-66,78] are produced by photochemical reactions. Photochemistry also produces species that are not volatile enough to be emitted to the atmosphere, such as complex hydrocarbons,[69] but these will modify snow and ice core chemistry and this aspect therefore also has interest. The rate of photochemical production is proportional to the light flux inside the snowpack. As detailed above, this will be inversely proportional to $SSA^{1/2}$. Similarly to the discussion in section 3.2, we speculate that a temperature increase without change in metamorphic regime will increase the rate of SSA decrease and therefore increase the light flux in the snow and photochemical production in the snowpack. Warming with a concomitant change of metamorphic regime will on the contrary decrease snowpack photochemical production.

4.5 Emissions of species from interstitial air

Species emitted to snowpack interstitial air must be transported out of the snow into the atmosphere to have an atmospheric impact. The most efficient mechanism is ventilation by wind, which depends on wind speed, snow surface structure and permeability. Hansen et al.[48] predict little surface wind change in polar regions. Snow surface structure depends mostly on wind, but speculating on its evolution is beyond our scope. We discussed that snow permeability will in general decrease with warming. The residence time of species produced in interstitial air will therefore increase under a warmer climate, reducing the chances of reactive species to escape before they react in the snowpack.

For example, HCHO and H_2O_2 can photolyze and the HO_x radicals produced can react with organic species contained in aerosols that are always present in snow. Other species will be produced,[64] that can also be released to the atmosphere, so that the nature of snowpack emissions are highly dependent on the residence time of primary products in the snowpack. Complex snow photochemical models, inexistent today, are needed to predict the effect of warming on snowpack emissions, caused by changes in snow permeability.

4.6 Overall climate-chemistry interactions

It is then clear that we lack critical data and understanding to predict with confidence how warming will affect snowpack chemical emissions to the atmosphere. It does appear, however, that in regions where both warming and a change in metamorphic regime will take place, most processes will lead to reduced emissions : reduced emissions of adsorbed species because of a slower SSA decrease, reduced photochemical activity because of the reduced light flux caused by the slower SSA decrease, slower release of dissolved species because of a reduced metamorphic intensity and slower transfer from the snowpack air to the atmosphere due to a decrease in permeability, although the impact of this last aspect is far from clear. Following previous discussions, such a scenario may take place at the southern edge of the taiga and in warmer Alpine snowpacks in the fall.

In regions where warming will not lead to a change in metamorphic regime, many antagonistic effects will take place. SSA will decrease faster, leading to enhanced desorption and photochemical activity. On the contrary, the reduced metamorphic intensity due to decreased temperature gradients will lead to lower emissions of dissolved species and slower transfer from snowpack air to the atmosphere, so that the net effect is at present unpredictable. This scenario will probably be seen in the northern taiga, in the colder Alpine areas and in the tundra that is not transformed into taiga by warming and the northern migration of vegetation zones.

If the southern tundra turns into shrub tundra or taiga[19] the transformation of the tundra windpack into faceted crystals or depth hoar of lower SSA, longer e-folding depth and higher permeability will lead to enhanced release of adsorbed and dissolved species, greater photochemical activity (modulated by tree shading) and more efficient release to the atmosphere, so that emissions may in this case increase.

In summary, the spatially-resolved prediction of how warming will affect chemical emissions by snow will probably need the coupling of a general circulation model to a photochemical model and to a vegetation model. Building a snow photochemical model would need new developments and more importantly, an improved understanding of the location of species in the snowpack. Quantifying the chemistry-climate feedback will furthermore require the coupling of these models with an atmospheric chemistry model.

5 CONCLUSION

The ideas proposed here and the data shown indicate that snow will exert many physical and chemical feedbacks on climate change. Most snow-climate feedbacks considered so far, such as the decrease in the areal extent of snow coverage,[4] the increase of soot in snow[18] and climate-vegetation-snow interactions,[19] are positive feedbacks. Here we investigate other feedbacks, such as the snow albedo-climate feedback and the snow heat conductivity-climate feedback, to conclude that their sign and magnitude will show large spatial and temporal variations, so that complex modeling is needed to quantify them. Our considerations, that have a large speculative character, suggest that significant negative climate feedbacks may take place in the southern taiga and in the warmer Alpine snowpack in the fall, while the tundra and Northern taiga will experience mostly positive feedbacks. The case of sea ice is interesting. A very simplified 1-D model suggests than in some areas, changes in the heat conductivity of snow may contribute to increased ice growth. Quantifying the overall effect of warming on ice growth must however also take into account many other complex effects.

The complexity of snow-climate feedbacks is illustrated by the observations that increases in atmospheric temperatures on the one hand and of snowpack and sea ice areal extents on the other, are not well correlated. Temperature has risen almost continuously, although at variable rates, since the beginning of the 20th century.[79] However, until the early 1980's, snow areal and temporal coverages have increased in most areas in the Northern hemisphere and a significant decrease has started to be observed only after that date.[80-82] Regarding sea ice, the signal is complex, with no significant trend in Antarctica, while the Arctic has seen a recent decrease.[4] These observations confirm that poorly identified and complex feedbacks may be at work. Elucidating them will require new model developments, the coupling of several models and perhaps more importantly, new observations that will produce time series of several physical and chemical properties of the snowpack, averaged over large areas. This is necessary to parameterize and test complex models. Setting up snow observatories would greatly assist in this enterprise.

Acknowledgements

F. Domine thanks the organizers of PCI 2006 for their kind invitation. Snow physics data obtained in Alaska was supported in part by the Chapman chair (Pr. Norbert Untersteiner) and by the International Arctic Research Center. The formaldehyde profiles were obtained by the FAMAS program funded by the French Polar Institute (IPEV) to S. Houdier. The Spitzbergen observations that inspired Figure 5 were funded by IPEV through the OSMAR program to F. Domine. Helpful comments by M. Sturm are gratefully acknowledged.

References

1   T. Zhang, *Rev. Geophys.*, 2005, **43**, RG4002. doi: 10.1029/2004RG000157.
2   M. Sturm, D.K. Perovich, and J. Holmgren, *J. Geophys. Res.*, 2002, **107**, 8047, doi: 10.1029/2000JC000409.
3   F. Dominé and P.B. Shepson, *Science*, 2002, **297**, 1506.
4   R.A. Pielke, G.E. Liston, W.L. Chapman and D.A. Robinson, *Climate Dynamics*, 2004, **22**, 591.
5   S.C. Colbeck, *Rev. Geophys. Space Phys.*, 1982, **20**, 45.
6   M. Legrand and P. Mayewski, *Rev. Geophys.* 1997, **35**, 219.
7   F. Dominé, E. Thibert, E. Silvente, M. Legrand, and J.-L. Jaffrezo, *J. Atmos. Chem.*, 1995, **21**, 165.
8   K. Nakamura, M. Nakawo, Y. Ageta, K. Goto-Azuma and K. Kamiyama, *Bull. Glaciol. Research*, 2000, **17**, 11.
9   Perrier, S., Houdier, S., F. Dominé, A. Cabanes, L. Legagneux, A.L. Sumner, P.B. Shepson, *Atmos. Environ.*, 2002, **36**, 2695.
10 M.A. Hutterli, J.R. McConnell, G. Chen, R.C. Bales, D.D. Davis and D.H. Lenschow, *Atmos. Environ.*, 2004, **38**, 5439.
11 W.R. Simpson, M.D. King, H.J. Beine, R.E.Honrath and X. Zhou, *Atmos. Environ.*, 2002, **36**, 2663.
12 A.L. Sumner and P.B. Shepson, *Nature*, 1999, **398**, 230.
13 R. Honrath, M.C. Peterson, Y. Lu, J.E. Dibb, M.A. Arsenault, N.J. Cullen and K. Steffen, *Atmos. Environ.*, 2002, **36**, 26290.
14 R.A . Schmidt, *Boundary-layer Meteorol.*, 1986, **34**, 213.
15 J.W. Pomeroy and L. Li, *J. Geophys. Res.*, 2000, **105**, 26619.
16 T. Aoki, T. Aoki, M. Fukabori, A. Hachikubo, Y. Tachibana and F. Nishio, *J. Geophys. Res.*, 2000, **105D**, 10219.
17 F. Dominé, R. Sparapani, A. Ianniello and H. J. Beine, *Atm. Chem. Phys.*, 2004, **4**, 2259.
18 J. Hansen and L. Nazarenko, *Proc. Nation. Acad. Sci.* 2004, **101**, 423.
19 M. Sturm, J.P. McFadden, G.E. Liston, F.S. Chapin, C.H. Racine and J. Holmgren, *J. Climate*, 2001, **14**, 336.
20 M. Kanakidou et al. *Atmos. Chem. Phys.*, 2005, **5**, 1053.
21 S.C. Colbeck, *J. Geophys. Res.*, 1983, **88**, 5475.

22   M. Sturm and C.S. Benson, *J. Glaciol.*, 1997, **43**, 42.

23   F. Dominé, T. Lauzier, A. Cabanes, L. Legagneux, W.F. Kuhs, K. Techmer and T. Heinrichs, *Microsc. Res. Tech.*, 2003, **62,** 33.

24   M. Sturm, J. Holmgren, and G. Liston, *J. Climate*, 1995, **85**, 1261.

25   F. Dominé, A. Cabanes and L. Legagneux *Atmos. Environ.*, 2002, **36**, 2753.

26   J.B. Brzoska, B. Lesafre, C. Coléou, K. Xu and R.A. Pieritz, *Eur. Phys. J. AP.*, 1999, **7**, 45.

27   D. Marbouty, *J. Glaciol.*, 1980, **26**, 303.

28   E. Brun, P. David, M. Sudul and G. Brunot, *J. Glaciol.*, 1992, **128**, 13.

29   R.A. Schmidt, *Boundary-layer Meteorol.*, 1982, **23,** 223.

30   F. Domine, R. Salvatori, L. Legagneux, R. Salzano, M. Fily and R. Casacchia, *Cold Regions Sci. Technol.*, 2006, **46**, 60.

31   L. Legagneux, A. Cabanes and F. Dominé, *J. Geophys. Res.*, 2002, **107D**, 4335, doi:10.1029/2001JD001016.

32   T.C. Grenfell, S. G. Warren and P. C. Mullen, *J. Geophys. Res.*, 1994, **99,** 18669.

33   S.G. Warren, *Rev. Geophys. Space Phys.*, 1982, **20**, 67.

34   T.C. Grenfell, B. Light, and M. Sturm, *J. Geophys. Res.*, 2002, **107**, C8032, doi:10.1029/2000JC000414.

35   W.J. Wiscombe, *The delta-Eddington approximation for a vertically inhomogeneous atmosphere*, NCAR Technical Note, NCAR/TN-121+STR, 1977, pp.1-66.

36   K. Stamnes, S.C. Tsay, W. Wiscombe and K. Jayaweera, *Applied Optics*, 1988, **27**, 2502.

37   T.C. Grenfell and S.G. Warren, *J. Geophys. Res.*, 1999, **104,** 31697.

38   S.P. Neshyba, T.C. Grenfell and S.G. Warren, *J. Geophys. Res.*, 2003, **108D**, 4448. doi: 10.1029/2002JD003302.

39   W.R. Simpson, G.A. Phillips, A.-S. Taillandier and F. Domine, in preparation.

40   F. Domine, A.-S. Taillandier and W.R. Simpson, *J. Geophys. Res.*, submitted.

41   A. Cabanes, L. Legagneux and F. Dominé, *Atmos. Environ.*, 2002, **36**, 2767.

42   A. Cabanes, L. Legagneux and F. Dominé, *Environ. Sci. Technol.*, 2003, **37**, 661.

43   L. Legagneux, T. Lauzier, F. Dominé, W.F. Kuhs, T. Heinrichs and K. Techmer, *Can. J. Phys.*, 2003, **81,** 459.

44   L. Legagneux, A.-S. Taillandier and F. Domine, *J. Appl. Phys.*, 2004, **95,** 6175.

45   A.-S. Taillandier; F. Domine, W.R. Simpson, M. Sturm and T.A. Douglas, *J. Geophys. Res.*, submitted.

46   L. Legagneux and F. Domine, *J. Geophys. Res.*, 2005, **110**, F04011. doi: 10.1029/2004JF000181.

47   M.G. Flanner and C.S. Zender, *J. Geophys. Res.*, 2006, **111**, D12208. doi:10.1029/2005JD006834.

48   J. Hansen et al., *J. Geophys. Res.*, 2005, **110**, D18104. doi:10.1029/2005JD005776.

49   X Zhou, S Li and K. Stamnes, *J. Geophys. Res.*, 2003, **108**, 4738. DOI: 10.1029/2003JD003859.

50   M. Sturm, J. Holmgren, M. König and K. Morris, *J. Glaciol.*, 1997, 43, 26.

51   M.L. Goulden, J.W. Munger, S.-M. Fan, B.C. Daube and S.C.Wofsy, *Science*, 1996, **271**, 1576.

52   M. Stieglitz, A. Ducharne, R. Koster and M. Suarez, *J. Hydromet.*, 2001, **2**, 228.

53   T. Fichefet, B. Tartinville and H. Goosse, *Geophys. Res. Lett.*, 2000, **27**, 401.

54   E.L. Andreas, R.E. Jordan and A.P. Makshtas, *J. Hydromet.* 2004, **5**, 611.

55   D.J. Pringle, H.J. Trodahl and T.G. Haskell, *J. Geophys. Res.* 2006, **111**, C05020. doi:10.1029/2005JC002990.

56   M.R. Albert and E. Shultz, *Atmos. Environ.*, 2002, **36,** 2789.

57 H. Shimizu, *Air permeability of deposited snow*, Institute of low temperature science, Sapporo, Japan, 1970 Contribution N° 1053. English translation.
58 J.E. Dibb, R.W. Talbot and M.H. Bergin, *Geophys. Res. Let.*, 1994, **21**, 1627.
59 M. Legrand and M. De Angelis, *J. Geophys. Res.*, 1996, **101**, 4129.
60 S. Houdier, S. Perrier, F. Dominé, A.M. Grannas, C. Guimbaud, P.B Shepson, H. Boudries and J.W. Bottenheim, *Atmos. Environ.*, 2002, **36**, 2609.
61 H.-W. Jacobi, M.M. Frey, M.A. Hutterli, R.C. Bales, O. Schrems, N.J. Cullen, K. Steffen and C. Koehler, *Atmos. Environ.*, 2002, **36**, 2619.
62 S. Villa, M. Vighi, V. Maggi, A. Finizio and E. Bolzacchini, *J. Atmos. Chem.*, 2003, **46**, 295.
63 J.-L. Jaffrezo, M.P. Clain and P. Masclet, *Atmos. Environ.*, 1994, **28**, 1139.
64 A.M. Grannas, P.B. Shepson and T.R. Filley, *Global. Biogeochem. Cycles*, 2004, **18**, GB1006.
65 H. J. Beine, R.E. Honrath, F. Dominé, W. R. Simpson and J. D. Fuentes, *J. Geophys. Res.*, 2002, **107D**, 4584, doi:10.1029/2002JD002082.
66 A.L. Swanson, N.J. Blake, D.R. Blake, F.S. Rowland and J.E. Dibb, *Atmos. Environ.*, 2002, **36**, 2671.
67 F. Dominé and L. Rey-Hanot, *Geophys. Res. Lett.*, 2002, **29**, 1873, doi:10.1029/2002GL015078.
68 G.L. Daly and F. Wania, *Environ. Sci. Technol.*, 2004, **38**, 4176.
69 J. Klanova, P. Klan, J. Nosek, I. Holoubek, *Environ. Sci. Technol.*, 2003, **37**, 1568.
70 F. Dominé and E. Thibert, *Geophys. Res. Lett.*, 1996, **23**, 3627.
71 E. Thibert and F. Dominé, *J. Phys. Chem. B.*, 1997, **101**, 3554.
72 E. Thibert and F. Dominé, *J. Phys. Chem. B.*, 1997, **102**, 4432.
73 S. Perrier, P. Sassin and F. Dominé, *Can. J. Phys.*, 2003, **81**, 319.
74 Y.D. Lei and F. Wania, *Atmos. Environ.*, 2004, **38**, 3557.
75 S. Harder, S.G. Warren and R.J. Charlson, *J. Geophys. Res.*, 2000, **105D**, 22,825.
76 C.Magano, F. Endoh, S. Ueno, S. Kubota and M. Itasaka, *Tellus*, 1979, **31**, 102.
77 F. Dominé and C. Rauzy, *Atm. Chem. Phys.*, 2004, **4**, 2513.
78 A.M. Grannas, P.B. Shepson, C. Guimbaud, A.L. Sumner, M. Albert, W. Simpson, F. Dominé, H. Boudries, J.W. Bottenheim, H.J. Beine, R. Honrath and X. Zhou, *Atmos. Environ.*, 2002, **36**, 2733.
79 IPCC, Climate Change 2001: The Scientific Basis J. T. Houghton, Y. Ding, D. J. Griggs, P. J. van der Linden, X. Dai, K. Maskell and C. A. Johnson, Eds., Cambridge University Press, 2001, 881 pp.
80 D.G. Dye, *Hydrol. Process.*, 2002, **16**, 3065.
81 H. Ye, *Geophys. Res. Lett.*, 2001, **28**, 551.
82 M. Laternser and M. Schneebeli, *Int. J. Climatol.*, 2003, **23**, 733. doi: 10.1002/joc.912

ICE ADHESION AND ICE FRICTION MODIFICATION USING PULSED THERMAL POWER

V.F. Petrenko

Dartmouth College, Hanover, NH 03755, USA

1 INTRODUCTION

Two physical properties of ice, its strong adhesion and its low dynamic friction, have driven the interest in ice research for decades. Since both properties are ice/solid-interface phenomena, significant attention has been paid to the structure and properties of the ice-solid interfaces.

Ice adhesion in particular brings dangerous and costly problems. There have been numerous attempts to reduce ice adhesion by developing a durable ice-phobic coating. All of these attempts have failed to decrease ice adhesion to a sufficiently low level where ice can be easily removed from the coating. There are three major physical mechanisms of ice adhesion: electrostatic interactions, hydrogen bonding, and Van der Waals interaction [1]. While the first two mechanisms can be significantly reduced or even totally eliminated, the third one, Van der Waals interaction, is strong enough to keep ice in place and cannot be cancelled.

A wide variety of deicing methods have been suggested in the past. The methods can be organized into three groups: mechanical ice removers, chemical methods, and thermal ice melters. Of those three, only the thermal melters could clean protected structures well without damaging either the structure or polluting the environment.

Although thermal deicing cleans surfaces well, it has a very serious disadvantage that limits its applications: the high energy requirement. For instance, even in still air it typically takes 2 MJ/m^2 or more to deice a solid surface. So much energy is needed, because an ice-structure interface is thermally connected to the ice bulk and to an engineering structure. Thus, while wanting to heat and melt only the interface, one must inevitably heat those large heat masses as well. When heat transfer by air-convection is involved, even more energy is needed for deicing. Thus, a large airplane may use up to 25% of its fuel to keep the wings ice-free.

A recently invented Pulse Electro Thermal Deicer (PETD) [2], successfully overcomes that high-energy requirement limitation by effectively isolating an ice-solid interface from the environment. When optimized, a pulse deicer requires only 1% of the energy conventionally used in thermal deicing and can deice surfaces in less then one second. This technology has been applied to deice airplanes, bridges, car windshields, to release ice from icemakers, and to defrost refrigeration evaporator coils.

It was also found that a technique similar to PETD technique applied to an ice-slider interface can increase static and low-velocity ice friction up to one hundred times [3]. The method was tested on cross-country skis, non-slip shoes, and prototypes of automotive tires. The method was named Pulse Electro-Thermal Brake (PETB).

2 THEORY OF PETD AND PETB

De-icing is a process in which interfacial ice attached to an engineering structure is either broken or melted. Some sort of external force (gravity, wind-drag) then removes the ice from the surface of the structure. Mechanical de-icers take less energy, but they don't clean structures well, leaving behind significant amounts of ice fragments. The mechanical deicers also can damage the structures they are supposed to protect and accelerate wear on materials. A conventional thermal de-icer does the ice-cleaning job well, but takes too much electric (or other) energy. In a typical thermal de-icer the heat loss exceeds by orders of magnitude the amount of "useful" heat that is used to melt the interfacial ice. The following example illustrates this statement.

Let us consider the de-icing of an airplane wing at T= -20°C and air speed of 720 km/hour. For a relatively thin wing (10 cm thick), a convective heat exchange coefficient on the leading edge of the wing is about h_c=415 watt/K·m². The power it takes to keep the wing surface at +5°C is then:

$$W = 25 \cdot h_c = 10 \; kW / m^2 \tag{1}$$

Keeping such power for just 3 minutes would take an enormous amount of energy:

$$Q = W \cdot t = 1.8 \cdot 10^6 \cdot \frac{Joule}{m^2} \tag{2}$$

But in fact, what it takes to remove ice is to melt about 10-µm thick layer of ice on the ice/wing interface. Estimates show that this would take only Q_{min}:

$$Q_{min} = d \cdot q \cdot \rho_i = 4 \cdot 10^3 \frac{Joule}{m^2} \tag{3}$$

where d is the thickness of the melted layer, ρ_i is ice density, and q is the ice latent heat of fusion. The energy of eqn. 3 is 450 times less than that in eqn.2. Moreover, because of heat drainage into the wing, in practice, jet-plane thermal de-icers currently use three to five times more power than that of Eq.1. So, the difference between an "ideal" de-icer and what is in use is even greater.

In contrast with conventional ice-melters, PETD and PETB melt a very thin layer of interfacial ice while significantly limiting the above-mentioned heat-drainage mechanisms. To limit heat lost to heat diffusion into ice and a substrate, in PETD the heating is applied as a very short pulse, rather than continuously. Shortening the heating-pulse duration minimizes the thickness of heated layers in both materials, thus decreasing the thermal mass of those layers. Short pulsing also reduces heat loss into the environment.

Because in pulse deicing (braking) the heat diffusion lengths are typically much shorter than the thickness of the ice and that of a substrate, the result would not depend on the materials thickness at all, to simplify boundary conditions we can consider infinitely thick layers of the materials:

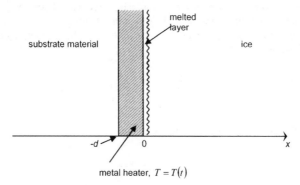

Figure 1 *Ice-heating film-substrate schematics.*

We can also break the mathematical problem into two halves:
1) Heating the ice/heater interface from an initial temperature, T_0, to the ice-melting point, $T_m=0°C$.
2) Melting interfacial ice. In this case, due to the large latent heat of ice melting, q, we can assume that the interfacial temperature remains almost constant at $T = 0°C$, see figure 2.

Mathematically, the problem in #2 is similar to the problem of a "constant surface temperature" time-dependent heat-conduction problem, but with a different set of initial conditions.

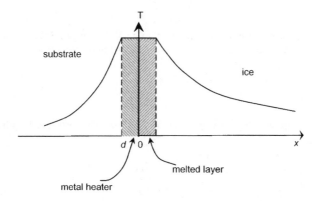

Figure 2 *Temperature distribution near the interface during formation of the melted layer. This picture implies constant temperature in the metal and in a thin water layer.*

After the heating power is "off", the melted layer refreezes.

2.1 Solving Problem #1

Let us first consider a simpler problem of a semi-infinite layer of ice, $0 \leq x < +\infty$. The surface at $x = 0$ is heated with a power density W (W/m^2) starting at time $t=0$. Let us first neglect the small heat capacity of a thin-film heater. The temperature in the ice is $T(x,t)$ and it obeys the heat diffusion equation:

$$\frac{\partial^2 T}{\partial x^2} = \frac{1}{\alpha_i}\frac{\partial T}{\partial T}, \qquad x > 0 \tag{4}$$

Where α_i is the thermal diffusivity coefficient of ice:

$$\alpha_i = \frac{k_i}{\rho_i C_i} \tag{5}$$

Where k_i is the ice's thermal conductivity, ρ_i is ice density, and C_i is the specific heat capacity of ice.

Boundary condition:

$$\dot{Q} = -k_i \frac{\partial T}{\partial x} = W, \qquad x = 0 \tag{6}$$

Initial condition:

$$T(x,0) = T_0 \tag{7}$$

Let us first consider a case of an infinitely thin heating element: d=0. The solution to eqn. (4) with the boundary condition given by eqn. (6) and the initial condition of eqn. (7) is:

$$T \cdot (x,t) = T_0 + \frac{W}{k_i}\left[2\sqrt{\frac{\alpha_i t}{\pi}} \cdot e^{-\frac{x^2}{4\alpha_i t}} - x\, erfc\left(\frac{x}{2\sqrt{\alpha_i \cdot t}}\right)\right] \tag{8}$$

Where *erfc* is a complimentary error function, *erfc=1-erf*. The diffusion-length of heat in the ice is :

$$L_{Di}(t) = \sqrt{\frac{k_i \cdot t}{\rho_i C_i}} \tag{9}$$

Let us introduce an equivalent thickness of the heated layer of ice, $L_{eq,i}$, as:

$$L_{eq,i} = \frac{1}{(T-T_0)} \int_0^\infty \left(\left(T(x,t) - T_0 \right) \right) dx \tag{10}$$

Then:

$$L_{eq,i} = \frac{\sqrt{\pi}}{2} \cdot L_{Di} = \frac{1}{2} \sqrt{\frac{\pi \cdot k_i \cdot t}{\rho_i C_i}} \tag{11}$$

Let us now consider an infinitely thin heater placed at x=0 in between two semi-infinite media, the ice ($0 < x < +\infty$) and the substrate material ($-\infty < x < 0$).

To heat the interface by $\Delta T = T_m - T_0$ degrees using constant power W:

$$W = W_1 + W_2 \tag{12}$$

where W_1 and W_2 are heat fluxes in the ice and in the substrate respectively, would take time t:

$$t = \frac{\pi \cdot \Delta T^2}{4W^2} \left[\sqrt{k_i \cdot \rho_i \cdot C_i} + \sqrt{k_s \cdot \rho_s \cdot C_s} \right]^2 \tag{13}$$

where ρ_s, k_s, and C_s are substrate density, thermal conductivity, and specific heat capacity respectively.

The total energy requirement, Q, it takes to heat the interface to the ice melting point is:

$$Q = W \cdot t = \frac{\pi \cdot \Delta T^2}{4W} \left[\sqrt{\rho_i \, k_i \, C_i} + \sqrt{\rho_s \, k_s \, C_s} \right]^2 \tag{14}$$

Notice, that in eqns. (13) and (14) W and Q are given per unit area.

One can also obtain the result of eqn. (14) by applying the energy conservation law:

$$Q = W \cdot t = \Delta T \cdot \left[C_i \, \rho_i \, L_{eq,i}(t) + C_s \, \rho_i \, L_{eq,s}(t) \right] \tag{15}$$

where

$$L_{eq,i}(t) = \frac{\sqrt{\pi}}{2} \cdot \sqrt{\frac{k_i \cdot t}{\rho_i \cdot C_i}} \tag{16}$$

and

$$L_{eq,s}(t) = \frac{\sqrt{\pi}}{2} \cdot \sqrt{\frac{k_s \cdot t}{\rho_s \cdot C_s}} \tag{17}$$

are equivalent lengths of heat diffusion in ice and the substrate correspondingly.

The total energy requirement, Q, can then be found by solving eqn. (15) for pulse duration, t, and by multiplying the t by the power W. The result obtained in that way will, of course, be the same as in eqn. 14.

The simplest way to take into account a finite heat capacity of the heater-film and the latent heat of melting the interfacial ice is to add them to eqn. (14):

$$Q = W \cdot t = \frac{\pi \cdot \Delta T^2}{4W} \left[\sqrt{k_i \, \rho_i \, C_i} + \sqrt{k_s \, \rho_s \, C_s} \right] + d_h \cdot \rho_h \cdot C_h \cdot \Delta T + d_i \, \rho_i \cdot q \qquad (18)$$

where d_h is the heater-film thickness, ρ_h and C_h are correspondingly the heater film density and specific heat capacity, d_i is thickness of melted layer of interfacial ice, and q is latent heat of ice melting. Notice that the first term in eqn (18) is typically much bigger than the second and the third ones. Because of this, in a typical deicing operation the total energy used by PETD, Q, is approximately inversely proportional to the density of heating power, W. This allows almost unlimited possibilities for the thermal deicing of large areas at a very low energy required "per unit area". Due to its counter-intuitive nature, the idea that one has to apply higher power to save overall energy was missed by thousands of practical engineers and designers of deicing systems.

Equation (18) assumes that temperature varies with time and a coordinate in accord with eqn. (8) (constant heating power applied to an interface). This assumption is not totally accurate, due to the fact that during the heating of the interface from T_0 to the ice melting point, the heat flux to ice and to substrate is slightly reduced by the heat flux stored in the metal heater. Moreover, on reaching the melting point, the interfacial temperature becomes tied to T_m, thus changing the "constant-interfacial-power" problem to the "constant-interfacial-temperature" problem. In the later case, an equivalent heat diffusion length in ice is:

$$L'_{eq,i}(t) = \frac{2}{\sqrt{\pi}} \cdot \sqrt{\frac{k_i \cdot t}{\rho_i \cdot C_i}} \qquad (19)$$

which is $\dfrac{2^2}{\pi} = 1.273$ times larger than that in equations (16-17). This difference can be used to estimate the maximum error of the used approximations. Because the "diffused" heat is just a part of the total heat flux, the maximum error is always less than 27% and typically less than 10%, which provides sufficient accuracy for most practical applications.

When the latent heat of melting interfacial ice and the heat capacity of the heater require an amount of energy comparable to or bigger than what it takes to heat the ice and the substrate, the energy conservation law states that:

$$W \cdot t = [C_s \cdot L_{eq,s}(t) + C_{eq,i} \cdot L_{eq,i}(t)] \cdot \Delta T + d_h \cdot \rho_h \cdot C_h \cdot \Delta T + d_i \, \rho_i \cdot q \qquad (20)$$

One can easily solve the quadratic equation in (20) for t to find the de-icing energy $Q = W \cdot t$:

$$Q = \left[\frac{b}{2} + \sqrt{\frac{b^2}{4} + c} \right]^2 \tag{21}$$

where:

$$b = \frac{\sqrt{\pi}}{2} \cdot \Delta T \cdot \frac{\sqrt{k_i \cdot \rho_i \cdot C_i} + \sqrt{k_s \cdot \rho_s \cdot C_s}}{\sqrt{W}} \tag{22}$$

$$C = C_h \cdot \rho_h \cdot d_h \cdot \Delta T + \rho_i \cdot d_i \cdot q$$

2.1.1 Freezing Time Calculations. After the heating pulse is over, the heat continues to diffuse from the interface. This heat flux is taken from the melted layer, unless it refreezes. Assuming heat diffusion lengths of eqns. 16-17 and applying the energy conservation law, we can easily arrive at the following result for the interface re-freezing time, t_f''

$$t_f = \frac{4 Q_L^2}{\left(\sqrt{k_i \, \rho_i C_i} + \sqrt{k_s \, \rho_s C_s} \right)^2 \pi} - \frac{4 \cdot Q_L \cdot \sqrt{t}}{\left(\sqrt{k_i \cdot \rho_i \cdot C_i} + \sqrt{k_s \cdot \rho_s \cdot C_s} \right) \sqrt{\pi}} \tag{23}$$

where:

$$Q_L = \rho_i \, d_i \cdot q \tag{24}$$

and where t is the heating-pulse duration:

$$t = \frac{Q}{W} \tag{25}$$

In deicing applications the refreezing time should at least allow for the ice to slide off of the surface before the interface refreezes. However, in a pulse electro-thermal brake, this refreezing time should be as short as possible to prevent slippage of the slider before the interface refreezes.

Note that in the case where heat-diffusion length is larger than or comparable to either the ice thickness or the substrate thickness, Q, t and t_f can be easily calculated using numerical methods. In our work we have successfully used FemLab software by Comsol for that purpose. Such numerical calculations have also demonstrated the great advantage of using shorter but higher-density power pulses for deicing surfaces.

3 EXPERIMENTS

3.1 PETD: Laboratory Experiments

Numerous experiments have been conducted on PETD. In those experiments we have tested as substrates the materials glass, concrete, wood, several polymers such as ABS, polyethylene,

HDPE, composite materials such as epoxy-glass fiber and epoxy-carbon fiber, and laminate structures such as polymer films on bulk metal substrates. Thin-conductive film heaters were made of electrically conductive paints, electrically conductive polymer films, metal foils (stainless-steel, copper, titanium, and titanium alloys), and thin metal films sputtered on a ceramics or glass. The samples' dimensions varied from a few centimeters to 5m.

The experiments were conducted in cold rooms and in an icing wind tunnel in a temperature range of -40°C to -1°C and with wind velocity up to 100 m/s. The range of heating density, W, varied from 500 W/m^2 to 250 kW/m^2. In this paper we present results obtained on a prototype of an automotive windshield deicer. The deicer was made of 356 mm x 356 mm x 6mm soda-lime glass sheet coated with 0.3 μm layer of indium tin oxide (ITO), a common transparent conductor. The ITO layer had a sheet-resistance of 8 ohm and was scratch-protected with an approximately

2-μm layer of Al$_2$O$_3$, which also played the role of anti-glare coating. The windshield had two electric buses on its sides (figure 3). The 1-cm wide buses were made of either a thin copper deposited electrochemically, or they were painted with silver-based conductive paint. Ice blocks 2-cm in thickness were frozen to the glass windshield by surface melting/refreezing for testing.

Short pulses of 60-Hz AC power were applied to the electric buses to heat the ice-glass interface and, thus, to deice the glass surface. When interfacial ice was melted, the ice block slid off of the windshield under the force of gravity. Figure 4 depicts deicing time, t, and deicing energy, Q, as functions of the density of heating power, W.

Figure 3 *General view of the glass windshield deicer with an ice block attached.*

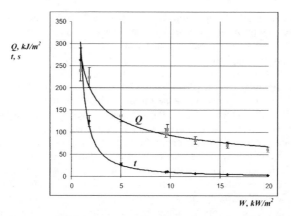

Figure 4 *Deicing time, t, and deicing energy, Q, versus the density of heating power, W. T = -10°C.*

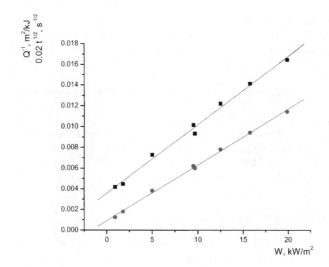

Figure 5 *Inverse deicing energy, 1/Q, and inverse square root of deicing time, $50 \cdot \sqrt{t}$, versus the density of heating power, W. T = -10°C. The same data as in figure 4 were used. The upper line relates to Q, and the lower line relates to t.*

Figures 4&5 show the results for Q and t obtained at T = -10°C. As can be seen from figure 5, at a heating pulse duration short compared with the heat diffusion time through the ice and glass sheets, $t \leq 30s$, Q and t follow correspondingly the simple inverse and inverse-quadratic dependencies on W, as predicted by eqns. 13&14.

3.2 PETB: Laboratory Experiments

While the theory of a pulse de-icer (PETD) and that of a pulse brake (PETB) is essentially the same, the eventual result of a heating pulse on an ice/substrate interface is opposite in these two devices. While in PETD ice slides off the substrate using the melt as a lubricant, in PETB the thin melted layer almost instantly refreezes, thus gluing the slider to the ice.

Numerous laboratory and field experiments have been conducted on pulse electro-thermal brakes. In these experiments we have tested as "sliders" plates made of glass, wood, polymers such as ABS, polyethylene, HDPE, composite materials such as epoxy-glass fiber and epoxy-carbon fiber, rubber, shoe soles, 1/3-scale rotating automotive tires, *etc*. Thin-conductive film heaters were made of electrically conductive paints, electrically conductive polymer films, metal foils (stainless-steel, copper, titanium, and titanium alloys), and thin metal films sputtered on a ceramics or glass. The samples' dimensions varied from a few centimeters to 2m. The tests were conducted on both snow and ice in cold rooms in a temperature range of -40°C to -1°C. The range of heating density, W, was from 100 kW/m^2 to 4 MW/m^2. As a single illustration we are presenting here results obtained on a prototype of a cross-country ski equipped with 30cm x 4 cm PETB made of a PCB (printed-circuit-board, FR-4). A 17-μm in thickness copper foil at the bottom of the PCB was used as the pulse-action heating element. A typical duration of the heating pulses varied from 50 μs to 300 μs, depending on snow temperature. Each time a pulse of energy sufficient to melt the snow surface was applied, the static friction on the snow increased to a level of the strength of snow. For instance, on an artificial ski trail at -10°C and normal load of 70 kg a 100 μs pulse of $W = 4$MW/m^2 increased the friction force from 45N to 270N, or by a factor of 6. When the same type of PETB was tested on solid ice, the friction force increased by a factor of up to 100, which corresponded to the strength of the copper/ice interface. A pair of such skis was also successfully tested on real ski trails in the Austrian Alps and in New Hampshire, USA. That work was supported by Fisher-Gesellschaft mbH.

4 FIELD TESTS AND PRACTICAL APPLICATIONS

PETD technology has already found several practical applications and a significant number of new applications are in current development. Goodrich Co. has acquired exclusive rights to use the technology for deicing airplanes, windmill turbines, and sea vessels. Together with the Thayer School of Engineering, Goodrich has developed and successfully tested the first airplane deicer seen in figure 6 on the leading edges of the Cessna-plane wings. That deicer was tested extensively during the 2003/2004 winter season and demonstrated almost instant ice cleaning and an extremely low average power consumption [4]. Goodrich is currently developing PETD for several large airplane manufacturers.

Figure 6 *Cessna 303T business plane with PETD in flight, Winter 2003/2004.*

The PETD deicer in usage thus far was installed on several cables and one pylon of the Uddevalla bridge in Sweden (figure 7). The longest of the bridge's cables are over 200 m in length and 25 cm in diameter. The bridge PETD is made of 0.3-mm thick stainless steel foil and is powered by a battery bank. The deicer cleans one cable at a time with a pulse several seconds in length.

One of the most promising applications of this new deicing technology is harvesting ice from commercial and residential icemakers. Because an ice-harvest cycle includes heating and re-cooling massive hardware, icemakers consume comparable amounts of electric energy to first grow the ice and then to release it. PETD can release the ice from icemakers using very little heat, thus reducing the energy needed for the ice harvest and re-cooling cycles. Our prototype icemakers demonstrated a savings of up to 50% of the energy used to produce 1 kg of ice.

Other applications of the technology under development are an automotive windshield deicer, non-slip shoes, a pulse brake for automotive tires, and a building-roof deicer.

Figure 7 *General view of the Uddevalla-bridge in Sweden. The PETD deicer has been installed and tested on several cables and one pylon-top during the 2004/2005 and 2005/2006 winters.*

Acknowledgements

The author thanks Drs. M. Starostin, M. Higa and V. Kozliouk, J. Chen and F. Petrenko for their help with the experiments described in the paper.

References

1 V. Petrenko and R. Whitworth, *Physics of Ice*, Oxford University Press, New York. 1999.
2 V. F. Petrenko, *System and method for modifying ice-to-object interface.* US Patent # 6, 870, 139, March 22, 2005.
3 V. F. Petrenko, *System and method for modifying friction between an object and ice/snow.* US Patent 7, 034, 257, April 25 2006.
4 G. Botura, D. Sweet, and D. Flosdorf, *Development and demonstration of low-power electrothermal de-icing system,* AIAA-2005-1460, 43rd AIAA Aerospace Sciences Meeting and Exhibit, Reno, Nevada, Jan. 10-13, 2005.

IMPROVING OUR UNDERSTANDING OF GAS HYDRATE FORMATION PROCESSES: THE IMPORTANCE OF MULTI-TECHNIQUE APPROACHES

J.A. Ripmeester

Steacie Institute for Molecular Sciences, National Research Council Canada, Ottawa, ON, Canada K1A 0R5

1 INTRODUCTION

1.1 The major issues

Gas hydrate should be considered as a material which directly impacts our planet's population, in many ways as yet to an unknown extent.[1] Although hydrates and their structures have been studied for many years[2,3], the understanding of hydrate formation processes is still in its infancy but is of critical importance today for a variety of reasons: understanding the nature of natural gas hydrate deposits and how they accumulate[4], the prevention and control of pipeline hydrates[5], and the development of hydrate-based processes such as gas storage[6].

Natural gas hydrates, ubiquitous[1] both offshore on the continental margins[7] and under the permafrost[8], are considered to have potential as a major energy resource, they may act as agents of climate change by liberating greenhouse gas methane[9], and they also may present a considerable geohazard as decomposing hydrate may cause seafloor instability[10]. Natural gas hydrate, as a mineral, should be expected to display the complexities of most other natural materials in that they reflect conditions during the initial hydrate formation, as well as the conditions during subsequent changes in the local environment. That is, depending on the gas source, natural hydrate is unlikely to be a pure material like samples synthesized in the laboratory. Also, since hydrate is labile, its properties will change on fairly short timescales in response to changes in ambient conditions (P, T, gas composition), [11] for instance if a gas or hydrocarbon fluid of a different composition comes in contact with the hydrate. The study of hydrate is further complicated by the fact that recovery of intact hydrate is quite difficult.

Hydrate formed in pipelines has been identified as a problem by the gas and oil industry since the 1930's, there being both hazards and economic factors in actual pipeline blockage, and the prevention and control of hydrate formation in pipelines is of major importance[5]. Although emphasis has been placed on finding practical solutions in this area by inhibiting hydrate formation with various additives it is clear that modern practice requires better alternatives to those in common use today that are safe and environmentally benign even for the ever more challenging deep water wells. A fundamental understanding of hydrate formation can only be helpful in formulating solutions.

For quite some time now, hydrates have also had the promise of being useful materials both for gas storage[6] and separation processes[12], however, usually a number of problems are identified that prevent a technology from developing or from becoming competitive. Certainly there are a number of barriers to overcome that will address the improvement of both the kinetics and the yield of hydrate formation before a practical technology can emerge.

1.2 Specific issues

Probably the main hydrate property that is of interest is the compressed state in which trapped gas molecules find themselves in gas hydrate [2, 3]. Each volume of gas hydrate can hold some 160 volumes of gas (STP), and for methane gas this represents a compression factor of ~9 comparing hydrate kept in its stability zone under a stabilizing pressure at $0°C$ to gas under the same pressure. In terms of understanding hydrate formation, especially in the context of hydrocarbon guest materials or other insoluble species, we see that a large volume of gas has to react with water, and neither component is soluble in the other to the extent that a homogeneous reaction (or simple freezing) can take place. Thus, hydrate formation depends critically on heat and mass transport and reactions in bulk require efficient mixing of the two components, often in stirred reactors[13]. One result is that it is very rare that kinetic parameters for hydrate formation are transferable from one apparatus to another. Often simplified systems are studied in order to eliminate some of the extrinsic parameters eg by reacting gas with water in quiescent systems, gas reacting with powdered crystalline ice[14, 15] or in frozen amorphous water – guest systems at low temperatures[16].

This contribution will examine some recent experimental results that challenge a number of ideas that often were taken as fundamental assumptions in the hydrate research community. Some of these ideas concern the predictability of phase equilibria [17-19], the existence of metastable phases [15, 20, 21] and the labile nature of hydrate structures [11, 21]. As well, very distinct notions of hydrate nucleation and growth exist among hydrate researchers.

Much of the new information has arisen because of novel and improved methods for the study of hydrate structures and processes [20 – 23]. Of course, modeling of hydrates, especially phase equilibria, is a well-developed area, but results from modeling calculation do need to be closely tied to experimental verification.[17-19] Over the last ten years or so hydrate researchers have come to the conclusion that in order to understand hydrate phase equilibria, the assumption that one can accurately predict the solid phase structures that one produces is no longer valid [17, 18, 24]. The seeds for these ideas were sown already in the 1980's when small molecules such as O_2, N_2, Kr, Ar were shown to form sII hydrate, instead of sI, as had been previously assumed[25,26]. The report of sH hydrate in 1987 further complicated matters in that heavier hydrocarbons also were shown to be able to form hydrates at pressures considerably less than the small "help gas" guests themselves [27, 28]. The recent work on the ethane-methane system [18,24] also recalls the much earlier work of von Stackelberg [29] who observed that some small guests that formed sI hydrate individually were able to form sII hydrate when used together to form hydrate under certain conditions.

As well, such important concepts as hydrate nucleation and the often-observed memory effects are not well understood [30-33]. The memory effect, occurs when a hydrate is melted or decomposed and hydrate reformation upon cooling is much faster than the original hydrate formation[30-33]. This effect often has been assumed to be structural and therefore, intrinsic[23]. However, it should then be observable in the realm of homogeneous nucleation and, as discussed later, this is not so.

Much of the information on hydrate processes has come from macroscopic studies, that is, from the observation of gas consumption, pressure drop, particle size measurements, or crystal morphology observations. However clathrate hydrates in many ways are unique materials that make it imperative that studies on the molecular scale are also carried out. For instance, several structures of hydrate may coexist, and often this is not obvious from phase equilibrium studies[17, 24]. As well, since the hydrates are clathrates, and as such, non-stoichiometric, the cage occupancies must be measured to define the system and to provide important insights into how well the various modeling parameters used to predict the structure, phase equilibria and composition actually fit experimental data [34-36]. Fortunately, instrumental methods that operate on a molecular scale often can be used to obtain both bulk properties as well as structural information[37].

Much of the new experimental information has depended on the ability to obtain structural information from specific experimental markers as a function of time. This has allowed the observation of some structural features that appear during the early stages of hydrate formation and how these evolve. Information on nucleation will require yet further experimental refinements, as much of the information available has been inferred indirectly.

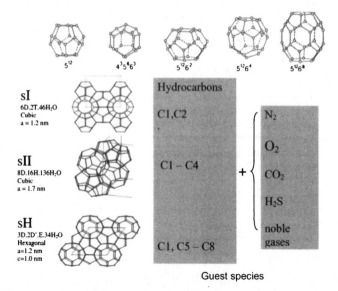

Figure 1 *The common hydrate cages and structural families, giving the unit cell compositions, lattice symmetry and parameters. The guests indicated are those typically found in natural gas hydrates of each structure. The guests in the last column usually are a minor component and may be present in any of the structures since they fit in the small D and D' cavities. The actual hydrate structure formed depends on the partial pressure of each gas component and the pressure and temperature.*

2 THE LABORATORY CHARACTERIZATION OF NATURAL GAS HYDRATES

Since ultimately it is necessary to understand hydrate formation processes, it is of interest to examine the various techniques that have been valuable in characterizing hydrate structures and compositions. Natural gas hydrates recovered from offshore or under the permafrost tend to be far more complex than laboratory-prepared materials where the combination of guest gases and their purity can be well defined. Natural hydrate minerals belong to three structural families known as cubic sI, cubic sII and hexagonal sH with unit cells $2D.6T.46H_2O$, $16D.8H.136H_2O$ and $3D.2D'.E.40H_2O$, where the D, T, H and E refer to the cages present in the structure ($D=5^{12}$, $T = 5^{12}6^2$, $H = 5^{12}6^4$, $D' = 4^35^66^3$ and $E = 5^{12}6^8$; X^y refers to y, the number of faces with X sides, that constitute each cage [3,36]. In most cases each cage holds a single guest (but note ref. 15 and recent work on high pressure hydrates), and the overall stoichiometry of the hydrate depends on the guest distribution and occupancy of each guest type[2]. So, in order to specify a material exactly one needs to know the structure (space group, lattice parameters and the absolute occupancy of each cage by each guest type). All of the reported diffraction work carried out on natural gas hydrates has been done on powders so far [38-41]. It is difficult to specify hydrate structure and composition from powder diffraction alone even in the case of a single guest species as it is not straightforward to obtain absolute cage occupancies without some

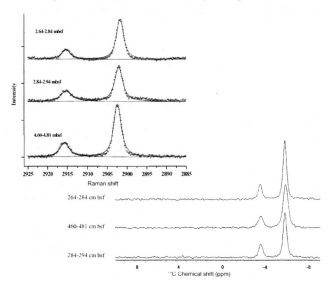

Figure 2 *Partial Raman (top) and ^{13}C MAS NMR spectra (bottom) from hydrate samples recovered from Cascadia, taken at 10K and 173K, respectively. The spectral signatures indicate that methane is the principal guest for both large (major peak) and small cages(minor peaks). The line intensities show that the populations are in an approximate 3-4:1 ratio (for large:small cages), confirming the powder X-ray data assignment that it is a sI hydrate. mbsf = meters below sea floor, cm bsf = cm below sea floor. From ref. 41*

assumptions or model calculations. Spectroscopic measurements have provided additional information which often also gives information on the structure type [37-40]. Specifically, ^{13}C NMR spectroscopy is able to distinguish methane in D, T and H cages[33], and sometimes even can distinguish D from D' cages. The C-H symmetric stretch frequencies in the Raman spectrum of methane also are diagnostic in that they can distinguish methane in D from methane in T and H cages [34]. For each guest such cage-specific features in the spectrum can be determined from studies on the pure model systems. For instance, for the hydrate minerals, the guest materials are confined to hydrocarbons of C1- C8, CO_2, N_2, O_2 and H_2S and, likely, traces of the rare gases (Fig. 1). For many of these the various cage-dependent signatures are known, although an exhaustive catalogue is not as yet available. Hydrate samples recovered from Cascadia [42, 43] offer a good example of the diversity that can be found as both thermogenic and biogenic hydrates are found in close proximity. The ^{13}C NMR and Raman spectra for a sI methane hydrate recovered from a cold vent field, Cascadia, are shown in fig.2., and as also confirmed by X-ray powder diffraction. The patterns are diagnostic of sI hydrate with the large cages essentially full and the small cages ~ 80-90% occupied. Application of the van der Waals and Platteeuw[44] statistical model allows an estimate of the hydration number, which gives hydration numbers of 6.1 +/-0.1 waters/guest.

The powder diffraction pattern for a second sample, obtained from near a hot vent in Barkley Canyon [45] is more complex and can be indexed in terms of sII hydrate and a smaller amount of sH hydrate[45]. The ^{13}C NMR spectrum confirms that a large number of hydrocarbons are present with most falling in the C1 – C4 range, but also heavier hydrocarbons. Especially with mixed gas species there is little hope to obtain guest distributions from powder data. However, since many natural samples are well annealed, some on a geological timescale, there is a chance of finding single crystals in natural deposits, and indeed this was done successfully for a hydrate sample recovered from the hot vent area[43]. The quality of the diffraction data for these well annealed samples is extraordinary and populations of mixed guests in a sII hydrate could be determined from the single crystal data[45]: C1 -66.6%, C2 - 22.2%, C3 - 11.1%; from gas analysis of a polycrystalline decomposed hydrate sample – C1 72.1 %, C2 - 14.8%, C3 - 7.4 %, C4 and iC4 – 4.5%, C5 – 0.67%, C6- 0.19%, C7- 0.18%)[45]. The polycrystalline sample contains a small quantity of sH hydrate which contains mainly methane and heavier hydrocarbons (C5-C8). C4 and iC4 are present in the sII crystal in quantities that are too small too resolve separately. The structure type and guest distribution determine the stability zone of the hydrate uniquely[45].

3 HYDRATE FORMATION

3.1 Stable and metastable states

We now have encountered a number of techniques that can be used to determine hydrate structure, either directly or indirectly. In order to study processes we must add some capability of time resolution so that the process can be followed, preferably from nucleation to some level of conversion of the aqueous or ice phase. The usual method by which hydrate has been made is to use a stirred reactor – often the method of choice in phase equilibrium studies.[13, 46] (fig. 3). However, as mentioned previously, it is difficult to obtain kinetic parameters that are system

Physics and Chemistry of Ice

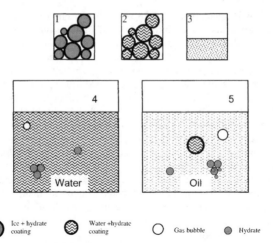

Figure 3   *Different methods of hydrate formation (1-5); each can give a distinct perspective on certain aspects of hydrate formation: 1-ice + gas at T< 273K; 2- ice +gas with the temperature ramped above T = 273K; 3 – amorphous ice +gas at T < ~ 130K; 4 and 5, either quiescent or with agitation are usually used for phase equilibrium studies and the effect of inhibitors. Simulation of natural gas hydrate requires the addition of sediment at various levels of water-sediment content.*

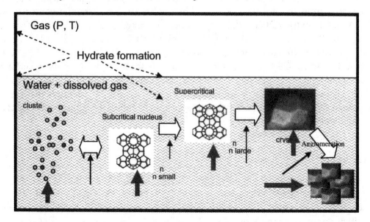

Figure 4   *A commonly used conceptual model for the formation of gas hydrate from water and gas. Since gas is only poorly soluble in water, hydrate forms on surfaces as well as in the bulk. Standard nucleation theory suggests that water-gas clusters form sub-critical nuclei/clusters of the thermodynamic product which at a certain minimum size become large enough to survive. Once crystals form they will grow and/or agglomerate.*

independent because of the unique experimental set up in terms of heat and mass transfer that each researcher has assembled. Simpler approaches include the use of quiescent systems without agitation, the reaction of powdered ice with gas, the reaction of powdered ice with gas followed by ramping the temperature up above the ice point, and the mimicking of hydrate formation in a pipeline which has oil, water and gas phases and which may or may not use agitation (fig. 3). All of these systems are different and have quite distinct control parameters. We can examine a general model for hydrate formation and then see what kind of relevant information is available. The model shown in fig. 4 suggests that we have to deal with a starting state[23], a final product and two identifiable intermediate states – so, five material states, and four processes. Some measurements on the initial and final crystal states are available; however, much about the intermediate states has to be inferred by examining the final product as a function of time after nucleation. Both powder diffraction and spectroscopic measurements have been made to follow the conversion of powdered ice and liquid water to hydrate.

As solid – gas reactions are slow, information on complex products, for instance in the case of the formation of N_2, CH_4 and CO_2 hydrates[15, 22, 47] can often be made by diffraction, with the caution that if non-crystalline intermediates exist these may be difficult to identify. One of the earliest attempts to follow hydrate formation from the study of guest distribution as a function of time came from the application of a new technology that made Xe an easily observed nucleus in NMR spectroscopy, allowing information to be obtained on local order [14]. Figure 5 shows that for the reaction of xenon with powdered ice to give sI Xe hydrate, after an induction period there is a period of rapid growth. During this time the large cage to small cage occupancy ratio, which is close to four for the equilibrium crystalline product, is achieved only after a time of ~ 100 seconds or so. At earlier times, the ratio is very much in favour of there being many more small cages than required for sI hydrate, indicating an intermediate quite different from the equilibrium structure.

A second example comes from the reaction of Xe with powdered THF hydrate[21]. It is easy to see that a hydrate of sII forms quickly on the surface as it is not necessary to form to nucleate a new phase. Since the activity of the Xe guest is much greater than that of THF under the initial experimental conditions, Xe replaces THF in the surface layers of the hydrate. This metastable hydrate persists for several hundred seconds before it converts completely to sI Xe hydrate (which does have to nucleate) plus a mixed THF-Xe hydrate. Such experiments reveal that hydrates are extremely labile at temperatures above ~200°C, and that metastable hydrates may be important in the early stages of hydrate formation.

Spectacular evidence to that effect was offered also by recently published work by Staykova et al[22] and Schicks et al [20]. In the first instance, diffraction showed that below the ice point, sII CO_2 hydrate formed as well as the stable sI hydrate, and that the metastable sII phase disappeared only slowly. In the second case, a Raman spectroscopic study of methane hydrate [20] formed under quite moderate conditions showing that sII methane hydrate formed along with sI and that it was quite persistent (fig. 6). One may well postulate that in the first Xe experiment mentioned above [14] and in the CO_2 and methane hydrate experiments [15, 20] the small cavities played a major role in determining which structure appeared first. Since there are many more small cavities in sII (S:L = 2:1) than sI (S:L = 1:3) hydrate, sII may well be the favoured kinetic product for small guests. Using a similar argument, one may in fact suggest that under certain conditions a sH hydrate or other novel hydrate may well appear during the early stages of a reaction although sI hydrate is the stable phase.

We can conclude even from the limited information at hand that hydrates are extremely labile materials, especially at surfaces where hydrate responds almost instantaneously to changing thermodynamic conditions [11, 21]. Away from the surface, mass transport becomes a limiting factor [11,22,47]. It is quite clear that in order to understand hydrate formation mechanisms it becomes very important to understand the structure and composition of the product during the various stages of the formation process. An as yet little understood area is the relationship between kinetics (mass and heat transfer) and crystal growth, leading to a variety of crystal morphologies.[48-51].

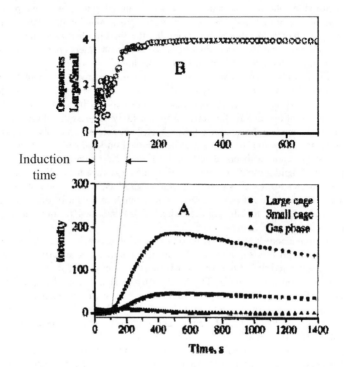

Figure 5   *Formation of sI Xe hydrate on the surface of powdered ice as a function of time. The bottom figure shows the hydrate cage occupancies as well as the gas line. There is an induction time of ~ 100s, followed by a period of rapid growth. Hydrate growth stops after ~ 400 sec. The decay in intensity after that is due to spin-lattice relaxation of the Xe NMR signal. The top figure shows the cage occupancy ratio as a function of time. After ~100 s. the ratio ghas a value of ~4, close to that expected for Xe hydrate made under equilibrium conditions. During the first 100s, the ratio decreases to values of 1+/- 0.8, indicating that during the induction time the product is quite different from sI hydrate with many more small cages than expected for sI. From ref. 14.*

3.2 Nucleation and the memory effect

Hydrate nucleation has often been treated as the early part of a hydrate forming reaction that precedes rapid growth[13]. As there is a strong random component that determines nucleation (depending on driving force)[52] it is difficult to study experimentally. It was established many years ago that in bulk water, ice almost always nucleates heterogeneously[53] upon freezing. Homogeneous nucleation takes place at temperatures below 240K, and then only in suspensions of micron-sized droplets where the vast majority of the droplets are impurity-free. Only recently has it been shown that hydrate nucleation is similar in that homogeneous nucleation likely does not take place in any situation commonly encountered[54]. Measurements of the effect of some hydrate inhibitors on homogeneous nucleation shows that these change the nucleation temperature very little, so that the memory effect and kinetic inhibition are not primarily a structural effect and must operate by inhibiting growth or by interfering with heterogeneous nucleation [54]. The former mode of operation can be studied by examining crystal morphology

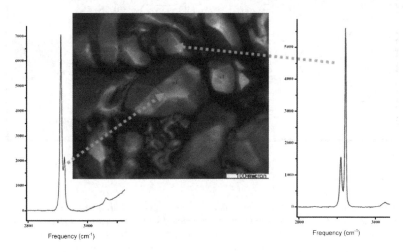

Figure 6 *The CH₄ C-H symmetric stretch region from confocal Raman spectroscopy for methane in hydrate cages for the product of the reaction of methane and water. The spectra suggest that crystals of both sI and sII hydrate form, the latter being metastable. From ref. 20*

and growth as well as by measuring the rate of gas uptake after nucleation, the latter by measuring induction times for nucleation [54,55.] A model for the memory effect, where impurities are imprinted by hydrate formation that makes them more effective nucleators when hydrate is reformed has been presented[54]. It has also been suggested that additives meant to act as kinetic inhibitors of hydrate formation should be reclassified as to their ability to a) inhibit nucleation, b) inhibit growth, and c) eliminate the memory effect.[56] Modeling [57-59] has been used to address the various aspects of nucleation and kinetic inhibition by adsorption, and it is hoped that by linking the models to experiment this difficult area will become better understood.

4 SUMMARY

As can be ascertained from this summary of the state of the art of understanding hydrate formation, it is very much a work in progress. Figure 7 shows another conceptual hydrate formation model that incorporates the various observations that have been made in the paper as summarized below. New insights that are becoming part of current thinking, or that need to be considered can be stated as follows.

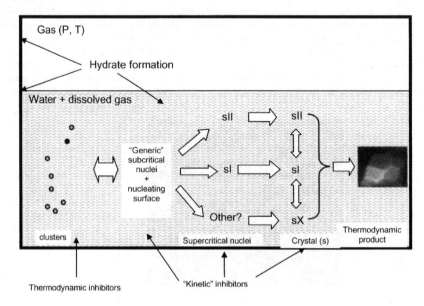

Figure 7   *A conceptual model that takes into account recent experimental data. It shows that subcritical nuclei should always be associated with a nucleating surface and that after nucleation, a variety of structures may result. The metastable structures, likely including kinetic products, will eventually convert to the thermodynamically stable product. The arrows underneath show where additives for inhibition are active. The "Kinetic" inhibitors may interfere with both nucleation and growth at any stage of crystal growth.*

1) Nucleation of hydrate in all commonly encountered situations is a heterogeneous process. Hence the prevention of hydrate formation, including nucleation, growth and elimination of memory effects, should be considered as involving impurities.
2) Nucleation of hydrates with small guests (Xe, CH_4, CO_2) appear to involve mainly small cages. As well, the hydrate structures for CH_4 and CO_2 guests that appear after nucleation include not only the thermodynamically stable sI hydrate, but also sII hydrate, which may be a kinetic product. This implies that it is not possible to predict the nature of the initial product after nucleating hydrate phases. Another point that can be made is that if the nucleation of hydrates with small guests involves small cages, there must be a

guest-size dependent effect on nucleation – how do hydrates made with larger guests nucleate?

3) Hydrate formation can appear to be a homogenous or quite inhomogeneous process depending on the measurement technique used. Measurements made on bulk samples tend to show an apparently homogeneous reaction as judged by gas uptake curves which can be fit to a variety of models. However, measurements on a local scale show that conversion of water to hydrate can be fast in one region of a sample and much slower in a different region with sudden increases in reaction rate that are characteristic of renewed cycles of nucleation. In that case the reaction only looks homogeneous because of averaging over many local environments.

Acknowledgments

Much of the work was contributed by my co-workers and former co-workers at NRC: Chris Ratcliffe, Igor Moudrakovski, Konstantin Udachin, Hailong Lu, Huang Zeng, Yu-Taek Seo, Jong-won Lee, Lee Wilson, Chris Tulk, John Tse, Dennis Klug, Graeme Gardner, as well as collaborators elsewhere: Virginia Walker (Queen's University), Huen Lee (KAIST), Judith Schicks (GFZ Potsdam), all gratefully acknowledged. I also would like to acknowledge Scott Dallimore (GSC) Fred Wright (GSC), Roy Hyndman (GSC), Michael Riedel (GSC), Tim Collett (USGS), Ross Chapman (University of Victoria), George Spence (University of Victoria), Rick Coffin (Naval Research Labs) for the provision of well-preserved natural hydrate samples recovered from a variety of locations.

References

1   K. A. Kvenvolden, *Proc. Nat. Acad. Sci USA* 1999, **96**, 3420
2   D. W. Davidson in *Water. A Comprehensive Treatise*, Ed F. Franks, Plenum, New York, 1972, Vol. 2
3   G. A. Jeffrey *Comprehensive Supramolecular Chemistry* 1996, **6**, 763.
4   R. Kleinberg, P. Brewer, *Am. Sci.* 2001, **89**, 244
5   E.D. Sloan, Jr, *Hydrates of Natural Gas*, 2nd ed'n, Marcel Dekker, New York, 1998
6   J. S. Gudmundsson, V. Anderson, O. I. Levik, M. Mork, *Ann. N. Y. Acad. Sci.* 2000, **912**, 403
7   G. Ginsberg and V. A. Soloviev, *Submarine Gas Hydrates*, VNIIOkeangeologia, St. Petersburg,1998; A. Milkov, *Earth Sci. Rev.* **2004**, 66, 183
8   Scientific Results from JAPEX/JNOC/GSC Gas Hydrate Research Well, Mackenzie Delta, Northwest Territories, Canada, *Geol. Survey of Canada Bull.* 1999, **544**
9   W. Xu, R. P. Lowell, E. T. Peltzer, *J. Geophys. Res.* 2001, **106**, 26413
10 W. P. Dillon, J. W. Nealon, M. H. Taylor, M. W. Lee, R. M. Drury, C. H. Anton, In: *Natural Gas Hydrates: Occurrence, Distribution, and Dynamics.*, eds. C. K. Paull and W. P. Dillon, AGU monograph
11 H. Lee, Y. Seo, Y.-T. Seo, I. L. Moudrakovski, J. A. Ripmeester, *Angew. Chem. Int'l Ed'.* 2003, **41**, 5048
12 A. J. Barduhn, *Chem. Eng. Proc.* 1967, **63**, 98
13 P. R. Bishnoi, V. Natarajan, N. Kalogerakis, *Ann. N. Y. Acad. Sci.* 1994, **715**, 311; P. Servio, F. Lagers, C. Peters, P. Englezos, *Fluid Phase Equil.* **1999**, 158, 795

14 I. L. Moudrakovski, A. A. Sanchez, C. I. Ratcliffe, J. A. Ripmeester, *J. Phys. Chem. B* 2001, **105**, 12338

15 W. F. Kuhs, B. Chazallon, P. G. Radaelli and F. Pauer, *J. Incl. Phenom.* 1997, **29**, 65; B. Chazallon, W. F. Kuhs, *J. Chem. Phys.* 2002, **117**, 308.

16 H. Nakayama, D.D.Klug, C. I. Ratcliffe and J. A. Ripmeester, *Chem. Eur. J.* 2003, **9**, 2969

17 G. D. Holder, D. J. Mangiello, *Chem. Eng. Sci,* 1982, **37**, 9

18 E. M. Hendriks, B. Edmonds R. A. S. Moorwood, R. Szczepanski, *Fluid Phase Equil.* 1996, **117**, 193

19 L. Ballard, E. D. Sloan, *Fluid Phase Equil.* 2004, **216** , 257

20 J. M. Schicks, J. A. Ripmeester, *Angew. Chem. Int'l. Ed'n.* 2004, **43**, 3310

21 I. L. Moudrakovski, C. I. Ratcliffe, J. A. Ripmeester, *Angew. Chem. Int. Ed'n,* 2001, **40**, 3890.

22 D. K. Staykova, W. F. Kuhs, A. N. Salamantin T. Hansen, *J. Phys.Chem. B,* 2003, **107** 10299; W. F. Kuhs, D. K. Staykova and A. N. Salamantin, *J. Phys. Chem. B*, 2006, **110** 13283

23 P. Buchanan, A. K. Soper, H. Thompson, R. E. Westacott, J. L. Creek, G. Hobson. C. A. Koh, *J. Chem. Phys.* 2005, **123**, 164507

24 S. Subramanian R. A. Kini, S. F. Dec E. D. Sloan, *Chem.Eng. Sci,* 2000, **55**, 1981

25 D. W. Davidson, Y. P. Handa, C. I. Ratcliffe, J. S. Tse, B. M. Powell, *Nature*, 1984, **311**, 142

26 J. S. Tse, Y. P. Handa, C. I. Ratcliffe, B. M. Powell, *J. Inclusion Phenom.* 1986, **4**, 235

27 J. A. Ripmeester, J. S. Tse, C. I. Ratcliffe, B. M. Powell, *Nature*, 1987, **325**, 135

28 J. A. Ripmeester, C. I. Ratcliffe, *J. Phys. Chem.,* 1990, **94**, 8773

29 M. von Stackelberg, W. Z. Jahns, *Elektrochem. Z*, 1954, **58**, 162

30 R. Ohmura, M. Ogawa, K. Yasuoka, Y. H. Mori, *J. Phys. Chem. B* 2003, **107** 5289

31 S. Takeya, A. Hori, T. Hondoh, T. Uchida, *J. Phys. Chem. B*, 2000, **104**, 4164

32 P. M. Rodger, *Ann. N. Y. Acad. Sci.* 2000, *912*, 474.

33 H. Zeng, L. D. Wilson, V. K. Walker and J.A. Ripmeester, *Can. J. Phys.* 2003, **81**, 17

34 J. A. Ripmeester, C. I. Ratcliffe, *J. Phys. Chem.* 1988, **97**, 337.

35 A. K. Sum, R. C. Burruss, E. D. Sloan, *J. Phys. Chem.* 1997, **101**, 7371

36 K. A. Udachin, C. I. Ratcliffe, J. A. Ripmeester, *J. Supramol. Chem.* 2002, **2**, 405.

37 I. L. Moudrakovski, G. E. McLaurin, C. I. Ratciffe, J. A. Ripmeester, *J. Phys. Chem.* 2004, **108** 17591

38 38. D. W. Davidson, S. K. Garg, S. R. Gough, Y. P. Handa, C. I. Ratcliffe, J. A. Ripmeester, J. S. Tse, W. F. Lawson, *Geochim. Cosmochim. Acta* 1986, **50**, 619

39 C. A. Tulk, C. I. Ratcliffe, J. A. Ripmeester, *Geol. Survey of Canada Bulletin*, 1999, **544**, 251

40 40. Scientific Results from the Mallik 2002 Gas Hydrate Production Research Well Program, Mackenzie Delta, Northwest Territories, Canada. Ed. S. R. Dallimore, T. S. Collett, *Geol. Survey of Canada Bulletin* 2005, **585**

41 41. H. Lu, I. L. Moudrakovski, M. Riedel, G. Spence, R. Dutrisac, J. A. Ripmeester, F. Wright, S. R. Dallimore, *J. Geophys. Res.* 2005, **110**, B10, 204

42 M. Riedel, G. D. Spence, N. R. Chapman, R. D. Hyndman, *J. Geophys. Res.* 2002, **107**, B9, 1.

43 R. Chapman, J. Pohlman, R. Coffin, J. Chanton, L. Lapham, *EOS*, 2004, **85**, 361

44 J. H. van der Waals, J. C. Platteeuw, *Adv. Chem. Phys.*, 1959, **2**, 1

45 45. H. Lu, I. L. Moudrakovski, J. A. Ripmeester, K. A. Udachin, R. Chapman, R. Coffin, J. Pohlman, unpublished results.

46 46. J. D. Lee, R. Susilo and P. Englezos, *Energy and Fuels* 2005, **19**, 1008

47 47. R. W. Henning, A. J. Schultz, V. Thieu and Y. Halpern, *J. Phys. Chem A,* 2000, **104**, 5066

48 48. P. Servio, P. Englezos, *AIChE Journal* 2003, **49**, 269

49 49. C. A. Knight, K. Rider, *Phil. Mag. A*, 2002, **82**, 1609

50 50. R. Ohmura, W. Shimada, T. Uchida, Y. Mori, S. Takeya, J. Nagao, H. Minagawa, T. Ebinuma, H. Narita, *Phil. Mag.* 2004, **84**, 1.

51 51. L. A. Stern, S. H. Kirby, S. Circone, W. B. Durham, *Am. Mineral.* 2004, **89** 1162.

52 52. P. Englezos, *Rev. Inst. Francais Petr.* 1996, **51**, 789

53 53. N. H. Fletcher, *The Chemical Physics of Ice*, Cambridge University Press, 1970; P. V. Hobbs, *Ice Physics*, Oxford University Press, Oxford, 1974, F. Franks, *Biophysics and Biochemistry at Low Temperatures*, Cambridge University Press, 1985

54 H. Zeng, L. Wilson, V. K. Walker, J. A. Ripmeester, *J. Am. Chem. Soc.* 2006, 128, 2844

55 R. Larsen, C. A. Knight and E. D. Sloan, *Fluid Phase Equil*, 1998, 150, 353

56 H. Zeng, V. K. Walker, J. A. Ripmeester, Proc. 5th Int'l Conf. Gas Hydr., June 13-16, 2005, Trondheim, Norway, 1295

57 C. Moon, P. C. Taylor, P. M. Rodger, J. Am. Chem. Soc. 2003, 125, 4706

58 B. J. Anderson, R. Radhakrishnan, J. W. Tester, B. L. Trout, Prepr. Div'n Petr. *Chem. ACS*, 2005, 50 , 50

59 G. Tegze, T. Pusztai, G. Toth, L. Grasy, A. Svandal, T. Buanes, T. Kuznetsova, B. Kvamme, *J. Chem. Phys.* 2006, 124, 234710

FAST THERMAL DESORPTION SPECTROSCOPY AND MICROCALORIMETRY: NEW TOOLS TO UNCOVER MYSTERIES OF ICE

V. Sadtchenko

Chemistry Department, The George Washington University, 725 21st Str. NW, Washington, DC 20052, USA

1 INTRODUCTION

Due to the crucial role of aqueous chemistry in a variety of environmental, biological, and industrial processes, experimental studies of ice remain an important field of modern physical chemistry. Examples of applied areas, which require knowledge of the physics and chemistry of various solid forms of water, include: atmospheric chemistry and climate change, soil chemistry, planetary and interstellar chemistry, cryopreservation, and research into alternative energy sources. Furthermore, ice and water are uniquely accessible to computational modelling, due to the availability of extensive information on the intermolecular interactions in water-containing systems. Therefore, ice is often considered to be a model system for studies of fundamental properties of condensed molecular phases.

Despite centuries of scientific inquiry many fascinating properties of ice still await a comprehensive explanation. The lack of fundamental understanding of physics and chemistry of condensed aqueous phases is due to the numerous challenges involved in laboratory studies of such systems. Many of the standard analytical methods that have been developed to investigate reactions and dynamics in solids and liquids are not sufficiently sensitive, fast, or selective to provide detailed characterization of processes of interest in varying aqueous environments. The failure of contemporary experimental science to address many fundamental questions about the nature of aqueous phases is reflected in the current jargon, which includes a notion of "No Man's Land", i.e., an "experimentally inaccessible" temperature region in studies of properties of amorphous ice. Without a doubt, the development of new experimental techniques for studies the dynamics and reactions in various condensed aqueous phases is long overdue, and, therefore, has been the focus of our research program at the George Washington University. Over the past five years, we have developed a set of novel experimental techniques, which we collectively refer to as Fast Thermal Desorption Spectroscopy - Ultrafast Scanning Calorimetry (FTDS-USC), and which we have already used to obtain data crucial for resolving several long-standing controversies in the field. The following sections of this article contain a description of our apparata, and the main venues of the research in our laboratory.

2 OVERVIEW OF EXPERIMENTAL TECHNIQUES

Over the past decade Thermal Desorption Spectroscopy (TDS), in its various modifications, has been widely used to study adsorption, desorption, diffusion, and reactions in thin (0.01-1 μm) aqueous films under high vacuum conditions [1-14]. In these types of experiments, thin ice films of various phase composition, saturated with a variety of chemical dopants are grown on a flat substrate, at cryogenic temperatures under vacuum using effusive or supersonic molecular beam sources. Because of the low temperatures at which the films are deposited reactions usually do not proceed during film deposition. After the film growth, the temperature of the substrate is raised in a controlled fashion, and vaporization of the film (accompanied by diffusion and reactions of dopant species) is initiated. During film's vaporization, a sensitive and selective analytical technique, or a combination of several such techniques are used to monitor chemical and physical phenomena in the film. For example, in Temperature Programmed Reaction Mass Spectrometry (TPRMS) experiments, the volatile chemical species, which evolve from the film, are monitored with a sensitive mass spectrometer.

There are several factors, which make the TDS-based approach particularly suitable for experimental studies designed to gain *molecular-level* information on ice properties. First of all, a TDS approach allows precise control of the film's morphology and phase composition. Indeed, amorphous microporous, amorphous solid, crystalline cubic, or crystalline hexagonal films can be grown on the variety of substrates [1-13]. Second, deposition of ice at cryogenic temperatures makes it possible to create intricate, non-equilibrium, spatial and chemical distributions of dopant species in the film. For example, a sandwich-like film consisting of various isotopes of water can easily be grown. The isotopic scrambling in such a film at higher temperatures has been used to determine transport properties of ice on the nanoscale [6,8]. Third, a wide variety of analytical techniques can be applied in order to monitor chemical transformations in thin ice films. Examples of analytical techniques used to study the aqueous surface at cryogenic temperatures are numerous and include mass-spectrometry, vibrational spectroscopes, photoelectron spectroscopy, atomic force microscopy, proton scattering etc. Finally, due to high vacuum conditions characteristic of TDS experiments, surface and condense phase phenomena are not obscured by processes in the gas phase.

In spite of all these powerful features, the TDS approach has its obvious limitations. In the case of ice, chemical phenomena can be studied with the TDS technique only at temperatures below -80 ^{0}C, where the volatility of this molecular solid is relatively low. One of the particularly nasty practical problems, arising during attempts to extend the temperature range of the TDS studies, is the formation of a gas boundary layer above the ice free surface. Such a layer forms as a result of rapid vaporization of the macroscopic ice sample and is due to collisions and back scattering of water and other desorbing molecules. The vapor layer above the surface of ice samples in high temperature TDS experiments eliminates the main advantage of this approach, i.e., its ability to exclude the gas phase phenomena from interpretation of the experimental results. Aqueous phase phenomena at temperatures below -80 ^{0}C are, indeed, of great fundamental, as well as applied significance. Nevertheless, due to the likelihood of fundamental changes in the nature of aqueous interfaces near ice melting point [14-16], the results of low temperature studies may not be automatically extrapolated to higher temperatures typical for many natural environments.

Several years ago, we found a solution to the problem described above. The central idea of our experimental approach was inspired by the pioneering work of Faubel and Kisters [17], who reported TOF spectra of H_2O and CH_3COOH molecules evaporating at near ambient temperatures from the surfaces of 10-50 μm thick jets of water injected into a vacuum chamber. These TOF measurements showed that when the diameter of the water jet was reduced from 50 to 10 μm, the TOF distributions changed from that characteristic of a supersonic molecular beam to a Maxwellian distribution representative of the collision-free molecular flow. Using this important result, we have achieved a critical improvement of the TDS technique by changing the substrate geometry[18,19]. Instead of using a flat substrate for ice film deposition, we grow ice on a thin (~ 10 μm) tungsten filament. A number of advantages instantly follow:

First, because desorbing molecules undergo only a few gas phase collisions the formation of a dense boundary vapor layer over the ice sample surface does not occur at temperatures up to 0 °C. Second, the small surface area of ice films grown on a thin wire results in a small net flux of desorbing molecules, which makes it possible to keep the load on a vacuum system at manageable levels. Third, due to small mass of thin wire, heating rates in excess of 10^5 °C·s⁻¹ can be easily achieved, making it possible to bring the temperature of a few micrometer thick aqueous films to a value as high as 0 °C *before* vaporization of a significant fraction the ice sample. Finally, because the mass of the microscopic filament is small and the heat capacity of tungsten is about an order of magnitude lower than the specific heat capacity of any condensed aqueous phase, scanning calorimetry can be used to measure the variations in the heat capacity of the aqueous film during rapid heating. As we discuss later in this article, high heating rates achieved in our experiments provide a unique opportunity for studies of molecular kinetics in amorphous ice.

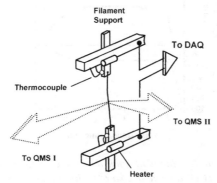

Figure 1 *The essentials of the Fast Thermal Desorption Spectroscopy apparatus*

A detailed description of apparatus and standard experimental procedures can be found in our publications[21]; therefore, only a brief account is given here. As illustrated in Figure 1, our ice films are grown on the surface of a tungsten filament (10 μm in diameter, 2 cm long), which is spot-welded to the supports of the filament assembly. The supports, while in thermal contact with the liquid nitrogen-cooled heat sink, are electrically isolated from the rest of the apparatus. The filament assembly includes a thermal control system capable of maintaining the filament supports at temperatures in a range from -160 °C to -120 °C. The assembly is surrounded by cryogenic shields (not shown in Fig. 1) and positioned in a vacuum chamber. The cryogenic shields significantly reduce the gas load on the pumping system of the apparatus, allowing it to maintain pressures below $3 \cdot 10^{-7}$ Torr, during all stages of FTDS experiments.

Water vapor and dopants are delivered from the vapor source to the filament via 12 effusive dosers (not shown in Fig. 1), which are equally spaced around the filament at a distance of approximately 5 cm from its centre. The intersecting vapor beams from the dosers create a relatively dense H_2O vapor cloud around the tungsten filament, facilitating deposition of uniform ice films with a neat cylindrical geometry[18,19]. After deposition, rapid isothermal vaporization of the film is initiated by applying a difference of a few volts across the filament. In the course of the entire FTDS experiment, the voltage across the filament and the current through the filament are recorded every four microseconds by a custom designed data acquisition (DAQ) system. The temperature of the filament is then calculated from the resistance data using the temperature coefficient of electrical resistivity of tungsten. During FTDS experiments, the kinetics of the vaporization of various chemical species from the film are monitored simultaneously with three detectors: a quadrupole mass spectrometer (QMS) positioned 10 cm away from the filament, a QMS positioned 95 cm away from the filament, and a Fast Ionization Gauge (FIG), positioned 6 cm away from the filament. The apparatus also employs a single-slit chopper disc, which facilitates Time-of-Flight characterization of the velocity distribution of the vaporization products.

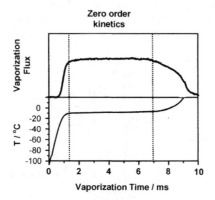

Figure 2 *H_2O flux from the filament and the corresponding temperature of the filament. The H_2O flux was measured with the fast ionization gauge (FIG), positioned 6 cm away from the filament.*

Figure 2 illustrates the basic thermal desorption experiment [18,19]. It shows the H_2O flux from the filament and the corresponding temperature of the filament, as determined from resistance data. The H_2O flux was measured with the FIG. As shown in the figure, applying a small (3.5 V) electrical potential across the filament results in rapid heating of the ice-filament system during the first millisecond of the experiment. Nevertheless, after approximately 2 ms, the filament achieves the steady-state vaporization regime, which is characterized by a nearly constant temperature and H_2O vaporization flux. At this stage of the experiment, the heat generated by the filament is carried away by vaporization of ice from its surface. The film vaporization follows zero-order kinetics for the next 5 ms, i.e., until 80% of the ice sample is gone, then the film breaks into clusters and islands on the filament surface resulting in non-zero kinetics, and the increase in the filament temperature.

Observation of zero-order vaporization kinetics leads to several important conclusions. First, during steady-state vaporization near 0 °C, the ice adlayer on the surface of the filament is a neat film free of *large* pores or ruptures, i.e., the film covers the entire filament surface. Second, the temperature gradient *along* the length of the filament is negligible, except within a fraction of a millimeter near the ends of the filament. Finally, since the temperature of the filament does not vary significantly with the film thickness, during vaporization, the temperature gradient in the film must be less than a few degrees [18,19]. Because the typical time scale of our desorption experiments is up to four orders of magnitude shorter than that in classical TDS studies, we have termed our technique, Fast Thermal Desorption Spectroscopy or FTDS.

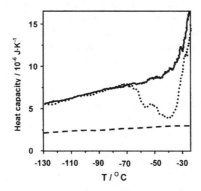

Figure 3 *Typical results of ultrafast calorimetry measurements. Dashed line: the effective heat capacity of the ice-free filament. Solid line: the effective heat capacity of the filament with an H₂O film vapor-deposited at – 120 °C. Dotted line: the effective heat capacity of the filament with an H₂O film vapor-deposited at -150 °C.*

Finally, we would like to explain the basics of our calorimetry technique, which is an integral part of the FTDS apparatus, and which makes it possible to determine mass and phase composition of our ice samples [18,19]. In a way similar to conventional scanning calorimetry studies, the combined effective heat capacity of the ice film and the filament is calculated in our experiments as the ratio of the power generated by the filament to the first time derivative of the temperature. However, the heating rates in our studies are more than six orders of magnitude greater. Taking into account the extremely high heating rates, and microscopic physical dimensions of our ice samples, we refer to our calorimetric technique as Ultrafast Scanning Calorimetry (USC), ultrafast microcalorimetry, or fast microcalorimetry. Figure 3 illustrates typical USC data. The dashed line shows the effective heat capacity of the ice-free filament; the solid line shows the effective heat capacity of the filament with an H_2O film vapor-deposited at – 120 °C; and the dotted line shows the effective heat capacity of the filament with an H_2O film vapor-deposited at -150 °C. In the case of deposition at -120 °C, the monotonic increase in heat capacity of the filament-film system, *i.e.*, the lack of exothermic transitions characteristic of crystallization, demonstrates that the ice sample is crystalline [18]. However, in the case of films deposited at -150 °C, the effective heat capacity of the film shows partially overlapping irreversible exotherms, which are due to crystallization of the initially amorphous ice samples [18]. Assignment of the exotherms shown in Fig. 3 to crystallization of amorphous ice is facilitated by the 2 kJ/mol enthalpy release, which is close to previously reported values[20]. These calorimetric experiments also show that non-crystalline ice films always undergo crystallization before steady state vaporization can occur at temperatures above -25 °C. In addition to providing information on the phase composition of ice samples, the data shown in Fig. 3 were used to determine the masses of the ice films. Assuming that the film has the specific heat capacity and density similar to that of hexagonal ice we can estimate the thickness of the ice films with accuracy of ± 10 % [18,19].

3 CURRENT RESULTS

In addition to the well-known benefits of thermal desorption spectroscopy, the apparata described in the previous section has another important advantage. Unlike any other technique, the FTDS-USC setup, in its full configuration, makes it possible to study ice in the entire temperature range from cryogenic to near ambient. While information on phase transitions, transport phenomena, and reactions near 0 °C can be studied with FTDS directly, USC can be employed to investigate phenomena in amorphous ice at cryogenic temperatures. Here, we provide a review of the current research projects. We begin with USC investigations of the molecular kinetics in amorphous ice.

3.1 USC Studies of Amorphous Solid Water

Water is the most important yet unusual liquid found in nature. The anomalous properties of water are most noticeably manifested when water is supercooled below its equilibrium freezing point. The constant-pressure, specific heat capacity, the coefficient of thermal expansion, and the isothermal compressibility of water begin to diverge as the temperature approaches 228 K (T_s). Development of realistic models of the structural transformations in water upon supercooling requires measurements of its properties (i.e. diffusivity, viscosity, heat capacity, etc.) at temperatures near and below T_s. However, such measurements have proven to be a great experimental challenge. Unless the cooling rate is exceedingly high (10^7 K·s^{-1}) and the sample dimensions are microscopic, crystallization becomes inevitable, as the temperature of supercooled water is lowered to the vicinity of T_s [22,23].

An alternative approach to the measurements described above utilizes glassy water samples grown by slow H_2O vapor deposition. The amorphous solid water (ASW) samples, grown at cryogenic temperatures and heated above the postulated glass transition, offer a way to obtain measurements of supercooled water properties at temperatures below T_s. Unfortunately, as the temperature of an amorphous sample is raised to about 150-160 K, the ASW rapidly crystallizes to form cubic ice[20]. Thus, the characteristic temperature of rapid crystallization of low density amorphous ice ($T_c \sim 160$ K) and the temperature of catastrophic crystallization of supercooled water ($T_s \sim 228$ K) mark the borders of a temperature region dubbed "No Man's Land", where experimental studies of non-crystalline states of water have been deemed to be impossible[21]. With the objective of gaining insights into the properties of deeply supercooled water, we utilized the capabilities of our ultrafast microcalorimetry technique to investigate thermodynamic properties of ASW, under conditions of ultrafast heating. Due to the high heating rates used in our experiments, we were able to conduct measurements of ASW heat capacity at temperatures as high as 205 K.

The thermogram in Figure 4 shows the most important results of our experiments. It presents the measured heat capacities of ASW ice at temperatures significantly higher than

T_c. The heat capacity of a crystalline (cubic) ice sample of *the same mass* is shown for comparison (dotted line). The exothermic peaks in the thermogram of the ASW sample are due to its crystallization and prove that our ASW samples contain a low fraction of crystalline ice. The endothermic upswing in heat capacity is due to the onset of rapid ice vaporization at temperatures above 220 K. The most striking feature of the thermogram shown in Fig. 4 is the lack of significant differences in the heat capacities of amorphous and crystalline aqueous phases at temperatures between 160 and 205 K, i.e., above the glass transition temperature of 136 K clamed in some previous studies[24,25]. According to the predictions of heat capacity dependence on temperature, which are based on the glass transition temperature of 136 K, the heat capacity of ASW must rapidly deviate from that of crystalline ice[26]. Our measurements show less than a few percent difference between heat capacities of ASW and cubic crystalline ice at temperatures up to 205 K, which is in direct contrast with the predictions.

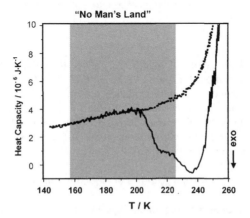

Figure 4 *Heat capacity of crystalline (dots), and ASW (line) ice films*

In order to demonstrate that it is actually possible to observe thermal phenomena characteristic of supercooled liquids with our microcalorimetry approach, we have conducted measurements of the heat capacity of ASW samples contaminated with acetic acid. Figure 5 compares the thermogram of pure ASW to that of ASW/Acetic Acid mixture (10:1). While the heat capacity of the ASW/Acetic acid mixture undergoes a rapid increase in heat capacity at 175 K, the heat capacity of pure ASW remains nearly equal to that of crystalline ice.

The data shown in Fig. 5 makes it possible to estimate the enthalpy relaxation time in the case of the ASW/Acetic acid mixture at 175 K (i.e. at the onset temperature of heat capacity rise). Taking into account the value of the heating rate in our experiments (10^5 $K \cdot s^{-1}$), we arrive at the conclusion that the relaxation time is on the order $10^{-5 \pm 0.5}$ s at 175 K for the ASW/Acetic acid mixture. Such an estimate is based on the assumption that a glass transition can be observed in a scanning calorimetry experiment only when the enthalpy relaxation time becomes shorter than a typical time of the experiment (note that the relaxation time decreases with temperature)[27]. Thus, because we do not observe any noticeable upswing in the heat capacity of the pure ASW sample, we conclude that even at temperatures as high as 205 K the enthalpy relaxation time for pure ASW must *exceed*

Physics and Chemistry of Ice

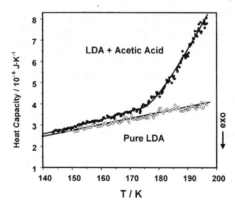

Figure 5 *Heat capacity of doped (solid circles) and pure (open circles) amorphous solid water*

$10^{-5\pm0.5}$ s, which is three orders of magnitude greater than that obtained assuming the existence of glass transition for pure ASW at 136 K[28]. In summary, our ultrafast microcalorimetry experiments provide evidence to support of the recent arguments by Angel[29] that the glass transition in ASW must occur at temperatures near 160 K, i.e., at temperatures consistent with rapid crystallization of ASW samples. Seemingly remarkable, this conclusion is in agreement with recent studies of water self diffusivity in ASW conducted Mullins and coworkers[13].

3.2 FTDS Studies of Vaporization Kinetics of Aqueous Films

In 1971 Davy and Somorjai presented their measurements of the desorption rate of ice up to – 40 ^{0}C [3]. These experiments have laid a foundation to the precursor mediated mechanism for ice desorption. The model predicts zero-order desorption kinetics in two limiting cases. At low temperatures, the mobile precursor's (MPs) desorption rate from the surface is low and the overall ice vaporization process is governed by pre-equilibrium between the MPs and the surface molecules. The effective activation energy is approximately the desorption enthalpy, which is approximately 50 kJ/mole. At higher temperatures, the rate of desorption of the MPs increases drastically destroying the equilibrium between the MPs and surface molecules. The formation of the mobile desorption precursors becomes the limiting step of the desorption process, which results in the effective activation energy of half the desorption enthalpy. Although many highly sophisticated measurements of ice vaporization rate were attempted in the past none of them were conducted at temperatures above - 40 ^{0}C.

Figure 6 shows the first direct measurements of absolute vaporization rate of crystalline D_2O and H_2O ice at temperatures above -40 ^{0}C obtained in our FTDS experiments[18]. Different symbols represent results from experiments with ice films of distinct thermal history and thickness. The solid lines in the Fig. 6 show the vaporization rate values calculated from ice equilibrium vapor pressure under the assumption that the mass accommodation coefficient is equal to unity, *i.e.* that the vaporization rate is equal to the maximum equilibrium rate given. The dotted lines show the range of possible desorption rate values predicted by the simple MP mechanism.

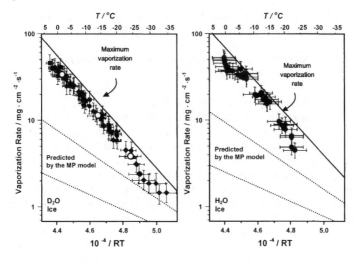

Figure 6 *Absolute vaporization rate of D_2O and H_2O ice. Different symbols represent the data obtained with polycrystalline ice samples of various thermal histories.*

Data shown in Fig. 6 lead to the following conclusions: at temperatures near 0 ^{0}C the desorption rate demonstrates Arrhenius behaviour with the effective activation energy of 50 ± 4 kJ/mol, which is much greater than the value predicted by the simple MP mechanism (25-35 kJ/mol). Also, at temperatures above -40 ^{0}C, the ice vaporization rate exceeds the desorption rate predicted by the simple MP model. In summary, the mobile precursor mechanism, as formulated by Somorjai and Davy, fails to describe the desorption kinetics of ice at temperatures near its melting point.

In the past we argued that the deviation of observed ice vaporization kinetics from a classical MP model can be explained by surface roughening or premelting transition at temperatures above -40 ^{0}C. Recent experimental studies have emphasized the complexity of water/vapor interface in respect to the dynamics of gas uptake and release which is now treated in terms of the nucleation theory[30]. Our work indicates that uptake and release of gases on *ice/vapor* interface at temperatures near ice melting point may be equally or even more complex as that on liquid surfaces. Because the uptake and release of gases on ice is likely to depend on ice surface morphology, theoretical description of these processes may require a new formalism that takes into account possible surface phase transitions at various temperatures. However, many questions remain unanswered. For example, it is still unclear at what temperature the onset of the premelting actually occurs. Is the formation of a liquid-like layer preceded by a roughening transition? How does it affect the ability of ice to trap and release various gases? Is the interface between the vapor and a quasi-liquid layer different from a neat water surface? These questions mark an exciting area for future studies and emphasize the importance of developing new experimental methods capable of probing the surface dynamics of volatile molecular solids near their melting points. Using Fast Thermal Desorption technique, we will continue investigation into vaporization kinetics of condensed aqueous phases. The next step in our work will be extensive investigations of vaporization kinetics of ice films doped with a variety of environmentally relevant impurities.

3.3 Studies of Transport Phenomena, Interfacial Phase Transitions, and Reactions in Polycrystalline Ice near Its Melting Point

In order to demonstrate that Fast Thermal Desorption spectroscopy can be used successfully to study reactions in volatile polycrystalline materials and to gain insights into nanoscale molecular transport in polycrystalline ice, we have conducted preliminary studies of H/D exchange kinetics near ice melting point. Figure 7 illustrates our approach.

As shown in the Figure 7, a thin layer of D_2O, initially positioned at a distance away from the film surface, evolves from the ice film only after vaporization of the overlaying H_2O^{16}. The time interval from the onset of H_2O vaporization to the appearance of the D_2O vaporization peak, i.e., the mean residence time of D_2O in the bulk of the H_2O film, gives the mean reaction time for H_2O/D_2O isotopic exchange. Positioning the D_2O layer further away from the surface, *i.e.*, increasing the mean reaction time, results in a gradual decrease of the magnitude of the D_2O peak. The decrease in the D_2O peak is accompanied by a gradual increase in the HDO yield, signifying rapid conversion of D_2O to HDO.

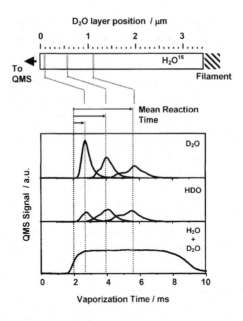

Figure 7 *Studies of H_2O/D_2O reaction. Top and middle panels: FTDS spectra of D_2O and HDO from 3.3 micrometer thick $H_2O/D_2O/H_2O$ ice films at -5 ^0C. Lower panel: vaporization kinetic of the entire film*

Figure 8 shows the H/D exchange kinetics between 50 nm thick D_2O layer and the surrounding H_2O^{16} polycrystalline ice at -5 ^0C. As shown in the figure, only half of the initial D_2O layer is converted into HDO on the time scale of our FTDS experiment. Analysis of the isothermal desorption spectra of HDO along with the observed reaction kinetics for D_2O layers of various thickness show that the H/D exchange is controlled by the rate of inter-diffusion of H_2O and D_2O in polycrystalline ice. According to our results, the H/D exchange reaction is limited to relatively narrow zone at H_2O/D_2O interface. Because the width of the reaction zone must be defined by the inter-diffusivity of H_2O and D_2O, and because the HDO and D_2O yields depend on the dimensions of the reaction zone, we can estimate the characteristic scale of inter-diffusion of D_2O and H_2O on the time scale of our experiments.

According to our estimates, the characteristic diffusion scale over 3 ms diffusion/reaction time is less that 50 nm at -2 ^0C. In the future, we will conduct similar experiments over the entire temperature range accessible with our Fast Thermal Desorption technique. Furthermore, we plan to investigate the influence of various environmentally relevant impurities on the water self-diffusivity in polycrystalline ice at various temperatures. Combining such data with an appropriate theoretical treatment will make it possible to gain valuable insight into interfacial phase transition and morphological dynamics in this important condensed phase system. In summary, the Fast Thermal Desorption Spectroscopy technique can be used to study the kinetics of reactions in polycrystalline ice near its melting point, and to obtain detailed information on nanoscale molecular transport in polycrystalline ice at temperatures near its melting point.

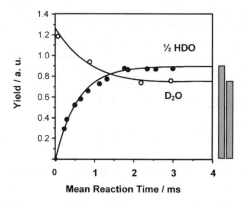

Figure 8 *Kinetics of H/D exchange between 50 nm thick D_2O layer and the surrounding polycrystalline H_2O ice at -5 ^0C. The ice film thickness was 3 μm. As shown in the figure, a significant fraction of initial D_2O film does not react with H_2O*

4 AFTERWORD

Several experiments reviewed in this article illustrate just a small fraction of opportunities for research into the fundamental physical and chemical properties of ice offered by our experimental approach. We emphasize that our core method, i.e., the FTDS can be combined with a variety analytical techniques. For instance, the existing apparatus can be equipped with an FTIR spectrometer for better initial characterization of phase and chemical composition of ice films during deposition. At the present time we are also developing an optical system that would allow us to conduct studies of photochemical reactions in polycrystalline ice. While our USC measurements were conducted with a fixed heating rate, further improvement of this particular component of our apparatus will make it possible to measure enthalpy relaxation times in doped polycrystalline ice for a wide range of temperatures. Finally, we are also in the process of designing a novel data acquisition system which would facilitate FTDS and USC measurements with ice films subjected to strong electrostatic fields (in excess of 10^9 V/m). We hope that at some point the FTDS-USC approach will become a standard tool for studies of reactions and dynamics in ice and other volatile molecular solids.

Acknowledgments

I gratefully acknowledge the talent and dedication of my students, H. Lu, S. McCartney, M. Chonde, and M. Brindza. I would like to thank the science committee of 11th PCI conference for the opportunity to present the results of our studies in Bremerhaven. The experiments described in this article were possible due to the generous support from the US National Science Foundation (award 0416091).

References

1   F. E. Livingston and S. M. George, *J. Phys. Chem.* 1998, **A 102**, 10280.
2   R. S. Smith, C. Huang, E. K. L. Wong, and B. D. Kay, *Phys. Rev. Lett.* 1997, **79**, 909.
3   J. Gordon Davy and G. A. Somorjai, *J. Chem. Phys.* 1971, **55**, 3624.
4   D. R. Haynes, N. J. Tro, and S. M. George, *J. Phys. Chem.* 1992, **96**, 8502.
5   N. J. Sack and R. A. Baragiola, *Phys. Rev.*, 1993, **B 48**, 9973.
6   D. E. Brown, S. M. George, C. Huang, E. K. L. Wong, K. B. Rider, R. S. Smith, and B. D. Kay, *J. Phys. Chem.* 1996, **100**, 4988.
7   R. J. Speedy, P. G. Debenedetti, R. S. Smith, C. Huang, and B. D. Kay, *J. Chem. Phys.* 1996, **105**, 240.
8   J. A. Smith, F. E. Livingston, and S. M. George, *J. Phys. Chem B.* 2003, **107**, 3871.
9   V. Sadtchenko, W. R. Gentry, and C. Giese, *J. Phys. Chem. B*, 2000, **104**, 9421 ().
10   J. D. Graham and J. T. Roberts, *J. Phys. Chem.* 1994, **98**, 5974.
11   A. Bar-Nun, I. Kleinfeld, and E. Kochavi, *Phys. Rev. B* 1988, **38**, 7749.
12   S. Haq, J. Harnett, and A. Hodgson, *J. Phys. Chem. B*, 2002, **106**, 3950.
13   S. M. McClure, D. J. Safarik, T. M. Truskett, and C. B. Mullins, *J. Phys. Chem. B*, 2006, **110**, 11033. References 1-13 are just examples of TDS approach, and do not represent not an extensive review of the field.
14   J. G. Dash, in *Ice Physics and the Natural Environment*, edited by J. S. Wettlaufer, J. G. Dash, and N. Untersteiner, NATO ASI Series I Vol. 56 (Springer-New York, 1999), p. 11.
15   J. S. Wettlaufer, Philos. Trans. R. Soc. London, Ser. A 1999 **357**, 1.
16   J. S. Wettlaufer, Dynamics of ice surfaces, *Interface Science*, 2001, **9**, 115.
17   M. Faubel, in *Photoionization and Photodetachment* edited by Cheuk-You Ng, Advanced Series in Physical Chemistry, 10A, Part. 1, (World Scientific: River Edge, N.J., 2000), pp. 634-690.
18   V. Sadtchenko , M. Brindza, M. Chonde, B. Palmor, R. Eom, *J. Chem. Phys.*, 2004, **121**, 11980, and references therein.
19   Haiping Lu, Stephanie A. McCartney, M. Chonde, D. Smyla, and Vlad Sadtchenko, *J. Chem. Phys.*, **2006**, 125, 044709, and references therein.
20   For review see R. A. Baragiola in *Water in Confining Geometries*, edited by V. Buch and J. P. Devlin, Springer Series in Cluster Physics (Springer-Verlag, Berlin Heidelberg, 2003), pp. 359-395.
21   For a review see Osamu Mishima & H. Eugene Stanley, *Nature*, 396, 329 (1998)

22 J. Huang, L. S. Bartell, *J. Phys. Chem.*, 1995, **99**, 3924

23 L. S. Bartell, J. Huang, *J. Phys. Chem.*, 1994, **98**, 7455

24 G. P. Johari, A. Hallbrucker, and E. Mayer, *Nature*, 330, 552 (1987).

25 A. Hallbrucker, E. Mayer, and G. P. Johari, *Philos. Mag.*, B 60, 179 (1989).

26 G. P. Johari, G. Fleissner, A. Hallbrucker, E. Mayer, *J. Phys. Chem.*, 98, 4719 (1994)

27 M. Chonde, M. Brindza, and Vlad Sadtchenko, *J. Chem. Phys,* 2006, **125**, 094501.

28 G. P. Johari , *J. Chem. Phys.*, 1996, **105**, 7079.

29 C. A. Angell, *Chem. Rev.*, 102 2627 (2002)

30 G. M. Nathanson, P. Davidovits, D. R. Worsnop, C. E. Kolb., *J. Phys. Chem.*, 1996, **100**, 13007.

SELECTION FOR 'ICE RESISTANCE', ANTIFREEZE PROTEINS AND ICE OR HYDRATE INHIBITION

V. K. Walker[1], S.L. Wilson[1], Z. Wu[1], E. Huva[1], G. R. Palmer[1], G. Voordouw[2], H. Zeng[3], and J. A. Ripmeester[1,3]

[1]Department of Biology, Queen's University, Kingston, ON, Canada K7L 3N6
[2]Biological Sciences, University of Calgary, Calgary, AB, Canada T2N 1N4
[3]Steacie Institute for Molecular Sciences, NRC, Ottawa, ON, Canada K1A 0R6

1 INTRODUCTION

The tenacity of existence under extreme conditions has long been of interest. This, coupled with our curiosity about the evolution of life, has lead to studies on thermophiles, organisms that survive at temperatures near boiling. Such environments, in places such as thermal vents, are thought to mimic conditions found on the earth's surface some 4 billion years ago, although there is no compelling evidence to support this particular hypothesis[1]. Nevertheless, these investigations inspired studies that have resulted in the discovery of a variety of gene products, such as thermophilic enzymes, which in turn, have allowed the development of new solutions for practical problems. For example, the discovery of a DNA polymerase from *Thermus aquaticus* (Taq polymerase) has facilitated the current revolution in genome analysis[2]. Other extremophiles are resistant to pH extremes, salts, desiccation and pressure[3]. Psychrophiles or psychrotolerant bacteria, which are hardy at low temperatures, have received comparatively less attention. However, there is some interest in prospecting for ice-associating proteins from psychrophiles that may find uses in the preservation of frozen products or for enzymes that are active at low temperatures. As a result, microbes with these characteristics have been sought at high altitudes, Arctic regions, glacial cores, ice accretions and in Antarctic lakes. Indeed, Antarctic expeditions have been mounted solely to identify bacteria with antifreeze proteins (AFPs), which bind to ice crystals and arrest their growth[4].

AFPs and antifreeze glycoproteins were first characterized and cloned from cold-water fish, and later from insects and plants that overwinter in temperate latitudes, as well as from a few bacterial species[5]. AFPs inhibit freezing in a non-colligative manner by binding to ice and making the addition of water molecules unfavorable, thereby resulting in the lowering of the equilibrium freezing point[6,7]. Since melting is affected in a colligative manner, the interaction between AFPs and the ice surface results in a separation of the freezing point and the melting point, a phenomenon termed thermal hysteresis. Inhibition of ice growth is thought to derive from local ice surface curvature effects induced by the adsorption of AFPs at the ice/solution interface (the adsorption-inhibition hypothesis[8]). Due to this action, small quantities of AFPs can have large effects on ice crystal growth[9,10]. Adsorption to certain ice faces by AFPs seems to be by a surface to surface complementarity of fit, made possible by regularly-spaced residues on the regularly spaced crystal lattice. Once adsorbed to ice, AFPs sit 'snugly' assisted by van der Waals and

Physics and Chemistry of Ice

hydrophobic interactions[5].

Organisms with AFPs survive by either not freezing, provided temperatures do not drop too low, or if freeze-tolerant, by preventing the growth of large, damaging ice crystals that may form at temperatures close to melting, a property known as ice recrystallization (IR). These ice-associating proteins have been known for over 30 years[5,11], and presumably are important adaptations to counter the threat of cellular damage associated with ice growth. Ironically perhaps, we still do not fully appreciate their biological role, nor even really understand their interaction with ice. This is partially due to the difficulties of obtaining quantities of AFPs for application studies since traditional sources are often impractical and animal AFPs expressed recombinantly often need to be processed or folded *in vitro* in order to have full activity[11].

In contrast to AFPs, ice nucleating proteins (INPs; known and cloned from only 3 genera of bacteria to date) nucleate solutions at temperatures close to freezing. The survival strategy of these organisms is presumably to freeze extracellular ice at high, subzero temperatures. In addition, since these microbes are considered plant pathogens, it has been speculated that INPs may serve to deliberately cause frost injury to plants, thereby increasing the fitness of the host bacteria[12]. Since the INP sequences cloned to date appear to be closely related although the host bacterial genera are divergent, it has been suggested that the genes have been horizontally transferred to several members of the microbial community living on leaves[13]. A few of these proteins have been recombinantly expressed and one strain is used commercially for snow making while a genetically-manipulated version of these same bacteria are used in agriculture[12].

Gas or clathrate hydrates are crystalline lattices of hydrogen-bonded water molecules that are stabilized by the inclusion of natural gas molecules such as methane, propane, ethane and carbon dioxide. These form at moderate pressures and low, but above zero temperatures. Gas hydrates represent a vast potential energy source, more than twice all the known coal, oil and gas deposits[14], and the large reserves off continental shelves and in the permafrost regions will undoubtedly be tapped in the future. However, hydrates can also spontaneously form in pipelines causing blockages, and are often troublesome during drilling operations, in gas lines and during fractionation[15]. In addition to posing economic concerns, hydrates can have severe environmental consequences, including the release of gas into boreholes and the potential for blowouts. Since the offshore industry is going to deeper waters, with consequent increased flow line pressures and decreased temperatures, hydrate formation is now a major concern. Currently methanol is used to 'melt' these plugs, but this and other preventative measures cost industry $500 million per year[16]. Due to environmental concerns the use of hydrate inhibitors is now restricted in the North Sea. As a result, there is interest in the development of new, more benign synthetic inhibitors[17]. Several commercial kinetic inhibitors have been developed that appear to function by slowing down hydrate growth so that there is a reduced probability of blockage.

We have recently reported that AFPs not only inhibit ice growth but also the formation of clathrate hydrates[18,19]. For example, under our conditions, the growth of a model gas hydrate, tetrahydrofuran (THF) hydrate, was decreased by a factor of 16 with AFP, compared to the decrease, by a factor of 6, in the presence of the kinetic inhibitor, polyvinyl pyrrolidone (PVP). Similarly, AFPs reduced propane hydrate growth rate 5-fold. It is possible that AFPs inhibit hydrates analogously to ice growth inhibition. Indeed, modelling calculations[20] and our preliminary *in silico* structure studies suggest that docking to ice and hydrate lattice repeats may depend on the position and orientation on the crystal structures. In addition, the use of hydrates as an alternative substrate may provide insight into the mechanisms of protein-ice interactions, but also be useful for the discovery of future 'green' kinetic inhibitors for the gas and oil industry. In order to further our investigations, we need access to quantities of active and inexpensive ice-associating proteins.

2 MATERIALS AND METHODS

2.1 Microbe Selection and Identification

The search for novel, ice-associating proteins in extreme locations can be expensive and bureaucratically cumbersome. In contrast, our investigations are based on the assumption that ice-associating proteins are encoded by microbes that also inhabit more easily sampled environments. Although these microbes may not be abundant if conditions are warm, presumably all that is required for their discovery is a method of selection for rare microbes with these properties. Two selection regimes have been developed for this purpose.

2.1.1 Freeze-thaw Treatments using a Cryocycler. An instrument for automatically subjecting cultures to programmed freeze-thaw treatments has been fabricated and dubbed a 'cryocycler'[21] (Figure 1). This instrument achieves maximum cooling and heating rates by switching between two fixed temperature baths containing 40% ethylene glycol using solenoid-activated valves powered by a timer. Multiple freeze thaw cycles were used to provide sufficiently powerful selection so that only a few freeze-resistant microbes would survive in thousands of different microbes without that property. Soil samples were literally obtained from 'backyards' at several temperate locations and cultured in dilute tryptic soy broth (10% TSB; vol/vol). Sample vials (3 ml) were placed in a jacketed beaker filled with 40% ethylene glycol, which was stirred. Samples were then subjected to the freeze-thaw regime consisting of a two hour cycle at −18 and 5°C (with warming rates of 0.5°C/ min and cooling rates of 1.0°C/ min). Aliquots were removed from each of the triplicate vials periodically during cycling up to 48 cycles. Survivors were determined as colony forming units (CFU) per ml on 10% TSB plates, and normalized to a common

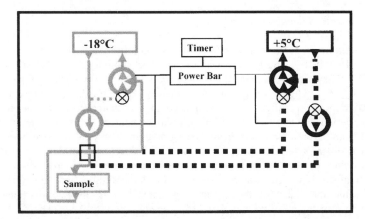

Figure 1 *Diagram of the cryocycler designed to automatically subject microbial cultures to freeze-thaw cycles. Grey lines show the flow pattern in the 'power-off' state with the valves in position to circulate ethylene glycol at -18°C through the jacketed sample chamber. Dotted lines show the partial 5°C flow pattern during the 'power-on' state.*

starting count of 1 x 10^8 CFU/ ml. Colony morphologies were also noted. Laboratory cultures of the gut bacterium, *Escherichia coli* TG2, as well as *Pseudomonas chlororaphis* were routinely used as controls.

Bacteria that remained viable after 48 freeze-thaw cycles were used to initiate new cultures and these were subjected to additional freeze-thaw cycles as described. Individual isolates, which had been previously identified, as well as the control cultures, were also subjected to freeze-thaw cycles. Occasionally, single isolates in 10% TSB would supercool rather than freeze at temperatures close to 0°C, and to ensure that all samples froze at the same temperature, a few sterilized AgI crystals were added to the vials at the start of the experiments. For some experiments, cells were harvested by centrifugation (6,000 xg for 10 min) and the cell pellet washed with 10% TSB and kept at 0°C until analysis. Spent media was obtained by centrifuging, as above, and filtering (0.45 μm).

2.1.2 Ice-affinity Selection. A second method of microbe selection was devised that takes advantage of the principle that when ice grows slowly it excludes solutes. A technique that allows for the ice-affinity purification of AFPs[22] has been adapted for the selection of microbes that are not excluded from ice. The apparatus consists of circulating temperature-controlled ethylene glycol solution through a brass cylinder immersed in the sample of interest. By holding the sample just at its freezing point and initiating freezing using seed ice crystals, an ice ball or 'Popsicle' was formed. Cultured diluted soil samples from three locations, obtained as described for the cryocycler experiments, were placed in a beaker and ice was grown slowly (about 0.01 ml/ min) over a 24-48 h period. After approximately 50% of the culture was frozen, the ice was rinsed and then melted to obtain the microorganisms that had partitioned into the ice phase. Individual colonies were obtained after culturing on 10% TSB plates.

After both selective regimes, microbes were isolated as single colonies, DNA isolated and primers for 16S rRNA genes were used to amplify fragments with polymerase chain reactions as previously described[21]. Comparisons of the DNA sequence from the various colonies with those in DNA sequence data bases allowed the identification of the recovered microbes.

2.2 Microbe and Molecule Analysis

2.2.1 Inhibition of Ice Recrystallization. IR inhibition assays were conducted as previously described[23]. Briefly, samples are placed in capillary tubes and snap-frozen at about -35°C. They were then placed in the cryocycler chamber, with the temperature held at -6°C for ≥16 h. Digital images through crossed polarizing filters were taken of the capillaries immediately after they were placed in the chamber and at the end of the incubation period. Crystal size in the images was compared. Samples were classified as showing IR inhibition if there was no apparent growth in the crystal size over time.

2.2.2 Crystal Morphology. Microscopic analysis of ice crystal morphology was done using a nanolitre osmometer. After freezing, the temperature was increased until a single crystal remained. The temperature was then lowered slightly and subsequently held constant while the shape of the crystal was noted[24]. THF hydrate morphology was examined by growing a crystal on the end of a pipette in a test tube containing THF:water (1:15 molar ratio) and then either allowing the crystal to continue to grow in this THF solution or by transferring the crystal to THF solutions containing AFP or PVP additives.

2.2.3 Ice Nucleation and Hydrate Crystallization Assays. Ice nucleation assays were modified from previously described assays[25-27]. Briefly, samples were loaded into capillary tubes, which were placed in the cryocycler chamber, and the temperature was lowered from -1°C to -15°C (at 0.1-0.2°C/ min). The temperature at which 90% of the capillaries (typically 30-50 capillaries were used for each experiment) were frozen was taken as the ice nucleation temperature of the most active fraction of the sample. Clathrate hydrate crystallization assays were similar except that the capillary tubes contained a THF solution as well as bacterial culture sample or bacterial medium (see figure legends) to make a final molar ratio of THF/water of 1:15. Digital images were captured and the temperature of crystallization was recorded.

    2.2.4 Thermal Profiles. A 16-channel data acquisition apparatus that records temperatures from thermistors was used as a thermal analyzer. The heat of fusion of first-order freezing and thawing phase transitions[28] were monitored in 2-3 ml samples that were chilled from approximately 5°C to -15°C at 1°C/ min.

3 RESULTS

3.1 Freeze-thaw Resistant Microbes

The viability of the various soil consortia was dramatically affected by multiple freeze-thaw cycles, with CFU/ ml typically decreasing by five orders of magnitude after 48 cycles (Figure 2). This was then repeated for a second 48 cycles after reculturing the survivors (not shown). The overall decrease in viability did not simply result from a reduction in numbers since the appearance of the colonies indicated that there was a shift in the complexity of the populations. Evidence for this differential susceptibility to freeze-thaw treatments is shown by the lack of *E. coli* colonies after 24 cycles and the absence of *P. chloroaphis* after 48 cycles (Figure 2). After isolating DNA from 16 colonies that survived this rigorous freeze-thaw selection in one series of experiments, five distinct sequences were identified. The bacteria from which these five sequences were isolated were then used

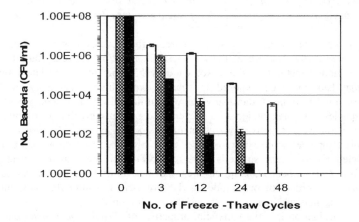

Figure 2 *Decay of viability of microbial cultures as a function of the number of freeze-thaw cycles in the cryocycler. Viable cells (CFU/ ml) in cultures derived from one of the soils (first light bar in each cycle set) were compared to pure cultures of P. chlororaphis (grey, second bar in each cycle set) or E. coli (black, third bar in each cycle set). In this series of experiments only the soil culture survived 48 freeze-thaw cycles. Error bars represent standard deviations.*

to initiate cultures so that individual freeze-thaw resistance could be determined. Of these, the bright yellow *Chryseobacterium* sp. cultures appeared to be unaffected by freeze-thaw cycling since no decrease in viability was seen even after 48 cycles (Figure 3). In contrast, the small pale yellow *Enterococcus* sp. samples were more freeze-thaw susceptible and lost about six orders of magnitude in viability over the course of the experiments.

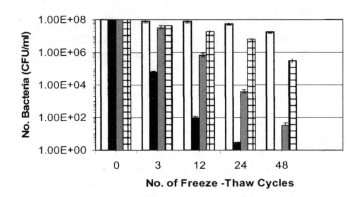

Figure 3 *Decay of viability of single colony isolates after serial freeze-thaw cycles. Isolates surviving 2 x 48 freeze-thaw cycles (see text) were used to initiate cultures and these were subjected to further freeze-thaw treatments. Single isolates: Chryseobacterium sp. (first light bar in each set), E. coli (black, second bar in each cycle set) and Enterococcus sp. (grey, third bar in each cycle set). A fraction of the Enterococcus sp. culture was also separated from the culture medium and resuspended in the spent medium derived from the Chryseobacterium sp. cultures prior to freeze-thaw cycles (hatched, fourth bar in each cycle set). Error bars represent standard deviations.*

Therefore, *Enterococcus* sp. survived freeze-thaw cycles as part of a consortium, but was, unexpectedly, less freeze-thaw resistant in isolation. Indeed, viability in these isolates was more affected than the mixed cultures that originated from late summer soil samples that had not been exposed to subzero temperatures for many months. An investigation to determine if their survival was influenced by the presence of another bacterial species was then undertaken. Several experiments involving combinations of different bacteria were negative, but when *Enterococcus* sp. and *Chryseobacterium* sp. were cultured together, both showed freeze thaw resistance (not shown). Furthermore, when pelleted *Enterococcus* sp. cells were resuspended in the cell-free, spent medium obtained from *Chryseobacterium* sp. cultures, the survival of *Enterococcus* sp. improved, so that only three orders of magnitude of viability were lost after the same number of treatments. Thus their survival improved a 1000-fold (Figure 3).

An investigation into the beneficial properties of *Chryseobacterium* sp. that might confer freeze-thaw protection to itself, as well as to other bacteria, showed that these cultures had IR inhibition activity; no large ice crystals could be seen in capillary tubes after the frozen solutions containing *Chryseobacterium* sp. cultures had been left overnight at -6°C (Figure 4). Furthermore, this IR inhibition activity appeared to be conferred by a protein since treatment with a protease, proteinase K, abolished this effect ([21]; not shown).

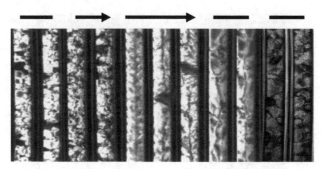

Figure 4 *Inhibition of ice recrystallization. Samples in 10 μl microcapillaries (740 μm diameter) were frozen and placed at -6°C and examined between crossed polarizing filters. Images were taken prior to, and after overnight incubation, but only the latter images are shown. From left to right samples include sample buffer controls (1, 2), bovine serum albumin, a control protein diluted to 0.2 and 0.02 mg/ ml, respectively (3,4), serial dilutions of 0.2, 0.02 and 0.002 mg/ ml Type I fish antifreeze protein in buffer (5-7), Chryseobacterium sp. cultures (8, 9), and E. coli cultures (10, 11). Note that only the fish AFP and the Chryseobacterium sp. cultures have crystals too small to be detected at this magnification and the overlying "feathery" pattern, typical of snap frozen samples, is apparent. Bacterial cultures were at ~2 x 10⁸ CFU/ ml. Lines indicate duplicate samples and arrows indicate samples that were diluted.*

3.2 Microbes Recovered by Ice Affinity

Similar to the results described after selection for freeze thaw resistance, there were reductions in microbial diversity and abundance after ice affinity selection. A total of 19 isolates were recovered from the three locations, but many of these were duplicated, so that only five distinct genera were identified by DNA sequencing. Included in this group were five isolates of *Chryseobacterium* sp. Again, there was some evidence of IR inhibition activity as seen with the *Chryseobacterium* sp. strain obtained by freeze-thaw resistance selection. When ice crystals were formed in the presence of this bacterium, the morphology was changed from a round disc-like shape seen with buffer solutions, culture media, or *E. coli* (Figure 5) to an oval-shaped crystal (not shown).

Another microbe that was recovered by ice affinity, *P. borealis,* showed some IR inhibition and when tested for ice nucleation activity (INA), it approached the activity of commercial 'snow making' preparations derived from *P. syringae*[12]. Again, protease treatments indicated that this activity was mediated by protein, likely an INP (not shown). Thermal analysis profiles showed that the addition of *P. borealis* cultures essentially prevented the supercooling of solutions, whereas control *E. coli* cultures did not (Figure 6). Moreover, this INA was concentration-dependent. The ability of this strain to inhibit supercooling also extended to the crystallization of a model gas hydrate, THF hydrate (Figure 7). Like some other isolates recovered by ice affinity, when *P. borealis* was used in ice morphology assays, ice crystals were distinct from controls, often taking a rectangular shape (Figure 5). Control, fish Type I AFP also changed the morphology of ice crystals, but in the presence of this protein, the crystals were shaped like diamonds (Figure 5).

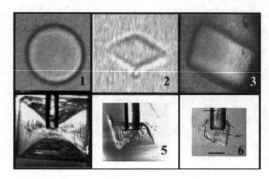

Figure 5 *Crystal morphology assays. Formed ice crystals (1-3) were examined under a microscope (40x). Representative crystals were grown in the presence of (1) control solutions such as buffer (shown), E. coli or medium, (2) Type I fish AFP and (3) P. borealis cultures. THF hydrate crystals (4-6) were grown at a pipette tip for about the same amount of time and transferred to solutions of (4) THF:solution (15:1 molar ratio), (5) THF:solution containing PVP and (6) THF:solution containing Type I AFP. The bar represents 1.3 mm.*

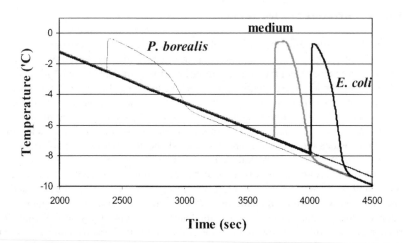

Figure 6 *Representative thermal profiles showing the heat of fusion in 2 ml samples. A reference thermister recording (straight, thin black line) shows the temperature of a control sample containing ethylene glycol. P. borealis cultures (approximately ~1 x 10^{-8} CFU/ ml) show freezing beginning at -2.5°C. Medium and E. coli cultures at ~1 x 10^{-8} CFU/ ml show freezing at -7°C and -8°C, respectively.*

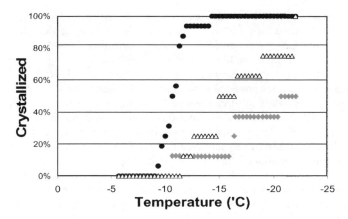

Figure 7   *Crystallization of a model hydrate as a function of temperature. THF:solution (15:1 final molar ratio) hydrate formation was examined in the presence of P. borealis cultures (~1 x 10⁻⁸ CFU/ ml; ●), control medium (Δ), and E. coli cultures (~1 x 10⁻⁸ CFU/ ml; ◊). Circulating ethylene glycol was cooled with a programmable bath and the temperature of crystallization in each microcapillary was taken as the temperature of the bath at the time of clathrate hydrate formation. Crystallization was recorded using time lapse photography. Each point represents a single microcapillary tube and thus the crystallization spectrum was normally determined with 30-50 microcapillaries for each sample, as above.*

5 DISCUSSION

The development of apparatus and methodology to mimic the rigors of high latitude terrestrial environments has resulted in the selection of a group of bacteria with a high proportion of ice-associating properties. Although the analysis of the selected cultures is still in progress, a total of half of the recovered, identified bacteria have known ice-associating activities. Two of these ten genera were recovered in both selective regimes. Those ten that have been previously described as associating with ice have been recovered from samples obtained from glacial cores, Antarctic lakes or sea ice[4,29,30]. It should be noted that ice-associating activities are not common properties amongst bacteria. To date, AFPs have been described in half a dozen genera, but these sequences have been characterized or cloned from only a few of these bacteria[31-34] Similarly, the production of INPs has been described in several plant-associating bacteria, and they appear to be homologous proteins[13]. Thus the selection and recovery of these ice-active microorganisms from late summer-collected soil samples is remarkable. This success demonstrates that sampling in extreme or highly remote areas is not required for the isolation of strains with these presumably rare properties.

Of the ten recovered isolates, some showed protein-mediated IR inhibition as well as ice-shaping activities, properties that together are suggestive of AFPs. One of these,

Chryseobacterium sp. showed IR inhibition, which has not been reported in this genera previously. It was more than 1000-fold more freeze-thaw tolerant than the original soil consortium, and was also recovered by ice affinity selection. In addition, these bacteria changed the shape of ice crystals (Figure 5). *Chryseobacterium* sp. appears to have a commensal relationship with an *Enterococcus* species because the latter bacterium survives multiple freeze-thaw cycles provided it undergoes these treatments with *Chryseobacterium* sp. (not shown) or with the cell-free medium from *Chryseobacterium* sp. cultures (Figure 3). AFPs may allow survival under freeze-thaw conditions since these proteins impede water mobility at the ice crystal surface, preventing the growth of large, damaging crystals seen in IR. It is interesting to hypothesize that a *Chryseobacterium* sp. AFP could help confer freeze-thaw resistance to closely-associating susceptible microbes such as *Enterococcus* sp. that show no IR inhibition activity themselves (not shown). Indeed, the development of this equipment as a selective tool to identify rare, but most interesting bacterial species, offers hope that insights into the overwintering survival of communities may come from these studies.

Strikingly, other recovered isolates had INA. One of these, a strain of *P. borealis*, appeared to have some IR inhibition but also significantly changed the morphology of ice crystals (Figure 5). This species has not been previously reported to have an INP, but its activity seen in the thermal analysis profile (Figure 6) approached that of a commercial ice nucleator. Partial sequencing of the gene encoding its INP shows that it is repetitive with ~75% identity with other reported INPs (not shown). We speculate that the repetitive structure of this membrane protein complements the reiterative structure of ice, and this coupled with its ability to change the shape of ice crystals, likely explains this strain's retrieval after ice-affinity selection.

Previously, we have shown that AFPs can inhibit gas hydrate growth and change the morphology of a growing THF hydrate crystal[18,19] (Figure 5). Since clathrate hydrates are crystalline ice-like cages of water that house small guest molecules, we hypothesized that *P. borealis* might also have some affinity for hydrates, either through the bacterium's INA or its ice-shaping activity. When THF hydrate crystallization was monitored in microcapillaries, approximately 60% of the samples containing *P. borealis* cultures were crystallized by -10°C, whereas at that temperature none of the capillaries containing THF and *E. coli* cultures had solidified (Figure 7). It was not until -20°C that even 50% of the capillaries with *E. coli* cultures showed crystallization. In addition, the microcapillaries containing THF and buffer or *E. coli* cultures showed that hydrate crystallization was a stochastic phenomenon, in that crystallization occurred over a much wider temperature range, with 25-50% of the samples remaining liquid even at temperatures below -20°C. Thus, whatever the mechanism, THF hydrates formed more quickly in the presence of *P. borealis*, presumably by preventing extreme supercooling, than with a variety of other control cultures or solutions (Figure 7; not shown). This is a significant finding since it represents a glimmer of hope that an affordable, environmentally-benign method to control and inhibit hydrate formation may be found in a microbe.

As mentioned, hydrate formation is a serious problem for the petroleum industry with hydrate plugs forming during production and transportation[15]. Although traditionally-used chemical treatments such as methanol can 'melt' or prevent the formation of hydrates, they are no longer favored for environmental reasons. New groups of synthesized chemicals such as the polymer kinetic inhibitors are being explored for their utility in retarding hydrate formation. Other groups of inhibitors such as the anti-agglomerants allow the formation of hydrate particles, but keep them dispersed so that they do not develop into large aggregates that would block pipelines[17].

The difficulty in expressing well-folded, active AFPs has constrained our ability to develop larger-scale laboratory tests of these protein hydrate inhibitors. Our observations that bacterial cultures influenced hydrate formation in these experiments offers hope that technologies to inhibit or control the growth of gas hydrates may come from such studies. Perhaps all that is required are robust selection methods to discover bacteria with these properties in the millions of microbes that inhabit our planet. Indeed, we already have two potential contenders. *Chryseobacteria* sp. and its putative AFP may find use both as an ice and hydrate inhibitor. As well, the two ice-associating activities of *P. borealis* may find use as a new ice nucleator, but as well, also as an inducer of gas hydrate crystallization. In the latter case, we envision a kind of 'green' anti-agglomerate, which would allow the formation of multiple small hydrate crystals and thus flow assurance.

5 CONCLUSION

To recapitulate, challenging and expensive expeditions to polar regions, glaciers and high altitudes may not be required to prospect for microbes with 'ice resistance'; all that may be necessary is the development of strong selective techniques to recover those with useful properties. After toiling for many years with the challenges of expressing sufficient amounts of active AFPs to investigate ice- and hydrate-associating properties it is humbling to know that answers to many of our quests may have been quietly awaiting discovery in our own 'backyards' all along.

Acknowledgements

This research was funded by a Discovery and Equipment grants to V.K. Walker by the Natural Sciences and Engineering Research Council (Canada). We thank our colleague, P.L. Davies, for his continued enthusiasm for our work.

References

1   A. Lazcano and S.L. Miller, *Cell,* 1996, **85**, 793.
2   R.K. Saiki, D.H. Gelfand, S. Stoffel, S.J. Scharf, R. Higuchi, G.T. Horn, K.B. Mullis and H.A. Erlich, *Science,* 1988, **239**, 487.
3   A. Reysenbach and E. Shock, *Science,* 2002, **296**, 1077.
4   J.A. Gilbert, P.J. Hill, C.E. Dodd and J. Laybourn-Parry, *Microbiol.,* 2004, **150**, 171.
5   P.L. Davies, J. Baardsnes, M.J. Kuiper and V.K. Walker, *Philos. Trans. R. Soc. Lond. B. Biol. Sci.* 2002, **357**, 927.
6   A. DeVries, *Science,* 1971, **172**, 1153.
7   Y. Yeh and R.E. Feeney, *Chem. Rev.,* 1996, **96**, 601.
8   J.A. Raymond and A.L. DeVries, *Proc. Natl. Acad. Sci. USA,* 1977, **74**, 2581.
9   K. Harrison, J. Hallett, T.S. Burcham, R.E. Feeney, W.L. Kerr and Y. Yeh, *Nature,* 1987, **328**, 241.
10   D.B. DeOliveira and R.A. Laursen, *J. Am. Chem. Soc.,* 1997, **119**, 10627.
11   V.K. Walker, M.J. Kuiper, M.G. Tyshenko, D. Doucet, S.P. Graether, Y-C. Liou, B.D. Sykes, Z. Jia, P.L. Davies and L.A. Graham, 'Surviving winter with antifreeze proteins: studies on budworms and beetles', in *Insect Timing: Circadian Rhythmicity to Seasonality,* eds., D.L. Denlinger, J.M. Giebultowicz and D.S. Saunders, Elsevier, Amsterdam, 2001, pp.199-211.

12 S.S. Hirano and C.D. Upper, *Microb. and Molec. Biol. Reviews*, 2000, **64**, 624.

13 P.K. Wolber and G.J. Warren, 'Evolutionary perspective on the ice nucleation gene-encoded membrane protein' in *Microbial Ecology of Leaves*, eds., J.H. Andrews and S.S. Hirano, Springer-Verlag, New York, 1991, pp.315-330.

14 K.A. Kevenvolden, *Proc. Nat. Acad. Sci. USA*, 1999, **96**, 3420.

15 B. Edmonds, R.A.S. Moorwood, and R. Szezepanski, *Offshore Magazine (Ultradeep Engineering Supplement, March)*, cited in P.F. Pickering, B. Edmonds, R.A.S. Moorwood, R. Szczepanski, and M.J. Watson, M. J. *IIR Conference*, Aberdeen, Scotland, 2001, http://www.feesa.net/research.htm, pp.1-15.

16 E.D. Sloan, *Gas Research Institute Topical Report GRI-91/0302 June 1*; 1992, Gas Research Institute: Chicago.

17 J.P. Lederhos, J.P. Long, A. Sum, R.L. Christiansen and E.D. Sloan, *Chem. Eng. Sci.*, 1996, **51**, 1221.

18 H. Zeng, L.D., Wilson, V.K. Walker and J.A. Ripmeester, *J. Amer. Chem. Soc.*, 2006, **128**, 2844.

19 H. Zeng, I.L. Moudrakovski, J.A. Ripmeester and V.K. Walker, *AIChE J.*, 2006, in press.

20 B. Wathen, M. Kuiper, V. Walker and Z. Jia, *J. Am Chem. Soc.*, 2003, **125**, 729.

21 V.K. Walker, G.R. Palmer and G. Voordouw, *Appl. Environ. Microbiol.*, 2006, **72**, 1784.

22 M. Kuiper, C. Lankin, S.Y. Gauthier, V.K. Walker and P.L. Davies, *Biochem. Biophys. Res. Commun.*, 2003, **300**, 645.

23 M.M. Tomczak, C.B. Marshall, J.A. Gilbert and P.L. Davies, *Biochem. Biophys. Res. Commun.*, 2003, **311**, 1041.

24 A. Chakrabartty and C.L. Hew, *Eur. J. Biochem.*, 1991, **202**, 1057.

25 G. Vali, *J. Atmos. Science*, 1971, **28**, 402.

26 L.R. Maki, E.L. Galyan, M.M. Chang-Chien and D.R. Caldwell, *Appl. Environ. Microbiol.*, 1974, **28**, 456.

27 L.M. Kozloff, M.A. Schofield and M. Lute, *J. Bacteriol.*, 1983, **153**, 222.

28 H.J. Borchardt and F. Daniels, *J. Amer. Chem. Soc.*, 1957, **79**, 41.

29 B.C. Christner, E. Mosley-Thompson, L.G. Thompson and J.N. Reeve, 'Recovery and identification of bacteria from polar and non-polar glacial ice' in *Life in Ancient Ice*, eds., S.O. Rogers and J. Castello, Princeton University Press, Princeton, 2005, pp. 209-227.

30 S.L. Wilson, D.L. Kelley and V.K. Walker, *Environ. Microbiol.*, 2006, in press.

31 N. Muryoi, M. Sato, S. Kaneko, H. Kawahara, H. Obata, M.W. F. Yaish, M. Griffith and B.R. Glick, *J. Bacteriol.*, 2004, **186**, 5661.

32 Y. Yamashita, N. Nakamura, K. Omiya, J. Nishikawa, H. Kawahara and H. Obata, *Biosci. Biotechnol. Biochem.*, 2002, **66**, 239.

33 J. A. Gilbert, P.L. Davies, and J. Laybourn-Parry, *FEMS Microbiol. Lett.*, 2005, **245**, 67.

34 S. D'Amico, T. Collins, J-C. Marx, G. Feller and C. Gerday, *EMBO Reports*, 2006, **7**, 385.

Contributed Papers

RAMAN SCATTERING STUDY OF PROTON ORDERED ICE-XI SINGLE CRYSTAL

K. Abe, Y. Ootake and T. Shigenari

Department Applied Physics and Chemistry, The University of Electro-Communications, Tokyo, 182-8585, Japan, E-mail:abe@pc.uec.ac.jp

1 INTRODUCTION

It is well known that protons in ice Ih crystal are disordered. The disorder remains even at zero temperature and shows the residual entropy. The existence of the residual entropy predicts the existence of the proton ordered state but it would take a geologic scale of time to transform into the ordered phase. Several decades ago, however, it was found that KOH doping to Ih crystal makes the proton ordering in Ih much easier and the transformation to the proton ordered state at ambient pressure (XI phase) at T_c=72K was confirmed by calorimetric[1] and dielectric measurement[2]. By this transition, the symmetry changes from D_{6h}-6_3/mmc in Ih to C_{2v}-Cmc2_1 in XI. In XI phase, dipole moments of water molecules are uniformly aligned in b-c plane (see Fig. 4(a)). Their directions along b-axis change oppositely for alternative a-b planes. So the net (ferroelectric) polarization appears along c-axes[3,4].

Some variations in the phonon density of states (DOS) in translational and librational mode regions were studied by inelastic incoherent neutron scattering (IINS)[5] and also studied by molecular dynamics (MD)[6,7]. For the spectroscopic studies of phonons in XI phase, only a few studies have been reported[8,9]. We have reported several studies on Raman spectra in the proton ordered XI phase[10,11]. However, detailed mode assignment has not been made successfully yet because of the difficulties to get a single crystal which undergoes the Ih-XI transition with enough transparency suitable for Raman measurement.

In this paper we present the recent results which give more reliable assignments for the translational and librational modes.

2 EXPERIMENTAL

Ice Ih single crystals were grown using the modified Bridgman method developed by Kawada's group[2]. In order to obtain optically semi-transparent good quality samples, the concentration of KOH solution is thinner (about 0.01M) than those used in other researches, e.g. neutron, x-ray diffraction and dielectric studies. The growth rate is about 1cm/hour and a sample grows in a direction perpendicular to the c-axis in most case. Sample preparation was done in a large refrigerator kept at 243K. Typical sample size is 6x6x8 mm^3. The c-axis of crystal was confirmed by an optical polarized microscope in the refrigerator and a-

and b-axes were determined by x-ray Laue method under the cooled N₂ gas. For Raman measurement, all surfaces were carefully polished on a paper filter with a small amount of methanol. A sample was set into an airtight holder and then the holder was loaded quickly into the cryostat with GM-refrigerator under the atmosphere of cooled N₂ purge to avoid the frost-formation on sample surfaces. The temperature stability is less than 0.2K/hour. Raman spectra were measured using an Ar-ion laser (514.5nm, 300mW) and a single monochromator (with 300 and 2400gr/mm for low and high resolution) equipped with a two dimensional detector (ICCD). A notch filter with bandwidth of about 5nm was used to suppress the intense elastic scattering, which masked the low frequency spectra below 170cm⁻¹. Spectra were measured in a right angle scattering geometry for various polarization configurations. Raman spectrum is specified using the conventional notations, e.g. as x(z,z)y, where x and y stand for a direction of incident and scattered light, respectively and the (z,z) in the parenthesis stand for the polarization of incident light and scattered light, respectively. The latter pair (i,j) corresponds to the Raman tensor $\alpha_{i,j}$.

In order to induce the proton ordered XI phase, some thermal treatments are required. Typically we kept the sample at 45K for 12 hours for nucleation and then it was annealed at 65K for a few days. Even after the heat treatment, samples do not always undergo the proton ordering transition. So, before the measurements of Raman spectra, we measured the temperature dependence of the dielectric constant at several low frequencies (20, 30, 60,100, 200,400 Hz). In addition to the dispersions around 160 and 230K(Fig. 1(a)), if a sample shows a clear dispersion around 90K as shown Fig. 1(b), it has enough high proton mobility and we found empirically that the proton ordering phase transition occurred with much higher probability.

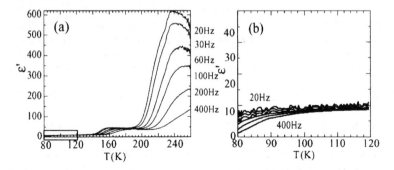

Figure 1   *(a) Temperature dependence of the dielectric constant at several frequencies. The square in (a) for 80-120K is magnified in (b), which shows an additional dispersion around 90K.*

3   RESULTS AND DISCUSSIONS

3.1   Wide Range Raman Spectra in Ih and XI Phase

Figure 2(a) shows Raman spectra in a wide frequency range in proton disordered Ih (thin spectra) and ordered XI phase (thick spectra). In spite of the low resolution (Δν~15cm⁻¹), changes of spectra by the transition can be seen clearly in the lattice (translational, librational) modes and also in the stretching bands of water molecules. Particularly, peaks

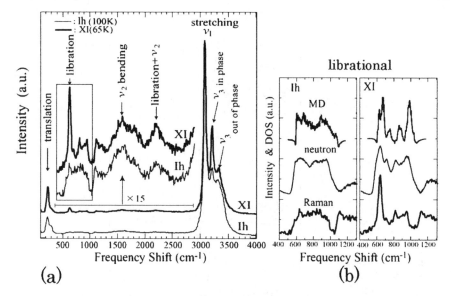

Figure 2 (a) *Raman spectra in a wide frequency range. Thin and thick lines represent the spectra in Ih (100K) and in XI (65K) phases, respectively. (b) The comparison of the librational mode region between Raman spectra (the rectangular part in (a)) and the DOS spectra obtained by neutron[5] and MD-calculation[6]. Note the frequency scale of DOS by MD is shifted for comparison with Raman.*

at 230cm[-1] in the translational region, at 630 cm[-1] in the librational region and at 3200cm[-1] become remarkably stronger in XI phase.

In Ih phase, the selection rules for D_{6h} are violated due to the proton disorder. So the observed band shape represents mostly the phonon density of states (DOS). As shown in Fig.2 (b), observed Raman spectra in the librational region in Ih phase surprisingly agrees with results of the neutron scattering (IINS)[5] and MD calculation[6]. In XI phase, Raman spectra show mostly the first order scattering around the Γ–point but qualitative agreement with neutron and MD studies is also seen in XI phases.

3.2 Translational region

Figure 3 shows Raman spectra of translational mode region observed with higher resolution ($\Delta\nu \sim 1.5$cm[-1]) in a geometry a(,)b. In Ih phase, two peaks at 230 and 315cm[-1] and a plateau between them are observed. It shows almost no polarization dependence. In contrast, clear polarization dependences are observed in XI phase indicating that the selection rules for C_{2v} become applicable by the proton ordering.

Translational modes for four rigid water molecules in a primitive cell are classified as $E_{1g}+E_{2g}+A_{1g}+E_{2u}+B_{2u}+B_{1g}$ in Ih phase (D_{6h}). According to the dispersion curves obtained from neutron scattering for ice Ih and also by MD calculation by Tse et al.[7], the peaks at 230cm[-1] is assigned as $E_{1g}+E_{2g}$ (with almost same frequency) and a peak at 315cm[-1] as A_{1g} in D_{6h}. The frequencies of B_{2u} could appear at around 270 cm[-1], but it is not seen clearly in Fig.3. The frequencies of E_{2u} and B_{1g} are lower than 150cm[-1]. In XI phase

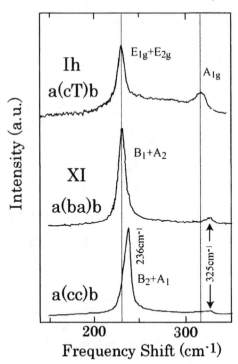

Figure 3 *Raman spectra in the translational region observed in the geometry a(,)b. The spectrum in Ih was measured at 100K and those in XI at 65K. (c,T) for Ih phase means scattered light includes both c and a polarization. Two thin lines at 230cm^{-1} and 315cm^{-1} are guides for eyes.*

(C_{2v}), symmetries change to $3A_1+2A_2+B_1+3B_2$ [10,13] and all modes are Raman active.   It is expected that E_{1g} splits to B_1+B_2 and E_{2g} splits to A_1+A_2.

The spectra a(b,a)b and a(c,c)b in Fig.3 show that the 230cm^{-1} peak in ice Ih splits into 230cm^{-1} and 236cm^{-1} in ice XI. We assigned the 230cm^{-1} as B_1+A_2 and 236cm^{-1} as B_2+A_1 from the following reason. As shown in Fig. 4(a), in contrast to Ih phase, dipole moments of all water molecule in XI phase are aligned in the bc plane and their directions (tilted from c and b) vary alternately between layers perpendicular to c-axes. In other words, it is ferroelectric along c-axes and anti-ferroelectric along b-axes. In C_{2v}, symmetries of a polar vector (a,b,c) are B_1(a), B_2(b) and A_1(c). Therefore, A_1 and B_2 modes of which displacements are in the bc plane would shift to higher frequency due to the static electrical field (depolarization field) caused by the ordering of the dipole moment, while B_1+A_2 remain at 230cm^{-1} because their displacements are perpendicular to the bc plane and are not influenced by the depolarization field.

From the same reason, A_1 mode with displacement parallel to c-axis shifts higher from 315cm^{-1} in Ih to 325cm^{-1} in XI. It is not clear, however, why 325 cm^{-1} mode is observed not only in the polarized configuration a(c,c)b but also in the depolarized spectra a(b,a)b.

Besides these shifts of the modes, we note that the intensity of the plateau (broad band) between 230cm^{-1} and 315cm^{-1} mode in Ih phase significantly decrease in XI phase. This indicates that the origin of the broad band is the density of states which is locally

Table 1 *Assignment of translational modes in XI*

Ih D_{6h} $\Delta v[cm^{-1}]$	assignment	XI C_{2v} $\Delta v[cm^{-1}]$	assignment	Raman tensor
230	E_{1g}	230	B_1	ac
		236	B_2	bc
230	E_{2g}	230	A_2	ab
		236	A_1	aa, bb, cc
315	A_{1g}	325	A_1	

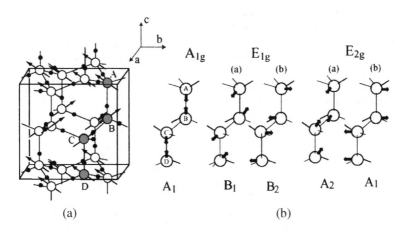

(a) (b)

Figure 4 *(a) Structure of ice XI [4]. Four water molecules (A to D) consist of a primitive cell. Arrows represent the direction of dipole moment of a water molecule. (b) Displacement of molecules for translational mode [12]. Symmetries in Ih and XI phases are given at upper side and at lower side, respectively.*

Raman activated due to the disorder in Ih phase. Thus Raman spectra observed in XI phase in the translational region are assigned quite successfully as summarized in Table1.

3.3 Librational Region

Hindered rotations (librational modes) of a rigid water molecule in ice are observed in the frequency range of 600-900cm^{-1}. In Ih phase, as shown in the top of Fig. 5, weak and broad peaks are observed at about 770cm^{-1} and 910cm^{-1}(shown by arrows) and the spectra show almost no polarization dependence. In addition to these peaks, a step-wise jump is seen at 1080cm^{-1} and it becomes more clear in c(a,a)b spectrum in XI phase, the origin of which has not been known yet.

As shown in Fig. 6(a), for an isolated water molecule there are three types of librational modes, twist(T), wag(W) and rock(R). Using the factor group analysis for the four molecules in a primitive cell, 12 librational modes in XI phase are divided into

$$2A_1 + 4A_2 + 4B_1 + 2B_2 \qquad (1)$$

Most significant change by the transition to XI phase is seen in the polarized spectrum a(c,c)b (bottom of Fig.5). A strong and sharp peak at 630cm^{-1} gradually appears during the annealing process.

It is interesting to note that another polarized spectrum c(a,a)b is quite different from a(c,c)b. It resembles the depolarized c(b,a)b spectrum. The resemblance can be attributed to the appearance of structural domains in XI phase, since the three equivalent b-axes in the hexagonal Ih phase would produce three different orthorhombic structures in XI phase with the unique b-axis rotated by 120degrees. Then, even if the incident light is a- or b-polarized, in some domains it would be the mixture of a- and b- polarization (e.g., $a \rightarrow (\sqrt{3}a \pm b)/2$). Thus, in the c(b,a)b not only the non-diagonal α_{ba} but also the diagonal α_{aa} and α_{bb} contribute to the spectrum.

The large difference in the intensity of the 630cm^{-1} mode between a(c,c)b and c(a,a)b indicates that Raman tensor α_{cc} for this mode is significantly enhanced by the ferroelectric alignment of dipoles parallel to c, while the tensors α_{aa} and α_{bb} are little enhanced by the anti-ferroelectric ordering in the a-b plane. Since in the a(c,c)b spectrum there is no mixing of Raman tensor due to the domains, two peaks at 630 and 810cm^{-1} in a(c,c)b can be assigned as totally symmetric A$_1$ modes.

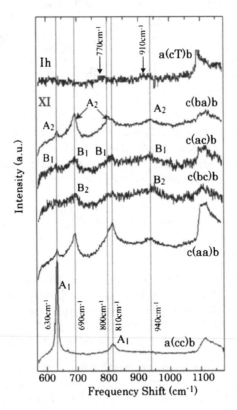

Figure 5   *Polarization dependence of Raman spectra in the librational region. Spectrum for Ih was measured at 100K and those for XI at 65K. Configuration (c,T) for Ih means, no analyzer was used for the scattered light.*

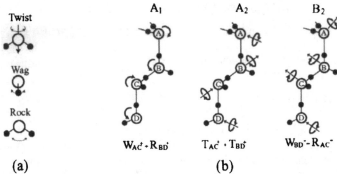

Figure 6 *(a) Three types of librational mode of a single water molecule. (b) Typical mode pattern of librational mode for four water molecules in a primitive cell of ice XI.*

In contrast, in other depolarized spectra, c(a,c)b and c(b,c)b, which are very similar as they should be, no diagonal tensor would be mixed due to the domains. Therefore, modes observed in these spectra, are assigned as one of A_2, B_1 or B_2.

As a lattice mode in crystal, librational motions should be expressed by the combinations of the four water molecules (A to D in Fig.4(a)). Among them, A and C (also B and D) are connected by the symmetry operations $\{C_2|\tau\}$ and $\{\sigma_v|\tau\}$, ($\tau = (0,0,1/2)$) in the space group $Cmc2_1$. So, the displacements are the linear combination of the two molecules, for example, by $T_{AC}^{\pm} \equiv (T_A \pm T_C)$, where $\pm$ represents in-phase and out of phase motion. Furthermore, the coupling between the pairs, (A,C) and (B,D) may exist. Thus we classify the librational modes in ice XI by the notations such as $T_{AC}^{\pm} + T_{BD}^{\pm} \equiv (T_A \pm T_C) + (T_B \pm T_D)$. Typical examples are shown in Fig. 6(b).

Separation of the observed modes to Twist, Wag and Rock, however, are not easy since in a crystal these modes would not be independent. Our assignments are based on the following assumptions. (1) Referring the MD calculation for ice Ih[6], the modes in the middle part of spectra in Fig.5 are the twist modes. (2) Twist motions are rather independent, where as Wag and Rock are coupled each other. It is because, as shown in Fig.6 (b), twist of B molecule would not significantly affect the motion of A and C. (3) Out of phase vibration (e.g., $T_{AC}^- - T_{BD}^-$) has higher frequency than that of in phase vibration ($T_{AC}^- + T_{BD}^-$) [11,13]. Then the librational motions in eq.(1) would be classified as follows

Table 2 *Tentative assignment of librational modes in XI in terms of Wag (W), twist (T) and rotation(R).*

$\Delta \nu$ [cm^{-1}]	a(cc)b, c(aa)b A_1	c(ba)b A_2	c(ac)b B_1	c(bc)b B_2
630	$W_{AC}^+ + R_{BD}^+$	$W_{BD}^+ + R_{AC}^+$	$W_{BD}^- + R_{AC}^-$	
690		$T_{AC}^+ + T_{BD}^+$	$T_{AC}^- + T_{BD}^-$	$W_{AC}^- + R_{BD}^-$
800		$T_{AC}^+ - T_{BD}^+$	$T_{AC}^- - T_{BD}^-$	
810	$W_{AC}^+ - R_{BD}^+$			
940		$W_{BD}^+ - R_{AC}^+$	$W_{BD}^- - R_{AC}^-$	$W_{AC}^- - R_{BD}^-$

$$2A_1(W_{AC}^+ \pm R_{BD}^+) + A_2(W_{BD}^+ \pm R_{AC}^+, T_{AC}^+ \pm T_{BD}^+)$$
$$+ 4B_1(W_{BD}^- \pm R_{AC}^-, T_{AC}^- \pm T_{BD}^-) + 2B_2(W_{AC}^- \pm R_{BD}^-). \qquad (2)$$

and the assignments of 12 librational modes are given in Table 2.

4   CONCLUSION

Polarized Raman spectra of the translational and librational modes in Ih and XI phases of ice single crystal were measured. Changes in the spectra due to the proton ordering phase transition were detected clearly. In the translational region, a peak at 230cm^{-1} in Ih phase splits into the intense peaks at 230 and 236cm^{-1} and the peak at 315cm^{-1} in Ih phase shifts to 325cm^{-1}. Assignments of the translational modes, except for the 325cm^{-1} were done successfully. In the librational region, an intense peak at 630cm^{-1} appears in XI phase. Clear polarization dependence, in particular the strong anisotropy in Raman tensor ($\alpha_{cc} \gg \alpha_{aa}, \alpha_{bb}$) of 630cm^{-1}, indicates the appearance of the depolarization field due to the ferroelectric ordering of water molecules along the c-axis. Taking into account the combination of the twist, rock and wag motion of the four water molecules in a primitive cell, the tentative assignments of librational modes as the lattice modes in ice XI phase were given.

Acknowledgements

We thank K.Ishii, M.Nakajima, H.Yamamoto, K.Nakano and T.Miasa for their collaboration in the preparation of samples and Raman experiments.

References

1   Y. Tajima, T. Mastuo and H. Suga, *Nature*, 1982, **229**, 810.
2   S. Kawada, *J. Phys. Soc. Jpn.*, **32**, 1442 (1972), ibid., 1988, **57**, 54.
3   R. Howe and R.W. Whitworth, *J.Chem.Phys.*, 1989, **90**, 4450.
4   S.M. Jackson, V.M. Nield, R.W. Whitworth, M. Oguro and C.C. Wilson, *J.Chem. Phys.*, 1997, **101**, 6142.
5   J-C. Li, *J.Chem.Phys.*, 1996, **105**, 6733.
6   H. Itoh, K. Kawamura, T. Hondoh and S. Mae, *J.Chem.Phys.*, 1998, **109**, 4894.
7   J.S. Tse, M.L. Klein and I.R. McDonald, *J. Chem. Phys.*, 1984, **81**, 6124.
8   M. Kobayashi, *Solid State Physics* (in Japanese), 1996, **31**,369.
9   J-C. Li, V.M. Nield and S.M. Jackson, *J.Chem.Phys. Lett.*, 1995, **241**, 290.
10 K. Abe, K. Ishi, M. Nakjima, H. Fukuda and T. Shigenari, *Ferroelectrics*, 2000, **239**, 871.
11 K. Abe, T. Miasa, Y.Ootake, K.Nakano, M. Nakajima, H. Yamamoto and T. Shigenari, *J.Korean Phys. Soc.*, 2004, **46**, 300.
12 P.T.T. Wong and E. Whalley, *J. Chem. Phys.* 1976, **65**, 829, ibid.,1975, **62**, 2418.
13 K. Nakano, *Master's theses*, The University of Electro-Communications *(in Japanese)*, 2003.

ON THE PERFORMANCE OF SIMPLE PLANAR MODELS OF WATER IN THE VAPOR AND THE ICE PHASES

A. Baranyai[1] , A. Bartók[1], and A.A. Chialvo[2]

[1]Institute of Chemistry, Eötvös University, Budapest, Hungary
[2]Oak Ridge National Laboratory, Tennessee, USA

1 INTRODUCTION

Interaction potential models of water developed for computer simulations are typically fitted to the properties of the liquid phase. The most frequently used experimental data to be matched are the heat of vaporization (or the configurational internal energy), the structure at the level of pair correlations and the density. In the case of the most popular models tests have been carried out for further properties to check their performance[1]. Thanks to their classical, nonpolarizable character rigid planar models of water like the SPC/E[2] and TIP4P[3] are inexpensive to implement in computer simulations. In the following table we present some alternative parametrizations suggested in the literature recently.

Table 1 *Geometry and potential parameters of the water models, where q is the partial charge on the hydrogen in the unit of elementary charge; L(O-H) is the distance between the oxygen and the hydrogen in Angstroms; α is the angle between the two O-H bonds in degrees; L(q) is the position of the negative partial charge from the oxygen on the bisector of the HOH bond ; ε(LJ) is the Lennard-Jones energy parameter in kJ/mol; σ(LJ) is the Lennard-Jones size parameter in Angstroms; μ is a dipole moment in Debyes.*

Model	q	L(O-H)	α	L(q)	ε(LJ)	σ(LJ)	μ
SPC/E	0.4238	1.0	109.47	0.0	0.6504	3.166	2.35
TIP4P	0.52	0.9572	104.52	0.15	0.6492	3.153	2.18
average	0.4761	0.9786	107.0	0.073	0.6499	3.160	2.33
TIP4P-ice	0.5897	0.9572	104.52	0.1577	0.8821	3.1668	2.42
TIP4P-EW	0.52422	0.9572	104.52	0.125	0.6809	3.164	2.32
TIP4P-m	0.5474	0.9572	104.52	0.15	0.6809	3.164	2.29
TIP4P-2005	0.5564	0.9572	104.52	0.1556	0.7803	3.163	2.345

The *average* model is a combination of the TIP4P and SPC/E interactions[4]. The TIP4P-ice was devised to reproduce correct melting temperature for hexagonal ice[5]. The TIP4P-EW is a recent reparametrization of the original TIP4P model to provide better estimations for liquid properties[6], while the TIP4P-m[7] and TIP4P-2005[8] were devised to give better properties for ice.

Using these models we wanted to find answers to several questions. Which of the above models is the best parametrization if we predict not only the liquid but the ice and the vapour phase properties as well? What is the performance of these models in general, or to put it differently, what are the inherent limitations of these simple models in terms of estimations of several properties simultaneously? As for simulations of ice phases the question is whether such a simple classical model is able to grasp the major characteristics of these systems, i.e., the energy, density and structure variations in terms of the pressure and temperature? The question of stability for the proton-ordered, proton-disordered polymorph pairs is a subtle problem. Can one obtain any useful information about these problems by using classical models?

2 METHOD AND RESULTS

2.1 Studies in the ice phases and in the vapour phase

We used a standard Monte Carlo code with Ewald summation and, if needed, Rahman-Parinello type volume variation (N,p,T ensemble) to simulate the 13 polymorphs of ice[4,9]. We studied the SPC/E and the TIP4P models at fixed volume[9] and the TIP4P-EW and the *average* model at fixed pressure[4]. Simulations for the TIP4P-2005 were carried out by Abascal and Vega[8]. It is obvious that the configurational energies of the ices depend sensitively on the potential parameters. A larger amount of calorimetric measurement should be accomplished to decide about the accurate internal energy to be reproduced by the simulations.

The most important finding was that potential models (not only the ones studied here) fitted to the density of liquid water overestimate the energy of the high pressure phases (ice VII, VIII) by 4-6 kJ/mol[9]. This was argued to be caused by the fact that in liquid water and in the low pressure ice phases the water molecule has 4 neighbours, while in the high pressure phases the number of neighbours is 8, and only 4 of them is connected by hydrogen bonds. The fixed repulsive part of the interaction is unable to adjust itself to this situation, in contrast to real molecules.

Table 2 *Critical conditions, melting and maximum density temperatures as predicted by the rigid planar models*

Model	$T_c(K)$	$\rho_c(g/cm^3)$	$P_c(bar)$	Ref.	$T_m^{(\#)}$	T_{md}	Ref.
TIP4P	558	0.313	112.0	4,9,14	232.0	253, 258	12'13
SPC/E	638	0.273	139.0	10	---	---	
average	609	0.326	120.0	4,9,14	---	---	
TIP4P-ice	705	0.305	184.0	4,9,14	272.2	295	13
TIP4P-2005	617	0.309	121.0	4,9,14	252.1	278	8
TIP4P-EW	639	0.315	118.0	4,9,14	---	----	----
TIP4P-m	623	0.342	128.0	4,9,14	---	---	---
Exp.	647.1	0.322	220.6	11	273.15	277	11

We calculated the temperature dependence of the second virial coefficient for some models and compared them to the experimental data[4].(See Figure 1) Each model underestimated the correct value. The size of the discrepancies varied by the size of their dipole moment, i.e., TIP4P (2.18D), TIP4P-m (2.29D), TIP4P-EW (2.32D), *average*

(2.33D), and SPC/E (2.35D). We also calculated the temperature dependence of the vapour pressure for the same set of models[4]. The TIP4P overestimates, the rest of the models underestimate the vapour pressure values[4].

The augmented and fixed dipole moments of the models could give good estimates for the vapour-liquid equilibrium only as a result of fortuitous cancellation of forces. In Table 2 we present the critical values of the state variables for the different models. None of the models predicts all the three state variables with acceptable accuracy. Both the pressures and in most cases the temperatures are smaller than the experimental values. For further details see Refs.4, 9,14.

The other properties shown in the table are the melting temperatures and the temperatures of maximum density. In addition to the literature data shown in the table, we also carried out some pilot calculations and experienced that these temperature values increase with the magnitude of the dipole moment. However, the difference of the melting temperature and the temperature of maximum density remains roughly the same: $\sim 20K$. It seems that there is no optimal value of the dipole moment which could bring this difference to the correct value of 4K.

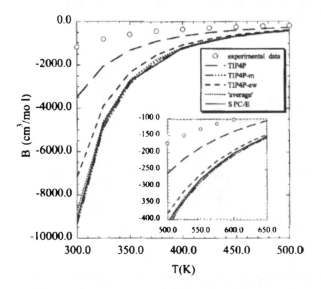

Figure 1 *Temperature dependence of the second virial coefficient of water as predicted by six models in comparison with experimental data*

2.2 Cis/trans isomerism of proton order in ice phases

Two water molecules connected by hydrogen bond in hexagonal ice can have four possible configurations. These configurations are distinguished by the relative orientations of the two molecules and termed for obvious reasons as c-cis, h-cis, c-trans, and h-trans. (See Figure 2) Occurrence of symmetry permitted dimer orientations is a characteristic feature of each ice polymorph. In the proton-ordered structures the occurrence of orientations is strictly determined, while in the proton-disordered structures within the limitations of the ice rule it can vary within certain limits.[15,16]

We performed Monte Carlo simulations using the TIP5P-EW[17], TIP4P-EW, and TIP5P-2005 models to study this isomerism in the phases of ice. First we tested these models for the ice VII (disordered) ice VIII (ordered) transition. To ensure equilibrium conditions we created the input of the ice VIII phase by introducing its proton order into the starting configuration of ice VII. The cubic system transformed into a tetragonal one, manifesting the properties of the equilibrated ice VIII model phase. Such a way we had identical number of particles. The ice VII phase to be compared was represented by several hundred samples selected from different realizations of proton disorder. Following the nomenclature of the hexagonal phase, in this structure there are only two orientations, h-cis and h-trans. According to the available experimental evidence[18,19] the transition takes place in the pressure range of 2 – 12 GPa around 263-273 K. The TIP5P-EW model used by us could not reproduce these results. In the calculations for the ice VII and ice VIII polymorph pairs using the TIP4P-EW and TIP4P-2005 potentials the energy of ice VIII was lower for both models by 0.05-0.15 kJ/mol, at each state point studied[20]. This means that even these simple planar, nonpolarizable models are able to capture the ice VII to ice VIII phase transition correctly. These transitions could be an additional test of performance for empirical potentials.

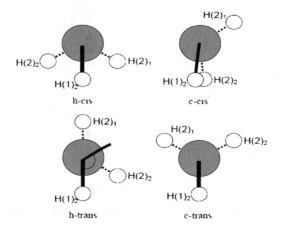

Figure 2   *A schematic diagram of the four possible arrangements of hydrogen bonded dimers in hexagonal ice. The oxygen molecules (gray circle) perfectly overlap. The numbers 1 and 2 in parentheses mark atoms of the donor and the acceptor molecules, respectively. On the diagram of the c-cis configuration the $H(1)_2$ and $H(2)_2$ overlapping atoms are shifted deliberately for the sake of clarity. On the diagram of the h-trans configuration we show the angle used for defining the relative orientation of the two molecules.*

The identification of the isomers is unequivocal only in the tetrahedral structures (Ih, Ic, VII, VIII, X). Nevertheless, in some phases we can replace the unequivocal formalism of the tetrahedral structures with a looser definition of dimer orientations.[20] In Figure 3 we show the distribution of angles for ice III and ice IX at 5 K. These dimer angles (see Figure 3) represent the input orientations created from the measurement data. At high temperatures the peaks of the ice III distribution smooth out and only four peaks remain, similarly in character to the dimer angle distribution of the hexagonal phase[20]. This high temperature (250 K) distribution shows a reasonable amount of order, in contrast to the ice IV or ice V phases, which justifies our following calculations. So, we divided up the dimer angle distribution of ice III into four ranges. (In Figure 3 dotted vertical lines).

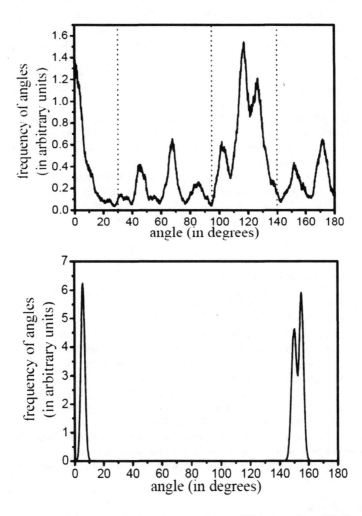

Figure 3 *Frequency of dimer angle distributions for ice III (up) and ice IX (down) at 5K.*

We created several input arrangements by different relative occurrences of orientations in these four ranges by reforming the hydrogen bond network in the structure while maintaining the ice rule and the zero net dipole moment of the phase. We expressed the configuration energy as a combined energy of energy values related to specific dimer orientations. In fact, we solved an over-determined linear equation where the ratio of dimer angle occurrences in the four ranges varied. The densities of the two phases at 5 K are not very different. So, we could estimate the energy of ice IX combining the corresponding energy values of ice III. For TIP5P-EW using the weights of the corresponding ice IX peaks we obtain $-56.0 \cdot \frac{1}{3} - 58.5 \cdot \frac{2}{3} \cong -57.7$ kJ/mol. The value obtained directly from the ice IX simulation is -57.95 kJ/mol. The energy values for TIP4P-EW at 10 K following the ranges of angles from 0° to 180° were -68.5, -38.1, -62.1, and -57.5 kJ/mol, respectively. From these data the calculated energy for ice IX is -61.2 kJ/mol, while the simulated energy is -61.2 kJ/mol. The energy values for TIP4P-2005 at 10 K following the ranges of angles from 0° to 180° were -70.4, -38.6, -63.6, and -58.6 kJ/mol, respectively. From these data the calculated energy for ice IX is -62.5 kJ/mol, while the simulated energy is -62.3 kJ/mol. These calculations revealed a remarkable agreement. While energy values are functions of several factors (system size, calculation of long range interactions, etc.), the leading term of the energy difference for a proton ordered-disordered pair is the occurrence of certain orientation angles of hydrogen bonded dimers. This proves that our method of energy assignation by orientation angles is a reasonable approximation for the energy content of different orientations.

It is important to mention by passing that the imperfect tetrahedral order of ice III and ice IX has no impact on our calculations because their imperfections are identical in character and cancel out any distortions influencing the results.

3 CONCLUSION

In summary, our simulation results provide strong support to the contention that rigid-planar nonpolarizable models of water suffer from an inherent transferability problem due to their inability to adjust their interaction strength to the actual polarizing environment. None of this type of models is capable of predicting correct critical data, vapour pressure or second virial coefficient. None of the models tested so far predicts the difference of the melting and the liquid density maximum temperature accurately.

Recent parametrizations achieved only partial improvements. The TIP4P-EW and the TIP4P-2005 in addition to provide better match of liquid properties[6,8], give more accurate estimates of densities and internal energies of ice phases. Results using these potentials support the ice VII – ice VIII phase transition. Using these potentials we could show that the energy difference of an ordered and disordered pair of ice phases is determined primarily by the occurrence of certain orientation angles in their hydrogen bonded dimers.

It seems that the method to fix the size parameter of the Lennard-Jones potential to match the density of ambient water is unable to predict the internal energy of the high density phases of ice.

Acknowledgements

A.B. acknowledges OTKA grant No. T043542. Research at ORNL was supported by Contract No. DE-AC05-00OR22725.

References

1 B. Guillot, *J. Mol. Liquid,*, 2002, **101**, 219.
2 H.J. Berendsen, J.E. Grigera,and T.P Straatma, *J. Phys. Chem.*, 1987, **91**, 6269.

3   W.L. Jorgensen, J. Chandrashekhar, J.D. Madura, R.W. Impey, and M.L. Klein, *J. Chem. Phys.*, 1983, **79**, 926.

4   A Baranyai, A. Bartók, and A.A. Chialvo, *J. Chem. Phys.*, 2006, **124**, 074507.

5   J.L.F. Abascal, E. Sanz, R.G. Fernandez, and C. Vega, *J. Chem. Phys.*, 2005, **122**, 234511.

6   H.W. Horn, W.C. Swope, J.W. Pitera, J.D. Madura, T.J. Dick, G.L. Hura, and T. Head-Gordon, *J. Chem. Phys.*, 2004, **120**, 9665.

7   E. Sanz, C. Vega, J.L.F. Abascal, and L.G. MacDowell, *J. Chem. Phys.*, 2004, **121**, 1165.

8   J.L.F. Abascal and C. Vega, *J. Chem. Phys.*, 2005, **123**, 234505.

9   A. Baranyai, A. Bartók, and A.A. Chialvo, *J. Chem. Phys.*, 2005, **123**, 054502.

10 J.R. Errington and A.Z. Panagiotopoulos, *J. Phys. Chem. B*, 1998, **102**, 7470.

11 H.Flyvbjerg and H.G. Petersen *J. Chem. Phys.*, 1989, **91**, 461.

12 W.L. Jorgensen and C. Jenson, *J. Comp. Chem.*, 1998, **19**, 1179.

13 C. Vega and J.L.F. Abascal, *J. Chem. Phys.*, 2005, **123**, 144504.

14 A.A. Chialvo, A. Bartók, and A. Baranyai, *J. Mol. Liquids,* in press.

15 V.Buch, P.Sandler, and J. Sadlej, *J. Phys. Chem., B* 1998, **102**, 8641.

16 S.W. Rick, *J. Chem. Phys.*, 2005, **122**, 94504.

17 S.W. Rick, *J. Chem. Phys.*, 2004, **120**, 6085.

18 G.P. Johari, A. Lavergne, and E. Whalley, *J. Chem. Phys,.* 1974, **61**, 4292; *ibid*, 1980, **73**, 4150(E).

19 W.F. Kuhs, J.L. Finney, C. Vettier, and D.V. Bliss, *J. Chem. Phys.*, 1984, **81**, 3612.

20 A. Bartók and A. Baranyai, *Phys. Rev. B*, submitted for publication.

PHASES OF SUPERCOOLED LIQUID WATER

I.Brovchenko and A.Oleinikova

Physical Chemistry, University of Dortmund, Germany

1 INTRODUCTION

Three different amorphous ices are seen in experiment: low-density amorphous ice (LDA), high-density amorphous ice (HDA) and very-high density amorphous ice (VHDA).[1,2] Observation of the distinct forms of amorphous ices corroborates an old idea that anomalous behaviour of the properties of liquid water upon cooling may be explained, if water is considered as a mixture of molecules having quite different local ordering (mixture model of liquid water).[3] Transformations between amorphous ices upon varying pressure or temperature occurs in a way, which looks similarly to the first-order phase transition.[1,4] This experimental finding was supported by the simulation studies of the liquid-liquid transitions using various water models.[5-9] Existence of a phase transition between different phases of supercooled water provides physical justification for a mixture model of liquid water. Besides, critical point of such phase transition may have distant effect on water properties in a wide thermodynamic range. So, to understand anomalous water properties, it is important to know location of the phase transitions between various phases of supercooled water.

Experimental studies of supercooled water at zero pressure indicate a singular-like behaviour of liquid water properties with approaching 228 K,[10] interrupted by crystallization at 235 K. This anomaly may be caused by the phase transition from normal liquid water (with a density of about 1 g/cm^3) to low-density water upon cooling. The location of the critical point of this transition is unknown. If it is located at negative pressures,[6,7] liquid water should undergo first-order transition upon cooling along the liquid-vapour coexistence curve. Alternatively, a singular-like behaviour of water at 228 K should be attributed to the crossing of a Widom line, which is defined as the locus of the correlation length maxima emanating from the liquid-liquid critical point located at positive pressures.[5,9] Note, that some properties of liquid water at room and higher temperatures change qualitatively at some positive pressure (when water density is about 1.1 g/cm^3),[11,12] that may originate from another liquid-liquid transition of supercooled water. Crystallization prevents direct experimental studies of the transitions between liquid phases of water. However, observation of the phase transitions between various glassy states of water (LDA, HDA and VHDA) evidences possibility of the liquid-liquid transitions at higher temperatures, where liquid water is unstable with respect to crystallization. Crystallization can be suppressed by confinement of water in pores. Recent experimental studies show the first-order phase transition between two liquid phases of water confined in Vycor glass at zero pressure and about 240 K.[13] This transition is accompanied by sharp changes of the structural and dynamical water properties. Liquid water, confined in mesoporous silica and carbon nanotubes, shows qualitative change of dynamics at about 225 K[14] and 218 K,[15] respectively. This dynamic transition was attributed to crossing a Widom line.[9]

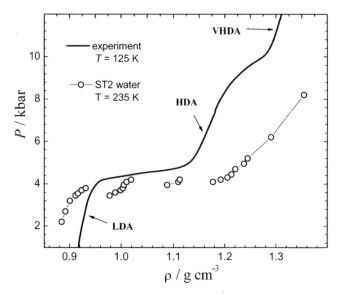

Figure 1   *Isotherms of the ST2 water model (shifted by +3.2kbar).[7] Solid line shows the experimental transformation LDA → HDA → VHDA under isothermal compression.[4]*

Simulations essentially extend possibility to study supercooled liquid water, as crystallization may be suppressed. However, there is no water model, which adequately reproduces phase diagram of water and its properties even in the thermodynamic region, where experimental data are available. In such situation, only comparative analysis of the results, obtained for various water models, can give information, relevant for the behaviour of real water in supercooled region. Additional complication appears due to the necessity to use sophisticated simulation methods, appropriate for the studies of the phase transitions, such as Monte Carlo simulations in the grand canonical or in the Gibbs ensemble (see Refs.7,16 for more details). Note, that simulations in the simple constant-volume or constant-pressure ensembles, widely used in the studies of supercooled water (see, for example Refs.17,18), are not appropriate for the location of the phase transitions.

In this paper we compare isotherms of supercooled water, obtained in simulations of various water models,[6,7,16] with available experimental data. Special attention is focused on the first (lowest-density) liquid-liquid transition, which influences properties of liquid water at zero pressure. This liquid-liquid transition of water is considered in terms of increasing concentration and percolation transition of the four-coordinated water molecules upon cooling.

2 FOUR PHASES OF SUPERCOOLED WATER

Various water models show multiple liquid-liquid transitions in supercooled region.[6-8] For example, there are three liquid-liquid transitions in ST2 water and two liquid-liquid transitions in TIP4P, TIP5P, SPCE, ST2RF (ST2 water with reaction field applied to account for long-range Coulombic interactions)[6,7] and in polarizable BSV water.[8] Accordingly, there are four water phases in the ST2 model and three in the other models. The phase transitions observed in simulations are usually the transitions between two liquid phases. In some cases, however, we observe transition between two amorphous phases below the glassy transition temperature.

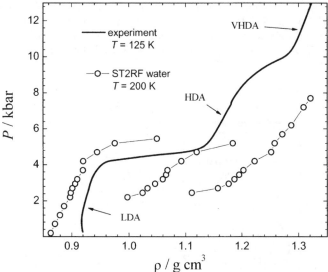

Figure 2 *Isotherms of the ST2RF water model (shifted by +1.2kbar).[7] Solid line shows the experimental transformation LDA → HDA → VHDA under isothermal compression[4].*

The pressure-density isotherms of various water models in supercooled region, obtained in simulations,[6,7] can be directly compared with the available experimental isotherms, showing transformations between amorphous ices upon compression.[4,19,20] Experimental measured isotherms are unavoidably affected by the transformation kinetics and by the strong hysteresis. These two effects may be reduced by using slow compression rates and higher temperatures, respectively. Therefore, the equilibrium isotherms obtained in simulations we compare with the experimental pressurization curves,[4] which were obtained under the slowest compression rate and at highest temperature.

In Figs.1-3 the simulated isotherms[7] of supercooled ST2, ST2RF and TIP5P water models are compared with the experimental pressurization curve, which shows transformations between LDA, HDA and VHDA at 125 K.[4] As the temperatures of the simulated isotherms differ from 125 K, they are shifted in figures in pressure, taking into account that LDA→HDA transition shifts by about +4 kbar upon cooling on 100°.[19] The extension of the branches of the isotherm in the pressure range strongly depends on temperature and becomes extremely large, when the system is deeply below the glassy transition temperature (TIP5P water, Fig.3). Comparison of simulated and experimental data in Figs.1,2 shows that ST2 and ST2RF models more or less adequately reproduce the density interval, where transitions between various phases of amorphous water occur. This is not the case for the TIP5P water model (Fig.3) and also for the SPCE and TIP4P water models.[16] Density interval, where these three models show transitions, is much narrower in comparison with the experimental one. Besides, contrary to ST2 and ST2RF models, TIP5P, SPCE and TIP4P models reproduce a low-density water phase (analogue of LDA) at strongly negative pressures only. Note also, that SPCE and TIP4P models do not reproduce even qualitatively the noticeable increase of the heat capacity of liquid water upon cooling.[7] So, among five considered water models, ST2 and ST2RF provide the most adequate description of liquid (amorphous) phases of real water in supercooled region. Contrary to ST2 water, ST2RF water gives unreasonably low density of liquid water at ambient conditions (about 0.93 g/cm³). Therefore, below we analyse four phases of supercooled ST2 water, which phase diagram in supercooled region is the closest to the one of real water.

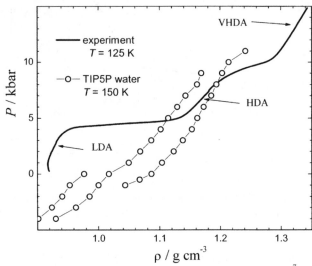

Figure 3   *Isotherms of the TIP5P water model (shifted by +1.0kbar).[7] Solid line shows the experimental transformation LDA → HDA → VHDA under isothermal compression[4].*

There are four phases of supercooled ST2 water,[17] which densities at corresponding liquid-liquid coexistence in the temperature interval from about 200 to 235 K are: $\rho^I \approx$ 0.90, $\rho^{II} \approx 1.00$, $\rho^{III} \approx 1.10$ and $\rho^{IV} \approx 1.20$ g/cm³. Obviously, water phase I is an analogue of LDA. It is important, that phase I of ST2 water and LDA exist at zero pressure. Most of water molecules in this phase are four-coordinated and show perfect tetrahedral ordering of the nearest four neighbours. First and second coordination shells are well defined in phase I. Phase II seems to be an analogue of "normal" liquid water, as its density is close to the density of liquid water at ambient conditions. There are no analogues of this phase among amorphous ices. This may reflect the absence of crystalline ices with the densities of about 1 g/cm³. We can not exclude that phase II disappears in deeply supercooled region. Phase

II is enriched with molecules, which have tetrahedrally ordered four nearest neighbours and up to 6 molecules in the first coordination shell. Phase III seems to be an analogue of HDA. Contrary to the phases I and II, first and second coordination shells of molecules in phase III is not clearly divided. Phases I, II and III are enriched with tetrahedrally ordered molecules. There is a noticeable drop of tetrahedral order in phase IV and it consists mainly of molecules with highly anisotropic distribution of the nearest neighbours. Phase IV may be considered as analogue of VHDA.

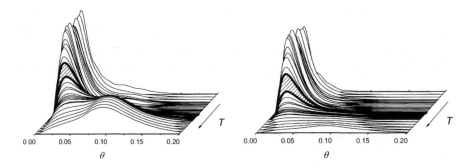

Figure 4 *Distribution of the tetrahedricity measure θ in liquid water along the liquid-vapour coexistence curve (from 100 to 450 K). Thick lines indicate two coexisting liquid phases. Left panel - total distributions; right panel – distributions for four-coordinated molecules.*

3 MIXTURE MODEL OF LIQUID WATER

The separation of a one-component fluid into two liquid phases should be attributed to the existence of molecules with qualitatively different local ordering. At least two components of water should be imposed in case of a single liquid-liquid phase transition. To apply mixture model for the analysis of water properties, it is necessary to determine the local ordering, which dominates in the considered water phase. In the phase I (analogue of LDA), tetrahedrally ordered four-coordinated water molecules seems to be the natural choice of the dominating component.

The tetrahedral arrangement of water molecule may be characterized by the tetrahedricity measure:

$$\theta = \frac{6\sum_{i\neq j}(l_i - l_j)^2}{15\sum_i l_i^2}, \tag{1}$$

where l_i are the length of the six edges of the tetrahedron formed by the four nearest neighbours. For an ideal tetrahedron, θ is equal to zero. Distributions of the tetrahedricity measure in liquid water at various temperatures along the liquid-vapour coexistence curve are shown in Fig.4 (left panel). There is a pronounced two-peak structure of these distributions in a wide temperature range. The left and right peaks correspond to tetrahedral and essentially isotropic distributions of the four nearest neighbours, respectively. These distributions were used to establish criterion ($\theta < 0.06$) for tetrahedrally ordered water molecules. Distributions of the tetrahedricity measure for the four-coordinated water molecules (with four neighbours within the coordination shell of 3.5 Å radius) are shown in the right panel of Fig.4. Thick lines denote the distributions, observed in two coexisting phases of liquid water at $T = 270$ K.

The obtained distributions of the tetrahedricity measure were used for estimation of the concentration C of the four-coordinated tetrahedrally ordered water molecules. Temperature dependence of this concentration along the liquid-vapour coexistence curve is shown in the upper panel of Fig.5. There is only slight increase of C upon cooling from the liquid-vapour critical temperature to about 350 K (due to the temperature mismatch of ST2 water and real water, about 30 to 35° lower temperature should be expected for real water). The drastic increase of C is evident at lower temperatures, when approaching the liquid-liquid phase transition. At $T = 270$ K, concentrations of the tetrahedrally ordered four-coordinated water molecules in two coexisting phases was found to be about 28% and 46.5%. Such step increase of C is related to a step decrease of density from 0.97 to 0.91 g/cm^3.

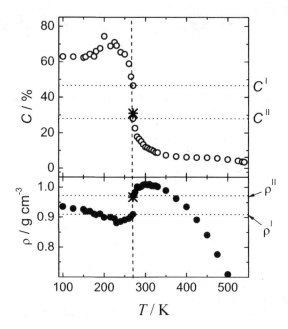

Figure 5   *Temperature dependence of the concentration C of the tetrahedrally ordered four-coordinated water molecules (upper panel) and of the liquid water density (lower panel) along the liquid-vapour coexistence curve. Vertical dashed line indicates the temperature of the liquid-liquid transition. Dotted lines indicate the densities and concentrations of the coexisting phases. Stars indicate percolation transition of the tetrahedrally ordered four-coordinated molecules.[18]*

The values of C in the coexisting phases indicate that critical concentration may be expected between 30 and 40%. More precise estimation may be obtained from the percolation analysis of the clustering of the tetrahedrally ordered four-coordinated molecules, as the line of percolation transitions should pass through the critical point. In fact, the concentration C at the percolation threshold was found slightly temperature dependent and equal to $C \approx 31\%$ (Fig.5).[21]

Drastic increase of C with approaching the liquid-liquid phase transition indicates close proximity of the liquid-liquid critical point, located at negative pressures. Similar increase of C (without the step increase, however) we may expect also for the ST2RF water model, where the liquid-liquid critical point is located at slightly positive pressure.[5,7,9] Strong temperature dependence of C leads to the strong increase of the isobaric heat capacity upon cooling,[22] which is clearly seen both in experiment[10] and in simulations.[7,9]

As we have mentioned in the Introduction, the location of the critical point of the lowest density liquid-liquid transition of real water is unknown and both scenarios (critical point at positive or at negative pressure) can qualitatively explain water anomalies. Recent simulation studies of confined water[23] show the way, how to locate the liquid-liquid critical point of water. Confinement in hydrophobic pores shifts the temperature of the liquid-liquid transition to lower temperatures (at the same pressure), whereas effect of confinement in hydrophilic pores is opposite. If the liquid-liquid critical point in real water is located at positive pressure, in hydrophobic pores it may be shifted to negative pressures. Alternatively, if the liquid-liquid critical point in real water is located at negative pressure, it may be shifted to positive pressures by confinement in hydrophilic pores. Interestingly, that it may be possible in both cases to place the liquid-liquid critical point at the liquid-vapour coexistence curve by tuning the pore hydrophilicity. We expect, that the experiments with confined supercooled water should finally answer the questions, concerning existence of the liquid-liquid phase transition in supercoleed water and its location.

References

1   O. Mishima, L.D. Calvert and E. Whalley, *Nature*, 1985, **314**, 76.

2   T. Loerting, C. Salzman, I. Kohl, E. Mayer and A. Hallbrucker, *Phys. Chem. Chem. Phys.*, 2001, **3**, 5355.

3   H. Whiting, *A New Theory of Cohesion Applied to the Thermodynamics of Liquids and Solids*, 1884, Harvard Physics PhD Thesis.

4   T. Loerting, W. Schustereder, K. Winkel, C.G. Salzmann, I. Kohl and E. Mayer, *Phys. Rev. Lett.*, 2006, **96**, 025702.

5   P.H. Poole, F. Sciortino, U. Essmann and H. E. Stanley, *Nature*, 1992, **360**, 324.

6   I. Brovchenko, A. Geiger and A. Oleinikova, *J. Chem. Phys.*, 2003, **118**, 9473.

7   I. Brovchenko, A. Geiger and A. Oleinikova, *J. Chem. Phys.*, 2005, **123**, 044515.

8   P. Jedlovszky and R. Vallauri, *J. Chem. Phys.*, 2005, **122**, 081101.

9   L. Xu, P. Kumar, S.V. Buldyrev, S.-H. Chen, P.H. Poole, F. Sciortino and H.E. Stanley, *PNAS*, 2005, **102**, 16558.

10   C.A. Angell, J. Shuppert and W.J. Sichina, *J. Phys. Chem.*, 1982, **86**, 998.

11   M. Krish, P. Loubeyre, G. Ruocco, F. Sette, M. DÁstuto, R. LeToulec, M. Lorenzen, A. Mermet, G. Monaco and R. Verbeni, *Phys. Rev. Lett.*, 2002, **89**, 125502.

12   F. Li, Q. Cui, Z. He, T. Cui, J. Zhang, Q. Zhou, G. Zou and S. Sasaki, *J. Chem. Phys.*, 2005, **123**, 174511.

13   J. M. Zanotti, M.-C. Bellissent-Funel and S.-H. Chen, *Europhys. Lett.*, 2005, **75**, 91.

14   L. Liu, S.-H. Chen, A. Faraone, C.-W. Yen and C.-Y. Mou, *Phys. Rev. Lett.*, 2005, **95**, 117802.

15   E. Mamontov, C. J. Burnham, S.-H. Chen, A. P. Moravsky, C.-K. Loong, N. R. de
     Souza and A. I. Kolesnikov, *J. Chem. Phys.*, 2006, **124**, 194703.
16   I. Brovchenko and A. Oleinikova, *J. Chem. Phys.*, 2006, **124**, 164505.
17   R.Martoniak, D.Donadio and M.Parinello, *Phys. Rev. Lett.*, 2004, **92**, 225702.
18   N.Giovambattista, H.E.Stanley and F.Sciortino, *Phys. Rev. E*, 2005, **72**, 031510.
19   O. Mishima, *J. Chem. Phys.*, 1994, **100**, 5910.
20   E.L.Gromnitskaya, O.V.Stalgorova, V.V.Brazhkin and A.G.Lyapin, *Phys. Rev. B*,
     2001, **64**, 094205.
21   A. Olenikova and I. Brovchenko, *J. Phys.: Cond. Matt.*, 2006, **18**, S2247.
22   E.G. Ponyatovsky, V.V. Sinitsyn and T.A. Pozdnyakova, *J. Chem. Phys.*, 1998, **109**,
     2413.
23   I. Brovchenko and A. Oleinikova, 2006, cond-mat/0606207.

USING GAUSSIAN CURVATURE FOR THE 3D SEGMENTATION OF SNOW GRAINS FROM MICROTOMOGRAPHIC DATA

J.B. Brzoska[1], F. Flin[2] and N. Ogawa[3]

[1]Météo-France/CNRM, Snow Research Center, Grenoble-St Martin d'Hères, France.
[2]Institute of Low Temperature Science, Hokkaido University, Sapporo, Japan.
[3]Hokkaido Institute of Technology, Sapporo, Japan.

1 INTRODUCTION

The study of snow on the ground - the snowpack - is traditionally based on the notion of grains considered as fully representative of the microstructure, with the aim of parametrizing small-scale effects in a continuous approach of snow physics and mechanics. [1, 2, 3] This lead to an extensive work of classification that takes presently the form of an official reference document. [4] Due to metamorphism and packing, any snow layer other than fresh snow is sintered at various levels [5] depending on the snowpack's history. Hence the main problem is to choose a geometric definition of a "grain" that is consistent with snow physics and mechanics.

The answer is trivial when collecting snow grains from the field: [6,7] a grain is the smallest element that can be mechanically detached from the snow structure. On numerical images of snow samples in 2-D (thin sections), standard algorithms based for instance on the distance map [8] or region propagation [9] allow an efficient segmentation that is consistent with the mechanical meaning of a grain separation - that is - a region of mechanical weakness; [10] they are already implemented in commercial software.

The picture is however different in 3-D because of the many possible shapes and topologies of grain assemblies, often hard to segment even manually. The standard strategy to solve such problems relies on the use of discrete [11] or vector [12, 13] distance maps that allow to compute 1-D (wiry) skeletons. It provides a convenient segmentation of materials for which either the grain [14] or pore [15] phase consists in nearly spherical elements. However, these methods are not usable on 3D snow samples: Except for wet grains, snow grains and bonds are far from spherical, and require the generation of a homotopic medial surface (a 2-D skeleton that respects the object's topology), which still remains a difficult problem. [16, 17, 18]

Another strategy consists in using the Gaussian curvature G as a means of segmentation. Such a method has been recently proposed for CAD purposes by Zhang *et al.*[19] to detect different typical "parts" on triangulated objects. The aim of the present paper is to adapt this concept to large voxel data sets (here, snow images from X-ray tomography [20, 21]) known presently to "resist" triangulation for reasons of CPU time, memory and topology.

Discrete approaches allow computing G directly from the distance map [22], but generally give results that are prone to high digitization effects. Our approach takes advantage of a previously developed adaptive method of normal computation [23] that smoothes digitization effects while preserving significant details of the voxel surface. Once a smooth and accurate normal field **n** is obtained from raw voxel data, it can then be used to obtain a precise Gaussian curvature map.

In this paper, an original method to derive G directly from any normal field is first presented. The resulting curvature field is then used to drive an iterative process of "matter removal" able to separate geometrical grains by a connected region whose thickness is small as compared to its perimeter. In most cases, the desired separation surface can be extracted from the latter region by using standard algorithms. The relevance of this purely geometric procedure to the physics and mechanics of sintered materials is then discussed.

2 COMPUTING THE GAUSSIAN CURVATURE FROM THE NORMALS

The Gaussian curvature G is often defined from the principal curvatures $\kappa_1 \equiv 1/r_1, \kappa_2 \equiv 1/r_2$ as $G = \kappa_1\kappa_2$. Here, r_1 and r_2 are the principal radii of curvature of the surface Σ at point P. Unfortunately, this definition is not well suited to express G as a function of the normal field **n** that describes always a numerical surface. Just as the mean curvature $\kappa \equiv 1/2(\kappa_1 + \kappa_2)$ can take the form $\kappa = 1/2\,(\text{div }\mathbf{n})$, G can be defined directly in terms of normals from what is called the Gauss map [24].

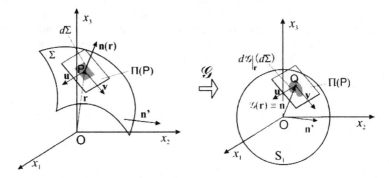

Figure 1 *The Gauss map $\mathcal{G}$ between a surface Σ around a point P of position vector* **r** *and the unit sphere S_1. Q on S_1 corresponds to P on Σ. The vector tangent plane Π is the same. $d\mathcal{G}$ transforms $d\Sigma$ into $dS_1 = d\mathcal{G}(d\Sigma)$, changing its shape and area.*

This mapping $\mathcal{G}$, defined from the surface Σ to the (abstract) unit sphere S_1, sends a point P of Σ whose position vector is **r** to its normal $\mathbf{n}(\mathbf{r})$ that is considered as the position vector of a point Q on the unit sphere (see Figure 1). In fact, $\mathcal{G}$ changes the point P of Σ into the point Q of S_1. Let $\mathcal{N}(P)$ be any neighborhood of P on Σ. By writing $a(X)$ to denote the surface area of X, the Gaussian curvature is defined by: [25]

$$G(P) \equiv \lim_{a(\mathcal{N}) \to 0} \frac{a\,[\mathcal{G}[\mathcal{N}(P)]]}{a\,[\mathcal{N}(P)]} \tag{1}$$

Its differential $d\mathcal{G}$ can be considered as a mapping on the vector tangent plane at P on Σ, or at Q on S_1 (by definition of $\mathcal{G}$, these tangent planes are identical). Let us choose a local frame $(\mathbf{u}\ \mathbf{v})$ such that $(\mathbf{u}, \mathbf{v}, \mathbf{n})$ be orthonormal direct. The surface element on Σ can then be written as $d\Sigma\ (\mathrm{P}) = dudv$. This element is a possible neighborhood of P, that is changed by $d\mathcal{G}$ into another oriented surface element $dS_1(\mathrm{Q})$ of the tangent plane such that $dS_1(\mathrm{Q}) = G\ d\Sigma\ (\mathrm{P})$. In the following, we will use the convention:

$$\mathbf{n}_x \equiv \frac{\partial \mathbf{n}}{\partial x} \tag{2}$$

By definition of $\mathcal{G}$, this surface element of the unit sphere is given by:

$$dS_1(\mathrm{Q}) = \mathbf{n} \cdot (\mathbf{n}_u \times \mathbf{n}_v)dudv \tag{3}$$

Hence, we obtain the following expression for G:

$$G = \mathbf{n} \cdot (\mathbf{n}_u \times \mathbf{n}_v) \tag{4}$$

This expression of G in the local frame $(\mathbf{u}, \mathbf{v})$ has now to be converted in the existing grid frame (x_1, x_2, x_3). For the following, it is convenient to orient locally the grid frame in a form (O, x, y, z) such that Σ can be represented around P as a function $z = f(x, y)$. In the configuration of Figure 2, this is achieved by identifying the x_2 axis to z (see Figure2), since:

$$\max\left(|n_{x_1}|, |n_{x_2}|, |n_{x_3}|\right) = |n_{x_2}| \tag{5}$$

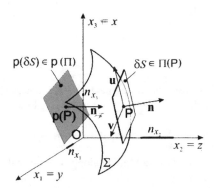

Figure 2 *Projection in 3-D of an element δS of the tangent plane to a surface onto the functional grid plane $\mathcal{F}$ ($\Pi(\mathrm{P})$), that is - the coordinate plane such that the surface equation can take the form $z = f(x, y)$. In the present configuration, $\mathcal{F}$ $[\Pi(\mathrm{P})] \equiv \mathsf{p}[\Pi(\mathrm{P})]$, with p being the projection mapping from $\Pi(\mathrm{P})$ to $\mathcal{F}$ $[\Pi(\mathrm{P})]$. This choice of orientation depends on $\Pi(\mathrm{P})$; it is done for each voxel by the algorithm before the computation of G*

To respect the orientation of frames, we will make also $x_1 \equiv y$ and $x_3 \equiv x$. By applying the chain rule for multivariate functions to the derivatives of Eqn. (4), we obtain:

$$\begin{aligned}
\mathbf{n}_u &= \mathbf{n}_x\,x_u + \mathbf{n}_y\,y_u \\
\mathbf{n}_v &= \mathbf{n}_x\,x_v + \mathbf{n}_y\,y_v
\end{aligned} \tag{6}$$

Then, we have:

$$\begin{aligned}
\mathbf{n}_u \times \mathbf{n}_v &= (\mathbf{n}_x\,x_u + \mathbf{n}_y\,y_u) \times (\mathbf{n}_x\,x_v + \mathbf{n}_y\,y_v) \\
&= (\mathbf{n}_x \times \mathbf{n}_y)\,(x_u\,y_v - x_v\,y_u)
\end{aligned} \tag{7}$$

The expression $x_u\,y_v - x_v\,y_u$ is the Jacobian $J(\mathbf{p}_z^{-1})$ of the backprojection mapping $\mathbf{p}_z^{-1}$ along the z-axis, from the plane $\mathcal{F}(\Pi)$ to Π at point P. By definition of the Jacobian, [26] m being the measure of the ball B_r of radius r, we have:

$$|J(\mathbf{p}_z^{-1})| = \lim_{r \to 0} \frac{m\left[\mathbf{p}_z^{-1}(B_r)\right]}{m(B_r)} \tag{8}$$

In two dimensions, Euclidean balls are disks whose measures are areas, so that $|J(\mathbf{p}_z^{-1})|$ is actually the backprojection ratio $\mathcal{R}$ that can be computed easily from the normal coordinates: [23]

$$\mathcal{R} \equiv \frac{\|\mathbf{n}_{\mathcal{F}}\|}{\|\mathbf{n}\|} = \max\left(|n_{x_1}|, |n_{x_2}|, |n_{x_3}|\right) \tag{9}$$

Once the grid frame is properly oriented around P, we obtain a practical formula to compute G from a discrete numerical normal field:

$$G = \mathbf{n} \cdot (\mathbf{n}_x \times \mathbf{n}_y)\,|n_z| \tag{10}$$

3 ALGORITHM

The practical aim of this work is to segment a 3-D numerical image into physically relevant "grains" by using the Gaussian curvature G as a criterion.

3.1 Computation of G

In the current implementation, G is computed along with normals from two-level (binary) voxel data of snow. For each surface voxel:
• d_f (P) being the background distance [8] at point P, an adaptive connected volume neighborhood $\mathcal{N}(P)$ that gathers vectors $\nabla d_f(Q)$ of similar orientations is constructed; $\mathbf{n}$ is then obtained by averaging $\nabla d_f(Q)$ for each point Q belonging to $\mathcal{N}(P)$. For more details, see the description of the ADGF algorithm. [23]
• The original coordinates (x_1, x_2, x_3) of the grid frame are identified to new coordinates (x, y, z) so that the surface can be expressed as $z = f(x, y)$ around P (see Figure 2). This is achieved by identifying the z-axis with the result given by Eqn.(5).

• Eqn.(10) can then be used by computing the derivatives $\mathbf{n}_x$ and $\mathbf{n}_y$ along with $\mathbf{n}$ in the adaptive neighborhood and provides the desired estimation of G. The figure 3 shows an example of computation on real snow data.

Figure 3 *Extract (width 1 mm) of the map of Gaussian curvature G as computed from a snow sample submitted to a long isothermal metamorphism (3 months at $-2\ °C$). Regions where G < 0 appear in dark gray.*

3.2 Segmentation

The segmentation procedure is based on a numerical model initially developed to simulate the isothermal metamorphism of snow. [27] The chosen strategy of segmentation consists in submitting the numerical sample to a non-physical metamorphism based on the local values of G instead of mean curvature $\kappa \equiv 1/2(\kappa_1 + \kappa_2)$ for the isothermal case, that "etches" the object iteratively at a rate proportional to G where $G < 0$ (see Figure 4). The medial surface between them is then reconstructed by using a "grass fire" algorithm. [28] The intersection of the ice volume and this medial surface is taken as the "numerical grain boundary". As can be seen from Figure5, a sharp, continuous and well-located grain boundary is generally obtained with this algorithm. However, for complicated necks where local differences in the Gaussian curvature field are poorly detectable (see the bottom neck on Figure 5.c), our method may fail in reconstructing fully connected grain boundaries. This delicate question will be addressed in more details in a further work.

Figure 4 *From left to right: Original object (2 joining spheres), the same during the "etching process".*

4 RELEVANCE TO SNOW PHYSICS AND MECHANICS

By definition of $G = \kappa_1\kappa_2$, connected regions of negative Gaussian curvature always describe one of the following local configurations of a curved surface:
• Indented regions of the surface, like notches, grooves or crevices.
• Narrow parts of the three-dimensional shape (the object) comprised inside the curved surface, such as necks or bonds.
• "Piercing" holes in the 3-D object like the inner region of a torus (here the surface is arcwise connected).

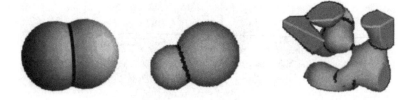

Figure 5   *Contours of grain separation surfaces (in dark gray) as found by the algorithm. From left to right: On the original object of Figure 4, on two spheres whose centers are not aligned on the grid, on a small snow sample.*

On the other hand, pits and craters are not concerned since both principal curvatures are negative in this case. Connected regions where $G < 0$ on a 3-D object of arbitrarily complicated shape describe indeed regions of mechanical weakness with respect to tensile or shear stress, providing that the object's material is considered homogenous (or with low strength regions like GB located in concave parts). When snow grains are collected on the field for observation, they are mostly cut off in such regions: what is called a "snow grain" is presumably well segmented by using an algorithm that uses Gaussian curvature. This observation applies to other sintered materials too.

Besides, there may exist a link between regions where $G < 0$ and grain boundaries. In the cloud, all convex parts ($\kappa > 0$) of low-rime snow crystals are monocrystals. Practically, a fresh snow layer that does not contain graupel is made of a collection of convex shapes separated either by a true GB (the limit between two contacting crystals) or by a concave neck already present in the initial snowflake; this neck may be either convex or concave in the sense of κ. Due to the so-called "destructive metamorphism" (isothermal metamorphism that acts on fresh snow crystals), convex necks commonly evaporate in a few hours, whereas concave necks within a crystal disappear because they are located between convex shapes; the resulting disjunctions are immediately followed by a geometric rearrangement of grains by gravity packing. The final result is then a collection of packed convex monocrystals whose shape is allowed to evolve by metamorphism. If this is isothermal metamorphism, concave regions evolve towards $\kappa \sim 0$ as mentioned above, so that the parameter κ is not convenient for segmentation purposes. However, G will stay negative in those "saddle" regions that include the grain boundaries. The morphologic evolution is different in case of a subsequent temperature gradient (TG for short) metamorphism, because the inhomogeneous vapor field generated by the TG displaces matter from warm to cold parts regardless of the location of grain boundaries. [5] Hence, there are no reason to expect GB where $G < 0$ in a snow sample that was previously submitted to a sustained TG metamorphism. Real "grains" of depth hoar as collected from the snowpack are indeed mostly polycrystalline, as can be seen for instance on a thin section between crossed polarizers.

5 CONCLUSION AND OUTLOOK

We presented a geometrical method that is able to segment 3-D images of sintered granular materials using Gaussian curvature as computed in an original way, directly from the raw field of normals. The resulting map of Gaussian curvature is used to drive a pseudo-metamorphism process of "matter removal". This process allows segmenting the digitized object at places that correspond to regions of structural weakness in the real object. On several types of snow layers, surface regions of negative Gaussian curvature are likely to include the contours of physical grain boundaries. Besides, the geometry of the "ablated regions" allows in most cases to reconstruct a thin separation surface without holes (simply connected) between grains. The robustness of the reconstruction is still to be checked on various topologies like multiple junctions. Once labelled, such numerical grains constitute a promising framework for further studies in micromechanics. One can for instance access directly the experimental coordination number that is considered as a key parameter in granular materials. [29] Moreover, a set of spatially located separation surfaces can be used as a realistic initialization of a discrete element model of sliding grain boundaries. [30] At least in the domain of high temperatures (like the "creep regime" close to $0°C$), this might open the road to the first explicit 3-D simulations of snow micromechanics.

Akcnowledgments

We thank Pierre-Jacques Font and Bernard Lesaffre for their valuable comments and fruitful discussions. Frédéric Flin is also grateful to the Japan Society for the Promotion of Science (JSPS) for its financial support.

References

1   R. Jordan, in *A one-dimensional temperature model for a snow cover: technical documentation for SNTHERM.89*. Technical Report 91-16, 1991.

2   E. Brun, P. David, M. Sudul, and G. Brunot, *J. Glaciol.*, 1992, **38**(128), 13.

3   P. B. Bartelt and M. Lehning, *Cold Reg. Sci. Technol.*, 2002, **35**, 123.

4   S. Colbeck, E. Akitaya, R. Armstrong, H. Gubler, J. Lafeuille, K. Lied, D. McClung, and E. Morris, in *The International classification for seasonal snow on the ground, International Commission on Snow and Ice of the International Association of Scientific Hydrology*, World Data Center A for Glaciology, University of Colorado, Boulder, 1990.

5   S. C. Colbeck in *A review of sintering in seasonal snow*, Technical Report 97-10, CRREL, 1997.

6   W. Good, in *International workshop on physics and mechanics of cometary materials*, ed. J. Hunt and T. Guyenne, ESA SP-302, 1989, 147.

7   B. Lesaffre, E. Pougatch, and E. Martin, *Ann. Glaciol.*, 1998, **26**, 112.

8   J.-M. Chassery and A. Montanvert, in *Géométrie discrète en analyse d'images*, Hermès, Paris, 1991.

9   J. Serra, in *Image analysis and mathematical morphology*, Vol. 1, Academic Press, London, 1982.

10   R. L. Brown and M. Q. Edens, *J. Glaciol.*, 1991, **37**, 203.
11   G. Borgefors, *Comput. Vision Graphics Image Process.*, 1986, **34**(3), 344.
12   P. E. Danielsson, *Comput. Graphics Image Process.*, 1980, **14**(3), 227.
13   J. A. Sethian, in *Level set methods and fast marching methods*, Cambridge University Press, Cambridge, 1999.
14   W. Lindquist and A. Venkatarangan, *Physics and Chemistry of the Earth, Part A: Solid Earth and Geodesy*, 1999, **24**(7), 593–599(7).
15   T. Dillard, F. NGuyen, S. Forest, Y. Bienvenu, J. Bartout, L. Salvo, R. Dendievel, E. Maire, P. Cloetens, and C. Lantuéjoul, in *Cellular metals and metal foaming technology*, 2003, pp. 1–6. Verlag MIT Publishing,.
16   G. Malandain and S. Fernandez-Vidal, *Image and Vision Computing*, 1998, **16**, 317
17   R. Z. M. Couprie, in *Discrete Geometry for Computer Imagery*, Lecture Notes in Computer Science, Springer-Verlag, 2005, pp. 216–225.
18   J. Toriwaki and K.Mori, in *Digital and Image Geometry*, ed. G. Bertrand, number 2243 in Lecture Notes in Computer Science, 2001, Springer-Verlag, pp. 412–429.
19   Y. Zhang, J. Paik, A. Koschan, and M. Abidi, in *Proc. Int. Conf. Im. Processing*, ed. K. M. Morton and M. J. Baines, Vol. III, pp. 273–276, 2002.
20   J.-B. Brzoska, C. Coléou, B. Lesaffre, S. Borel, O. Brissaud, W. Ludwig, E. Boller, and J. Baruchel, *ESRF Newsletters*, Apr 1999, **32**, 22.
21   M. Schneebeli and S. A. Sokratov, *Hydrological Processes*, 2004, **18**(18), 3655.
22   M. Jones, J. Baerentzen, and M. Sramek, *IEEE Trans. Visualization and Computer Graphics*, 2005, **12**(4), 581.
23   F. Flin, J.-B. Brzoska, D. Coeurjolly, R. A. Pieritz, B. Lesaffre, C. Coléou, P. Lamboley, O. Teytaud, G. L. Vignoles, and J.-F. Delesse, *IEEE Trans. Im. Proc.*, 2005, **14**(5), 585.
24   M. P. do Carmo, in *Differential geometry of curves and surfaces*, Prentice- Hall, Englewood Cliffs, New Jersey, 1976.
25   F. Rouvière, in *Petit guide de calcul différentiel,* 2nd ed., Cassini, Paris, 2003.
26   R. Courant and F. John, in *Introduction to calculus and analysis,* Vol. II/2, Springer-Verlag, Berlin, 1999.
27   F. Flin, J.-B. Brzoska, B. Lesaffre, C. Coléou, and R. A. Pieritz, *J. Phys. D. Appl. Phys.*, 2003, **36**, A49–A54.
28   I. Pitas, in *Digital image processing algorithms and applications*, Wiley, New York, 2000.
29   L. Sibille, F. Darve, and F. Nicot in Ed. Pande and Pietruszczak, in *Numerical models in geomechanics*, Balkema, Leiden, The Netherlands, 2004, pp. 91–98.
30   J. B. Johnson and M. A. Hopkins, *J. Glaciol.*, 2005, **51**(174), 432.

ETHANOL HYDRATES FORMED BY GAS-CONDENSATION: INVESTIGATIONS BY RAMAN SCATTERING AND X-RAY DIFFRACTION

B. Chazallon[1], C. Focsa[1], F. Capet[2], Y. Guinet[2]

[1]Laboratoire de Physique des Lasers, Atomes et Molécules (PhLAM), Université Lille 1, CNRS UMR8523, FR CNRS 2416, 59655 Villeneuve d'Ascq, France
[2]Laboratoire de Dynamique et Structure des Matériaux Moléculaires, Université Lille 1, CNRS UMR 8024, Villeneuve d'Ascq, France

1 INTRODUCTION

There is an increasing interest for the study of interaction of small oxygenated organics (e.g. formaldehyde, methanol, ethanol, etc) with ice due to their potential impact on the atmospheric and snowpack chemistry[1,2]. A highly reactive atmospheric compound like formaldehyde is known to hydrolyse rapidly in contact with water and generate hydrates during freezing of water droplets at low temperatures[3]. Conjointly, ethanol (EtOH) is predicted to be an important constituent of cometary's ices or dense cores where star formation occurs. In these media, condensed ethanol is found in abundance relative to H_2O (between 0.5 % and 5 %). It is many orders of magnitude in excess of predictions based on pure gas-phase chemistry which may partially be explained by surface grain reaction chemistry[4].

Previous works reported on the study of the freezing process of aqueous solutions of ethanol in relation with the phase diagram[5-7]. They were focused on the identification of different stable or metastable ethanol-hydrate structures using X-ray diffraction and differential scanning calorimetry (DSC). They also mentioned the formation of clathrate hydrates. In spite of the familiarity of alcohol-water systems and their importance in many fields of chemistry, cryobiology, astrophysics[8], there are still significant discrepancies in the literature with uncertainties' concerning the composition and the structure of the stable or metastable eutectic or peritectic hydrates[5-7,9].

On the other hand, a number of studies have been reported on the co-condensation of mixtures including water vapor and gas[10-13]. Clathrate hydrates can form after deposition at temperature below 100 K of an appropriate water:gas vapor ratio. A thermal treatment applied in the 110-150 K range promotes their crystallization[11,12]. These ice-like solids crystallize into distinct cubic structure (type I, with space group $Pm\overline{3}n$ and type II with space group $Fd\overline{3}m$) characterized by the presence of gas molecules (guests) stabilizing the host water framework[14]. A direct crystallization of the deposit may occur in the 100-150 K range if an appropriate amount of the occluded gas includes proton-acceptor molecules. These latter convey greater orientational mobility of the water lattice at low temperature by creating mobile defects that facilitate the trapping of non (or weakly) polar gas molecules and ensure the crystallization of the deposit[15]. Ethanol represents an interesting case in that the molecule is suggested to form occasionally H-bonds with the water lattice. Our previous Raman spectroscopic work suggests the formation of distinct hydrates that crystallize after annealing in the range 130 K -150 K for $0.23 < X_{EtOH} < 0.54$ and at ~ 170 K[16]. These hydrates are thought to reflect a molecular association (clusters) of different nature with respect to that encountered in frozen aqueous solutions.

In this work, detail Raman analysis of the EtOH internal vibrational modes and X-ray diffraction are performed to get insights on the structure of the EtOH hydrate phases formed by vapor deposition. The temperature and concentration effect are examined to determine how they affect the EtOH hydrate structure.

2 EXPERIMENTAL METHODS

2.1   Vapour deposition

Ethanol purchased from Verbièse (France) (purity > 99.77 %) is mixed with double distilled and deionized water (resistivity ~ 18 MΩ/cm) for the preparation of aqueous solutions of concentration: 0.6, 1, 2, 3, 4.5, 9 and 17 mol%. The solutions are placed in an insulated glass bulb connected to a vacuum line. The vapour-liquid equilibrium conditions are first obtained at 295 K before a contact between the gas phase in equilibrium above the liquid and the pre-cooled sample holder of a cryostage is established. The deposition takes place at ~ 10^{-1} Torr and 88 K in 3-5 min. Using thermodynamic modelling[16] and a condensation kinetic model (relation below), one derives the concentration (X_{EtOH}) of EtOH in the deposited solids as:

$$X_{Eth} = \left(1 + \frac{P_{H2O}\alpha_{H2O}\sqrt{M_{Eth}}}{P_{Eth}\alpha_{Eth}\sqrt{M_{H2O}}}\right)^{-1} \tag{1}$$

where P_i is the partial vapor pressure of a component i, α_i is the mass accommodation coefficient (assumed to be 1 at 88 K), M_i the molar mass of the component i. The corresponding values are then 9, 14, 23, 31, 40, 54 and 65 mol% respectively[16]. The temperature is then increased at rates of 1 K/min or 5 K/min in 10 or 5 K increment steps. To avoid evaporation of the sample, few mbar of high purity nitrogen gas is added at ~ 120 K. Further details of the deposition procedure can be found elsewhere[16].

2.2   Characterization tools

Micro-Raman spectroscopy is used to probe in situ the ice film grown by condensation. A modified LINKAM cryostage is attached to the BX 40 microscope of a XY-DILOR spectrometer. Working conditions are adjusted between 77 K and 273 K (± 0.1° K) and from 10^{-3} Torr to the atmosphere. The excitation radiation at 514.5 nm is produced by an Ar ion laser source (Coherent). A typical resolution of 1 cm^{-1} is chosen by adjusting the entrance slit width. To avoid the sample damage under laser irradiation, the laser power at the sample through the cell window is fixed at ~ 5 mW. Neon lines are used to calibrate the spectrometer.

X-ray diffraction is performed with a X-pert pro MPD powder diffractometer from Panalytical. This latter has been upgraded with the Panalytical X'celerator detector for high speed powder XRD measurements. The X-ray source consists in a high energy ceramic sealed tube (1.8 kW). The diffraction data are recorded with the Cu$_{K\alpha1, K\alpha2}$ (λ = 1.54056 Å, 1.5444 Å) radiation and the incident beam optics is alternatively used with a fixed divergence slit or a programmable divergence slit. This latter ensures that the irradiated length of the sample remains constant over full measured angular range thus improving the signal/noise ratio acquired at high θ. The sample stage consists in an Anton-Paar TTK 450 (80 K – 723 K) chamber. It is modified to allow the gas introduction through a membrane valve.

3 RESULTS AND DISCUSSION

3.1 Internal vibrational motions: the [2600-3720] cm^{-1} spectral range

Figure 1A and 1B display respectively the temperature dependent Raman spectra of condensed EtOH and the spectra of different EtOH-H$_2$O ice mixtures deposited at 88 K. Pure EtOH is investigated to get better insights on the behaviour of the mixtures. The bands observed at ~ 2882 cm^{-1}, 2930.8 cm^{-1} and ~ 2977 cm^{-1} in pure EtOH correspond respectively to the symmetric C-H stretching modes of the ethanol molecules $v_s(CH_2)$, $v_s(CH_3)$, $v'_s(CH_3)$ [17]. The corresponding bands in the EtOH-H$_2$O ice mixtures are found at somewhat higher frequencies (at 88 K): for e.g. ~ 2883 cm^{-1}, 2931.5 cm^{-1} and ~ 2977 cm^{-1} for the 65 mol%. The position of the intense band $v_s(CH_3)$ is determined with more precision and is significantly blue-shifted for mixtures of low concentration (Figure 1B).

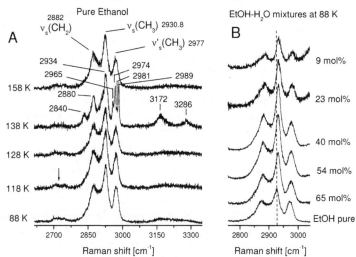

Figure 1   *(A) Raman spectra of vapour deposited pure EtOH and (B) EtOH-H$_2$O ice mixtures collected at 88K and different concentrations (C-H spectral region). The temperature cycle applied in (A) is read from the bottom to the top of the figure*

As reported in the case of the liquid mixtures[16,18], the softening of the C-H bonds as X_{EtOH} increases is explained by a partial and gradual electron transfer from the hydrogen and carbon to the oxygen of the EtOH molecule. This is consistent with a gradual increase with X_{EtOH} of the H-bond strengths of the EtOH molecules. This also highlights the role played by the EtOH oxygen acceptor or the end chain EtOH proton donor in H-bonds formation. The effect can also be observed for the modes $\nu_s(CH_2)$ and $\nu'_s(CH_3)$, thus indicating a complete participation of all hydrogen in the H-bonds formation. A weak contribution is observed at ~ 2729 cm⁻¹ (see arrow in Figure 1A) on the low frequency side of the band assigned to $\nu(CH_2)$. The energy is too low to correspond to a C-H stretching mode. It may be attributed to an overtone of a CH bending vibration occurring at ~ 1382 cm⁻¹, which is apparently slightly perturbed by Fermi resonance effect with a C-H stretching mode[19].

    3.1.1 Ethanol crystal. An important difference between pure and mixed EtOH solid occurs at 138 K. Additional components are observed at ~ 2840 cm⁻¹, ~ 3172 cm⁻¹ and ~ 3286 cm⁻¹ in pure EtOH, while the band at 2977 cm⁻¹ is split into multiple sharp bands with two intense peaks at ~ 2974 cm⁻¹ and ~ 2989 cm⁻¹ and weaker ones at ~ 2965 cm⁻¹ and ~ 2981 cm⁻¹.

    Such changes are attributed to the EtOH crystallization. Distinct solid phases of EtOH are known to exist. An "orientational glass" (OG) or a "structural glass" (SG) appear at T < 95 K and are obtained from the liquid state by cooling (at rate > 6 K /min for SG) or by applying a thermal treatment between T < 95 K and 110 K for the OG phase[20]. OG refers to a solid where the molecules are arranged on an ordered lattice with disordered orientations while SG corresponds to a solid where the molecules are disordered both in position and orientation. One deduces that our Raman spectra of condensed ethanol at 88 K correspond certainly to a SG phase. During heating at 108 K, the expected rotator crystal phase (RP) does not appear since our sample is maintained only a few minutes at this temperature. Several hours of equilibrium are necessary to observe the RP phase at annealing temperature of 105-110 K. The complicated band structure observed at 138 K (Figure 1) may in turn be ascribed to the stable monoclinic EtOH crystal[21], in which two crystallographically independent EtOH molecules are joined by H-bonds of different lengths. They form infinite chains of repeated units containing four EtOH molecules. The weak and broad band at ~ 3180 cm⁻¹ corresponding to O-H symmetric stretching in the amorphous phase at 88 K splits into two well resolved components in the crystal. This is also a characteristic spectral feature of a number of alcohols as reported by Jakobsen et al.[22] using infrared spectroscopy (IR). These authors proposed that the origin of the two bands comes from intermolecular interactions due to a coupling of the O-H stretching vibrations through nearest neighbours. The dual bands may thus represent the in-phase and out of phase vibrations respectively. However, a more recent IR treatment[23] of vapour deposited EtOH suggests that this alcohol behaves differently, in that the dual components assigned to the coupled O-H stretching split at low temperature. The same tendency is observed for the out of plane bending COH. Additionally, in the isotopic diluted ethanol crystals the uncoupled O-H stretching gradually split as temperature decreases down to 18 K and tends to average in one component at higher temperature (150 K). This is considered as a confirmation that the ethanol chains include molecules of distinct H-bond strengths and points to the existence of two independent sets of hydrogen bonds[23] as suggested previously[21].

    The different experimental conditions (high pressure, ambient temperature) in Jakobsen's work may have lead to the formation of a different allotropic form of solid ethanol as suggested from the polarized EtOH spectra reported by Mikawa et al.[24]. In our spectra, dynamic interactions may affect the coupled O-H vibration and may contribute to the presence of two components for $\nu(OH)$. Therefore, the origin of the two bands cannot be unambiguously assigned to the contribution of unequal H-bond strengths in the crystal.

    The C-H stretching mode spectral region is strongly perturbed in the EtOH crystal at 138 K. The bands at ~ 2965 cm⁻¹ and ~ 2974 cm⁻¹ may be associated with $\nu_s'(CH_3)$

symmetric stretching and the band at ~ 2989 cm⁻¹ may correspond to $v_a(CH_3)$. This attribution is based on a previous extensive IR analysis of liquid and solid ethanol[17]. The ~ 2840 cm⁻¹ band can correspond to a C-H symmetric stretching as deduced from the fact that symmetric stretching are assigned at lower frequencies than anti-symmetric. These results are consistent with the occurrence of two independent EtOH molecules in the monoclinic structure. The melting of the EtOH solid is observed at temperature slightly lower than 158 K, i.e. close to the melting point of EtOH[21].

3.1.2 Ethanol hydrates. It is remarkable that the EtOH-H₂O ice mixture exhibits relatively "soft" spectral changes between spectra collected at different temperatures. This is illustrated in figure 2 which displays the temperature dependency of $v_s(CH_3)$ for pure EtOH and EtOH-H₂O ice mixtures of different concentrations. The temperature and concentration dependency of $v_s(CH_3)$ can be qualitatively compared with that of $v(OH)$ reported previously[16].

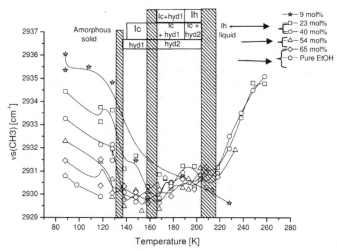

Figure 2 *Temperature dependence of the band assigned to $v_s(CH_3)$ for different EtOH-H₂O ice mixtures (see legend). The diagram at the top of the figure is based on both Raman and X-ray results.*

At 88 K, the molecules are frozen in random orientations and the characteristic decrease in frequency as X_{EtOH} increases merely reflects the gradual increase of H-bond strengths in which the CH₃ hydrogens participate (see above). The slight decrease of $v_s(CH_3)$ at any concentration from 88 K to ~ 120 K may be assigned to a change in the molecular organization around EtOH molecules or EtOH clusters. A remarkable drop in frequency is observed at ~ 138 K that is ascribed to the onset of crystallization. The C-H bond strengths are perturbed due to a modification of the local molecular environment in the growing crystalline phase. H-bonds formation between EtOH and water molecules may be responsible for the frequency shift observed. It should be noted that phase separation with ice and crystalline EtOH is not expected at this step as the characteristic spectra of pure crystalline EtOH do not appear during heating (Figure 1). Rather, an EtOH-hydrate 1 apparently stable between ~ 130 K and ~ 165 K is formed. Moreover, the higher the EtOH content in the deposit, the lower is the EtOH-hydrate formation temperature. This behaviour seems related to the nature of EtOH, which as an oxygen-containing molecule should generate an important number of defects of Bjerrum-type in the water lattice and induced extra mobility to water molecules. This provides a faster relaxation of water in the amorphous deposits[25] and subsequently promotes crystallization at lower temperature in comparison to ice. Hydrate formation is further facilitated by the added mobility induced by EtOH present in the crystal.

Further annealing leads to an additional change of frequencies. $v_s(CH_3)$ exhibits a blue shift (~ 1 cm^{-1}) at ~ 165 K (Figure 2) which may indicate that a structural rearrangement takes place. As confirms by X-rays, this corresponds to the formation of a new hydrate of EtOH (hydrate 2). The hydrate 2 formation temperature is apparently shifted to higher values at low X_{EtOH} (23 mol%). Above 210 K, the hydrate 2 dissociates (Figure 2) and a mixture of ice Ih and a viscous liquid of ethanol remain, just before the ice starts to melt[16]. The same scenario is assumed when considering the bands assigned to the skeletal EtOH vibrations (C-C and C-O stretching modes) with however less pronounced frequency variations between the different hydrate structures.

3.2   Hydrate structures

Our preliminary results obtained by X-ray diffraction confirm the occurrence of two independent EtOH hydrates. Figure 3 displays the diffraction pattern of a 54 mol% EtOH-H$_2$O ice mixture at different temperatures. The amorphous deposit shows up as broad background scattering especially at low temperatures (T $\leq \sim 128$ K). Progressively Bragg reflections begin to develop between 128 K and 138 K. This transformation is ascribed to the formation of the crystalline hydrate 1 phase, in agreement with the changes observed for the C-H vibrational motions (Figure 2). Diffraction patterns of ice Ih, ice Ic, pure ethanol RP and stable monoclinic EtOH are recorded in order to discriminate and recognize the presence of these phases. However, none of the reflections of the hydrate 1 (Figure 3) can be indexed on the basis of the aforementioned crystal patterns. This again confirms that phase separation (with crystallization of the separated phases) does not occur during the devitrification process of the deposit. 17 reflections can however be indexed in terms of the cubic Pm$\overline{3}$n space group of a type I clathrate. However, low angle reflections that usually characterize the type I are very weak or quasi absent from the pattern. Departure from the usual type I structure may happen due to H-bonds formation between EtOH and water molecules that may cause distortion of the lattice and form a semi-clathrate[26]. A refinement of the lattice parameters gives a unit cell dimension of a $=$ 12.006(3) Å at 143 K. The somewhat unusual large lattice may be explained by the size and shape of ethanol molecules (~ 6.28 Å) that is on the limit of fitting the large cage of type I. The lattice parameter is relatively close to the values reported in frozen EtOH aqueous mixtures for type I clathrate (a $= 11.88$ Å at 108 K)[26]. This hydrate should correspond to a composition of C$_2$H$_5$OH 5.75H$_2$O or most probably C$_2$H$_5$OH 7.6H$_2$O (empty small cavities). Using our condensation model[16], the estimated EtOH content corresponds rather to C$_2$H$_5$OH·1.2H$_2$O. This would mean that excess EtOH is apparently excluded from the crystalline phase and remains under a non-crystallized form (as suggested by the remaining broad background scattering) or is partially evaporated from the sample. The exclusion takes place in order to achieve the proper EtOH/H$_2$O ratio for hydrate formation. The Bragg reflections indicate that this hydrate is stable up to ~ 158 K, depending however on the EtOH content of the deposit. At ~ 158 K, additional peaks start to grow at 2θ $\sim 22.4°$, $\sim 25.7°$, $\sim 27.02°$ and $\sim 28.3°$, $\sim 28.9°$ while the hydrate 1 peaks decrease (Figure 3). This corresponds to the onset of hydrate 2 formation. 59 Bragg reflections can be distinguished using longer counting statistics on the 65 mol% EtOH, for which a good sample is obtained after transformation 1 → 2 at ~ 160 K.

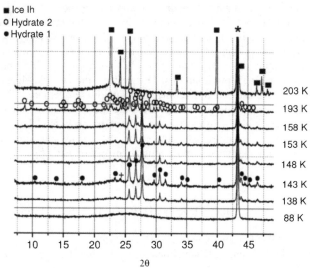

Figure 3   *Diffraction patterns of a mixed EtOH-H₂O deposit (54 mol%). Annealing temperatures are given on the right (bottom to the top). *parasitic reflections of the sample holder.+ ice Ih contamination.*

Unfortunately, neither gives the calculation of the peak indexing procedure an unambiguous result for the crystal system, nor can the observed Bragg lines be assigned to the reported lines of the other known EtOH hydrate structures[5,7]. Nevertheless, in light of our spectroscopic data, one infers that this crystal consists in somewhat weaker intermolecular interactions between water and ethanol molecules than in the hydrate I structure. Moreover, the hydrate 2 composition may approach 1/1 water/EtOH i.e. correspond to the monohydrate (intermediate higher hydrates showing distinct diffraction patterns[5,7]). It should be noted that excess EtOH (non-crystallized) and water excluded from the dissociated hydrate 1 may also contribute to hydrate formation. Further work is in progress to better characterize this step.

In lowering the concentration in ethanol (~ 9 or 14 mol%), ice Ic starts first to crystallize at ~143 K. The hydrate I begins to develop at ~ 163 K, i.e. at higher temperature compared with the 54 mol% EtOH deposit (Figure 3). Apparently, water molecules in excess need first to be excluded as ice Ic in order for the mixture to achieve the appropriate component ratio H₂O/EtOH for hydrate 1 formation. The hydrate 2 formation is shifted to 188 K (14 mol%), 193 K (31 mol%) or can be absent from the pattern (9 mol%). In this latter case, the (hydrated) alcohol clusters apparently remain disseminated in ice as the ratio water/EtOH required for hydrate 2 formation cannot be reached within the stability field of the phase. Further, its domain of existence seems to narrow as X_{EtOH} decreases. A more detail account will be given in a forthcoming paper.

4 CONCLUSION

Co-condensed EtOH-water mixtures reveal the formation of distinct EtOH hydrate phases in different temperature domains. A hydrate 1 appears in the ~ 130 K – 163 K range depending on the EtOH content. It is proposed to have a cubic lattice similar to that of the clathrate type I. Hydrate 2 is found to crystallize at ~ 158 K or 188 K-193 K in correlation with the absence or the presence of ice Ic and EtOH content. Its composition seems to correspond to the monohydrate. The deposited solids undergo crystallization ~ 10 K lower in comparison to frozen aqueous solutions. This reflects the remarkable ease with which water molecules initiate molecular rearrangement at low temperature. This seems most likely due to EtOH generating defects that facilitate the water reorientation[25]. This may also reflect the generation of clusters (in the vapour phase before deposition) having a different nature relative to those encountered in the liquid solutions. These unusual structures may have implications in atmospheric chemistry or astrophysics.

References

1   H. Singh, Y. Chen, A. Staudt, D. Jacob, D. Blake, B. Heikes, J. Snow, *Nature*, 2001, **410**, 1078.
2   F. Dominé, & P.B. Shepson, *Science*, 2002, **297**, 1506.
3   B. Chazallon, N. Lebrun, P. Dhamelincourt, C. Toubin, C. Focsa, *J. Phys. Chem. B*, 2005, **109**, 432.
4   N. Boudin, W.A. Schutte, J.M. Greenberg, *Astron. Astrophys.*, 1998, **331**, 749.
5   P. Boutron, A. Kaufmann, *J. Chem. Phys.*, 1978, **68**, 5032.
6   K. Takaizumi & T. Wakabayashi, *J. Solution Chem.*, 1997, **26**, 927.
7   T. Takamuku, K. Saisho, S.Nozawa, T. Yamaguchi, *J. Mol. Liq.* 2005, **119**, 133.
8   F. Franks & J. Desnoyers, In *Water Science Reviews*, vol. 1 *F. Franks, ed.* Cambridge University press (Cambridge), 1985.
9   S.S.N. Murthy, *J. Phys. Chem. A*, 1999, **103**, 7927.
10   K. Consani & G.C. Pimentel, *J. Phys. Chem.*, 1987, **91**, 289.
11   J.E. Bertie & J.P. Devlin, *J. Chem. Phys.*, 1983, **78(10)**, 6340.
12   F. Fleyfel & J.P. Devlin, *J. Phys. Chem.*, 1988, **92**, 631.
13   H.H. Richardson, P.J. Wooldridge and J.P. Devlin, *J. Chem. Phys.* 1985, **83**, 4387.
14   M. von Stackelberg, *Naturwissenshaften*, 1949, **36**, 327.
15   D.W. Davidson & J.A. Ripmeester, *Inclusion Compounds* vol. 3, *ed. J.L. Atwood, J.E.D. Davies and D.D. MacNicol*, Academic San Diego, Calif pp. 69, 1984.
16   B. Chazallon, Y. Celik, C. Focsa, Y. Guinet, *Vib. Spectrosc.* 2006 (in press).
17   J.P. Perchard and M.L. Josien, *J. Chim. Phys. Chim. Biol.*, 1969, **65**, 1856 (in French).
18   K. Mizuno, Y. Miyashita, Y. Shindo, H. Ogawa, *J. Phys. Chem.* 1995, **99**, 3225.
19   N.B. Colthup, L.H. Daly, S.E. Wiberley, *Introduction to infrared and Raman spectroscopy, 3rd Ed.*, Academic press San Diego, 1990.
20   M.A. Ramos, S. Viera, F.J. Bermejo, J. Dawidowski, H.E. Fischer, H. Schober, M.A. Gonzales, C.K. Loong, D.L. Price, *Phys. Rev. Lett.* 1997, **78**, 82.
21   P-G. Jönsson, *Acta Cryst.*, 1976, **B32**, 232.
22   R.J. Jakobsen, J.W. Brasch and Y. Mikawa, *J. Mol. Structure*, 1967-68, **1**, 309.
23   M. Rozenberg, A. Loewenschuss, Y. Marcus, *Spectrochim. Acta*, 1997, **53A**, 1969.
24   Y. Mikawa, J.W. Brasch and R.J. Jakobsen, *Spectrochim Acta*, 1971, **27A**, 529.
25   J.P. Devlin, *Int. Rev. Phys. Chem.*, 1990, **9**, 29.
26   D.W. Davidson, In *Water. A comprehensive treatise, F. Franks, ed.*, Plenum press (London), 1972 vol2, Chap 3.

DISLOCATION PATTERNING AND DEFORMATION PROCESSES IN ICE SINGLE CRYSTALS DEFORMED BY TORSION

J. Chevy[1,2], M. Montagnat[1], P. Duval[1], M. Fivel[2] and J. Weiss[1]

[1]Laboratoire de Glaciologie et Géophysique de l'Environnement, 38402 St Martin d'Hères, France.
[2]Génie Physique et Mécanique des Matériaux, ENSPG, 101 Rue de la Physique, BP 46, 38402 St Martin d'Hères, France.

1 INTRODUCTION

The early observation of macroscopic "slip lines" in deformed ice single crystals by Nakaya (1958)[1] or Readings and Bartlett (1968)[2] indicated the simultaneous and correlated motion of many dislocations. More recently, acoustic emission analyses performed on ice single crystals during deformation[3] revealed the scale-free intermittent motion of dislocations through dislocation avalanches.

In this paper we present synchrotron X-ray analysis of ice single crystals deformed in pure torsion. The observation of diffraction topographs revealed scale invariant arrangement of dislocations, independently of the macroscopically imposed deformation gradient[4]. Scale invariant patterning is thought to be induced by the long-range interactions between dislocations. This spatio-temporal heterogeneity of slip is induced by the motion of dislocations in a collective and self organized manner that can not be described in terms of local and uncorrelated relaxations of individual dislocation segments toward configurations of lower energy[5].

Preliminary Dislocation Dynamics (DD) simulations using the model developed by Verdier et al. [6] provide a plausible scenario for the dislocation patterning occuring during the deformation of ice single crystals based on cross-slip mechanism. The simulated dislocation multiplication mechanism is consistent with the scale invariant patternings observed experimentally.

2 EXPERIMENTAL OBSERVATIONS

2.1 Torsion deformation of ice single crystals

Pure torsion tests were performed on ice single crystals at a constant imposed external shear stress[4,7]. Softening was evidenced as the creep curves revealed a strain-rate increase, up to a cumulated plastic strain of 7%, see figure 1. Note that such a behaviour was also observed during compression and tension tests[8,9].

The initial dislocation density of the sample was estimated to be less than 10^8 m^{-2}. The orientation of the samples was chosen in order to align the torsion axis as close as possible with the c-axis (±1°). The maximum resolved shear stress is then applied on the basal planes. The plastic deformation is accommodated by the glide of screw dislocations on the three slip systems a/3<11-20>{0001} contained within the basal planes. We assume that dislocations can be freely nucleated at the surface of the sample, where the imposed shear stress is the highest.

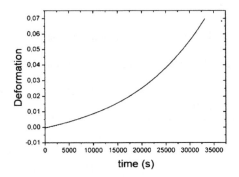

Figure 1 *Typical creep curve of an ice single crystal deformed in torsion up to 7% along the c-axis. The deformation is the measured shear strain on the surface.*

2.2 X-ray diffraction analyses

Specimens were extracted from the deformed samples at different radii from the cylinder axis, along the torsion axis direction. Some of the diffraction analyses were performed using a hard X-ray diffraction technique developped at Institut Laue Langevin, others were observed via synchrotron X-ray diffraction topography at ESRF (European Synchrotron Radiation Facilities) on the "ID19" beamline.

The hard-ray diffraction technique allows a fast characterization of the lattice distortions for sample size in the centimeter range or more[10]. The spatial resolution is low, but this technique showed the predominance of excess screw dislocations to accommodate the torsion deformation in all deformed crystals[7]. A negligible contribution of edge dislocations was evidenced, which is consistent with the loading conditions.

The ID19 beamline at ESRF is characterized by a spectrally and spatially homogeneous, highly coherent beam at the sample position. X-ray topography, an imaging technique based on Bragg diffraction, provides a two-dimensional intensity mapping of the beam(s) diffracted by the crystal. Topography analysis consists in the study of the fine structure of a Bragg spot which contains all the information about the defect structure. High energy (50-120 keV) white beam (with all wavelengths in the provided energy range) was used to allow the investigation of bulk ice samples in transmission[4]. Final images provide a final resolution of about 10μm. The topograph given in figure 2 is the diffraction pattern on the prismatic plane for a specimen extracted from a sample deformed at 2.69%.

The diffracted intensity records the orientation contrast and is, in our case, related to the dislocation density at the origin of the lattice distortion[4]. Although a direct relationship between diffracted intensity variations and dislocation density is difficult to assess, darker zones in the topographs correspond to higher dislocation density regions. We extracted the dislocation density variation signal along 1D profiles in the direction of the c-axis (figure 2) on [100] prismatic plane diffraction pattern. Indeed, as no extinction is expected for any of the 3 glide systems with a/3<11-20> Burgers vectors, the signal on this diffraction pattern is supposed to be complete.

Figure 2 *Diffraction pattern on the prismatic plane for a specimen taken from a sample deformed at 2.69%. The long dimension is the height of the specimen, while the width corresponds to the size of the diffracted beam through 0.1mm slit. The intensity profile is taken along the black line.*

Spectral analyses were performed along the 1D intensity profiles. The scaling properties of the profiles are independent on the position of the profile along the basal plane. The diffracted intensity along the 1D profile is spatially heterogeneous in the sense that it is not distributed around a mean value following a Gaussian distribution; instead it is characterized by a wide distribution and, as shown below, by spatial correlations. Figure 3 shows the power spectrum of the intensity record taken along a profile as represented in figure 2 (calculated from the fast Fourier transform of the signal). The power spectrum is characterized by a power law regime over the entire available scale range (bounded by the resolution of the analysis, and by the sample size), $E(f) \sim f^{-\mu}$. These scale ranges extend over more than 2 orders of magnitude, from 15 μm to about 7 mm. The power spectra obtained on 2 samples deformed at different levels are characterized by a same exponent $\mu = 1.3 \pm 0.1$. Such a power law reveals the scale invariance of the intensity records.

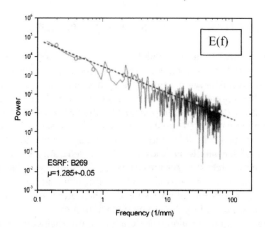

Figure 3 *Power spectrum of the intensity record taken from a 1D profile along the topograph represented in figure 2.*

Such scale invariant spectra reveal that the dislocation density variation is scale invariant, and put forward spatial correlations over very large distances with a correlation length of the order of the system (sample) size[3]. This is confirmed with the autocorrelation function of the records shown in figure 4, in which is also represented a signal taken from a non-deformed sample. The autocorrelation function obtained for the deformed sample slowly decreases with increasing distance Δx, to fall around nil values for distances of few millimetres, whereas the correlation falls off for much lower distances on the non-deformed sample. The increase in long-range correlations is due to the deformation processes. Such correlations are the fingerprints of strong interactions between dislocations. Thus, dislocations are neither distributed randomly, nor concentrated on well-defined "slip bands" more or less regularly spaced. Instead, they self-organize into a scale free pattern.

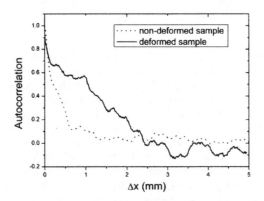

Figure 4   *Autocorrelation functions obtained for a deformed sample and a non-deformed sample.*

3 DISLOCATION DYNAMICS SIMULATIONS. Preliminary results

The Discrete Dislocation Dynamics code used here is based on the discretization of dislocations into edge and screw segments whose extremities are located on the nodes of a discrete network. Each dislocation segment displacement is due to an effective stress, which depends on the applied stress, the internal stress and the Peierls lattice friction. Cross-slip is allowed for the screw component of a dislocation line through a probabilistic law which involves the resolved shear stress in the deviate system. The code was originally developed for FCC metals[6]. In this study, it has been adapted to the simulation of the torsion deformation of ice single crystals mainly by selecting 3 over the 12 FCC glide systems to represent the ice slip systems (hexagonal symetry). The torsion moment is applied through a shear stress proportional to the radius superimposed to the internal stress fields. Dislocations sources are randomly spread at the outer radius of the specimen, where the applied shear stress induced by the moment is the highest.

When a moment is applied along the c-axis, screw dislocations are simultaneously activated in the 3 coplanar basal glide systems <11-20>{0001}. These dislocations are pushed toward the torsion axis inducing a kind of pile up in each glide system. Although the torsion loading does not induce any applied stress in the prismatic system, the internal stresses due to the dislocation arrangement create a shear stress component on the prismatic system that could favour cross slip events. Thus, when the shear stress component is higher in the prismatic plane than in the basal plane, a screw dislocation has a high probability to cross slip under the internal driving force. Then, at a given distance from the initial basal plane, the shear stress in the prismatic plane becomes lower than that of the basal plane and the dislocation cross slip back to the primary system. Such a double cross-slip creates a Koehler source that emits basal dislocations in a new basal plane.

Figure 5 represents a typical evolution of the dislocation pattern during the deformation. The simulation was performed in a 20 mm diameter crystal, with 2 initial basal planes activated (one system in each plane) at the beginning of the deformation. It clearly appears that the double cross-slip mechanism propagates the plasticity in many other basal planes. One can also notice the asymmetry in the plane expansion due to the dislocation interactions.

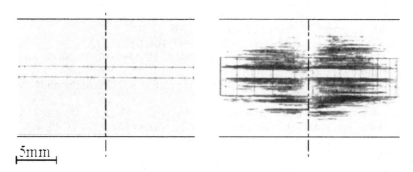

Figure 5 *Thickening of slip planes due to the double cross-slip of basal dislocations. Begining of the simulation (left), after deformation (right). Basal dislocations glide horizontally (horizontal lines), dislocations in the prismatic plane vertically (white vertical lines). The center of the cylinder is represented.*

4 DISCUSSION

During the torsion experiments performed in this study, only screw dislocations are activated on the basal planes by the imposed shear stress. If dislocations can easily nucleate at the free surface, such a random multiplication process cannot explain the spatial correlations observed by X-Ray diffraction topography. As suggested by the DD modelling, the double cross-slip mechanism appears to be an efficient mechanism to provide a dislocation multiplication process that strongly depends on neighbouring dislocation interactions. As previously modeled by Mendelson[11], cross-slip is enhanced by the repulsive stress of neighbouring screw arrays. Such a propagation of dislocation slip in other planes is influenced by dislocation interactions, and can lead to scale invariant patterning for dislocation arrangements with long range correlations.

The double cross-slip mechanism can then be considered as the most probable deformation process, complementary to the basal slip. Indeed, dislocation climb can hardly be invoked in this torsion loading conditions since most of the dislocations are of screw type.

Usually, creep deformation of ice single crystals is associated to a steady-state creep regime, with a stress exponent equal to 2 when basal glide is activated[8,9]. In the torsion experiments performed, the steady-state creep was not reached, but one would expect it to be achieved for larger strain when the immobilisation of the basal dislocations in the pile-ups is balanced by the dislocation multiplication induced by the double cross-slip mechanism.

In summary, torsion creep tests on well-oriented ice single crystals appear to be a pertinent experiment to try to understand and represent the fundamental mechanisms of deformation in ice single crystals. The presented evidence for the occurence of cross-slip as a rate-limiting process questions the role of dislocation climb as suggested by Louchet (2004) [12].

References

1   U. Nakaya, in Mechanical properties of single crystals of ice (US Army Snow Ice and Permafrost Res. Establish 1958) Report 28, 1958.
2   C.J. Readings and J.T. Bartlett. *J. Glaciol.*, 1968, **51**, 479.
3   J. Weiss and J.R. Grasso, *J. Phys. Chem.*, 1997, **101**, 6113-6117.
4   M. Montagnat. J. Weiss, J. Chevy, P. Duval, H. Brunjail, P. Bastie and J. Gil Sevillano, *Phil. Mag.*, 2006, **86**, 4259.
5   H. Neuhauser, in Dislocations in Solids, edited by F.R.N. Nabarro (North Holland, Amsterdam 1983) pp. 319.
6   M. Verdier, M. Fivel and I. Groma, *Mod. Simul. Mater. Sci. Eng.*, 1998, **6**, 755.
7   M. Montagnat, P. Duval, P. Bastie, B. Hamelin, *Scripta Mater.*, 2003, **49**, 411.
8   S.J. Jones and J.W. Glen, *J. Glaciol.*, 1969, **8**, 463.
9   A. Higashi, S. Koinuma and S. Mae, *Japanese J. Appl. Phys.*, 1964, **3**, 610.
10   B. Hamelin and P. Bastie, *J. Phys. IV France*, 2004, **118**, 27.
11   S. Mendelson, *Phil. Mag.*, 1963, **8**, 1633.
12   F. Louchet, *Phil. Mag. Let.*, 2004, **84**, 797.

FORMATION OF CARBON DIOXIDE GAS HYDRATES IN FREEZING SEDIMENTS AND DECOMPOSITION KINETICS OF THE HYDRATES FORMED

E.M. Chuvilin[1], S.Yu. Petrakova[1], O.M. Gureva[1] and V.A. Istomin[2]

[1] Geological Faculty, Moscow State University, Leninskie Gory, Moscow, 119992, Russia
[2] VNIIGAZ (Research Institute of Natural Gases and Gas Technologies), Moscow Region, p. Razvilka, 142717, Russia

1 INTRODUCTION

It is well known that permafrost regions provide favourable environments for gas hydrates accumulations [1]. The p-T conditions required for gas hydrate formation are created partly due to freezing processes, leading to a concentration of intra-sediment gas at increased pore pressures. Permafrost sediments differ from unfrozen sediments by their low value of gas permeability, promoting an accumulation of raising gas, in sub-permafrost sediment layers. It allows considering permafrost and sub-permafrost sediment layers as possible environments for greenhouse gases storage, mainly CO_2 in the form of gas hydrates. Special attention to the problem of carbon dioxide hydrates formation, stability and decomposition is given in connection with projects of liquid CO_2 disposal in sub-permafrost layers which appears less critical than a storage in sub-seafloor sediments [2,3]. It is supposed that the sequestration of liquid CO_2 in permafrost areas entails the formation of CO_2 hydrates within the sediment layers.

In this connection it is vitally important to carry out experiments to study the CO_2 hydrate formation in freezing soils and the behaviour of frozen hydrate saturated samples under non-equilibrium conditions.

This work presents experimental data on the CO_2 hydrate formation in gas saturated wet samples under cooling conditions as well as on the hydrate decomposition kinetics in frozen CO_2-hydrate saturated samples.

2 METHODS

Our approach for studies of gas hydrate formation and decomposition in sedimentary pore space consists of two steps. The first one is devoted to the hydrate accumulation kinetics in pore space of frozen soils to obtain frozen hydrate-saturated samples. The second one concentrates on the pore hydrate dissociation kinetics in frozen soils under non-equilibrium conditions.

Experimental modelling of CO_2 gas hydrate formation in pore space of wet gas saturated soils has been carried out using an experimental installation consisting of the following basic elements: a pressure chamber (about 420 cm^3), a refrigerator to provide temperature variability of the pressure chamber, a converter of electric signals of the temperature and pressure into digital data, and a computer [4,5]. This experimental installation allows automatic registration of temperature and pressure readings at any time step.

Various soils (Table 1) were used in the course of experiments: quartz sand and sand-clay mixes consisting of quartz sand with kaolinite or montmorillonite clay particles. The weight content of clay particles was chosen as 7 and 14%.

The initial water content of soil was set to 10% and 17%, ensuring a well-developed gas-water interfacial area in the pore space. For this experiment wet soil was compacted layer-by-layer in a metal container (internal diameter 4.5 cm, length 10 cm) and then placed in the pressure chamber. In its initial condition the soil samples featured a homogeneous distribution of water and has a known density established by experimental data [6].

The experimental pressure chamber filled with the model soil was pressurized at room temperature ($t \approx$ =+20-:-+22 $^{\circ}$C). Then the chamber was evacuated and subsequently saturated with carbon dioxide (99.995%) up to a pressure of 4 MPa and left for 24 hours until sample temperature and pressure became stable.

The study of hydrate and ice formation in gas saturated samples was carried out under conditions of cyclic temperature changes. At the beginning the temperature in a sample was lowered from room to low positive values ($\approx+1 \div +3^{\circ}$C) (cooling speed about 0.02°C/min), causing a hydrate accumulation in the pore space of soil, detected by a sharp pressure drop in the pressure chamber. Then, after the hydrate accumulation process, the chamber with the sample was cooled to sub-zero temperatures (-7 $\div$ -8°C). In this step the remaining pore water content froze consolidating further the sample. Subsequently, the pressure chamber with the sample was warmed up to room temperature in order to carry out the second cycle of hydrate and ice formation. As a rule two cycles of cooling were carried out in the course of our experiments.

As a result we have obtained frozen CO_2-hydrate-saturated samples. The analysis of the thermo-baric changes in the pressure chamber during the process of hydrate and ice formation allows us to localized any phase transition in the soil samples as well as to calculate the water content, the volumetrical hydrate content (H_v) and the hydrate coefficient K_H (fraction of liquid water transformed into hydrate).

In order to study the kinetics of pore hydrate dissociation in frozen soils, the pressure in the pressure chamber was lowered to atmospheric values at the second stage of cooling after the complete freezing of the sample. Then the pressure chamber was opened in a cold-room at a temperature of -8°C

The hydrated samples were taken out of the pressure chamber and subjected to a

Table 1 *Grain size of investigated sediments*

Type of sediment	Diameter of particles, mm							
	1 - 0.5	0.5 - 0.25	0.25 – 0.1	0.1 – 0.05	0.05 - 0.01	0.01- 0.005	0.005- 0.001	< 0.001
	Particle content of each fraction, %							
Quartz sand	6.5	6.5	79.6	2.2	2.4	0.4	0.3	2.1
Montmorillonite clay	0.0 (2.0)	0.0 (1.0)	0.2 (1.3)	0.1 (8.5)	18.8 (15.6)	7.3 (5.9)	20.1 (15.4)	53.5 (50.3)
Kaolinite clay	0.7 (1.5)	0.5 (0.2)	0.4 (0.2)	2.9 (1.2)	19.5 (32.2)	11.2 (24.0)	40.2 (34.5)	24.6 (6.2)

0.1 – content of certain grain size particle
(8.5) – content of certain diameter micro-aggregates

detailed petrophysical study. Normally it was made within 30 minutes after pressure release. This study included morphological observations, layer-by-layer analysis of the water content (accuracy 1%), density (accuracy 0.02 g/cm^3), gas content (accuracy 0.1 cm^3/g), as well as the calculation of porosity, hydrate coefficient (K_H), hydrate (G_h) and ice (G_i) saturation. The determination of the sample's gas content was carried out by means of measurements of gas released in the course of defrosting of the sample placed in a saturated NaCl solution. These data were used for the determination of the volumetric hydrate content (H_v) and hydrate saturation. The calculation of the volumetrical hydrate content (H_v) and the K_H coefficient was carried out in two ways: (1) under gas pressure – *via* the calculation of the p-T dependent CO_2 consumption during hydrate formation and (2) at atmospheric pressure by a recalculation of the gas content on the basis of the composition of CO_2 hydrate assuming a stoichiometry of $CO_2 \cdot 6.1 H_2O$ [5,6].

Then the remaining hydrate-containing soil was placed in weighing cups for long-term storage at various negative temperatures. To prevent sublimation the samples were covered by a snow crumb. After a certain period of time the sample stored at negative temperature were probed in order to control changes of their gas and hydrate content.

Apart from a direct observation of the gas content of frozen samples in the course of time, the study of gas releases from the frozen hydrate saturated samples at atmospheric pressure were carried out with the help of DC-1 gas meter.

3 EXPERIMENTAL RESULTS

3.1 Thermo-baric conditions and kinetics of CO_2 hydrate accumulation in pore space of soils

The analysis of thermo-baric changes in the wet soil samples saturated with CO_2 as a function of time under condition of cyclic cooling and heating permits to follow the kinetic and thermo-baric indicators of phase transitions within the pore space of the samples. On cooling of wet gas-saturated soils under gas pressures higher than the three-phase equilibrium line "gas – water – CO_2 hydrate", conditions for gas hydrates nucleation in pore space of soils are created. Pressure stabilization marks the end of the phase transition of water into hydrate. Upon further cooling below 0°C the remaining, untransformed liquid turns into ice.

The extent of hydrate formation depends on the supercooling of the system (ΔT_{sc}) with respect to the temperature of the hydrate stability limit; the supercooling of pore water reached from one up to several degrees (Table 2) which is comparable to earlier measurements for CH_4 hydrate formation in the same soils [5] (excluding the sample with 7% of kaolinite where the values are different). As shown in Table 2, the supercooling for the second cycle showed a marked increase in contrast to methane saturated samples where a typical decrease of the supercooling is seen.

The comparison of supercooling values in samples with different clay additives shows in samples with kaolinite values up by 2 degrees and more while supercooling in samples with montmorillonite do not exceed 1.3 degree. Evidently, the presence of montmorillonite as compared to kaolinite does not complicate the formation of gas hydrate crystals in pore water and even may favour it as was observed earlier [5,6,7].

The extent of hydrate accumulation in the pore space of soils as obtained from the CO_2 consumption rate is shown in Figure 1. The figure shows a rapid CO_2 consumption in the initial stage of hydrate formation, a slower consumption in the middle part and a

Table 2 *Supercooling (ΔT_{sc}, °C) of pore water for the formation of CO_2 and CH_4 gas hydrates in model samples*

Type of sediment	1 cycle		2 cycle	
	CO_2	CH_4	CO_2	CH_4
Sand (W_{in}=10%)	1.1	1.6	1.9	0.7
Sand with 7% montmorillonite clay (W_{in}=10%)	0.6	1.2	1.2	0.4
Sand with 7% montmorillonite clay (W_{in}=17%)	0.8	1.6	-	0.4
Sand with 7% kaolinite clay (W_{in}=10%)	3.9	0.3	3.9	0.2
Sand with 14% kaolinite clay (W_{in}=10%)	1.4	-	2.4	0.7

declining consumption at the terminal stage of hydrate formation. According to experimental data the quantity of absorbed CO_2 in the presence of clay particles decreases (Figure 1). The presence of montmorillonite causes significant changes of the chemical potential of pore water (increase of binding energy of the pore water with the mineral surface) and thus a larger decrease of CO_2 consumption. However, we can see that at the initial stage of hydrate formation the samples with kaolinite show a higher consumption than even in sand samples with no clay particles. This behaviour apparently is governed by the existence of a more developed gas-water interface in pore space of sand containing kaolinite.

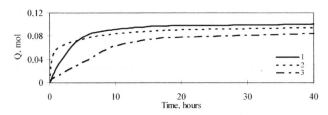

Figure 1 *CO_2 consumption rate (Q) during the hydrate accumulation for different model sediments (W_{in}=10%):1- sand, 2- sand with 7% of kaolinite clay, 3- sand with 7% of montmorillonite clay*

At repeated hydrate formation in the studied soils the amount of hydrate accumulation varies in different ways. During the second cycle of cooling sand samples with kaolinite show a decrease of hydrate accumulation as compared to the first cycle. On the contrary, samples with montmorillonite indicate a trend for an increased hydrate accumulation in the second cycle (Table 2). This behaviour may be explained by a response of the soil microstructure to phase transformations and concomitant changes in the pore space arrangement which is different for soils of different composition [7].

The influence of the soil water content on the extent of CO_2 hydrate formation can be traced on a sand sample with 7% of montmorillonite particles (Figure 2). According to our data samples with less water content (W_{in}=10%) show a more developed hydrate formation at the initial stage ceasing after 30 hours in contract to the sample with 17% water content. As a result the total CO_2 consumption in the 17% water sample becomes higher. The higher speed of the hydrate formation in the sample with W_{in}=10% is explained by a lower degree of water saturation and a bigger contact surface of gas and pore water.

A comparison of the extent of hydrate formation in the pore space of sand samples (W_{in}=10%) saturated with different hydrate-forming gases (CO_2 and methane) shows that at comparable conditions of hydrate formation the CO_2 hydrate accumulation is higher than that of methane hydrate.

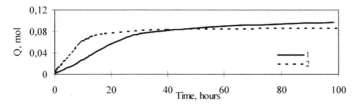

Figure 2 *The amount of CO_2 consumption (Q) during the hydrate accumulation for model sediments (sand with 7% of montmorillonite clay) with water content (W_{in}): 1- 17%, 2- 10%*

The examination of the kinetics of pore filling with CO_2 hydrate for the sand sample with 7% of kaolinite allows us to establish the slowing down of the hydrate accumulation. During the first 2 hours of hydrate formation hydrate the saturation was 0.13, while during the following 18 hours it increased by 0.08 and in further 20 hours it increase only by 0.01 (Figure 3).

The pore water content was normally decreasing and amounted by the end of experiment to 0.15. All experiments carried out gave an incomplete transition of pore water into hydrate. This is quantitatively reflected in the so called hydrate coefficient (K_H)[5]. For the sample with 7% kaolinite a fraction of 0.2 of the sample's water content has been transformed into hydrate in 2 hours after the start of hydrate formation. By the end of the run (40 hours) this value was 0.36.

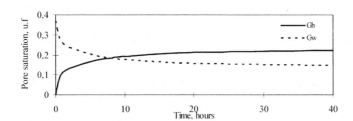

Figure 3 *Kinetics of CO_2 hydrate accumulation in pore media (sand with 7% of kaolinite clay, $W_{in}=10\%$), G_h – hydrate saturation, G_w – water saturation*

Under condition of cooling to negative temperatures (-8°C) there is a freeze-out of the remaining, untransformed water. According to experimental data the water-ice transition in hydrate containing samples is characterized by a lowering of the freezing point of water as compared to reference samples (with no hydrate). Thus, in the sand sample with 7% of montmorillonite and 17% of water content the temperature of the start of the freezing in the reference sample was -0.3°C, while in hydrate containing sample it was -3.0 °C. This can be explained by influence of gas pressure and soil water saturation [10].

The analysis of hydrate formation under cooling to sub-zero temperatures shows that there is an activation of the hydrate formation process during the freezing of the remaining pore water. Some 10-25 % of the total amount of CO_2 hydrate is formed during the water freezing process. This portion increases in samples with clay particles, in particular for kaolinite.

3.2 Decomposition kinetics of frozen hydrate-saturated samples at non-equilibrium pressure

In the course of the experiments described above frozen hydrate-saturated samples were obtained at temperatures of -8°C. According to our petrographic analysis they feature quite homogeneous structures and a uniform distribution of hydrate and ice in the pore space of samples. The porosity of these samples was 0.37-0.41. The hydrate saturation of the samples in equilibrium conditions before the pressure release was 0.36-0.46. The volumetric hydrate content was 15-17 %, and the hydrate coefficient between 0.7-0.75.

An vigorous CO_2 hydrate dissociation was observed in frozen hydrate saturated samples after the pressure release in the pressure chamber. The hydrate coefficient decreased 1.5-3.0 fold in 30 minutes after a pressure drop to atmospheric values. The maximum decrease was observed in the sand sample with 14% of kaolinite particles, the minimum decrease in the sand sample with 7% montmorillonite particles with 17% of initial water content. In the course of time the intensity of CO_2 hydrate dissociation in frozen samples dropped sharply with even a complete stop of the dissociation process as a consequence of gas the hydrates self-preservation effect at sub-zero temperatures [11,12,13]. A comparative analysis of the self-preservation effect in frozen soils with methane hydrates at similar test conditions (T=-7°C) demonstrated a higher resistance [8]. Thus, in the sample with 7% montmorillonite CO_2 hydrates decomposed in 200 hours and in the sample with methane hydrates they were present within the whole period of observation (500 hours) stabilizing in the last 400 hours at 3% of volume.

Temperature acts as the most important factor influencing the process of self-preservation of gas hydrates in pore space. The study of sand sample with 7% (W_{in}=10%) carried out at different temperature conditions shows that the time of CO_2 pore hydrate decomposition in frozen samples varies from 5 hours at -2°C to 60 hours at -13°C. According to our observations, the pore hydrate dissociation process at -20°C stopped in one hour with no further dissociation in the following 40 hours. At the end of the experiment (about 100 hours) the CO_2 hydrate content was about 7% in volume.

Pore ice plays a particular role for the self-preservation effect in frozen soils [8]. Initial hydrate preservation apparently is helped by frozen pore water (not transformed into hydrate). Additional ice formation in the form of a film on the surface of gas-hydrate forming due to hydrate surface dissociation is expected to take place upon gas pressure release. Thus in the sample with 7% of montmorillonite particles (W_{in}=17%), pore hydrate showed a higher stability after pressure release as a consequence of the greater ice content due to the freezing of remaining pore water (Figure 4).Our results clearly indicate that the hydrate content decreases on the expense of an increases of ice (Figure 5).

The mineral composition of the soil will also influence the kinetics of gas hydrates dissociation in frozen soils. Our results show, that gas hydrate formations in pore space of samples with montmorillonite particles dissociate less markedly as compared to the samples with kaolinite admixture. This influence may be explained by microstructural specificities of pore hydrate saturated samples but undoubtedly requires additional micro-morphological studies for a full understanding.

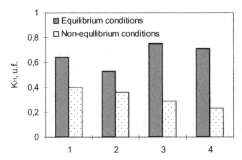

Figure 4 *Hydrate coefficient (K_H) in artificial hydrate-saturated samples (W_{in}=10%) at a temperature of -8°C in equilibrium and non-equilibrium conditions (30 min after pressure release): 1- sand; 2- sand with 7% montmorillonite clay (W_{in}=17%); 3- sand with 7% montmorillonite clay; 4- sand with 14% kaolinite clay*

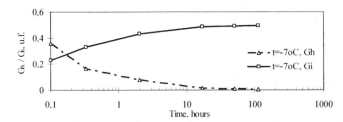

Figure 5 *CO_2 hydrate dissociation kinetic in frozen hydrate-saturated sample (sand with 7% montmorillonite clay, W_{in}=10%) after pressure release to 0.1 MPa, G_h – hydrate saturation, G_i – ice saturation*

4 CONCLUSION

Our experiments describe the thermo-baric conditions and CO_2 hydrate accumulation kinetics in model soil media under conditions of cooling. The influence of mineral composition on the supercooling of pore water, the conditions and rates of CO_2 hydrate crystal growth, and also on the hydrate coefficient has been shown. Samples with kaolinite particles exhibit a pore water super-cooling during hydrate formation that is much higher than in samples with montmorillonite particles. The final CO_2 hydrate saturation is decreasing in the order pure sand, sand with kaolinite, sand with montmorillonite. It is also found, that during the freezing of the remaining pore water there is an activation of the CO_2 hydrate formation process. The amount of hydrate formed during this stage amounts to 10-25 % of the total hydrate accumulation.

The frozen hydrate-saturated media formed during these experiments were used for a study of the CO_2 hydrate decomposition kinetics in the pore space. The influence of soil mineral composition, ice content and temperature on the CO_2 hydrate self-preservation effect was established. It is revealed, that a temperature decrease slows down the CO_2 hydrate dissociation; low negative temperatures (below -13°C) cause a complete stop of the CO_2 hydrate dissociation process. It is also shown that ice forming in the remaining pore space from freezing of unreacted water enhances the CO_2 hydrate self-preservation effect.

Acknowledgments

These investigations were supported by grants INTAS № 03-51-4259 and the RFBR № 04-05-64757 and № 05-05-39019. We thank the anonymous referees for various suggestions to improve the manuscript.

References

1   V.S. Yakushev, E.M. Chuvilin, *Cold Regions Science and Technology*, 2000, **31**, 189.
2   A. Yamasaki, M. Wakatsuki, H. Teng, Y. Yanagisawa, K. Yamada, *Energy*, 2000, **25(1)**, 85.
3   P.G. Brewer, G. Friederich, E.T. Peltzer, F.M. Orr, *Science*, **284**, 1999, 943.
4   E.M. Chuvilin, E.V. Kozlova, *Earth Cryosphere Journal*, 2005, **9 (1)**, 73 (in Russian).
5   E.M. Chuvilin, E.V. Kozlova, T.S. Skolotneva, in *Proccedings of the Fifth International Conference on Gas Hydrate,* Trondheim, Norway, 2005, v.5, 1540.
6   S.B. Cha, H. Ouar, T.R. Wildeman and E.D. Sloan. *J. Phys. Chem.*, 1988, **92**(23), 6492.
7   T. Uchida, S. Takeya, E.M. Chuvilin, R. Ohmura, J. Nagao, V.S. Yakushev, V.A. Istomin, H. Minagawa, T. Ebinuma, H. Narita. *J. Geophys. Res.*, 2004, **109**, 8.
8   E.M. Chuvilin, E.V. Kozlova, in *Proccedings of the Fifth International Conference on Gas Hydrate,* Trondheim, Norway, 2005, v.5, 1561.
9   E.M. Chuvilin, O.M. Yazynin, in *Proccedings of the Fifth International Permafrost Conference,* Trondheim, Norway, 1988, **1**, 320
10 E.M. Chuvilin T. Ebinuma, Y. Kamata, T. Uchida, S. Takeya, J. Nagao and H. Narita, *Can. J. Phys.*, 2003, **81**, 343.
11 E.D. Yershov, Yu.P. Lebedenko, E.M. Chuvilin, V.S. Yakushev, V.A. Istomin, *Doklady Akademii Nauk SSSR*, 1991, **321(4)**, 788 (in Russian).
12 S. Circone, L. Stern, S. Kirby, W. Durham, B. Chakoumakos, C. Rawn, A.Rondinone, Y. Ishii. *J. Phys. Chem. B*, 2003, **107**, 5529.
13 W.F. Kuhs, G. Genov, D.K. Staykova, T. Hansen. *Phys. Chem. Chem. Phys.*, 2004, **6**, 4917.

FIRST-PRINCIPLES STUDY OF BJERRUM DEFECTS IN ICE IH: AN ANALYSIS OF FORMATION AND MIGRATION PROPERTIES

M. de Koning[1], A. Antonelli[1], A.J.R. da Silva[2] and A. Fazzio[2]

[1]Instituto de Física "Gleb Wataghin", Universidade Estadual de Campinas, Caixa Postal 6165, 13083-970, Campinas, SP, Brazil.
[2]Instituto de Física, Universidade de São Paulo, Caixa Postal 66318, 05315-970, São Paulo, SP, Brazil.

1 INTRODUCTION

While the isolated water molecule is one of the simplest in Nature, the condensed phases of H_2O reveal many complex features that still elude complete understanding[1,2]. The proton-disordered hexagonal crystalline form of water, ice I_h, for which several aspects of structure-properties relationship remain unclear[3], is an example of such a phase.

A particularly important question involves the understanding of the role of crystal defects in the peculiar electrical behaviour of ice I_h. Upon the application of an electric field, the solid becomes polarized by the thermally activated reorientation of the molecular dipoles. Niels Bjerrum postulated the existence of orientational defects[4], which represent local disruptions of the hydrogen-bond network of ice I_h, to explain the microscopic origin of this phenomenon.

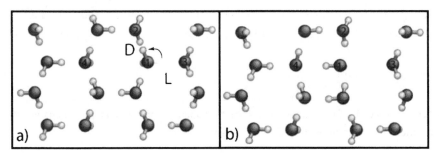

Figure 1   *a) Formation of a D-L Bjerrum defect pair. b) Corresponding relaxed DFT structure.*

Figure 1a) provides a schematic picture of the formation of a Bjerrum defect pair in ice I_h. Defect-free ice I_h obeys Pauling's two ice rules[4]: (i) each molecule offers/accepts two protons to/from two neighbouring molecules, and (ii) there is precisely one proton between each nearest-neighbour pair of oxygens. The proton-disordered character of the structure implies that there is no long-range order in the orientations of the molecules or hydrogen bonds. The Bjerrum defect pair constitutes a violation of ice rule (ii) and is obtained by the indicated rotation of molecule 1, after which there are two protons between molecules 1 and 2 (D defect) and none between molecules 1 and 3 (L defect). After the formation of this "embryonic" defect pair, the D and L defects can be further separated by successive thermally activated molecular rotations, eventually creating a pair of free D and L defects. Given that a jump of a D or L defect from one site to the next involves the rotation of a water molecule, their motion through the crystal provides a molecular mechanism for electrical polarization.

Even though from the conceptual point of view the role of Bjerrum defects is well established[4], their molecular structure and energetics, remains a subject of debate. Recent atomistic studies[5-7] based on empirical water potentials have provided important qualitative insight into the structure and dynamics of Bjerrum defects, although they have not yet attempted to make direct contact with experimental conductivity data. Furthermore, the few *ab initio* studies[8,9] involve clusters that are too small to reliably capture the properties of a defect embedded in bulk crystal.

In this contribution, we present a first-principles study of the structure and energetics of Bjerrum defects in ice I_h using a large supercell subject to periodic boundary conditions[10]. The results are interpreted in the context of experimental electrical conductivity data for doped ice Ih, using the framework of Jaccard's defect-based microscopic electrical theory of ice[11].

2 THEORETICAL BACKGROUND AND COMPUTATIONAL APPROACH

In both pure as well as doped ice, the electrical conductivity σ_{DL} due to Bjerrum defects is essentially controlled by L defects[4]. According to Jaccard's electrical theory of ice[11] it takes the form

$$\sigma_{DL} = q_{DL} n_L \mu_L , \tag{1}$$

where q_{DL} is the effective charge carried by the Bjerrum defects, n_L is the concentration of free L defects and μ_L is their mobility. Its temperature dependence is described by[3]

$$\sigma_{DL}(T) \sim \frac{q_{DL}}{T} \exp(-E / k_B T), \tag{2}$$

where T is the absolute temperature, k_B is Boltzmann's constant and E is a characteristic activation energy. In *pure* ice, the latter is given by[4]

$$E = \frac{1}{2} E_{DL} + E_{Lm} , \tag{3}$$

where E_{DL} is the formation energy of a pair of free D and L defects which controls the thermal equilibrium concentration n_L, and E_{Lm} is the migration energy characterizing the free mobility μ_L.

While the activation energy can be determined experimentally by measuring σ_{DL} as a function of T and fitting it to Eq. (2), it is not possible to independently measure the components E_{DL} and E_{Lm}. To achieve this, additional measurements need to be carried out on *doped* ice samples[3], in which an extrinsic concentration of L defects is injected. This indirect procedure, however, is subject to large uncertainties and has revealed incompatibilities between different experiments[3]. The purpose of the present contribution is to use a density-functional theory (DFT)[12] first-principles approach to explicitly and independently compute the formation and migration energies E_{DL} and E_{Lm}, providing new theoretical insight into the parameters central to Jaccard's defect-based microscopic electrical theory of ice[11].

All the calculations were based on the 96-molecule proton-disordered ice I_h cell labelled $3 \times 2 \times 2$ in Ref. 12. The DFT calculations were carried out using the state-of-the art VASP package[14,15] using the Perdew-Wang 1991 generalized-gradient approximation[12] and the projector-augmented-wave[16] approach. Brillouin-zone sampling was limited to the Γ-point and we used a plane wave cut-off of 700 eV. The effects of spurious image dipole interactions due to the periodic boundary conditions were found to be negligible for all investigated structures[17]. To account for quantum-mechanical zero-point contributions, known to be relevant in ice[3], we also evaluated the change in the local inter and intramolecular vibrational modes of the molecules in the immediate vicinity of the defect structure compared to those in the defect-free crystal, using the local harmonic approximation[18].

3 CALCULATIONS AND RESULTS

As a first step, we relax the defect-free crystal supercell, allowing both the atomic and the supercell degrees of freedom to relax at zero stress. The resulting hexagonal lattice parameters are in good agreement with experimental values, with a=4.383 Å and c=7.16 Å, about 2% below the values measured[3] at T=10K. The average intramolecular oxygen-hydrogen separation of 1.01 Å is in excellent agreement with the experimental value of ~1.006-1.008 Å. As appears typical of DFT calculations on ice I_h[19] our calculations slightly overbind the solid, giving a sublimation energy of 0.69 eV compared to the experimental value 0.61 eV.

Next, we create an ``embryonic'' Bjerrum defect pair according to Figure 1a) and minimize the Hellman-Feynman forces at constant volume. The resulting structure is shown in Fig. 1b) and is qualitatively similar to those observed in a recent molecular dynamics (MD) study[7]. Compared to the initial geometry, one can no longer recognize a D defect in the schematic sense of Figure 1a) due to the large electrostatic repulsion between the two hydrogen atoms[3,6,7]. The total-energy part of the formation energy of this structure was found to be 0.55 eV, while the zero-point contribution lowers it by about 10%, giving a formation energy of 0.50 eV. The corresponding formation entropy at the melting point is $S_f = 2.2\, k_B$.

To estimate the formation energy E_{DL} of an independent defect pair, we move the L defect through the crystal by a series of molecular rotations, separating it from the D defect, followed by structural relaxation. To estimate the effect due to the proton disorder, we carried out this procedure for two different embryonic D-L defect pairs.

Because of their effective charges q_{DL}, the formation energy is expected to be of the form

$$E_{\text{form}}(r) = E_{DL} - \frac{q_{DL}^2}{4\pi\varepsilon_0\varepsilon_\infty r}, \qquad (4)$$

where r is the distance between the point charges q_{DL}, and ε_∞ is the high-frequency dielectric constant of ice I_h^3. We define the centers of charge such that the charge $+q_{DL}$ is positioned on the dangling proton of the D defect, while the charge $-q_{DL}$ is centered midway between the two oxygen atoms of the L defect. Assuming that the zero-point contribution is the same as for the first embryonic configuration, the formation energy as a function of the inverse distance between the D and L defects for two distinct embryonic D-L defect pairs, in which the D defect, respectively, points along the c axis and the ab plane[6,7], is shown in Figure 2.

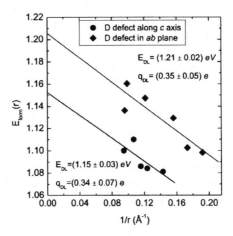

Figure 2   *Formation energy as a function of the inverse distance between D and L defects for two distinct embryonic D-L defect pairs.*

The results for both defect pairs are consistent with the $1/r$ behaviour of Eq. (4), with the scatter of the data points showing fluctuations ~ 0.01 eV due to the changing environment of the L defect along their displacement paths. As expected, both D-L displacement curves are essentially parallel and, assuming the experimental value $\varepsilon_\infty = 3.2$ for the high-frequency dielectric constant, yield an estimate for the effective charge of $q_{DL} = (0.35\pm0.07)e$, which is in excellent agreement with the well-established experimental value $q_{DL} = 0.38e$ [3]. The intercepts of both straight lines give estimates for the independent D-L pair formation energy E_{DL}, which, because of the different local environments of the D defects, differ by about 4%, giving an average estimate $E_{DL} = (1.18\pm0.05)$ eV.

Next we compute the migration energy barrier E_{Lm}. Because of the proton disorder, this barrier is also expected to fluctuate depending on the local environment of the L defect. For this purpose, we computed 6 distinct barriers at 3 different L-defect sites by starting from the relaxed D-L configuration and rigidly rotating one of the 2 molecules hosting the L defect as shown in Figure 3a) in 5° angular increments, followed by a constrained relaxation in which the rotating proton is allowed to move only in the plane of the molecule, while allowing full relaxation of all other degrees of freedom. The maxima of the relaxed energies as a function of rotation angle give transition states in which $\angle ABC$ is essentially bisected by the coplanar rotating OH bond, as shown in Figure 3b). The resulting energy barriers (neglecting zero-point effects) vary in the range $E_{Lm} \approx 0.10 - 0.14\,\text{eV}$, indicating an overall high mobility of free L defects.

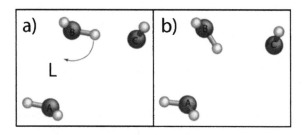

Figure 3 *Identification of transition state for independent L-defect motion. a) Equilibrium structure. b) Typical transition state.*

4 DISCUSSION

Before analyzing these results in the context of experimental data, we need to assess the extent to which the obtained results might depend on the particular disordered hydrogen-bond configuration of the used supercell. The fact that the electrostatic $1/r$ behaviour in Figure 2 sets in for distances as short as $r \approx 6\,\text{Å}$ indicates that the disorder effect operates on a length scale that is significantly smaller than the dimensions of the cell. This suggests that the influence of disorder is local in nature, rather than a property of the cell as a whole. Furthermore, these results for E_{Lm} are consistent with recent MD findings[6] based on cells with different sizes and hydrogen-bond arrangements. If we compute the average migration times corresponding to our E_{Lm} data using an attempt frequency of 24 THz, which is typical for the librational modes in ice I_h[3], we obtain values between 4 and 35 ps at $T = 230K$. This is consistent with the MD values for the same temperature, which were reported to range between a few ps to several tens of ps[6].

The experimental estimates for the formation and migration energies are obtained from conductivity measurements on doped ice[3], in which a substitutional concentration of HF[11,20] or HCl[21] molecules is introduced into the crystal. Since each molecule has only one proton, they introduce an extrinsic, temperature-independent concentration of L defects. Measurements of the conductivity σ_{DL} as a function of T then typically yield an Arrhenius-type plot with 3 distinct activation energies, as shown schematically in Figure 4.

The high-temperature regime I is attributed to intrinsic behaviour, characterized by the activation energy Eq. (3). The lower-temperature regimes II and III are believed to be dominated by the extrinsic L defects, where the former has been attributed to the *free* motion of extrinsic L defects[3] with $E_{II} = E_{Lm}$. In regime III the temperature is so low that the extrinsic L defects are not completely dissociated from their dopant molecules so that the activation energy involves an additional dissociation energy, giving[3] $E_{III} = \frac{1}{2}E_{\text{diss}} + E_{Lm}$.

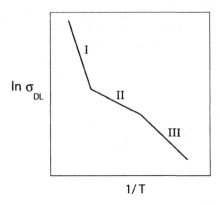

Figure 4   *Schematic representation of a characteristic Arrhenius plot of the conductivity*
σ_{DL} *as a function of temperature in a doped ice I_h sample[3].*

Considering the experimental results[11,20,21] reproduced in Table 1, the activation
energy for regime I is quite well established, even for different dopant species, showing a
dispersion of less than 0.05 eV among the different experiments. The individual
components E_{DL} and E_{Lm}, however, show considerably larger deviations. The migration
barrier values vary between 0.19 eV and 0.315 eV, leading to formation energies E_{DL}
ranging between 0.66 eV and 0.79 eV.

Table 1   *Activation energies and their components E_{DL} and E_{Lm} determined from doped
ice experiments and the present DFT calculations.*

	E	E_{Lm}	E_{DL}
Ref. 10 (HF)	0.575	0.235	0.68
Ref. 19 (HF)	0.625	0.315	0.664
Ref. 20 (HCl)	0.585	0.19	0.79
DFT	0.71±0.05	0.12±0.02	1.18±0.05

Comparing our DFT calculations to these experimental results, we notice that both
components of the activation energy deviate significantly from the experimental values.
The DFT result for E_{DL} is more than 46% larger than the largest experimental value,
whereas E_{Lm} is about 37% lower than the lowest experimental estimate. In this light, it is
quite striking that, despite the discrepancies for the individual components, the DFT
estimate for the net activation energy of Eq. (3) agrees quite well with the experiments,
deviating about 13% from the highest experimental estimate. This seems to be a direct
demonstration of the difficulties involved in the interpretation of conductivity experiments
in doped ice samples under conditions not controlled by intrinsic ice properties.
Specifically, the fact that the DFT estimate for the migration barrier in pure ice is
systematically and significantly lower than all experimental estimates indicates that, as
suggested earlier[3], the regime interpreted as being controlled by free extrinsic L defects,
may in fact involve the activity of traps that obstruct their motion, leading to the higher
effective migration barriers deduced experimentally.

5 CONCLUSIONS

We have conducted a first-principles DFT study of the energetics of Bjerrum defects in ice I_h and compared the results to experimental data for doped ice samples. The results provide new insight into the parameters in Jaccard's microscopic electrical theory of ice. While the DFT value for the net activation energy is in good agreement with experiment, we find that its two components have quite different values from those inferred from experiment. In particular, our results predict an L-defect migration barrier that is significantly lower than the lowest experimental estimate, hinting at the presence of traps in the regime usually interpreted as being controlled by the free migration of extrinsic L defects.

Acknowledgements

The authors gratefully acknowledge financial support from the Brazilian agencies FAPESP and CNPq. Part of the calculations were performed at CENAPAD-SP.

References

1   I.F.W. Kuo and C.J. Mundy, *Science*, 2004, **303**, 658.
2   B.J. Murray, D.A. Knopf, and A.K. Bertram, *Nature*, 2005, **434**, 202.
3   V.F. Petrenko and R.W. Whitworth, *The Physics of Ice*, 1999, Oxford University Press, Oxford.
4   N. Bjerrum, *Science*, 1952, **115**, 385.
5   R. Hassan and E.S. Campbell, *J. Chem. Phys.*, 1992, **97**, 4326.
6   R. Podeszwa and V. Buch, *Phys. Rev. Lett.*, 1999, **83**, 4570.
7   N. Grishina and V. Buch, *J. Chem. Phys.*, 2004, **120**, 5217.
8   M.D. Newton, *J. Chem. Phys.*, 1983, **87**, 4267.
9   P.L. Plummer, 'Structural studies and molecular dynamics simulations of defects in ice' in *Physics and Chemistry of Ice*, 1992, edited by N. Maeno and T. Hondoh, Hokkaido University Press, Sapporo.
10 M. de Koning, A. Antonelli, A.J.R. da Silva and A. Fazzio, *Phys. Rev. Lett.*, 2006, **96**, 075501.
11 C. Jaccard, *Helv. Phys. Acta*, 1959, **32**, 89.
12 R.M. Martin, *Electronic Structure: Basic Theory and Practical Methods*, 2004, Cambridge University Press, Cambridge.
13 J.A. Hayward and J.R. Reimers, *J. Chem. Phys.*, 1997, **106**, 1518.
14 G. Kresse and J. Hafner, *Phys. Rev. B*, 1993, **47**, R558; *Phys. Rev. B*, 1994, **49**, 14251.
15 G. Kresse and J. Furthmüller, *Comput. Mater. Sci.*, 1996, **6**, 15; *Phys. Rev. B*, 1996, **54**, 11169.
16 G. Kresse and D. Joubert, *Phys. Rev. B*, 1999, **59**, 1758.
17 L.N. Kantorovich, *Phys. Rev. B*, 1999, **60**, 15476.
18 R. LeSar, R. Najafabadi, and D.J. Srolovitz, *Phys. Rev. Lett.*, 1989, **63**, 624.
19 P.J. Feibelman, *Science*, 2002, **295**, 99.
20 G.C. Camplin, J.W. Glen, and J.G. Paren, *J. Glaciol.*, 1978, **21**, 123.
21 Takei and N. Maeno, *J. Phys. (Paris)*, 1987, **48**, Colloque C1, 121.

FIRST-PRINCIPLES STUDY OF MOLECULAR POINT DEFECTS IN ICE IH: INTERSTITIAL VS. VACANCY

M. de Koning[1], A. Antonelli[1], A.J.R. da Silva[2] and A. Fazzio[2]

[1]Instituto de Física "Gleb Wataghin", Universidade Estadual de Campinas, Caixa Postal 6165, 13083-970, Campinas, SP, Brazil
[2]Instituto de Física, Universidade de São Paulo, Caixa Postal 66318, 05315-970, São Paulo, SP, Brazil

1 INTRODUCTION

Ice I_h, the proton-disordered hexagonal phase of water, has been by far the most studied phase of ice, however, a complete understanding of many of its properties still remains elusive.[1] The role played by crystal defects is among the important issues requiring better comprehension. It is in fact quite remarkable that not even the most basic disruptions of crystalline order, the molecular vacancy and self-interstitial, are well understood.[1]

Molecular defects are involved in diffusion and possibly affect the electrical properties of ice I_h,[1,2] the central issue revolves around the questions which of the two is the primary molecular point-defect, and what is its structure.[1] From the experimental side, positron annihilation experiments[3,4] first indicated that the vacancy should be overall dominant, but a more recent series of X-ray topographical studies of dislocation loops[5,6,7,8] provided convincing evidence that for temperatures above −50° C the self-interstitial should take over as the principal point defect. The structure of this self-interstitial, however, remains unknown. In addition to a surprisingly high formation entropy of ~ 4.9 k_B, the X-ray studies inferred a formation energy (~ 0.4 eV) below the sublimation energy of ice I_h, which suggests that its structure might involve bonding to the surrounding hydrogen-bond network.[5] This idea, however, is inconsistent with the established consensus[1] that the relevant self-interstitial structures in ice I_h involve the two cavity-center positions, i.e. the capped (Tc) and uncapped (Tu) trigonal sites[1], for which such bonding is not expected.[1,9] The theoretical insight from classical molecular dynamics (MD) simulations[10,11,12] has also been largely inconclusive, mostly due to the sensitivity of the results to the choice of water model, as well as the limited availability of formation energetics.

In this work we present a density-functional-theory (DFT) *ab initio* study in which we compare the structure and formation free energies of three self-interstitial configurations and the molecular vacancy in ice I_h as a function of temperature. The results strongly suggest that a structure different from the established Tc and Tu configurations is the preferred self-interstitial structure in ice I_h. This finding lends support to the experimental indications in that, unlike the Tc and Tu configurations, this interstitial involves the formation of hydrogen bonds with the surrounding water molecules. Furthermore, a comparison between the vacancy and this interstitial reveals that, although the former has the lowest formation free energy for T ≤ 200 K, a crossover at a temperature below the melting transition, as indicated experimentally,[5] is conceivable.

2 CALCULATION METHODS

In our calculations we used an approach that has been successfully applied to study ice I_h,[13,14] which was implemented through the VASP package,[15,16] employing the Perdew-Wang 91 generalized-gradient approximation and the projector-augmented-wave[17] technique. We utilize a 96-molecule proton-disordered supercell[18], restrict Brillouin-zone sampling to the Γ-point and adopt a plane-wave cut-off of 700 eV. The free-energy calculations are carried out within the local harmonic approximation.[19] To consider effects related to thermal expansion we carry out the calculations for three sets of lattice parameters, chosen to correspond to their zero-pressure experimental values[1] at 10, 205 and 265 K, respectively. Furthermore, to estimate the influence of the proton-disordered character of ice I_h we create a number of different replicas of the defects by placing them at different positions within the cell.

3 RESULTS AND DISCUSSION

Fig. 1 depicts relaxed molecular structures of the three investigated interstitial geometries at T=10 K. Panels a) and b) show the Tc and Tu interstitial molecules, respectively. Both defect conformations are similar in that neither significantly disrupts the surrounding hydrogen-bond network and both closely resemble isolated water molecules, characterized by covalent O-H bond lengths ~ 0.97 Å significantly shorter than the typical hydrogen-bonded value ~ 1.00 Å in ice I_h. The structure of the third interstitial configuration, as shown in panel c), is fundamentally different from these two. Its formation involves the breaking of a hydrogen bond between the bulk molecules 1 and 2, followed by the creation of two new ones in which the interstitial molecule accepts and donates one O-H bond to/from the molecules 1 and 2, respectively, incorporating it into the surrounding hydrogen-bond network. This structure, referred to as the bond-center (Bc) interstitial, seems to resemble a configuration observed in one of the recent MD simulations,[12] although no detailed structural nor energetics data were reported. Considering the hydrogen bonds in the vicinity of the defect, the two that link the interstitial molecule to the molecules 1 and 2, as well as the six that connect the latter two to their three regular neighbours, are significantly distorted with respect to the defect-free crystal. The first two are considerably compressed, with nearest-neighbour O-O distances of 2.61 Å and 2.65 Å compared to the typical value ~ 2.75 Å in defect-free ice I_h. The other six hydrogen bonds are also mostly compressed, five of them with O-O distances between 2.70 Å and 2.74 Å and only one stretched bond with an O-O distance of 2.77 Å. The creation of a molecular vacancy (not shown) involves the breaking of four hydrogen bonds and its relaxed structure involves two dangling covalent O-H bonds with bond lengths ~0.97, as well as the distortion of the 12 hydrogen bonds involving its four nearest-neighbour molecules. In contrast to the Bc interstitial, however, the majority of these are stretched with respect to the defect-free crystal, with eight O-O distances ranging between 2.80 Å and 2.84 Å and only four compressed bond lengths between 2.65 Å and 2.67 Å.

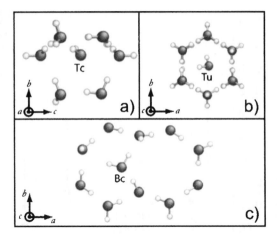

Figure 1 *Investigated molecular interstitial structures in ice I_h. Arrows indicate crystallographic directions (c.f. Ref. 1). a) Capped trigonal (Tc) intestitial. b) Uncapped trigonal (Tu) interstitial. c) Bond center (Bc) interstitial.)*

In order to obtain quantitative insight into the relative thermodynamic stability of these defects we compute their formation free energies as a function of temperature. Using the relaxed structures for the defect-free and defected cells, we first compute the DFT total-energy contributions to the formation free energies. The results are shown in Fig. 2, in which the data points and error bars correspond to the average values and their variation, respectively, as obtained from 2 replicas for the Tu interstitial and 3 for the vacancy and Bc interstitial, respectively. In view of its elevated DFT formation energy, only one replica was considered for the Tc interstitial. The solid curves represent second-order polynomial fits to the data points. If we examine the DFT formation energy curves of the individual replicas (not shown), we find that they are essentially parallel to the shown average curves, with a vertical spread indicated by the error bars. This suggests that the effect of the proton disorder on the formation energetics involves only constant energy shifts of the order of ~0.01, consistent with previous findings.[14]

The DFT formation energy of the Tu interstitial is approximately 0.2 eV lower than that of the Tc structure, probably rendering the latter insignificant as an equilibrium point defect in ice I_h. The Tu interstitial is not expected to play a significant role either, given that its DFT formation energy is at least ~ 0.1 eV higher than that of the Bc interstitial and molecular vacancy. Comparing the latter two, although having essentially equal formation energies at low temperatures, it is found that the Bc interstitial is particularly sensitive to thermal expansion, showing a reduction of ~ 7 % upon a linear lattice dilatation of only ~0.5 %. This effect is visibly less pronounced for the vacancy, which shows a decrease of only ~ 2 %. A comparison between the structures of both defects suggests that the strengthened bonding of the Bc interstitial is associated with the relief of compressed hydrogen bonds upon thermal expansion. Since almost all the hydrogen bonds in the vicinity of the Bc interstitial are compressed, this effect is expected to be stronger than for the vacancy, for which only a third of the affected hydrogen bonds are compressed. Indeed, the thermal-expansion-driven bond strengthening for the Bc interstitial leads to a DFT formation energy that is lower than the DFT sublimation energy of 0.69 eV.[14]

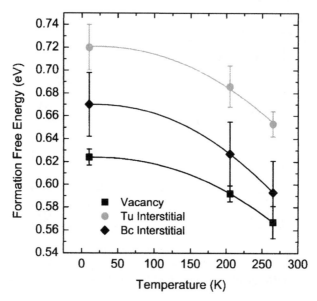

Figure 2   *DFT formation energies as a function of temperature for the Tc interstitial (triangles), Tu interstitial (circles), Bc interstitial (diamonds), and vacancy (squares), averaged over different replicas. Error bars denote the magnitude of the fluctuations due to proton disorder. Lines represent polynomial fits to the average results.*

To compute the *total* formation free energies of the defects, we need to include zero-point and finite-temperature vibrational contributions. To this end, we apply the local harmonic approximation[19], in which these effects are taken into account by comparing the local vibrational modes of the molecules in the vicinity of the defect to those that characterize their vibration in the defect-free crystal. In the case of ice I_h, this requires the analysis of both inter- as well as intramolecular vibrational frequencies. We compute them by numerically estimating the force and torque constants associated with the translational and librational degrees of freedom of the rigid molecules as well as the force constants involving the internal degrees of freedom of the molecules, followed by the diagonalization of the corresponding dynamical matrices. Given the high DFT formation energy of the Tc interstitial and the elevated computational cost we restrict this analysis to single replicas of the molecular vacancy and the Tu and Bc interstitials.

The results for T=10 K are displayed in Fig. 3. Each data point represents a particular vibrational mode ν, whose energy in the defect-free cell is plotted on the horizontal axis. The vertical axis then describes how it changes in the presence of the defect. The perfect-crystal vibrational modes can be divided into four groups, which, respectively, correspond to the molecular translational, molecular librational, intramolecular bending and intramolecular stretching modes. The energies of the four groups are in excellent agreement with experimental inelastic neutron scattering data.[20] Considering the influence of the point defects, their presence mostly affects two groups, causing softening of molecular librational modes and stiffening of intramolecular stretching modes while the molecular translational and intramolecular bending modes remain largely unaltered. These frequency shifts are related to the partially hydrogen-bonded molecules in the defect region, with the stiffening revealing free-molecule-like O-H bonds and the softening indicative of weakened hydrogen bonds. The temperature-dependence of the frequencies, investigated by comparing the results obtained for the 10 K and 265 K supercells, was found to be negligible.

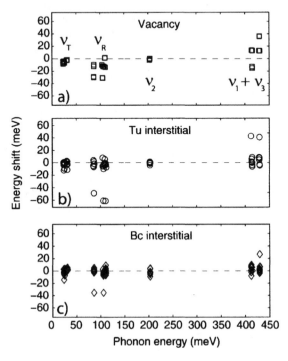

Figure 3 *Vibrational frequency shifts due to the presence of the three point defects. Horizontal axis describes the energy $h\nu$ of each vibrational mode in the defect-free crystal. Vertical axis describes the energy shift in the presence of the defect. Panel a) Vacancy, b) Tu interstitial, c) Bc interstitial. Symbols ν_T, ν_R, ν_2 and $\nu_1 + \nu_3$ denote, respectively, groups of molecular translational, molecular librational, intramolecular bending and intramolecular stretching modes, c.f. Ref. 1.*

Using the data of Figs. 2 and 3 we now compute the full formation free energy as a function of temperature within the local harmonic approximation. The results are shown in Fig. 4. A comparison between both interstitial configurations reveals that, coherent with the results in Fig. 2, the formation free energy of the Bc interstitial is found to be lower than that of the Tu structure across the entire temperature interval. This supports the view that a structure different from the established Tc and Tu interstitials should be the preferred interstitial configuration in ice I_h and, given the structural properties of the Bc interstitial, is consistent with the experimental suggestion of a bound self-interstitial.[5] Moreover, in light of its similarity to the Bc interstitial, the present findings may renew interest in an interstitial/Bjerrum defect complex that has been proposed to explain the equality of the relaxation times and activation energies associated with dielectric/mechanical relaxation and diffusion.[2]

Regarding the relative stability of the vacancy and interstitial, the results of Fig. 4 indicate that the vacancy should be the dominant point-defect species below T = 200 K, with its formation free-energy being lower than that of the Bc and Tu interstitials beyond the limits of the error bars. Compared to Fig. 2, the downward shift of the vacancy curve with respect to the Bc interstitial is due to zero-point energy contributions, which lower the DFT vacancy formation energy by ~ 0.08 eV compared to a reduction of ~ 0.02 eV for the Bc interstitial. This implies that the vibrations of the Bc interstitial are more ``bulk-like'' than the vacancy, containing only one molecule that is not fully hydrogen bonded, compared to four in the case of the vacancy. As the temperature increases, the formation free-energy difference between the defects decreases because of the high formation entropy of Bc interstitial. At T = 265 K, for instance, the latter reaches a value of ~ 7 k_B compared to ~ 5 k_B for the vacancy. The origin of this elevated formation entropy value, which is compatible with the experimental indications,[5] is mostly associated with the appreciable temperature-dependence of the DFT total-energy contribution to the formation free energy, as shown in Fig. 2. As a result of this formation-entropy difference, within the precision of the error bars, the formation free energies of the vacancy and Bc interstitials become essentially equal for T ≥ 200 K. This is consistent with the crossover scenario suggested in Ref. 5, in which the interstitial is assumed to dominate for temperatures above –50° C whereas the vacancy becomes the principal thermal equilibrium point defect at lower temperatures.

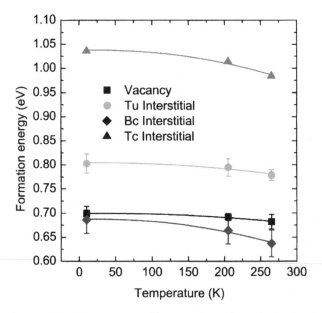

Figure 4   *Formation free energies as a function of temperature for the Tu interstitial (circles), Bc interstitial (diamonds) and vacancy (squares), averaged over different replicas. Error bars denote the magnitude of the fluctuations due to proton disorder. Lines represent polynomial fits to the average results.*

4 CONCLUSIONS

In conclusion, we present a DFT-based study of the structure and formation energetics of a series of molecular point defects in ice I_h. The results suggest that, due to its bonding to the surrounding hydrogen-bond network, the Bc interstitial is the favoured molecular interstitial structure in ice I_h. Considering the formation free energies as a function of temperature, the molecular vacancy is found to be the favoured molecular point defect in ice across most of the considered temperature interval. Due to the high formation entropy of the Bc interstitial, however, a crossover scenario in which the Bc interstitial becomes favoured before reaching the melting point is conceivable.

Acknowledgements

The authors gratefully acknowledge financial support from the Brazilian agencies FAPESP and CNPq.

References

1 V. F. Petrenko and R. W. Whitworth, *The Physics of Ice*, Oxford University Press, Oxford, 1999.

2 C. Haas, *Phys. Lett.*, 1962, **3**, 126.

3 O. E. Mogensen and M. Eldrup, *J. Glaciology*, 1978, **21**, 85.

4 M. Eldrup, O. E. Mogensen, and J. H. Bilgram, *J. Glaciology*, 1978, **21**, 101.

5 T. Hondoh, T. Itoh, and A. Higashi, 'Behavior of point defects in ice crystals revealed by X-Ray topography' in *Point Defects and Defect Interactions in Metals*, eds. J. Takamura, M. Doyama, and M. Kiritani, University of Tokyo Press, 1982, p. 599.

6 K. Goto, T. Hondoh, and A. Higashi, 'Experimental determinations of the concentration and mobility of interstitials in pure ice crystals' in *Point Defects and Defect Interactions in Metals*, eds. J. Takamura, M. Doyama, and M. Kiritani, University of Tokyo Press, 1982, p. 174.

7 K. Goto, T. Hondoh, and A. Higashi, *Jpn. J. Appl. Phys.*, 1986, **25**, 351.

8 T. Hondoh, K. Azuma, and A. Higashi, *J. Phys., Paris*, 1987, **48**, C1.

9 N. H. Fletcher, *The Chemical Physics of Ice*, Cambridge University Press, London, 1970.

10 H. Itih, K. Kawamura, T. Hondoh, and S. Mae, *J. Chem. Phys.*, 1996, **105**, 2408.

11 A. Demurov, R. Radhakrishnan, and B. L. Trout, *J. Chem. Phys*, 2002, **116**, 702.

12 T. Ikeda-Fukazawa, S. Horikawa, T. Hondoh, and K. Kawamura, *J. Chem. Phys*, 2002, **117**, 3886.

13 P. J. Feibelman, *Science*, 2002, **295**, 99.

14 M. de Koning, A. Antonelli, A. J. R. Silva, and A. Fazzio, *Phys. Rev. Lett.*, 2006, **96**, 075501.

15 G. Kresse and J. Hafner, *Phys. Rev. B*, 1993, **47**, R558.

16 G. Kresse and J. Futrthmüller, *Comp. Mat. Sc.*, 1996, **6**, 15.

17 G. Kresse and D. Joubert, *Phys. Rev. B*, 1999, **59**, 1758.

18 J. A. Hayward and J. R. Reimers, *J. Chem. Phys.*, 1997, **106**, 1518.

19 R. LeSar, R. Najafabadi, and D. J. Srolovitz, *Phys. Rev. Lett.*, 1989, **63**, 624.

20 J.-C. Li and D. K. Ross, 'Neutron Scattering studies of ice dynamics. Part I – Inelastic incoherent neutron scattering studies of ice I_h (D_2O, H_2O and HDO)' in Physics and Chemistry of Ice, eds. N. Maeno and T. Hondoh, Hokkaido University Press, 1992, p.27.

FROM ICE TO CO_2 HYDRATES AND BACK – STUDY OF NUCLEATION AND INITIAL GROWTH USING SCANNING ELECTRON MICROSCOPY

A. Falenty[1], G. Genov[1, 2], W.F. Kuhs[1]

[1]GZG, Abt. Kristallographie, Universität Göttingen, Goldschmidtstr. 1, 37077 Göttingen, Germany
[2]Kjemski Institutt, Universitetet i Bergen, Realfagbygget, Allegaten 41, 5007 Bergen, Norway

1 INTRODUCTION

Gas hydrates (GH) form crystalline, non-stoichiometrid compounds belonging to the clathrate hydrate structural family in which gas molecules are trapped in hydrogen bonded water cages[1]. While studies of the formation of GH from liquid water are numerous[2] the formation from ice is less well covered and will be one of the subjects of the present study; the latter has considerable importance for hydrate formation and decomposition in Solar system bodies[3,4], in particular for Mars, on which hydrates could play an important role in geomorphological processes[5]. Two main formation stages can be distinguished: (1) an initial nucleation and growth limited stage and (2) a later diffusion-controlled stage[6-9]. The latter can be easily investigated by gas consumption methods, X-ray or neutron diffraction, which give a good quantitative account of the reactions in a bulk sample. Unfortunately, for the initial phase, the application of these techniques is somewhat limited by the insufficient sensitivity of the methods (with neutron diffraction still performing best); moreover, they do not resolve information on the ongoing nucleation and growth processes like nucleation sites density, shape, size of the formed crystals and their mutual arrangement. The application of an imaging technique like field-emission scanning electron microscopy (FE-SEM)[10-12] can provide the key to understand these processes[7,8], as has been demonstrated in our earlier work in which we developed a shrinking core model for gas hydrate growth[7-9]. Equally interesting is the microstructural arrangement of ice crystals formed upon clathrate decomposition, in particular for a better understanding of the phenomenon of anomalous- (or self-) preservation[13-15]. In this contribution we will discuss recent results concerning nucleation and initial growth processes on the ice / hydrate surface at various p-T conditions and various degrees of transformation. Cryo-SEM techniques will be mainly used for this purpose, performed on samples quenched to liq.N_2 temperatures and recovered at various stages of the transformation process (so-called "interrupted runs"). Neutron and X-ray diffraction will be used in addition to control the averaged transformation degree as a function of time.

2 EXPERIMENTAL SECTION

A custom-build low-pressure cryo system[7,8] adapted to the high-intensity 2-axis neutron diffractometer D20 at the Institute Laue-Langevin, Grenoble, France, allowed us to study transformation processes *in-situ* with a time resolution of a few seconds. Unfortunately, the limited beam time allocations make this technique inadequate for longer experiments. Such runs, up to several months, require a supplementary system[9] based on measurements of the gas consumption (formation) or release (decomposition) in a closed system.

To investigate transformation phenomena at the surface of the solid reactant we use an *ex-situ* FE-SEM imaging method, which provides a frozen-in visualization of the reaction fronts with a maximum resolution of a few tens of nm. The FE-SEMs used (LEO Gemini 1530, QUANTA 200F) are designed for work at low acceleration voltages of less than 2keV, which reduces to a minimum the destructive power of the electron beam and undesired charging effect. During the measurements, the uncoated samples are held in a cryo stage, cooled to about 90K with liq.N_2 at a pressure of about 0.1 Pa (1×10^{-6} bar).
For all clathrates formation experiments discussed in this paper, we used ice spheres with a diameter of <350µm formed by quenching sprayed, demineralised water droplets in liq.N_2. The decomposition runs were carried out on powders ~250µm, prepared by crushing and sieving of CO_2 hydrate through a set of 200 and 300 µm meshes.

3 NUCLEATION AND GROWTH OF CO_2 HYDRATE

There is not much information in literature on the processes of nucleation and initial growth of gas hydrates coating ice particles[16]. At temperatures of ~190K, it is assumed to be a relatively slow process, which needs several hours to transform the free ice surface exposed to the reacting gas[17]. According to Schmitt[17] below 200K a so-called incubation period often delays the onset of the nucleation. Whether this is truly an inhibition of nucleation or just a very slow start following a sigmoidal reaction pattern remains largely an open question. The neutron diffraction data shows no increase in the hydrate fraction during the first few hours. The shape of the reaction curve suggests rather sigmoidal

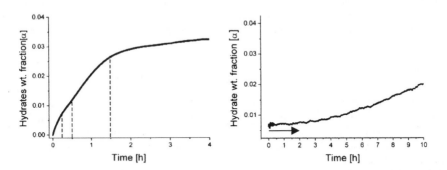

Figure 1   *The initial part of the CO_2 hydrates formation reaction at 193K and 80kPa in the gas consumption system (left). The complete coverage of the surface is expressed by a clearly visible bend of the reaction curve after 90min. Dashed line marks the position of each interrupted run on the reaction curve: 15, 30 and 90 min. The neutron diffraction experiment of the CO_2 hydrate formation at 185K and 36kPa (right) starts significantly only after 2 hours.*

growth but do not rule out the possibility of an incubation period either (Figure 1). The recent formation experiments, performed on gas consumption system, at low temperatures and pressure (193K 80kPa) with CO_2 gas suggest that the time for the nucleation and total coverage could be much shorter. From our SEM work (discussed below) we can infer that this part of the reaction is most probably related to the formation of a hydrate layer on the surface of the ice spheres. With a transformation degree of about 2.5% (after 90 min), we have observed a significant slowing down of the formation process possibly caused by the depletion of ice surface exposed directly to the gas molecules and a corresponding switching to the diffusion controlled growth. It is worthwhile to notice that the reaction described above is still in the time frame of induction period, observed during the neutron diffraction experiments. To better understand the role of microstructure evolution, we have tracked down the surface changes in a series of interrupted runs with steps: 15 min, 30 min, and 90 min.

For the run interrupted after 15 min, the SEM images reveal an ice surface covered with clearly distinguishable nucleation sites and spreading fronts with a well visible sub-micron size porous structure (Figure 2D), which is characteristic for GH[10,12,18,19]. The crystallites show a characteristic "jackfruit pattern" (Figure 2B). The average size of the observed seemingly radially grown features is estimated as 10-15 μm (Figure 2C, D, E, F). The roughly estimated density of nucleation centres is about $4.9\text{-}8.8 \cdot 10^7$ nuclei/m² (assigning one centre to one "jackfruit" area); the distribution of the nucleation centres is rather inhomogeneous. The shape of individual crystallites is "pancake like", which is often related to heterogeneous nucleation[20] (Figure 2D, F). Most of the GH crystallites were developed along the cracks caused by shock freezing or other defective zones (e.g. grain boundaries, small frost particles electrostaticly attached to the spheres), which apparently promote the formation of the new phase (Figure 2E, G). It is interesting to note that at some places nucleation did not occur while other areas were completely covered by GH. Yet, the free surfaces do not resemble typical properties for ice-like high resistance to etching by electron beam. EDX analyses show only the traces of carbon in measured areas, which additionally complicates identification. A striation pattern commonly seen on the ice starting material (Figure 2A) is also not present; instead, one could observe flat and smooth surfaces (Figure 2C, D, E). After 30 min. the reaction is approaching complete coverage. In those regions, where the surface transformation had been already completed, the grain boundaries between separate hydrate crystallites starts to disappear (Figure 2H). In spite of this, it is still relatively easy to see the original "jackfruit patterns" on the ice spheres as well as regions with sub-micron porous structures.

A quite different picture was observed after 90 min. The ice surface is completely replaced by growing clathrates (Figure 3A-D). All borders between crystallites are strongly blurred

Figure 2 *SEM images of the initial ice spheres and after a reaction with CO_2 gas at 193K and 80kPa for 15 and 30 min. (A) surface of a polycrystalline ice sphere before the reaction with clearly visible grain boundaries and striation patterns (B)"jackfruit pattern", (C) nucleation sites with clearly visible uneven surface, (D, F) one of the seemingly radially grown "pancake", nucleation sites, (E) defect and cracks promote GH crystallization, (G) still well visible porous structure after 30min. reaction, (H) slow blurring of the borders between separate GH crystallites.*

START

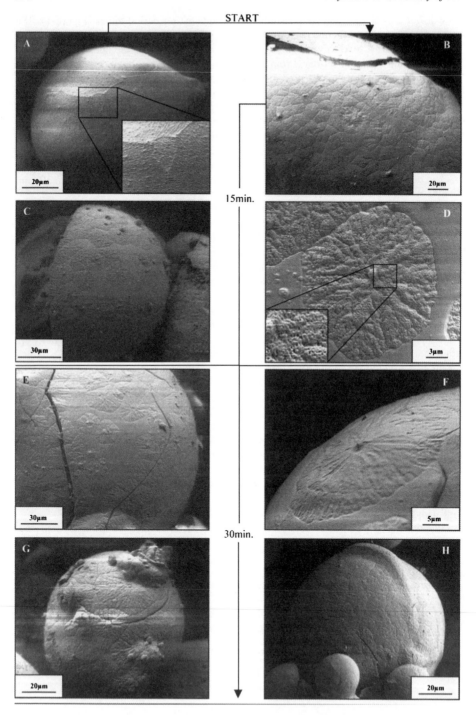

15min.

30min.

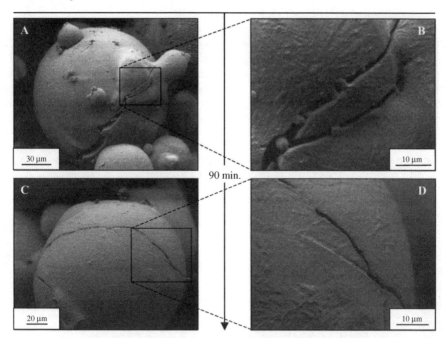

Figure 3 *The SEM images after 90 min. of GH formation at 193K and 80kPa. (A-D) The borders between individual clathrate crystallites are barely visible or have completely vanished. Figure B visualizes the initial and figure D the advanced crack bridging and filling process as enlarged portions of figures A and C respectively.*

and the sub-micron porous structures, previously easily visible, are not present. Instead, one could observe surfaces, which may be easily mistaken as water ice striation patterns (Figure 3B, D). Undulated pseudo-dendritic features building each crystallite, commonly observed after 15-30 minutes, become smoother; yet the differences between elevations and depressions are big enough to be mistaken with the ice striation patterns. A careful analysis of the SEM images reveal smoother and rounded edges (Figure 3B), which are untypical for ice surfaces. At this transformation stage, we have observed the growth of the hydrates in cracks of the initial ice spheres (Figure 3D). At higher temperatures[9] this crack-filling process was assumed to be relatively fast and promote hydrate crystallization due to slightly smaller activation energy necessary to initiate the nucleation. The results at lower temperature presented here suggest the opposite situation: The bridging and crack filling appears to be a consequence of the surface transformation and thickness increase of the GH coating which covers whole surface. (Figure 3B, D). Apparently, the combination of volume expansion with the growth creates an effective mechanism for closing of the smaller cracks at this p-T condition.

4 DECOMPOSITION OF CO$_2$ HYDRATE

The decomposition of gas hydrates is a multi-stage[15], still not well-understood, process. In the initial stage different structures appear depending on the speed of growth related to the p-T conditions and the duration of the decomposition. Above 240K, the considerable mobility of water molecules increases the importance of stacking-fault annealing[15] and Ostwald ripening process, which leads to additional surface changes. In this complex system, two different types of microstructures can be distinguished.

Firstly, the series of the interrupted runs at low temperature (<200K at 0.6kPa) or close to the thermodynamic stability field (220K at 90kPa) reveal surfaces of the hydrate particles, which are covered with relatively well-developed subhedral ice crystals (Figure 4A-D). The slow nucleation and growth from decomposing hydrates leads in certain cases even to the development of clearly visible prismatic and basal planes (Figure 4B, E).

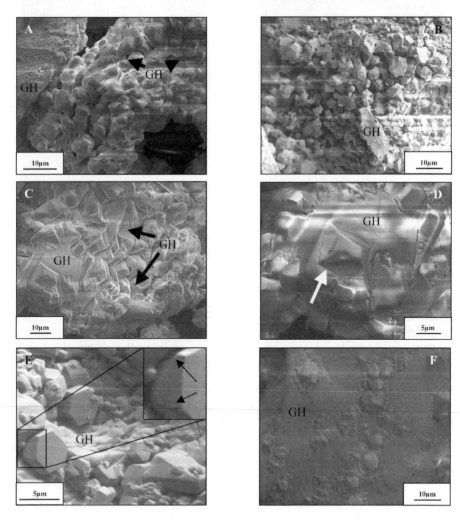

Figure 4 *The SEM images exhibiting typical GH surface evolution during slow decomposition: (A) subhedral ice crystals after 5min at 220K and 90kPa, (B) euhedral and subhedral ice agglomerate on the GH surface after 12h at 180K and 0.6kPa, (C) Between densely packed ice crystals one can still see GH in the close-up (arrows), (D) Magnification of the previous image. The ice crystals look like some sort of melting into the hydrate surface due to the inward growth, (E) Euhedral ice crystals with very well visible prismatic and basal planes at 195K and 0.6kPa, (F) Ice crystals surrounded by a transition zone after 10 min. at 220K and 90kPa. Some of them have clearly visible kinks in the prismatic faces[15] (arrow) indicative for stacking faults.*

The average size of the ice crystals is 5-10μm and increases with the temperature (up to 20μm at 220K and 90kPa). The commonly observed small depressions surrounding ice crystals (Figure 4D-white arrow) suggest an inward growth. The distribution of the newly formed crystals is rather inhomogeneous and in the empty spaces between them one could still observe initial material (Figure 4A, C). It is worth noticing that at 220K 90kPa some of the ice crystals are surrounded by the undulated rim of a "transition zone" (Figure 4F). The thickness of this zone is not larger than few micrometers; typically it is about 1-2μm broad. The second type is related to a rapid growth of ice on the clathrate surface. The fast 5 to 10 minutes long decomposition at 220K and 60kPa (i.e. far from the field of thermodynamic stability of GH) leads to a nucleation of "pancake crystallites" (Figure 5A, B, D). Each of the disks is formed from ice growing in a dendritic manner starting from a central point (Figure 5C). The typical size of the observed features is 15-20μm.

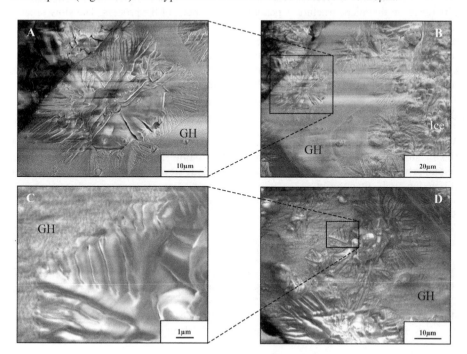

Figure 5 *The surface transformation during fast decomposition (220K 60kPa): (A, C) Inhomogeneously distributed "pancake" ice crystallites covering decomposing GH surface, (A) The magnification of (B) showing well visible transition zone surrounding ice crystallite, (C) The magnification of (D) on dendritic like spreading front from the centre of the nucleation site.*

Many crystals are surrounded by a transition zone (Figure 5A-white arrow). It is somewhat similar to the one observed for the case of slow growth. The distribution of nucleation sites appears to be rather inhomogeneous. In certain places one can still observe the primordial hydrate surface, where other areas exhibit an advanced transformation (Figure 5B). One possible explanation of this state might be the hindered gas outflow, which locally increases the fugacity of CO_2 gas and therefore slows down reaction.

5 CONCLUSIONS

Ice and gas hydrate surfaces in the initial stage of transformation to the other show rich variations in microstructures. Undoubtedly, the microstructures of the newly formed phases at the reacting surface influence the kinetics of these phase transformations. In most cases the observed growing GH exhibits a pancake-like shape; such structures are inhomogeneously distributed. The transformation at the investigated p-T conditions is characterised by a fast outward expansion. The total surface coverage, linked to the slowing down of the reaction curve, marks the transition to the development of a new surface structure, which persists also at later stages. Decomposition of GH shows clear differences in the microstructure depending on the speed of the process. The slow process is characterized by the development of rather well shaped ice crystals, sometimes with clearly visible basal and prismatic planes. In the fast decomposition process pancake-like ice crystallites are formed. The ice coating, developed with time, is almost free from voids. Consequently, the ice cover formed provides a good sealing and could well be related to the appearance of the phenomenon of self-preservation.

Acknowledgements

We thank the Deutsche Forschungsgemeinschaft for financial support (Grant Ku 920/11), the Institut Laue-Langevin in Grenoble for beam time and support, Kirsten Techmer (Göttingen) for her help in the SEM sessions and Steve Kirby (USGS) for valuable discussion on the manuscript.

References

1   E. D Sloan, *Clathrate hydrates of natural gases*. 2nd edn, Marcel Dekker Inc., New York, 1998.
2   Y.F.Makaogon, *Hydrates of Hydrocarbons*, PennWell Books, Tulsa, Oklahoma, 1997.
3   O. Prieto-Ballesteros, J. S. Kargel, M. Fernandez-Sampedro, F. Selsis, E. S. Martinez, D. L. Hogenboom, *Icarus*, 2005, **177**, 491.
4   G. Notesco and A. Bar-Nun, *Icarus*, 2000, **148**, 456.
5   J. S. Kargel, *Mars: A Warmer Wetter Planet*, Springer, Berlin, 2004.

6   R. W. Henning, A. J. Schultz, V. Thieu and Y. Halpern, *J. Phys. Chem. A.*, 2000, **104**, 5066.

7   D. K. Staykova, W. F. Kuhs, A. N. Salamatin, and T. Hansen, *J. Phys. Chem. B.*, 2003, **107**, 10299.

8   W. F. Kuhs, D. K. Staykova and A. N. Salamatin, *J.Phys.Chem. B.*, 2006, **110**, 13283.

9   G. Genov, W. F. Kuhs, D. K. Staykova, E. Goreshnik, and A. N. Salamatin, *Am. Miner.*, 2004, **89**, 1228.

10 W. F. Kuhs, A. Klapproth, F. Gotthardt, K. Techmer, and T. Heinrichs, *Geophys. Res. Lett.*, 2000, **27** (18), 2929.

11 S. Circone, L. A. Stern, S. H. Kirby, W. B. Durham, B. C. Chakoumakos, C. J. Rawn, A. J. Rondinone and Y. Ishii, *J.Phys.Chem. B.*, 2003, **107**, 5529.

12 L. A. Stern, S. H. Kirby, S. Circone and W. B. Durham, *Am. Miner.*, 2004, **89**, 1162.

13 J. S. Takeya, W. Shimada, Y. Kamata, T. Ebinuma, T. Uchida, J. Nagao and H. Narita, *J.Phys. Chem. A.*, 2001, **105**, 9756.

14 L.A. Stern, S. Circone, and S.H. Kirby, *J. Phys. Chem. B.*, 2001, **105**, 1756.

15 W. F. Kuhs, G. Genov, D. K. Staykova, and T. Hansen, *Phys. Chem. Chem. Phys.*, 2004, **6**, 4917.

16 I. L. Moudrakovski, A. A. Sanchez, CH. I. Ratcliffe and J. A. Ripmeester, *J. Phys. Chem. B.*, 2001, **150**, 12338.

17 B. Schmitt, PhD Thesis, Laboratoire de Glaciologie et Géophysique de L'Environnement, Grenoble, 1986.

18 W. F. Kuhs, G. Y. Genov, E. Goreshnik, A. Zeller, K. Techmer, and G. Bohrmann *Intern. J. Offshore Polar Eng.*, 2004, **14** (4), 305.

19 K. Techmer, T. Heinrichs and W. F. Kuhs, *Geological Survey of Canada Bulletin*, 2005, **585** (12 p.)

20 A. C. Lasaga, *Kinetic theory in earth science*, Princeton University Press, Princeton, New Jersey, 1997.

THE TEMPERATURE GRADIENT METAMORPHISM OF SNOW: MODEL AND FIRST VALIDATIONS USING X-RAY MICROTOMOGRAPHIC IMAGES

F. Flin[1], J.-B. Brzoska[2], R. A. Pieritz[3], B. Lesaffre[2], C. Coléou[2] and Y. Furukawa[1]

[1] ILTS, Hokkaido University, Kita-19, Nishi-8, Kita-ku, Sapporo, Japan
[2] CEN, Météo-France, 1441 rue de la piscine, 38406 Saint Martin d'Hères CEDEX, France
[3] ESRF, 6 rue Jules Horowitz, BP220, 38043 Grenoble CEDEX, France

1 INTRODUCTION

During a snowfall, the snow crystals accumulate on the ground and gradually form a complex porous medium constituted of air, water vapour, ice and sometimes liquid water. This ground-lying snow transforms with time, depending on the physical parameters of the environment. This process, called *metamorphism*, can be divided into three main types: the *wet snow* metamorphism, the *isothermal* metamorphism, and the *temperature gradient* (TG) metamorphism.

Among these different kinds of metamorphisms, the last one is probably the most interesting. Typically occurring by cold and clear nights, when the TG between the top and the bottom of the snow layer is high, this metamorphism is characterized by the formation of facets at the bottom of the grains, while upper parts remain rounded.[1,2]

Since the TG metamorphism may be the source of weak layer formation in the snow cover, its study has major issues in avalanche sciences,[3] and is an active research field in snow and ice community (see the introduction of Sommerfeld,[4] for a detailed review until 1983). Despite of this interest, the TG metamorphism remains quite poorly understood. In particular, two fundamental questions have not really been solved. First, what is the driving force of the matter exchange in the ice matrix and what are the associated mechanisms? Second, what determines practically whether well-rounded or faceted shapes can appear?

These two questions have been addressed and partly solved by Colbeck[2] twenty years ago, but the results where based on 2D observations and very simple approximations on the snow geometry. In our approach, we would like to take advantage of X-ray microtomographic techniques and revisit these questions by using high-resolution 3D images.

We first present a simple model that estimates the matter fluxes in a snow sample submitted to temperature gradients, and address the faceting issue by using standard concepts of crystal growth. We then describe a TG experiment, followed by X-ray microtomographic measurements that give 3D images of the metamorphosed snow structures. Then, we apply our model to one of the obtained snow samples, in order to check its validity.

2 PHYSICAL MODEL

2.1 Origin of the Matter Transfer inside the Snow Matrix

The basic idea behind this model is the following: snow crystals grow (respectively decrease) by the condensation (sublimation) of vapour on their bottom (top) portions because they are colder (warmer) than the average snow temperature at the considered height (Figure 1).

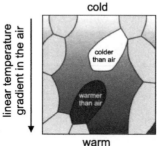

Figure 1   *Simplified model of TG metamorphism: the upper (respectively lower) grain is colder (warmer) than the surrounding vapour. It leads to condensation (sublimation) on the grain surface.*

In this model, we assume that the main limiting mechanism is the sublimation-condensation at the ice/pore interface, as supposed in a previous work.[5] This assumption may be subject to debate, as generally the vapour diffusion mechanism is directly assumed to be predominant (see for instance the recent work of Flanner and Zender[6]), in spite of numerous other proposed mechanisms such as, for example, grain boundary effects[7] or surface diffusion.[8] The usual argument to prefer the vapour diffusion to the sublimation-condensation as being a possible limiting process consists in using a sphere model to estimate the contributions from the different mechanisms.[9] By considering then that the condensation coefficient is close to 1, it is straightforward to conclude that the predominant mechanism is the vapour diffusion in the air. However, there are two main objections to this kind of reasoning:

-1) the geometrical models based on spheres or assemblies of spheres, which are used to represent the snow microstructure, are clearly oversimplified, and one could really wonder about the physical relevance of such models for real structures of snow.

-2) the experimental values commonly found in the literature for the condensation coefficient α actually range from 1 to 10^{-4}.[10] This large scatter can be partly explained by the fact that α is probably depending of both the temperature and the supersaturation ranges.[11] Thus, the condensation coefficient cannot be considered as being equal to 1 a priori, but needs to be estimated according to the imposed experimental conditions.

In our approach, we address these two problems (1) by taking into account the real 3D geometry of the snow structure by using X-ray microtomographic images and (2) by assuming a value of alpha around 10^{-3}, as provided by a previous study on isothermal metamorphism in nearly similar conditions.[12] This value is in good accordance with the results given by Libbrecht[11] for extremely low supersaturations (in our snow samples, the supersaturation is lower than 0.01% as the temperature gradient do not exceed 10 K m^{-1}) provided that the Libbrecht's available data near -10°C can be extrapolated at -3°C.

Consequently, there is no obvious reason to discard the sublimation-condensation as a possible predominant mechanism, at least for small samples of deposited snow undergoing low supersaturations. Finding out the limiting mechanism is a difficult task, and both the diffusion and the surface phenomena may actually coexist in the same metamorphic process. In this work, we choose to consider only the sublimation-condensation mechanism, which is much easier to implement in a 3D numerical model, and gives realistic results (see sections 4 and 5). Taking into account vapour diffusion in a 3D real-structure model and comparing the obtained results in both cases would probably allow learning more about the rate limiting mechanisms, and will be the subject of a further study.

Besides, we did also the following hypotheses in our modelling work:
-the latent heat is quickly removed through the highly conductive ice;[2]
-the convection effects in the pore space are negligible;[13, 14]

In order to quantify the matter transfers involved at the ice/pore interface, we can express the incoming matter flux j [mol m^{-2} s^{-1}] on a point of the ice surface by using the Langmuir-Knudsen formula: [15, 16]

$$j(C,C_a,T,T_a) = \alpha \frac{P(C_a,T_a) - P(C,T)}{(2\pi MRT_a)^{1/2}} \tag{1}$$

where R is the universal gas constant and M is the molar mass of water. C, T, P are respectively the local mean curvature of the ice/pore interface, the local temperature of the interface, and the local equilibrium vapour pressure at the interface. The a-subscript designs the corresponding averaged quantities that define the ambient surrounding vapour pressure. The condensation coefficient α is supposed to be the same for both condensation and sublimation.

In a preceding work on snow isothermal metamorphism,[5] we introduced a simple numerical model based on this local equation (1). However, $T = T_a$ was a constant and $P(C,T)$ was just the Kelvin equation. In our present case, we would like to obtain a similar expression, which also takes into account the temperature dependency. By combining the Clausius-Clapeyron and Kelvin equation, we have thus:

$$P(C,T) = P_0 \exp\left[\frac{1}{RT}\left(\Delta h_{sub}\frac{T-T_0}{T_0} + 2\gamma\Omega C\right)\right] \tag{2}$$

$$P(C_a,T_a) = P_0 \exp\left[\frac{1}{RT_a}\left(\Delta h_{sub}\frac{T_a-T_0}{T_0} + 2\gamma\Omega C_a\right)\right] \tag{3}$$

where P_0 is the reference vapour pressure at some temperature T_0, Δh_{sub} is the molar enthalpy of sublimation, γ is the surface tension of the interface and Ω the molar volume of ice. By inserting (2) and (3) into (1), we can determine an expression of j that just depends on the parameters C, C_a, T, T_a and numerical constants:

$$j(C,C_a,T,T_a) = Z(C,C_a,T,T_a) \cdot k \tag{4}$$

where k is a positive numerical constant and Z is defined by:

$$Z(C,C_a,T,T_a) = \Delta h_{sub} \left(\frac{1}{T} - \frac{1}{T_a} \right) + 2\gamma\Omega \left(\frac{C_a}{T_a} - \frac{C}{T} \right) \tag{5}$$

In our model, the ambient vapour pressure is considered as uniform on any horizontal plane of the snow sample (because the macroscopic TG can be supposed constant across the sample). Consequently, on such a plane, $P(C_a,T_a)$ is given by the values of the curvature and temperature that have been averaged on the considered plane. By knowing the temperature and the curvature fields on the ice/pore interface, one can then use the equation (5) to compute the local matter fluxes occurring in each point of this interface.

2.2 Faceting

To address the problem of faceting of grains, we use a classical atomistic approach by reasoning with a Kossel crystal representation,[17, 18] which is commonly used in crystal growth theories. The main idea is summarized as follows:

a) During the growth of a Kossel crystal, some sites (such as step or kink sites) are energetically more likely to adsorb atoms than flat surfaces. It results in a layer by layer process and a production of facets (Figure 2a).

b) On the contrary, the sublimation of such a crystal will begin by removing the step and kink sites. This will produce new kink and step sites, and finally lead to an atomically rough surface (Figure 2b), that is, a macroscopically rounded interface.

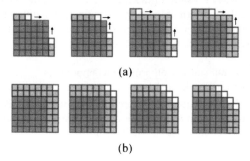

(a)

(b)

Figure 2   *Simple faceting model based on Kossel crystal explanation. Condensation leads to faceting (a) while sublimation generates a rounding of the grains (b).*

Thus, growth of grains triggers faceting while sublimation leads to rounding of shapes. As a consequence, the sign of $Z(C,C_a,T,T_a)$ can predict the faceting or the rounding of shapes. In order to check this model, we submitted snow to a temperature gradient and imaged the obtained samples.

3 EXPERIMENT OF TEMPERATURE GRADIENT METAMORPHISM

3.1 Sample Preparation

A three week long experiment of TG metamorphism was run at Col de Porte, French Alps, in order to provide precise microtomographic 3D data for the validation of TG metamorphism models.

Snow was sieved into a specially designed cool box, which is able to generate several TG in a single experiment (Figure 3). This insulated cool box uses two thick copper plates to force a quasi vertical TG, whose values vary linearly along the horizontal direction. A fluid circulation maintains the copper plates at the desired temperature. Temperature probes were used in order to monitor the temperature in different points of the snow layer. The snow, whose density was about 280 kg m^{-3}, was submitted to a permanent temperature gradient ranging from 3 K m^{-1} at one side of the device to 16 K m^{-1} at the other side. After three weeks of experiment at an average temperature of -3°C, samples of 4 cm in diameter were cored and impregnated at -8°C with 1-chloronaphtalene. The cold room was then set to -25°C in order to freeze completely the samples and allow further machining. Smaller cores of 9 mm in diameter were then extracted with a precision hole-saw, sealed inside thin PMMA boxes and kept ready for tomography acquisition at -50°C.

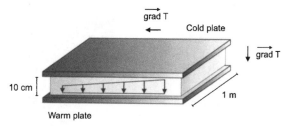

Figure 3 *TG experimental setup. Several TG, from 3 to 16 K m^{-1}, were generated in a single experiment.*

3.2 Image Acquisition by X-ray Microtomography

3D images of snow samples were obtained at the ID19 beamline of the European Synchrotron Radiation Facility (ESRF), Grenoble, France by X-ray absorption microtomography using a specially designed refrigerated cell.[19, 20] All the images were obtained at 18 keV, with a voxel (volume element) size of 4.91 µm. The grey-level images obtained, reconstructed at the ESRF, were contoured using a semi-automatic procedure. More information about core preparation and image processing can be found in a previous paper.[5] To allow fast processing times for our image analysis algorithms, the resolution of the 3D images was reduced by a factor 2 in the 3 axes. Thus, in all the 3D snow images presented in this article, one voxel corresponds to 9.82 µm.

4 MODEL VALIDATION

As shown in part 2, the sign of Z can predict the faceting or the rounding of the shapes. So, having a 3D image of snow offers to check the validity of our model. We tested the model on a sample submitted to a low TG of 3 K m^{-1}. To compute Z, we need to estimate both the curvature and the temperature in each point of the ice/pore interface:

-Curvature maps (Figure 4, left hand side) can be easily obtained by using image analysis algorithms previously developed.[21, 22]

-Temperature maps (Figure 4, right hand side) require modelling the heat transfer into the 3D snow structure. As mentioned in section 2.1, convection and latent heat effects in snow are negligible for the considered TG (3 K m^{-1}). Consequently, the temperature field was computed by considering only conduction effects in both the ice and the pores of the snow structure. This was accomplished by using a Finite Volume scheme.[23] Because of the large meshes used, the solver was written following an iterative approach. The initial boundary conditions were fixed according to the experimental measurements: the temperatures on the top and the bottom of the 3D image were estimated from the experimental temperature data by linear interpolation. For the model validation, a cube of 3x3x3 mm^3 was extracted from the centre of the wider (6x6x6 mm^3) 3D field initially obtained, in order to remove any possible side effect.

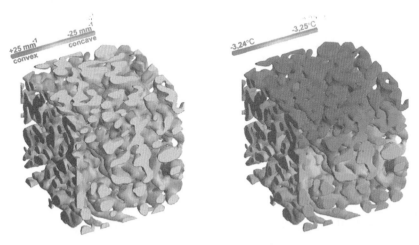

Figure 4   *Computed curvature map (left hand side) and temperature field (right hand side) of a sample submitted to a TG of 3 K m^{-1}. Edge size of the volume: 300 voxels ≈ 3 mm.*

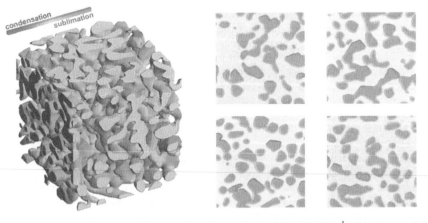

Figure 5   *Computed Z map of a sample submitted to a TG of 3 K m^{-1}. Edge size of the volume: 300 voxels ≈ 3 mm. Some 2D cuts extracted from the 3D volume are represented on the right side of the figure. The top (respectively, the bottom) of the images corresponds to the lowest (highest) and warmest (coolest) side of the physical sample.*

For a better visualisation of the facetted shapes, all the images of the Figures 4 and 5 are presented "upside down": the top (respectively, the bottom) of the images corresponds to the lowest (highest) and warmest (coolest) side of the physical sample. From Figure 5, it appears clearly that predicted condensation sites ($Z > 0$) correspond to faceted shapes while rounded shapes are defined by sites where $Z < 0$ (sublimation). This model validation is confirmed by the graph of the Figure 6, which displays the curvature distribution depending on the sign of Z: the curve for $Z < 0$ is typical of rounded shapes while the curve for $Z > 0$ is a signature of faceted shapes.[20]

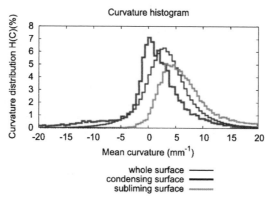

Figure 6 *Curvature distribution for the whole surface as well as the condensing and subliming parts. The histogram of the condensing (respectively subliming) surface is typical from faceted (rounded) shapes.*[20]

5 CONCLUSION

A simple physical model describing the TG metamorphism of snow has been presented. This model, based on Kelvin and Langmuir-Knudsen equations, takes into account the variation of the saturating vapour pressure with temperature. It can determine locally whether the ice is condensing or subliming, just depending on both the temperatures in the snow matrix and the local mean curvatures of the ice/pore interface. This model can also explain the formation and location of facets that appear during the metamorphism. It is consistent with several observations,[1, 2] suggesting that, during the TG metamorphism, grains sublime at their top parts by generating rounded shapes while their bottom parts grow into faceted shapes. More recent time-lapse observations of grain motion also strongly support this interpretation.[24]

Our experimental results also points out that faceting may occur at TG that are much lower than the common accepted values available in the literature (about 10 K m^{-1})[2, 4]. This discrepancy with previous studies may be due to the two following reasons:

-our investigating method used a TG experimental setup able to generate at a same time, a lot of different TG for the same snow type. This specific setup, combined with high resolution tomographic image acquisition, probably allowed tracking the faceting occurrence more precisely.

-we submitted the snow samples to a particularly high average temperature (around -3°C) and a long metamorphism time (3 weeks), which enhance the possibility of observing facets that are characterized with very low growth rates.

The validity of the model has been successfully checked on a snow 3D image obtained from a sample submitted to a low TG. The choice of the surface kinetics mechanism over the generally more accepted diffusion mechanism seems pertinent to determine which parts of the snow structure will undergo faceting or rounding, for the considered TG. Of course, one cannot discard the diffusion as a possible limiting mechanism directly from these results. Indeed, either the diffusion or the surface kinetics, or any combination of these two mechanisms may result in quite similar snow morphologies. Further investigations, requiring to model the vapour diffusion in 3D snow structures seems mandatory to solve this complicated question effectively.

The model should now be tested on a larger set of temperature gradients, by using the other images provided by X-ray microtomography. If confirmed, it offers really interesting outcomes for 3D time-lapse numerical simulations of the TG metamorphism.

Acknowledgments

This work was partially supported by the Japan Society for the Promotion of Science (JSPS). The authors are also grateful to all the staff of the ESRF ID19 beamline, and especially X. Thibault and J. Baruchel.

References

1   Z. Yosida, H. Oura, D. Kuroiwa, T. Huzioka, K. Kojima, S. Aoki and S. Kinosita, *Contrib. Inst. Low. Temp. Sci.*, 1955, **7**, 19.

2   S. C. Colbeck, *J. Geophys. Res.*, 1983, **88**, No. C9, 5475.

3   J. Schweitzer, J. B. Jamieson and M. Schneebeli, *Rev. Geophys.*, 2003, **41**, No. 4, 1016.

4   R. A. Sommerfeld, *J. Geophys. Res.*, 1983, **88**, No. C2, 1484.

5   F. Flin, J.-B. Brzoska, B. Lesaffre, C. Coléou and R. A. Pieritz, *J. Phys. D: Appl. Phys.*, 2003, **36**, No 10, A49.

6   M. G. Flanner and C. S. Zender, *J. Geophys. Res.*, 2006, **111**, No. D12208, doi: 10.1029/2005JD006834.

7   S. C. Colbeck, *Journal of Applied Physics*, 2001, **89**, No. 8, 4612.

8   N. Maeno and T. Ebinuma, *J. Phys. Chem.*, 1983, **87**, 4103.

9   L. Legagneux and F. Dominé, *J. Geophys. Res.*, 2005, **110**, No. F04011, doi: 10.1029/2004JF000181.

10   P. W. Hobbs, in *Ice Physics*, Oxford University Press, London, 1974, pp. 441-443.

11   K. G. Libbrecht, *Journal of Crystal Growth*, 2003, **247**, 530.

12   F. Flin, PhD Thesis, Université Grenoble1, 2004 (partially in French).

13   E. Brun and F. Touvier, *Journal de Physique*, 1987, **C1**, supplement to No. 3, Tome 48, 257 (in French).

14   T. U. Kaempfer, M. Schneebeli and S. A. Sokratov, *Geophys. Res. Lett.*, 2005, **32**, L21503.

15   B. Mutaftschiev, in *The Atomistic Nature of Crystal Growth*, Springer Series in Material Science, Berlin, 2001, Chapter 16.1, p. 292.

16   I. W. Eames, N. J. Marr and H. Sabir, *Int. J. Heat Mass Transfer.*, 1997, **40**, No. 12, 2963.

17   I. V. Markov, in *Crystal Growth for beginners*, World Scientific Publishing Co. Pte. Ltd., Singapore, 1995, Chapter 1, p. 32.

18   B. Mutaftschiev, in *The Atomistic Nature of Crystal Growth*, Springer Series in Material Science, Berlin, 2001, Chapter 2, p. 46.

19   J.-B. Brzoska, C. Coléou, B. Lesaffre, S. Borel, O. Brissaud, W. Ludwig, E. Boller and J. Baruchel, *ESRF Newsletter*, 1999, **32**, 22.

20   C. Coléou, B. Lesaffre, J.-B. Brzoska, W. Ludwig and E. Boller, *Ann. Glaciol.*, 2001, **32**, 75.

21   F. Flin, J.-B. Brzoska, B. Lesaffre, C. Coléou and R. A. Pieritz, *Ann. Glaciol.*, 2004, **38**, 39.

22   F. Flin, J.-B. Brzoska, D. Coeurjolly, R. A. Pieritz, B. Lesaffre, C. Coléou, P. Lamboley, O. Teytaud, G. L. Vignoles and J.-F. Delesse, *IEEE Trans. Image Process.*, 2005, **14**, No. 5, 585.

23   S. V. Patankar in *Numerical Heat Transfer and Fluid Flow (Series in Computational Methods in Mechanics and Thermal Sciences)*, Taylor and Francis, 1980.

24   A. Sato and Y. Kamata, *Proceedings of the 2000 Cold Region Technology Conference*, 2000, **16**, 143 (in Japanese).

SUM FREQUENCY GENERATION ON SINGLE-CRYSTALLINE ICE I$_h$

H. Groenzin, I. Li, and M.J. Shultz

Pearson Laboratory, Tufts University, 62 Talbot Ave., Medford, MA, 02155 USA

1 INTRODUCTION

Ice is arguably one of the most important solids in the Universe. Despite the importance and prevalence of ice, much remains unknown about the atomic-level arrangement at the surface. The atomic-level arrangement, however, affects chemical interactions at the surface. In interstellar space, these interactions make ice a veritable chemical factory, which is of major importance to the composition of comets. Closer to terra firma, generation and interparticle transfer of charges on ice particles are responsible for most lightening storms.[1] In the stratosphere, ice surfaces play a key role in enabling and catalyzing reactions responsible for the now famous ozone hole. Throughout the troposphere, ice and aqueous particles transport and cycle countless species as well as provide a staging site for reactions that are either forbidden or very slow in the gas phase. Our world would literally be a different place without this marvelous solid. Yet, until recently, the surface of ice was a mysterious material at the molecular level. Glancing angle X-ray has investigated the oxygen atom arrangement, examining the question of surface premelting.[2,3] LEED measurements[4] of a thin film of ice on a Pt substrate have shown that the surface either reconstructs into a somewhat disordered material or undergoes large amplitude motion at 90 K – a large amplitude motion that is predicted to persist even at 0 K. More recent He atom scattering experiments[5,6] affirm *ab initio*

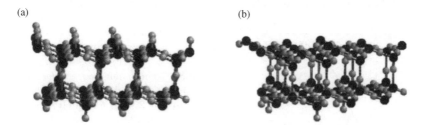

Figure 1 *Cartoon rendition of an ice surface with (a) an ideal hydrogen versus (b) lone pair termination. A so called 'proton disordered' surface consists of a mixture of hydrogen and lone-pair termination.*

calculations that the surface has a full bilayer termination with well-ordered oxygen atoms, completing the picture of the oxygen atoms at low temperatures (<130 K). The hydrogen atoms are more elusive. FTIR measurements on low temperature (<120 K) ice[7,8] suggest that crystalline samples have a low density of free-OH groups whereas amorphous samples have abundant dangling-OH groups. Given the insensitivity of linear spectroscopy to the surface, the location of the surface hydrogen atoms is still uncertain. For example, it is difficult for bulk techniques to distinguish dangling bonds on the surface from those in micropores. In particular, location of dangling OH groups remains unexplored at higher, atmospherically relevant temperatures.

For a mechanistic understanding of reactions at the ice surface however, the location and accessibility of the hydrogen atoms is all important. As illustrated by the cartoon rendition in Figure 1, hydrogen termination provides a donor for hydrogen bond formation with incoming adsorbates, while the oxygen terminated surface features lone pairs that are hydrogen bond acceptors. Since ice is not a rigidly bonded solid, the atomic arrangement can change with interactions and most certainly does change with temperature.[9,10] Little is known about the dynamics of the ice surface in the very relevant 190-273 K range.

To address this gap in understanding, the surface specific nonlinear vibrational spectroscopy, sum frequency generation (SFG) has been applied to study the hexagonal ice surface, I_h. The I_h crystalline form is chosen as the focus of this work because it is the stable form of ice for ambient conditions on Earth. Further, the most abundant exposed face is the hexagonal or basal face.[11,12] The second most abundant face consists of the cylinder sides, also called the prism face. This work examines both of these common faces.

2 EXPERIMENTAL

This section contains a brief description of the probe spectroscopy as well as information about surface preparation and crystal orientation. The combination of SFG and selected orientation provides exquisite detail about the surface.

2.1 Probe

SFG is a vibrational spectroscopy. As such, it is an ideal probe for locating hydrogen atoms since the OH oscillator is sensitive to the hydrogen bonding environment. In addition, due to the polarization information contained in SFG, this vibrational spectroscopy provides orientation information.

Only the major features of this technique are described here since excellent reviews exist.[13-18] Two input beams, one visible and the other infrared, mutually overlap on the surface. Since the surface is noncentrosymmetric, these beams combine as described by the second-order, nonlinear susceptibility, $\chi^{(2)}$, to produce a nonlinear polarization. This nonlinear *polarization* is proportional to the *spatially averaged product* of the infrared and Raman transition moments and the number of scattering centers. The scattered *intensity* is proportional to the nonlinear polarization squared. The spatial average over the coherence length of the two probe beams

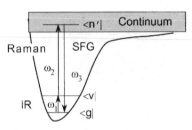

Figure 2 *Energy-level diagram for sum frequency generation.*

results in the interference that makes the method both surface sensitive and a function of the molecular orientation on the surface.[19-23] Polarization measurements provide further information about orientation on the surface. The plane of incidence is defined by the incident angle and the surface normal: *p* polarization is in the plane of incidence, *s* polarization is perpendicular to the plane. Polarization is specified in the order sum frequency, visible, and infrared.

An energy-level diagram for SFG is shown in Figure 2; scanning the infrared

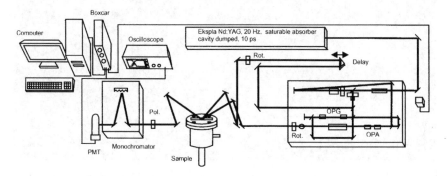

Figure 3 *Schematic of the experimental set up for sum frequency generation.*

wavelength produces a vibrational spectrum of the interface. A schematic of the experimental set up is shown in Figure 3. Both the visible and infrared beams are generated from a *ps* YAG (Ekspla PL2143A/20). The 1064 nm output of the YAG is fed into an optical parametric generator/optical parametric amplifier OPG/OPA (LaserVision Custom KTP/KTA/AgGaSe$_2$ OPG/OPA) where the beam is split. One beam is frequency doubled to provide the 532 nm pump light for the first OPG down conversion stage, as well as to provide the visible portion of the beams used for sum frequency generation. The other beam is sent through a delay before recombining with the idler output of the first stage and proceeding to the second OPA converter stage where a difference frequency process generates a tunable (1.35 to 4 μm) IR beam. The infrared wavelength is selected by angle tuning the OPG/OPA crystals producing 4 cm^{-1} bandwidth infrared. The tunable infrared and fixed visible beams overlap spatially and temporally on the surface producing the sum frequency, which is filtered, sent through a monochromator to a photomultiplier tube and the signal processed through a box car averager. The beams impinge on the surface at angles of 50° (vis) and 60° (IR). This determines the direction of the SF beam to be just over 50°.

2.2 Single Crystal Generation and Verification

Surface preparation is a key ingredient for successful characterization. For this work, single crystals were grown in a modified Bridgman apparatus. A capillary is attached to the growth tube to seed ice growth. Slow growth results in the *c*-axis oriented perpendicular to the growth tube axis. Careful control of growth conditions (Figure 4) results in crypto-morphological growth of single crystals that are 2.5 cm diameter by 7-8 cm long. These large crystals greatly facilitate locating the *c*-axis and probing a single face, either the prism (Figure 4e) or basal face (Figure 4f), as desired.

Characterization of the crystal structure uses the birefringence of ice (n_e = 1.3105, n_o = 1.3091). Between crossed polarizers, only a crystal oriented with the *c*-axis along the light

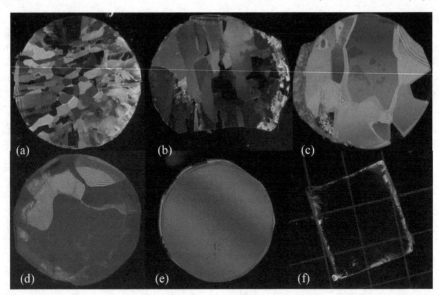

Figure 4 *Crystal cross sections as seen on the Rigsby stage. (a) A growth rate of 2*
μm/s produces a multicrystalline sample. (b) Slowing growth to 1 μm/s
produces larger domains. (c,d) Careful control of growth conditions and
slowing growth to 0.75 μm/s produces yet larger domains that nonetheless
show signs of hoppering (stepped structures). Hoppering indicates that
growth is insufficiently slow. (e) A single domain cut along the c-axis *(the*
prism face) shows alternating light and dark as the crystal is rotated on the
Rigsby stage. (f) A cut perpendicular to the c-axis *(along the basal face)*
remains dark as the crystal is rotated.

propagation direction remains dark as the ice is rotated (Figure 4f). The procedure for
production of the basal face is as follows. The ice crystal is mounted in an orientation
stage and a section perpendicular to the ice growth direction is cut from the crystal. This
section is placed on a Rigsby stage[24,25] on top of a light box. The rotation angles required
to orient the c-axis along the light propagation direction are recorded. The angles are
transferred to the orientation stage and the crystal rotated so that the c-axis is
perpendicular to the cutting plane. A thin section is cut from the ice plug and mounted on
a glass slide. The c-axis orientation is verified with the Rigsby stage. The combination
enables production of the basal face to within 1°. A prism face sample is cut by rotation of
the ice plug so that the c-axis is parallel to the cutting direction.

After cutting the appropriate face, the surface is smoothed by shaving several tens to
hundreds of microns from the surface with a microtome. This produces a very flat surface.
To produce an optically flat surface, the microtomed surface is allowed to self anneal in a
freezer at – 15 °C for several hours. Since the freezer walls are slightly colder than the ice
surface, annealing removes asperities on the surface. Optical flatness is verified by
illuminating the surface with a green laser pointer and examining the reflection at a meter
distance. The absence of interference from microtome produced striations indicates nm
scale flatness. The resulting surfaces are sufficiently reflective for SFG characterization.

3 RESULTS AND SIGNIFICANCE

The SFG spectrum of the basal face of single crystalline hexagonal ice in the *ssp* polarization combination (Figure 5) is dominated by a peak at 3100 cm^{-1}. This peak is commonly associated with the bonded OH-stretching modes in the hydrogen bonding lattice[26-28] and often referred to as the "ice peak." It can also be observed in the Raman spectrum of ice.[26,27] The main peak is accompanied by shoulders on either side of 3100 cm^{-1}, indicating more than one oscillator type in the hydrogen-bonded region. More interestingly, the *ppp* polarization combination reveals strong oscillators blue-shifted from 3100 cm^{-1} that rival the main "ice peak" in strength. Similar peaks are observed in the IR spectrum of ice.[29] No less than five distinct features are observable in the hydrogen-bonded region of the *ppp* spectrum of the single-crystalline ice surface: 3100, 3150, 3200, 3250 and 3400 cm^{-1}. The designation *ssp* and *ppp* refer to the polarizations of the input and output beams in the order of SF, vis, IR. The peaks in the 3100-3400 cm^{-1} range have not been shown with this clarity or intensity in prior work. Figure 5 further shows that the peak at 3100 cm^{-1}, which is very sharp (in particular in comparison to the hydrogen bonded region of the water surface), continues to red-shift with decreasing temperatures. In fact, the peak that is identified in this work by its position at 3100 cm^{-1} was measured in previous works at 3150 cm^{-1} at 173 K.[28] Our spectrum at 178 K is in good agreement with these measurements. Note in particular the increasing intensity of the entire hydrogen-bonded region and the switch in relative intensity between the 3100 and 3200 cm^{-1} peak with decreasing temperature.

Figure 6 shows a temperature cycling of the sample from 128 K up to 178 K and back down to 153 K and reveals that the peaks in the 3150-3300 cm^{-1} range greatly diminish. Clearly the surface undergoes a reconfiguration, likely due to differential pumping

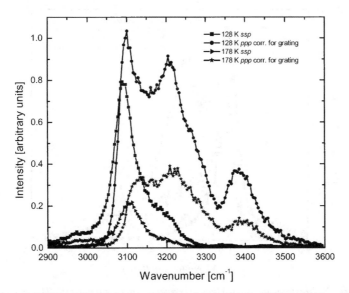

Figure 5 *SFG spectra of the ice surface of a single-crystalline species with the c-axis oriented along the surface normal (the basal face) to within 2°. The spectrum is shown at two temperatures and two polarization combinations.*

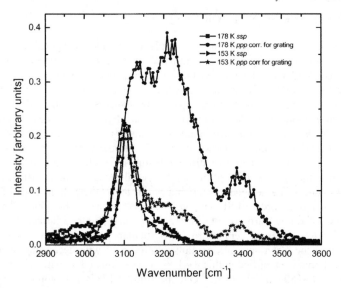

Figure 6 *Spectra after warming ice from 128 K to 178 K then cooled to 153 K.*

between warmer and colder parts in the cell during cycling.

Figure 7 shows that reconfiguration of the surface is not permanent. Warming the sample to freezer temperature (263 K), annealing overnight, and recooling to 128 K

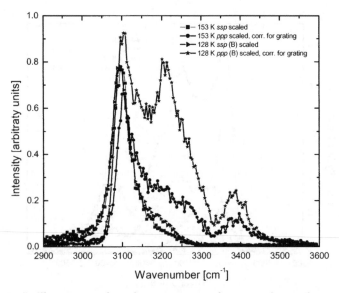

Figure 7 *Illustration of surface recovery after annealing. The spectrum compares the modified surface at 153 K and the recovered surface at 128 K. The two temperature spectra are scaled to the same intensity for the 3100 cm⁻¹ peak to aid comparison.*

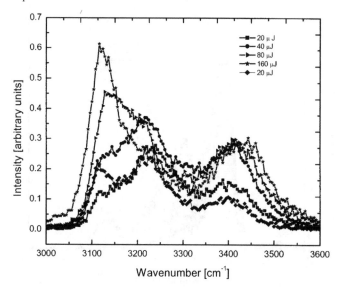

Figure 8 *The* ppp *spectrum at approximately 193 K of the surface single-crystalline ice, cut at approximately 7° to the normal, at different IR beam intensities. The spectra were run in the order they are listed in the legend (top to bottom).*

almost fully restores the spectral features. During the annealing the ice sample is kept in the sealed sample cell to avoid significant mass transport from or to the ice surface. The robustness of the features at 3100 cm^{-1} contrasted with the fragility and dynamics of the peaks in the 3150-3300 cm^{-1} range suggests that the former originate from deeper bi-layers in which the inversion symmetry of the bulk is still broken due to the proximity of the surface while the latter are reflective of the topmost bilayer. The robustness of the 3100 cm^{-1} peak, particularly in the *ssp* polarization combination, makes it ideally suited as a reference point: It has been used (Figure 7) to scale the absolute intensities for better comparability as the absolute intensity is dependent on the reflection efficiency of the sample, which varies from sample to sample and spot to spot.

The above-mentioned findings are further corroborated by the beam heating results shown in Figure 8. Excessive beam intensities induce a reconstruction of the surface which is again evidenced by diminished oscillator strength in the 3150-3300 cm^{-1} region. The features are mostly restored by reducing the intensity of the input IR beam. The relative speed of the recovery (no overnight annealing required) is likely due to the small size of the affected area (the diameter of the IR beam is about 1 mm), which is surrounded by a large surface area without reconfiguration. The general increase of the peak at 3100 cm^{-1} is mainly a function of increased beam penetration.

One of the consequences of SFG's ability to distinguish surface layers is connected with the debate concerning surface premelting. Premelting temperatures have been variously identified as 261 K (prism face), 260K (basal face)[2] and ~200 K:[30] a considerable discrepancy. We suggest resolution of this issue as follows. A liquid has no surface anisotropy in the absence of an applied field. Ice, however, is anisotropic: The prism face can be oriented either with the *c*-axis in the plane of incidence (*p*) or

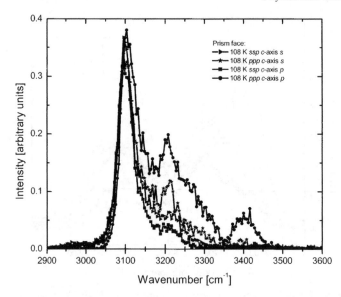

Figure 9 *Spectra of the prism face (within 2°) of single-crystalline ice with the c-axis oriented parallel (p) and perpendicular (s) to the plane of incidence at two polarization combinations.*

perpendicular to it (*s*). Figure 9 shows the first published SFG spectrum of the prism face of single-crystalline ice. The *ppp* spectrum clearly shows the surface anisotropy – in the *s* orientation, the 3400 cm⁻¹ resonance is practically extinct but has significant intensity in the *p* orientation. Thus, monitoring the 3400 cm⁻¹ oscillator in *p* vs. *s* orientation enables determination of the temperature at which the surface reaches a liquid-like mobility: it is the temperature at which these spectra are indistinguishable. Experiments to determine this temperature are in progress.

4   CONCLUSIONS

The presented data shows that SFG is capable of offering unique and novel insights into the structure of ice surfaces. This work has shown that the spectral range between 3150 and 3400 cm⁻¹ is most sensitive to the top bi-layer of the ice surface. Carefully prepared and characterized single-crystal samples enable clear detection of the spectral features of this range. The assignment and clarity of these features suggest that increasing polar ordering continues at least to sample temperatures as low as 128 K. The identified spectral features enable detection of pre-melting as the temperature at which the surface of ice loses its anisotropy.

Acknowledgements

We acknowledge support for this work from the United States National Science Foundation (CHE0240172) and the Petroleum Research Fund of the American Chemical Society (Grant #37517-AC5).

References

1   J. Wettlaufer, J. G. Dash, and N. Untersteiner, *'Ice Physics and the Natural Environment'*, Springer, 1997.
2   A. Lied, H. Dosch, and J. H. Bilgram, *Phys. Rev. Lett.*, 1994, **72**, 3554.
3   H. Dosch, A. Lied, and J. H. Bilgram, *Surf. Sci.*, 1995, **327**, 145.
4   N. Materer, U. Starke, A. Barbieri, M. A. V. Hove, G. A. Somorjai, G. J. Kroes, and C. Minot, *J. Phys. Chem.*, 1995, **99**, 6267.
5   J. Braun, A. Glebov, A. P. Graham, A. Menzel, and J. P. Toennies, *Phys. Rev. Lett.*, 1998, **80**, 2638.
6   A. Glebov, A. P. Graphan, A. Menzel, J. P. Toennies, and P. Senet, *J. Chem. Phys.*, 2000, **112**, 11011.
7   J. E. Schaff and J. T. Roberts, *J. Phys. Chem.*, 1996, **100**, 14151.
8   M. A. Zondlo, T. B. Onasch, M. S. Warshawsky, M. A. Tolbert, G. Mallick, P. Arentz, and M. S. Robinson, *J. Phys. Chem. B*, 1997, **101**, 10887.
9   L. Delzeit, J. P. Devlin, and V. Buch, *J. Chem. Phys.*, 1997, **107**, 3726.
10 V. Buch, S. Bauerecker, J. P. Devlin, U. Buck, and J. K. Kazimirski, *Int. Rev. Phys. Chem.*, 2004, **23**, 375.
11 F. M. Geiger, A. C. Tridico, and J. M. Hicks, *J. Phys. Chem. B*, 1999, **103**, 8205.
12 F. M. Geiger, C. D. Pibel, and J. M. Hicks, *J. Phys. Chem. A*, 2001, **105**, 4940.
13 Y. R. Shen, *Solid State Comm.*, 1999, **108**, 399.
14 P. B. Miranda and Y. R. Shen, *J. Phys. Chem. B*, 1999, **103**, 3292.
15 Q. Du, E. Freysz, and R. Shen, *Science*, 1994, **264**, 826.
16 Q. Du, R. Superfine, E. Freysz, and Y. R. Shen, *Phys. Rev. Lett.*, 1993, **70**, 2313.
17 Y. R. Shen, *Nature*, 1989, **337**, 519.
18 Y. R. Shen, *Ann. Rev. Phys. Chem.*, 1989, **40**, 327.
19 C. Hirose, N. Akamatsu, and K. Domen, *J. Chem. Phys.*, 1992, **96**, 997.
20 N. Akamatsu, K. Domen, and C. Hirose, *Appl. Spec.*, 1992, **46**, 1051.
21 C. Hirose, N. Akamatsu, and K. Domen, *Appl. Spec.*, 1992, **46**, 1051.
22 N. Akamatsu, K. Domen, C. Hirose, and H. Yamamoto, *J. Phys. Chem.*, 1993, **97**, 10064.
23 H. Yamamoto, N. Akamatsu, A. Wada, and C. Hirose, *J. Elect. Spec. Relat. Phenom.*, 1993, **64/65**, 507.
24 L. A. Wilen, *J. Glaciology*, 2000, **46**, 129.
25 C. J. Langway, *'Ice fabrics and the universal stage'*, Technical Report 62, U. S. Army snow ice and permafrost research establishment, Wilmette, Il, 1958. Special thanks for assistance with the Rigsby stage from Rachel Obbard, Dartmouth College.
26 J. R. Scherer and R. G. Snyder, *J. Chem. Phys.*, 1977, **67**, 4794.
27 P. T. T. Wang and E. Whalley, *J. Chem. Phys.*, 1975, **62**, 2418.
28 X. Wei, P. B. Miranda, C. Zhang, and Y. R. Shen, *Phys. Rev. B*, 2002, **66**, art. no. 085401.
29 V. Buch and J. P. Devlin, *J. Chem. Phys.*, 1999, **110**, 3437.
30 X. Wei, P. B. Miranda, and Y. R. Shen, *Phys. Rev. Let.*, 2001, **86**, 1554.

MODELLING ICE Ic OF DIFFERENT ORIGIN AND STACKING-FAULTED HEXAGONAL ICE USING NEUTRON POWDER DIFFRACTION DATA

T.C. Hansen[1,2], A. Falenty[2] and W.F. Kuhs[2]

[1]Institut Max von Laue-Paul Langevin, BP 156, 38042 Grenoble Cedex 9, France
[2]Geowissenschaftliches Zentrum der Universität Göttingen, Abteilung Kristallographie, Goldschmidtstraße 1, 37077 Göttingen, Germany

1 INTRODUCTION

1.1 Background

1.1.1 Formation of ice Ic. Ice Ic (named cubic ice by König[1]), known as a further solid crystalline phase at ambient pressure since 1943[1], can be obtained only in the form of very small crystallites[1-12]. The formation of bulk ice Ic requires special approaches such as relaxation of recovered high-pressure polymorphs of ice (e.g., II, V, XII, IX, VII, VIII and HDA)[4, 7, 13-15] upon warming at ambient pressure, condensation of water vapour at temperatures from 130 to 150 K[16], hyper-quenching liquid water at 170 to 190 K[2, 3], freezing of pore water[17], or by homogeneous freezing of micrometer sized aqueous droplets[18, 19].

1.1.2 Ice Ic in nature. The shape of snowflakes made crystallographers anticipate that ordinary ice would show a hexagonal symmetry. However, studies of the topology of a snowflake and the ice-crystal growth patterns suggested later[20] that some dendrite snow flakes grow from a nucleus of cubic ice. In the 1980s, Whalley[21] reasoned that water droplets may freeze to cubic ice crystals of substantial size and deduced that the 28° halo around the Sun, which had been first described in Christopher Scheiner's 20 March 1629 notes[22], may have been formed by refraction of light passing through the octahedral crystals of cubic ice in the atmosphere. Further evidence of ice Ic in the atmosphere has been provided more recently[23]. Murray et al.[18, 19] reproduced atmospheric conditions in laboratory experiments and thus confirmed the atmospheric presence of ice Ic when micrometer-sized droplets freeze below 190 K, corresponding to conditions in the polar stratosphere and tropical tropopause. As a metastable phase, ice Ic has a higher vapor pressure than ice Ih, and as a result together with the transformation to ice Ih, ice crystals in clouds at 180 K to 200 K will be larger. These crystals can more efficiently dehydrate air and thus enhance the dehydration of the tropopause[24].

1.1.3 Metastability of ice Ic. Above 150 K ice Ic is unstable with respect to hexagonal ice Ih. The transition takes place over a large temperature range, slowly at 150 K, faster at 190 and 210 K. It depends, again, on the way the cubic ice was produced. Even at these temperatures, the transition is not complete but leaves a stacking faulted hexagonal ice, which transforms further to pure ice Ih above 240 K as shown by an *in situ* neutron powder diffraction study[25] and an *ex situ* X-ray diffraction investigation[18]. Moreover, ice Ic and defective ice Ih is formed at temperatures well above 200 K upon decomposition of gas hydrates as observed in time-resolved neutron diffraction experiments[25]. It appears that ice Ic is stable up to the melting point if it crystallizes in sufficiently small mesopores (a few nm)[26]. A number of papers discuss the energetic differences of ice Ih/ice Ic in their pure forms, e.g. based on dipole energies[27] or on MD[28].

1.1.4 Diffraction pattern of ice Ic. None of the obtained ice Ic phases show a diffraction pattern corresponding to well-crystallized cubic ice[2, 3, 6-9]. Moreover, the deviations from this idealized structure are different for different starting materials[1-3, 5-11, 17]. The appearance of broad reflections in neutron diffraction experiments with intensities not[19] in agreement with a simple cubic structure was first explained by anisotropic particle size effects[8]. A closer peak shape investigation of the main cubic reflections and the appearance of broad peaks at Bragg angles typical for ice Ih lead to the assignment of the underlying defects as deformation stacking faults[11].

The powder pattern of ice Ic (e.g., Figure 2) shows the reflections 111, 220, 222, 331, 422, 511, 333, 440, 531, 442. Considering isotropic spherical particles, the evaluation of the 111, 220, 222 and 440 peaks according to the Scherrer formula leads to a diameter of 160 Å for ice Ic from ice II.[11] The skewness of the 111 and 222 reflections (shoulder at the high-angle, respectively at the low-angle side) indicate qualitatively the presence of so-called deformation-faults in contrast to so-called growth-faults[29]. The first diffraction peak, best visible in ice Ic samples from ice IX, cannot be indexed with a cubic symmetry and corresponds to the 100 peak of hexagonal ice Ih. It's presence (and the absence of further hexagonal reflections) cannot be explained by a high concentration of growth faults, but with either the existence of thin hexagonal sequences or by the occurrence of regular stacking sequences (polytypes).[11]

Here, a *fault* is defined as a break in the sequence of alternation of close-packed layers. A *growth* fault in a cubic close packed lattice (f.c.c.) is the introduction of one hexagonal packing: ABCABCABCAB becomes ABCABA**CBACB**, the whole layer sequence is mirrored from the faulty plane onwards. A *deformation* fault corresponds to the formation of two adjacent hexagonal packing sequences: The layer sequence becomes AB-CAB**AB**CABC. Starting from a hexagonal close packed lattice (h.c.p.) a *growth* fault corresponds to the formation of one cubic packing sequence: ABABAB becomes AB**C**BCB (structures of 2nd type).[30] A *deformation* fault corresponds to the introduction of two (structures of 3rd type) or three cubic packing sequences (structures of 1st type).[30] In all cases, more complex types of faults can be defined.

1.2 Motivation

To quantify the amount of the resulting ice phase from diffraction data a proper description of these defective ice phases is needed. Here we present a way to fit the diffraction peak profiles with different models of stacking disorder[30]. This may allow us to distinguish different samples with different histories in terms of their stacking distributions. An adapted range of sufficiently large unit cells with different, non-random stacking sequences is created for each sample and a linear combination of these replicas is fitted to the measured neutron diffraction data. In this way a quantitative determination of the total amount of ice present can be obtained.

2 EXPERIMENTAL

2.1 Samples and measurements

The D_2O ice samples investigated in this context were prepared inside vanadium sample cans inside an Orange Cryostat on the constant wavelength neutron powder diffractometers D2B (wavelength 1.6 Å) and D20[31, 32] (wavelength 2.4 Å), based at the high-flux reactor of the Institut Laue-Langevin (ILL). Precursors were high-density amorphous ice (HDA), and the crystalline high-pressure ice phases V and IX, recovered at ambient pressure and 77 K. Ice Ic was obtained by warming the high-pressure phases at ambient pressure. All neutron diffraction data were collected *in situ*.

2.2 Computing

2.2.1 Model generation. First we have to define the basic layer of hexagonal and cubic ice, which will be stacked afterwards following rules for hexagonal or cubic stacking. With the layers as defined subsequently, the terminology of stacking faults as used for close packed metals can be applied correspondingly to ice Ih/Ic. As one layer we shall regard here the two one-fold oxygen atom positions, which lay in a lattice of hexagonal metric (subgroup P3m1 of the space groups P6$_3$/mmc of ice Ih) on the same fractional x and y co-ordinates, but at different z. There are three different value pairs (xy) possible: (00), (⅓⅔) and (⅔⅓). The fractional heights (z) in a hypothetical one-layer cell (half the height of the ice Ih cell in direction of the c axis) are – idealized for the purely cubic case – ⅛ and ⅞ respectively. The three different possible layers shall be called A, B and C respectively. Two half-occupied disordered deuterium position is on the bonding lines between two oxygen atoms of same fractional coordinates x and y in one layer, and two three-fold half occupied ones are between two neighbouring oxygen atoms of adjacent layers. The layers are equivalent but two same layers cannot follow each other. Hexagonal stacking (h) occurs when a new layer is similar to the layer preceding the previous one, e.g., (A-)B-A (the hexagonal structure 2H of ice Ih). Cubic stacking (c) occurs when the third layer is different from the first one, e.g., A-B-C (the cubic structure 3C of ideal ice Ic). A given stacking sequence, e.g., *hhc*, may be realizable only in a unit cell with a number of layers being a multiple of the periodicity of the stacking sequence (here 3) to fulfil the periodic boundary conditions of a crystal, here (A-B-)A-B-C-B-C-A-C-A-B, the so-called 9R lattice[33] (Figure 1). The stacking variants can thus be equally characterised by the sequence of layers A, B and C or by the sequence of cubic or hexagonal stacking, c and h.

For stacking sequences of a certain height the number of possible models soon exceeds a number of distinct models, which can be treated one by one, and we have to create models by using probabilities. In the simplest case $(s=2)$, one probability is used to determine if a stacking is cubic or hexagonal (no interactions between two layers). In a more complex case $(s=3)$[34] two probabilities are used, one (α) to determine, if a cubic stacking follows a hexagonal one, and another (β) to determine, if a cubic stacking follow a cubic one (interaction only with the directly adjacent layer). In the investigation of ice Ic as resulting from relaxation of ice IX it revealed useful to go a step further $(s=4)$[35] and to work with four probabilities $(\alpha, \beta, \gamma$ and $\delta)$, determining the probability that a cubic stacking (c) follows one of four possible combinations of hexagonal and cubic stacking $(hh, hc, ch$ and $cc)$ in the two previous layers. This interaction range is required to obtain, e.g., the 9R lattice, of which the prototype is samarium metal[33]. It was not considered worthwhile to go for an even higher interaction range.

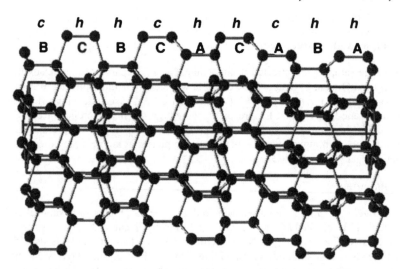

Figure 1   *Structure model of hhc-ice (9R), only the oxygen positions are shown*

Thus, using the programming package Igor[36], a routine was written which, for a given number of layers in a stacking variant of ice Ic/Ih, creates possible stacking sequences based upon the four probabilities for $s=2$ (or less for $s<4$). Each new sequence is forwarded either into a routine computing neutron powder diffraction patterns or as a starting model into a Rietveld refinement program (FullProf[37]). Within this, a limited number of parameters (e.g., size and strain peak broadening, scale factor) can be refined to fit the modelled structure to the diffraction data. Instrumental parameters (i.e. the instrument resolution dependent peak widths), peak shape parameters and other parameters related to the sample (size broadening, atomic thermal displacement), were previously refined from a pure ice Ih sample. This sample was measured in exactly the same conditions as it results from the transition of ice Ic to Ih through heating in the same experimental set-up. The resulting figures of merit of a fit to the given powder diffraction data set is stored in a table, together with the effective stacking probabilities for this particular model. The table is kept sorted after a figure of merit, and the probabilities used to create new models are taken from the computed values for the model having given the best fit so far. Once no more new models can be created with the given number of layers (for a low number of layers), or if the number of tries to obtain a new model from the probability-driven, random-generator dependent model generator exceeds a given value, the number of layers is increased by one.

In a first approximation we assume that the displacements of the lattice nodes due to the presence of a stacking "fault" are negligibly small, although there is a small difference of the interlayer distance in ice Ih compared to what one would expect in a purely cubic ice Ic, due to the non-ideal c/a ratio in ice Ih[38]. Note that the term "fault" is ambiguous in the case of cubic ice since it cannot strictly be described as belonging to either the cubic or the hexagonal system. Therefore, one intra-layer oxygen-oxygen distance in ice Ih differs from the three symmetry-equivalent inter-layer oxygen-oxygen distances.

2.2.2 Analytical approach.   An alternative to this Monte-Carlo like procedure leading to individual crystalline models fitting only relatively well the observed data, is an analytical description of diffraction patterns based on the stacking fault probabilities and thus on the resulting layer-layer correlation probability distribution, which has been tried for the description of ice in mesopores[26]. Analytical approaches have been widely used for the description of one-dimensional stacking disorder of close-packed metals since half a century[29]. They have been mostly based on single-crystal diffraction data, allowing for a more detailed and direct observation of the kind of stacking disorder than powder diffraction data would permit. Such an approach has been successfully applied to the structure of lithium metal at different temperatures[35]. However, one deals in that case with low stacking-fault probabilities where a simple classification of faults in twin, growth and deformation faults is appropriate. In this investigation, this approach has not been used, except for getting a qualitative hint of how to interpret the diffraction patterns' particularities, just as it had already been done in the first attempt to explain the defect nature of ice Ic by stacking faults: Kuhs et al.[11] took the indication for deformation faults in a basically cubic ice (double hexagonal stacking ...*chhc*..., in the $s=4$ case, this would correspond to $\alpha=1$, $\beta \approx 1$, $\gamma=0$, $\delta \approx 1$) from the *hkl*-specific broadening and skewness of higher angle peaks of ice Ic from ice II. Morishige and Uematsu[26] used a related approach to interpret the skewness of the hexagonal 100-peak in ice Ic confined in mesopores as growth faults in a basically hexagonal ice (single cubic stacking ...*hhchh*..., in the $s=4$ case, this would correspond to $\alpha \approx 0$, $\beta=0$, $\gamma \approx 0$, $\delta=0$). However, a quantitative full-pattern analysis has not been presented yet.

Inspired by this analytical approach we performed an alternative modelling approach no longer based on a single Monte-Carlo-created single structure model. The later fits reasonably well to the pattern, but does not necessarily represent the average structure very well. With different stacking probabilities α, β, γ and δ (varying each from 0.1 to 0.9 in steps of 0.2) a set of 10 ice supercells of 66 layers have been created. 10 corresponding powder patterns were calculated using Fullprof[39] with instrumental parameters previously refined on an ice Ih reference sample and added together, representing thus a rough general pattern for one of the 5^4 combinations of probabilities (if computing time permits, the steps should be smaller, the supercells larger and the number of individual models higher). One can easily determine the pattern fitting the data best and thus the representative stacking probabilities. The disadvantage of this method is the impossibility at this stage to refine further parameters like the anisotropic size-broadening using a standard Rietveld refinement program such as Fullprof[39].

3   RESULTS

The diffraction patterns of ice Ic from ice V and HDA, measured on D2B, show, compared to ice Ic from ice IX, a relatively weakly expressed 100 Bragg peak of hexagonal ice Ih. A simple modelling with $s=2$ (one simple probability of cubic or hexagonal stacking) delivers a reasonable fit which cannot be distinguished from the case $s=3$ (probabilities for a given stacking depending on previous stacking, thus two probabilities).

The best fitting models for ice Ic as obtained from the relaxation of ice IX revealed some similarity to a 9R lattice combined with large sequences of only cubic stacking. Lower interaction range models ($s=2$ or $s=3$) are not sufficient to model the particularities of the pattern. This alone is already an important qualitative result in itself: Ice Ic formed by the decomposition of ice IX is not just a cubic phase with statistically distributed faults in form of hexagonal stacking sequences, but there are characteristic patterns of how the hexagonal stacking sequences appear in a matrix of cubic stacking. Ice Ic as obtained from decomposition of ice II or HDA has a higher degree of cubic stacking and therefore a lower interaction range ($s=2$) is able to produce models, which are relatively well fitting the observed data.

From D2b data we obtained a structure model through the iterative generation of stacking variants delivering best possible Rietveld fits from a single structural model. The stacking probabilities for the best fitting models converges slowly to $\alpha=0.7$, $\beta=0.1$, $\gamma=0.1$, $\delta=0.8$ which is close to the 9R and the 3C lattice. The ratio of cubic packing versus hexagonal packing converges very clearly to 45% versus 55%. The structure models obtained from D20 data on a different ice Ic sample from recovered ice IX converge to slightly different models with $\alpha=0.5$, $\beta=0.2$, $\gamma=0.2$, $\delta=0.8$ with the same ratio of cubic packing of 45%.

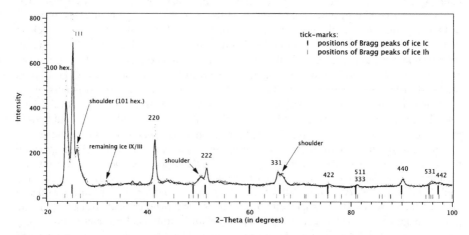

Figure 2   *Diffraction pattern (markers) of ice Ic from ice IX measured at D2b with $\lambda=1.6$ Å and Rietveld fit (solid line) of best fitting model; the ticks mark theoretical peak positions for ice Ic (above) and Ih (below), with lattice constants calculated from those of the refined model*

4   DISCUSSION

The drawback of modelling the structure of bulk samples with a complex stacking sequence is that even a structure with 50 layers or more cannot be considered as a unique model to describe the bulk sample. At the given instrumental resolution of the diffractometers there is not visible influence to the diffraction pattern of two layers, which are at a distance of about 100 Å from each other. However, the modelled patterns are far from being as smooth as the measured ones. The structure is thus best modelled by an average of a reasonable number of individual structure models, characterized by the stacking probabilities as given. This could be achieved by a linear combination of patterns from a set of reasonable models and then extracting the stacking probabilities from the result. However, the data do not allow a stable refinement yet. Equivalent to a stacking probability description of the structure, the layer-layer correlation probability can be used as a fingerprint of a given Ic structure.

A second difficulty is the anisotropic size and strain broadening of the diffraction peaks, which is considerable as well, but highly correlated with the broadening due to stacking disorder. A de-convolution is not possible without determining the anisotropic size distribution, e.g., with electron microscopy, which is difficult to perform due to technical temperature and pressure constraints making it impossible to observe a stable sample of ice Ic.

5   CONCLUSIONS

An improved modelling of the average structure of ice Ic includes linear combination of stacking-probability driven structure models and anisotropic size broadening. It allows for a quantitative modelling of neutron diffraction data of different ice Ic samples. It will serve as well for the description of ice Ih samples with stacking faults.

5.1   Outlook

5.1.1   Layer-layer correlation probabilities.   The analytical approach described by Berliner and Werner[35] is worth to be pursued further. From the four independent stacking probabilities corresponding to an interaction range of s=4, layer-layer correlation probabilities can be readily computed by a numerical approach[35] for any finite crystallite size. It is then less trivial to compute the precise diffraction pattern corresponding to a correlation probability, and then to refine the parameters – the stacking probabilities – to fit an observed pattern. In principle, even the anisotropic size broadening can be taken into account in this approach as well as deviations from the layer positions.

5.1.2   Small-angle neutron scattering (SANS).   Due to the ambiguities of the described modelling of ice Ic one should consider further experiments to distinguish possible scenarios. It is not yet shown, that we deal with a homogeneous sample or with a heterogeneous mixture of two or more phases with a given dispersion around different structure models. Small angle scattering will give helpful hints here as it did in the investigation of HDA[40] Additionally, SANS will reveal the size of suggested blocks of purely cubic stacking and of the 9R-like stacking in ice Ic as obtained from ice IX.

5.1.3   In situ transformation of ice Ic.   It is planned to follow in detail the evolution of the stacking probabilities during the slow transformation of ice Ic to ice Ih. It gives a key to the exact process of rearranging the layers and thus to the kinetics observed, which is different for samples of different origin.

5.1.4   Stacking-faulty ice Ih.   In this context it is of particular interest to describe the stacking-faulted ice Ih as it results from the decomposition of ice Ic but as well of gas-hydrates below 240 K[25]. It is still not entirely understood why these faults stay stable below 240K. The knowledge of the exact nature of the stacking faults will help to determine the amount of bond-breaking necessary to obtain a stacking fault free sample. This determines the activation energy and thus the transition kinetics at different temperatures.

5.1.5   Reconstructive phase transitions leading to ice Ic.   One could gain insight from a quantitative description of the stacking faults in terms of a first or better understanding the reconstructive phase transitions that lead to the formation of different cubic ices, i.e. whilst minimizing the breaking of H-bonds upon restructuring.

Acknowledgements

The presenting author (T.C. Hansen) thanks the University of Göttingen for the opportunity to perform a sabbatical stay including pleasure of teaching crystal chemistry during the long shutdown of the ILL's high-flux reactor for major upgrading.

References

1   H. König, *Z. Krist.*, 1943, **105**, 279.
2   E. Mayer and A. Hallbrucker, *Nature*, 1987, **325**, 601.
3   I. Kohl, E. Mayer, and A. Hallbrucker, *Phys. Chem. Chem. Phys.*, 2000, **2**, 1579.
4   E. F. Burton and W. F. Oliver, *Proc. R. Soc. London, A*, 1935, **153**, 166.

5   N. D. Lisgarten and M. Blackman, *Nature*, 1956, **178**, 39.
6   F. V. Shallcross and G. B. Carpenter, *J. Chem. Phys.*, 1957, **26**, 782.
7   L. G. Dowell and A. P. Rinfret, *Nature*, 1960, **188**, 1144.
8   G. P. Arnold, E. D. Finch, S. W. Rabideau, and R. G. Wenzel, *J. Phys.*, 1968, **49**, 4354.
9   A. Elarby-Aouizerat, J. F. Jal, J. Dupuy, H. Schildberg, and P. Chieux, *J. Phys. C1*, 1987, **48**, 465.
10   J. E. Bertie and S. M. Jacobs, *J. Chem. Phys.*, 1977, **67**, 2445.
11   W. F. Kuhs, D. V. Bliss, and J. L. Finney, *J. Phys. C1*, 1987, **48**, 631.
12   J. Huang and L. S. Bartell, *J. Phys. Chem.*, 1995, **99**, 3924.
13   J. E. Bertie, L. D. Calvert, and E. Whalley, *J. Chem. Phys.*, 1963, **38**, 840.
14   S. Klotz, J. M. Besson, G. Hamel, R. J. Nelmes, J. S. Loveday, and W. G. Marshall, *Nature*, 1999, **398**, 681.
15   D. D. Klug, Y. P. Handa, J. S. Tse, and E. Whalley, *J. Chem. Phys.*, 1989, **90**, 2390.
16   M. Blackman and N. D. Lisgarten, *Adv. Phys.*, 1958, **7**, 189.
17   D. C. Steytler, J. C. Dore, and C. J. Wright, *J. Phys. Chem.*, 1983, **87**, 2458.
18   B. J. Murray and A. K. Bertram, *Phys. Chem. Chem. Phys.*, 2006, **8**, 186.
19   B. J. Murray, D. A. Knopf, and A. K. Bertram, *Nature*, 2005, **434**, 202.
20   T. Kobayashi, Y. Furukawa, T. Takahashi, and H. Uyeda, *J. Cryst. Growth*, 1976, **35**, 262.
21   E. Whalley, *J. Phys. Chem.*, 1983, **87**, 4174.
22   P. Gassendi, 'Phenomenum rarum Romae observatum 20 martij, et ejus causarum explicatio', Amstelodami. Henr. Guerard, 1629.
23   M. Riikonen, M. Sillanpää, L. Virta, D. Sullivan, J. Moilanen, and I. Luukkonen, *Appl. Opt.*, 2000, **39**, 6080.
24   D. M. Murphy, *Geophys. Res. Lett.*, 2003, **30**, 2230.
25   W. F. Kuhs, G. Genov, D. K. Staykova, and T. Hansen, *Phys. Chem. Chem. Phys.*, 2004, **6**, 4917.
26   K. Morishige and H. Uematsu, *J. Chem. Phys.*, 2005, **122**, 044711.
27   D. A. Huckaby, R. Pitis, R. H. Kincaid, and C. Hamilton, *J. Chem. Phys.*, 1993, **98**, 8105.
28   H. Tanaka and I. Okabe, *Chem. Phys. Lett.*, 1996, **259**, 593.
29   M. S. Paterson, *J. Appl. Phys.*, 1952, **23**, 805.
30   Z. Weiss and P. Capkova, *IUCr Monographs on Crystallography*, 1999, **10**, 318.
31   P. Convert, T. Hansen, A. Oed, and J. Torregrossa, *Physica B*, 1998, **241-243**, 195.
32   T. C. Hansen, *Mater. Sci. Forum*, 2004, **443-444**, 181.
33   L. S. Ramsdell, *Am. Mineral.*, 1947, **32**, 64.
34   H. Jagodzinski, *Acta Cryst.*, 1949, **2**, 208.
35   R. Berliner and S. A. Werner, *Phys. Rev. B: Condens. Matter*, 1986, **34**, 3586.
36   Igor Pro 5.04B, Wavemetrics Inc., Lake Oswego, Oregon, USA, 2005.
37   J. Rodriguez-Carvajal, *Physica B*, 1993, **192**, 55.
38   K. Röttger, A. Endriss, J. Ihringer, S. Doyle, and W. F. Kuhs, *Acta Crystallogr., Sect. B: Struct. Sci.*, 1994, **50**, 644.
39   J. Dore, *Chem. Phys.*, 2000, **258**, 327.
40   M. M. Koza, F. Czeschka, H. Schober, T. Hansen, B. Geil, K. Winkel, C. Köhler, and M. Scheuermann, *Phys. Rev. Lett.*, 2005, **94**, 1.

FUNDAMENTAL STUDIES FOR A NEW H_2 SEPARATION METHOD USING GAS HYDRATES

S. Hashimoto, S. Murayama, T. Sugahara and K. Ohgaki

Division of Chemical Engineering, Department of Materials Engineering Science, Graduate School of Engineering Science, Osaka University, 1-3, Machikaneyama, Toyonaka, Osaka 560 - 8531, Japan

1 INTRODUCTION

Gas hydrates, one of clathrate compounds, are stabilized by the enclathrated relatively small guest molecules in the cavity of cages composed of hydrogen-bonded water molecules. Three types of hydrate cages are well known, D-cage (pentagonal dodecahedron, 5^{12}), T-cage ($5^{12}6^2$) and H-cage ($5^{12}6^4$). The structure-I (s-I) hydrate is composed of two D-cages and six T-cages, and the structure-II (s-II) hydrate is sixteen D-cages and eight H-cages.

H_2 has become the object of attention as a clean and promising energy resource. The steam reforming of hydrocarbons is a well-known technique in the H_2 production processes. The gas mixture including CO, CO_2 and H_2 is generated by the steam reforming. At present, the Pressure Swing Adsorption (PSA) is usually used as one of the H_2 separation methods for such mixtures. In our future, it is expected to develop a simple and potential H_2 separation method. A new technique for separating H_2 from (H_2 + CH_4) gas mixtures using hydrates has been reported[1]. Pure H_2 hydrate is generated only in extremely high-pressure regions[2]. Mao et al.[3] and Mao and Mao[4] reveal that the mixtures of H_2 and water generate the s-II hydrate at high pressure of 200 MPa or the low temperature of about 80 K, where the hydrate cages are multiply occupied with a cluster of two H_2 molecules in the D-cage and four H_2 molecules in the H-cage. On the other hand, CO_2 generates s-I hydrate easily in moderate conditions[5, 6]. That is, gas hydrates can be available as a medium of H_2 separation from such gas mixtures. As a fundamental study, the isothermal phase equilibrium (p-y) relations for the H_2, CO_2 and water ternary system in the presence of gas hydrate phase have been measured[7].

Tetrahydrofuran (hereafter, THF) generates s-II hydrate[8] below atmospheric pressure. THF can be stabilized in the H-cage and cannot be enclathrated in the D-cage[9]. Therefore, the chemical formula of THF hydrate is given by THF·17H_2O. THF has been used widely as an additive that reduces the equilibrium pressure for other gas hydrate systems (for example, CH_4, N_2 hydrate)[10] and this effect of THF on hydrate formation is the highest among several additives (for example, acetone, 1, 4-dioxane)[11]. That is, the addition of THF could make the equilibrium pressure much lower. The separation process using gas hydrates is feasible in the much moderate condition. Then, we have measured the isothermal phase equilibrium (p-y) relations for the H_2, CO_2, THF and water quaternary system under the three-phase (gas + aqueous solution + hydrate) coexisting conditions. Finally, the cage occupancy of the H_2 molecules has been discussed briefly by use of Raman spectroscopy.

2 METHOD AND RESULTS

2.1 Method

2.1.1 Experimental apparatus. A schematic illustration of the experimental apparatus for the phase equilibrium measurements is shown in Figure 1. The inner volume and maximum working pressure of the high-pressure cell were 150 cm^3 and 10 MPa, respectively. The cell had a set of windows for visually observing the phase behavior in the high-pressure cell. All of them were immersed in a temperature-controlled water bath. The contents were agitated using an up-and-down mixing bar driven by an exterior permanent magnetic ring.

The inner volume and maximum working pressure of the high-pressure optical cell for the Raman spectroscopic analysis were 0.2 cm^3 and 400 MPa, respectively. The cell had a pair of sapphire (or quartz) windows on both the upper and lower sides. The thermostated water was circulated constantly in the exterior jacket of the high-pressure optical cell. A ruby ball was enclosed to agitate the contents by the vibration from outside.

The system temperature was measured within a reproducibility of 0.02 K using a thermistor probe (Takara D-632), which was inserted into a hole in the cell wall. The probe was calibrated with a Pt resistance thermometer. The system pressure was measured by the pressure gauge (Valcom VPRT) calibrated by RUSKA quartz Bourdon tube gauge with an

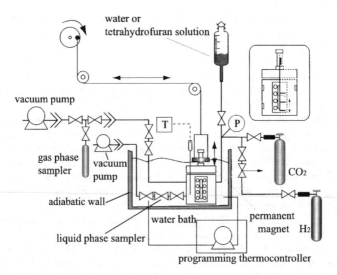

Figure 1 *Schematic illustration of experimental apparatus for the phase equilibrium measurement.*

estimated maximum uncertainty of 0.01 MPa.

2.1.2 Experimental procedure. The distilled water or the THF solution prepared at a desired composition was introduced into evacuated high-pressure cell. The contents were pressurized up to a desired pressure by supplying the H_2 and CO_2 gas mixture at a desired composition and then continuously agitated using a mixing bar. After the formation of gas hydrates, the system temperature was kept constant to establish the three-phase coexisting state of hydrate + aqueous solution + gas. The phase behavior was observed directly through the window. After establishing the equilibrium state of three-phase coexistence, a small volume (~0.5 cm^3) of gas phase was sampled for composition analysis. The equilibrium composition of gas phase was analyzed for H_2 and CO_2 by the TCD-Gas Chromatography (TCD-GC, Shimadzu GC-14B) as the water and THF composition of gas phase is negligibly small under the present experimental conditions.

The single crystal of gas hydrate prepared from H_2 + CO_2 and H_2 + CO_2 + THF mixtures was analyzed by *in situ* Raman spectroscopy using a laser Raman microprobe spectrophotometer with multichannel CCD detector. In the present study, the "single crystal" was defined as the gas hydrate crystal for which the Raman peak of the intermolecular O-O vibration mode can be detected. The argon ion laser beam (wavelength: 514.5 nm, power: 100 mW) or He-Ne laser beam (wavelength: 632.8 nm, power: 35 mW) condensed to 2 μm in spot diameter was irradiated to the sample through the upper sapphire (or quartz) window. The backscatter of the opposite direction was taken in with the same lens. The spectral resolution was about 1 cm^{-1}.

2.1.3 Materials. Research grade H_2 (mole fraction purity 0.999999) was obtained from the Neriki Gas Co., Ltd. The maximum impurity was 0.2 ppm of nitrogen. Research grade CO_2 (mole fraction purity 0.9999) was obtained from the Takachiho Tradings Co., Ltd. Research grade THF (mole fraction purity 0.997) was obtained from Yashima Pure Chemicals Co., Ltd. The distilled water was obtained from the Yashima Pure Chemicals Co., Ltd. All of them were used without further purifications.

2.2 Results and Discussions

2.2.1 H_2 + CO_2 + water ternary system[7]. Figure 2 shows the isothermal phase equilibrium (*p-y*) relations for the H_2, CO_2 and water ternary system in the presence of gas hydrate phase at various temperatures. The three-phase equilibrium pressure increases monotonically with the mole fraction of H_2. The variation of equilibrium pressure with the H_2 composition gives similar behavior in the whole temperature range of the present study. The equilibrium pressure for the pure CO_2 hydrate (left vertical axis in Figure 2) is good consistent with the previous ones[5, 6]. At a given temperature and gas-phase composition, the total pressure is estimated by numerical calculation (various lines in Figure 2) in order to give the equilibrium fugacity of pure CO_2 hydrate, which is calculated from the experimental data obtained in the present study. The fugacity coefficient of CO_2 in the gas mixture is calculated by the Soave-Redlich-Kwong equation of state[12] with the ordinary mixing rule (k_{12} = -0.3426)[13]. The critical constants of CO_2 and H_2 and other parameters in the reference[14] were adopted. As shown in Figure 2, the estimated values agree well with the experimental equilibrium pressures. This indicates that the gas hydrate generated from the H_2 + CO_2 + water mixtures is pure CO_2 hydrate. The CO_2 hydrate cannot be generated until the CO_2 fugacity in the gas mixture exceeds the equilibrium fugacity of pure CO_2 hydrate system. That is, H_2 behaves only like a diluent gas.

Raman spectra corresponding to the intramolecular vibration of H_2 and CO_2 were measured at 274.6 K and 15.2 MPa. In order to avoid the optical effect of sapphire window

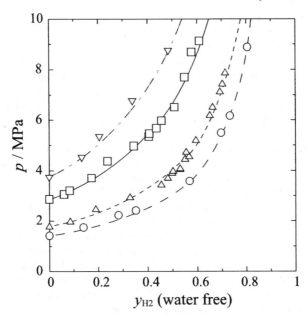

Figure 2 *Isothermal phase equilibrium (pressure-composition) relations for H_2 +*
CO_2 + water system in the presence of gas hydrate phase; $\bigcirc$*: 274.3 K,*
$\triangle$*: 276.5 K,* $\square$*: 280.1 K,* $\bigtriangledown$*: 281.9 K. The lines are estimated values.*

used in the present study, the spectra of CO_2 are obtained by the irradiation of the Ar ion
laser. On the other hand, the spectra of H_2 are given by the He-Ne laser. The Raman peaks
corresponding to the intramolecular symmetric C=O stretching vibration mode of CO_2 are
detected in both gas and hydrate phases and the spectra exhibit the double peaks due to
Fermi resonance effect. This result is consistent with the Raman shifts of pure CO_2
hydrate[15] at the same pressure. On the contrary, the Raman peaks corresponding to the H-H
stretching vibration mode of H_2 is detected only in the gas phase, while it is not in the
hydrate phase. Kim and Lee[16] have reported that the H_2 molecule can be entrapped by
hydrate cages generated from H_2 + CO_2 + water mixtures at extremely low temperature
(123.15 K), which is much lower than that of our study. At extremely low temperature
condition (~100 K), H_2 can be easily entrapped by hydrate cages without additives[3]. H_2
may be enclathrated in the hydrate cage competing with CO_2 at extremely low temperature.
We claim that H_2 cannot occupy hydrate cages and the hydrate generated from H_2 + CO_2 +
water mixtures is pure CO_2 hydrate in a temperature region higher than 274.3 K and a
pressure region up to 10 MPa.

2.2.2 H_2 + CO_2 + THF + water quaternary system[17]. The isothermal phase equilibrium
(p-y) relations for H_2 + CO_2 + THF + water (the mole fraction of THF in water is 0.030,
0.056 and 0.080) mixed system containing gas hydrate at 280.1 K are shown in Figure 3. In
the present study, it is assumed that the mole fraction of THF in the gas phase can be
neglected. In comparison with the H_2 + CO_2 + water mixed system[7], the most remarkable
change is a large depression of equilibrium pressure which is caused by the addition of a
small amount of THF. This pressure depression is the largest at the THF mole fraction of
0.056, which is stoichiometric for the pure THF hydrate[18]. The experimental equilibrium

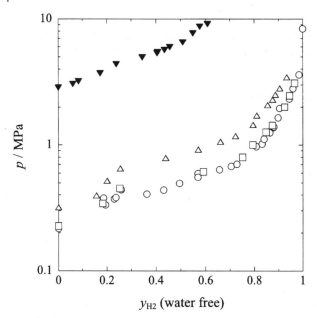

Figure 3 *Isothermal phase equilibrium (pressure-composition) relations for the H_2 + CO_2 + THF + water mixed system in the presence of hydrate phase at 280.1 K at the THF composition of 0.030, 0.056 and 0.080; ▼: no additive[7], △: $x_{THF} = 0.030$, ○: $x_{THF} = 0.056$, □: $x_{THF} = 0.080$.*

pressure for the H_2 + THF mixed hydrate is 8.30 MPa at 280.1 K, which is consistent with that of Florusse *et al.*[19] The degree of pressure depression depends on the additive composition, that is, the additional THF depresses the equilibrium pressure till the THF composition comes up to the stoichiometric mixture. The THF composition exceeding the stoichiometric ratio does not depress the equilibrium pressure any more.

It is also notable characteristic that the unusual behavior comes into existence on the *p-y* curve for the quaternary mixtures including THF. A stepwise increase in the equilibrium pressure appears around 0.2 in the H_2 mole fraction of gas phase (y_{H2}) for every aqueous solution of different THF mole fraction. It is reasonable to guess that the H_2 molecule starts to occupy the D-cage of structure-II in the CO_2 + THF mixed gas hydrate around $y_{H2} = 0.2$, although we could not obtain a clear evidence of non-occupation with H_2 in the composition range of $y_{H2} < 0.2$. It would be the further subject to make sure a physical meaning of the above discontinuous point by use of X-ray and/or neutron diffraction analyses. In order to confirm the existence of H_2 in the hydrate phase, the single crystals of H_2 + CO_2 + THF mixed gas hydrate were prepared in the higher composition region than $y_{H2} = 0.2$. The Raman spectroscopy for the above single crystal gives the characteristic signals of THF, H_2 and CO_2. The Raman shifts obtained in the present study are shown in Figure 4. The mole fraction of THF is 0.056, which is stoichiometric for the pure THF hydrate[18] and mole fraction of H_2 in the gas mixture is 0.90. As shown in Figure 4 (a), the peaks observed at 919 and 1035 cm^{-1} correspond to the enclathrated THF molecules[20] in hydrate cages and the quadruplet peaks corresponding to the H_2 rotation in the gas phase are observed at about 350, 585, 815 and 1036 cm^{-1}. In the hydrate phase, the weak and broad peaks are also detected at the similar position except for the peaks at 815

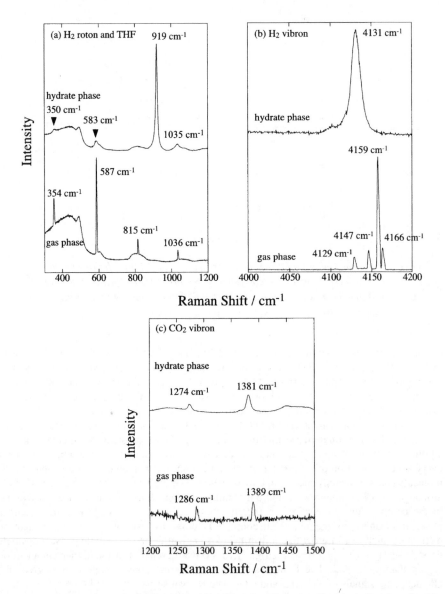

Figure 4 *Raman spectra of rotation for H_2 (a), and intramolecular vibration for H_2 (b) and CO_2 (c) molecules in the gas and hydrate phases at 280.1 K and 4.3 MPa. The mole fraction of THF is 0.056. The high base line less than 520 cm^{-1} and the broad peaks at 600, 810 and 1060 cm^{-1} are due to the quartz window of high-pressure optical cell.*

and 1036 cm^{-1}. The H_2-rotation peaks at about 815 and 1036 cm^{-1} in the hydrate phase may overlap with those of quartz window and THF, respectively. The peaks corresponding to H_2 rotation in the both gas and hydrate phases are consistent with those of reference data[3, 19, 21]. Figure 4 (b) also shows that the peaks corresponding to H-H stretching vibration mode of H_2 molecules are detected in the both gas and hydrate phases. The Raman peaks corresponding to the H-H stretching vibration mode of the H_2 molecule in the gas and hydrate phases are consistent with the reference data[19]. The shoulder is also observed at about 4100 cm^{-1}, however, it is not clear what this shoulder is and the detail about this shoulder cannot be discussed here at this stage. In addition, Figure 4 (c) indicates that the doublet peak by Fermi resonance effect corresponding to C=O symmetric stretching vibration mode of CO_2 are detected in the both gas and hydrate phases. It is reasonable to guess that the THF completely occupies the H-cage because the mole fraction of THF is stoichiometric for pure THF hydrate formation. Therefore, the H_2 and CO_2 can be competitively enclathrated in the D-cage.

We have claimed that the hydrate generated in the ternary system of H_2, CO_2 and water (without THF) can be regarded as the pure CO_2 hydrate crystal (s-I) from Raman spectroscopic study[7]. The gas hydrate crystal generated in H_2 + CO_2 + THF + water mixed system is the s-II. The D-cage of s-II is somewhat smaller than that of s-I[22]. Therefore, it is possible that the H_2 molecule occupies the D-cage to generate H_2 + CO_2 + THF mixed hydrate crystal.

3 CONCLUSION

The present study has revealed the following findings;
1. H_2 cannot be enclathrated in the hydrate cages of CO_2 hydrate under the present experimental conditions.
2. The addition of THF induces the large pressure reduction from the equilibrium pressure without THF. The largest pressure reduction is obtained when the THF mole fraction is 0.056, which is the stoichiometric mole fraction for the pure THF hydrate.
3. With the existence of THF, H_2 can be enclathrated in the hydrate cages depending on the composition of gas mixtures. In the H_2 + CO_2 + THF mixed hydrate, it is likely that THF occupies all of the H-cage while H_2 and CO_2 can occupy the D-cage of s-II hydrate competitively.

Notation

f: fugacity [Pa]
p: pressure [Pa]
T: temperature [K]
x: mole fraction of aqueous phase [-]
y: mole fraction of gas phase [-]

Acknowledgement

This work was financially supported by New Energy Industrial Technology Development Organization (NEDO). The participation in the conference on the "Physics and Chemistry of Ice" (PCI2006) is financially supported by Kansai Research Foundation for technology promotion (KRF). One of the authors (S. H.) shows his gratitude for the center of excellence (21th century COE) program "Creation of Integrated EcoChemistry of Osaka

University". We gratefully acknowledge the Division of Chemical Engineering, Graduate School of Engineering Science, Osaka University for the scientific support by "Gas-Hydrate Analyzing System (GHAS)".

References

1   G. -J. Chen, C. -Y. Sun, C. -F. Ma and T. -M. Guo, *Proceeding of the Fourth International Conference on Gas Hydrates*, 2002, **2**, 1016-1020.
2   Y. A. Dyadin, E. G. Larionov, E. Y. Aladko, A. Y. Manakov, F. V. Zhurko, T. V. Mikina, V. Y. Komarov and E. V. Grachev, *J. Struct. Chem.*, 1999, **40**, 790-795.
3   W. L. Mao, H. Mao, A. F. Goncharov, V. V. Struzhkin, Q. Guo, J. Hu, J. Shu, R. J. Hemley, M. Somayazulu and Y. Zhao, *SCIENCE*, 2002, **297**, 2247-2249.
4   W. L. Mao and H. Mao, *Proc. Natl. Acad. Sci. USA*, 2004, **101**, 708-710.
5   K. Ohgaki, Y. Makihara and K. Takano, *J. Chem. Eng. Japan*, 1993, **26**, 558-564.
6   D. B. Robinson and B. R. Mehta, *J. Can. Petr. Tech.*, 1971, **10**, 33-35.
7   T. Sugahara, S. Murayama, S. Hashimoto and K. Ohgaki, *Fluid Phase Equilib.*, 2005, **233**(2), 190-193.
8   R. E. Hawkins and D. W. Davidson, *J. Phys. Chem.*, 1966, **70**, 1889-1894.
9   S. R. Gough and D. W. Davidson, *Can. J. Chem.*, 1971, **49**, 2691-2699.
10 S. -P. Kang, H. Lee, C. -S. Lee and W. -M. Sung, *Fluid Phase Equilib.*, 2005, **185**, 101-109.
11 Y. -T. Seo, S. -P. Kang and H. Lee, *Fluid Phase Equilib.*, 2001, **189**, 99-110.
12 G. Soave, *Chem. Eng. Sci.*, 1972, **27**, 1197-1203.
13 H. Knapp, R. Doring, L. Oellrich, U. Plocker and J. M. Praushnitz, *Vapor - Liquid Equilibria for Mixtures of Low Boiling Substances, Chemistry Data Series Vol. VI*, Dechema, Frankfurt, 1982.
14 R. C. Reid, J. M. Prausnitz and B. E. Poling, *The Properties of Gases and Liquids, 4th Ed.*, McGraw-Hill, New York, 1986.
15 S. Nakano, K. Ohgaki and M. Moritoki, *J. Chem. Eng. Data*, 1998, **43**, 807-810.
16 D. -Y. Kim and H. Lee, *J. Am. Chem. Sci.*, 2005, **127**, 9996-9997.
17 S. Hashimoto, S. Murayama, T. Sugahara and K. Ohgaki, *J. Chem. Eng. Data*, 2006, **51**(5), 1884-1886.
18 T. Makino, T. Sugahara and K. Ohgaki, *J. Chem. Eng. Data*, 2005, **50**(6), 2058-2060.
19 L. J. Florusse, C. J. Peters, J. Schoonman, K. C. Hester, C. A. Koh, S. F. Dec, K. N. Marsh, E. D. Sloan, *SCIENCE*, 2004, **306**, 469-471.
20 A. Y. Manakov, S. V. Goryainov, A. V. Kurnosov, A. Y. Likhacheva, Y. A. Dyadin and E. G. Larionov, *J. Phys. Chem. B*, 2003, **107**, 7861-7866.
21 U. Fink, T. A. Wiggins and D. H. Rank, *J. Mol. Spectrosc.*, 1965, **18**, 384-395.
22 S. Subramanian and E. D. Sloan Jr., *J. Phys. Chem. B*, 2002, **106**, 4348-4355.

SEGREGATION OF SALT IONS AT AMORPHOUS SOLID AND LIQUID SURFACES

O. Höfft[1], U. Kahnert[1], S. Bahr[1], V. Kempter[1], P. Jungwirth[2] and L. X. Dang[3]

[1]Institut für Physik und Physikalische Technologien, TU Clausthal, Leibnizstr. 4 38678 Clausthal-Zellerfeld, Germany
[2]Institute of Organic Chemistry and Biochemistry, Academy of Sciences of the Czech Republic and Center for Biomolecules and Complex Molecular Systems, Flemingovo nám. 2, 16610 Prague 6, Czech Republic
[3]Chemical Science Division, Pacific Northwest National Lab, Richmond, Washington, USA

1 INTRODUCTION

Traditionally, the surfaces of aqueous electrolytes are described as inactive and practically devoid of ions [1, 2]. Indeed, this has turned out to be true for non - polarizable ions, as alkali cations and small anions, as fluoride as well. However, due to polarization interactions singly charged anions, with the heavy halides as particular examples, exhibit a propensity for the water / air (vacuum) interface. This was first suggested in order to rationalize the occurrence of chemical reactions on aqueous interfaces, sea - salt particles, ocean surfaces etc. This initiated MD calculations using polarizable potentials. They suggest that highly polarisable anions can indeed be preferentially adsorbed at the outermost liquid layer. In this description, the ions are polarized by the anisotropy of the interface, creating an induced dipole that is stronger than in the bulk. The interaction between the polarized ions and the surrounding water molecules compensates for the reduced solvation available at the surface. This has triggered a number of laboratory studies, applying mainly non - linear optical probes.

Additional insight into hydration and segregation phenomena would be gained from adsorption experiments on the surface of amorphous solid water (ASW), provided that ASW could be considered as a supercooled extension of liquid water, an issue which is however a matter of active research and intense scientific debate at present [3, 4, 5]. In this case, the entire scope of well – developed vacuum based methods could be applied to the study of the solvation and segregation phenomena at ASW during its annealing above the glass transition temperature T_g. Hopefully, the results would be applicable to surfaces of liquid water as well (where the application of vacuum - based techniques is difficult because of the high vapour pressure of water). Clearly, the study of processes with ASW offers the opportunity to contribute to our understanding of the properties of ASW and its relation to liquid water [6]. First steps into this direction have already been undertaken:
The interaction of water, around T_g (136K), -i.e. when the molecular motion in the solvent becomes important, with some polar and nonpolar molecules was studied with TOF - SIMS [7]. It was found that non - polar molecules, hexane as a specific example, are fully

incorporated into the bulk while polar molecules, formic acid for instance, stay at the surface.

Kim et al. [8] examined the migratory motion of Na$^+$ and Cl$^-$ ions on ASW surfaces by low energy sputtering (LES). The Na$^+$ and Cl$^-$ populations at the surface were measured as a function of the ASW temperature (100 to 140K). The Na$^+$ intensity decreased in the range 110 to 130K while the Cl$^-$ intensity remained practically constant. This indicates that the inward migration of Na$^+$ ions takes place at temperatures where solvent diffusion becomes important, in contrast to surface residence of Cl$^-$. This surface propensity of certain halides is characteristic for liquid water [9, 10].

The combination of Metastable Impact Electron Spectroscopy (MIES) and Ultraviolet Photoelectron Spectroscopy, UPS(HeI) has been applied to the study of the interaction between halides (CsI, CsF, and NaI) with solid water and methanol around their respective T_g values [11,12]. Surface segregation of iodide, but not of fluoride or Cs ions, took place from ASW, exposed to CsI or CsF vapour, during annealing [11]. The same behaviour was also derived from molecular dynamics (MD) simulations of the corresponding aqueous salt solutions. In contrast, no appreciable surface segregation of ions was observed in methanol under similar conditions, neither in the experiment nor in the simulation of the corresponding liquid solution. It was pointed out that, as far as solvation phenomena are concerned, water and methanol ices, when heated above their respective T_g values, behave remarkably similar to the corresponding liquid solutions. The surface propensity of iodide is also seen when Cs is replaced by Na [12].

In the present report, we present a more through-out analysis of the data for the interaction of halides with solid water and methanol films, in particular those of ref. [12]. This leads to a deepened understanding of the analogy observed between amorphous solids above their T_g and their respective liquid counterparts.

2 EXPERIMENT

The present analysis is based on MIES and UPS (HeI) results presented in [11,12]. The apparatus, applied to the study of the salt - ASW and MeOH interactions, has been described in detail previously [13, 14, 15] (see also [11,12]). It is important for the present analysis that the electrons detected with MIES result exclusively from the outermost layer of the film under study, in contrast to UPS(HeI) which averages over 3 layers, typically. Detailed introductions into MIES and its various applications in molecular and surface spectroscopy) can be found elsewhere [16, 17].

The TPD experiments are carried out using a differentially pumped quadrupole mass spectrometer (QMS), connected to the UHV apparatus employed for the MIES/UPS studies. The ramping time in TPD is considerably shorter (1K/s) than in MIES and UPS (1K/min). For this reason, the maximum desorption rate in MIES/UPS occurs about 15K earlier than in TPD.

The Au sample is cooled with liquid nitrogen to 100 K and exposed to water or methanol by backfilling the chamber. Exposures are stated in Langmuirs (L) (1 L=10^{-6} Torr * s). It is expected that amorphous solid water and methanol are formed during the slow deposition of the respective vapours at temperatures below 120 K (see also ref. [11]).

3 RESULTS AND DISCUSSION

NaI and Cs halide layers were produced on ASW (MeOH) films at about 105K and annealed up to about 200K, and the electron emission from the ionization of the highest - lying states of H_2O, CH_3OH, and of iodide was studied [11,12]. **Fig. 1** presents (in arbitrary units) the variation of the probability for ionization of the states $5p_{3/2}I$ and $1b_1$ H_2O ($M_{3;4}$ MeOH; notation as in [12]) for NaI interacting with amorphous solid water (MeOH) films as a function of the film temperatures. For NaI/ASW, a gradual decrease of I(5p) becomes noticeable above 125K. An intensity minimum occurs at 134 K, close to the accepted water glass transition ($T_g = 136$ K) [7]. In the same region, water desorption becomes noticeable as manifested by the decrease of $1b_1$ H_2O. As for CsI / ASW, I(5p) remains visible throughout the entire annealing procedure, although the intensity passes through a minimum. This is at variance with NaI / MeOH: here, I(5p) does disappear almost completely when annealing beyond 117 K ($Tg = 120K$ [18]), -i.e. full solvation takes place. The same behaviour has been noticed for CsI / ASW and MeOH [11]. In ref. [11] this was attributed to the segregation of iodide at the ASW, but not at the solid MeOH surface. Surface segregation of heavy halide ions in liquid water, but not in liquid MeOH, has been predicted by classical MD calculations [2, 19] performed for NaI aqueous and MeOH solutions, and is subject of two recent reviews [9, 10]. Besides an increase of I(5p), we notice a decrease of the $5p_{3/2}I$ binding energy by about 0.6 (0.4)eV for ASW (MeOH), respectively. This decrease occurs around 135K, i.e. when, according to MIES, most of the water has already desorbed.

The portion of the total surface area occupied by iodide species (surface area accessed (ASA) by iodide [11]) can be accessed by both theory and experiment [11,12]. We can then estimate that the ASA of iodide is roughly 20 percent for both CsI and NaI. Theory predicts a value of ASA of 16 % for 1.2 M aqueous CsI solution [11]. For NaI we get a similar value, albeit smaller by ~2 %. The experiment and calculation are thus in a semi-quantitative agreement (note that the value of the experimental salt concentration is hard to obtain very accurately). For illustration, **Fig. 2** show, for illustration, the surface of a 1.2 M aqueous NaI solution, as obtained from a MD simulation [2]. Iodide anions are clearly seen at the surface, while sodium cations are absent (being, however, present in the subsurface).

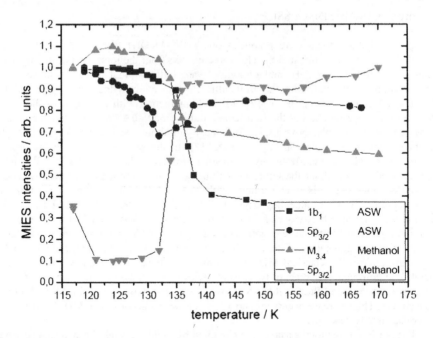

Figure 1   *Probablities for the emission of electrons from the $5p_{3/2}I$ and $1b_1$ H_2O states from iodide and water species, located in the top layer of the film produced by depositing 0.5ML NaI on amorphous solid water ASW and methanol (MeOH) films at 105K. Shown is the variation of the probabilites with the film temperature between 115 and 165K.*

Figure 2   *A snapshot from a MD simulation of 1.2 M aqueous solution of NaI showing the structure of the surface. Large dark grey balls: iodide, small light grey balls: water oxygen, and white balls: water hydrogen.*

Summarizing, the comparison of experiments and computations suggests that viscous liquids, namely supercooled NaI (CsI) solutions (water and methanol), are obtained when depositing the salt molecules on amorphous solid films of the solvent molecules, and annealing the films above the respective T_g's. As to solvation, these supercooled solutions behave similar as the corresponding liquid solutions. The formation of the supercooled solution precedes water desorption. These findings suggest that the local environment of the solute species, iodide and alkali, is similar to that seen by the same species in the respective normal liquids. The obvious consequence of this conclusion is that the water (MeOH) - ion interaction energies (binding energies), I^- - H_2O (MeOH) in particular, and the electronic properties, ionization energies and the optical band gap, should follow from cluster – type calculations that simulate the local environment of the salt ions. For NaI / water such computations are available for room temperature solutions [20]. In the following, it will be demonstrated that this argumentation is indeed reasonable; in the following a comparison will be made between the MIES / UPS and TPD results [12] and the DFT results available for $NaI(H_2O)_n$ complexes [20].

TPD spectra recorded on mass 18 (H_2O) for neat water (20L) on Au (a), as well as for 0.7ML NaI deposited on ASW films of various thicknesses, and analogous data for the interaction of NaI with MeOH can be found in ref. [12]. The desorption of the neat "solvent", ASW, produces a narrow structure around 158K which we attribute to multilayer desorption. These peaks develop tails towards higher temperatures as the consequence of the NaI deposition. This was interpreted as follows: the NaI - solvent interaction gives rise to a broad range of H_2O binding energies, extending from about 42 kJ mol^{-1} (close to the values for desorption from the neat multilayer) to more than 60 kJmol^{-1} for the strongest - bound species. It was shown that the remarkable broadening of the TPD spectra for NaI/ASW (as compared to neat water) towards higher temperatures under the influence of the supplied NaI molecules can be understood when adopting the following scenario: the preparation produces hydrated NaI structures, partially or fully dissociated in the sense of ref. [20]. When annealing, the increase of T first removes the weakest - bound water molecules, -i.e. those not involved into the hydration of NaI species. This is followed by the removal of H_2O molecules from the largest hydrated structures present in the solution; further increase of the temperature leads to the removal of water molecules from increasingly smaller structures. This process can be visualized as the transition from $NaI(H_2O)_n$ to $NaI(H_2O)_{(n-1)}$ whereby n decreases with increasing temperature; decreasing values of n lead to increasingly stronger bound hydrated structures. A comparatively steep increase can be noticed for n = 3 $\rightarrow$ 2; it stems from the fact that below n = 4 all water molecules are in direct contact with the Na - I complex [20].

We proceed now to a discussion of the electronic properties of the NaI/ASW films as they emerge from the MIES / UPS spectra [12]; we adopt the same scenario as discussed above: we assume that the electronic properties of solvated iodide are determined by those of $NaI(H_2O)_n$ complexes of the type studied in [20]. As all energies in fig.1 refer to the Fermi level, the vertical IPs with respect to the vacuum are obtained by adding WF (4.8eV for the top spectrum in fig.1(a)), as derived from the high – binding energy cut - off of the spectrum. For neat water (MeOH) this yielded 10.2 (8.9) eV for the distance between the valence band maximum and the vacuum level. In the same way we can determine ionization energy of 5pI. In order to discuss the observed solvation shift of I(5p) in ASW, -i.e. the decrease of the 5pI IP's during the anneal process from 8.9 eV by about 0.5eV, we rely on the electronic properties of $NaI(H_2O)_n$ (n=1 to 6) clusters as computed by ref. [20]. The computed vertical IPs increase from 7.88eV for unsolvated NaI to values between 8.3 and 8.61eV for various n = 6 conformers; the increase is essentially up to n = 3. Thus, we are led to the conclusion that the solvation shift of the experimental IPs takes place when the coordination in the first solvation shell changes (for n < 4 while for larger n the IP will change comparatively little), -i.e. the observed shift of IP reflects the degree of solvation onto the electronic properties of the NaI - H_2O complexes. Thus, the essential aspects of the electronic properties can also be explained under the assumption of a supercooled solution whose local structure ressembles that of the corresponding room temperature solution.

4 SUMMARY

We have studied the interaction of NaI and CsI with amorphous solid water (ASW) and methanol films, deposited on Au substrates. The deposition of the halides on the amorphous solid films (105K), followed by heating above the glass transition point, leads to the formation of supercooled halide solutions. For water, but not for methanol, we find a surface propensity of the iodide species of the solution. The same behaviour is predicted for the corresponding liquids by classical MD calculations. For NaI / ASW we have analyzed the ionization energy of $I(5p_{3/2})$ as well as the solvent desorption energies as a function of temperature on the basis of cluster DFT calculations, available for NaI - H_2O complexes. Our results suggest that the similarities observed for the iodide interaction with ASW (MeOH) and the corresponding liquids are based on the presence of similar NaI - solvent complexes in both cases.

Acknowledgements

Support from the Czech Ministry of Education (grants LC512 and ME644) and via the NSF - funded Environmental Molecular Science Institute (grants CHE 0431312 and 0209719) is gratefully acknowledged.

References

1   P.B. Petersen, R.J. Saykally, *Ann. Rev. Phys. Chem.,* 2006, **57**, 333.
2   P. Jungwirth, D.J. Tobias, J. Phys. Chem. B, 2002, **106**, 6361.
3   R.S. Smith and B.D., Kay, Nature, 1999, **398**, 788.
4   V. Buch, J.P. Devlin, Eds. "Water in Confining Geometries" (Springer, Berlin, 2002)
5   S. M. McClure, D.J. Safarik, T.M. Truskett, C.B. Mullins, J. Phys. Chem. B, 2006, **110**, 11033
6   H. Kang, Acc. Chem. Res., 2005, **38**, 893.
7   R. Souda, Phys. Rev. B, 2004, **70**, 165402.
8   J. - H. Kim, T. Shin, K.-H. Jung, H. Kang, ChemPhysChem., 2005, **6**, 440.
9   P. Jungwirth, D.J. Tobias, Chem. Rev., 2006, **106**, 1259.
10 T.M. Chang, L.X. Dang, Chem. Rev., 2006, **106**, 1305.
11 O. Höfft, A. Borodin, U. Kahnert, V. Kempter, L.X. Dang, P. Jungwirth, J. Phys. Chem. B, 2006, **110**, 11971.
12 O. Höfft, U. Kahnert, S. Bahr, V. Kempter, J. Phys. Chem. B, **110**, 17115
13 A. Borodin, O. Höfft, S. Krischok, V. Kempter, Nucl. Instr. Meth. B, 2003, **203**, 205.
14 A. Borodin, O. Höfft, S. Krischok, V. Kempter, J. Phys. Chem. B, 2003, **107**, 9357.
15 A. Borodin, O. Höfft, U. Kahnert, V. Kempter, A. Poddey, P. Blöchl, J. Chem. Phys., 2004, **121**, 9671.
16 H. Morgner, Adv. Atom. Mol. Opt. Phys., 2000, **42**, 387.
17 Y. Harada, Y., S. Masuda, H. Osaki, Chem. Rev., 1997, **97**, 1897.
18 R. Souda, Phys. Rev. B, 2005, **72**, 115414.
19 L.X. Dang, Phys. Chem. A, 2004, **108**, 9014.
20 A.C. Olleta, H.M. Lee, K.S. Kim, J. Chem. Phys., 2006, **124**, 024321.

THEORETICAL STUDY ON GASES IN HEXAGONAL ICE INVESTIGATED BY THE MOLECULAR ORBITAL METHOD

A. Hori* and T. Hondoh

Institute of Low Temperature Science, Hokkaido University, N19 W8 Kita-ku, Sapporo 060-0819, Japan
*present address: Kitami Institute of Technology, Kouen-cho 165, Kitami, 090-8507, Japan. E-mail: horiak@mail.kitami-it.ac.jp

1 INTRODUCTION

In the Antarctic ice sheets, gases, for example, air and greenhouse gases, are trapped, and the variations of these gas concentrations have been measured in order to reconstruct the paleoclimate over the past over 300,000 years . CO_2 and CH_4 and other constituents in the ice sheet have been intensively studied to reveal the history of the climatic variation of the earth.[1-3] In the deep sections of the polar ice sheets, air exists in the form of clathrate hydrate as found in nature.[4] An interesting phenomenon of the air hydrate in polar ice sheets based on Raman spectroscopy is that its composition depends on the depth; O_2-rich hydrates are found at around a 700 m depth, and the ratio of N_2 increases with the depth, finally a reaching the present N_2/O_2 ratio of 3.7.[5] This fractionation of N_2 and O_2 was observed in air bubbles and clathrate hydrates found in the deep ice cores and was attributed to the difference in the diffusion constants for N_2 and O_2 in ice. However, there is no experimental data to directly support this result, and the diffusion constants are expected to be very low and difficult to measure.[5-7] The Solubility of gases in ice is also necessary in order to understand the behaviour of gases in ice sheets, but the experimental solubility data have been obtained only for a few gases.[6]

Recently, the diffusion constant of various gases in ice have been estimated by molecular dynamics (MD) simulations.[8-10] The authors reported that the N_2 molecule hops in the ice by breaking hydrogen bonds in the ice lattice while it was considered to diffuse through the bottleneck of the ice lattice from one interstitial site (T_u site) to the adjacent interstitial site (T_u site).[9] They also reported that the diffusion constants estimated from the MD simulations were almost similar for various gases larger than Ar and the mechanism is the same as in the case of N_2.[10] From our previous study[11] using the ab-initio molecular orbital method, the calculated barrier energy for the diffusion of Ne in ice through the bottleneck was in line with the experimental results[12]. It would be better to examine the above-mentioned breaking bond mechanism using quantum chemical calculations since, in general, the formation and breaking of chemical bonds cannot be discussed by classical simulations.

In this paper, we conducted molecular orbital calculations for gases in ice in order to investigate the solubility of the gas in ice and the diffusion paths of gases in ice by the molecular orbital method mainly using a semiempirical molecular orbital calculation program (MOPAC2002[13]).

2 METHOD

To investigate the characteristic features of gases in hexagonal ice Ih, molecular orbital calculations of the ice structure containing various gas molecules were carried out using the semiempirical quantum mechanics package, MOPAC2002[13], with the PM3 parameterization[14] under the periodic boundary conditions. The calculations were conducted for the model structures of ice which were composed of 96, 288, and 360 water molecules in the unit cell to check the unit cell size dependence of the results. Since it is difficult to obtain the structure of proton-disordered ice for any size of the unit cell, we used the structure of proton-ordered ice in this paper. Furthermore, we obtained the proton-ordered ice structure for any size by repeating a unit cell composed of eight water molecules as shown in Figure 1. The ice structures of the different unit cell sizes containing a gas molecule were optimized under the periodic boundary condition. For the ice structure containing an O_2 molecule, Unrestricted Hartree-Fock (UHF) calculations were carried out.

We also conducted first principle calculations for a N_2 diffusion in an ice lattice consisting of 96 water molecules using the DFT program package Dmol3 solid.[15,16] in order to check the validity of the semiempirical calculations. We used this program package to reduce the computer time although the lattice constants were fixed due to the limitations of the program at present.

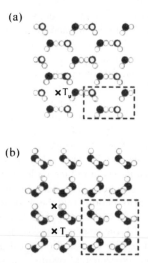

(a)

(b)

Figure 1 *The ice lattice viewed from the top of the $(H_2O)_{96}$ super cell as seen from the top (a) and from the side (b). The basic unit cell of eight water molecules is shown in the dashed line rectangle. One of the two interstitial sites of the hexagonal ice, T_u, is also shown.*

3 RESULTS AND DISCUSSION

3.1 Solution of Gas Molecules in Ice

The ice structures containing various gas molecules were fully optimized under the periodic boundary conditions. The gas molecules in the optimized structures were almost all located at the interstitial T_u site. The heat of solution of the gases into ice lattice was estimated as the difference in the cohesive energies between the ice with a gas molecule and that without it. The cohesive energies, E_{coh}, were defined as the difference between the heats of formation, ΔH, of the ice structure containing a gas molecule and those of its constituents: $E_{coh} = \Delta H$ (ice containing gas) – n x ΔH (water) – ΔH (gas), where n is the number of water molecules.

First, we conducted the calculations for the ice structures of different sizes, i.e., the number of water molecules in the structure, to evaluate the effect of the size of the ice structure on the results. Table 1 shows that the energy of solution for various gases in ice containing 96 and 288 water molecules (and 360 water molecules only for N_2). All values are positive and those for N_2, CO_2, O_2, and CH_4 decrease in this order. Particularly, CH_4 was far more stable than the others in the ice. For O_2 and N_2, the changes in energy between the ice structure consisting of 96 water molecules and that of 288 water molecules are lower than 0.2 kJ/mol, while in the case of CH_4 and CO_2, they are higher than 0.4 kJ/mol. This is because the sizes of the molecules in the latter are larger than those in the former, and larger structures are required for the relaxation.

Table 1 *Energy of solution for various gases in the ice structures of various sizes*

Gas	Energ of solution (kJ/mol)		
	Number of water moelules		
	96	288	360
O_2	44.87	44.72	-
N_2	92.13	92.10	92.09
CH_4	6.12	5.71	-
CO_2	76.19	75.48	-

3.2 Diffusion of Gas Molecules in Ice

In our previous study, we calculated the barrier energy of the diffusion of various gases in ice by the semiempirical molecular orbital method using the model structures of the ice lattice consisting of 378 water molecules under the constraint that the atomic positions of the water molecules on the exterior of the cluster were fixed.[11] For these calculations, we assumed the diffusion path to be along the c-axis from a T_u site to the neighbouring T_u site through the bottleneck of the 6-memberd water ring. We first calculated the barrier energy for the ice structure containing 96 water molecules with O_2, and N_2 molecules for both cases in which the lattice constants were fixed and optimized. The first principle DFT calculations were carried out for the N_2 molecule in the ice structure containing 96 water molecules with the lattice constants fixed under the periodic boundary conditions. Hence these calculations were conducted in order to evaluate the effect of the lattice relaxation. Furthermore, only for the N_2 molecule, were the calculations conducted for the ice structure containing 360 water molecules under the constraint that the lattice was not fixed in order to evaluate the lattice size effect on the results.

The results are shown in Figure 2 and Table 2. For the ice structure composed of 96 water molecules, the barrier energies of the O_2 diffusion hardly depended on the relaxation of the lattice constants while that of the N_2 diffusion for the structure with the fixed lattice constants was about 0.01 eV greater than that with the optimized lattice constants. This indicates that the effect of the lattice relaxation on the barrier energy is negligible for the O_2 and N_2 diffusions in the ice structure $(H_2O)_{96}$. Therefore, we carried out the first principle DFT calculations of N_2 for the ice structure $(H_2O)_{96}$ with the fixed lattice constants. As described later, the barrier energy estimated from the DFT calculations was about 0.48 eV. This is almost consistent with the result calculated using MOPAC2002, which indicates that the semiempirical calculations is almost valid at least

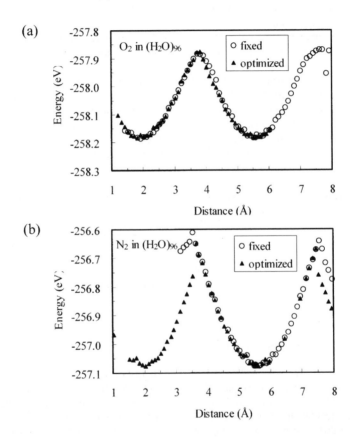

Figure 2   *Energy change by the diffusion of gas molecules in the ice lattice along the c-axis. The constraint that the lattice constant are fixed and optimized is imposed in order to evaluate the effect of the lattice relaxation of the ice structure.*

for the N_2 diffusion in the ice structure $(H_2O)_{96}$. To investigate the effect of the lattice size, we also conducted calculations for the N_2 diffusion in the ice structure containing 360 water molecules, with the optimized lattice constants. As shown in Table 2, the barrier energy is greater than that in $(H_2O)_{96}$. The displacement of water molecules decreases with the increasing distance from the gas molecule and it causes a local distortion which results in the increase of total energy. Hence, the total energy with the gas molecule at the saddle point will increase with the increasing size of the lattice, i.e., number of water molecules, until it is saturated although the energy change will decrease. In the present study, we did not confirm the saturation due to limited computer resources and computation time.

Table 2 *Comparisons of barrier of the diffusion of gas molecules in the ice lattice along the c-axis in order in order to investigate the effect of the lattice relaxation and the model size*

Gas	Number of H_2O	Constraint on the lattice constants	Barrier energy (eV)
O_2	96	fixed	0.301
		–	0.307
N_2	96	fixed	0.436
		–	0.428
	360	–	0.468

For the diffusion path of the gases in ice, Ikeda-Fukazawa and others carried out classical MD simulations and proposed a new diffusion mechanism for gases larger than Ne.[10] According to their new mechanism, gases diffuse between the neighbouring water molecules by breaking the hydrogen bond between them, and the diffusion constants of the gases grater than Ne are of almost the same order. However, it is not evident that the breaking of chemical bonds can be treated by the classical MD simulations. Therefore, we conducted quantum chemical calculations for this mechanism or diffusion path, first by the semiempirical molecular orbital method using MOPAC2002 and second by the first principle DFT using Dmol3 solid. The energy barrier of the N_2 diffusion in ice calculated by the semiempirical method was in line with that by the first principle DFT method under the constraint that the lattice constants of the ice structure were fixed.

Table 3 shows the differences in the barrier energy between the diffusion path from a T_u site to the neighbouring T_u site through the bottleneck (path 1) and the diffusion path between the neighbouring water molecules by breaking the hydrogen bond (path 2) for O_2, N_2, and CH_4. These results agree with those by Ikeda-Fukazawa and others. The change in energy (converted from the heat of formation calculated by the program) for the diffusion of CH_4 in the ice lattice along the two different paths is shown in Figure 3. The snapshots are shown in Figure 4 and a pair of water molecules appears to come apart during the diffusion of the CH_4 molecule, and finally both water molecules are isolated in the ice lattice. The barrier energies of path 2 were almost comparable to those of path 1 although the diffusion by path 2 required the breaking of bonds and the energy due to the greater distortion of the lattice. Therefore, the present study supports the result by Ikeda-Fukazawa *et al.* although our calculations are static and mostly based on the semiempirical and molecular orbital method.

Table 3 *Barrier energy of diffusion of gases in the ice lattice of $(H_2O)_{96}$ by two different paths*

Gas	Barrier energy (eV)	
	path 1	path 2
O_2	0.307	0.326
N_2	0.428	0.343
CH_4	0.344	0.370

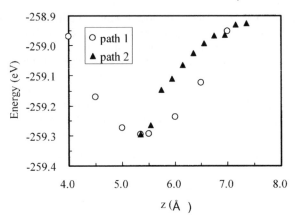

Figure 3   *Energy change by the diffusion of of the diffusion of CH₄ molecule in the ice lattice (H₂O)₉₆ along two different paths: path 1 (○) and path 2(▲).*

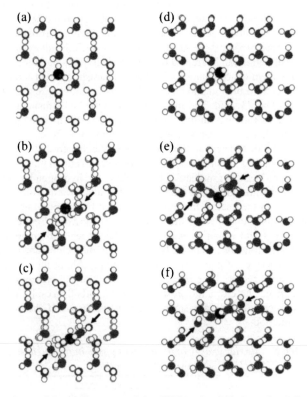

Figure 4   *Snapshots of the ice lattice and the CH4 in the diffusion along the path 2 as seen from the top ((a),(b),(c)) and from the side ((d),(e),(f). A pair of water molecules isolated in the ice lattice are denoted by the arrows. (a) and (d), (b) and (e), and (c) and (f) correspond to the snapshots at z=5.3, 6.5, and 7.5 in Figure 3, respectively.*

4 SUMMARY

We conducted molecular orbital calculations for the solution of O_2, N_2, CH_4, and CO_2 in the ice structure under the periodic boundary conditions using the semiempirical program package, MOPAC2002. Based on the calculations, the values of energy of solution for N_2, CO_2, O_2, and CH_4 in the ice were positive and they decreased in this order. Particularly, CH_4 was far more stable than others. The lattice size dependence of solution energy was found to be small particularly for O_2 and N_2.

We also calculated the barrier energy of the diffusion by two different paths, i.e., a conventional one and one based on a recently proposed mechanism. The calculated values are very similar in spite of the disadvantage of breaking bonds in the latter. Therefore, our results are at least qualitatively in agreement with those of Ikeda-Fukazawa *et al.* although our calculations were static and semiempirical.

Acknowledgements

This study was partially supported by the Japanese Ministry of Education, Science, Sports and Culture, Grant-in-Aid for Creative Scientific Research, 14GS0202, 2002-2006.

References

1 J. R. Petit, J. Jousel, D. Raynaud, N. I. Barkov, J.–M. Baarnola, I. Basile, M. Bender, J. Chappellaz, M. Davis, G. Delaygue, M. Delmotte, V. M. Kotlyakov, M. Legrand, V. Y. Lipenkov, C. Lorius, L. Pepin, C. Ritz, E.Saltzman, and M. Stievneard, *Nature*, 1999, **399**, 429.

2 O. Watanabe, J. Jpusel, S. Johnsen, F. Parrenin, H. Shoji, and N. Yoshida, *Nature*, 2003, **422**, 509.

3 EPICA community members., *Nature*, 2004, **429**, 623.

4 H. Shoji,and C. C. Langway Jr., *Nature*, 1982, **298,** 548.

5 T. Ikeda, H. Fukazawa, S. Mae, L. Pepin, P. Duval, , B. Champagnon, V.Y. Lipenkov, and T. Hondoh, *Geophysical Research Letters*, 1999, **26**, 91.

6 K. Satoh, T. Uchida, T. Hondoh, and S. Mae, Proc. NIPR Symp. Polar Meteorol. Glaciol., 1996, **10**, 73.

7 T. Ikeda, A. N. Salamatin, V. Y. Lipenkov and T. Hondoh, *in Physics of Ice Core Records*, edited by T. Hondoh (Hokkaido Univ. Press, Sapporo, 2000), p. 393.

8 T. Ikeda-Fukazawa, S. Horikawa, T. Hondoh, and K. Kawamura, *J. Chem. Phys.* , 2002, **117**, 386.

9 T. Ikeda-Fukazawa, K. Kawamura, and T. Hondoh, *Chem. Phys. Lett.*, 2004, **385**, 467.

10 T. Ikeda-Fukazawa, K. Kawamura, and T. Hondoh, *Molecular Simulation*, 2004, **30**, 973.

11 A. Hori, and T. Hondoh, *Can.a J. Phys.*, 2003, **81**, 251.

12 P. V. Hobbs, in *ICE PHYSICS* (Clarendon Press, Oxford, 1973), p. 742.

13 J.J.P. Stewart, *MOPAC 2002*, Fujitsu Limited, Tokyo, Japan, 2001.

14 J.J.P. Stewart, *Journal of Computational Chemistry*, 1989, **10**, 209.

15 B. Delley, *J. Chem. Phys.,* 1990, **92**, 508.

16 B. Delley, *J. Chem. Phys.*, 2000, **113**, 7756.

DEVELOPMENT OF *IN SITU* LOW TEMPERATURE INFRARED SPECTROSCOPY FOR A STUDY OF METHANE HYDRATE

K. Ishikawa[1], A. Tani[1], and S. Nakashima[1]

[1] Department of Earth and Space Science, Graduate School of Science, Osaka University, 1-1 Machikaneyama, Toyonaka, Osaka, 560-0043, Japan

1 INTRODUCTION

Methane hydrate is a clathrate compound of water molecules surrounding a methane molecule. Natural methane hydrate is found in permafrost and deep-sea sediment,[1,2] and has recently attracted much attention as a potential new resource because of the large amount of deposits. Methane hydrate is also expected as new materials for gas storage and transportation due to its unique properties called anomalous preservation, quite slow dissociation from -40 to -10°C at atmospheric pressure,[3] despite of its dissociation over -80°C.[4]

Decomposition mechanism of methane hydrate has been investigated with various methods to understand this effect. For example, Takeya et al. (2002) studied decomposition of methane hydrate to ice and methane gas by X-ray diffraction (XRD), and suggested that the slow dissociation might be caused by a layer of ice on the surface of methane hydrate which prevents methane gas from diffusing outward.[5] On the contrary, Takeya et al. (2005) proposed that methane hydrate dissociated into liquid water and gaseous methane in temperature range of -38 to -13°C because the activation energy of methyl radical decay in gamma-irradiated methane hydrate was close to the enthalpy change at methane hydrate dissociation to water and methane gas.[6] Behaviour of H_2O molecules forming hydrate cages is important to understand decomposition of methane hydrate.

One of the most popular analytical methods to study properties and structures of water molecules is infrared (IR) spectroscopy, because it is highly sensitive to hydrogen bonding nature of water molecules. Bertie and Whalley (1964) measured IR spectra of polymorphic ice and showed their minute change of IR spectrum due to the difference of hydrogen bonding.[7,8] Therefore, we developed here *in situ* IR spectroscopy at low temperatures under atmospheric pressure using commercial thermostatic stage and nitrogen gas flow system to avoid frost formation.

In this paper, we introduce the developed *in situ* low temperature IR spectroscopy and report preliminary IR spectra of thin methane hydrate film.

2 EXPERIMENTAL METHODS

2.1 Sample Preparation

Methane hydrate formation was conducted by the following procedures, introduced by Hachikubo et al. (2004).[9]

(1) A thin water film was sandwiched between two CaF_2 plane crystals with a plastic film spacer of about 50 ×m. A water-ice film was prepared by freezing those plates in liquid nitrogen (-196°C). One CaF_2 plate was carefully removed to keep the ice film on the other plate. This sample was called water-ice film.

(2) The plate with water-ice film was set in a high-pressure vessel immersed in ethanol coolant cooled by dry ice at -80°C. Methane gas was supplied in the vessel up to 8 MPa at this temperature.

(3) Methane hydrate was formed on the surface of the ice film by keeping the vessel in a freezer (-22°C) for 36 hours. Methane gas pressure in the vessel was initially 10 MPa and decreased by formation of the hydrate.

(4) The vessel was immersed in liquid nitrogen and the sample plate was taken out. Before IR measurements, the sample plate was kept under -196°C to avoid decomposition.

A CaF_2 plate (diameter: 10 mm, thickness: 1 mm) was employed as a substrate in this study. CaF_2 is transparent to IR light for wavenumbers higher than 1000 cm^{-1}, permitting measurements of IR absorption bands for CH bending and stretching or OH bending and stretching vibrations. In addition, a water film sample was prepared by sandwiching a water droplet by two CaF_2 plates.

2.2 Thermostatic Stage at Low Temperature

We have developed an *in situ* cooling system to perform IR measurements at very low temperatures under atmospheric pressure by combining a commercial thermostatic stage (Linkam FT-IR 600) and nitrogen gas flow system to avoid frost formation (Figure 1).

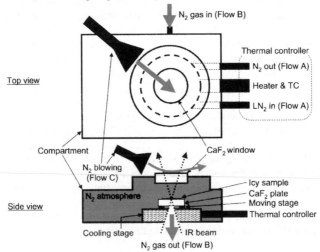

Figure 1   *Schematic illustration of the compartment and the cooling stage. Icy sample on a CaF₂ plate was set on the center of the cooling stage. Moving stage was used to move the CaF₂ plate horizontally. Gray arrows indicate N₂ flow directions.*

For thermal control, the cooling stage uses a Pt-Rh heater and vaporized liquid nitrogen pumped from its container (Flow A in Figure 1) to regulate the sample temperature from room temperature down to -180°C. A Pt-Rh thermocouple is placed just beside the sample position to monitor the sample temperature.

A CaF$_2$ window covers the upper part of the compartment in order to prevent water vapor from entering through the window (Figure 1). N$_2$ gas was flowed from the side through the interior of the compartment to the entrance of the lower part (Flow B in Figure 1) to avoid frost on the sample surface. Another N$_2$ gas flow unit was employed to blow out frost on the top CaF$_2$ window of the compartment (Flow C in Figure 1).

2.3 IR Measurements

For IR measurements of the film samples upon cooling, the above cooling system is placed on the sample stage of the IR micro-spectrometer (Jasco FT/IR-620 Plus + IRT30).

A ceramic light source, a KBr beam splitter and a Mercury Cadmium Telluride (MCT) detector were used to measure IR spectra. An aperture of the IR light was $100 \times 100 \times m^2$. All spectra were obtained by collecting 100 scans with a spectral range from 7000 to 700 cm^{-1} at 4 cm^{-1} resolution. The precision of peak position is within 2 cm^{-1}.

A representative methane hydrate film position was selected for IR measurements by observing magnified images (160×) of the sample texture with a Cassegrainian mirror objective of 16× and an ocular of 10×.

All IR measurements were performed by a transmission method. IR spectra through a CaF$_2$ plate without any samples were first measured from 25 to -120°C to clarify how the IR spectra depend on measurement temperatures. The IR spectrum at 25°C was used as a background spectrum I_0 to obtain absorption spectra ($Abs = -\log(I/I_0)$) for all IR measurements. The spectra for the methane hydrate and water-ice films were taken at every 20°C after keeping for 10 minutes at temperatures from -120 to -20°C. An IR spectrum for the liquid water film was performed at 25°C.

3 RESULTS AND DISCUSSION

3.1 IR Absorption Spectra of a CaF$_2$ Plate without Any Samples from 25 to -120°C

Figure 2 shows IR absorption spectra through only a CaF$_2$ plate at different temperatures. Although these spectra showed variations by water vapor and atmospheric CO$_2$, no other significant absorption bands were recognized. Since the IR spectrum at 25°C was used as a background spectrum I_0, IR absorption spectrum at 25°C is almost zero except three bands due to water vapor and gaseous CO$_2$. The absorbance variation in different temperatures was generally small and less than 0.02 absorbance unit (Figure 2). Therefore, all spectra at different temperatures were divided by the background spectrum at 25°C for further discussion.

3.2 IR Absorption Spectra of Methane Hydrate, Water-ice and Water Films

An IR absorption spectrum of the methane hydrate sample at -120°C is shown in Figure 3. For comparison, absorption spectra for the water-ice sample at -120°C and the water sample at 25°C are also presented.

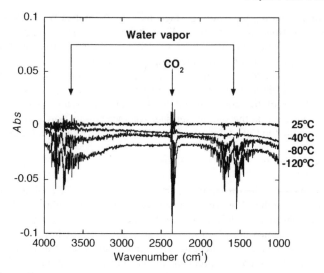

Figure 2   *IR absorption spectra at different temperatures for a CaF₂ plate without any*
samples. Background spectra, I₀ was measured at 25°C and absorbance at
each temperature, T was obtained by the formula Abs = -log(I(T)/I₀(25°C)).
Difference of absorbance between at 25°C and -120°C was mostly 0.02, except
variations due to reduction of water vapor and gaseous CO₂ at low
temperature.

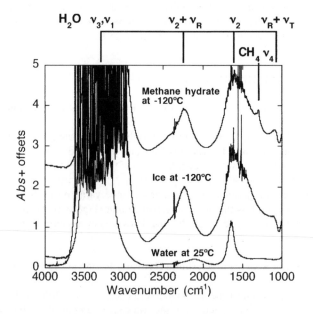

Figure 3   *IR absorption spectra of the methane hydrate, water-ice and liquid water films.*

A broad band around 3400 cm^{-1} corresponding to fundamental OH stretching modes (ν_3, ν_1)[7] is saturated due to the thick samples (Figure 3). Bending vibrations of H$_2$O molecules are observed around 1650 cm^{-1} (ν_2)[7] and the band is wider for methane hydrate and ice than for liquid water. Association band of H$_2$O molecules $(\nu_2 + \nu_R)$ observed at a band around 2200 cm^{-1} is due to the combination band of bending vibration (ν_2) and librational vibration (ν_R).[7] Another association band of H$_2$O molecules $(\nu_R + \nu_T)$ may be observed around 1100 cm^{-1} due to the combination band of libration (ν_R) and translation (ν_T). These bands are weak for liquid water but become stronger for the methane hydrate and water-ice films. A small band at 1300 cm^{-1} corresponding to CH bending mode (ν_4) of CH$_4$ is observed only for the methane hydrate film.[10,11]

Two IR absorption spectra for the methane hydrate and water-ice films are quite similar except for the CH bending band at 1300 cm^{-1}. According to IR absorption spectra of a mixture of CH$_4$ and H$_2$O molecules with a molar ratio of 1:20, peak height at 1650 cm^{-1} in HOH bending mode is close to that at 1300 cm^{-1} in CH bending mode.[11] Therefore, the methane hydrate film is considered to be a mixture of methane hydrate and water-ice. The thickness of the film was estimated to be about 12 ×m using H$_2$O peak height at 1650 cm^{-1}.[12] Assuming that methane occupancy in the hydrate was 100%, the thickness of the methane hydrate layer was roughly evaluated to be about 0.6 ×m using peak height ratio at 1300 to 1650 cm^{-1} based on the IR spectrum of methane and ice mixture.[11] The methane hydrate sample is considered to be a 0.6 ×m thin methane hydrate layer on 11.4 ×m water-ice film.

3.3 IR Absorption Spectra of the Methane Hydrate Sample from -120 to -20°C

IR absorption spectra of the methane hydrate sample from -120 to -20°C are shown in Figure 4. The peak intensity at 1300 cm^{-1} corresponding to CH bending of methane appeared to decrease with increasing temperature. In order to represent quantitatively the spectral changes, peak height at 1300 cm^{-1} was calculated with a liner baseline from 1270 to 1315 cm^{-1}. The obtained peak heights at 1300 cm^{-1} for temperatures of -120 to -20°C are

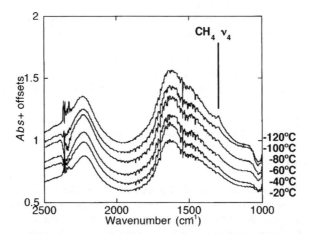

Figure 4   *IR absorption spectra of the methane hydrate sample from -120°C to -20°C. Spectra were measured at every 20°C after keeping for 10 minute at each temperature.*

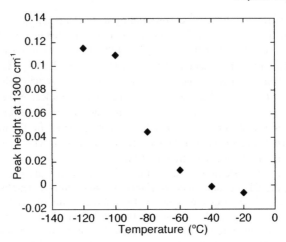

Figure 5   *Changes in the peak height at 1300 cm[-1] for the methane hydrate sample as a function of temperature. A measurement error in peak height is within the marker size.*

shown in Figure 5. The peak height slightly decreases at -100°C due to small amount of decomposition and decreases drastically at -80°C where methane hydrate is unstable around this temperature.[4] These are consistent with the results of XRD experiment.[5] The details of the decomposition behavior of methane hydrate should be further studied using this *in situ* low temperature IR spectroscopy.

4 CONCLUSIONS

Our conclusions are listed as follows.

(1) We have developed an *in situ* low temperature IR spectroscopy using a commercial thermostatic stage and nitrogen gas flow system to avoid frost formation on experimental system.

(2) Transmission IR spectra for methane hydrate film were measured by this method from -120 to -20°C under atmospheric pressure.

(3) Obtained IR spectra indicate that the sample film is a mixture of methane hydrate and water-ice.

(4) The methane peak at 1300 cm[-1] decreased drastically around -80°C in accordance with the literature results on the decomposition of methane hydrate.

Acknowledgements

We thank Prof. A. Tsuchiyama, Mr. K. Takeya, Mr. T. Otsuka, Ms. M. Igisu, and all the members of Tsuchiyama laboratory and Nakashima laboratory at Osaka University for their help and advice. We appreciate Dr. A. Hachikubo for the experimental data of temperature and pressure conditions for forming methane hydrate under ice point. We also appreciate Mr. R. Yoshida for developing cooling stage with nitrogen gas flow system. This study was supported by the Granted-in-Aid for the 21st Century COE program for "Towards a New Basic Science; Depth and Synthesis".

References

1   T.S. Collett and S.R. Dallimore, in *Natural Gas Hydrate in Oceanic and Permafrost Environments*, ed. M.D. Max, Kluwer Academic Publishers, Dordrecht, 2000, ch. 5.

2   W.P. Dillon and M.D. Max, in *Natural Gas Hydrate in Oceanic and Permafrost Environments*, ed. M.D. Max, Kluwer Academic Publishers, Dordrecht, 2000, ch. 6.

3   L.A. Stern, S. Circone, S.H. Kirby and W.B. Durham, *J. Phys. Chem. B*, 2001, **105**, 1756.

4   E.D. Sloan, Jr., in *Clathrate Hydrates of Natural Gases*, 2nd Edn., Marcel Dekker Inc., New York, 1998, ch. 6, p. 319.

5   S. Takeya, T. Ebinuma, T. Uchida, J. Nango and H. Narita, *J. Cryst. Growth*, 2002, **237-239**, 379.

6   K. Takeya, K. Nango, T. Sugahara, K. Ohgaki and A. Tani, *J. Phys. Chem. B*, 2005, **109**, 21086.

7   J.E. Bertie and E. Whalley, *J. Chem. Phys.*, 1964, **40**, 1637.

8   J.E. Bertie and E. Whalley, *J. Chem. Phys.*, 1964, **40**, 1646.

9   A. Hachikubo, K. Yamada, T. Miura, K. Hyakutake, K. Abe and H. Shoji, *Ocean Polar Res.*, 2004, **26**, 515.

10   C. Chapados and A. Cabana, *Can. J. Chem.*, 1972, **50**, 3521.

11   D.M. Hudgins, S.A. Sandford, L.J. Allamandola and A.G.G.M. Tielens, *Astrophys. J. Suppl. Ser.*, 1993, **86**, 713.

12   S.Y. Venyaminov and F.G. Prendergast, *Anal. Biochem.*, 1997, **248**, 234.

A MECHANISM FOR PHOTOCHEMICAL REACTIONS IN THE QUASI-LIQUID LAYER OF SNOW CRYSTALS IN POLAR REGIONS

H.-W. Jacobi[1], T. Annor[1], B. Kwakye-Awuah[1, 2], B. Hilker[1, 3], and E. Quansah[1]

[1] Alfred Wegener Institute for Polar and Marine Research, Am Handelshafen 12, 27570 Bremerhaven, Germany
[2] Now at Research Centre in Applied Sciences, University of Wolverhampton, Wolverhampton WV1 1SB, UK
[3] Now at Institute of Solid State Physics, University Bremen, 28334 Bremen, Germany

1 INTRODUCTION

Photochemical processes in snow have recently attracted considerable scientific interest.[1] Field and laboratory experiments have been reported regarding the influence of photochemical processes in the snow on concentrations of species like nitrogen oxides (NO_x), hydrogen peroxide (H_2O_2), and formaldehyde (HCHO). Field measurements have studied the following processes in the surface snow: 1. the production of NO_x in the snow and the release to the atmosphere, 2. the bi-directional exchange of H_2O_2 and HCHO between the atmosphere and the snow, and 3. the impact of the emission on the composition of the atmospheric boundary layer in snow-covered regions.[1-15]

Among the manifold reactions in snow the photolysis of nitrate (NO_3^-) is now the best characterized reaction due to a range of laboratory experiments. These studies have been used to extract information about the absorption coefficients and quantum yields of NO_3^- in ice as a function of wavelength,[16] the formation of products like the hydroxyl radical (OH)[16, 17] and nitrite (NO_2^-),[17-19] and the release of NO_x to the gas phase.[18, 20-23] Limited information regarding the absorption coefficients and the formation of OH radicals can also be found for the photolysis of H_2O_2.[24] Investigations regarding the photolytic decomposition of further compounds with a potential relevance for photochemical processes in surface snow are limited to studies, in which reactions of HCHO and NO_2^- have been examined.[19, 25, 26]

These results have demonstrated that snow photochemistry involves several complex chemical processes. Such a system can only be accounted for with a comprehensive reaction mechanism. Therefore, we used available experimental and field data to assemble a comprehensive reaction mechanism for surface snow. In addition, rate constants for the reactions involved for typical polar conditions in summer at snow temperatures of –20 °C are presented together with typical initial concentrations of the stable compounds involved in the mechanism. First simulations are performed for typical summer conditions in Greenland and results of these calculations are presented.

2 LABORATORY STUDIES OF PHOTOCHEMICAL REACTIONS IN ARTIFICIAL SNOW

We developed an experimental set-up to investigate single photochemical reactions in artificial snow samples. Details of the experiments and the preparation of the snow samples have previously been described.[19, 25] Snow samples were prepared by spraying solutions of purified water containing single impurities into liquid nitrogen. Photolytic reactions of H_2O_2, HCHO, NO_3^- and NO_2^- have been investigated under comparable experimental conditions.[19, 25] The results demonstrated that under the applied conditions NO_2^- was produced during the NO_3^- photolysis. However, the photolysis of NO_2^- also led to the formation of significant amounts of NO_3^-. The photolysis of H_2O_2 in the snow was observed for a wide range of initial H_2O_2 concentrations in the snow. In contrast, a first-order decrease of HCHO was only observed with high initial concentrations well beyond the range of HCHO concentrations observed in natural snow samples. The obtained experimental photolysis rates are summarized in Table 1.

3 PHOTOCHEMICAL MECHANISM FOR SNOW

Jacobi et al.[26] presented a reaction mechanism for the transformation of NO_3^- and NO_2^- in snow using several series of laboratory experiments investigating the photolysis of NO_3^- and NO_2^- in artificial snow samples. Using the experimental results, rate constants were determined for the involved photolysis reactions of NO_3^- and NO_2^- and the transfer of both compounds from the snow to the gas phase for the applied experimental conditions. The calculations were performed with the assumptions that all of the impurities were located in the so-called quasi-liquid layer (QLL)[27] at the surface of the snow crystals and that the reactions occur in this liquid-like medium. Although it is known that the formation of QLL appears close to the melting point,[27] impurities can enhance the disorder in the surface layer leading to a thicker QLL also at much lower temperatures. For example, the results of Cho et al.[28] indicate that the presence of sodium chloride in the snow with typical concentrations found in coastal polar snow samples leads to the formation of a significant QLL even at temperatures as low as -20 °C.[26] Since further ions are omnipresent in snow even in polar regions, the presence of a QLL on the snow crystals at these low snow temperatures is likely.

Subsequently, the obtained photolysis rate constants were adjusted for Arctic summer conditions as observed on the Greenland ice sheet. Further calculations with the adjusted rate constants demonstrated that under natural conditions the formation of nitrogen oxides

Table 1 *Experimental photolysis rates in artificial snow for several investigated compounds. Rates were obtained for -20 °C.*

Compound	Photolysis rate h^{-1}	Reference
NO_3^-	1.2	26
NO_2^-	30	26
H_2O_2	0.48	19
HCHO	0.1	19

in the snow is dominated by NO_2, which is either generated directly by the photolysis of NO_3^- or by the reaction of NO_2^- with the hydroxyl radical (OH). Due to the quick transformation of NO_2^- to NO_2 the NO_2^- concentrations remain low. Consequently, even if the pH of the QLL would be low enough to favor the formation of nitrous acid (HONO), the HONO generation in the QLL remains probably negligible.

The full and simplified mechanisms for the transformation of NO_3^- and NO_2^- and the production of NO_x in snow presented by Jacobi et al.[26] take only into account reactions of N-containing compounds. However, the importance of the role of the OH radical for the reactions in the QLL is well known. Further laboratory studies indicated that the photolysis of H_2O_2 is probably the most important OH source in the QLL.[24] Thus, a comprehensive reaction mechanism needs to consider a complete set of OH sources and sinks.

The importance of the OH radical for the chemical reactions in the QLL becomes obvious if the calculated QLL concentrations are compared to levels calculated for tropospheric cloud droplets. For example, Herrmann et al.[29] demonstrated that OH levels in the tropospheric aqueous phase reach maximum values between 1 and $2 \cdot 10^{-12}$ M (= mol dm^{-3}) depending mainly on the concentrations of organic compounds. In contrast, the OH levels in the calculations presented by Jacobi et al.[26] increase to levels on the order of $1 \cdot 10^{-9}$ M in the QLL for conditions in surface snow in Greenland. This number represents an upper limit for the OH concentrations since a rage of reactions, which are significant OH sinks in the tropospheric aqueous phase, are not included in the previously presented mechanism. Among these reactions are mainly the reactions with organic compounds. However, knowledge about concentrations of single organic compounds in the snow is still very limited.[e.g. 7] Currently, only formaldehyde concentrations in snow have been investigated in detail at several polar locations.[5, 10, 11, 30-32] The reported concentrations range from 0.05 to $1.2 \cdot 10^{-6}$ M. A more comprehensive characterization of the organic content is available also for Summit station. Measurements of total organic carbon and

Table 2 *Reported concentrations of organic components in snow samples from Summit station collected in June 2000.[7]*

Compound or class of compounds	Concentration	
	$\mu g\ L^{-1}$	μM
Total organic carbon (TOC)	1850	3.7[a]
Inorganic Carbon	1080	17.7[b]
Formaldehyde	35.7	1.19
Acetate	21.8	0.369
Propionate	5.9	0.080
Formate	3.9	0.087
Methanesulfonate	0.6	0.006
Lactate	0.4	0.004

[a] Since Grannas et al.[33] reported that the molecular mass of the organic carbon in snow can reach values of higher than 1 kDa, we used an estimated average molecular mass of 500 Da to translate the measured TOC concentration into μM.

[b] Assuming that the inorganic carbon is dominated by carbonate.

several individual components were performed using surface snow samples collected in June 2000.[7] The reported concentrations are summarized in Table 2.

An updated reaction mechanism with additional reactions and estimated rate coefficient adjusted to conditions encountered at Summit station in Greenland in June 2000 is presented in Table 3. In addition to the chemistry of N-containing compounds, we included further sources and sinks of the OH radical.

The reaction rate for the photolysis rate of H_2O_2 (R8) was calculated using data from previously published laboratory experiments of photolysis reactions of NO_3^- and H_2O_2 in artificial snow for comparable experimental conditions (Table 1). Therefore, the obtained experimental rate constant of 0.48 h^{-1} for the H_2O_2 photolysis was divided by a factor of 400 similar to the procedure for the photolysis rate of NO_3^- as described in Jacobi et al.[19, 26] The HCHO photolysis reaction in snow is probably negligible under natural conditions and is not included in the reaction mechanism.

According to modeling studies regarding the tropospheric aqueous phase the reaction of hydrated formaldehyde ($CH_2(OH)_2$) with OH represents an important OH sink.[e.g. 29] Using temperature dependent kinetic data for this reaction,[29] a rate constant of $5.4 \cdot 10^8$ M^{-1} s^{-1} is calculated for –20 °C. Since the used kinetic data was obtained in bulk aqueous solutions, the reaction (R9) involves the attack of the OH radical on the hydrated formaldehyde $CH_2(OH)_2$. It was suggested that formaldehyde in snow is mainly present in the non-hydrated form HCHO.[14] Nevertheless, we include the aqueous phase rate constant since it seems likely that the hydration of the HCHO molecules can occur in the QLL of the snow crystals. Due to the presence of the high concentration of organic compounds in the snow[7] the reaction of these compounds with OH are also included (Table 3). Since most of the individual organic compounds are not identified, reactions with single organic reactants cannot be included. We rather use a general reaction (R10) with an estimated rate constant on the order of $1 \cdot 10^8$ M^{-1} s^{-1}. This rate constant represents in average for the reaction of the OH radical with several organic compounds in the aqueous phase.[28]

Table 3 *Recommended reactions with rates estimated for -20 °C for a comprehensive mechanism for reactions occurring in natural surface snow.*

	Reaction	Rate constants	Reference
(R1)	NO_3^- $(+ H^+)$ $\xrightarrow{h\nu}$ $NO_2 + OH$	$8.3 \cdot 10^{-7}$ s^{-1}	26
(R2)	NO_3^- $\xrightarrow{h\nu}$ $NO_2^- + O$	$1.7 \cdot 10^{-7}$ s^{-1}	26
(R3)	$NO_3^- + O \rightarrow NO_2^- + O_2$	$2 \cdot 10^8$ M^{-1} s^{-1}	26
(R4)	$O (+ O_2) \rightarrow O_3$	$1.2 \cdot 10^6$ s^{-1}	26
(R5)	$NO_2^- + OH \rightarrow NO_2 + OH^-$	$1 \cdot 10^{10}$ M^{-1} s^{-1}	26
(R6)	$NO_2 + OH \rightarrow H^+ + NO_3^-$	$5 \cdot 10^9$ M^{-1} s^{-1}	26
(R7)	$NO_2 \rightarrow NO_2$ gas	9.7 s^{-1}	26
(R8)	H_2O_2 $\xrightarrow{h\nu}$ 2 OH	$3.3 \cdot 10^{-7}$ s^{-1} [a]	See text
(R9)	$CH_2(OH)_2 + OH \rightarrow$ prod.	$5.4 \cdot 10^8$ M^{-1} s^{-1}	29
(R10)	ORG + OH $\rightarrow$ prod.	$1 \cdot 10^8$ M^{-1} s^{-1}	See text

[a] The photolysis rate is extrapolated relative to the total NO_3^- photolysis rate using the ratio of the experimental photolysis rates reported by Jacobi et al.[19, 26]

To start simulations of reactions in surface snow initial concentrations of the involved stable species are needed. Table 4 gives a summary of concentrations of NO_3^-, H_2O_2, HCHO, and organic compounds (ORG) observed in surface snow samples collected at Summit station in the summer of the year 2000. Assuming that all photochemical reactions take place in the QLL these concentrations need to be translated into QLL concentrations. As described by Jacobi et al.[26] the impurities are confined to the very small volume of the QLL leading to significantly higher concentrations. Such an enrichment factor representing the summer conditions at Summit was used to obtain the initial NO_3^- concentration in the QLL for the simulation of processes in natural snow.[26] Applying the same factor of $1.94 \cdot 10^{-5}$ representing the ratio of the QLL to the total volume of the snow, the QLL concentrations for the further compounds can be calculated (Table 4). In the case of H_2O_2 this value possibly represents only an upper limit. Previous studies have shown that larger fractions of the H_2O_2 incorporated in snow can also be located within the snow crystal.[5]

4 MODEL CALCULATIONS

We performed simulations for the QLL with the photochemical mechanism summarized in Table 3 using the commercial software FACSIMILE,[34] which uses an implicit integration scheme for stiff differential equations[35] with a self-adjusting time step. This software is able to translate a reaction mechanism into a set of dependent differential equations for the concentrations of all involved species. Further input parameters include reaction rate constants and initial concentrations if necessary. Additional details about the software can be found on the AEA Technology website (www.aeat.co.uk/mcpa/areas/software/facsimil.htm).

Calculations were started with initial concentrations as shown in Table 4. Initial concentrations of all further compounds were set to zero. Results of the calculations are shown in Figure 1. We present numbers obtained after a simulation period of 100 min with constant reaction rates. Although concentrations of the stable compounds H_2O_2, NO_3^-, HCHO, and ORG steadily decrease due to the photolysis reactions or the reactions with OH, calculated concentrations remain relatively constant over longer periods. For example, concentrations of the stable compounds changed less than 1% after the simulation of 100 min (Figure 1) compared to the initial concentrations (Table 4).

Table 4 *Concentrations of stable compounds included in the reaction mechanism as observed in the snow at Summit station in the summer of the year 2000.*

Compound	Observed Concentration μM	QLL concentration mM	Reference
NO_3^-	4.4	230	26
NO_2^-	0	0	26
H_2O_2	18	930	9
HCHO	1.19	61	7
ORG	2.5 [a]	130	7

[a] The concentration for the sum of organic compounds was calculated as the difference of the total organic content and the formaldehyde measurements shown in Table 1.

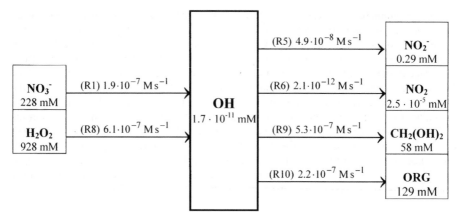

Figure 1 *Sources and sinks of OH calculated for the QLL of surface snow. Numbers are calculated after a simulation period of 100 minutes. Numbers in brackets refer to reactions as shown in Table 2.*

The calculations clearly demonstrate the strong impact of the organic compounds on the OH levels in the QLL. The reactions of OH with HCHO and ORG clearly dominate the OH sinks. The additional OH production due to the photolysis of H_2O_2 cannot outweigh this OH sink. As a result the OH concentrations are significantly lower as compared to the calculations with the N-containing compounds alone.[26]

Nevertheless, the production of OH in the QLL is dominated by the photolysis of H_2O_2. The OH source strength is more than a factor of three higher than the OH production due to the NO_3^- photolysis. This is in agreement with the results presented by Chu and Anastasio.[24]

5 CONCLUSIONS

The experiments and model calculations regarding photochemical processes in surface snow clearly demonstrate that photochemical transformations in the snow are very diverse. As in the atmospheric gas and liquid phase, the OH radical plays a critical role in these transformations. However, the sinks of this radical are not well defined. The reactions with organic compounds are probably the most important OH destruction reactions. However, due to the limited information of the concentrations of single organic compounds in snow it is currently impossible to assemble a detailed mechanism for snow chemistry. Therefore, we decided to introduce a class of compounds, which represents organic material. Additional investigations of organic components in snow can be used to further refine the mechanism.

The comprehensive reaction mechanism presented here includes only ten reactions. Assuming that the conditions at Summit are typical summer conditions for both polar regions we suggest using these reactions together with the recommended rates as a basic set of reactions for further modeling studies of photochemical processes in surface snow. Stable compounds include NO_3^-, H_2O_2, HCHO, and organics in snow. For the first three compounds snow concentration measurements for several polar locations are available in the literature. Therefore, this mechanism can be introduced into one-dimensional models to investigate the impact of photochemical reactions in the snow for a given location in detail.

A further introduction into regional or global models will be helpful for the evaluation of the impact of the surface snow reactions on a larger spatial scale.

The applicability of the recommended mechanism is possibly limited by different snow properties. For example, since the reactions occur in the QLL of the snow crystals, changes in the specific surface area (SSA) during the metamorphosis of the snow can also alter concentrations of the impurities. A strong interaction between chemical and physical properties of the snow can be expected. Therefore, further studies taking also into account physical snow properties like density and SSA are needed to verify the proposed transformations by photochemical reactions in the snow.

References

1   F. Dominé and P.B. Shepson, *Science*, 2002, **297**, 1506.
2   A.E. Jones, R. Weller, E.W. Wolff and H.-W. Jacobi, *Geophys.Res.Lett.*, 2000, **27**, 345.
3   A.E. Jones, R. Weller, P.S. Anderson, H.-W. Jacobi, E.W. Wolff, O. Schrems and H. Miller, *Geophys.Res.Lett.*, 2001, **28**, 1499.
4   J. Yang, R.E. Honrath, M.C. Peterson, J.E. Dibb, A.L. Sumner, P.B. Shepson, M. Frey, H.-W. Jacobi, A. Swanson and N. Blake, *Atmos.Environ.*, 2002, **36**, 2523.
5   H.-W. Jacobi, M.M. Frey, M.A. Hutterli, R.C. Bales, O. Schrems, N.J. Cullen, K. Steffen and C. Koehler, *Atmos.Environ.*, 2002, **36**, 2619.
6   H.-W. Jacobi, R.C. Bales, R.E. Honrath, M.C. Peterson, J.E. Dibb, A.L. Swanson and M.R. Albert, *Atmos.Environ.*, 2004, **38**, 1687.
7   T.M. Dassau, A.L. Sumner, S.L. Koeniger, P.B. Shepson, J. Yang, R.E. Honrath, N.J. Cullen, K. Steffen, H.-W. Jacobi, M. Frey and R.C. Bales, *J.Geophys.Res.*, 2002, **107**, 4394, doi: 10.1029/2002JD002182.
8   M.A. Hutterli, R. Röthlisberger and R.C. Bales, *Geophys.Res.Lett.*, 1999, **26**, 1691.
9   M.A. Hutterli, J.R. McConnell, R.W. Stewart, H.-W. Jacobi and R.C. Bales, *J.Geophys.Res.*, 2001, **106**, 15395.
10 M.A. Hutterli, R.C. Bales, J.R. McConnell and R.W. Stewart, *Geophys.Res.Lett.*, 2002, **29**, doi: 10.1029/2001GL014256.
11 A.L. Sumner and P.B. Shepson, *Nature*, 1999, **398**, 230.
12 A.L. Sumner, P.B. Shepson, A.M. Grannas, J.W. Bottenheim, K.G. Anlauf, D. Worthy, W.H. Schroeder, A. Steffen, F. Domine, S. Perrier and S. Houdier, *Atmos.Environ.*, 2002, **36**, 2553.
13 H. Boudries, J.W. Bottenheim, C. Guimbaud, A.M. Grannas, P.B. Shepson, S. Houdier, S. Perrier and F. Dominé, *Atmos.Environ.*, 2002, **36**, 2573.
14 A.M. Grannas, P.B. Shepson, C. Guimbaud, A.L. Sumner, M. Albert, W. Simpson, F. Domine, H. Boudries, J. Bottenheim, H.J. Beine, R. Honrath and X. Zhou, *Atmos.Environ.*, 2002, **36**, 2733.
15 S. Perrier, S. Houdier, F. Dominé, A. Cabanes, L. Legagneux, A.L. Sumner and P.B. Shepson, *Atmos.Environ.*, 2002, **36**, 2695.
16 L. Chu and C. Anastasio, *J.Phys.Chem.A*, 2003, **107**, 9594.
17 Y. Dubowski, A.J. Colussi, C. Boxe and M.R. Hoffmann, *J.Phys.Chem.A*, 2002, **106**, 6967.
18 Y. Dubowski, A.J. Colussi and M.R. Hoffmann, *J.Phys.Chem.A*, 2001, **105**, 4928.
19 H.-W. Jacobi, T. Annor and E. Quansah, *J.Photochem.Photobiol.A*, 2006, **179**, 330.
20 R.E. Honrath, S. Guo, M.C. Peterson, M.P. Dziobak, J.E. Dibb and M.A. Arsenault, *J.Geophys.Res.*, 2000, **105**, 24183.

21 E.S.N. Cotter, A.E. Jones, E.W. Wolff and S.J.-B. Baugitte, *J.Geophys.Res.*, 2003, **108**, 4147, doi: 10.1029/2002JD002602.

22 C.S. Boxe, A.J. Colussi, M.R. Hoffmann, J.G. Murphy, P.J. Wooldridge, T.H. Bertram and R.C. Cohen, *J.Phys.Chem.A*, 2005, **109**, 8520.

23 C.S. Boxe, A.J. Colussi, M.R. Hoffmann, I.M. Perez, J.G. Murphy and R.C. Cohen, *J.Phys.Chem.A*, 2006, **110**, 3578.

24 L. Chu and C. Anastasio, *J.Phys.Chem.A*, 2005, **109**, 6264.

25 H.-W. Jacobi, B. Kwakye-Awuah and O. Schrems, *Ann.Glaciol.*, 2004, **39**, 29.

26 H.-W.   Jacobi   and   B.   Hilker,   *J.Photochem.Photobiol.A*,   doi: 10.1016/j.jphotochem.2006.06.039, in press.

27 V.F. Petrenko and R.W. Whitworth, *Physics of ice*, Oxford University Press, Oxford, UK, 1999.

28 H. Cho, P.B. Shepson, L.A. Barrie, J.P. Cowin and R. Zaveri, *J.Phys.Chem.B*, 2002, **106**, 11226.

29 H. Herrmann, B. Ervens, H.-W. Jacobi, R. Wolke, P. Nowacki and R. Zellner, *J.Atmos.Chem.*, 2000, **36**, 231.

30 T. Staffelbach, A. Neftel, B. Stauffer and D. Jacob, *Nature*, 1991, **349**, 603.

31 R.W. Gillett, T.D. van Ommen, A.V. Jackson and G.P. Ayers, *J.Glaciol.*, 2000, **46**, 15.

32 S. Houdier, S. Perrier, E. Defrancq and M. Legrand, *Anal.Chim.Acta*, 2000, **412**, 221.

33 A.M. Grannas, P.B. Shepson and T.R. Filley, *Global Biogeochem.Cycles*, 2004, **18**, doi: 10.1029/2003GB002133.

34 A.R. Curties and W.P. Sweetenham, *Facsimile / Chekmat User's Manual*, Computer Science and System Division Harwell Laboratory, Oxford, UK, 1987.

35 C.W. Gear, *Numerical initial value problems in ordinary differential equations*, Prentice-Hall, Englewood Cliffs, New Jersey, 1971.

TOPOLOGICAL TRANSITIONS BETWEEN ICE PHASES

S. Jenkins[1], S.R. Kirk[1] and P.W. Ayers[2]

[1]Dept. of Technology, Mathematics and Computer Science, University West, P.O. Box 957, Trollhättan, 461 29 SWEDEN
[2]Department of Chemistry, McMaster University, Hamilton, ON L8S 4M1, CANADA

1 INTRODUCTION

Thermal and statistical physics[1] have served well in the elucidation of the features of ice's famously complex phase diagram. But the microscopic origins of the transitions between ice phases are rather difficult to understand fully, largely due to the phase boundaries between low temperature phases possessing large metastability fields (large hysteresis). The literature has many in-depth theoretical studies of the phase transitions between ordered and disordered ice phases such as the ice VII - ice VIII proton ordering transition[2] or ice VII, ice VIII and ice X[3], to name but a few, which involve sophisticated statistical treatments and/or entropic equation of state approaches within density functional theory (DFT). An alternative formulation involves symmetry consideration to elucidate these phase transitions[4]. Our work lies somewhere in the middle of these two antipodes: dynamic (i.e., computational 'experiment') and static (purely mathematical). This is because we use the theory of atoms in molecules (AIM) (see Section 3) based on topological features (critical points) in the gradient vector field of the charge density. This theory is also relevant for experimental techniques (e.g. X-ray diffraction), since they yield a measurable real space charge density distribution. In this work we show that the gradient vector field of the charge density contains information about preferred directions of motions of atoms and combine this with bond instabilities and structural compressibilities (all of which are straightforward to calculate from AIM), and present possible phase transition paths for a selection of ice phases, ice IX, ice II, ice IX, ice VIII and ice X. The phases studied in this work were chosen because they are share a common pressure at which they can exist. With the exception of ice XI we consider the zero pressure and temperature structures. We choose an artificially high pressure (50 kbar) that is within the stability field of ice VIII because ice XI at zero pressure and temperature is not close to a phase boundary, as evidenced by the lack of O—O bonding interactions, but this is not the case when the pressure is increased.

In Section 2 the computational methods are outlined, and in Section 3 the AIM theory that is used in this work, and some extensions, are included. The results of this work are presented in Section 4 and the conclusions made in Section 5.

2 COMPUTATIONAL METHODS

The generalized gradient approximation (GGA) to density functional theory is used, with a plane-wave basis set, as embodied in the Quantum-ESPRESSO code[5] Ultrasoft pseudopotentials were used to describe the both the oxygen and hydrogen cores. All of the AIM results were obtained using the code Integrity[6], in which the locations of the critical points were calculated using a three-dimensional Newton–Raphson search, with the Hessian matrix and $\nabla\rho$ calculated at each step using Lagrangian interpolation.

3 ATOMS IN MOLECULES (AIM)

The charge density $\rho(\mathbf{r})$ is a scalar field, and its topological properties are reducible to a description of the number and type of its *critical points,* where $\nabla\rho(\mathbf{r}) = 0$[7]. Critical points may be further characterized by the (rotationally invariant) Laplacian, $\nabla^2\rho(\mathbf{r})$, and by the principal axes and corresponding curvatures derived from the eigenvectors ($\mathbf{e}_1$, $\mathbf{e}_2$ and $\mathbf{e}_3$) and corresponding ordered eigenvalues ($\lambda_1 < \lambda_2 < \lambda_3$) produced in the diagonalisation of the 3x3 Hessian matrix (partial second derivatives with respect to the conventional Cartesian 3D coordinate system.) The critical points can be characterised by their *signature*, an ordered pair of values (number of distinct eigenvalues, arithmetic sum of eigenvalue signs) - signatures of (3,+3), (3,+1), (3,-1) and (3,-3) correspond to cage critical points (CCPs), ring critical points (RCPs), bond critical points (BCPs) and nuclear critical points (NCPs). Here we focus mainly on the charge density at bond critical points $\rho(\mathbf{r}_b)$, where the subscript *b* refers to a BCP. The eigenvalues λ_1 and λ_2 describe the plane perpendicular to the bond path which passes through the BCP, and the $\mathbf{e}_3$ eigenvector defines the direction of the bond path. The ellipticity is denoted by $\varepsilon = \lambda_1/\lambda_2 - 1$, and the eigenvectors associated with λ_2 and λ_1 provide a measure of the extent to which charge density is maximally and minimally accumulated respectively in the given plane locally perpendicular to the bond path. Large (e.g. greater than 0.1) values of ellipticity also indicate π character as well as bond instability[8]. In such situations of large ellipticity the λ_2 eigenvector $\mathbf{e}_2$ has previously been linked to the direction in which atoms most easily slide[9], and more recently the direction in which a defect hydrogen atom moves[10]. Further to this $\mathbf{e}_1$, $\mathbf{e}_2$ and $\mathbf{e}_3$ and linear combinations were found to be very closely related to the the motion of the normal modes in an exploratory study on ice Ic[11].

4 RESULTS AND DISCUSSION

4.1 Topological transformation of ice XI to ice II

From Figure 1(a), depicting ice XI at 50 kbar, it can be seen that there is a relative twisting motion about the basal plane, since the upper row of hydrogen bonds (i.e. parallel to the c-axis) have associated $\mathbf{e}_2$ eigenvectors that are perpendicular to those of the equivalent hydrogen bonds in the lower layer (compare BCPs 1 and 4 respectively). This was not the case for ice XI at 0 kbar, where all of the $\mathbf{e}_2$ directions of the c-axis orientated hydrogen bonds were parallel to the [010] direction. The former may be a display of ice II structural changes being imposed on ice XI at the artificially high pressure of 50 kbar: further evidence of this is provided by the O—O BCP (Figure 1(a) complete with bondpath).

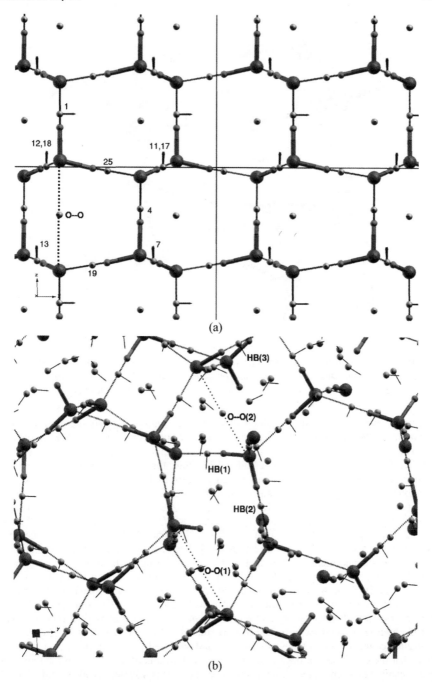

(a)

(b)

Figure 1 *The topological transformation of ice XI (at 50 kbar) to ice II with* **e**$_2$, *'easy' directions displayed as arrows for hydrogen and O—O BCPs, (a) ice XI at 50 kbar, (b) supercell of ice II (most of O—O BCPs removed for clarity).*

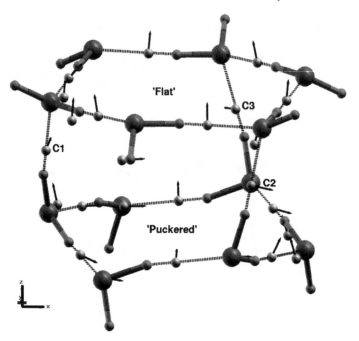

Figure 2 *The topological transformation of ice XI (at 50 kbar) to ice II with e_2, 'easy' directions displayed as arrows for hydrogen and O—O BCPs, showing flat and puckered rings of ice II (most of O—O BCPs removed for clarity).*

Conversely we examined the topology of ice II (see Figure 1(b)) to see evidence of its origins from ice XI. Half of the hexagonal rings are still present, in the form of stacks of alternating flat and puckered rings (see Figure 2, where most of the O—O BCPs have been removed for clarity). Examination of Figure 1(b) shows that between these columns of rings there is a region of denser bonding, which appears to be the remnants of the crushed hexagonal rings. Using AIM we can see further evidence of ice XI character; the hydrogen bond (shown as HB(1) in Figure 1(b)) that connects the flat ring and the puckered ring is somewhat unstable, having an ellipticity, ε of 0.09 and D_{rb} (distance between a BCP and its nearest RCP) value of 1.260 Å. Further to this, notice that HB(2) denotes a hydrogen bond BCP with e_2 eigenvector in the direction parallel to the c-axis, which is parallel to the direction of displacement of the stacks of hexagonal rings[12]. In order to complete the plausibility study of the phase change ice II to ice XI, there need to be hydrogen bonds forming with BCPs at the sites which currently possess O—O BCPs, shown as O—O(1) and O—O(2) in Figure 1(b). The formation of a hydrogen bond at position O—O(2) could happen as indicated, since the e_2 vector of the hydrogen bond BCP marked HB(3) suggests that the hydrogen could swing round towards the site of the O—O(2) BCP. The hydrogen bonds that connect the two types of rings (flat and puckered)) of ice II, parallel to the c axis, show that there is indeed a degree of squashing downwards of ice XI and a rotation of the rings relative to each other about a common axis. This rotation is also evident in the directions of e_2 eigenvectors of the hydrogen bonds parallel to the c-axis (marked as C1, C2 and C3 in Figure 2).

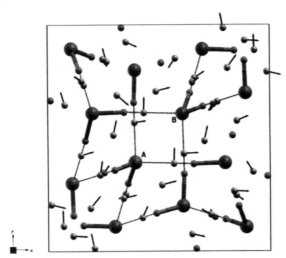

Figure 3 *The topological transformation of ice II to ice IX. - twelve molecule unit cell of ice IX*

4.2. Topological transformation of ice II to iceIX

Careful examination of the bonding topology of the region between the stacked rings of ice II shows a region which is dense, with unstable hydrogen bonds and O—O bonding interactions. The denser packed region is also the site of the crushed ice XI rings. It is not surprising that in this area we found an ice IX-like structure with 12 associated molecules. In Figure 3 the hydrogen bond and O—O BCPs and their e_2 eigenvectors are displayed. Figure 4(a) highlights the ice IX-like structure to show its location in ice II; see also Figure 1(b). In Figure 4(a) the two molecules displayed with a 'ball and stick' notation refer to molecules marked A and B in Figure 3. Figure 4(b) is a reorientation of the ice II unit cell highlighted in Figure 4(a) to emphasise the ice IX-like topology. The creation of the ice IX-like structure, if such a pathway were possible, would require the breaking of some 7 hydrogen bonds although 10 of the original ice II hydrogen bonds would remain unbroken. The reason that it is possible to identify ice IX or at least an ice IX-like structure within ice II may simply be because both phases may exist at the same pressure.

4.3. Topological transformation of ice VIII to ice X

The transition from the tetragonal unit cell of ice VIII(see Figure5(a)) to the cubic (bcc) ice X, (see Figures 5(b-c)), involves the hydrogen bond becoming centrosymmetric. From Figure 5(a) it can be seen that for ice VIII, the general trend of the preferred direction of motion of the sub-lattices is parallel to the c-axis. Ice VIII is most compressible parallel to the c-axis; this is explainable by the existence of linear chains of O—O BCPs (denoted by a G in Figure 5(a)) alternating with cage critical points (CCPs) (shown by the 'X' in Figure 5(a)) parallel only to the c-axis, which are local minima in $\nabla\rho$ and indicate regions of a structure that are maximally crushable. In all other directions there are either continuous linear chains of O—O BCPs, referred to as H-type O—O BCPs[13], or alternating O—O BCPs/hydrogen bonds are present(E and F O—O BCPs).

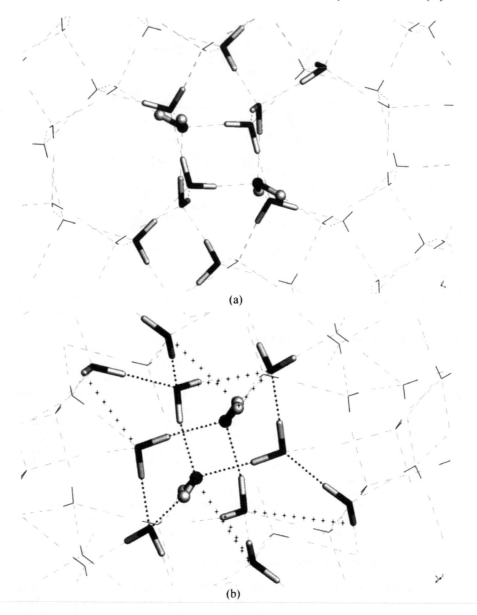

(a)

(b)

Figure 4 *The topological transformation of ice II to ice IX. (a), supercell of ice II highlighted ice IX-like structure (b) Ice II unit cell rotated to emphasize ice IX-like structure.*

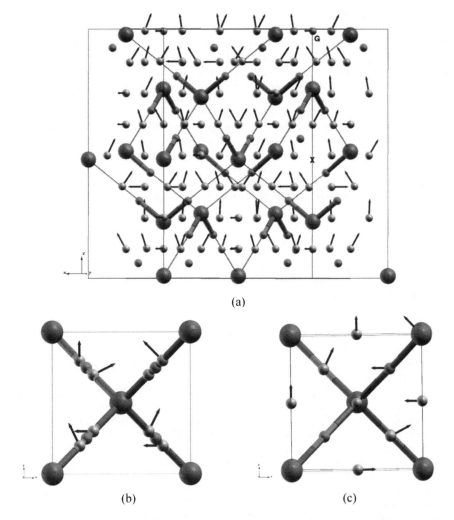

(a)

(b)   (c)

Figure 5 *The topological transformation of ice VIII to ice X. (a) Sixteen molecule unit cell of ice VIII with hydrogen and O—O e_2 directions and BCPs. (b) and (c) show the ice X O-H and O—O BCP easy directions respectively.*

In Figure 5(b) the easy direction eigenvectors are displayed for the high density BCPs for ice X, it can be seen that there are two BCPs either side of the hydrogen atom, all pressures being in excess of 1000 kbar. This shows that the orientation of e_2 rotates by 90 degrees traversing in either direction from the hydrogen along its connecting bondpath.

After the transition to ice X, the E and G interactions become equivalent, as do the E and F interactions(see Figure 5(c)); note the e_2 is attached to a BCP positioned behind the hydrogen in this view along the [010] direction.

5 CONCLUSIONS

We have found it possible to probe the microscopic origins of phase transitions close to a phase boundary above and beyond considerations of symmetry. One of the reasons this is successful in ice in particular is because the hydrogen atom is so light it responds in a manner more in accordance with predictions from the easy direction eigenvector e_2 than might be the case for a compound not containing hydrogen. A hypothesis for the physical meaning of the O—O bonding interactions is that they indicate the onset of soft phonon modes which are already known to accompany structural changes. In future a more comprehensive study of the ice phase diagram could be undertaken to include ice VI and proton disordered phases, as well as other phases with ill-defined phase boundaries. To this end, comparisons could be made with molecular dynamics simulations and experiment could be made. The easy direction eigenvectors e_2 could then be used as reaction coordinates in future work to determine the location phase boundaries which are currently either not known or are uncertain.

Acknowledgments

This research was supported by KK Grant number 2004/0284 and made possible by the facilities of the Shared Hierarchical Academic Research Computing Network (SHARCNET:www.sharcnet.ca).

References

1 V.F. Petrenko and R.W. Whitworth, *Physics of Ice*, Oxford, 1998.
2 S.J. Singer, J-L. Kuo, T.K. Hirsch, C. Knight, L. Ojamäe and M.L. Klein, *Phys. Rev. Lett.*, 2005, **94(13)**, 135701
3 D. D. Klug, J. S. Tse, Z. Liu, X. Gonze, and R. J. Hemley, *Phys.Rev. B*,2004, **70**, 144113
4 V. P Dmitriev, S.B. Rochal and P. Toledano, *Phys. Rev Lett,*, 1993, **71(4)**, 553
5 S. Baroni, A. Dal Corso, S. de Gironcoli, P. Giannozzi, C. Cavazzoni, G. Ballabio, S. Scandolo, G. Chiarotti, P. Focher, A. Pasquarello, K. Laasonen, A. Trave, R. Car, N. Marzari and A. Kokalj, *Quantum-ESPRESSO,* http://www.pwscf.org
6 C. Katan, P. Rabiller, C. Lecomte, M. Guezo, M., V. Oison and M. Souhassou, *J. Appl. Cryst.*, 2003, **36**, 65
7 R.F.W. Bader, *Atoms in Molecules: A Quantum Theory* ,Oxford University Press, Oxford, 1990; R.F.W. Bader, *J. Chem. Phys.*, 1980, **73**, 2871
8 U. Koch and P.L.A. Popelier, *J. Chem. Phys.*, 1995, **99**, 9747
9 R.G.A. Bone and R.F.W. Bader, *J. Chem. Phys.*, 1996, **100**, 10892
10 S. Jenkins and M.I. Heggie, *J. Phys.: Condens. Matter*, 2000, **12**, 10325
11 S. Jenkins, S.R. Kirk, A.S. Coté, D.K. Ross, and I. Morrison, *Can. J. Phys.*, 2003, **81**, 225
12 A. D. Fortes, I. G. Wood, J.P. Brodholt and L. Vočadlo, *J. Chem. Phys.*, 2003, **119(8)**, 4567
13 S. Jenkins and I. Morrison, *J. Phys. Chem. B*, 1999, **103**, 11041

THE IMPORTANCE OF O—O BONDING INTERACTIONS IN VARIOUS PHASES OF ICE

S. Jenkins[1], S.R. Kirk[1] and P.W. Ayers[2]

[1]Dept. of Technology, Mathematics and Computer Science, University West, P.O. Box 957, Trollhättan, 461 29 SWEDEN
[2]Department of Chemistry, McMaster University, Hamilton, ON L8S 4M1, CANADA

1 INTRODUCTION

In this work we begin by exploring the importance of O—O bonding in ice VIII[1-3]. We choose ice VIII because of its high symmetry(space group I4$_1$amd) and anti-ferroelectric ordering. We then extend this to the general case of a selection of ice phases of ice[4,5,6]. In particular Ice VIII is ideal for analyzing O—O bonding interactions because the zone-centre (zc) modes of ice VIII with symmetry $v_{Tz}A_{1g}$ and $v_{Tx}, v_{Tz}E_g$ involve the hydrogen bonded sublattices move as rigid units without any distortion in the hydrogen bonded networks, so effectively isolating the O—O bonding interactions. We attempt to determine the degree to which the hydrogen bonding and O—O bonding networks couple by quantifying the O—O interactions such as bond stiffness and relative charge movement.

The relative movement of charge density is found by considering the ratio of the charge density to the Laplacian of the charge density at the bond critical point (BCP). The Laplacian is a known measure for quantifying the tendencies for charge density to move away from a BCP (i.e $\nabla^2\rho(r_b) > 0$) or contract towards a BCP, the former being the case for all of the bonding interactions considered in this study.

The low temperature anti-ferroelectrically ordered polymorph of ice VI with a ten molecule unit cell, and a sixteen-molecule super-cell for ice VIII, were also used. Both ice VII(A) and ice VII(B) consist of identical eight molecule sub-lattices anti-parallel to one another. Ice VII(A) consists of six C and two D type-hydrogen bonds[1] respectively, while ice VII(B) contains four C and D type hydrogen bonds. The structure of these phases is considered in detail[8].

In Section 2. the computational methods used are outlined. The theory of atoms and molecules (AIM)[9] is outlined in Section 3. in the context of this work including some extensions. The results of this work are presented in Section 4; in subsection 4.1. the effect of cooperative polarization and coupling of bonds on the PDOS is quantified. The findings are discussed in section 5 and the conclusions made in Section 6.

2 COMPUTATIONAL METHODS

In a previous publication[1] the dynamical properties of the zc modes of seven phases of ice were analyzed within the harmonic approximation via a finite-difference evaluation of dynamical matrices from atomic forces. The generalized gradient approximation (GGA)[10] to density functional theory is used. A norm conserving pseudopotential was used to describe the oxygen core and a bare Coulomb potential to describe the proton. All of the AIM results were obtained used a suite of scratch-built programs, in which the locations of the critical points were calculated using a three dimensional Newton-Raphson search, with the Hessian matrix and $\nabla\rho$ calculated at each step using an explicit sum over plane waves[11].

3 ATOMS IN MOLECULES (AIM)

The charge density $\rho(r)$ is a scalar field and its topological properties are reducible to a description of the number and type of its *critical points* (where $\nabla\rho$ ($r_{critical}$) = 0). Critical points may be further characterized by the (rotationally invariant) Laplacian $\nabla^2\rho(r)$, and by the principal axes and corresponding curvatures derived from the eigenvectors and corresponding ordered eigenvalues ($\lambda_1 < \lambda_2 < \lambda_3$) produced in the diagonalisation of the 3x3 Hessian matrix of $\rho(r_b)$ (partial second derivatives with respect to the conventional Cartesian 3D coordinate system.) The critical points can also be characterised by their *signature*, an ordered pair of values (number of distinct eigenvalues, arithmetic sum of eigenvalue signs) - signatures of (3,+3), (3,+1), (3,-1) and (3,-3) correspond to cage critical points (CCPs), ring critical points (RCPs), bond critical points (BCPs) and nuclear critical points (NCPs) respectively. The existence of a BCP is a necessary condition for the formation of a bound state and implies the existence of an *atomic interaction line* — a line linking the nuclei along which the charge density is a maximum with respect to any neighbouring line. If the forces resulting from the accumulation of electronic charge in the binding region are sufficient to exceed the anti-binding forces over a range of separations to yield an equilibrium configuration, then the state is bound and the atomic interaction line is called a *bond path*.

At a BCP (defined by a position vector r_b), the eigenvalues λ_1 and λ_2 describe the plane perpendicular to the bond path which passes through the BCP, and the e_3 eigenvector defines the direction of the bond path. Two curvatures are negative and ρ is a maximum in the plane defined by their corresponding axes; ρ is a minimum at r_b along the third axis, perpendicular to this plane. The BCPs are found to join some but not all of the nuclear critical points. For instance in the phases of ice that possess sub-lattices there are only O—O bonding interactions between sub-lattices even though in ice VI there is a non bonded O—O separation shorter than that of bonded O—O separations.

Using AIM all interactions are characterized as one of two types, where there is a continuum of chemical character between the two types. These two types of interactions are designated according to the sign of the Laplacian at the BCP, $\nabla^2\rho(r_b)$: it is positive for a 'closed-shell' and negative for a 'shared-shell' interaction, a positive value of $\nabla^2\rho(r_b)$ indicating a greater tendency for charge to move away from the BCP along the bond path into the two atomic basins connected by the bond path. A negative value of the Laplacian means there is a greater tendency for the charge density to remain in the inter-nuclear region and away from atomic basins. Examples of the later are the strong 'covalent' carbon–carbon bonds in diamond[12], and those of the former type include hydrogen bonds[13,14]. All of the bonding interactions in this study are of the closed-shell type.

The inequality $|\lambda_1|/\lambda_3 < 1$ holds for all closed shell bonding interactions, and $|\lambda_1|/\lambda_3$ is related to the rigidity of the bond path. This can be seen from the fact that these interactions (all the bonding interactions in this paper are closed shell) are dominated by the contraction of charge away from the inter-atomic surface towards each of the respective atomic basins.

The larger the value of $|\lambda_1|/\lambda_3$ at a BCP, the 'softer' or fuzzier a bond is. This idea of bond softness is related to metallic character, so the softer a bond the more metallic it is. Though none of the interactions in this study are metallic, (ice VI shows evidence of semi-metallic character), it is useful to be able to observe the comparative insulating character of the different interactions in this study. We can define:

$$\xi(r_b) = \rho(r_b)/\nabla^2\rho(r_b), \qquad \text{for } \nabla^2\rho(r_b) > 0 \qquad (1)$$

where $\rho(r_b)$ and $\nabla^2\rho(r_b)$ are the values of the real space charge density and the Laplacian respectively at the BCP. If the ratio $\xi(r_b)$ is in the region of unity, then the BCP can be described as being semi-metallic in character; if less than this then the interaction is described as insulating. In previous work values for $\xi(r_b)$ were found to indicate metallicity[16].

4 RESULTS

4.1. Investigations of the relationship between the translational frequencies of the zc (zone-centre) modes and coupling of hydrogen and O—O bonds.

The symmetry coordinates of all the zc modes in ice VIII have been analyzed experimentally using Raman[17], infrared[18] and neutron scattering techniques[19] and theoretically[3] The theoretical work[3] contains diagrams of the vibrations for all of the symmetry coordinates in Table 1.

In Table 1 T^1 and T^2 refer to water molecules in one sub-lattice and T^3 and T^4 the opposing sub-lattice. We can refer to Figure 1 in conjunction with these symmetry coordinates to understand how the O—O interactions were being distorted in these zc modes, and to see the relative motion of the oxygen atoms comprising these interactions. In Figure1 the letters E, F, G and H correspond to the bonds formed by the pairwise orientation of water molecules that occur in ice VIII. From Table 1 and Figure 1 it can be seen that both the E and F O—O interactions form a linear chain with the hydrogen bonds since $\underline{e}_{OO} \cdot \underline{e}_{HB} = 1.0$ for both types of O—O interactions. This means that both the E and F O—O interactions can couple very well to the hydrogen bonding network. This also means that for modes where these interactions couple well to hydrogen bonds the frequencies will tend to be higher, if the O—O interaction couples poorly with the zc vibration, e.g. compare the vibration $\nu_{Tz}B_{1g}$ to $\nu_{Tx}, \nu_{Tz}E_g$ where from Table 1, the former has a frequency of 240.8cm^{-1} compared with 195.5cm^{-1}.

The zc mode that has the highest frequency is $\nu_{Tz}B_{1g}$ because the F O—O interaction couples well with the hydrogen bonded network and the O—O interaction couples poorly with the zc vibration. The $\nu_{Tz}B_{2u}$ mode has the second highest frequency as the G O—O interaction doesn't couple as well to the hydrogen bonds since $\underline{e}_{OO} \cdot \underline{e}_{HB} = 0.6$. The $\nu_{Tx}, \nu_{Tz}E_g$ mode has a lower frequency still because the O—O interaction couples very well with the zc vibration since $\underline{e}_{OO} \cdot \underline{zc}_{vib} = 1.0$.

The three optic types modes ($\nu_{Tz} A_{1g}$ and $\nu_{Tx} E_g$) the latter mode $\nu_{Tx} E_g$ is doubly degenerate, only involve distortions to the O—O bonding interactions i.e. the sublattices move as rigid units. The frequencies of these modes are in effect the 'spring-constants' of the O—O interactions involved which can be related to the bond stiffness $|\lambda_1|/\lambda_3$.

The $\nu_{Tx}, \nu_{Tz}E_u$ zc mode does involve the distortion of hydrogen bonding and yet has the lowest frequency in Table 1. This is because only the H O—O interaction is involved in the vibration; this does not couple well ($\underline{e}_{OO} \cdot \underline{e}_{HB} = 0.6$) to the hydrogen bond and has the lowest stiffness ($|\lambda_1|/\lambda_3 = 0.0799$) of all the O—O interactions.

Table 1 *Relationship between translational frequencies of the zc (zone-centre) modes and coupling of hydrogen and O—O bonds, see Figure 1 and caption for geometries of E, F, G and H type O—O interactions.*

Vibration	Symmetry coordinate[a]	ω (cm^{-1})	$(\varrho_{OO} \cdot \varrho_{HB}, \varrho_{OO} \cdot zc_{Vib})$[b,c]
$v_{Tz}B_{1g}$	$\frac{1}{2}(T^1_z - T^2_z - T^3_z + T^4_z)$	240.8	**F(1.0, 0.6)**, E(1.0, 0.6)
$v_{Tz}B_{2u}$	$\frac{1}{2}(T^1_z - T^2_z + T^3_z - T^4_z)$	226.5	**G(0.6, 1.0)**
$v_{Tx}, v_{Tz}E_g$	$\frac{1}{2}(T^1_x - T^2_x - T^3_x + T^4_x)$	195.5	**F(1.0, 0.8)**, E(1.0, 0.8)
$v_{Tx}, v_{Tz}E_g$	$\frac{1}{2}(T^1_x + T^2_x - T^3_x - T^4_x)$	158.4	**F(1.0, 0.8)**, H(0.6, 0.7), E(1.0, 0.8)
$v_{Tz}A_{1g}$	$\frac{1}{2}(T^1_z + T^2_z - T^3_z - T^4_z)$	158.0	**G(0.6, 1.0)**, E(1.0, 0.6)
$v_{Tx}, v_{Tz}E_u$	$\frac{1}{2}(T^1_x - T^2_x + T^3_x - T^4_x)$	89.2	H(0.6, 0.7)

[a] T_x and T_z represent the translations along the crystal axes.

[b]   Fonts are used as a guide for the eye to show relative values of $|\lambda_i|/\lambda_3$ (see Figure 2).

[c] $(\varrho_{OO} \cdot \varrho_{HB}, \varrho_{OO} \cdot zc_{Vib})$ represents the magnitude of the dot product of unit vectors in the direction of the hydrogen bond and the particular O—O interaction given. The second number is for the dot product O—O interaction with the zc vibration[c].

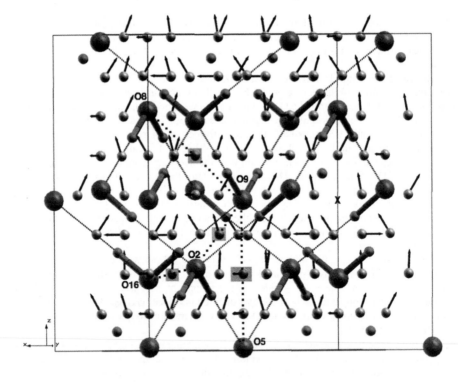

Figure 1. *The sixteen molecule supercell of ice VIII. All bond critical points are shown as small spheres, with the associated e_2 eigenvectors shown using arrows. O2—O9, O9—O8, O9—O5 and O2—O16 dotted lines display the E, F, G and H interactions respectively. The G interaction is parallel to the c-axis.*

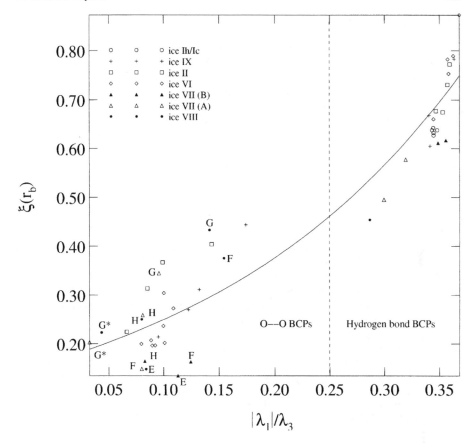

Figure 2 *A plot of* $\xi(r_b)$ *against the bond softness* $|\lambda_1|/\lambda_3$ *the data points to the right of the vertical dashed line all correspond to hydrogen bond critical points, whilst those to the left are for O—O interaction BCPs. The curve was fitted to:* $\xi(r_b) = A \exp[B(|\lambda_1|/\lambda_3)]$ *where A = 0.165 and B = 4.105 and the correlation was 0.96.*

5 DISCUSSION

5.1 Relationship between $\xi(r_b)$ and $|\lambda_1|/\lambda_3$ in ice VIII for the zone centre modes.

In ice VIII the $\nu_{Tz} A_{1g}$ mode the motion is always along the BCP of the G O—O interaction since $\varrho_{oo} \cdot zc_{vib} = 1.0$ from Table1, and so causes maximal distortion of the bond. This may well explain the large values of $\xi(r_b)$ and $|\lambda_1|/\lambda_3$ present in this interaction (see Figure 2 and Table 1). In both the $\nu_{Tx} E_g$ modes the motion is almost parallel to the direction of the F interaction, explaining why its values of $\xi(r_b)$ and $|\lambda_1|/\lambda_3$ are large ($\varrho_{oo} \cdot zc_{vib} = 0.8$) but not as large as those of the G interaction.

5.2 Comparison of $\xi(r_b)$ and $|\lambda_1|/\lambda_3$ in ice VIII, iceVII (A) and ice VII(B).

Ice VII(A), ice VII(B) and ice VIII have sublattice dipole moment vectors $\underline{P}$ of magnitude (for eight molecule sublattices) of 4, $(32)^{\frac{1}{2}}$ and 8 respectively. It is clear that the anti-ferroelectric ordering in ice VIII has an effect on the electronic properties such as the larger values of $\xi(r_b)$ and $|\lambda_1|/\lambda_3$ as compared with either of the ice VII structures. Presumably this is because the sublattices in ice VIII have the largest possible value of $\underline{P}$, From the scattering of the data in Figure 2 the effects of the value of $\underline{P}$ which may be causing a cooperative polarization are most evident in ice VIII, followed by ice VII(A) then ice VII(B). This may well explain why the G and F interactions in ice VIII possess much larger values of $\xi(r_b)$ and $|\lambda_1|/\lambda_3$ (see Figure 2, where O—O and hydrogen bond data for five other phases is given for comparison), than either of the ice VII structures considered in this study.

6 CONCLUSIONS

It was a straightforward matter to put forward a consistent explanation as to the orderings of the zc translational modes in ice VIII. This demonstrates that AIM goes beyond symmetry in its use as a descriptor. Using AIM provides a description of the O—O interactions as entities which possess properties that allow coupling and bond stiffness $|\lambda_1|/\lambda_3$ and charge movement $\xi(r_b)$ to provide a full picture of the vibrational characteristics. this picture would be absent if the only available descriptors were 'repulsion' or 'attraction'. In particular, it proved possible to probe the zc modes v_{Tx}, v_{TzEg} and v_{TzAlg} where the only bonding interactions that were distorted were O—O bonding interactions.

We have been able to compare the difference that full anti-ferroelectric ordering present in ice VIII has with the incomplete anti-ferroelectric ordering in two ice VII structures in terms of bond stiffness $|\lambda_1|/\lambda_3$ and charge movement $\xi(r_b)$. The ability of the O—O interactions to couple to hydrogen bonds may explain why the water ice phase diagram is rather complicated[7].

Acknowledgments

This research was supported by KK Grant number 2004/0284 and made possible by the facilities of the Shared Hierarchical Academic Research Computing Network (SHARCNET:www.sharcnet.ca).

References

1 S. Jenkins and I. Morrison, *J. Phys.: Condens. Matter*, 2001, **13**, 9207
2 S. Jenkins and I. Morrison, *J. Phys.: Condens. Matter*, 2000, **12**, 10263
3 S. Jenkins, I. Morrison and D.K. Ross, *J. Phys.: Condens. Matter*, 2000, 12, 815
4 S. Jenkins and I. Morrison, *J. Chem. Phys. A*, 1999, **103**, 11041
5 S. Jenkins and I. Morrison, *Chem. Phys. Lett.* , 2000, **317**, 97.
6 J.D. Bernal and R.H. Fowler, R H., *J. Chem. Phys.*, 1933, **1**, 515
7 V.F. Petrenko and R.W. Whitworth, *Physics of Ice*, Oxford University Press, Oxford 1999
8 P.V. Hobbs, *Ice Physics*, Clarendon, Oxford, 1974
9 R.F.W. Bader, *Atoms in Molecules: a Quantum Theory (The International Series of Monographs on Chemistry 22)*, Oxford University Press, Oxford, 1998
10 J.P. Perdew and Y. Wang, *Phys. Rev. B*, 1992, **46**, 12947
11 S. Jenkins, Ph.D. thesis, University of Salford, 1999
12 Yu A. Abramov and F.P. Okamura, *Acta Crystallogr. A*, 1997, **53**, 187

13 E. Espinosa, E. Molins C. Lecomte, *Chem. Phys. Lett.* , 1998, **285**, 170

14 U. Koch and P.L.A. Popelier, *J. Chem. Phys.*, 1995, **99**, 9747

15 D. Cremer and E. Kraka, *Croatica Chem. Acta*, 1984, **57**, 1259

16 S. Jenkins, *J. Phys.: Condens. Matter*, 2002, **14**, 10251

17 P.T.T. Wong and E. Whalley, *J. Chem. Phys.*, 1976, **64**, 2359

18 S.P. Tay, D.D. Klug and E. Whalley, *J. Chem. Phys.*, 1985, **83**, 2708

19 A. Kolesnikov, J-C. Li, D.K. Ross, V.V. Sinitzin, O.I. Barkalov, E.L. Bokhenkov and E.G. Ponyatovskii, *Phys.Lett. A*, 1992, **168**, 308

THE CHEMICAL CHARACTER OF VERY HIGH PRESSURE ICE PHASES

S. Jenkins[1], S.R. Kirk[1] and P.W. Ayers[2]

[1]Dept. of Technology, Mathematics and Computer Science, University West, P.O. Box 957, Trollhättan, 461 29 SWEDEN
[2]Department of Chemistry, McMaster University, Hamilton, ON L8S 4M1, CANADA

1 INTRODUCTION

The high-pressure phase (cuprite structure) labeled as ice X has been determined experimentally above 40 GPa[1]. This phase has a symmetric hydrogen-bonded structure and oxygen atoms take the body-centred cubic (BCC) arrangement. In relation to this, it was proposed that at very high pressure the oxygen sublattice becomes the denser face-centred cubic (fcc) form and the phase X is transformed into the anti-fluorite structure becoming metallic at 1.76 TPa[2] via an hexagonal close-packed HCP structure and after this at even higher pressures a pressure-induced insulator-metal transition occurs[3].

The purpose of the present work is investigate the effect of pressure on the topology and mechanics of the electron density. In addition we explore the definition of metallicity introduced earlier[4], based only on the charge density, rather than the determination of a zero band gap.

In this work we explore using only properties derived from the gradient vector field of the charge density, using the theory of atoms in molecules (AIM)[5]. In this way we demonstrate the usefulness of the charge density to understand such properties as metallicity when ice is in such extreme conditions (> 100 GPa, since above this pressure the tunneling effect of the proton ceases). The pressure is applied via an external isotropic force of compression.

In Section **2.** we give the procedures for obtaining the charge density distributions and how the AIM quantities are obtained from them, while in Section **3** we provide the necessary AIM theory for this work. In Section **4** we present the results and a discussion, followed by the conclusions in Section **5**.

2 COMPUTATIONAL METHODS

The generalized gradient approximation (GGA) to density functional theory is used, with a plane wave basis set, as implemented in the Quantum-ESPRESSO code[6]. Ultrasoft pseudopotentials were used to describe the both the oxygen and hydrogen cores, along with a plane-wave basis set with a 108.85 Ry as plane-wave cut-off using an FFT grid of 40x40x40. The number of **k** points in the irreducible wedge of the Brillouin zone that have been used for all calculations on ice X and antifluorite structures is ten. All-electron basis sets were tried, but completely failed to cope with the high pressures in this work. All of the results from examination of the topology of the charge density distribution were obtained using the code Integrity[7], in which the locations of the critical points (see next section) were calculated using a three-dimensional Newton–Raphson search, with the Hessian matrix and $\nabla\rho$ calculated at each step using Lagrangian interpolation.

In this work the cutoff radius for the oxygen and hydrogen ultrasoft pseudopotential were 0.74Å and 0.58Å respectively. For the antifluorite structure the hydrogen and oxygen core radius interferes (i.e. exceeds the nuclear BCP separations) at pressures of greater than 4000 and 2800kbars respectively. The presence of metallic character in the BCPs of the antifluorite structure stops above 4000kbars, presumably due to the presence of the overlapping pseudopotentials.

3 ATOMS IN MOLECULES (AIM)

3.1 Characterization of atomic interactions

The charge density distribution $\rho(r)$ is a scalar field and its topological properties are reducible to a description of the number and type of its *critical points* (where $\nabla\rho(r) = 0$). Critical points may be additionally characterized by the (rotationally invariant) Laplacian ∇^2 $\rho(r)$, and by the principal axes and corresponding curvatures derived from the eigenvectors (e_1, e_2 and e_3) and corresponding ordered eigenvalues ($\lambda_1 < \lambda_2 < \lambda_3$) produced in the diagonalisation of the 3x3 Hessian matrix (partial second derivatives with respect to the conventional Cartesian 3D coordinate system.) The critical points can also be characterised by their *signature*, an ordered pair of values (number of distinct eigenvalues, arithmetic sum of eigenvalue signs); signatures of (3,+3), (3,+1), (3,-1) and (3,-3) correspond to cage critical points (CCPs), ring critical points (RCPs), bond critical points (BCPs) and nuclear critical points (NCPs). Here we focus mainly on the charge density at bond critical points $\rho(r_b)$, where the subscript b refers to a BCP. The eigenvalues λ_1 and λ_2 describe the plane perpendicular to the bond path which passes through the BCP, and the e_3 eigenvector defines the direction of the bond path (the curve through the BCP connecting NCPs along which the charge density is maximal with respect to any neighbouring curve). Previously in ice the authors found O—O bonding interactions and showed that they are not caused by mere steric hindrance; nor are they always present between oxygen atoms that are close together. At a BCP (defined by position vector r_b) $\lambda_1 < \lambda_2 < 0, \lambda_3 > 0$. The existence of a BCP is a necessary condition for the formation of a bound state and implies the existence of a bondpath.

The interaction between bonded atoms is characterized by the values of $\rho(r)$, $\nabla^2\rho(r)$, $G(r)$ and $V(r)$ at the bond critical point. $G(r)$ is the positive definite kinetic energy density and $V(r)$ is the potential energy density. At a bond critical point, the kinetic and potential energy densities are related to the Laplacian by the local form of the virial relation:

$$(\hbar^2/4m)\nabla^2\rho\,(r_b) = 2G(r_b) + V(r_b) \tag{1}$$

In a shared interaction, electron density is both accumulated and concentrated along the bond path between the nuclei. The degree of accumulation is measured by the value for $\rho(r_b)$, and the extent of concentration of the density by the magnitude of $\nabla^2\rho(r_b)$, which is negative because of the dominance of the perpendicular contractions of r.   Shared interactions achieve their stability through the lowering of the potential energy resulting from the accumulation of electronic charge between the nuclei, an accumulation that is shared by both atoms. The opposite extreme to a shared interaction occurs when two closed-shell systems interact, as found in ionic, hydrogen bonded, Van der Waals and repulsive interactions. In those cases the, positive curvature of $\rho(r_b)$ along the bond path is dominant, the Laplacian $\nabla^2\rho(r_b) > 0$ and, since the density contracts away from the surface, the interaction is characterized by a relatively low value of $\rho(r_b)$ with density being concentrated separately in each of the atomic basins.

3.2 Metallicity in bonding interactions

The Laplacian for closed shell interactions is always positive, a larger magnitude indicating a greater tendency for charge to move away from the BCP along the bond path into the two atomic basins connected by the bond path. A smaller value of the Laplacian means there is a greater tendency for the charge density to remain in the inter-nuclear region and away from atomic basins. This situation leads to bonding interactions with more metallic character. We can therefore define a metallicity measure[4]

$$\xi(r_b) = \rho(r_b)/\nabla^2\rho(r_b), \quad \text{for } \nabla^2\rho(r_b) > 0 \quad (2)$$

where $\rho(r_b)$ and $\nabla^2\rho(r_b)$ are the values of the real space charge density and the Laplacian respectively at the BCP. This relation holds for $\nabla^2\rho(r_b) > 0$, which is the case for metallic interactions. If the ratio $\xi(r_b)$ is of the order unity or less, then the BCP can be described as being non-metallic in character, or insulating. Values of $\rho(r_b)$ and $\nabla^2\rho(r_b)$ are listed in previous work, making it possible to calculate the metallicity. (See Table 7 in previous work[8] under the heading interaction C4–C5, diamond. This is a previous example of a metallic bond, where for a metastable state (during kink-pair formation) in the unreconstructed core of the 90° partial in diamond, a closed shell C–C interaction has a $\xi(r_b) > 1$. This bond can be judged to be metallic, since it is responsible for the existence of a half-filled band.)

3.3 Kinetic and potential energy densities

Since $G(r) > 0$ and $V(r) < 0$, the lowering of the potential energy dominates the total energy in those regions of space where electronic charge is concentrated, where $\nabla^2\rho(r) < 0$, while the kinetic energy is dominant in regions where $\nabla^2\rho(r) > 0$. We can define the amount of covalent character $H(r_b)$ where

$$H(r_b) = G(r_b) + V(r_b) \quad (3)$$

Covalent character is present at a BCP if $H(r_b) < 0$. Previously covalent character was found in the hydrogen bond BCPs of many of the phases of ice[9]. Given the extreme pressures involved in these calculations there will be debate about the repulsive or attractive nature of the interactions found in these structures, but as was the case for the O—O bonding interactions in ice VIII[9], providing a description for such interactions is still useful.

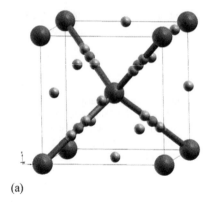
(a)

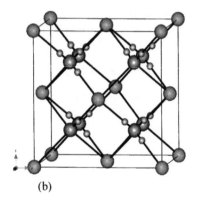
(b)

Figure 1 *The sites of the bond critical point (BCP) data shown for ice X and the antifluorite structure in (a) and (b) respectively.*

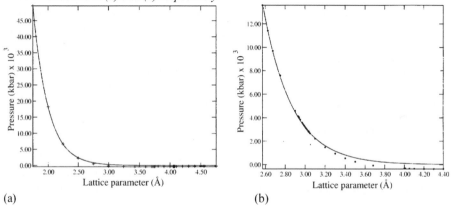

(a)   (b)

Figure 2 *The relationship between the lattice parameter L(in Ångstroms), and pressure P with the same simple exponential relation for comparative purposes: P=A exp(- BL) given for ice X and the antifluorite structure in (a) and (b) respectively, where in (a) A = 5.80x10⁷, B = 4.04 and correlation = 0.9999 and (b) A = 8.95x10⁷, B = 3.41, correlation = 0.998.*

4 RESULTS AND DISCUSSION

In Figure 1 we see the bond critical points (BCPs) for ice X and the antifluorite structure in (a) and (b) respectively. The large spheres (in Figure 1a represent the oxygens, the hydrogens sit midway between the oxygens, the remaining spheres show the sites of BCPs, O-H close to and either side of the hydrogens and the rest are O—O BCPs. The antifluorite has no O—O interactions; this might seem surprising given how dense the phase is, but it shows that BCPs do not always exist as a result of steric hindrance. We postulate that this is because for the antifluorite structure there doesn't appear to be denser phases left for antifluorite ice to transform to. The O—O BCP interactions appear to exist as a mechanism for phases to respond to (for instance) increasing pressure, for example ice VIII transforms to

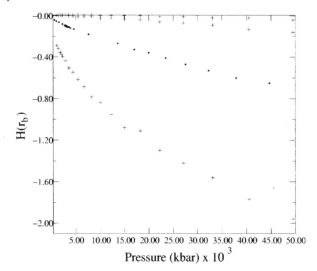

Figure 3 *The variation of covalent character; $H(r_b)$ with pressure for ice X (crosses) and the antifluorite structure(dots).*

ice X, then HCP ice, then the antifluorite structure. Ice X has two types of O—O BCPs, a symmetry increase from the four types ice VIII possesses. The ice X O-H BCPs were shared shell ($\nabla^2\rho(r_b) < 0$) interactions for all pressures (i.e. greater than 1000 kbar), and both of the two distinct types of O—O BCPs were closed shell ($\nabla^2\rho(r_b) > 0$). In the antifluorite structure the O-H BCPs were shared shell above a pressure of 2890.47 kbars and closed shell below it, presumably this is due to the oxygen psuedopotential overlapping or interfering in some way with the O-H BCP. Figure 2 shows the variation with lattice parameter with pressure, with a noticeably better fit for ice X than the antifluorite structure. This may be because of the pseudopotentials not being optimized for the use of AIM. In Table 1 is shown a selection of the AIM properties for ice X and antifluorite structures, in particular notice that for the antifluorite structure $\xi(r_b) > 1$ demonstrating metallic character.

In Figure 3, the O-H BCP of the antifluorite structure had a lower degree of covalent character, $H(r_b)$ (defined by Equation (3)) than the ice X O-H BCPs. This initially seems a bit surprising but can be explained by the increased metallic character (see Equation (3)) of antifluorite ice, since from Equation (1) and (2) the Laplacian must be positive (for metallic character), so the $G(r_b)$ contribution to Equation (1) must dominate $V(r_b)$ and so reduce the covalent character. In ice X the O—O interactions are very largely covalent in character, which was not the case in ice VIII. The metallic character of antifluorite ice can be seen in Figure 4: no results are presented for ice X since the maximum value of the metallicity $\xi(r_b)$ was found to be 0.2, well below the threshold of unity required. The onset of metallic character happens to coincide with the point where curves (i.e to the left hand side of the minimum in two parabolic fits) for the calculated EV data for ice X and antifluorite structures[10].

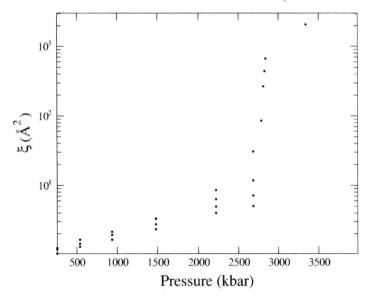

Figure 4 *The variation of metallicity; ξ(r$_b$) with pressure for the antifluorite structure. NB the multiple values of ξ(r$_b$) are due to there being four distinct BCPs for the antifluorite structure due to variations in ∇²ρ(r).*

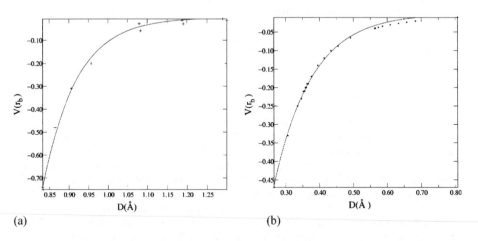

(a)   (b)

Figure 5 *The relationship between D(O-H BCP and oxygen atom separation) and local potential energy density V(r$_b$) with the simple exponential relation: V(r$_b$)=A exp(-BD) given for ice X and the antifluorite structure in (a) and (b) respectively, where in (a) A = -11824, B = 11.62 and correlation = 0.998 and (b) A = -4.90, B = 8.88, correlation = 0.999.*

Table 1 *Variation of properties of ice X (p = 3550.19 kbar and lattice parameter a = 2.40 Å) and antifluorite ice (p = 3588.89 kbar, a = 2.97 Å) see Figure 6 for associated charge density maps at the quoted pressures.*

Ice phase	interaction	$\rho(r_b)$	ε	$\xi(r_b)$	$H(r_b)$
ice X					
	O-H	0.2851	0.04	-0.158	-0.505
	O—O	0.0598	0.01	0.181	0.001
	O—O	0.0354	0.01	0.174	0.006
antifluorite ice					
	O-H	0.1411	0.00	14.24	-0.110

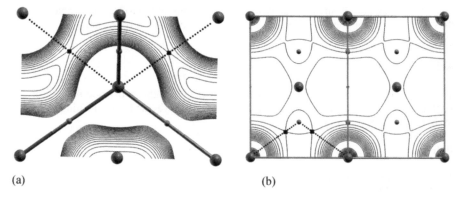

(a)   (b)

Figure 6 *Charge density maps for ice X and antifluorite structures in (a) and (b) respectively, note the O—O BCPs and bondpaths in (a) and O-H BCPs in (b) denoted by small filled squares, see Table 1.*

In Figure 5 we present the results for the variation of $V(r_b)$ with D (separation between the O-H BCP and hydrogen atom measured in Ångstroms) for ice X in (a) and the antifluorite structure in (b). In Figure 5(a) a value of $V(r_b)$ = -0.4 corresponds to a D value of 0.88 Å and a pressure of 1366.27 kbar, and in Figure 5(b) for the antifluorite structure the same $V(r_b)$ yields D = 0.28 Å and a pressure of 10,000 kbar. The charge density maps for ice X and the antifluorite structures are given in figures 6 with the same values of pressure as given in Table 1. The antifluorite structure shown in Figure 6(b) and Table 1 is metallic, but that of ice X is not. This comparison highlights the differences between these two very high pressure structures in terms of the response of their chemical character to pressure. It would appear (from these calculations) that metallization of antifluorite ice occurs at a much lower pressures than the 1.76 TPa given from experiment. Direct comparison with experiment is not possible since the value of D (see Figure 5(b)) at this pressure is much less than the core radius of the hydrogen pseudopotential (1.2 Å). This needs further investigation with more robust computational technology

5 CONCLUSIONS

For ice X and the (fcc) antifluorite structure the quantification of various aspects of their structural and chemical character and their dependence with pressure was found using AIM in a novel approach. Metallic character was found to be present in the antifluorite structure, but did not persist with increased pressure since the BCPs then fell within the pseudopotential core radii. In future studies on the antifluorite structure it will be necessary to replace the core with the true all electron distribution. In addition we present a hypothesis for the physical meaning of the O—O bonding interactions, namely that they indicate the onset of, soft phonon modes that are known to accompany structural changes. The fact that there are no O—O interactions in the antifluorite structure is consistent with this hypothesis, since to date there aren't any higher pressure phases of ice than antifluorite ice and so no pressure induced phase change can occur in this structure. Thus our hypothesis would explain why there are no O—O interactions in the antifluorite structure.

Since calculation of the metallicity measure $\xi(r_b)$ only depended on being able to obtain the gradient vector field of the charge density, this quantity it may be useful to experimentalists in this field.

Acknowledgments

This research was supported by KK Grant number 2004/0284 and made possible by the facilities of the Shared Hierarchical Academic Research Computing Network (SHARCNET:www.sharcnet.ca).

References

1 B. Schwager, L Chudinovskikh, A. Gavriliuk and R. Boehler, *J. Phys: Condens. Matter*, 2004, **16**, 1177

2 J .Hama, Y. Shiomi, and K. Suito, J. Phys. Condens. Matt, 1990, **2**(40), 8107

3 M. Benoit, M. Bernasconi, P. Focher and M. Parinello, *Phys. Rev. Lett.*, 1996, **76(16)**, 2934

4 S. Jenkins, *J. Phys.: Condens. Matter*, 2002, **14**, 10251.

5 R.F.W. Bader, *Atoms in Molecules: A Quantum Theory* ,Oxford University Press, Oxford, 1990; R.F.W. Bader, *J. Chem. Phys.*, 1980, **73**, 2871

6 S. Baroni, A. Dal Corso, S. de Gironcoli, P. Giannozzi, C. Cavazzoni, G. Ballabio, S. Scandolo, G. Chiarotti, P. Focher, A. Pasquarello, K. Laasonen, A. Trave, R. Car, N. Marzari and A. Kokalj, *Quantum-ESPRESSO*, http://www.pwscf.org

7 C. Katan, P. Rabiller, C. Lecomte, M. Guezo, M., V. Oison and M. Souhassou, *J. Appl. Cryst.*, 2003, **36**, 65

8 S. Jenkins and M.I. Heggie, *J. Phys.: Condens. Matter*, 2000, **12**, 10325

9 S. Jenkins and I. Morrison, *Chem. Phys. Lett.*, 2000, **317**, 97

10 A. D. Fortes, PhD Thesis, Research School of Geological and Geophysical Sciences, Birkbeck College and University College London, 2003

REAL-SPACE STUDY OF MECHANICAL INSTABILITY OF ICE XI ON A 'BOND-BY-BOND' BASIS

S. Jenkins[1], S.R. Kirk[1] and P.W. Ayers[2]

[1]Dept. of Technology, Mathematics and Computer Science, University West, P.O. Box 957, Trollhättan, 461 29 SWEDEN
[2]Department of Chemistry, McMaster University, Hamilton, ON L8S 4M1, CANADA

1 INTRODUCTION

In this work we investigate by means of first principles computations the pressure induced behavior of ice XI and VIII[1]. Ice XI is the low temperature, low pressure, ferroelectric, and hydrogen-ordered form of common ice Ih, and ice VIII is the low temperature, high pressure, anti-ferroelectric, and hydrogen-ordered form of ice VII. We show that ice XI transforms continuously into ice VIII under compression in a manner that retains the ice XI electric dipole ordering. The phase found here, 'ice VIII-like', has not been observed experimentally yet and it is unlikely to, but both ice XI[2-4] and ice VIII[5-7] are expected to amorphize inside each others stability field. Therefore, the present study is relevant to understanding amorphization itself. It provides a method for interpreting the high pressure amorphization process as intermediate phases produced in a stepwise transformation process between 1HBN and 2HBN hydrogen ordered structures. We followed the procedure of Umemoto and Wentzcovitch[1] (see Figure 1), for one leg of a hysteresis loop from 0-90 kbar and our results agree with theirs. (In addition we obtain an additional data point at 85 kbar that is consistent with Fig. 1 of Umemoto and Wentzcovitch's work[1], Summarizing the relevant results of Umemoto and Wentzcovitch's work[1]; near 35 kbar, an incommensurate phonon instability develops, followed by the collapse of the lowest acoustic branch. This multitude of unstable phonons, each one leading to a slightly different metastable structure, is what is referred to as amorphization. Up to 90 kbar, adjacent layers perpendicular to the [001] direction shift laterally with respect to each other along the [010] direction. Then, ice XI transforms to a 2HBN structure, i.e., an 'ice VIII-like' phase. Similarly to ice XI, 'ice VIII-like' is also ferroelectric, instead of antiferroelectric like ice VIII. This transition is parallel to the case of disordered ice Ih; paraelectric ice Ih transforms under pressure to paraelectric ice VII preceded by an intermediate amorphous phase[8,9]. The conclusions of Umemoto and Wentzcovitch's work[1] were that a more drastic molecular reorientation is required to produce an anti-ferroelectric ice VIII structure which must involve a higher energy barrier than the ice XI-ice VIII-like transition.

Our goals are to understand the changes in the topology of the gradient vector field of the charge density when the reconstruction leading to amorphization takes place, in addition to the formation of an 'ice VIII-like' structure, and to find consistency with the conclusions of work of Umemoto and Wentzcovitch[1]. In addition we wish to explore the association of O—O bonding interactions with structural instability and hence phonon softening in the process of amorphization.

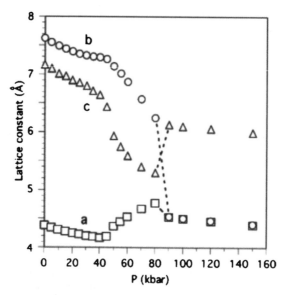

Figure 1 *Pressure dependence of lattice constants of ice XI taken from the work of Umemoto and Wentzcovitch[1]*

2 COMPUTATIONAL METHODS

The generalized gradient approximation (GGA) to density functional theory is used, with a plane wave basis set, as implemented in the Quantum-ESPRESSO code[10]. Ultrasoft pseudopotentials were used to describe the both the oxygen and hydrogen cores. Structural searches and optimizations under pressure were performed using variable cell shape molecular dynamics. The unit cell of ice XI consist of eight molecules. The six k points in the irreducible wedge of the Brillouin zone were used in the calculations. All of the results from examination of the topology of the charge density distribution were obtained using the code Integrity[11], in which the locations of the critical points (see next section) were calculated using a three-dimensional Newton–Raphson search, with the Hessian matrix and $\nabla \rho(\mathbf{r})$ calculated at each step using Lagrangian interpolation.

3 ATOMS IN MOLECULES (AIM)

The charge density distribution $\rho(\mathbf{r})$ is a scalar field and its topological properties are reducible to a description of the number and type of its *critical points* (where $\nabla \rho(\mathbf{r}) = 0$). Critical points may be characterized by the (rotationally invariant) Laplacian $\nabla^2 \rho(\mathbf{r})$, and by the principal axes and corresponding curvatures derived from the eigenvectors ($\mathbf{e_1}, \mathbf{e_2}$ and $\mathbf{e_3}$) and corresponding ordered eigenvalues ($\lambda_1 < \lambda_2 < \lambda_3$) produced in the diagonalisation of the 3x3 Hessian matrix (partial second derivatives with respect to the conventional Cartesian 3D coordinate system.) The critical points can also be characterised by their *signature*, an ordered pair of values (number of distinct eigenvalues, arithmetic sum of eigenvalue signs) - signatures of (3,+3), (3,+1), (3,-1) and (3,-3) correspond to cage critical points (CCPs), ring critical points (RCPs), bond critical points (BCPs) and nuclear critical points (NCPs). At a

BCP (defined by position vector r_b) , $\lambda_1 < \lambda_2 < 0$, $\lambda_3 > 0$. The existence of a BCP is a necessary condition for the formation of a bound state and implies the existence of a bondpath. Here we focus mainly on the charge density at bond critical points $\rho(r_b)$, where the subscript b refers to a BCP. The eigenvalues λ_1 and λ_2 describe the plane perpendicular to the bond path which passes through the BCP, and the e_3 eigenvector defines the direction of the bond path (the line through the BCP connecting NCPs along which the charge density is maximal with respect to any neighbouring line). The ellipticity is denoted by $\varepsilon = \lambda_1/\lambda_2 - 1$, and the eigenvectors associated with λ_2 and λ_1 provide a measure of the extent to which charge density is maximally and minimally accumulated respectively in given plane locally perpendicular to the bond path. Large (e.g. greater than 0.1) values of ellipticity also indicate π character as well as bond instability[12]. In such situations of large ellipticity the λ_2 eigenvector e_2 has previously been linked to the direction in which atoms most easily slide[13], and more recently the direction in which a defect hydrogen atom moves[14]. Further to this e_1, e_2 and e_3 and linear combinations were found to be very closely related to the the motion of the normal modes in an exploratory study on ice Ic[15]. In addition, $|\lambda_1|/\lambda_3$ defines the bond stiffness, typically taking values of 0.30 and 0.27 for ice Ih and ice VIII hydrogen bond critical points (BCPs) respectively. Previously in ice the authors found O—O bonding interactions and showed that they do not result from mere steric hindrance, nor are they always present between oxygen atoms that are close together.

4 RESULTS AND DISCUSSION

It can be seen from Figure 2(a) that the values of the ellipticities of the hydrogen bonds 'oscillate' as the pressure increases; notice at 20 kbar a dip, showing an increase in stability. This is also the pressure at which the first O—O BCP appears, with ellipticity value of 4.9, see Figure 3(a). At 55 and 60 kbar there are no O—O BCPs, perhaps due to the strain having been relieved by the sudden change in the structure; see Figure 4 to compare the change in topology of ice XI with pressure. From 70 to 90 kbar the general trend is a decrease in the hydrogen bond ellipticity values. To summarize, the ellipticities of the hydrogen bond BCPs increase with increasing structural distortion, and then are temporarily reduced by sudden changes of topology, in particular at 55 and 85 kbar. Notice also that after 60kbar the spread of the ellipticities and bond stiffness values of the hydrogen bond BCPs increases noticeably. The O—O BCP ellipticities are at a minimum at 50 kbar when the hydrogen bond stiffness | $\lambda_1|/\lambda_3$ reaches a maximum value(see Figure 3(a), with the exception of 55 and 60 kbar (where there are no O—O BCPs) the O—O bond stiffness values climb with pressure, as can be seen in Figure 3(b). Therefore the abrupt change in structure of ice XI at 55kbars, in particular is reflected in the distribution of the values of the hydrogen bond stiffness $|\lambda_1|/\lambda_3$.

Table 1 shows the comparison of ice VIII and 'ice VIII-like' structures; notice that the instability of the ice VIII structure is not apparent from the hydrogen bond BCP data as the values are similar for both structures. The difference is very apparent in the O—O BCP data, especially in the values of the ellipticity ε, for the F and F-like, G and G-like BCPs. The F and G O—O interactions feature very prominently in the zone centre modes with symmetry coordinates v_{Tx}, $v_{Tz}E_g$ and $v_{Tz}A_{1g}$ [16].

An attractive electrostatic force in the form of the G-type O—O interaction between these networks (in ice VIII) induces a symmetry-allowed displacement of these networks along the [001] direction. Since in the 'ice VIII-like' phase both networks have all dipoles oriented along the same direction, this attractive interaction and the relative displacement of the networks vanish. The BCP of the G-type O—O interaction in the 'ice VIII-like' structure, however, has much higher ellipticity than for actual ice VIII; compare the value 2.79

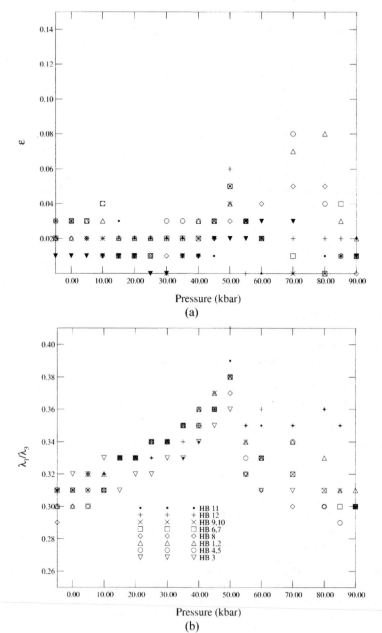

Figure 2 *The variation with pressure for ice XI of the AIM quantities (a) ellipticity ε, and (b) bond stiffness λ₁/λ₃, of the hydrogen bonds (BCPs). The numbering scheme in the legend matches that in **Figure 4**, also the legend in Figure 2(b) is appropriate for Figure 2 (a).*

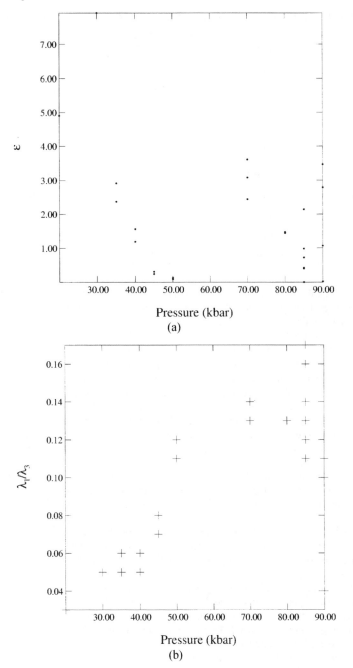

Figure 3 *The variation with pressure for ice XI of the AIM quantities (a) ellipticity* ε, *and (b) bond stiffness* λ_1/λ_3, *for the O—O interactions. The numbering scheme in the legend matches that in* **Figure 4**.

Table 1 *Stability of all inequivalent hydrogen bond and O—O bond critical points. Ice VIII values previously calculated by the authors[17].*

	interaction	$|\lambda_1|/\lambda_3$	ε
ice VIII	O-H	0.29-0.31	0.02
	E O2--O9	0.08	0.38
	F O6--O10	0.15	0.26
	G O3—O16	0.14	0.00
	H O3--O9	0.08	0.44
iceVIII-like	O-H	0.30-032	0.01-0.03
	E-like	0.11	0.02
	F-like	0.04	3.47
	G-like	0.10	2.79

(extremely unstable, as much greater than 0.1) to 0.00 for an ice VIII G-type O—O BCP. This is because of the differences in the dipoles orientations in 'ice VIII-like' and ice VIII, i.e. ferroelectric and anti-ferroelectric respectively.

From Figure 4(a) we see the results for the AIM analysis on the ice XI structure at 0 kbar: the hydrogen bond BCPs and e_2 eigenvectors are displayed. Note that the e_2 eigenvectors are defined to within 180° i.e. [0 1 0] = [0 -1 0] for BCP 11 and 12. As the pressure is increased above 0 kbar insignificant structural or topological changes are apparent until 20 kbar, when O—O BCPs emerge, albeit very unstable ones, in the same position as for ice XI at 35 kbar in Figure 4(b), where the O—O BCP is shown with its bondpath denoted with a dashed line. This O—O BCP is present as a consequence of the increased structural instability relative to the ice XI at 0 kbar pressure. The appearance of these O—O bonding interactions at 35kbar also coincides with the pressure at which an incommensurate phonon instability develops, followed by the collapse of the lowest acoustic branch[1].

In Figure 4(b), comparing the orientation of e_2 for the BCPs labeled 9 and 10 with same numbered BCPs in figure 2(a) shows rotation about the [0 1 0] direction. Note that in Figure 4(b) the presence of an O—O BCP with its bondpath denoted with a dashed line. From 40 to 50 kbar the O—O BCPs take on increasingly stable (i.e. lower) values of ellipticity until the structure yields somewhat to the pressure at 55 kbar. (See Figure 4(c), where a considerable degree of distortion is apparent, though the structure is still recognizable as ice XI) Ice XI remains in a situation of unstable equilibrium until a major rearrangement occurs at 85 kbar, see Figure 4(d). In Figure 4(d) the O—O BCPs labeled 3 and 10 are marked with a heavy dashed line. Notice also the long hydrogen bonds labeled 8 and 9.

The hydrogen bonds labelled 8 and 3 break and these BCPs transform into O—O bonding interactions; the bondpath for BCP 3 is marked with a heavy dashed line. New, rather unstable hydrogen bonds form (BCPs labelled 9 and 8) and additional O—O BCPs form, such as at the BCP site labelled 10. On moving to 90 kbar the hydrogen bond labeled 9 breaks to form an O—O BCP, see Figure 4(e).

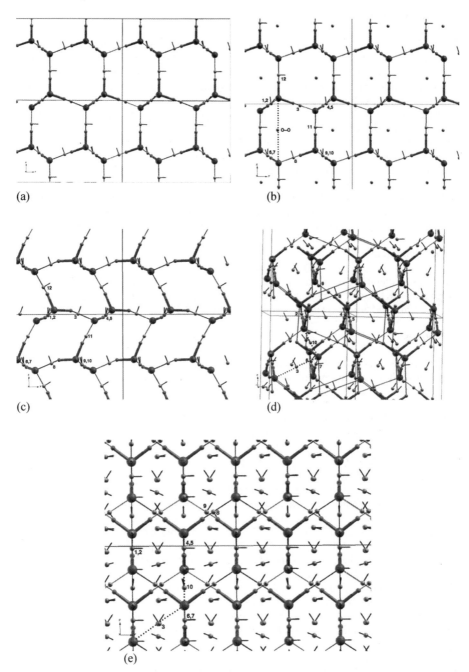

(a)

(b)

(c)

(d)

(e)

Figure 4 *The ice XI hydrogen bond BCPs with their 'easy direction' i.e. the e_2 eigenvectors are shown:(a) 0 kbar, (b) 35 kbar, (c)55 kbar, (d) 85 kbar, and (e) 90 kbar.*

5 CONCLUSIONS

The multitude of unstable phonons in ice XI under compression, each one leading to a different metastable structure, may interpreted using AIM by the appearance of O—O interactions, since they are markers of structural instability, as bondpaths switch easily between such bonding interactions. The e_2 eigenvectors provide a novel tool for predicting structural change, though they may be most successful when they are sued to describe the rearrangement of hydrogen atoms due its extreme lightness. The findings of this work are consistent with the conclusions of previous work[1], namely that a more drastic molecular reorientation is required to produce an anti-ferroelectric ice VIII structure, which must involve a higher energy barrier. This we demonstrate by showing that the ice VIII-like structure is much more unstable than ice VIII, since the values of the BCP ellipticities of the O—O interactions are much higher in the former, as can be seen from Table 1. Further work underway is to understand how the O—O bonding interactions are related to soft phonon modes by using a supercell approach[3] and to develop a theory of pressure amorphisation through, for example, displacive disorder as inspired by the work of Morrel Cohen[18].

Acknowledgments

We thank Koichiro Umemoto for useful discussions on this topic. This research was supported by KK Grant number 2004/0284 and made possible by the facilities of the Shared Hierarchical Academic Research Computing Network (SHARCNET: www.sharcnet.ca).

References

1 K. Umemoto and R. M. Wentzcovitch, *Chem. Phys. Lett.*, 2005, **405**, 53
2 V.F. Petrenko and R.W. Whitworth, *Physics of Ice*, Oxford, 1998.
3 K. Umemoto, R.M. Wentzcovitch, S. Baroni, S. de Gironcoli, *Phys. Rev. Lett.*, 2004, **92**, 105502.
4 O. Mishima, L.D. Calvert, E. Whalley, *Nature*, 1984, **310**, 393.
5 D.D. Klug, Y.P. Handa, T.S. Tse, E. Whalley, *J. Chem. Phys.*, 1989, **90**, 2390.
6 A.M. Balagurov, O.I. Barkalov, A.I. Kolesnikov, G.M. Mironova, E.G. Ponyatovskii, V.V. Sinitsyn, V.K. Fedotov, *JETP Lett.*, 1991, **53**, 30.
7 K. Umemoto, R.M. Wentzcovitch, *Phys. Rev. B*, 2004, **69**, R180103.
8 J.S. Tse, M.L. Klein, *Phys. Rev. Lett.*, 1987, **58**, 1672.
9 R.J. Hemley, L.C. Chen, H.K. Mao, *Nature*, 1989, **338**, 638.
10 S. Baroni, A. Dal Corso, S. de Gironcoli, P. Giannozzi, C. Cavazzoni, G. Ballabio, S. Scandolo, G. Chiarotti, P. Focher, A. Pasquarello, K. Laasonen, A. Trave, R. Car, N. Marzari and A. Kokalj, *Quantum-ESPRESSO*, http://www.pwscf.org
11 C. Katan, P. Rabiller, C. Lecomte, M. Guezo, M., V. Oison and M. Souhassou, *J. Appl. Cryst.*, 2003, **36**, 65
12 U. Koch and P.L.A. Popelier, *J. Chem. Phys.*, 1995, **99**, 9747
13 R.G.A. Bone and R.F.W. Bader, *J. Chem. Phys.*, 1996, **100**, 10892
14 S. Jenkins and M.I. Heggie, *J. Phys.: Condens. Matter*, 2000, **12**, 10325
15 S. Jenkins, S.R. Kirk, A.S. Coté, D.K. Ross, and I. Morrison, *Can. J. Phys.*, 2003, **81**, 225
16 S. Jenkins, I. Morrison and D.K. Ross, *J. Phys.: Condens. Matter*, 2000, **12**, 815
17 S. Jenkins and I. Morrison, *J. Phys. Chem. B*, 1999, **103**, 11041
18 M.H. Cohen, J. Íñiguez, and J.B. Neaton *Eur. Phys. J. E*, 2002, **9**, 239–243.

WATER-VAPOR TRANSPORT IN SNOW WITH HIGH TEMPERATURE GRADIENT

Y. Kamata[1] and A. Sato[2]

[1] Railway Technical Research Institute, Disaster Prevention Technology Division, Hikari 2-8-38, Kokubunji, Japan
[2] National Research Institute for Earth Science and Disaster Prevention, Snow and Ice Research Center, Suyoshi, Nagaoka, Japan

1 INTRODUCTION

Snow crystals change their shapes and states of bonding to each other, after depositing and accumulating on the ground. There have been many basic research studies of snow metamorphism because the dynamic and thermal properties change drastically through it. Water vapor transport in the snow is the cause of dry snow metamorphism, like an isothermal process and temperature gradient process.[1, 2]

Depth hoar grows under a temperature gradient. There have been much research about depth hoar since it can be a major factor in avalanche initiation. It was reported that the growth rate of depth hoar snow was proportional to the temperature gradient and the crystal size was a decreasing function of initial density.[3, 4, 5, 6] Those researches were about depth hoar in the temperate region and the effect of the temperature gradient was considered important. Depth hoar snow also exists widely in the Arctic region during winter period.[7] Such Arctic region snow is important as a cold source, when the heat balance of the climate is considered. From observations in Arctic regions such as in Alaska, or Finland, during the severe winter, the lowest temperature is about -40 °C and depth hoar forms under a large temperature gradient even at these low temperatures.[8, 9, 10]

It was reported that depth hoar crystals developed mostly near the ground although temperature gradient near the ground was smaller than near the surface layer.[8, 11] Such tendency was inconsistent with previous reports and was thought to be due to the difference in growth conditions. It was reported that depth hoar in the Arctic region was affected firstly by temperature and secondarily by temperature gradient due to water vapor concentration.[12] Sturm and Benson (1997) were pioneers, who studied the relationship between water vapor transport and depth hoar development in the subarctic snow with field experiment.[11] Our previous study also presents the possibility that water vapor transfer affected density change, snow temperature distribution, and crystal growth.[13] However, sufficient experimental confirmation of mass transfer have not occurred.

In this study, to clarify the effect of water vapor, we chose a wide temperature range (-65 °C to -12 °C), for which the water vapor concentrations differed by a factor 1000 between the lowest and highest temperatures, and carried out snow metamorphism experiment under high temperature gradient in a cold room. The density change with water vapor was measured and the effects of water vapor transport on crystal growth, and on density change were examined experimentally.

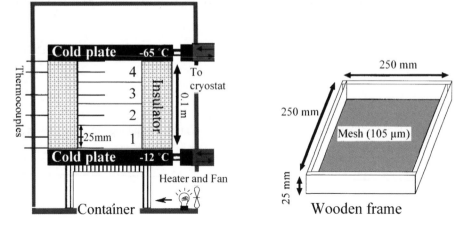

Figure 1   *Schematic experimental equipment and wooden frame affixed nylon mesh.*

2 EXPERIMENTAL

Figure 1 presents schematic of the experimental equipment. A constant-temperature container (0.6×0.7×1.2 m) was installed in a -15 °C cold room. The inside temperature was maintained at -10±1 °C by heater and fan connected to a thermo regulator. A sample box was made of 0.1 m thickness insulator, blocking heat from the outside. In order to measure the change of mass due to water vapor transfer, wooden frames affixed with a nylon mesh were used; the frame size was 250×250×25 mm, the frame thickness was 10 mm and the mesh opening is 105 µm. The nylon mesh can be penetrated by water vapor, but it prevents the transfer of snow particles. The frames were superimposed in four layers, in order from the bottom, layer 1, layer 2, layer 3, and layer 4. The snow sample was sieved in the frame and then installed in the sample box. We used lightly compacted snow (Class 2dc) for the initial stage; the calculated average diameter was 2.7×10^{-4} m, and the density was 165 kg m^{-3} (Figure 2(a)).[14] The top and bottom ends of the box were fully covered with cold plates, to prevent inflow of water vapor from the outside and then the box was installed in the container. The plates were connected to a circulating cryostat, so that temperatures can be individually controlled. Copper-Constantan thermocouples were installed at the positions 0, 12.5, 37.5, 62.5, 87.5, and 100 mm from the bottom. The thermocouple has been connected to a digital multimeter using zero-temperature standard. The temperatures were recorded at one-minute intervals.

The snow sample was kept at -10 °C as an initial state. Next, the cooling plate was set at a fixed temperature (bottom end: -12 °C, top end: -65 °C). Therefore, the snow sample was subjected to a 530 K m^{-1} temperature gradient condition for 133 hours.

3 ANALYSIS

Crystal shapes of each layer were observed and classified into three types: the initial crystal type (lightly compacted snow (Class 2dc)), solid type depth hoar (Class 3mx, 4a) and skeleton type depth hoar (Class 5cp).[14] Snow crystal photomicrographs of each layer were taken and then the projected area of the crystals were obtained.

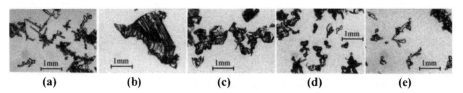

(a)   (b)   (c)   (d)   (e)

Figure 2 *Photograph of snow crystal at initial and final stage of this experiment.
(a) initial snow: lightly compacted snow (Class 2dc), (b) layer 1: skeleton type
depth hoar (Class 5cp), (c) layer 2: solid and skeleton type depth hoar (Class
3mx, 4a, 5cp), (d) layer 3: solid type depth hoar (Class 3mx, 4a), (e) layer 4:
lightly compacted snow (Class 2dc).*

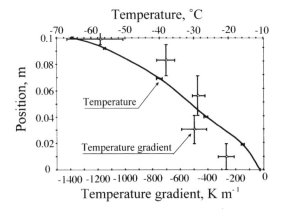

Figure 3 *Quasi-steady state temperature distribution and calculated temperature
gradient. Solid circles are measured temperature and are connected by smooth
solid line. Open triangle is calculated temperature gradient.*

The circle diameter equivalent to the projected area was calculated. Such diameter of the
initial snow crystal was also calculated using the same method. We calculated 250 ~ 400
crystals in each layer. In this paper, we use the statistically averaged diameter and the
growth rate, which is the amount of change in the diameter per unit time.

The density change of each layer was measured in the following method. First, we
measured the weight of each wooden frame. Second, initial snow was sieved and initial
mass of each layer was weighed. Third, after the experiment, we confirmed there were no
voids between the layers and mass of each layer was weighed. The density of each layer
was calculated from the volume and mass of the sample.

The mass difference between layers was divided by the cross-sectional area of the
snow sample and the duration of the experiment, and the measured water vapor flux f_m was
obtained. The measured mass flux M_m was calculated from the density change divided by
the duration of the experiment.

The calculated water vapor flux f_c was calculated from the quasi-steady state
temperature distribution in the snow sample. Fick's law was used to model the vapor flux.
The calculation was performed as follows:

$$f = \partial D_e \frac{\partial C}{\partial T} \frac{\partial T}{\partial z} = \partial D_e \frac{\partial C}{\partial z} \tag{1}$$

Here, f is the water vapor flux, C is the saturated water vapor concentration, D_e is the water vapor diffusion coefficient in air, T is the temperature, and z is a position in the sample. In this study, it was assumed that the water vapor in the snow sample was saturated. Although there were some arguments about water vapor diffusion coefficient in snow and tortuosity dependence, we did not use those values because we have no data on how those values change during dry snow metamorphism.[15, 16] In reality, we think those values probably affect the water vapor flux.

The saturated water vapor concentration C was calculated from the temperature at each position. The distribution of the water vapor flux was then calculated using the water vapor diffusion coefficient D_e. Finally, we compared f_c and f_m and examined the relationship between the water vapor flux and the crystal growth rate.

It is possible to require a change of mass per unit volume and unit time, by differentiating equation (1), that is, the calculated mass flux M_c at position z.

$$\frac{d\partial}{dt} = \frac{df}{dz} = \partial D_e \frac{\partial^2 C}{\partial z^2} \tag{2}$$

Here, ρ is density of each layer, and t is time. M_c and M_m were compared and the relationship between the density change and the mass flux were examined.

4 RESULTS AND DISCUSSION

4.1   Snow type and temperature distribution

The snow type of each layer after metamorphism is given in Figure 2. Layer 4 consisted entirely of lightly compacted snow (Class 2dc) that had not changed from its initial stage. It changed to solid (Class 3mx, 4a) and skeleton type (Class 5cp) depth hoar snow in layers 3, 2, and 1. Especially, layer 1 had many greatly developed skeleton type (Class 5cp) depth hoar snow crystals and was very fragile.

Next, the distribution of temperature and temperature gradient of the quasi steady state is depicted in Figure 3. The snow sample was initially kept at -10 °C, and was cooled from the top cooling plate to -65 °C, reaching a quasi steady state in about 6 hours. The temperature distribution showed upward convex curve. Therefore, the temperature gradient was not uniform in the sample. This tendency is same as a typical temperature profile in subarctic snow.[11] In addition, the nonlinearity of the snow temperature distribution resulted from water vapor transport in the snow.[17]

The results of the crystal growth rate in each layer, along with the temperature and the magnitude of the temperature gradient, are listed in Table 1. The growth rate is not always proportional to the temperature gradient. Layers with a higher mean temperature yielded larger particles, as has been reported for subarctic snow.[8, 11, 12]

Table 1   *Temperature, temperature gradient, and growth rate of each layer. Temperature was measured value in the layer. Temperature gradient was inside value of each layer. Growth rate was averaged value of each layer.*

layer	Temperatue (°C)	Temperature gradient (K m^{-1})	Growth rate (x10^{-10} m s^{-1})
1	-16	-270	10.3
2	-26	-500	4.9
3	-40	-480	0.8
4	-56	-700	-0.5

4.2 Water vapor flux and growth rate

Water vapor transport develops depth hoar crystals. It has been reported that the growth rate agrees well with the calculated water vapor flux.[12] Figure 4 shows the relationship between the water vapor fluxes (f_m and f_c) and the growth rate of each layer. The distribution of the water vapor fluxes and the growth rate agreed well. f_m and f_c also agreed well. This indicates the validity of f_c and the temperature distribution in the snow largely reflected crystal growth.

Figure 5 re-expresses the relationship between f_c and the growth rate seen in Figure 4. It was demonstrated that depth hoar crystals could not grow very large in a layer with low water vapor flux, even if the temperature gradient is large. We can see this tendency when the temperature is very low. In the figure, it also accounts for the result at 290 kg m^{-3} which was obtained under conditions similar to those reported in Kamata et al. (1999).[12] A high proportion coefficient was required when the initial density was low. In short, depth hoar crystal was able to grow extensively when the initial density was low, since spaces in the snow were large. This agreed with previous studies, in which depth hoar snow developed when the initial density was low.[3, 5, 6]

In this study, the range of water vapor flux was from 0.3 to 8 $\times 10^{-7}$ kg m^{-2} s^{-1}. Sturm and Benson (1997) observed temperature profiles in snow and calculated layer-to-layer vapor flux using Fick's law.[11] They obtained an average water vapor flux about 2.5 $\times 10^{-7}$ kg m^{-2} s^{-1}, with peak values at 15 $\times 10^{-7}$ kg m^{-2} s^{-1}. Our results fell within their value range with the exception of the extremely low temperature layer (less than about -40 °C).

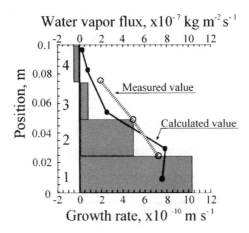

Figure 4 *Relationship between the water vapor fluxes and the crystal growth rate in each layer. Solid circle is the calculated water vapor flux. Open circle is the measured water vapor flux. Columns are the average growth rate of each layer.*

4.3 Mass flux and density change

The water vapor flux seems to affect significantly the growth rate of depth hoar snow. It seemed that water vapor transportation also caused a density change in snow. The initial and final density of each layer is illustrated in Figure 6. Before and after the experiment, the lost mass was only 0.5 %. The density of layer 1 (warmest) decreased,

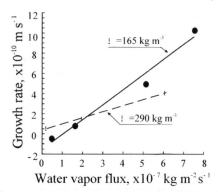

Figure 5   *Re-expressed relationships between the water vapor flux and the growth rate seen in Figure 4. Previous result reported in Kamata et al. (1999) was also shown to examine the density dependence.*

while the densities of layers 2, 3, and 4 increased. We calculated the density change subtracting initial density from final one. Figure 7 shows the relationship between the mass fluxes (M_m and M_c) and the density change. M_m and M_c were of the same order, and the distribution of both mass fluxes showed the same tendency as the density change. In the layer 1, M_c showed a negative value where the density decreased. This result shows that it might be possible to estimate the amount of density change from the temperature distribution within the snow.

An average mass flux reported in Sturm and Benson (1997) was about 50×10^{-7} kg m^{-3} s^{-1}, with peak values at 500×10^{-7} kg m^{-3} s^{-1}.[11] The range of our mass fluxes was within their values.

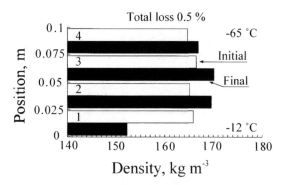

Figure 6   *Density change over the duration of the test. Blank columns show initial density of each layer. Black columns show final density of each layer.*

4.4   Snow metamorphism under high temperature gradient

Water vapor transport in snow not only greatly affected the crystal growth, but also caused density change. We tried to examine depth hoar growth for a wide range of temperature and temperature gradient from the viewpoint of water vapor flux (Figure 8). The isogram line of water vapor flux was ideally obtained using equation (1) in the following way. The value of water vapor flux was fixed and then temperature gradient was calculated at certain temperature value. We connected those values on the graph.

The hatched area is the region where observations and experiments of previous studies on depth hoar snow have been carried out. This area is sensitive to the temperature gradient even if the temperature gradient changes only slightly, since the isogram line indicates that the water vapor flux changes significantly. This accounts for previous studies reporting that the growth rate of depth hoar is proportional to the magnitude of the temperature gradient. [3, 4, 5, 6] In contrast, the experiment in this study was carried out in the circled region. The isogram line stands in this region. Therefore, the amount of change in the water vapor flux is slight even if the temperature gradient changes significantly. However, the water vapor flux changes greatly, when the temperature changes. In this way, by analyzing the water vapor flux distribution, we could understand temperature gradient metamorphism over a wide temperature range.

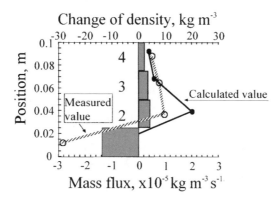

Figure 7 *Relationship between density change and mass flux. Solid circle is the calculated water vapor flux. Open circle is the measured water vapor flux. Columns are the density change of each layer.*

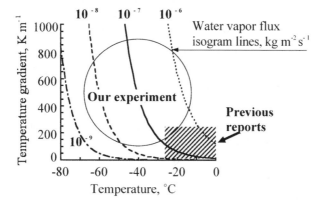

Figure 8 *Water vapor isogram line on temperature and temperature gradient graph. Isogram lines were ideally calculated using formula (1). Hatched area is sensitive to temperature gradient. Large circle area is sensitive to temperature.*

5 CONCLUSIONS

Using equipment that can measure water vapor transfer, to examine the effect of water vapor transport on growth rate of grains and on density changes over a wide range of temperatures and temperature gradients, an experiment under high temperature gradient was carried out. The growth rate of depth hoar snow was found to be proportional to the water vapor flux. The measured water vapor flux and the water vapor flux calculated from the temperature distribution agreed well. In addition, the calculated mass flux and the measured mass flux also agreed well. These results helped us understand the relationship between temperature gradient metamorphism and water vapor flux over a wide temperature range. In the future, the experiments would be carried out under various conditions exactly, and these kinds of data might be accumulated.

References

1   Brown, R.L., Edens, M.Q., and Sato, A., *Ann. Glaciol.*, 1994, **19**, 69.
2   Sato, A., Adams, E.E., and Brown, R.L., *Proc. ISSW*, 1994, 176.
3   Akitaya, E., *Contrib. Inst. Low Temp. Sci.*, 1974, Ser.A **26**,1.
4   Fukuzawa, T. and Akitaya, E., *Ann.Glaciol.*, 1995, **18**, 39.
5   Marbouty, D., *J. Glaciol.*, 1980, **26**(94), 303.
6   Perla, R., and Ommaneney, C.S.L., *Cold Reg. Sci. Tech.*, 1985, **11**, 23.
7   Sturm, M., J., Holmgren and Liston, G.E., *J. Climate*, 1995, **8**(5), Part2: 1261.
8   Kamata, Y., Kosugi, K., Abe, O., Sato, T., Shimizu, M., Yoshikawa, K., Hosono, M., and Sato, A., *"Tohoku no Yuki to Seikatsu"*, 1998, **13**, 8, in Japanese.
9   Sato, T., Nakamura, W., Abe, O., Kosugi, K., Kamata, Y., Shimizu, M., Yoshikawa, K., Hosono, M., and Sato, A., *"Tohoku no Yuki to Seikatsu"*, 1998, **13**, 4, in Japanese.
10 Sturm, M., and Johnson, J.B., *J. Geophys. Res.*, 1991, **96**(B7), 11657.
11 Sturm, M., and Benson, C.S., *J. Glaciol.*, 1997, **43**(143), 42.
12 Kamata, Y., Sokratov, S.A., and Sato, A., *Advances in Cold Regions Thermal Engineering and Sciences*, In: Hutter, K., Wang, Y., and Beer, H. (eds), Springer Verlag, Part IV, 1999, 395.
13 Sokratov, S.A., Sato, A., and Kamata, Y., *Ann.Glaciol.*, 2001, **32**, 51.
14 Colbeck, S., Akitaya, E., Armstrong, R., Gubler, H., Lafeuille, J., Leid, K., McClung, D., and Morris, E., *NTIS, 5285 Port Royal Rd., Springfield, Virginia 22161, Issued by: The International Commission on Snow and Ice of the International Association of Scientific Hydrology and Co-Issued by; International Glaciological Society*, 1990.
15 Colbeck, S.C., *Water Resour. Res.*, 1993, **29**(1), 109.
16 Maeno, N., Narita, H. and Araoka, K., *Natl. Inst. Polar Res. Mem.*, 1978, Special Issue 10, 62.
17 Sokratov, S.A., Kamata, Y., and Sato, A., *Advances in Cold Regions Thermal Engineering and Sciences*, In: Hutter, K., Wang, Y., and Beer, H. (eds), Springer Verlag, Part IV, 1999, 409.

EXPERIMENTAL GEOSCIENCE IN A FREEZER: ICE AND ICY COMPOUNDS AS USEFUL EDUCATIONAL ANALOGUES FOR TEACHING EARTH AND PLANETARY MATERIALS SCIENCE AND THE PHYSICAL SCIENCES

S.H. Kirby[1]

[1]U.S. Geological Survey MS 977, 345 Middlefield Road, Menlo Park, California 94025 USA

1 INTRODUCTION AND BACKGROUND

As our societies become ever more technologically complex, citizens and decision makers need to become more knowledgeable about the science of materials if they are not to become victims of technology rather than the masters of it. Many recent examples of failures of public-works projects, such as collapses of airport terminals, tunnels, and aircraft losses, have issues regarding materials selection and properties. Decision makers have learned that they cannot passively accept expert opinion and need to be knowledgeable enough to ask the right questions concerning complex public-works, transportation and other large and costly projects. Moreover, many science students and professionals generally have but a rudimentary knowledge of the science of materials and hence are often unprepared for research involving knowledge of material properties and processes.

The purpose of this paper is to stimulate interactive discussion on how to exploit what we have learned about the physics and chemistry of ice in the PCI community by developing engaging, inexpensive, and safe laboratory experiments on ices for use in classrooms at the primary school to lower-division undergraduate university level. The behavior of terrestrial glacial ice (plastic flow, brittle behavior, development of crystallographic preferred orientations and property anisotropy, etc.) has long been recognized by our community as instructive to students of Earth's interior and the icy planets and moons and the physical sciences in general. Education for students is most effective when it uses a balance of classroom instruction, personal reading, and "hands on" laboratory experience in working with real materials. This process of doing science rather than just reading about it is often called "active learning" and experience has shown that it is an effective tool in demonstrating scientific principles and also showing how science is actually done. In the geosciences, lab experience of beginning students is usually limited to field trips and classroom identification of rocks and minerals. What is typically lacking is lab experience using experiments on mineral systems in which the fundamental material properties and governing processes are investigated. To do such experiments on silicate mineral systems, particularly at the high-school or undergraduate level, is often impractical, given the high-temperatures required to investigate thermally-activated processes and melting and the high pressures that are often necessary to investigate metamorphic reactions.

Thus not only do ice and icy compounds occur on Earth, in polar planetary regions, and in the moons of the outer solar systems, they are excellent analogues for silicate mineral systems. Many of the processes that occur in Earth's interior, such as large-scale thermal convection, also occur in these icy bodies. The hydrogen and van-der-Waals bonding in their structures confer a number of useful properties, including low melting temperatures, low elastic moduli, high rates of thermally-activated processes at low temperatures, low pressures required for polymorphic reactions, usually relatively low ductile strengths, and low fracture toughness. These materials are also typically optically transparent and their internal structures may be investigated by sectioning and by transmission optical microscopic techniques in optical cells. Experiments involving ice and icy compounds evoke the wonder and fascination of crystals, how they grow from their liquids and how they respond to the environmental demands of this planet and those of the frigid worlds in our outer solar system. Some of the potential of such experiments may be found in the semi-popular book in Japanese by Norikazu Maeno titled *Kohori no Kagaku Ice Science*[1] and in the technical book on ice physics by Petrenko and Whitworth.[2]

Although isolated examples of science teaching through experiments on ice do exist and may be found, for example, on the world wide web or in books, to the author's knowledge, no attempt has been made to develop a comprehensive curriculum that exploits what has been learned by the research community about ice and icy compounds. Accordingly, the intended readership of this review of lab opportunities is not the science teaching community itself, but rather the ice research community as represented by PCI, a group that has centuries or perhaps even millennia of experience in working with ice as an experimental material. It is the view of the author that the teaching of science could be enriched and strengthened if this extensive experience held by our community could be tapped. Science teachers interested in reading this article may want to consult the glossary provided in the appendix and also glossaries of scientific terms or scientific dictionaries to find additional definitions of the technical terms used in this paper.

2 SOLAR SYSTEM MATERIALS AND PROCESSES

2.1 Earth Materials and First-Order Internal Processes

Earth science has greatly benefited from advances in numerical modeling of complex processes in Earth's interior that are not subject to direct observations, such as thermal convection in the mantle, magma generation and ascent and volcanic eruptions, and earthquake slip. The speed of seismic waves based on measurements of travel times of seismic waves, surface variations in the acceleration of gravity, Earth's natural electrical currents, GPS measurement of surface deformation rates, and the shape of Earth's surface are among the observations that can be used to constrain these models. Broadly speaking, the interrelated processes to be understood (and especially how they link to the foregoing observables) are:

2.1.1 Solid-state time-dependent plastic flow (creep) and the development of elastic anisotropy through flow-induced crystallographic preferred orientations (CPOs). All rock-forming minerals of Earth's mantle and crust are to varying degrees elastically anisotropic (*i.e.*, having properties that vary with the orientation of measurement, such as thermal expansion, thermal conductivity). Crystalline materials develop CPO's in connection with plastic flow and hence the resulting rocks develop anisotropy in properties. For example, the velocities of seismic waves (low-frequency elastic waves) vary with the direction of the ray paths in the mantle and shear waves speeds often vary with the direction of polarization. Other important issues for creep flow include the effects of mineral grain size and the effects of fluids, such as water and CO_2.

2.1.2 Mineralogical changes of state with or without the involvement of fluids. Rocks and minerals change their crystalline structures due to changes in pressure or temperature or both and with exposure to or release of fluids, such as water and CO_2. Such changes can have profound effects on the rheology of the mantle and crust (in both the brittle and ductile regimes) and are routinely mapped as discontinuous changes in seismic wave speeds.

2.1.3 Melting of the mantle and changes in the resulting magma composition with mixing with the crust and by fractional crystallization. The primary magmas of the Earth derive from melting of the mantle to produce what are called mafic melts (e.g., basalts and gabbros). The crystallization products of multi-component silicate melts generally are more dense less mobile than their melts, and tend to sink, and separate chemically from each other. This process, called fractional crystallization, combined with mixing with crustal rocks, leads to magma differentiation with resulting compositions ranging from mafic (such as basalts and gabbros with low SiO_2, K_2O, Na_2O, Al_2O_3 and water of hydration and high in MgO and CaO) to siliceous (such as granites, pegmatites, and rhyolites that are rich in SiO_2, Al_2O_3, K_2O, Na_2O and low in MgO).

(D) Multi-phase flow in silicate melts, involving silicate melts with dissolved water or CO_2, fluid-filled bubbles of free water and/or CO_2, and crystals that form during initial crystallization as the magma cools and depressurizes during ascent. Such media are rheologically and chemically complex and their pathways to Earth's surface to be expressed as volcanic eruptions, such as through porous media or along fractures, are poorly known.

Although the materials of the Earth are multi-component and involve a complex suite of processes that govern their origins and evolution, experiments with simpler model materials such as ice can help isolate, clarify, and give insight into how important physical processes can lead to well-defined physical behavior.

2.2 Planetary Ices

Although the basic chemical and material building blocks for the planets and their satellites were fairly uniform during the initial formation of the solar nebula from inter-stellar cloud materials, chemical differentiation, and segregation occurred over time during accretion of the planets, and their moons such that the volatile chemical components of the solar nebula ended up as present day near-surface ice on Earth, and ice plus solid CO_2 on Mars, and as ice and other molecular solids and fluids (such as hydrocarbons and ammonia) on most of the moons of Jupiter and Saturn, and as water ices and increasingly volatile species such as nitrogen in the outermost solar system.

Added to this mix of molecular chemistries are the water-soluble salts, such as sulfates and chlorides, that likely form salt hydrate compounds and eutectic intergrowths with water ice. These salts are probably leached out from the accreted rocky components of primitive rocky solids that make up chondritic meteorites that rain down on Earth[3]. Some of the larger icy moons of the outer solar system are big enough to have likely formed some of the high-pressure polymorphs of ice and to have segregated their denser silicate components to rocky cores[4]. Thermo-physical models of many of these larger bodies suggest that thermal convection has occurred during their histories and may be occurring today.[4]

Space exploration by planetary missions from Earth has revolutionized our view of these moons and planets. Europa, Jupiter's second large moon, is easily among the most fascinating bodies. Its surface features suggest that it is a tectonically active body that is driven by large solid-body tides from its huge planetary parent and by related internal convection. Results from the Galileo mission show that it has a magnetic field that is consistent with a briney (and hence electrolytic) subsurface ocean that has electrical currents driven by the magnetic field of Jupiter.[5] Moreover, the NIMS imaging spectrometer on board spacecraft showed clear evidence from the solar relectance spectra that sulfate hydrates exist on the surface of Europa and that their distribution show variations that correspond to boundaries between geological terranes.[6, 7, 8, 9, 10, 11, 12]

Evidence for abundant non-ice molecular species and active planetary surfaces and interior was also obtained from the ongoing Cassini-Huygens mission to the moons of Saturn, including the surfaces of Titan and Enceladus. The Voyager II mission also investigated the surface of Triton, Nepune's largest moon, showing evidence for nitrogen in solid and vapor form on this moon. These observations raise the possibility that these molecular components may form clathrate hydrates by reacting with water and water ices. Multi-component ices and molecular compounds are therefore expected to be the norm, rather than the exception, in the outer solar system.

3 WATER ICE AS A MODEL MATERIAL

3.1 Ice as an Ideal One-component Model Crystalline Material

There are a number of properties of water ice that make it an ideal one-component model material:
1. It is easy to obtain in high-purity both as a solid and a liquid.
2. It is completely non-toxic.
3. Simple methods exist for making isotropic nearly pore-free polycrystalline aggregates, and also for growing large nearly strain-free single crystals.
4. It is optically transparent. This makes possible to do transmission optical microscopy to explore internal processes, such as the creation of brittle fractures.
5. Sublimation rates of ice surfaces are highest at grain boundaries and at intersections of crystal dislocations with ice surfaces, making it easy to reveal grain boundaries as grooves and dislocations as etch pits using optical reflection microscopy. Rates of selective etching of these features by sublimation also may be increased by sublimating under vacuum.
6. Weak hydrogen bonding leads to low T_m (the temperature in Kelvins) and low elastic moduli, shear and bulk (μ and K). Thus high T/T_m (the ratio of the experiment temperature to the melting temperature, both in Kelvins), high σ/μ (the ratio of the applied stress σ to the shear modulus μ), and high P/K (the ratio of the applied pressure, P, to the bulk modulus, K) are easy to accomplish under freezer conditions. Such a range of conditions produce a wide range of deformation mechanisms and pressure-induced polymorphic phase changes (changes from one crystalline form to another without changes in chemistry).
7. It is also easy to produce conditions favorable for brittle fracture by reducing T/T_m and by operating at atmospheric pressure or under tensile stresses.
8. Most of the fundamental physical properties of ice are already known, thus making it Earth's best characterized rock-forming mineral and a rich source of supporting information for student scholarship.

3.2 A tour of physical processes accessible in single-component water ice: Melting and freezing

3.2.1 *Supercooling* liquid water. It is easy to demonstrate the fundamental difference between freezing water and melting ice using a cheap digital kitchen thermistor. Supercooling of water by several degrees may be observed in very clean test tubes as well as the sudden temperature rise associated with the latent heat of fusion. The importance of nucleation in controlling the degree of undercooling of the melting line is also easily shown by sprinkling powdered ice on the supercooled liquid. The sudden crystallization of ice on the surface of the liquid is always impressive.

3.2.2 *Superheating of ice.* Significant superheating does not ordinarily occur due to the ease of nucleation of the liquid as a more disorder phase. This may be shown by freezing in the tip of a bulb thermometer or a thermistor in a surface plate of ice immersed in water and then observing the temperature at which the tip of the thermistor begins to melt. Fascinating localized internal melting of impure ice interiors at temperature just below bulk melting may also be observed under strong illumination by light, such as the famous Tyndall figures[13], (also known as ice flowers or negative crystals).[13,14]

3.2.3 *Expansion associated with the freezing of water.* This phenomenon is well known and may be easily measured in a clear plastic graduated cylinder. Care must be taken to lightly lubricate the inner surface in order to avoid breaking the cylinder. More advanced high-school and lower division undergraduates are capable of demonstrating that negative pressure effect on T_m is related to changes in volume, ΔV, and enthalpy, ΔH, via the Clapeyron equation:

$$dT_m/dP = T\Delta V/\Delta H \tag{1}$$

Direct observations of T_m (P) and ΔV may be made in a sapphire optical cell with simple screw-press pump by measuring the offset in the pressure *versus* volume curve. ΔH can be measured at room pressure using a simple differential calorimeter comprised of two paper nut cups outfitted with kitchen thermistors and containing water in one and a standard solid material in the other for which the heat capacity curve is known. Direct observations of pressure-release freezing of water (as compared to pressure-release melting in silicates) may be observed in such an optical pressure cell by sudden release of pressure.

3.3 A tour of physical processes accessible in single-component water ice: brittle fracture

Optical transparency also permits direct observation of the nucleation and growth of cracks at low pressures. Several examples are cited in Petrenko and Whitworth.[2]

3.3.1 *Lab observations of brittle fracture in clear ice.* This may be observed under compressive loading using a small arbor press by drilling a hole in the ice. Compressive loading of such samples produces tensile stresses that nucleate and grow tensile cracks in ice approximately parallel to the direction of loading. Such fractures after unloading can also slowly heal with time through vapor transport and crystallization driven by fracture-surface-energy reduction, essentially a surface-tension effect.

3.3.2 *Tensile fracture by thermal shock.* Ice may be also be cracked by sudden increases in temperature of ice surfaces. The resultant thermal expansion of the near-surface region of the ice sample puts the interior of the sample in tension and nucleates tensile fractures. Cold ice cubes, for example, fracture in such a manner in the ice-cube interior when they a placed in warm water. The geometry of the fractures depends on temperature distribution and hence on the thermal boundary conditions (specimen shape) as well as the presence of pre-existing flaws, such as air bubbles. Natural examples fractures produced by thermal-shock may be observed in alpine lake ice when cold lake ice is suddenly exposed to sunlight as the sun rises, producing loud, booming "ice quakes" and, over time, offsets in the lake-ice surface.

3.3.3 Applications of fracture mechanics to ice. There are many natural examples, such such as the fracture in lake ice and in ice shelves (ice sheets that extend over water), crevasse formation in glaciers, and fracture in sea ice and these may be used as natural examples in terrestrial ice.[2,15]

3.4 A tour of physical processes accessible in single-component water ice in experiments in a freezer: ductile flow and deformation-induced melting and refreezing

A number of important physical processes and laws may be potentially investigated in an ordinary freezer using pure ice:

3.4.1 Ductile flow of ice by slow creep. This is easily investigated by compressive loading of molded ice in a freezer using weights. Measuring the changes in sample length may be done using a dial gauge or calipers. By doing successive runs using different weights and at different temperatures, Glen's law for the rates of steady-state plastic flow of ice ($d\varepsilon/dt = B\,\sigma^n \exp(-Q/RT)$ where $d\varepsilon/dt$ is the rate of shortening strain, σ is the applied stress, R is the gas constant, and T is the temperature in Kelvins, and B, n and Q are material constants and R is the gas constant) may be reproduced.[16,17,18] This law has been successful in modeling the first-order internal flow of ice sheets and alpine glaciers and its success shows that flow law scales up from lab specimen with dimensions of about 1 cm to 10^7 cm (100 km) in these natural settings. High-temperature steady-state creep of all crystalline inorganic materials also obey this fundamental flow law (usually called by other names), including metals, ceramics, and minerals.

3.4.2 Dynamic recrystallization and associated flow softening. This phenomenon was first unequivocally demonstrated in deformed ice of glaciers and in lab experiments on ice in the 1950's. These profound changes in the grain size and textures of ice deformed at high temperatures may be demonstrated by sublimation etching of the grain boundaries of specimens deformed, as described above.

3.4.3 Deformation by intracrystalline slip and dynamic recrystallization. These are important mechanisms for the development of crystallographic preferred orientation (CPO) and property anisotropy. Slip bands associated with intracrystalline flow may easily be seen in optical reflection microscopy of previously flattened and polished surfaces of specimens that are subsequently deformed as described above. CPO developments are less easily demonstrated, because it required the making of thin sections of deformed ice and the use of a simple universal stage to determine the orientation of the crystallographic **c** axis.

3.4.4 Pressure melting (regelation). This is an important process in accommodating bed irregularities ('bumps") in the bed slip of glaciers. As has been demonstrated for more than a century, this process is easily shown by crossing a wire weighted at both ends through a block of clear ice and observing the slow melting of the ice below the wire and re-freezing of the resulting liquid above the wire.

4 MULTI-COMPONENT ICES

4.1 A tour of physical processes in multi-component ices: Melting and freezing at room pressure

The mixing of other soluble chemical components with water produces fascinating departures from the simple freezing behavior observed in pure water. A prominent natural example is in the freezing of seawater to form sea ice. A huge literature exists on this process and on the structure, mineralogy, and thermal evolution of sea ice. Similar effects of salts on melting of planetary ices have already been discussed. [Compositions of such multicomponent ices and liquids are now easily measured using modern digital refractometers (e.g., the Palm Abbe^{TM}) that require only a drop of liquid. Refractive index of such liquids generally varies linearly with solute concentration, so solute concentrations can be inferred from these index of refraction measurements.] In addition to solute depression of the melting temperature and the broad temperature interval over which melting occurs, a number of interesting phenomena occur in these multi-component systems:

4.1.1 Eutectic phase relations in water-salt systems These systems usually involve ice and salt hydrates.[19,20,21] Water-soluble salts typically show eutectic phase relations involving both water ice and salt hydrates. These phase relations are a consequence of the well-known property of ice of rejecting solutes during freezing and the property of salts of hydrating in the presence of water. In terrestrial and planetary systems, these salts are generally sulfates and chlorides, but also may include acids, such as sulfuric acid. A whole suite of different hydrate structures have been described from desert and polar environments in addition to those in very cold sea ice. Depending on the composition of the starting solution, X, in relation to the eutectic composition X_E, either ice or the hydrate will crystallize first and the solution will move toward X_E as crystallization proceeds during cooling. In theory, simultaneous crystallization of ice and the hydrate will occur at temperature T_E when the liquid reaches the eutectic composition X_E. In practice, the eutectic solution composition X_E may routinely be undercooled below T_E by many degrees and crystallization may take minutes to days to complete. Again, such undercooling may be arrested and crystallization may occur very rapidly if granular "seeds" of the eutectic crystallization product are sprinkled on the solution. The crystallized eutectic materials are fine-scaled, intimate intergrowths of ice and hydrates and are always white (compared to the translucent or clear ice or hydrate crystals) because they scatter light and hence are easy to recognize.

The NaCl-H$_2$O system is an excellent example of such a eutectic system. Its eutectic temperature, T_E is 252.4 K and composition X_E is 23.3 weight percent. The results of a freezing experiment are shown in Figure 1 that shows the progress of a freezing. A saturated salt solution was held at 247 K (below T_E) for about 1.5 hours before initial crystallization of large, clear hydrohalite began to occur at the bottom of the tube. Hydrohalite is denser than the remaining solution and hence crystallizes at the bottom (Fig. 1A and F). After 1.7 hours, crystallization of a eutectic intergrowth of ice and hydrohalite (the thermodynamically stable assemblage) began to occur at the top of the tube on the tube walls and grew downward and more slowly inward (Fig. 1A and E). The eutectic mixture is less dense than the solution and always nucleate at the top of the tube. This assemblage is always white because it is very fine grained (Fig. 1E inset). The eutectic intergrowth exhibits radial cellular growth from the original points of nucleation. Completion would have taken several more hours, but we interrupted this freezing history by carefully inserting a glass tube through the annulus of eutectic material without touching it (Fig. 1B). Nucleii immediately began to grow in the middle of the test tube and within a few minutes, the tube was filled with eutectic "colonies" that rapidly grew and slowly floated upward in the solution (Fig. 1C and D). Not only did the solution demonstrate considerable undercooling of T_E, and sinking and floating buoyancy of the crystallization products, but we interpret the last phase of the experiment as indicating that eutectic nuclei were also stratified by density to the top of the tube. Evidently the liquid entrained by insertion of the glass tube displaced some of these nuclei to the supercooled liquid below and these started to immediately grow.

Figure 2 Photographs documenting the freezing history of a saturated NaCl solution (A-D) at 247 K with closeups of eutectic ice (E) and hydrohalite intergrowths (E) at the top of the test tube. A cryogenic SEM micrograph of that assemblage is shown as an inset in Figure 1E. Figure 1 F is a close-up image of the large monoclinic hydrohalite crystals at the bottom of the tube in Figure 1A. Tube diameter is about 31 mm. See details in text. The images had variable backgrounds, hence the differences in the gray levels and textures.

4.1.2 Fractional crystallization during freezing. Such freezing phenomena occur when crystals and remaining melt of differing composition separate by gravity due to density differences between the solids and the remaining solution. This phenomenon is easily seen in the eutectic systems just discussed. The hydrates are always denser than the solution and hence nucleate and always grow near the bottoms of test tubes and dilute the remaining solution. On the ice sides of these eutectics, ice forms first and always nucleates and grows at the top of cooling test tubes. Eutectic intergrowths are somewhat less dense than the remaining solution and hence slowly float. Subsampling of the frozen products of these freezing experiments always shows such stratification with the bottoms of the samples always having a greater solute concentration than in the starting solution. As mentioned earlier, such component separation by the densities of the silicate crystallization products is important in creating much of the compositional and mineralogical diversity of terrestrial lavas and plutons on Earth.

4.1.3 Solute depression of the melting point can also occur due to the dissolution of gases in water. Ordinarily the solubility of ordinary gases is so small that this depression is not easily measurable. However, some gases, such as CO_2, have a far higher solubility in water and solute depression can be easily measured with a thermistor.

4.1.4 Clathrate hydrates. Chemical systems involving dissolved gases can also form *clathrate hydrates* at modest pressures and temperatures in a simple low-pressure optical cell. Hydrocarbon clathrate hydrates occur in marine and polar regions where hydrocarbons react with cold water. CO_2 hydrates have been reported in seafloor environments near cold CO_2 vents. Natural air hydrates occur in ice cores taken from polar ice sheets. Air is trapped and sealed in snow during compaction to form low-porosity ice.

4.1.5 Air precipitation during freezing of air-water solutions. This can be observed during the freezing of ordinary tap water to make ice cubes. Air-filled bubbles form after initial crystallization of clear ice on the cube periphery has enriched the concentration of air in the remaining water, raising that concentration to above its solubility limit in water and leading to air-bubble nucleation on the freezing front. Subsequent precipitation of air occurs at these same bubble sites, producing air tubes. These features may be seen in the last-crystallized cores of ice cubes where the trajectories of the bubble trails and air tubes track the direction of the ice freezing front.

4.1.6 Direct observation of pressure-release freezing of water-gas solutions. Pressure release boils out dissolved gas, such as CO_2, and raises the melting point over and above that due to pressure release not involving as. This may be directly observed during the pressure release of such solutions in optical cells.

4.1.7 Schlieren production and motions. This is a fascinating phenomenon that can easily be observed during ice-cube melting in brines or sugar water solutions. Thin-layer schlieren of dense solution represent cold, dense boundary layers that are shed from ice cubes due to their sinking buoyancy. Such observations are small-scale counterparts of thermohaline circulation in the oceans and the fluid counterpart of subduction of the cold, dense boundary layer of the Earth (the lithosphere) into hot, less-dense mantle below.

4.2 A tour of physical processes in multi-component ices: Mechanical properties

Water ice, being hydrogen bonded, is generally mechanically weaker than other compounds grown by freezing of simple inorganic aqueous solutions. Many ionic salts, such as soluble alkali halides and alkali and alkaline-earth sulfates, form complex salt-hydrate structures that are ionically or covalently bonded and hence are usually stronger than and structurally dissimilar from water ice and show little or no variability in composition due to component mixing with ice. As such, these solutions exhibit *eutectic* freezing behavior and intimate eutectic intergrowths of ice and hydrates, as explained above. Such composite materials involving ice high much higher fracture toughness than ice based on the difficulty of mechanically cleaving synthetic samples in my laboratory[19,20,21] and unpublished USGS Ice Physics Laboratory observations, 2002-2006) and the higher applied forces required to initiate and grow cracks (J. Pape, USGS Ice Physics Laboratory, unpublished data, 2006). This type of strengthening in the brittle regime of deformation is caused by the same mechanism that strengthens *pykrete*[22], a synthetic mixture of ice and sawdust: an intimate intergrowth of a material with ice makes it impossible for a fracture to grow along weaker grain boundaries or wholly within ice; fractures are therefore forced to propagate across the grains of the tougher material as well as ice. Some of the hydrate-ice systems discussed above are stronger in the ductile regime at higher pressures since the harder phase must also be deformed if it constitutes 50% or more of the eutectic mixture[23].

5 DEMONSTRATION EXPERIMENTS IN A FREEZER, THE DEVELOPMENT OF TEACHING PLANS FOR PRE-UNIVERSITY PUBLIC-SCHOOL EDUCATION, AND CONCLUSION

5.1 The technical challenges

The technical challenges of creating inexpensive and safe demonstration apparatus that make possible evocative observations that effectively teach the principles of materials science falls to scientists themselves. Meeting these challenges is not trivial and it may require innovative apparatus design and component selection. The responsibility of getting the principles and ideas across to students, on the other hand, falls to teachers in consultation with scientists.

Laboratory scientists are accustomed to using sophisticated and generally expensive apparatus that may not necessarily be robust enough to withstand the rough and tumble of a classroom environment. Also, instruments should be multi-purpose so that the costs of equipment can be kept modest. One of the desirable goals is to make visible the results of experiments in the large classroom (*i.e.,* seeing is believing). Technological advances have opened a number of new opportunities for such visualization:

5.1.1 Cheap and capable digital web-cameras and microscopes. These make possible surveillance of experiments and their display on computer monitors. Simple and intuitive analogue instruments, such as bulb thermometers and bourdon-tube pressure gauges, may also be monitored web cameras.

5.1.2 Programmable digital cameras. Such equipment also can be used for surveillance purposes and have the capability of time-lapse photography in order to monitor slow processes.

5.1.3. Simple digital data acquisition. For more advanced students, PC-based or PDA-based data acquisition and analysis of data are possible. This are skills that students in technical fields need to learn.

5.1.4 Optical cells. Another desirable capability is to achieve elevated fluid pressures in an optical-cell. Simple flat-plate designs exist using silica or sapphire windows and O-ring seals and have the potential of routinely achieving pressures to 100 MPa. A challenge would be to find ways of making these optical cells inexpensive and safe for use in the classroom.

5.2 Developing demonstration experiments and instructional materials

The teaching profession is becoming more open to nontraditional "hands-on" laboratory activities as supplements to more conventional classroom science instruction. Moreover there is growing interest in ices in nature, including interest in glaciers and sea ice as sensitive indicators of climate change and in planetary ices stimulated by missions to the outer planets and their moons. Such interest spans the educational system from grade school to advanced undergraduate instruction. Teaching the teachers is vital. This involves developing well-conceived teaching plans and other instructional materials that address science standards that are required in most public-school systems. "Hands-on" workshops and tutorials for teachers would also be an important part of preparing them to use freezer-based systems for scientific discovery in the classroom. Such activities are probably best developed by science-education professionals in consultation with research scientists who study ice in the lab and in nature.

5.3 Conclusions – What can ice scientists do?

Suites of demonstration experiments have been discussed in this paper involving lab investigations on the deformation, melting, and freezing of water ice and multi-component icy systems in a freezer. The results of these experiments can individually be interesting, engaging, thought provoking, and accessible in a simple deep (top-door) freezer. Taken together, it is suggested that they have the potential for becoming a key part a curriculum of experimental geoscience and material science in public schools and at the undergraduate level.

The Physics and Chemistry of Ice Conference (PCI) participants have several centuries of collective experience in conducting experiments on ice and icy systems as well having read a vast literature on such materials. This research community is invited to share their experiences and citations of related literature by contacting the author at: **skirby@usgs.gov** or mailing him at the U.S. Geological Survey, 345 Middlefield Road, Menlo Park, California 94025 USA or phoning him at +1 650 329-4847. I also suggest that this effort may be an appropriate activity for PCI, including teacher participation in a short course sponsored by PCI at future conferences, and encouraging PCI participants to write popular or semi-popular books on ice such as the one by Norikazu Maeno cited earlier.[1]

5.4 Additional reading

Additional reading on Earth's interior and dynamics may be found in references [4, 25, 26, 27, 28, and 29] and on icy planets and moons in the references [6, 12, 24, and 30] and on clathrate hydrates in [31].

Acknowledgements

I thank Christine McCarthy and Karen Rieck as NASA Undergraduate Interns for the enthusiasms that they showed in doing freezing experiments on salt solutions. I also thank Sergio H. Faria for his very helpful reviews of this paper.

References

1   N. Maeno, *Kohori no Kagaku (Ice Science)*, Revised 1st Edition (in Japanese), Hokkaido University Press, Sapporo, Japan, 2004, 234 p. (ISBN4-8329-7371-1).

2   V.F. Petrenko, and R.W. Whitworth, Physics of Ice, Oxford University Press, 1999, 373 p.
3   J.S. Kargel, Icarus, **94**, 1991, 368.
4   G. Schubert, D.L. Turcotte, and P. Olsen, *Mantle Convection in the Earth and Planets*, Cambridge University Press, 2001, 940 p.
5   M. Kivelson, F. Bagenal, W.S. Kurth, F. Neubuerger, C. Paranicas, and J. Sauer, "Magnetospheric Interactions with Satellites" in F. Bagenal, T.E. Dowling, and W.B. McKinnon, eds., *Jupiter – The Planet, Satellites, and Magnetosphere*, Cambridge University Press, 2004, pp. 513-536.
6   Greeley and 7 others in in F. Bagenal, T.E. Dowling, and W.B. McKinnon, eds., *Jupiter – The Planet, Satellites, and Magnetosphere*, Cambridge University Press, 2004, pp. 329-362.
7   T.B. McCord, G.B. Hansen, F.P. Fanale, R.W. Carlson, D.L. Matson, J.V. Johnson, W.D. Smythe, K.K. Crowley, P.D. Martin, A. Ocampo, C.A. Hibbitts, J.C. Granahan, and the NIMS Team, *Science*, 1996, **280**, 1242.
8   T.B. McCord, G.B. Hansen, D.L. Matson, T.V. Johnson, J.K. Crowley, F.P. Fanale, R.W. Carlson, W.D. Smythe, P.D. Martin, C.A. Hibbitts, J.C. Granahan, and A. Ocampo, A., *J. Geophys. Res*, 1999, **104**(E5), 11827.
9   M.Y. Zolotov and E.L. Shock, Journal of Geophysical Research, 2001, **106**(E12), 32815.
10 J.S. Kargel, J.Z. Kaye, J.W. Head, G.M. Marion R. Sassen, J. K. Crowley, O.P. Ballesteros, S.A. Grant D. L. Hogenboom, Icarus, 2000, **148**(1), 226-265.
11 J.B. Dalton, O. Prieto-Ballesteros, J.S. Kargel, C.S. Jamieson, J. Jolivet, and R. Quinn, *Icarus*, 2005, **177**, 472.
12 R. Greenberg, *Europa – The Ocean Moon and Search for an Alien Biosphere*, Springer and Praxis Publishing, 2005, 380 p.
13 J. Tyndall, *The Forms of Water in Clouds and Rivers, Ice and Glaciers*, Henry S. King & Company, 1872, 192 p.
14 P.A. Shumskii, *Principles of Structural Glaciology*, English translation from the Russian by David Kraus, Dover Publications, 1964, 497 p.
15 W.S.B. Paterson, *The Physics of Glaciers*, 2nd edn., Pergamon Press, 1981, 380 p.
16 J.W. Glen, Journal of Glaciology, 1952, **2**, 111.
17 J.W. Glen, Nature, 1953, **172**, 721.
18 J.W. Glen, Proc. Royal Society Ser. A, 1955, **228**, 519.
19 C. McCarthy, R.F. Cooper, S.H. Kirby, and W.B. Durham, *Lunar and Planetary Science Conference* XXXVII, 2006a, 2467.
20 C. McCarthy, R.F. Cooper, S.H. Kirby, and W.B. Durham, R.D. Rieck, and L.A. Stern, *American Mineralogist* (Submitted July 2006).
21 C. McCarthy, K.D. Rieck, S.H. Kirby, W.B. Durham, L.A. Stern, and R.F. Cooper, Royal Society of Chemistry UK, 2007, Proceedings of the 11th Conference on the Physics and Chemistry of Ice, Bremerhaven, 2007, (this volume).
22 M. F. Perutz, "A Description of the Iceberg Aircraft Carrier and the Bearing of the Mechanical Properties of Frozen Wood Pulp upon Some Problems of Glacier Flow", *The Journal of Glaciology*, 1948, **1** (3): 95.
23 W.B.Durham, L.A. Stern., K. Tomoaki, and S.H.Kirby, Journal of Geophysical Research Planets, 2005, 110, E12010, doi:10.1029/2005JE002475.
24 F. Bagenal, T.E. Dowling, and W.B. McKinnon, eds., *Jupiter – The Planet, Satellites, and Magnetosphere*, Cambridge, 2004, 719 p.
25 G.F. Davies, Dynamic Earth: Plates, Plumes, and Mantle Convection, Cambridge 1999, 458 p.

26 P.C. Hess, *Origins of Igneous Rocks*, Harvard University Press, 1989, 336 p.
27 J.I. Lunine, *Earth – Evolution of a Habitable World*, Cambridge University Press, 1999, 319 p.
28 F. Press and R. Siever, *Understanding Earth*, 3rd edn., Text and CD, 1994, 682 p.
29 M.P. Ryan, ed., *Magmatic Systems*, Academic Press, 1994, 398 p.
30 P.D. Nicholson, ed., "Special Issue on Europa", Icarus (Elsevier), 2005, 177, pp. 293-576.
30 P.R. Weissman, L.-A. McFadden, and T.V. Johnson, *Encyclopedia of the Solar System*, Academic Press, 992 p.
31 M.D. Max, ed., *Natural Gas Hydrate in Oceanic and Permafrost Environments*, Kluwer Dordrecht, 2000, 415 p.

Appendix

Glossary of technical terms used in this paper

[c] axis	A unique direction in an ice crystal that shows hexagonal rotational symmetry (and leads to identical properties along six hexagonal directions perpendicular to the [c]-axis and the familiar hexagonal symmetry of snowflakes).
chondritic meteorite	A stony meteorite, i.e., one that is composed primarily of silicate minerals rather than iron-nickel metal alloy of metallic meteorites (sometimes called "irons").
clathrate hydrate	A class of hydrates (compounds containing water) with crystal structures composed of a molecular-water framework that encloses (or *enclathrates*) other molecules, such as gases. An example is the compound methane hydrate, $CH_4 \cdot nH_2O$ (where n is about 6) that occurs in abundance on Earth in marine sediments and under arctic permafrost.
creep	The slow ductile (non-brittle) deformation of a material under stress. Creep in ice in glaciers occurs by *intracrystalline slip* (see above).
crystallographic preferred orientation (CPO)	A preferred orientation of crystallographic directions in the component crystals of an aggregate that is acquired during freezing or during deformation. For example, the crystallographic [c] direction in ice often shows a preferred orientation with respect to the vertical in glaciers.
dynamic recrystallization	A phenomenon that occurs when a compound is deformed at high temperature where new grains can nucleate and grow at the expense of pre-existing grains. Ice extensively recrystallizes in terrestrial glaciers under stresses produced by gravity.
elastic moduli	The ratio of an applied stress to the resultant elastic strain.

	If the stress and strain are both shear, that ratio is the shear Modulus, μ, and if they are pressure and volume strain, that ratio is called the bulk modulus, K.
eutectic intergrowth	An intimate intergrowth of two different compounds or crystal structures as a consequence of their simultaneous crystallization from a melt.
geological terrain	A geographic region that is characterized by a rock type or a suite of rock types that have a common origin (e.g., glacial terrain).
intracrystalline slip	The block-like slip produced by the shear or offset of a crystal between layers of atoms that maintains that the same crystal structure and cohesion before and after slip occurs. Metals commonly plastically deform by intra-crystalline slip.
mafic melt	A silicate melt that has the composition of basalt, a common igneous rock.
multi-component ice	An icy compound composed of one or more chemical components besides water.
non-ice molecular species	A chemical component of an icy compound that is not molecular (*i.e.*, it is not composed of discrete molecules, but an ensemble of atoms). An example is the ionic common table salt, NaCl, in the compound $NaCl_2\,H_2O$, the mineral hydrohalite.
pluton	A body of igneous rock that has not reached Earth's surface and erupted as a lava or fragmental ejecta ("ash").
polymorph of ice	A crystalline form of molecular water ice. There are at least eleven polymorphs of ice that can exist under certain pressure-temperature conditions or can be made under different pressure-temperature-time pathways.
regelation	Melting under pressure in ice that occurs because pressure reduces the melting temperature of ice. Regelation can occur at the base of moving glaciers due to the pressures produced at irregularities of the rocky bed on which the glacier moves.
rheology	The mathematical law that governs how stress, temperature, grain size affect the rate of deformation in a material.
slip band	An offset of a pre-existing surface of a material produced by intracrystalline slip. Such bands may be visible if the amount of slip and hence the offset is large.

slip dislocation | A closed-loop line crystal defect, the movement of which on a slip plane produces a shear offset (intracrystalline slip).

solar relectance spectrum | The intensity of visible, infrared, and ultraviolet light produced by the sun and reflected by a solar-system body as a function of wavelength. Such spectra can be characteristic of the chemical compounds that occur on the surface of that body as well as surface temperature.

tensile stress | A stress that tends to pull a material apart.

texture | First meaning: The geometrical relationships between crystals (and crystal shapes) and pores in a porous aggregate.
Second meaning: The nature of a *crystallographic preferred orientation (CPO)* in a material.

thermohaline circulation | Circulation of seawater in the ocean basins on Earth caused by density differences due to differences in water temperature and/or differences in salt content.

water of hydration | Water that incorporated in the crystal structure of a compound as discrete and identifiable molecules.

CLASSIFICATION OF LOW-ENERGY CONFIGURATIONS OF POLYHEDRAL WATER CLUSTERS FROM CUBE UP TO BACKMINSTERFULLERENE

M.V. Kirov

Institute of the Earth Cryosphere SB RAS, 625000, Tyumen, Russia. E-mail: kirov@ikz.ru

1 INTRODUCTION

Polyhedral water clusters (PWCs) possess a large variety of features that make them extremely convenient and attractive for theoretical analysis of unusual properties of water and ice.[1] One important feature of PWCs is the essential dependence of their properties on the proton subsystem structure. The extensive theoretical analysis of PWCs energy profile was provided by Singer and co-authors[2,3] and also by Anick.[4-6] In a number of studies, statistical correlation between the stabilization energy of PWCs and structural characteristics was investigated,[7-9] using modern quantum-chemical methods.

In this article an alternate approach based on the analysis of specific features of inter-molecular interactions is presented. The strict tetrahedral alignment of H-bonds in the majority of PWCs limits severely the polarizability of their molecules and, thus, increases the significance "additive cooperativity".[10] At the same time, rigid molecular potentials are growing in importance. In order to study the energetics of PWCs we developed two discrete models of molecular interaction. First of them is the topological nearest-neighbor model of strong and week H-bonds.[11] The second model allows for consideration the Coulomb interaction between the second- and third-neighbor molecules.[12] The main aim of this paper is to demonstrate the effectiveness of the simple discrete models of inter-molecular interactions and exact combinatorial optimization methods for classification of low-energy PWCs configurations.

2 METHOD AND RESULTS

2.1 Residual Entropy of Polyhedral Water Clusters

The total number of various proton configurations in ice-like systems is rather large. By analogy with well-known Pauling's formula[13] we obtained the following expression for residual entropy of PWCs:[14]

$$S_0 = k\ln(M) = k\ln\left(\left(\frac{6}{8}\right)^N \cdot 2^{3N/2}\right) = Nk\ln(3/\sqrt{2}) = 0.752Nk, \qquad (1)$$

where M is the total number of configurations satisfying Bernal–Fowler's rules, N the number of molecules and k the Boltzmann constant. We will prove below that formula (1) can be used as a reasonable estimate of the total number of the Bernal–Fowler configurations M. For prismoidal and fullerene-like clusters, the exact values of the numbers have been obtained using the transfer-matrix method.[15] The estimate of residual entropy (1) is suitable for ice monolayer too.[16,17] It is based on an assumption that all proton configurations have the same energy. However, there are considerable energy differences between the various proton arrangements.[2,3]

2.2 Topological Strong and Weak H-bond (SWB) Model

In earlier works, computer simulations of PWCs with ST2 potential[18] revealed that the cluster energy and some local energetic characteristics are highly dependent on the actual arrangement of protons; and that the distribution of H-bonds according to their energy is essentially bimodal.[14] This is caused by energy non-equivalence of H-bonds with different conformational orientation. Within the framework of the approximate additive model of molecular interaction determined by a rigid five-site ST2 and TIP5P potentials,[19] the bimodality is caused by the interaction of non-H-bonded effective positive and negative charges participating in the formation of adjacent H-bonds (Fig. 1, at the left). If the positive and negative charges on the adjacent bonds of H-bonded molecules line up opposite to each other, then the interaction energy of H-bonded dimer will be much higher.[14] One of the three conformational orientations of the mirror-symmetrical (staggered) H-bond corresponds to a stronger *trans*-conformation whereas two remaining orientations correspond to a weaker *cis*-conformation. From five types of H-bonds on polyhedral surface (Fig. 1, at the right) only two configurations are *trans*-conformation.[11] The classification of H-bond types is the base of the topological SWB model.

Certainly, the discrete SWB model is rather crude. It does not take into account the long-range nature of Coulomb interaction, although it is of great significance for aqueous systems. Moreover, this model does not consider molecules' polarizability and other effects related to the non-additivity of intermolecular interaction.[20] Therefore it is quite surprising to observe a close correlation between the theoretical predictions based on the SWB model and contemporary quantum-chemical calculations. Some of the coincidences are listed below.

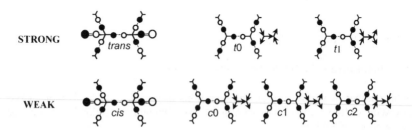

Figure 1 *At the left side: H-bonded dimers with their 3-dimensional surroundings according to ST2 and TIP5P models. At the right side: Types of H-bonds on the polyhedron surface; the effective charges of outward-directed non-realized (dangling) bonds are not shown.*

First of all, the most stable configurations of planar water cycles[20] contain the maximum number of strong H-bonds.[12] Secondly, according to statistical analysis of quantum chemical calculations, many of the lowest energy configurations of PWCs[2,4,5] again contain the maximum numbers of strong H-bonds.[21] Thirdly, enhanced stability of optimal configurations in even PWCs (with square and hexagonal faces)[5] is in complete agreement with our predictions.[11] In this case all H-bonds are strong. At last, global minimum energy structures for $(H_2O)_N$, $N = 12, 14, 16$[22] with modern TTM2-F potential,[23] are in agreement with the SWB model too; they have the maximum number of strong H-bonds.[21] This is in spite of the fact that some of molecules take part in four H-bonds. The main reasons of high predictability of the SWB model are discussed in the following section.

2.3 Strong and Weak Effective H-bond (SWEB) Model

Singer and co-authors found that the number of nearest neighbor double-acceptors ($c1$-bonds) was the most important factor determining the energy of PWCs.[2] It was previously noted by Smith and Dang[24]. On the other hand, Anick shown that the number of only $t1$-bonds very closely correlates with energy of PWCs.[4,5] As well, Anick derived the following formula which resolves the contradiction:

$$c1 + t1 = N / 2, \tag{2}$$

where N is the number of molecules. Furthermore, he proved that $c1 = c2$.[4] Thus, the most stable configurations have the maximum number of $t1$-bonds and the minimum number of $c1$- and $c2$-bonds. In the framework of statistical approach,[2-6] these two statements are absolutely equivalent. Our new discrete model enables to separate the cause from the effect.

Some explicit qualitative regularities that determine the PWCs energy can be seen in Fig. 1. First of all, it is noteworthy that for $t1$-bond the arrangement of effective electric charges is extremely favorable in energy. In this case both the Coulomb interaction of H-bonded molecules (strong bond) and Coulomb interaction of second neighbors with participation of dimer composing molecules are energy-favored. In second-neighbor interaction only those pairs where one of molecules forms a particular H-bond have been taken into account. The interaction of third neighbors also appears to be energy-favored. This simple qualitative analyses indicates a particular energetic preference of $t1$-bond.

A quantitative analysis of relationship between the local structure and local energetics of PWCs is based on a special decomposition of stabilization energy within the framework of additive models of interaction. In this approximation, the stabilization energy of PWCs is equal to the sum of all pair interactions. We define the effective pair interaction in PWCs (effective energy of H-bond) as the following:[12]

$$U_{eff} = U_1 + U_2 / 2 + U_3 \tag{3}$$

Here U_1 is the potential energy of two molecules forming the specified H-bond; U_2 is the energy of interaction of second neighbors, one of which participates in formation of this H-bond. The weighting factor $1/2$ is introduced to avoid counting the second neighbor interactions twice when calculating the total energy through the sum of pair effective interactions; U_3 is the energy of third neighbor interactions. We directly take into consideration the interaction of diagonal molecules only. In this case general Coulomb interaction of molecules is obviously evaluated in accordance with signs of the most close

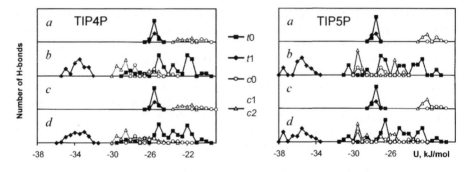

Figure 2   *The distribution of different types of H-bonds in dodecahedral cluster for optimal configurations according to SWB model. A local geometry optimization performed using TIP4P and TIP5P potentials: (a, b) – 21 strong H-bonds, (c, d) – 20 strong H-bonds; (a, c) – Energy values of H-bonds, as such; (b, d) – Effective energy values of H-bonds.*

charges. An interaction of lower and upper molecules (Fig. 1) in square and pentagonal cycles has been already taken into account in U_1 and U_2. Note that the majority of cycles in unstrained PWCs is pentagonal. It was shown that the unaccounted third neighbor interaction in hexagonal cycles additionally increases the accuracy of the model.[21]

Fig. 2 represents the distributions of various H-bond types according to their energy and effective bond energy. It was shown that the distributions for $c1$ and $c2$-bonds are very similar. The total distribution of H-bonds over their energy is bimodal for both TIP4P[25] and TIP5P potentials (Fig. 2, a, c). In our works, the bimodality lies at the basis of defining strong and weak hydrogen bonds[11,26], or more correctly the strong and weak nearest-neighbor pair interactions. At the same time, the effective H-bonds energy distribution (Fig. 2, b, d) is bimodal too. This allows us to conclude that just the $t1$-bonds determine the stability of PWCs. Note, that formula (2) is a consequence of purely topological correlations caused by Bernal-Fowler rules. From physical point of view, the number $c1$-bonds is a destabilization factor as well (weak bond). However, the number of $t1$-bonds is preferable taking into account the second and third neighbor interaction.

The strong and weak effective H-bond (SWEB) model do not takes into account the energy differences within each H-bond types. An evident advantage of $t1$-bonds form the basis of the simple discrete model. Though decomposition (3) allows to take into account the differences caused by different orientations of the surrounding molecules. The more complex model for gas hydrate frameworks suggested in our companion article.[27]

2.4 Classification of Low-Energy Configurations

Exact results of combinatorial optimization of PWCs structure are presented in the Table. The data were obtained by different methods. For small and middle-size clusters ($N \leq 28$), we used a complete enumeration method.[11,26] For large clusters ($N = 36, 60$), the data were obtained using symbolic calculations and *max-plus* algebra.[28] Characteristics of two lowest energy levels for the SWEB model are presented. For water fullerene ($N = 60$) on the base of SWEB model, the data obtained only for ground state level without isomorphism analysis. Values in italics correspond to ground sates of the SWB model. For example, the

Table *Combinatorial characteristics of polyhedral water clusters*

PCW	M, S_0	k	N_0	n	m	A
C(4^6)	450	2	216	6	6	6
8	0.7637	4	18	2	2	2
		12	6	1	1	1
D(5^{12})	3600000	6	184320	1541	803	65
20	0.7548	7	12720	106	64	22
		21	1440	12	9	6
D˙($4^3 5^6 6^3$)	3660000	7	24816	2068	1108	148
20	0.7556	8	1872	171	109	47
		24	132	14	11	8
T($5^{12} 6^2$)	73041408	8	354408	14795	7492	189
24	0.7544	9	7464	321	176	31
		27	200	13	9	5
T˙($4^6 6^8$)	75400162	10	10488	224	152	80
24	0.7558	12	338	17	17	17
		36	6	2	2	2
H($5^{12} 6^4$)	1482080256	10	434376	18099	9089	79
28	0.7542	11	1152	48	27	6
		31	24	1	1	1
E($5^{12} 6^8$)	610754032000	14	450984	18883	9742	601
36	0.7538	15	5184	228	130	32
		42	1504	77	66	55
F($5^{12} 6^{20}$)	4279050327→	24	360788736	>3006573	-	-
60	→8182400000					
	0.7534	78	32000	280	142	4

M is the total number of configurations; S_0 is the non-dimensional residual entropy (formula 1); k the number of energetically preferable bonds (energy level); n_0 and n the total number and the number of symmetry-distinct configurations for a given energy level; m the number of symmetry-distinct configurations with allowing of antisymmetry; a the number of antisymmetrical configurations. Italic data are calculated based on SWB model and the others calculated using SWEB model.

number of symmetry-distinct SWEB optimal configurations in cluster D is equal to 106. All of these structures have a maximum number of $t1$-bonds 7. Among them there are 22 antisymmetric configurations and 42 pairs of ordinary configurations with opposite H-bond directions. Therefore, the number of configurations allowing for antisymmetry is 64.

First of all, one can see that exact numbers of Bernal-Fowler configurations M give approximately the same residual entropy close to 0.7520 (1). On the other hand, the total numbers of configurations (M, n_0) give a good approximation for the number of symmetry-distinct structures. For example, in cluster D there are 30026 symmetry-distinct structures.[2] It is clear that most of them are asymmetrical. This gives the following estimate: 3600000/120 = 30000, where 120 is the order of symmetry group. At the same time, the exact numbers M and n_0 can give an absolute guarantee of completeness of the calculated configuration classes. It is especially important for large clusters. A set of symmetry-distinct optimal configurations can be obtained by simulated annealing method.[29] The total numbers n_0 is equal to the sum of all equivalent configurations for each of symmetry-distinct structures. The total number of equivalent configurations is determined by the symmetry of the configurations. This is not a particular problem. We can compare the total numbers n_0 with values, which are calculated by linear matrix methods. This enumerating techniques is alternate to Polya theorem and to cyclic group index.[2] This method allows to calculate the total number of symmetry-distinct structures as well as the number of discrete optimized structures.

It was established that all SWB optimal configurations (maximum number of strong H-bonds) are contained in more SWEB optimal classes, with the exception of cluster E.[30] The main reasons for this are: 1) As noted by Anick, there is evident positive correlation between the numbers of strong H-bond (*t0*+ *t1*) and the number of *t1*-bonds.[6] 2) There is a direct correlation between the numbers of *t0* and *t1*-bonds. In clusters *D* the correlation coefficient is 0.45.[12] 3) An interaction of opposite molecules in hexagonal cycles both for SWB and SWEB optimal configurations usually are stabilizing, i.e. the unaccounted factor raises the accuracy of the discrete models.[21] Two last reasons are the manifestation of purely topological correlations caused by Bernal Fowler rules in PWCs.

A good upper estimate of the maximum number of *t1*-bonds was obtained by Anick:[4]

$$t1 \le N/2 - f_{odd}/4 \tag{4}$$

Here f_{odd} is the number faces with odd numbers of sides. Our results (Table) indicate that the number of *t1*-bonds reaches the upper bound in all clusters with the exception of cluster *F*. It may be related with relatively large quantity of hexagonal faces. For cluster *F* the upper estimate of *t1*-bonds is 27.

An important result of the present study is the splitting of the energy between isomers belonging to the two lowest energy levels in according to SWEB model. It is especially visible when one takes into account the magnitude of total dipole moment (Fig. 3). This fact enhances a significance of the single-factor optimization. A completeness of classes of discrete optimized configurations is of fundamental importance for this classification. For the TTM2-F potential, a subset of second level configurations revert to non-polyhedral structures upon geometry optimization. Additional a priori ranking of configurations of the same energy level was carried out with taking into account the magnitude of the total dipole moment in idealized polyhedral geometry. Dipole moment was calculated as a vector sum of H-bond dipoles. Relatively isolated groups of configurations in Fig. 3 correspond approximately to the same value of the total dipole moment. Optimal configurations with the smallest dipole moment are of most importance to the energy optimization. It is interesting to note that there are no such structures in clusters H and E.

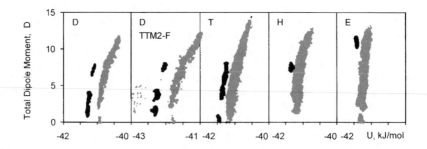

Figure 3   *Potential energies of PWCs versus total dipole moments for complete classes of discrete optimized configurations. Mainly the values are calculated using TIP4P potential.*

2.5 Supramolecular Asymmetry of Water Systems

A reversal of direction of all H-bonds can be considered as a special operation of generalized symmetry (antisymmetry)[26]. Energies and total dipole moments of the "anti-configurations" are approximately equal.[12] At the same time, the appreciable energy distinction is of fundamental importance. It was shown that all proton configurations of smallest water associates (isolated molecule, *trans*- and *cis*-configurations of H-bonded dimer, cyclic trimer and cyclic tetramer) are antisymmetrical;[31,32] all "anti-configurations" are related by ordinary symmetry operations. Energetically non-equivalent antipodal configurations emerge only in cyclic pentamer. Thus we can say about the presence of "supramolecular asymmetry" of water systems. Note that a majority of 14 symmetry-distinct configurations of cubic cluster are antisymmetrical too. In this cluster, there are only two non-antisymmetric configurations related by mirror symmetry.[32] The discrete models discussed above can not evaluate energy difference of antipodal configurations. However, they are quite suit for studying the dependence between the energy and antisymmetry statistics of regular ice-like systems.

3 CONCLUSION

The comparison with results of high level quantum-chemical calculations proves the utility of the simple discrete models of molecular interaction for predicting the most stable topologies of water cycles and PWCs. Based on these discrete models an effective enumerating techniques was developed for hierarchical classification of proton configurations. In spite of the fact that PWCs are very complex systems with complicated interactions, the discrete models of inter-molecular interaction help us "to see the wood for the trees" (Fig. 3).

Acknowledgements

The author thanks S.S. Xantheas and G.S. Fanourgakis for useful discussion and the examination of the SWEB model for pentagonal dodecahedron cluster. This work was supported by Russian Foundation for Basic Research (Grant 06-03-32321).

References

1   D. Eisenberg and W. Kauzmann, *The Structure and Properties of Water*, Clerendon Press, Oxford, 1969.
2   S. McDonald, L. Ojamäe and S.J. Singer, *J. Phys. Chem. A*, 1998, **102**, 2824.
3   J.-L. Kuo, C.V. Ciobanu, L. Ojamäe, I. Shavitt and S.J. Singer, *J. Chem. Phys.*, 2003, **118**, 3583.
4   D.J. Anick, *J. Mol. Struct. (Theochem)*, 2002, **587**, 87.
5   D.J. Anick, *J. Mol. Struct. (Theochem)*, 2002, **587**, 97.
6   D.J. Anick, *J. Chem. Phys.*, 2003, **119**, 12442.
7   S.D. Belair and J.S. Francisco, *Phys. Rev. A*, 2003, **67**, 063206.
8   V. Chihaia, S. Adams and W.F. Kuhs, *Chem. Phus.*, 2004, **297**, 271.
9   A. Lenz and L. Ojamäe, *Phys. Chem. Chem. Phys.*, 2005,7, 1905.
10   J.J. Dannenberg, *J. Mol. Struct.*, 2002, **615**, 219.
11   M.V. Kirov, *J. Struct. Chem.*, 1996, **37**, 84.
12   M.V. Kirov, *J. Struct. Chem.*, 2005, **46**, S188.

13   L. Pauling, *J. Am. Chem. Soc.*, 1935, **57**, 2680.
14   M.V. Kirov, *J. Struct. Chem.*, 1993, **34**, 557.
15   M.V. Kirov, *J. Struct. Chem.*, 1994, **35**, 126.
16   M.V. Kirov, *J. Struct. Chem.*, 1996, **37**, 920.
17   P.J. Feibelman and A. Alavi, *J. Phys. Chem. B*, 2004, **108**, 14362.
18   F.H. Stillinger, A. Rahman, *J. Chem. Phys.*, 1974, **60**, 1545.
19   M.W. Mahoney, W. L. Jorgensen, *J. Chem. Phys.*, 2000, **112**, 8910.
20   S.S. Xantheas, *Chem. Phys.*, 258 (2000) 225.
21   M.V. Kirov, *J. Struct. Chem.*, 2006, **47**, 701.
22   B. Hartke, *Phys. Chem. Chem. Phys.*, 2003, **5**, 275.
23   C.J. Burnham and S.S. Xantheas, *J. Chem. Phys.*, 2002, **116**, 5115.
24   D.E. Smith and L.X. Dang, *J.Chem Phys.*, 1994, **101**, 7873.
25   M.W. Mahoney, W. L. Jorgensen, *J. Chem. Phys.*, 2000, **112**, 8910.
26   M.V. Kirov, *J. Struct. Chem.*, 2002, **43**, 790.
27   M.V. Kirov, *Physics and Chemistry of Ice*, Royal Society of Chemistry, Cambridge, 2007, p.
28   M.V. Kirov, *Preprint deposited at VINITI*, 2005, N 982-B2005, [in Russian].
29   M.V. Kirov, *J. Struct. Chem.*, 2003, **44**, 420.
30   M.V. Kirov, *J. Struct. Chem.*, 2006, **47**, 708.
31   M.V. Kirov, *Preprint deposited at VINITI*, 2005, N 983-B2005, [in Russian].
32   M.V. Kirov (in press).

ENERGY OPTIMIZATION OF GAS HYDRATE FRAMEWORKS ON THE BASIS OF DISCRETE MODELS OF INTER-MOLECULAR INTERACTIONS

M.V. Kirov

Institute of the Earth Cryosphere SB RAS, 625000, Tyumen, Russia. E-mail: kirov@ikz.ru

1   INTRODUCTION

Gas hydrates are inclusion compounds formed from water and gas molecules. Guest molecules are distributed into cages of water framework. Although empty hydrate lattice is unstable under ambient conditions, the water-water interaction gives the main contribution to the total binding energy of gas hydrates.[1] An important feature of ice and clathrate hydrates is a disorder in the proton positions. However, there are energy differences between various proton arrangements. The problem of proton ordering in ice has been the subject of much theoretical investigation.[2,3] From the structural point, gas hydrate framework is composed entirely of eclipsed arrangement of hydrogen bonding. In this regard, gas hydrate frameworks are more structurally homogeneous than hexagonal ice.

Earlier, we have studied proton ordering in gas hydrate frameworks on the base of the strong and weak H-bond (SWB) model.[4] We analyzed isolated small and large gas hydrate structures using the cluster variation method. Now we present some results in order to evaluate the role of the size of the model cell and the effect of periodic boundary conditions. Furthermore, we present a new discrete model of inter-molecular interaction that take into consideration the interaction between the second- and third-neighbor molecules. This model is based on a special energy decomposition procedure for bulk frameworks. New discrete model is a generalization of the strong and weak effective H-bond (SWEB) model for polyhedral water clusters[5,6] to bulk gas hydrate frameworks. But, there is a radical departure between the two models. The new model is numerical, whereas SWEB model is parameter-free and purely topological.

2 METHOD AND RESULTS

2.1 Computational Methods

We developed an algorithm for fast generating proton configurations in a such way that the Bernal-Fowler rules (ice rules) are totally satisfied everywhere in the infinite framework created by periodic boundary conditions. This algorithm is intended for two types of structure: structure I (sI) and hexagonal structure (sH). It allows to change the size of model cell. The set of Bernal-Fowler configurations can be used for statistical analysis and for screening of proton configurations using standard computer programs.

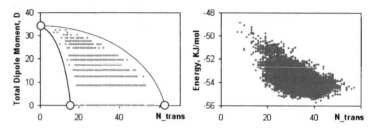

Figure 1   *The left plot represent total dipole moment versus a number of strong H-bonds for minimal unit cell of structure I.   The right diagram shows the energy distribution of local optimized configurations (TIP4P potential).*

A procedure of Monte Carlo simulation annealing method was developed for combinatorial structure optimization on the base of discrete models of inter-molecular interactions. Note that ice-like systems are strongly correlated. It is rather difficult to construct and optimize H-bond topology in accordance with Bernal-Fowler rules. In this regard, periodic boundary conditions are an additional problem. Simulated annealing is an excellent method for the solution of the difficult combinatorial problems. It allows for the generating proton configurations with zero total dipole moment. In this case, total dipole moment is estimated as a sum of all H-bond dipoles. Generally speaking, any single- and multi-factor discrete optimization can be carried out.

2.2 Combinatorial Optimization Using the SWB Model

Earlier, we have studied the proton ordering of gas hydrate frameworks on the base of SWB model[4]. According to the model, the energy of water framework is defined by the number of strong H-bonds corresponding to *trans*-configurations of H-bonding dimers. Note that it is more correctly to use the terms "strong and weak nearest-neighbour pair interactions" in this context. It was established that according to the model, gas hydrate frameworks are frustrated systems. It is not possible to build a clathrate structure with only strong H-bonds. In any pentagonal ring at least one of H-bonds is weak. We estimated the minimum number of strong H-bonds as zero, the average as 34% and the maximum close to 70%. In the earlier work we do not used the periodic boundary conditions. Here we present the results of combinatorial optimization under periodic boundary conditions.

It is well known that the sI is a cubic structure with the space group *Pm3n* and a lattice constant of about 12 Å. There are two types of cages in the structure. The 12-sided cavity has twelve pentagonal faces (5^{12}) and is a pentagonal dodecahedron. The 14-sided cavity is a tetrakaidecahedron consisting of 12 pentagonal and 2 hexagonal faces ($5^{12}6^{2}$). Unit cell of the structure contains 46 water molecules, which form 92 H-bonds. First of all, we present results for the minimal unit cell. Fig. 1 shows the correlation between the number of *trans*-configurations of H-bonded dimers (*nt*) and the total dipole moment (TDM). As pointed out above, TDM is estimated as a vector sum of all H-bond dipole moments according to idealized geometry of gas hydrate framework. Extreme points of the distribution are of the most interest. In Fig. 1 the points are indicated by circles. At the same time, search of these configurations is a challenging task. It was shown that the upper bound of *nt* is 62 (67.4%). There are five such configurations. All of them have zero TDM. There exist a unique "weak" configuration *nt* = 0. Note that this proves the presence of "weak" configurations in a cell of any size. TDM of this configuration is great. At the same time, configurations with small (near zero) TDM are of most interest. We have established that among zero TDM configurations (D0-configurations) the low bound of nt is 13 (14.1%). There are four such configurations.

Similar results were obtained for sH. The type sH is a hexagonal structure with the space group *P6/mmm*. The lattice constants are about $a=12$ Å and $c=10$ Å. It is formed by three different types of cages: three pentagonal dodecahedra (5^{12}), one 20-sided cavity, the icosahedron ($5^{12}6^8$), consisting of 12 pentagonal and 8 hexagonal faces, and two 12-sided irregular dodecahedra ($4^35^66^3$). Unit cell of the structure contains 34 water molecules, which form 68 H-bonds. Under periodic boundary condition, the upper bound of nt is 46 (67.6%). There are seven such configurations. All of them have zero TDM in idealized structure. For D0-configurations, the low bound of nt is 10 (14.7%). We have found only one such configuration. It can be seen that the structure H has approximately the same characteristics of optimized structures.

An effect of the size of the model cell on characteristics of optimized structures was also studied. We have carried out an optimization of large cubic unit cell sI that contains 368 water molecules. In this case, the upper bound of nt is 528 (71,7%). The result agrees closely with our previous results for large multi-cage fragments. We have found one D0-configuration that contains only 15 (2.0%) *trans*-configurations per unit cell. Thus the size of the model cell can have a pronounced effect on the structure and properties of ground state configurations. The limitations caused by periodic boundary conditions are more important in the minimal unit cells. It is clear that our results are approximate because they are based on simulated annealing method. At the same time, it is necessary to stress that coincidentally with extreme configurations of the minimal unit cells, we have found all equivalent structures for each mentioned configurations. The equivalent structures are related with each other by symmetry operations.

The results of local optimization for minimal unit cell sI using TIP4P potential are presented in Fig. 1 (at the right). For simplicity we used TIP4P potential and minimum image method with a cutoff distance of 7.5 A for gas hydrate structure I. One can see an obvious correlation between the energy and the number of strong H-bons nt. But the correlation mainly is due to the considered correlation between TDM and nt. It was shown that for the set of D0-configurations, the correlation is rather weak. Indeed, one can see (Fig. 1) that SWB optimal configurations have not the lowest energy. The situation is totally different from the one that exists in isolated polyhedral water clusters. Recall that all SWB optimal configurations of polyhedral clusters are very stable.[6]

2.3 Energy Decomposition Analysis of Inter-Molecular Interaction

Recently we have proposed a decomposition scheme of inter-molecular interaction energy in polyhedral water clusters.[5,6] Central to the scheme is the concept of effective pair interaction energy (effective H-bond energy). In additive approximation, the interaction energy of any molecular system is equal to the sum of all pair interactions. The effective energy of pair interaction (effective energy of H-bond) is

$$U_{eff} = U_1 + U_2/2 + U_3 \qquad (1)$$

Here U_1 is the potential energy of two molecules forming the specified H-bond; U_2 is the energy of interaction of second neighbors, one of which participates in formation of this H-bond. The weighting factor $1/2$ is introduced to avoid counting the second neighbor interactions twice when calculating the total energy through the sum of pair effective

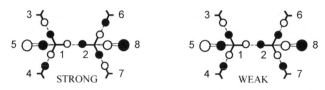

Figure 2   *H-bonded dimers with their 3-dimensional surroundings according to ST2 and TIP5P models.*

interactions. Here U_3 is the energy of third neighbor interactions of surrounding molecules, which do not belong to the same H-bonding cycles. In this case general Coulomb interaction of molecules is easily evaluated regarding the signs of the most close charges. The interactions 3-6, 4-7 and 5-8 (Fig. 2) in square and pentagonal cycles has been already taken into account in U_1 and U_2. Note that the majority of cycles in unstrained PWCs are pentagonal. Summing the effective energies of all H-bonds allows for an approximate evaluation of total interaction energy. In this way we take into account all the nearest, second and third neighbor interaction in water network, though the interaction of opposite molecules in hexagonal cycles should be considered separately.

The lowest values of U_{eff} correspond to more favorable H-bonds taking into account their surroundings. Two different types of H-bonded dimers in gas hydrate frameworks are shown in Fig. 2 using 5-point molecular model. The simple model offers a clearer view of the peculiarities both of the nearest and non-H-bonding Coulomb interaction. Based on purely electrostatic arguments, i.e. considering the interaction between only the positive charges on the hydrogen atoms (filled circles in Fig. 2) and the lone pairs on the oxygen atoms (open circles), the *cis*-conformation is energetically less stable than the *trans*-conformation because of the repulsion between peripheral atoms located opposite to each another. Essentially, this is the basis for discussed above SWB model.

Now let us consider the contributions of the second and third neighbors to total energy of three-dimensional gas hydrate frameworks. On the qualitative level, preferability of the second and third molecules can be estimated by the signs of nearest effective charges. If the nearest charges have different signs (colors in Fig. 2), then the interaction of these molecules is attractive and vice versa. One can readily see that an interaction between the second neighbors are equally attractive for both types of H-bonds. For strong H-bond the interactions 1-6, 1-8, 2-3 and 2-5 are attractive. Whereas only 1-8 and 2-4 interactions are repulsive. It can easily be checked that the same situation pertains to weak H-bond. Again there are four attractive and two repulsive inter-molecular interactions. Thus the second neighbor interactions has little or no effect on the energy non-equivalence of the two H-bond types.

In an analogous way consider the contributions from third neighbors. It is important to point out that mutual interactions of opposite molecules (3-6, 5-8 and 4-7) in square and pentagonal cycles have been already taken into account in U_1 and U_2. Recall that in gas hydrate frameworks, the majority of cycles are pentagonal. Strictly speaking, an interaction of opposite molecules in hexagonal cycles should be considered separately. Thus we have the following. The interactions 3-7, 4-6, 4-8 and 5-7 are repulsive whereas only 3-8 and 5-6 interactions are attractive for strong H-bond. Conversely, there are four attractive and two repulsive interactions for weak H-bond. The evident competition between U_1 and U_3 contributions significantly reduces the accuracy of SWB model for bulk gas hydrate frameworks (see Fig. 1, at the right).

2.4 Numerical Model of Effective Bonds (NEB model)

Using the concept of effective pair interactions we have developed a new discrete model of inter-molecular interactions in gas hydrate frameworks. Contrary to H-bond types on a surface of polyhedral clusters, in this case it is beyond reason to divide additionally the *trans-* and *cis-*configurations of distinct H-bonded dimer. It is necessary to take into consideration an interactions of all surrounding molecules. We have found that there are 1098 symmetry-distinct types of H-bond surroundings (extended H-bonds). Among them 369 types (33.6%) correspond to strong H-bonds. The rest of them contain weak H-bonded dimer. Taking into account antisymmetry,[6] the number of different extended H-bonds is 567.

An extended H-bond in graph theory representation is shown in Fig. 3. Dotted line in upper part of Figure (at the left) means the possible presence of hexagonal cycle. At the right, a configuration of the opposite molecules in hexagonal cycle is shown. Here we used again 5-point molecular model. In such model, four effective charges are located in tetrahedral directions.

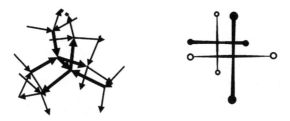

Figure 3 *Extended H-bond in graph theory representation (at the left); bold arrows determine the configuration of H-bonded dimer. At the right: Opposite molecules in hexagonal cycles (ST2 and TIP5P models).*

New discrete numerical model uses 1098 averaged values of effective bond energies without taking the opposite molecule interactions into account. In order to calculate these values we have optimized 5000 random configurations of minimal unit cell of structure I. For local optimization we used TIP4P potential and steepest descent method. In order to take into account the opposite molecule interactions in hexagonal cycles we used a square 6×6-matrix, because for each molecule there are 6 different ways to arrange hydrogen atoms. The elements of the matrix are equal to averaged values of opposite molecules interaction in hexagonal cycles. By reason of symmetry of diametrically opposite molecule pairs, the number of independent matrix elements is 9.

Firstly, we calculated the vector of averaged effective H-bond energies $U_{eff}(k)$ and the matrix of averaged interactions of opposite molecules in hexagonal cycles $U_{opp}(m,n)$. Total energy per unit cell to third neighbor interaction accuracy is

$$U = \sum_{i=1}^{N} U_{eff}(k_i) + \sum_{j}^{N_0} U_{opp}(m_j, n_j) \tag{2}$$

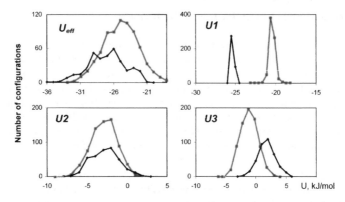

Figure 4   *Different contributions to effective H-bond energy for strong (black curves)*
and weak (grey) H-bonds.

Were $N=92$ is the number of H-bonds, $N_0=15$ the number of diametrically opposite
molecule pairs per unit cell. What we really need to know are the types of all extended H-
bonds and all mutual arrangements of opposite molecules in hexagonal cycles. Note that
the first term in (2) is considerably large

Before proceeding to results of combinatorial optimization, let us consider more
exactly different contributions to the effective energy. Fig. 4 represents the contributions
separately for strong and weak H-bonds. One can see that an advantage of the "strong"
extended H-bonds (U_{eff}) over "weak" extended bonds is not great. At the same time, the
nearest neighbor interactions U_1 are grouped into two distinct energy ranges. The
bimodality lies at the basis of defining strong and weak hydrogen bonds in our works. It's
necessary to point out that this is the distribution of average values. Average-free
distribution of U_1 has small overlap for the strong and weak H-bonds. For strong and weak
bonds, the second neighbor interactions U_2 fall into the same range of energies. Recall, that
the number of "weak" extended H-bond types is approximately twice greater than the
number of the "strong" types. In both cases interactions U_2, as a rule, are attractive. Of
special interest is the third neighbor interactions. On the average, interactions U_3 for
"strong" extended H-bonds are less attractive. The negative correlation between U_1 and U_3
is in reasonably good agreement with the results of our simple qualitative analysis (see
previous section).

The wide-range distribution of U_{eff} is of great importance to energy optimization.
However, because of overlapping neighboring extended H-bonds it is impossible to
construct a configuration consisting only of favorable extended bond types. There is a
strong mutual competition of the neighboring extended H-bonds. The frustration is
considerably stronger than in two-level nearest neighbor SWB model.[4]

Energy distribution of 5000 random D0-configurations are shown in Fig. 5. More
stable configurations were found during the process of combinatorial optimization on the
base of new discrete model. Two iterations were used in this optimization procedure. In the
beginning, using the set of random configurations, we calculated all model coefficients, i.e.
vector $U_{eff}(k)$ and matrix $U_{opp}(m,n)$. Then we found the large set of discrete optimized
configurations. After that, we re-calculated all model coefficients. As a result of the two-
step optimization we obtained the most stable configurations. These two configurations

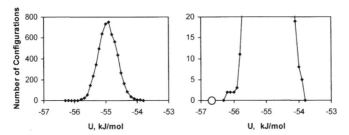

Figure 5   *The energy distribution of random configurations. Right side shows a part of the plot on an enlarged scale. Open circle indicates the energy of discrete optimized structure.*

(one circle in Fig. 5) have 44 strong H-bonds and close values of energy. It is interesting to note that we could not obtain a new less stable configurations in similar manner. In all probability, this means that the most stable configurations are more difficult to access.

3   CONCLUSION

In many respects the new discrete model is crude. It based on additive decomposition scheme of molecular interactions. In this work we don't take into account long-range Coulomb interactions. At the same time, this work has provided fresh insight into molecular interactions in gas hydrate frameworks. Really, the effective energies of extended H-bonds could differ in two times. It is clear that U_2 and U_3 interactions are mainly electrostatic. It necessary to point out that this model takes into account an additive part of H-bond cooperativity.[4] For example, the second neighbor interactions U_2 (Fig. 1) change a configuration of the H-bonded dimer. Hence, it is changed an interaction of the H-bonded molecules.

The NEB model is a natural outgrowth of the nearest neighbor SWB model.[4] Both these models use a graph representation of H-bond network and special algorithms of combinatorial optimization. But the NEB model is numerical. A vector of effective energies and a matrix of interaction of opposite molecules in hexagonal cycles essentially depend on the type of pairwise potential and on the size of unit cell. In this work the NEB model was applied to study only minimal cell of gas hydrate frameworks with TIP4P potential. These results are mainly of methodical significance. To obtain more general results we are currently studying the effect of other water model potentials and of the size of the model cell.

Acknowledgements

This work was partly supported by Russian Foundation for Basic Research (Grant 06-03-32321).

References

1   D. Sloan, *Clathrate Hydrates of Natural Gases*, Marcel Dekker Inc. Pub., New York, 1997.
2   C. Knight, S.J. Singer, J.-L. Kuo, T.K. Hirsch, L. Ojamäe and M.L. Klein, *Phys. Rev.* 2006, **E73,** 056113.
3   J.-L. Kuo and W. F. Kuhs, *J. Phys. Chem. B*, 2006, **110**, 3697.
4   M.V. Kirov, *Proc. IVth International Conference on Gas Hydrates*, 2002, Yokohama. Japan, pp. 649-654.
5   M.V. Kirov, *J. Struct. Chem.*, 2005, **46**, S188.
6   M.V. Kirov, *Physics and Chemistry of Ice*, Royal Society of Chemistry, Cambridge, 2007, p.

MICROSTRUCTURE OF GAS HYDRATES IN POROUS MEDIA

A. Klapproth, K.S. Techmer, S. A. Klapp, M. M. Murshed and W. F. Kuhs

GZG Abt. Kristallographie, Universität Göttingen, Goldschmidtstr. 1, 37077 Göttingen

1 INTRODUCTION

The vast deposits of natural gas hydrates found in permafrost regions and deepwater basins around the world are commonly confined in a porous sedimentary matrix. The large amount of methane stored in sediments has promoted gas hydrate deposits to be considered as potential energy resources, geohazards as well as agents for climate change. High-resolution seismic methods are important tools for identifying and quantifying gas hydrate deposits in the upper seabed. In this context, the understanding of the effect of gas hydrates on physical properties of sediments, like seismic velocity and attenuation, is essential. The relation between the hydrate fraction in the sediment and the elastic properties of the hydrate-sediment composite must be known. Seismic velocities depend on the bulk elastic moduli of the system, which are controlled by the grain-scale arrangements of hydrates and sediments.[1] Different models for the distribution of pores filled with gas hydrates to varying degrees within the sediment are proposed, which show different impact on elastic properties of gas hydrates bearing sediment. They can be basically classified as cement contact models and pore filling models. Cement contact models assume hydrates forming only at the grain contacts and by surrounding and cementation of sediment grains[1,2]. Numerical modelling of cementation on grain contacts provides a conspicuous increase of the system's elastic moduli.[1,3] Pore filling models assume hydrates forming either in the pore volume, which have no contact with sediment grains, or as load-bearing sediment (as a part of the sediment frame). For a given hydrate saturation, hydrates floating in pore fluid have a small influence on the sediment's elastic properties primarily by increasing the moduli of the pore fluid.[1,2]

Most authors on hydrate distribution in sediments favour pore filling models over cement contact models.[4,5,6] With rock physics modelling on the base of a load-bearing model Helgerud et al.[4] predicted hydrate concentrations consistent with estimates obtained from resistivity, chlorinity and gas content measured by decomposition. Tohidi et al.[5] showed that methane hydrates form initially in the centre of pores in water-rich systems. Based on Scanning Electron Microscope (SEM) imaging of a synthesized and compacted gas hydrate bearing quartz sand sample, Stern et al.[6] assumes that the hydrate phase forms very strong load-bearing cement that occurs between sediment grains and often surrounds them. They discussed the comparability of natural hydrates formed in a sand package recovered from a sub-permafrost environment and the synthesized and compacted samples in laboratory. They reported resemblances of the sample structure and phase distribution as observed by SEM investigations. Waite et al.[1] measured the compressional wave speed through synthesized, partially water-saturated, methane hydrate-bearing Ottawa sands and assumed that hydrates surround and cement sediment grains. The effect of hydrate formation in partially saturated, unconsolidated sand was derived from comparison of measured wave speeds and predicted. Based on rock physics models they calculated bulk and shear moduli, which were combined with the sample density to obtain the predicted compressional wave speed. For methane hydrate-bearing sand and its effect on seismic wave attenuation in gas-saturated environment Priest et al.[7] suggested that viscous squirt flow of absorbed water or free gas within the pore space is enhanced by hydrate cement at grain contacts and by the porosity of the hydrate itself.

For any assessment of natural gas hydrates in hydrate-bearing sediments, the determination of microstructure, composition and physical properties, as well as of the underlying formation and possible decomposition processes (including the effect of self-preservation[8,9,10]) is essential. Likewise, it is not clearly proven whether gas hydrates nucleate along the surface of mineral grains or they nucleate spontaneously from the saturated water phase. Scanning Electron Microscopy (SEM) provides an insight into the microstructure of gas hydrate-bearing sediments at different stages of the formation. We present cryogenic SEM investigations of methane hydrates synthesised in different porous media by the transformation of water in a gas-saturated environment. We use both quartz and various quartz-clay mixtures (quartz admixed with kaolinite or montmorillonite) as sediment models.

2 EXPERIMENTAL METHODS

2.1 Preparation of Gas Hydrates-bearing Sediments

The synthesis of methane hydrates begins with the preparation of exemplary porous media. The constituents are natural minerals from Lubercy, Moscow area (quartz) and further regions in Russia (kaolinite, montmorillonite). Quartz samples consist of globular grains with mean grain sizes ranging from 200 to 300 μm, while kaolinite and montmorillonite are small plate-like particles, which form very frequently micro-aggregates. Montmorillonite aggregates have an average size of ~ 7 μm, while kaolinite has an average size of ~ 20 μm.

Table 1   *Composition (weight-%) and formation conditions of the prepared methane hydrate-bearing sediment*

	porous media	frost/ice	gas	p/T conditions
medium I	quartz	10 %(1), 14 %(1), 17 %(1)	CH_4	100 bar, 3°C
medium II	quartz + 7 %(1) kaolinite	10 %(1) 17 %(1)	CH_4	100 bar, 3°C
medium III	quartz + 7 %(1) montmorillonite	10 %(1) 17 %(1)	CH_4	100 bar, 3°C

The media constituents were dried in an oven at 105 °C for at least 48 hours and stored in a cold room at -10°C. The various media were obtained by mixing the appropriate amount of constituents in a mortar in the cold room. For the generation of synthetic gas hydrates, the pre-cooled media were subsequently admixed with a certain amount of frost/ice, which ensured the unsaturated concentration of water in the porous media. The porosity of the starting material varies between 35 and 45%, based on the mass and volume of the sample as well as the known mass densities of the constituents. The samples characteristics are shown in Table 1.

Frost was generated by spraying deionized water vapour with a heat gun on a pre-cooled rotating metal disk. The resulting precipitates (very fine droplets of average grain size < 10μm) were scraped from the disk into liquid nitrogen. The whole process was carried out in a glove box filled with nitrogen to avoid any air frost. Notably, our laboratory-prepared frost offers a better degree of homogeneity as well as a smaller averaged grain size than crushed and sieved ice in the above-mentioned admixing process. In order to ensure a minimal and precise fraction of ice/water (10-17%) in the porous media, it is simpler to weight the fine grained solid and admix it thoroughly than to micro-pipette water in it.

Finally, gas hydrates were prepared using our gas-pressure equipment.[11,12] The cooled mixtures were filled into a pressure cell layer by layer at −10°C. To form methane hydrate from these porous media the pressure cells were mounted onto the gas pressure rig at a constant temperature of 3°C. After 25 minutes of temperature equilibration and ongoing melting of the ice phase, 10 MPa of methane gas pressure was applied. The progress of the transformation from water into hydrate was followed by the observed pressure drop in the cells, continually recorded in a computer. Methane gas was refilled manually whenever the pressure had fallen below 9.6 MPa.

2.2 Field-Emission-Scanning Electron Microscope (FE-SEM) Investigation

The microstructure of the sediment-hosted methane hydrates has been investigated using a high resolution Field-Emission-Scanning Electron Microscope (FE-SEM) equipped with a cryo-stage and a cooled sample transfer chamber (LEO Gemini 1530 and a FEI Quanta 200F microscope with OXFORD and Polaron cryo-stages respectively).

The prepared samples were recovered within 5 seconds in liquid nitrogen temperatures to ambient pressure at various stages of the generation process. The cylindrical sediment-gas hydrate samples as well as the starting material were kept in liquid nitrogen all the time and then small pieces (~2 mm^3) of the sample were transferred via a pre-cooled interlock into the pre-cooled cryo-stage of the FE-SEM. To avoid condensation on the extracted sample surface, the sample was also kept in liquid nitrogen during crushing and transferring. Several pieces of the same sample were investigated. The SEM images were received at pressures of <10^{-5} mbar at temperatures in the range of 90 to 100 K with acceleration voltages in a range of 1.2 to 3 keV. A resolution in the range of a few tens of nanometres could be achieved with this set-up.

3 RESULTS

3.1 Kinetics of Methane Hydrate Formation in Various Porous Media

The gas consumption rate during the methane hydrate formation tends to be different for each porous medium, therefore suggesting different reaction kinetics (Figure 1a). The gas hydrate formation is initially fast and then slows down. It suggests an initial fast surface reaction with free water and gas access. Subsequently, the transformation from water into gas hydrates is slowing down limited by the gas and water transfer. Finally, no further uptake is observed. From former experience we can safely assume that no further gas uptake indicates the almost complete transformation of water into gas hydrate under these conditions; quantitative diffraction work to confirm this supposition is in progress. The observed gas uptake for hydrate saturation is different in various porous media. In medium II (containing kaolinite) the gas consumption is nearly twice as large as in montmorillonite-containing medium III (Figure 1a) and hydrate saturation occurred also faster for the latter case. A first and obvious explanation is that the previously dried montmorillonite takes up part of the water and protects it from any reaction with the gas. However, other particularities of the different clay minerals may also interfere. It is interesting to note the findings of Rogers et al.[13] that in-situ generated surfactin promoted gas hydrate formation on bentonite surfaces in comparison to sand and kaolinite. Moreover these authors propose for the hydrate growth in media without biosurfactant a larger formation rate on bentonite (essentially consisting of montmorillonite) than on kaolinite, which indirectly supports our initial guess. However, the remarkably different reaction kinetics in these clay-containing sediments may well, at least partly, also reflect the different surface activities of the minerals in combination with free water or interfacial water.

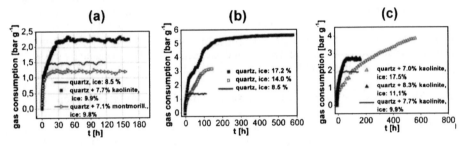

Figure 1   *Recorded gas consumption during methane hydrate formation at 3°C in various porous media (see also Table 1). Small pressure fluctuations are caused by the sensitivity of the pressure sensor to fluctuations of the room temperature.*

The reactions in media I (quartz) and II (quartz and kaolinite) show decreasing reaction rates as the concentration of frost/water is increased (Figure 1a,b). The different curve progressions for varying water content of the porous media suggest a multi-stage reaction process, as already proposed in the literature.[14,15] In the experiments with higher water content (Figure 1b, 14 % and 17 % ice) individual stages can not be identified easily, but after an initial surface reaction it seems that the formation of gas hydrates undergoes a transition to a diffusion controlled process. This could for instance occur if the gas hydrate cement seals some gas filled cavities in the porous matrix, allowing a subsequent gas transfer through the hydrate phase only by diffusion (Figure 1b, 17% ice). A similar tendency of transition to a diffusion controlled process by lowering the gas consumption was found in medium II containing kaolinite (Figure 1c). The reaction rate in these medium containing 17 % ice in the starting mixture shows no gas hydrate saturation. Only a slow removal of the gas consumption caused by the gas hydrate formation is observed.

3.2 Microstructure of Methane Hydrate in Various Porous Media

3.2.1 Starting Material. Scanning Electron Microscopy (FE-SEM) was used to study the starting porous material admixed with ice or water. The initial mixture of quartz and frost (medium I) consists of small aggregates of quartz particles with hexagonal-shaped ice grains well distributed along the surfaces of quartz (Figure 2).

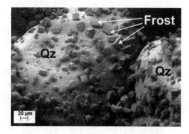

Figure 2 *SEM picture illustrating frozen quartz grains admixed with frost having mass content of 10 %.*

The ice grains are agglutinated along a rough surface of the quartz grains. Under the formation conditions, at 3°C above the melting point of ice, one could expect a water layer coating most of the quartz particles, with water necks at contact areas originating from capillary forces of the liquid and based on the migration of water with ongoing melting of ice. Subsequently, water in the porous medium was found concentrated between the quartz grains and in the valleys of the rough quartz surface. So far a complete uniform coating of the quartz grains by a thin water layer has not been observed up to a water content of 17%, which suggests hydrate cementation by surrounding the grains. However, a complete coating may well occur at higher water concentrations (or lower quartz surface areas).

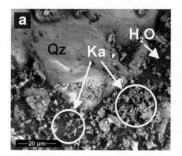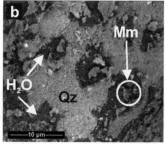

Figure 3 *SEM images of porous media admixed with frost after melting of the ice phase at 3°C and subsequent refreezing down to -196°C (frozen H₂O; dark grey background). a: medium II containing quartz (Qz) and kaolinite (Ka). b: medium III containing quartz (Qz) and montmorillonite (Mm).*

The mixture containing quartz and kaolinite (medium II) shows accumulation of clay particles mainly along the surface of the water phase, which is concentrated between the quartz grains (Figure 3a). It seems that the kaolinite particles form a filigree structure on water and even rarely on quartz surfaces. This behaviour might be explained by the ability of kaolinite to flocculate in an aqueous suspension driven by the negative charged plates and positive charged corners of these clay particles.[16] In both, medium III (quartz and montmorillonite) and medium II, clay particles/aggregates often appear on the surface of the water phase (Figure 3b). In comparison to kaolinite, montmorillonite shows a tendency to form dense flakes, like a frozen suspension on the water surface. The generation of dense flakes might be explained due to the aggregation of swelled montmorillonite particles originating from absorbed water molecules.[17]

3.2.2 Methane Hydrate-bearing Sediments. SEM images of methane hydrates in gas-saturated sediments display gas hydrates between the quartz grains like a glue or cement (Figure 4a). This cement should contain almost pure gas hydrate after the complete transformation from water (see Sect. 3.1).

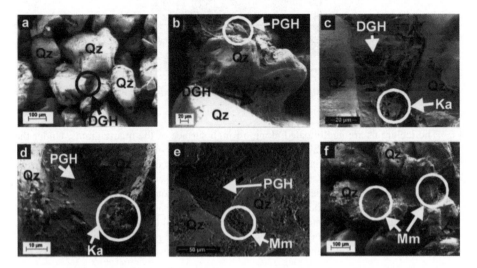

Figure 4   *SEM images of methane hydrate formed due to the transformation of liquid water in gas-saturated porous media. a, b: porous medium I (quartz (Qz) and frost having mass content of 10 %) showing dense (DGH) and porous gas hydrate (PGH) cement. c, d: medium II (Qz, kaolinite (Ka)) presenting kaolinite particles on the gas hydrate cement surface. d, e: medium III (Qz, montmorillonite (Mm)) showing montmorillonite flakes on porous gas hydrate cement between quartz grains. f: overview of medium III.*

The cemented quartz grains form agglomerates with increasing stiffness as the water concentration is increased. Thus, the cementation suggests stiffening of the sample, which is qualitatively well perceived while handling the samples, in particular during crushing. Nonetheless, at higher water concentration (media admixed with 17 % frost) there are still open cavities in the sediment, which ensure the gas transport through the reaction front. Free quartz grain surfaces have been observed in all investigated systems. This observation suggests the absence of a thin uniform bulk water layer around quartz grains within the instrumental resolution. On the other hand, the images could also display a fast sublimation of a thin gas hydrate layer promoted by the high-vacuum conditions and the erosive action of the electron beam. The alteration along the surface of the gas hydrate due to the accelerating beam is a limiting factor for observing first reactions. On the other hand not more than a few tens of nanometre sublimation should occur at accelerated electron beams of 1 kV before achieving the first image scan.

We learned from former investigations on gas hydrates while using the electron microscopy that they have either dense or porous morphological microstructures.[18,19] In medium I, methane hydrates are often dense and less frequently meso-porous (Figure 4b). Methane hydrates formed in medium II (quartz-kaolinite mixture) seem to be either porous or dense. On the surface of the cement-like gas hydrate, the platy kaolinite particles are located towards their flat sides, which are negatively charged. Moreover, these clay particles are seen to build a filigree network through the connection over their edges as a second layer (Figure 4c,d) mainly on the gas hydrate cement. Some bigger kaolinite particles appear perpendicular to the cement surface.

In contrast, when using montmorillonite (media III), individual clay particles are not visible. It seems as if they form flakes with different sizes and those are linked up to create a crust, which is more or less dense (Figure 4e,f). The appearance of the crust and big flakes resembles that of a frozen suspension of clay and water. The crust is mainly located between the quartz grains and affects the observation of methane hydrate. In the mixture containing around 10% water it is impossible to distinguish montmorillonite from gas hydrate or to separate the different crystals or particles apart from quartz (Figure 4f). At higher water concentrations, meso- to macro-porous gas hydrate cement was observed, which shows macro-cavities within the cement and at the mineral interface (Figure 4e, small black ovals). At the state of advanced hydrate growth these macro-cavities may, therefore, suggest that as the hydrate cements the quartz grains and reduces the gas and water transport through the reaction front, there may still be some contribution through the cavities.[7]

4 CONCLUSIONS

We have presented methane hydrate formation in gas-saturated-porous media by transformation of liquid water. Three types of media are used namely medium I (quartz), medium II (quartz + kaolinite) and medium III (quartz + montmorillonite). In the SEM investigations, methane hydrates appear between the quartz grains like cement. Kaolinite particles are observed as a filigree network on the surface of hydrate cement, while montmorillonite looks like flakes or crust. Each of the minerals, used in a specific media composition, may play individual/coupled interaction with water and gas hydrate, and thereby display characteristic configuration under SEM. Dissimilar kinetic features, using different porous media at the investigated conditions, lead to confirm that porous media directly function on gas hydrate formation. In contrast to medium I and II, medium III contains montmorillonite shows the fastest hydrate saturation. With varying water content of the porous media (medium I and II) the reaction seems to proceed through a multi-stage progression. The macro-pores in the cement, formed during the hydrate growth in medium III (along with 17% water), likely not to hinder for the transportation of water and gas to the reaction front. The results presented here may concern with some physical properties that are essential for economic geology, environmental science as well as the commercial exploration of gas hydrates for energy source.

Acknowledgments

We thank Dr. Chuvilin (Moscow State University) for providing the mineral samples. This is publication no. GEOTECH - 241 of the R&D-Programme GEOTECHNOLOGIEN funded by the German Ministry of Education and Research (BMBF) and German Research

Foundation (DFG), Grant G0605B. We also thank Dr. Sergio H. Faria and Dr. Marta Krzyzak for constructive discussions.

References

1   W. F. Waite, W. J. Winters and D. H. Mason, *American Mineralogist*, 2004, **89** 1202.

2   J. Dvorkin, M. B. Helgerud, W. F. Waite, S. H. Kirby, and A. Nur, *Natural Gas Hydrate in Oceanic and Permafrost Environments*, 2000, Ed. M. D. Max, Kluwer, Dordrecht, Netherlands, 245.

3   M. B. Helgerud, Doctoral Thesis, *Wave speeds in gas hydrate and sediments containing gas hydrate: A laboratory and modelling study*, 2001, Stanford University, Palo Alto California, U.S.A.

4   M. B. Helgerud, J. Dvorkin and A. Nur, *Annals New York Academy of Sciences*, 2000, **912**, 116.

5   B. Tohidi, R. Anderson, M. B. Clennell, R. W. Burgass and A. B. Biderkab, *Geology*, 2001, **29**, 867.

6   L. Stern, S. Circone, S. Kirby and W. Durham, *Proceedings of the Fifth Conference on Gas Hydrates*, June 12-16, 2005, Trondheim, Norway, Abstr. **1046**, 300.

7   J. A. Priest, A. I. Best and C. R. I. Clayton, *Geophys. J. Int.*, 2006, **164**, 149.

8   W. F. Kuhs, G. Genov, D. K. Staykova and T. Hansen, *Phys. Chem. Chem. Phys.*, 2004, **6**, 4917.

9   L. A. Stern, S. Circone, S. H. Kirby and W. B. Durham, *J. Phys. Chem. B*, 2001, **105**, 1756.

10 S. Takeya, W. Shimada, Y. Kamata, T. Ebinuma, T. Uccida, J. Nagao and H. Marita, *J. Phys. Chem. A*, 2001, **105**, 9756.

11 G. Genov, W. F. Kuhs, D. K. Staykova, E. Goreshnik and A. N. Salamatin, *American Mineralogist*, 2004, **89**, 1228.

12 G. Genov, Ph.D. Thesis, *Physical processes of CO_2 hydrate formation and decomposition at conditions relevant to Mars*, 2005, University of Göttingen.

13 R. E. Rogers, G. Zhang, C. Kothapalli and W. T. French, *Proceedings of The Fourteenth International Offshore and Polar Engineering Conference*, May 23-28, 2004, Toulon, France.

14 D. K. Staykova, W. F. Kuhs, A. N. Salamatin, and T. Hansen, *J. Phys. Chem. B*, 2003, **107**, 10299.

15 W. F. Kuhs, D. K. Staykova and A. N. Salamatin, *J. Physical Chemistry B*, 2006, **110**, 13283.

16 R. F. Giese and C. J. van Oss, *Colloid and Surface Properties of Clays and Related Minerals, Surfactant Science Series*, 2002, Ed. M. J. Schick and A. T. Hubbard, **105**, 226.

17 A. Meunier, *Clays, Series Geosciences, France*, 2003, ED (French). Scientifiques GB and Société Géologique de France, translated by N. Fradin, Springer

18 W. F. Kuhs, A. Klapproth, F. Gotthardt, K. Techmer, and T. Heinrichs, *Geophys. Res. Lett.*, 2000, **27**, 2929.

19 W. F. Kuhs, G. Y. Genov, E. Goreshnik, A. Zeller, K. Techmer, and G. Bohrmann, *Intern. J. Offshore Polar Eng.*, 2004, **14**, 305.

TACKLING THE PROBLEM OF HYDROGEN BOND ORDER AND DISORDER IN ICE

C. Knight and S. J. Singer

Department of Chemistry, Ohio State University
100 W. 18th Ave., Columbus, OH 43210, USA

1 INTRODUCTION

The liquid phase of water is ringed by a series of hydrogen bond disordered ice phases (Fig. 1). How hydrogen bond disorder arises is illustrated in Fig. 2. The constraints of the "ice rules[2]" – each water molecule required to form two H-bonds as a donor and two as an acceptor – do not fix the orientation of the water molecules because each molecule has six ways to point its two hydrogens in two tetrahedral directions and lone pairs in two other directions. Fig. 2 illustrates several different unit cells for ice Ih that each satisfy the ice rules for an infinite lattice when periodically replicated. Going from lowest to highest pressure, ice-Ih, ice-III, ice-V, ice-VI and then

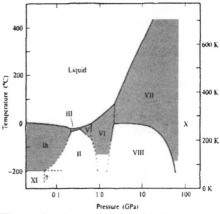

Figure 1 *Hydrogen bond disordered phases, the shaded regions, surround the liquid phase. (Adapted from Petrenko and Whitworth.[1])*

ice-VII are adjacent to the liquid phase. In all likelihood, every one of these phases gives way to a hydrogen bond ordered version of themselves at lower temperatures. Ice-Ih, when suitably doped with hydroxide, transforms to what is thought to be a ferroelectric phase, ice

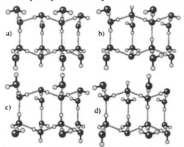

Figure 2 *Four possible arrangements of H-bonds within a unit cell of ice-Ih.*

XI, near $72K$ for H_2O and $76K$ for D_2O.[3–7] The oxygen atom positions in ice XI are very close to those of ice-Ih.[8–11] The transition principally involves selection of one particular H-bond arrangement in the low temperature phase. Ice-III, when cooled at about $1K$ per minute or faster, undergoes a similar transformation to a metastable H-bond ordered version known as ice-IX.[12–16] (Ice-IX, because it is metastable with respect to ice-II, is not shown in Fig. 2.) In the 2.1-12GPa range, the hydrogen bond ordered form of ice-VII, known as ice-VIII, appears near $0°C$.[17,18] The hydrogen bond ordered form of ice-VI has not been clearly identified, although some experiments give some preliminary indications of low temperature, fully ordered phase.[18–21] We will describe our prediction for the structure of the H-bond ordered version of ice VI in this work. Salzmann *et al.* recently observed ice XIII, the H-bond ordered version of ice V.[22] Those authors stated, "It would be a challenging test of the ability of modern day computational methods to reproduce our experimentally found lowest energy state." Indeed, in this paper we describe theoretical methods that predict the ground state and characterize the H-bond ordering transition for all the cases mentioned above, including ice V-XIII. and which are essential for understanding the behavior of defects moving in a disordered H-bond environment.

2 METHODOLOGY

Describing the statistical mechanics of hydrogen bond disorder immediately encounters several severe obstacles. Common empirical potentials do a poor job predicting the relative energies of H-bond isomers in ice, as documented by Buch *et al.*[23] Even relatively modest electronic density functional theory (DFT) methods perform much better, but these methods are too costly for straightforward simulation of phase transformations or defect motion in ice phases. Our strategy is to distill the in-

Figure 3 *Cis and trans H-bonds.*

formation we obtain from electronic structure calculations into a form in which it is feasible to perform the type of statistical mechanical calculations needed to predict phase transitions or describe defect motion. Since H-bond ordering transitions in ice involve condensation of the system into a single H-bond topology out of the myriad of disordered ones, we seek to link energy and other physical properties to the H-bond topology of the ice lattice.

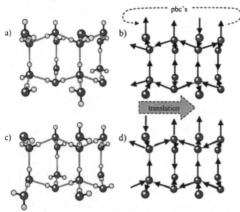

Figure 4 *H-bond topology of the unit cells (left) is described by directed graphs (right). This figure illustrates the process discussed in the text when the symmetry operation is a translation. Bonds moved outside the cell are translated back into the cell according to periodic boundary conditions (pbc's).*

Each different hydrogen bond arrangement in ice is a local minimum on the Born-Oppenheimer electronic potential energy surface. As an illustrative thought experiment, imagine exploring the Born-Oppenheimer surface of one of the periodically replicated unit cells shown in Fig. 2 using an electronic structure program, say Gaussian,[24] CPMD[25] or VASP.[26-30] A common exercise is to take an initial guess for the atomic positions and "optimize the geometry", meaning find the nearby local minimum of the potential energy surface. If the initial geometry was reasonable, the starting and optimized structures will have the same H-bond topology, which can be summarized in terms of directed or oriented graphs (Fig. 4). The topology only depends on the connectivity pattern, and is a distillation of the greater amount of information available from the full three-dimensional structure. Next, imagine taking the set of H-bond bond arrows that summarize the H-bond topology (Fig. 4b) and moving them to other bonds (Fig. 4d) using one of the group operations, then generating another starting structure from the new arrows (Fig. 4c), and finally optimizing the new structure. Of course, one would get exactly the same energy because energy is an invariant with respect to symmetry operations. Hence, if H-bond topology is a useful predictor of scalar physical properties like the energy, the prediction must originate from features of the H-bond topology that are invariant to symmetry operations.

A suitable way to capture the H-bond topology is to define bond variables b_r for H-bonds at each bond position r which take the values ±1 depending on the direction of the H-bond with respect to an arbitrarily defined reference direction. It is not obvious that the energy of the various H-bond arrangements of ice can be predicted from the H-bond topology as captured by a set of bond variables $\{b_r\}$, as illustrated by the arrows in Figs. 4b and d. The success of linking the energy to H-bond topology is evaluated below. However, the brief thought experiment of the previous paragraph illustrates that *if energy can be predicted from the H-bond topology, it must depend on combinations of the bond variables $\{b_r\}$ that are invariant to symmetry operations.* We call these symmetry invariant combinations of the H-bond variables *graph invariants*. These combinations are generated using the symmetry operations of the point group of a cluster, or the space group of an ice lattice. Linear combinations of bond variables are *first order graph invariants*, and are generated by application of a group theoretical projection operator on any bond variable.

$$I_r = \frac{1}{|G|} \sum_{\alpha=1}^{|G|} g_\alpha(b_r) \qquad (1)$$

In the above equation, G is the point or space group, and $|G|$ is the number of elements $\{g_\alpha\}$ in G. The normalization $|G|^{-1}$ makes graph invariants intensive. *Second order invariants* are quadratic polynomials in the bond variables,

$$I_{rs} = \frac{1}{|G|} \sum_{\alpha=1}^{|G|} g_\alpha(b_r b_s), \qquad (2)$$

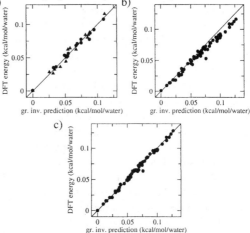

and so on for higher order invariants. In this way, we construct a hierarchy of basis functions for describing the dependence of scalar physical quantities on H-bond topology. In most cases, it is impossible to construct invariant linear functions of bond variables and the leading order invariants are second order.

The italicized statement in the previous paragraph began with a rather important "if" as to whether the abstracted H-bond

Figure 5 *a) Graph invariant fit of the form of Eq. (3) to the energies of the 14 H-bond isomers of a 12-water hexagonal($\bullet$) unit cell and the 16 H-bond isomers of an 8-water orthorhombic($\blacktriangle$) unit cell of ice Ih. If the fit was perfect, all points would lie on the diagonal. b) Calculated DFT energy of H-bond isomers of a 48-water hexagonal ice Ih unit cell plotted against energies predicted from graph invariant parameters derived from the small unit cells. c) Graph invariant fit to the energies of 63 H-bond isomers of a 48-water hexagonal unit cell of ice Ih.*

topology is sufficient to predict energy and other properties. However, we have demonstrated for water clusters[31–33] and for various forms of ice[34–37] that in all cases the leading order graph invariants through second order are sufficient to capture the dependence of energy and other scalar physical properties (i.e. squared dipole moment) on H-bond topology. Moreover, for our work to date on pure ice without defects, the only second order graph invariants needed are those containing products $b_r b_s$ for bonds which are very close to each other. Typical results are shown in Fig. 5. A simultaneous fit of the form

$$E(b_1, b_2, \ldots) = \underbrace{\sum_r \alpha_r I_r(b_1, b_2, \ldots)}_{\text{sum over 1st order invariants}} + \underbrace{\sum_{rs} \alpha_{rs} I_{rs}(b_1, b_2, \ldots)}_{\text{sum over 2nd order invariants}} + \underbrace{\ldots}_{\text{negligible}} \qquad (3)$$

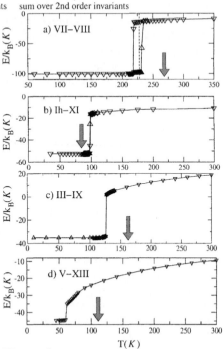

to the energy of H-bond isomers of two small unit cells of ice Ih is given in Fig. 5a. The energy is calculated using the CPMD program.[25,38] To achieve the level of precision shown in Fig 5, the fit required only three quadratic terms in Eq. (3), that is, three second order graph invariants, plus an overall constant. We have found that this procedure is indeed very rapidly convergent. In Fig. 5b, energies calculated for the H-bond isomers for a 48-water unit cell are compared with estimates using parameters from the fit for small unit cells shown in Fig. 5a. As can be seen in Fig. 5b, the prediction is nearly quantitative. The slight deviation between predicted and calculated results in Fig. 5b is actually not due to lack of convergence of the graph invariant parameters, but rather due to an artifact of the way we performed our calculations that causes the energies to be a moving target. Only the Γ-point was used in the electronic structure calculation for both the small cells and the 48-water cell, The same H-bond topology, when calculated for the larger cell, has a slightly different energy because the large cell effectively includes greater k-point sampling. We have verified that more extensive k-point sampling for small unit cells eliminates the slight discrepancy noticeable when only the Γ-point is used.[36,37]

Figure 6 *Energy vs. temperature calculated using graph invariant theory for a) the ice VII-VIII, b) Ih-XI, c) VI-VI′ and d) V-XIII systems. Upward and downward triangles indicate results of Metropolis Monte Carlo simulations ascending and descending in temperature. The dashed line in panel (a) indicates the transition point established by thermodynamic integration. The arrows indicate experimental transition temperatures (not well established for the III-IX system).*

3 KNOWN AND NEW ICE PHASES

Calculations using the graph invariant methodology described in section   predict that, of all the possible H-bond isomers, ice VII condenses into the correct low temperature ice VIII structure.[34,35] The calculated transition temperature of 228K (Fig. 6a) is in qualitative agreement with experimental reports in the range $263 - 274$[17,18] This case is an important benchmark because experimentally the VII-VIII system is the best characterized. For ice Ih (Fig. 6b), we indeed find a transition to the ferroelectric $Cmc2_1$ structure of ice XI.[34,35] Again, the calculated transition temperature of 98K is in qualitative agreement with the experiment result of 72K for H_2O and 76K for D_2O.[3-7] These results confirm the ferroelectric structure proposed by Leadbetter, Whitworth and their co-workers.[8-11] Our calculations for ice III-IX[37] (Fig. 6c) again predict the structure observed experimentally with a transition temperature in agreement with experiment.[12, 13, 15, 39] The ice V-XIII system is one of the most challenging theoretically because the unit cell is large[40] and there are many energetically low-lying isomers whose energy must be carefully converged. We find a transition to the same ice XIII structure observed by Salzmann *et al.*[22] Experimentally the transition is observed to occur between 110 and 120K. Our calculated transition temperature for the V-XIII transition is 61K.[41]

Because our work includes a statistical mechanical treatment of H-bond fluctuations, we can estimate partial hydrogen bond order in the high-temperature ordered phases and partial disorder in the low temperature ordered phases. To date, our calculated transition entropies are in good agreement when experimental comparisons are available,[34–37, 41] and for the ice III-IX case provide some insight regarding an apparent contradiction in the literature. When Lobban *et al.*[42] compared their partial ice-III order and the partial disorder in ice-IX measured by LaPlaca *et al.*,[14] they find that the entropy difference is incompatible with the entropy difference measured calorimetrically by Nishibata and Whalley.[15] Our calculations indicate which part of the picture needs revision.[37] To link diffraction data with calorimetric data, previous work has relied on mean field theories[43–47] that relate entropy and free energy (calorimetry) to order parameters which are fractional occupancies of certain hydrogen positions (diffraction). We can test how well these mean field theories relate thermodynamics to order parameters. The answer is that some do better than others for the low temperature phase, ice IX, and they all are in serious error for the high temperature phase. Entropy as a function of temperature is shown in Figure 7 (heavy solid line). The entropy of

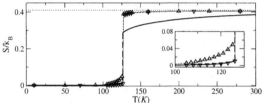

the low-temperature phase was calculated via thermodynamic integration from 0 K, and that of the high-temperature phase integrating from infinite temperature. The entropy at infinite temperature is taken to be Nagle's result for the Pauling entropy for a fully disordered arrangement of H-bonds in ice.[48] The calculated entropy at the transition is 67.7% of the ideal value for a fully disordered ice phase which is larger than the ex-

Figure 7 *Entropy vs. temperature from graph invariant theory (thick solid line). The horizontal line is the Pauling entropy for fully disordered ice. Entropy calculated using the occupational probabilities obtained from our simulations, is plotted using theexpressions of Nagle($\triangle$)[45] and Howe and Whitworth(∇).[46]*

perimentally observed 40%[15] of the ideal value. (Taking the lattice distortion that accompanies the phase transition into account would lower the calculated transition entropy.[37]) Also plotted in Figure 7 (curves with symbols $\triangle$ or ∇) is predicted entropy as a function of temperature calculated using various mean field theories with the occupational probabilities obtained *from our simulations* as input.

If the mean field expressions were accurate, the mean field entropy curves in Fig. 7, the curves with symbols, would coincide with the entropy from simulations. All of the mean field theories overestimate the entropy of the high-temperature phase, and Nagle's expression[45] also overestimates the entropy of the low-temperature phase. We can ask whether the mean field expressions are meaningful for our model, regardless of whether our model accurately corresponds to experiment. In actuality, our model is reasonably close to experiment. Two occupation probabilities for ice III and IX are traditionally called α and β. Our value for α, 38%, at the transition is close to the experimental values, 33%[16] and 35%.[42] Our β,

38%, is close to the value extracted from one diffraction experiment, 41%,[16] but rather low compared to a more recent report, 50%.[42]

A phase of ice that is an ordered version of ice VI has not been unambiguously characterized. In 1965, Kamb noticed an X-ray reflection at $77K$ that was incompatible with the $P4_2/nmc$ space group of ice VI and could have signaled the formation of a hydrogen bond ordered version of ice VI.[19] Later, he reported neutron diffraction data taken at $100K$ on a sample previously equilibrated at high pressure and $77K$ which indicated antiferroelectric ordering.[44] In 1976, Johari and Whalley predicted an ordering transition at $47K$ to a ferroelectrically ordered state in ice VI based on observed Curie-Weiss behavior of the low-frequency dielectric constant.[20] Later they concluded that the high frequency permittivity of ice VI at $0.9GPa$ indicated a very slow phase transition occurs in the temperature range 123-128K, but these experiments did not reveal the structure of the low temperature phase. Kuhs et al.[18] obtained neutron diffraction data on ice VI under temperature and pressure conditions where ice VI is stable, unlike earlier diffraction experiments where the diffraction experiments were performed at ambient pressure on samples recovered from high pressure cells. They took data at 225, 125 and 8K, but data was not sufficient to fully determine the structure. They found no evidence of the transformation observed by Johari and Whalley,[21] although the transition might be too slow compared to their experimental time scales.

Using electronic DFT calculations as input and graph invariant methods to extrapolate to the bulk limit, we find a transition to a proton-ordered phase at 108K.[36] Following Kamb,[19,44] we use primes to distinguish the hydrogen bond ordered phase and call this phase ice VI'.[49] The calculations indicate a large degree of pre-transitional order and post-transitional disorder: 29.8% of the Pauling entropy is lost as the transition is approached from above, 45.5% is lost at the transition, and 24.7% is lost below the transition.

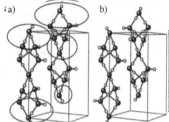

Figure 8 *(a) The lowest energy and (b) second lowest energy isomer, as determined from DFT calculations on a 40-water unit cell of ice VI. The H-bonds for each lattice in (b) are anti-ferroelectrically oriented, making the overall structure anti-ferroelectric. This arrangement of H-bonds has tetragonal symmetry and is assigned the space group $P2_12_12_1$. The H-bonds parallel to the a and b axes point in a counterclockwise fashion as one looks down the c-axis. The bonds that must be reversed to interconvert the two structures are circled on the left.*

We have extended our calculations to a 40-water unit cell of ice VI, and carefully converged the electronic DFT calculations with respect to k-point sampling and k-space cut off. We find two candidates for the ground state. The lowest energy structure, Fig. 8a, is ferroelectric. However, an anti-ferroelectric structure, Fig. 8b, is separated from the ground state by an energy gap per water molecule of $\Delta E/k_B = 4K$. While the electronic DFT calculations have been remarkably accurate for the energetic ordering of the H-bond isomers, we would recommend that experimentalists consider both structures as candidates for ice VI'. We find that energy and entropy as a function of temperature are not strongly affected by a change of the graph invariant parameters that reverse the energetic ordering of the two candidate structures.

4 ACCURACY OF METHODS

Energy differences between H-bond isomers in ice are on the order of 100K.[50] This difference is certainly far smaller than the absolute accuracy of even high-level *ab initio* quantum chemical methods, and even smaller than the expected accuracy for calculating, say, reactant-product energy differences in a chemical reaction. We most often use BLYP density functional theory[51,52] as implemented in the CPMD (Car-Parrinello molecular dynamics) program.[53] Because we perform these calculations for unit cells as large as 96 waters, this choice is dictated by practical concerns – we would rather be using periodic MP2 or CCSD(T) with a generously large basis set. It is certainly incumbent on us to document that a method as low-level as BLYP implemented with CPMD captures the energetic trends among the H-bond isomers of ice:

• Recall that common empirical potentials give wildly different energy differences among H-bond isomers.[23] However Fig. 9 demonstrates that CPMD implemented with a plane wave basis, DMol with a numerical basis, and CASTEP again with a plane wave basis give very similar results. CPMD and DMol used the BLYP functional,[51,52] while CASTEP employed PW91.[54-56] The solid curves near the top of Fig. 9 are calculations using three different DFT methods for the same geometry. The agreement between the methods is excellent. The lower dashed curves are for geometry optimized structures. In that case, the results using different methods are still consistent with each other, although the CASTEP

curves depart more from the other two because the unit cell was optimized along with the molecule geometry while for CPMD and DMol the unit cell was fixed.

These results suggest that the energy differences among H-bond isomers in the ices is *not* such a difficult problem after all, but that common empirical potentials are missing an important element.

• The relatively low level periodic DFT that we use correctly identifies the ground state H-bond configuration for the ice Ih-XI, VII-VIII, III-IX and V-XIII transitions. Moreover, it gives excellent structural parameters for ice VIII,[57] and for ice XI.[35,36]

• The transition temperatures we calculate for the ice Ih-XI, III-IX, VII-VIII and V-XIII transitions are all qualitatively correct. This would not happen if the energies of the higher-lying H-bond isomers were not estimated with at least qualitative accuracy by the electronic structure methods we use.

For pure water ice, we assume that

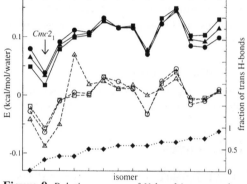

Figure 9 *Relative energy of H-bond isomers calculated by electronic DFT methods for 14 isomers of a 12-water hexagonal unit cell listed in order of increasing fraction of trans H-bonds. The lowest graph (dotted lines) gives the fraction of trans H-bonds associated with each isomer. Energies were calculated using the programs (●, ○)CPMD, (■,□)DMol, and (▲, △)CASTEP. Solid lines: energy of H-bond isomers before geometry optimization. Dashed lines: energies after optimization of the molecular coordinates, and for the CASTEP results cell dimensions as well. The data sets are plotted with their average taken as the zero of energy to facilitate comparison of the relative energies of the isomers. The $Cmc2_1$ isomer is noted and the optimized data sets are shifted by .06 kcal/mol.*

the vibrational free energy of the various H-bond isomers is nearly the same, and that quantum effects on free energy *differences* between isomers can be neglected. The excellence of these approximations is confirmed by the fact that H_2O/D_2O isotope effects have very little effect on the transition temperature of H-bond order/disorder transitions. Even the ice-Ih/XI transition, which occurs at very low temperature, shifts only by $4K$ from $72K$ to $76K$ upon replacing hydrogen by deuterium.[6] With these assumptions, the effective free energy surface for H-bond fluctuations is that of an equivalent spin-lattice model for the bond variables $\{b_r\}$, one that is easily treated by Monte Carlo methods[58,59] without introducing further approximations.

5 OUTLOOK

Electronic structure calculations coupled with graph invariant theory provides an effective treatment of hydrogen bond order-disorder phase transitions in ice. In the future, there are still more phase transitions that should be explored, such as possible differences in the low-temperature behavior between ice-Ih and ice-Ic,[60] and the ice XII-XIV system.[22] Hydrogen bond order-disorder phenomena in ice clathrates is also of great interest. Even though modest levels of electronic DFT seem to perform quite well for H-bond energetics in the ice phases, we are eager to benchmark our results against more powerful techniques as they become available. Perhaps the most exciting new horizons are found in the areas of dielectric phenomena and defect behavior. Ice responds to electric fields and conducts the passage of impurities and structural defects as a disordered medium. We have developed graph invariant theory for defects in water clusters,[33] and analogous work for defects in ice is on-going.

ACKNOWLEDGMENT

We gratefully acknowledge National Science Foundation for support of this work, and the Ohio Supercomputer Center for resources needed to perform the calculations reported here.

References

1 V. F. Petrenko and R. W. Whitworth _Physics of Ice;_ Oxford, 1999.

2 J. D. Bernal and R. H. Fowler _J. Chem. Phys._ 1933, **1**, 515.

3 S. Kawada _J. Phys. Soc. Japan_ 1972, **32**, 1442.

4 Y. Tajima, T. Matsuo, and H. Suga _Nature_ 1982, **299**, 810.

5 Y. Tajima, T. Matsuo, and H. Suga _J. Phys. Chem. Solids_ 1984, **45**, 1135.

6 T. Matsuo, Y. Tajima, and H. Suga _J. Phys. Chem. Solids_ 1986, **47**, 165.

7 T. Matsuo and H. Suga _J. Phys., Colloq._ 1987, **48**, 477.

8 A. J. Leadbetter, R. C. Ward, J. W. Clark, P. A. Tucker, T. Matsuo, and H. Suga _J. Chem. Phys._ 1985, **82**, 424.

9 R. Howe and R. W. Whitworth _J. Chem. Phys._ 1989, **90**, 4450.

10 C. M. B. Line and R. W. Whitworth _J. Chem. Phys._ 1996, **104**, 10008.

11 S. M. Jackson, V. M. Nield, R. W. Whitworth, M. Oguro, and C. C. Wilson _J. Phys. Chem._ 1997, **B101**, 6142.

12 B. Kamb and A. Prakash _Acta. Cryst._ 1968, **B24**, 1317.

13 E. Whalley, J. B. R. Heath, and D. W. Davidson _J. Chem. Phys._ 1968, **48**, 2362.

14 S. J. LaPlaca, W. C. Hamilton, B. Kamb, and A. Prakash _J. Chem. Phys._ 1973, **58**, 567.

15 K. Nishibata and E. Whalley _J. Chem. Phys._ 1974, **60**, 3189.

16 J. D. Londono, W. F. Kuhs, and J. L. Finney _J. Chem. Phys._ 1993, **98**, 4878.

17 G. P. Johari, A. Lavergne, and E. Whalley _J. Chem. Phys._ 1974, **61**, 4292.

18 W. F. Kuhs, J. L. Finney, C. Vettier, and D. V. Bliss *J. Chem. Phys.* 1984, **81**, 3612.

19 B. Kamb *Science* 1965, **150**, 205.

20 G. P. Johari and E. Whalley *J. Chem. Phys.* 1976, **64**, 4484.

21 G. P. Johari and E. Whalley *J. Chem. Phys.* 1979, **70**, 2094.

22 C. G. Salzmann, P. G. Radaelli, A. Hallbrucker, E. Mayer, and J. L. Finney *Science* 2006, **311**, 1758.

23 V. Buch, P. Sandler, and J. Sadlej *J. Phys. Chem.* 1998, **B102**, 8641.

24 M. J. Frisch *et al.*; Gaussian 03, Revision C.02; Gaussian, Inc., Wallingford, CT, 2004.

25 CPMD, Copyright IBM Corp. 1990-2001, Copyright MPI fur Festkörperforschung Stuttgart 1997-2004.

26 G. Kresse and J. Hafner *Phys. Rev. B* 1993, **47**, 558.

27 G. Kresse and J. Hafner *Phys. Rev. B* 1994, **49**, 14251.

28 G. Kresse and J. Furthmuller *Phys. Rev. B* 1996, **54** , 11169.

29 G. Kresse and J. Furthmuller *Comput. Mat. Sci.* 1996, **6**, 15.

30 G. Kresse and J. Furthmuller *Phys. Rev.* 1996, **54**, 11169.

31 S. McDonald, L. Ojamae, and S. J. Singer *J. Phys. Chem.* 1998, **A102** , 2824.

32 J.-L. Kuo, J. V. Coe, S. J. Singer, Y. B. Band, and L. Ojamae *J. Chem. Phys.* 2001, **114**, 2527.

33 S. D. Belair, J. S. Francisco, and S. J. Singer *Phys. Rev.* 2005, **A71**, 013204.

34 S. J. Singer, J.-L. Kuo, T. K. Hirsch, C. Knight, L. Ojamae, and M. L. Klein *Phys. Rev. Lett.* 2005, **94**, 135701.

35 C. Knight, S. J. Singer, J.-L. Kuo, T. K. Hirsch, L. P. Ojamae, and M. L. Klein *Phys. Rev.* 2006, **E73**, 056113.

36 C. Knight and S. J. Singer *J. Phys. Chem.* 2005, **B109**, 21040.

37 C. Knight and S. J. Singer *J. Chem. Phys.* 2006, **125**, 64506.

38 These calculations were performed using the BLYP functional[51,52], Troullier-Martins pseudopotentials[61] on oxygens, and a 70Ry plane wave cut-off.

39 B. Kamb and S. J. L. Placa *Trans. Am. Geophys. Union* 1974, **56**, 1202.

40 B. Kamb, A. Prakash, and C. Knobler *Acta Cryst.* 1967, **22**, 706.

41 C. Knight and S. J. Singer; Unpublished work.

42 C. Lobban, J. L. Finney, and W. F. Kuhs *J. Chem. Phys.* 2000, **112**, 7169.

43 I. Minagawa *J. Phys. Soc. Japan* 1981, **50**, 3669.

44 B. Kamb In E. Whalley, S. J. Jones, and L. W. Gold, Eds., *Physics and chemistry of ice*, Symposium on the Physics and Chemistry of Ice, page 28, Ottawa, (1973). Royal Society of Canada.

45 J. F. Nagle In E. Whalley, S. J. Jones, and L. W. Gold, Eds., *Physics and chemistry of ice*, Symposium on the Physics and Chemistry of Ice, page 70, Ottawa, (1973). Royal Society of Canada.

46 R. Howe and R. W. Whitworth *J. Chem. Phys.* 1987, **86**, 6443.

47 L. G. MacDowell, E. Sanz, C. Vega, and J. L. F. Abascal *J. Chem. Phys.* 2004, **121**, 10145.

48 J. F. Nagle *J. Math. Phys.* 1966, 7, 1484.

49 Actually, Kamb uses VI ′ to indicate a partially ordered phase, and VI″ to denote the fully ordered form of ice-VI.

50 That is, the energy difference is comparable to Boltzmann's constant k_B times $100K$.

51 A. D. Becke *Phys. Rev.* 1988, **A38**, 3098.

52 C. Lee, W. Yang, and R. G. Parr *Phys. Rev.* 1988, **B37**, 785.

53 R. Car and M. Parrinello *Phys. Rev. Lett.* 1985, **55**, 2471.

54 J. P. Perdew and Y. Wang *Phys. Rev.* 1986, **B33**, R8800.

55 J. P. Perdew In P. Ziesche and H. Eschrig, Eds., *Electronic Structure of Solids, '91*, page 11, Berlin, 1991. Wissenschaftsbereich Theoretische Physik, Akademie Verlag.

56 J. P. Perdew and Y. Wang *Phys. Rev.* 1992, **B45**, 13244.

57 J.-L. Kuo and M. L. Klein *J. Phys. Chem.* 2004, **B108**, 19634.

58 K. Binder, Ed. *Monte Carlo Methods in Statistical Physics*, Vol. 7 of *Topics in Current Physics;* Springer: New York, 2nd ed., 1986.

59 A. Rahman and F. H. Stillinger *J. Chem. Phys.* 1972, **57**, 4009.

60 H. Suga *Thermochimica Acta* 1997, **300**, 117.

61 N. Troullier and J. L. Martins *Phys. Rev.* 1991, **B43**, 1993.

THEORETICAL STUDY OF A HYDROXIDE ION WITHIN THE ICE-Ih LATTICE

C. Knight and S. J. Singer

Department of Chemistry, Ohio State University
100 W. 18th Ave., Columbus, OH 43210, USA

1 INTRODUCTION

One of the most difficult and intriguing aspects of ice physics is the behavior of defects.[1] Although defects have little effect on the statics of phase transitions, they are the key to understanding the dynamics and mechanism of phase transitions. For example, ice-Ih must be doped with hydroxide to catalyze the transition to ice-XI.[2-10] The mechanism by which hydroxide (OH^-) catalyzes the ice-Ih/XI transition is not understood.[11-13] In fact, it is not even clear that hydroxide defects have significant mobility in this temperature range near $70K$. The structure and transport properties of defects is relevant to problems in environmental and atmospheric science[14] and glaciology.[1] In this work, we introduce techniques applicable to the study of ionic defects, H^+ and OH^-, and neutral defects, such as the OH radical, in ice.

There is general agreement that ionic defects are immobilized on an accessible experimental time scale somewhere between 100 and $200K$. Beyond that, there is considerable disagreement, and seeming contradiction in the literature. Devlin and co-workers doped D_2O impurities in H_2O ice, or D_2O impurities in H_2O ice.[15-18] For the case of D_2O impurities in H_2O, passage of an ionic defect through the location of the impurity, will lead to the formation of HOD molecules.[19] Devlin's group spectroscopically monitors the formation of HOD as a function of temperature, providing information on the mobility of ionic impurities as a function of temperature. They observe that excess protons begin to actively diffuse at temperatures above $120K$, while hydroxide diffusion seems to be much slower.

The results of the Devlin group raise several issues. Recall that the ice-Ih/XI phase transition at $72K$ is catalyzed by hydroxide.[2-10] However, it appears that *no* ionic defects are actively diffusing at $72K$. Furthermore, even though hydroxide, not protons, catalyze this phase transition, the results of Devlin's group indicate that protons are initially the most active diffusing species as temperature is raised. Hence, the results of the Devlin group call out for a better understanding of the behavior of ionic defects in ice, and in particular, the mechanism by which hydroxide catalyzes the Ih/XI phase transition.

There are further gaps in our understanding of ionic defects in ice. Cowin and co-workers probed the diffusion of protons by a different method, by gently landing hydronium ions on the surface of a layer ice and subsequently observing the electric potential set up by the ionic layer.[20] In contrast to the Devlin group experiments, they found that the excess protons did not diffuse into the ice until the temperature reached $190K$. The difference between the Cowin and Devlin group experiments may originate from the fact that the former probes long range movement of the excess protons, while the latter may be more sensitive to local motions. Clearly, there is ample room for theoretical methods to make an impact on this field. However, this is a challenging system. Since ice-Ih is an H-bond disordered substance, it is incorrect to picture ionic defects as diffusing in a periodic potential. The H-bond disorder sets up a *random* medium, with potential traps for both ionic (and Bjerrum) defects in the disordered material. Characterizing the statistical properties of this random medium, particularly with interest in characterizing the passage of ionic defects, is one of the goals of this work.

We have previously shown that periodic electronic density function theory (DFT) is capable of describing the delicate energy differences between H-bond isomers in ice phases.[21-25] At this stage, we have accumulated a body of evidence for several different hydrogen bond order/disorder phase transitions (Ih-XI, VII-VIII, III-IX, V-XIII) that confirm that electronic DFT for small unit cells, combined with an analytic method we call *graph invariants*[26,27] to extrapolate to the bulk limit for statistical mechanical systems, can accurately describe hydrogen bond order-disorder phenomena in ice phases. The methodology is not as well tested for charged defects. Whereas we have verified that H-bond energetics in pure ice were rather insensitive to the DFT functional and basis set,[21,23,25] it is known that hydroxide in liquid water is more sensitive to the DFT method.[28-34] Therefore, we regard the results presented here for hydroxide ion in ice as a first step in the process of building a reliable model of charged defects in ice. We first present results on the structure of the hydroxide defect in section , and then describe graph invariant theory applied to defects in section . The outlook for further studies is given in section

2 STRUCTURE OF THE HYDROXIDE ION IN ICE-Ih

An ionic defect cannot be studied in isolation in a periodic system. For example, removal of a proton from one water molecule creates a hydroxide and L-defect pair, which can diffuse away from each other, as shown schematically in Fig. 1. Alternatively, a hydroxide/hydronium pair can be cre-

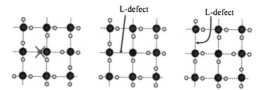

Figure 1 Creation of a hydroxide (OH$^-$) defect is shown schematically in the left panel. The result, center panel, is actually neighboring L and hydroxide defects, which can diffuse from each other as shown in the right panel.

ated. We used the process shown in Fig. 1, removing a proton and rearranging H-bonds in a unit cell containing 96 water molecules, to generate configurations with variable distance between the ionic and Bjerrum defect. The structural parameters given here are obtained from electronic DFT calculations using the BLYP functional,[35,36] Troullier-Martins pseudopotentials[37] on oxygens with a $70Ry$ plane wave cut-off.

Strong hydrogen bonds are characterized by a shortening of the distance between donor and acceptor oxygens, and a lengthening of the oxygen-hydrogen distance of the donor bond. Based on this criterion, the hydroxide ion in ice-Ih forms strong H-bonds as an acceptor, and weak H-bonds as a donor. The local structure near a hydroxide ion shows little variation depending on the hydrogen bonding arrangement of the water surrounding the defect, and even depends little on whether the hydroxide points along the c-axis or lies within an ab-bilayer. A generic local geometry near a hydroxide is shown in Fig. 2. Bond lengths variation from site to site is on the order of 0.01Å and bond angle variation is typically a degree or less. Compared to normal oxygen-oxygen bonds in ice,

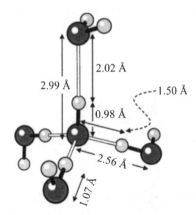

Figure 2 Local geometry near a hydroxide defect.

the distance from the oxygens of water molecules that donate to the hydroxide is shortened to about 2.56Å, while the distance from the hydroxide oxygen to the oxygen of the water that accepts a hydrogen bond from the hydroxide is lengthened to 2.99Å. The bond lengths are the only structural feature that deviate strongly from bulk ice. The hydrogen bonds to the hydroxide are all within 3 degrees of being linear and the water angles are close to 106°, both typical of bulk ice.

We now turn to the interaction between the hydroxide ion and the surrounding random H-bond disordered medium. To illustrate the effect, two configurations of a 96-water (95 waters and a hydroxide) unit cell are shown in Fig. 3. The hydroxide and L-defects in Fig. 3 are at the same lattice positions. The H-bond arrangement of the other water molecules differs, while, except for the defects, still maintaining the ice rules[38] (each water donates to two H-bonds, and accept two other H-bonds).

With the help of the graph invariant theory described in section , we were able to identify the most stable H-bond configurations surrounding the defect pair. In pure water ice, the $Cmc2_1$ arrangement of ice-XI proposed on the basis of diffraction data[7-10] is

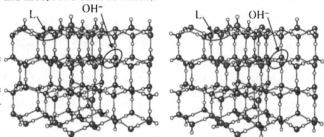

Figure 3 Two possible unit cells for 95 waters and a hydroxide. The L-defect, originally created when a proton was removed from the system, has diffused away from the hydroxide by a distance of roughly 8.5Å.

calculated to be the lowest energy arrangement. Now the question arises as to the degree to which these defects disrupt the lowest energy pure water arrangements. The answer is that the disruption is minimal: a hydroxide prefers to be surrounded by an ice XI structure. Of course, neither a hydroxide nor L-defect is compatible with a perfect ice XI structure. At least one of the surrounding bonds must be reversed. The family of low-energy H-bond arrangements preferred in our unit cell contain a path of bonds reversed from the ice XI structure that connects the two defects. The lowest energy structure located to date is shown in Fig. 4. In the $Cmc2_1$ ice XI structure, all c-axis H-bonds point in the same direction. The H-bonds in each ice XI ab-bilayer are aligned in a direction that alternates from layer to layer. The H-bonds in Fig. 4 are all in the ice XI structure except for a path of reversed H-bonds leading from the L-defect to the hydroxide. The reversed H-bonds are shown as thicker in the figure. Because of periodic boundary conditions (pbc's), the chain of reversed H-bonds leaves the unit cell at the top of Fig. 4 and then re-emerges at the bottom to point into the hydroxide defect.

These results are significant because they indicate that the hydroxide defect and ice XI

structure are highly compatible, and that the ice XI structure is disturbed only to the extent
necessary to maintain full hydrogen bonding.

3 GRAPH INVARIANT THEORY

It would not be feasible to
use electronic structure calculations to search through
the billions of possible H-
bond arrangements possi-
ble for the unit cell pictured
in Figs. 3 and 4. Nor would
it be possible to use only
electronic structure methods to describe the statistical mechanics of the hydrogen bond disordered ice
lattice. To sift through, or
thermally average over H-
bond fluctuations, we first
link the energy to H-bond
topology.   The configu-
ration of H-bonds is de-
scribed by a set of bond
variables. In pure water ice
the bond variable b_r takes
the values $+1$ or -1 depending on the orientation

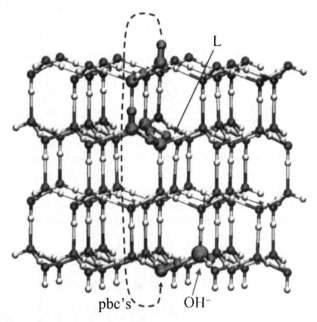

Figure 4 The most stable hydrogen bond arrangements in a
96-water unit cell containing a hydroxide and L-defect.

of the H-bond at bond position r. *If* the energy, or any other scalar physical quantity, can be
linked to H-bond topology (and this is indeed a significant "if" because the topology contains less information than the full set of atomic position coordinates), then it must depend
upon combinations of the full set of bond variables $\{b_r\}$ in special combinations that, like
the energy, are invariant to symmetry operations. We refer to these combinations, polynomials in the bond variables, as *graph invariants*. First order invariant polynomials in the $\{b_r\}$
are *first order graph invariants*, quadratic polynomials are *second order graph invariants*,
and so on. For most ice lattices, there are no invariant linear functions of the $\{b_r\}$ and the
leading order basis functions used to link energy to H-bond topology are the second order
graph invariants. This method is described in previous publications,[21–24,26,27] and an article in this volume.[25] The ability of graph invariants to capture the energetic trends among
H-bond arrangements in water clusters and in ice phases, and to predict H-bond ordering
phase transitions between ice phases, is documented in these previous works. Here we are
concerned with the extension to systems with defects.

To describe a system containing both an ionic and L-defect, as pictured in Figs. 3 or
4, the bond variables are also allowed to take the value of 0, which signifies the absence
of the OH bond of a neutral water molecule. In addition to the b-variables, we introduce a
variable c_r for each bond indexed by r. The c's are zero except at the hydroxide site where
the c-variable takes the value ± 1 to describe the orientation of the hydroxide. This scheme
is not unique, but it is sufficient to capture all possible H-bond arrangements in the system

pictured in Figs. 3 or 4.

Following our earlier work, polynomials in the b- and c- variables invariant to symmetry operations are generated using the projection operator for the totally symmetry representation of the appropriate symmetry group for the system under study.[21-27] For ice lattices, this is the crystal space group[27] while for finite water clusters it is the point group.[26] Because there are usually no invariant linear polynomials (first order graph invariants) and higher order polynomials have, to date, never been needed, the most important of these polynomials are the quadratic ones (second order graph invariants). Second order graph invariants involving only b-variables,

$$I_{rs} = \frac{1}{|G|} \sum_{\alpha=1}^{|G|} g_\alpha(b_r b_s) , \tag{1}$$

are used to describe pure water ice. In this equation $|G|$ is the order of the symmetry group and is used to make the invariants intensive functions. The sum is over the members of the symmetry group. The invariant is labeled by the bond pair, r and s, from which it is generated by the projection operator. With a single hydroxide ion present, invariants of the form

$$I_{rs(c)} = \frac{1}{|G|} \sum_{\alpha=1}^{|G|} g_\alpha(b_r c_s) , \tag{2}$$

involving a b- and a c-variable are also needed to describe interactions between the hydroxide and surrounding water molecules. In principle, if several hydroxides were present and interacting with each other, invariants involving two c-variables would be required. We have previously shown how this formalism can be applied to a hydroxyl radical defect in a water cluster.[39]

Using expressions like Eqs. (1-2) we can generate a complete set of invariant functions to parameterize the dependence of energy on H-bond topology in the following form.

$$E(b_1, b_2, \ldots, c_1, c_2, \ldots) = \underbrace{\sum_r \alpha_r I_r(b_1, b_2, \ldots, c_1, c_2, \ldots)}_{\text{sum over 1st order invariants}} \tag{3}$$
$$+ \underbrace{\sum_{rs} \alpha_{rs} I_{rs}(b_1, b_2, \ldots, c_1, c_2, \ldots)}_{\text{sum over 2nd order invariants}} + \underbrace{\cdots}_{\text{negligible}}$$

In practice, the α-coefficients are determined for small unit cells for which electronic structure calculations are feasible. Since we have shown that invariants for a small unit cell are automatically invariants for larger cells, the formalism provides a means to extrapolate an expression for the energy to unit cells large enough to be a simulation cell for statistical mechanical calculations. In practice, the expansion is truncated at second order, as indicated in Eq. (3).

To date, for pure water ice phases only second order invariants generated by projection on a small number of nearby bond pairs were needed. For example, for ice-Ih three second order invariant functions provided an accurate parameterization of the energy.[21,23,25] We used those same three invariant functions with identical α coefficients to describe the pure water portion of the system with an L- and ionic defect. We incorporated 6 additional invariants of the form given in Eq. (2) involving a b- and closeby c-variable. On physical grounds, we expect charge-dipole interactions to be important in the presence of ionic defects. The implication would be that invariants involving variables b_r and c_s where bonds r and s are not close to each other might also be significant. To account for this possibility, we included an interaction (also invariant!) involving a charge on the the hydroxide oxygen and oxygens bracketing the L-defect, and dipoles on water molecules. The sum of the hydroxide and L-defect charge was constrained to be -1. The energy expression with the charge-dipole interaction is still in the form of Eq. (3). The charge-dipole interaction amounts to slaving a large number of $b_r c_s$ terms together with a single coefficient proportional to the combination $\frac{q\mu}{\epsilon_{eff}}$, where ϵ_{eff} is an effective dielectric constant for the medium.

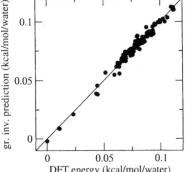

At the beginning of this section, we described the link between energy and H-bond topology as something that cannot be taken for granted and must be verified. The data in Fig. 5 is an important part of that verification process. If the fit using the graph invariant expression of Eq. (3) was perfect, all points would lie on the diagonal. The low-energy points in Fig. 5 are the ice XI-like structures described in section with a path of reversed H-bonds connecting the hydroxide and L-defect. In our fit to the data, the charge on the L-defect went to zero and the hydroxide had a charge of -1. (The charge of $-.38$ normally associated with an L-defect is a charge associated with a polarization current when a water molecule reorients, not a physical charge.)

Figure 5 Energy of H-bond isomers of a 96-water unit cell containing a hydroxide and L-defect is compared with a fit of the form of Eq. (3).

4 OUTLOOK

The information provided here, which concentrates on energetic and structural aspects of the problem, represents only the first steps toward understanding the behavior of a hydroxide defect in ice. The next stage is to analyze the dynamics of hydroxide motion, which will require both "microscopic" and "macroscopic" approaches. The microscopic event consists of proton transfer from one of the three waters pictured at the bottom of Fig. 2 to the hydroxide. It would not be surprising if calculation of transition states requires levels of electronic structure theory beyond the modest levels of DFT that suffice for equilibrium structures. The energy and free energy barrier of this process sets the time scale for local hopping of the hydroxide ion. The macroscopic level is a description of defect diffusion within the disordered H-bond network.

ACKNOWLEDGMENT

We gratefully acknowledge National Science Foundation for support of this work, and the Ohio Supercomputer Center for resources needed to perform the calculations reported here.

References

1 V. F. Petrenko and R. W. Whitworth *Physics of Ice;* Oxford, 1999.

2 S. Kawada *J. Phys. Soc. Japan* 1972, **32**, 1442.

3 Y. Tajima, T. Matsuo, and H. Suga *Nature* 1982, **299**, 810.

4 Y. Tajima, T. Matsuo, and H. Suga *J. Phys. Chem. Solids* 1984, **45**, 1135.

5 T. Matsuo, Y. Tajima, and H. Suga *J. Phys. Chem. Solids* 1986, **47**, 165.

6 T. Matsuo and H. Suga *J. Phys., Colloq.* 1987, **48**, 477.

7 A. J. Leadbetter, R. C. Ward, J. W. Clark, P. A. Tucker, T. Matsuo, and H. Suga *J. Chem. Phys.* 1985, **82**, 424.

8 R. Howe and R. W. Whitworth *J. Chem. Phys.* 1989, **90**, 4450.

9 C. M. B. Line and R. W. Whitworth *J. Chem. Phys.* 1996, **104**, 10008.

10 S. M. Jackson, V. M. Nield, R. W. Whitworth, M. Oguro, and C. C. Wilson *J. Phys. Chem.* 1997, **B101**, 6142.

11 M. J. Iedema, M. J. Dresser, D. L. Doering, J. B. Rowland, W. P. Hess, A. A. Tsekouras, and J. P. Cowin *J. Phys. Chem.* 1998, **B102**, 9203.

12 R. W. Whitworth *J. Phys. Chem.* 1999, **103**, 8192.

13 J. P. Cowin and M. J. Iedema *J. Phys. Chem.* 1999, **103**, 8194.

14 R. P. Wayne *Chemistry of Atmospheres;* Oxford University Press: Oxford, 1991.

15 G. Ritzhaupt and J. P. Devlin *Chem. Phys. Lett.* 1979, **65**, 592.

16 W. B. Collier, G. Ritzhaupt, and J. P. Devlin *J. Phys. Chem.* 1984, **88**, 363.

17 P. J. Wooldridge and P. Devlin *J. Chem. Phys.* 1988, **88**, 3086.

18 J. P. Devlin *Int. Rev. Phys. Chem.* 1990, **9**, 29.

19 Ionic defects diffusing through a D_2O molecule split that molecule into two neighboring HOD molecules,[15-18] designated as $(HOD)_2$ in the Devlin group's work. Both Bjerrum and ionic defects are needed for further diffusion of the deuterium contained in the $(HOD)_2$ pair, ultimately leading to isolated HOD molecules. D_2O, $(HOD)_2$ and isolated HOD molecules are spectroscopically identifiable as separate species.

20 J. P. Cowin, A. A. Tsekouras, M. J. Iedema, K. Wu, and G. B. Ellison *Nature* 1999, **398**, 405.

21 S. J. Singer, J.-L. Kuo, T. K. Hirsch, C. Knight, L. Ojamae, and M. L. Klein *Phys. Rev. Lett.* 2005, **94**, 135701.

22 C. Knight and S. J. Singer *J. Phys. Chem.* 2005, **B109**, 21040.

23 C. Knight, S. J. Singer, J.-L. Kuo, T. K. Hirsch, L. P. Ojamae, and M. L. Klein *Phys. Rev.* 2006, **E73**, 056113.

24 C. Knight and S. J. Singer *J. Chem. Phys.* 2006, **125**, 64506.

25 C. Knight and S. J. Singer (this volume, page xxx.).

26 J.-L. Kuo, J. V. Coe, S. J. Singer, Y. B. Band, and L. Ojamae *J. Chem. Phys.* 2001, **114**, 2527.

27 J.-L. Kuo and S. J. Singer *Phys. Rev.* 2003, **E67**, 016114.

28 B. Chen, J. M. Park, I. Ivanov, G. Tabacchi, M. L. Klein, and M. Parrinello *J. Amer. Chem. Soc.* 2002, **124**, 8534.

29 B. Chen, I. Ivanov, J. M. Park, M. Parrinello, and M. L. Klein *J. Phys. Chem.* 2002, **B106**, 12006.

30 Z. Zhu and M. E. Tuckerman *Journal of Physical Chemistry* 2002, **B106**, 8009.

31 M. E. Tuckerman, D. Marx, and M. Parrinello *Nature* 2002, **417**, 925.

32 D. Asthagiri, L. R. Pratt, J. D. Kress, and M. A. Gomez *Chem. Phys. Lett.* 2003, **380**, 530.

33 D. Asthagiri, L. R. Pratt, J. D. Kress, and M. A. Gomez *Proc. Nat. Acad. (USA)* 2004, **101**, 7229.

34 M. E. Tuckerman, A. Chandra, and D. Marx *Acc. Chem. Res.* 2006, **39**, 151.
35 A. D. Becke *Phys. Rev.* 1988, **A38**, 3098.
36 C. Lee, W. Yang, and R. G. Parr *Phys. Rev.* 1988, **B37**, 785.
37 N. Troullier and J. L. Martins *Phys. Rev.* 1991, **B43**, 1993.
38 J. D. Bernal and R. H. Fowler *J. Chem. Phys.* 1933, **1**, 515.
39 S. D. Belair, J. S. Francisco, and S. J. Singer *Phys. Rev.* 2005, **A71**, 013204.

ATOMIC FORCE MICROSCOPY OF REARRANGING ICE SURFACES

M. Krzyzak, K.S. Techmer, S.H. Faria, G. Genov*, and W.F. Kuhs

GZG, Abt. Kristallographie, Universität Göttingen, Goldschmidtstr. 1, 37077 Göttingen, Germany

1 INTRODUCTION

The study of crystallization and alteration processes of natural and artificial snow crystals has been a subject of interest for several decades. Nakaya[1] was the first who investigated the relations between growth forms and experimental conditions (temperature and water vapour saturation relative to ice). Although many laboratory[2-4] and theoretical studies[5-8] on ice crystals have been carried out since his seminal work, snow crystal growth is still not completely understood.[8] Experimental studies are challenged by the complexity of the physical processes influencing the crystal growth, which are further complicated by effects of chemical impurities.[8]

Ice crystallization phenomena and the properties of ice surfaces have been the subject of many experimental investigations using optical methods,[1-4,9] Scanning Electron Microscopy (SEM),[10,11] ellipsometry,[12,13] Nuclear Magnetic Resonance (NMR),[14] just to mention a few. Complementing these methods, Atomic Force Microscopy (AFM) offers the possibility to investigate dynamic changes of the ice surface under various conditions at a vertical resolution of a few nanometres. AFM enables studies of the surface kinetics of ice and allows for an imaging of surface features in three dimensions at variable water vapour pressures in various atmospheres. A further advantage of AFM is its ability to measure the physical properties of the ice surface (e.g., friction, adhesion, elastic and plastic deformation) and to determine the nature of the Quasi-Liquid-Layer (QLL) on ice. Several research groups applied AFM to investigate ice surfaces;[15-21] most of these studies focussed on the thickness of the QLL on ice. The thickness of this layer is calculated from measurements of force-versus-distance curves. In a typical force experiment, the normal force between the tip and the sample surface can be obtained as a function of the tip position. The interpretation of force curves is very complex, because they can be affected by elastic and plastic deformation of ice, capillary forces, adhesion, flow of the QLL under the tip, electrostatic as well as van der Waals forces.

The results of various research groups show strong differences in the thickness of the QLL. Pittenger et al.[16] calculated the thickness of the QLL assuming that it has the viscosity of supercooled water; they estimated the QLL thickness to be about 1 nm at -1°C and 0.2 nm at -10°C for a silicon tip. For a hydrophobically coated tip, the layer thicknesses were slightly smaller. Several authors have used the jump-in distance to estimate the thickness of the QLL (gradient of the tip-sample forces becomes greater than

* present affiliation: Department of Chemistry, University of Bergen, Allegaten 41, N-5007 Bergen, Norway

the spring constant of the cantilever and the tip jumps-in to the surface). With this method, Petrenko[15] obtained 2-16 nm of QLL at -10.7°C, Döppenschmidt and Butt[17] about 30-50 nm at -1°C and about 5-15 nm at -10°C. In another paper, Butt et al.[18] interpreted the force versus distance curve as produced by plastic deformation of ice and excluded the existence of an interfacial layer (between the moving tip and the ice surface) significantly thicker than a monolayer, still leaving an option for a QLL in the unperturbed system. Bluhm and Salmeron[19] estimated the thickness of the QLL by comparison of results obtained in various AFM operation modes. They obtained a thickness of about 5 nm between -20°C and -10°C.

In previous AFM work different solutions for temperature control were chosen: experiments in the cold chamber or freezer, where the temperature of the sample was determined by the environmental temperature;[15,17,18] experiments with a special sample holder, where the sample temperature was locally controlled while keeping the surrounding atmosphere below 0°C;[16,23] or experiments with the sample submerged in organic solvents.[15,23] In most cases, except the last one, the humidity was kept constant around 80-85% relative to the temperature of the ice sample. Most research groups used an optical filter to minimize the intensity of the laser beam to avoid the heating of the cantilever.

In general, AFM studies of the interface between ice and air seem to be tricky[15,22] and even the results obtained under similar experimental conditions show large differences.[19] Especially at temperatures above -20°C the ice surface is very mobile and changeable.[15,22] Whereas the thickness of the QLL and the ice surface properties have been extensively analysed through the interpretation of force curves, the in situ investigations of ice crystal growth depending on temperature and relative humidity have not been systematically performed. Most ice surfaces investigated by AFM are thin ice films grown on mica substrates. Zepeda et al.[23] have grown polycrystalline ice samples on mica from the vapour phase (through a pulse of N_2-gas with high saturation of water vapour). They obtained ice crystals of about tens of micrometres in diameter. The interface between two grain boundaries was tens or even hundreds of nanometres deep. About 3 min after nucleation the smaller grains with higher surface energy were incorporated into larger grains with lower surface energy and they observed sublimation of the sample. Bluhm and Salmeron[19] obtained an ice film of round, flat 0.2-0.3 nm high platelets on mica at temperatures below -30°C. On top of this film, supercooled water droplets formed with a height of max. 12 nm and 2 μm in diameter, which subsequently have been removed during scanning. Outside the scan area, the morphology of the ice film did not change within 30 min. After annealing to -17°C, the droplets were replaced by a contiguous liquid film with a height of ~5 nm. The images of 0.5 mm thick ice samples obtained from Döppenschmidt et al.[22] show a very rough ice surface, with "icebergs" and "valleys". The typical peak-to-valley distance was in the order of 100 nm.

The purpose of the AFM investigations presented here is to obtain information about crystallization and alteration processes of ice, in particular about the formation and metamorphosis of natural snow. The imaging of the ice surface combined with the possibility for an estimation of occurrences of the QLL on it at various conditions reveal new abilities for a better understanding of crystallization and alteration of ice. Nevertheless, the observation of ice crystal formation in situ is a very difficult task, seeing that this process is very sensitive to the conditions of pressure, temperature and relative humidity. We performed experiments under various conditions (sample thickness, temperature and relative humidity) to test the potentials of AFM for in situ investigations on changing ice surfaces. The sample was placed on a Peltier cooler working in the temperature range from 0 to -35°C and located in a glove box at a room temperature of about 20°C. This setup allows investigating the dynamic changes of the ice surface, which is quite unstable because of the changes in the ice surface exposed to room temperature and driven also by the large temperature gradient between the ice sample and the tip. Comparative surface investigations on samples recovered within a few seconds to liquid N_2 temperature were made by Cryo-SEM at -175°C for verifying the AFM images and to help identifying AFM scanning artifacts.

2 METHODS AND RESULTS

2.1 Experimental

For our studies we used a commercial Atomic Force Microscope (PicoSPM, Molecular Imaging) as it provided a large open access area for installing the low-temperature sample environment. Several technical problems needed to be solved before we could investigate the surface of ice in air at temperatures down to –35°C.

The steep slope of the ice/vapour coexistence line leads to large fluctuations in the equilibrium vapour pressure, which causes the surface to grow or sublimate rapidly. A constant temperature maintained in the vicinity of the measured ice surface is a key to obtain useful representative and reproducible surface images. For a precise control of the ice samples temperature a regulated sample stage was used (Fig.1); it consists of a stainless steel plate with a central hole for a copper plate, a Peltier cooler and copper block. The copper plate is located between the sample holder and the Peltier cooler, and the other copper block is mounted under the Peltier cooler to dissipate the heat. The lowest temperature of the sample stage is determined by the heat dissipation through the copper block, which is actively cooled via a circulating water-glycol mixture. The temperature of the sample stage was permanently controlled within the range of 0 to -35°C with 0.01°C precision using a commercially available temperature controller. The temperature can be held better than 0.1°C for several hours. The copper sample holder was fixed on the copper plate with a thermally conductive paste. To control the humidity in the atmosphere close to the sample surface and to avoid excessive ice growth the AFM investigations were performed inside a glove box at about 20°C, while the relative humidity was controlled by nitrogen flow. Before the experiment started, the chosen humidity of the chamber was initially adjusted. The relative humidity was measured with a commercially available relative humidity sensor (Lambrecht, HygroLog-D).

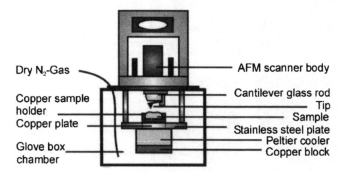

Figure 1   *Schematic diagram of the Atomic Force Microscope with a cooling stage.*

The samples were prepared in two different ways:

- Samples A: Obtained from freezing 0.5 ml of distilled water on the copper plate down to -30°C (cooling rate 1.5-2.0°C/min). The thickness of the bulk ice sample was about 2.8 mm.
- Samples B: Obtained from condensation of water vapour. The copper plate is cooled down to -30°C. Ice grows on the copper plate that is the coldest spot exposed to the water vapour in the chamber glove box. The thickness of the ice film was about 3-10 μm.

After freezing of the samples we wait 10-15 min to allow the ice sample to equilibrate. Primarily, the tip temperature is influenced by the temperature of the glove box of about 20°C. To avoid the alteration of the ice samples due to heating, the tip is brought in the vicinity of the cold sample so that the interaction between the sample and the tip could be observed. Subsequently, the tip is left there for 20 min for cooling before the start of the measurements. Then the samples were displaced to obtain images from "fresh" parts of the surface. Different commercially available types of uncoated cantilevers with hydrophilic tips were used, such as Si_3N_4 with spring constant <1 N/m, as well as silicon cantilevers with spring constant of 3 N/m in contact mode. For investigations in the alternating contact (AAC) mode we used a cantilever with spring constants between of 0.5-9.5 N/m.

2.2 Results

2.2.1 Samples A (obtained from freezing of distilled water). The AFM measurements of the surface of thick ice samples A are quite difficult. The ice surface features seem to be unstable and changeable from scan to scan and even the images obtained under the same experimental conditions show different ice surface characteristics. The appearance of the ice surface depends on the relative humidity (Fig.2).

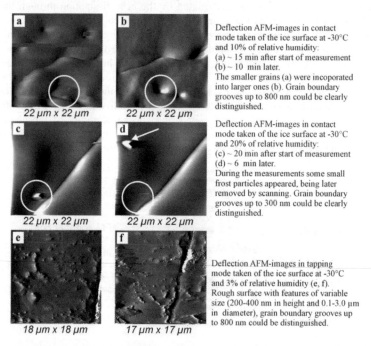

a

b

22 μm x 22 μm   22 μm x 22 μm

Deflection AFM-images in contact mode taken of the ice surface at -30°C and 10% of relative humidity:
(a) ~ 15 min after start of measurement
(b) ~ 10 min later.
The smaller grains (a) were incoporated into larger ones (b). Grain boundary grooves up to 800 nm could be clearly distinguished.

c

d

22 μm x 22 μm   22 μm x 22 μm

Deflection AFM-images in contact mode taken of the ice surface at -30°C and 20% of relative humidity:
(c) ~ 20 min after start of measurement
(d) ~ 6 min later.
During the measurements some small frost particles appeared, being later removed by scanning. Grain boundary grooves up to 300 nm could be clearly distinguished.

e

f

18 μm x 18 μm   17 μm x 17 μm

Deflection AFM-images in tapping mode taken of the ice surface at -30°C and 3% of relative humidity (e, f).
Rough surface with features of variable size (200-400 nm in height and 0.1-3.0 μm in diameter), grain boundary grooves up to 800 nm could be distinguished.

Figure 2   *AFM images along ice surface of samples A (prepared by freezing of distilled water) under different relative humidity conditions.*

AFM pictures taken 15 min after the start of the measurement at the humidity ≥10% showed often fine small crystallites (2-10 μm). Small grains were incorporated in ~10 min into larger ones (Fig.2a and 2b). The small crystals were detected at several occasions. Frequently the AFM-images show grain boundary grooves of a width of 100–800 nm (Fig.2). The large difference in the height of grain boundary grooves is influenced by the mutual orientation of the crystals and the sizes of the crystals. The sizes of individual crystals could hardly be defined, most likely because the crystals soon got larger than the scan area (30 μm^2). In the course of the measurements small angular frost particles appeared, being later removed during the scanning (Fig.2c and 2d). At a relative humidity of 3%, especially in tapping mode, one could recognize rough surfaces with features of variable size (2-400 nm in height and 0.1-3.0 μm in diameter) and grain boundary grooves (Fig.2e and 2f). In contrast to the ice surface observed at the humidity ≥10% (see Fig.2a and 2b), at the humidity of 3% the appearance and coarsening of small grains (<20 μm) were not observed.

Comparative SEM investigations show that large frost crystals appear on ice surfaces at relative humidity ≥10% (Fig.3a and 3b, RH=30%), while the ice surface at a relative humidity of ≤5% is flat and dense without secondary alteration (Fig.3c and 3d, RH=3%).

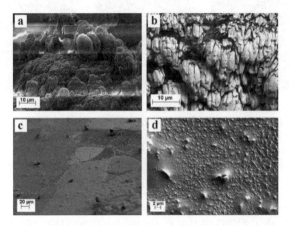

Figure 3 *SEM images of surface of samples A (prepared by freezing of distilled water) at -30°C and relative humidity of 30% (a and b) and 3% (c and d). The surface of samples prepared at 30% of relative humidity is covered with frost crystals that formed in the glove box chamber due to the high humidity. The surface of samples prepared at 3% of relative humidity is flat and dense, some grain boundaries are visible. Isolated crystals (dark dots) on the ice surface are frost crystals that formed during the transfer to the SEM-cryo-chamber (c).*

Comparison of the frost crystals observed via SEM at humidity of 30% (Fig.3a and 3b) prepared with those observed via AFM under similar conditions (Fig.2a, 2b, 2c and 2d) shows very little similarities. The temperature difference between the laser heated tip[24] and the ice surface has a large impact on the AFM imaging process. When the tip touches the surface covered with frost, melting of frost crystals immediately takes place, so that the tip loses contact with frost/ice (established via force-versus-distance curves). In this case the tip was approached further to the surface. The small size of the crystallites obtained at humidity of ≥10 % at the first minutes of AFM measurements (Fig.2a and 2b) suggests a rapid crystallization, probably by refreezing of melted frost. The melting process of the rough frost induced by AFM measurements can cause considerable differences in the results from AFM and SEM.

The investigations of ice surface made by these two techniques at relative humidity of 3% indicate a similar, flat and dense structure with grain boundary grooves without frost (Fig.2e and 2f, Fig.3c and 3d). Also in this case melting induced by the tip and refreezing of the ice surface take place, but at lower relative humidity we obtain AFM images showing etched flat ice surface similar to SEM investigations.

In the course of one AFM experiment, after the measurement at 5% relative humidity, we increased the relative humidity of the environment up to 50% during ~1 min. The increase of the relative humidity led to the formation of delicate frost on the flat surface of ice (Fig.4a). These individual frost crystals disappeared within a few minutes of scanning (Fig.4b and 4c) and from them remained just a few residuals (Fig.4c). It is however not clear whether the crystals were removed, crushed or melted by the AFM tip.

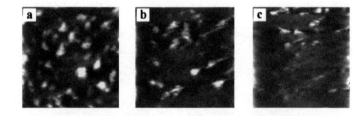

Figure 4   *AFM-image showing the nucleation of frost. The freezing process of distilled water took place on the sample stage at -30°C and at 5% of relative humidity; afterwards the relative humidity was increased up to 50%. The images were taken continuously with a scanning time of 3 min for each picture.*

Apparently, we were also able to capture the growth of an individual ice crystal. Fig.5a shows a topographic image of the ice surface obtained in contact mode at 30% relative humidity at -30°C after 30 min of start of the measurement. One could distinguish features on the ice surface that grow fast from one frame to the next within 3 min of scanning time (Fig.5b).

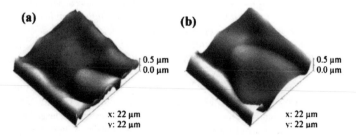

Figure 5   *AFM-image of Sample A (obtained from freezing of distilled water) showing the (re-)growth of ice crystals on the sample surface at 30% relative humidity in the glove box chamber and -30°C at the stage. Images were taken continuously with a scanning time of 3 min for a picture.*

The ice crystal growth could be observed in several of our AFM experiments at various values of relative humidity. The new individual grain grew in width and got rapidly larger than our scan area. The linear growth rate was estimated to be approximately 0.016-0.055 μm/s and is in principle consistent with extrapolated published results of basal face growth.[25] This comparison, however, is not fully conclusive and should be taken with some caution, since we were not able to determine the orientation of the growing ice crystals.

The instability of the AFM pictures obtained for Samples A (freezing of distilled water) can be explained by the possible variation of the thickness of liquid layer on the rough surface[22] and the lack of thermal equilibrium between tip and surface in the first minutes of measurements. The large roughness of frost crystals makes AFM investigations difficult, especially in tapping mode. Moreover, as soon as the tip comes into contact with the liquid layer at the ice/air interface, it is wetted by the liquid and pulled towards the sample surface, which potentially deforms the sample.

2.2.2   Samples B (obtained from condensation of water vapour). The AFM investigations of the samples formed from condensation of water vapour at 10-, 20- an 30% of relative humidity at -30°C show a polycrystalline ice structure (Fig.6). The ice crystals have diameters of approximately 5-15 μm and are easily identified by grain boundary grooves between 50-500 nm depth, (see Section 2.2.1). The thickness of the sample is 3-10 μm (measured by AFM via complete melting of the sample after investigation).

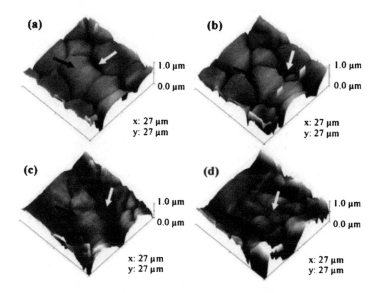

Figure 6   *AFM pictures of sample B at 30 % of relative humidity (obtained from condensation of water vapour). All images were taken in the course of one sample measurement. The images (a,b) show two snapshots (time interval of approx. 30 min) of small crystals growing at -30°C (marked by a white arrow). Images (c,d) show the same sample after annealing at -12.3°C (c) and -11.7°C (d) for 3 min. The images are noisy and the surface seems to be undulated.*

The AFM images are reproducible, since in this case the ice surface is stable in the course of the measurements with a typical duration of one hour (Fig.6a and 6b). In comparison with the samples A, prepared from distilled water under the same conditions (Fig.2), the size of ice crystals is clearly smaller. Melting and recrystallization of frost occur in similar way in both cases. In the case of samples A, the small crystals from frost (Fig 2a and 2b) are incorporated into underlying crystals from distilled water. In contrast, the grain coarsening of polycrystalline sample B could not be clearly observed, probably due to the considerably lower gradient of surface energy of the neighbouring ice grains.

Additionally, small crystals appeared in the course of the experiment (Fig. 6b). We can propose for their origin: exposition of existing grains by sublimation of the ice surface, or growth of underlying grains towards the ice-air interface. The simultaneous existence of these processes makes very difficult to distinguish between them. We also tried to investigate the impact of the temperature on the ice surface by annealing the sample. We could recognise that the "new" (small) crystals grew faster than the "older" (larger) grains, but the quality of the images became also much worse. Possible causes are either an increasing undulation or deformation[16] of the surface or some disturbance of the increasing liquid layer.[17]

3 CONCLUSIONS

We modified a commercial AFM to image ice surfaces in situ under various experimental conditions of temperature and relative humidity. The ice surfaces showed different structures (e.g. grain boundary grooves, defects of ice surface, frost crystals) depending on thickness of the ice sample, temperature and humidity.

At the temperature of -30°C the thick samples A and thin samples B showed different surface characteristics. In the case of thick samples (obtained from freezing of distilled water) the grain size of one crystal exceeded our scan area (30 μm^2). The ice surfaces were unstable and showed different features depending on humidity. Moreover we were able to capture the growth of a single ice crystal. In the case of thin ice films (grown from water vapour) at humidity of 10-, 20- and 30% and temperature of -30°C the images of the surfaces are reproducible. The grains are 5-15 μm large in base and the ice surface is stable during the course of measurement (1h).

The results presented here hold some promise that AFM work will give new insight for studying dynamic changes of ice surfaces on a nanometric scale which is relevant also for snow crystal growth as well as snow metamorphism.

Acknowledgements

We thank the Ministerium für Wissenschaft und Kunst of the Land Niedersachsen for a grant supporting this work. We also gratefully acknowledge the excellent technical support by Heiner Bartels and Eberhard Hensel.

References

1   U. Nakaya, *Snow Crystals, Natural and Artificial*, Harvard University Press, Cambridge, 1954, p. 510.
2   T. Kobayashi, *Physics of Snow and Ice*, Part 1, Institute of Low Temperature Science, Hokkaido University, Sapporo, 1967, p. 95.
3   N. Fukuta, T. Takahashi, *J. Atmos. Sci.*, 1999, **56**, 1963.

4   N.J. Bacon, M.B. Baker, B.D. Swanson, *Q. J. R. Meteorol. Soc.*, 2003, **129**, 1903.
5   B.J. Mason, Snow crystals, natural and man made, *Contemp. Phys.*, 1992, **33**, 227.
6   J. Nelson, *Phil. Mag. A*, 2001, **81**, 2337.
7   F.C. Frank, *Contemp. Phys.*, 1982, **23**, 3.
8   K.G. Libbrecht, *Rep. Prog. Phys.*, 2005, **68**, 855.
9   M. Elbaum, S.G. Lipson, J.G. Dash, *J. Cryst. Growth*, 1993, **129**, 491.
10   F. Domine, T. Lauzier, A. Cabanes, L. Legagneux, W.F. Kuhs, K. Techmer,
     T. Heinrichs, *Microsc. Res. Tech.*, 2003, **62**, 33.
11   E.F. Erbe, A. Rango, J. Foster, E.G. Josberger, C. Pooley, W.P. Wergin, *Microsc. Res. Tech.*, 2003, **62**, 19.
12   Y. Furukawa, M. Yamamoto, T. Kuroda, J. Cryst. Growth, 1987, **82**, 665.
13   D. Beaglehole, D. Nason, *Surf. Sci.*, 1980, **96**, 357.
14   J. Ocampo, J. Klinger, *J. Phys. Chem.*, 1983, **87**, 4325.
15   V. F. Petrenko, *J. Phys. Chem.*, 1997, **101**, 6276.
16   B. Pittenger, S.C. Fain, Jr., M.J. Cochran, J.M.K. Donev, B.E. Robertson,
     A. Szuchmacher, R.M. Overney, *Phys. Rev. B*, 2001, **63**, 134102-1.
17   A. Döppenschmidt, H.-J. Butt, *Langmuir*, 2000, **16**, 6709.
18   H.-J. Butt, A. Döppenschmidt, G. Hüttl, E. Müller, O.I. Vinogradova, *J. Chem. Phys.*, 2000, **113**, 1194.
19   H. Bluhm, M. Salmeron, *J. Chem. Phys.*, 1999, **111**, 6947.
20   C.R. Slaughterbeck, E.W. Kukes, B.Pittenger, D.J. Cook, P.C. Williams, V.L. Eden,
     S.C. Fain, Jr., *J. Vac. Sci. Technol. A*, 1996, **14**, 1213.
21   K. Ogawa, A, Majumdar, *Microscale Thermophys. Eng.*, 1999, **3**, 101.
22   A. Döppenschmidt, M. Kappl, H.-J Butt, *J. Phys. Chem. B*, 1998, **102**, 7813.
23   S. Zepeda, Y. Yeh, C.A. Orme, *Rev. Sci. Instrum.*, 2001, **72**, 4159.
24   T. Eastman and D.-M. Zhu, J. Colloid Interface Sci., 1995 **172**, 297.
25   D. Lamb, P.V. Hobbs, *J. Atmos. Sci.*, 1971, **28**, 1506.

IONISATION OF HCl ON ICE AT VERY LOW TEMPERATURE

C. Laffon and Ph. Parent

Laboratoire de Chimie Physique, Matière et Rayonnement, UMR 7614 Université Pierre et Marie Curie et CNRS, 11 rue Pierre et Marie Curie, 75231 Paris Cedex 05, France.

1 INTRODUCTION

The chemical state of HCl on ice can greatly influence the heterogeneous reactions leading to the release of active chlorine in the atmosphere.[1,2] A lot of studies have been recently dedicated to the HCl adsorption on ice,[3-16] each aiming at a better understanding of the dissociation/solvatation mechanism of HCl, but sometimes reporting contradictory conclusions.[3,4]

In this article, we summarize our recent results on this topic using Near Edge X ray Absorption Fine Structures (NEXAFS) spectroscopy at the Cl2p edge (200 eV),[3,5,6] with, precisely, a particular emphasis made on the description at a molecular level of the chemical state of hydrogen chloride.

NEXAFS probes the empty electronic levels of the system, which, in the case of HCl, directly reveal if it is dissociated or not (or ionised/molecular). Our measurements show unambiguously the ionisation of HCl on ice at 20 K. This facile ionisation at very low temperature is in good agreement with the idea of a barrier- free ionization of HCl on ice.[12] Furthermore, when measuring the NEXAFS spectra in Photon-Stimulated-Desorption mode (PSD), we greatly enhance the sensitivity to molecular HCl species that could eventually be adsorbed on the surface, especially at low temperature where HCl condensation can occur. Indeed, the PSD spectra show that at 20 K and at low coverage (0.1 L - 0.3 L) a small amount of molecular HCl is condensed at the side of the Cl⁻ anions.

Finally, these results also show that at 90 K *all* the HCl molecules adsorbed on the surface of ice are dissociated.

2 EXPERIMENTAL

The experimental set up consists in a special UHV chamber dedicated to the deposition of HCl and H_2O, which can be isolated from the main UHV chamber where the NEXAFS measurements are made. This avoids any further exposure during the measurements. This chamber is equipped with two separated dosing systems for HCl (Air liquide, electronic grade) and H_2O to avoid the hydratation of HCl prior to the deposition. The water ice films (100 ML thick) are dosed from the vapour pressure of purified degassed water. Further UPS experiments performed on these films show that they are well uniform (no signal from the Pt(111) is detected). In the results presented here, HCl was adsorbed at 20 K or 90

K at increasing exposures (0.1L-80L) on crystalline ice films obtained by water deposition on a Pt(111) substrate held at 150 K. The NEXAFS measurements were performed in the total electron yield (TEY) mode using the monochromatic soft X-ray radiation of SA22 beam line of Super ACO at the LURE facility (Orsay, France). The resolution was 0.02 eV at 190 eV. The PSD-NEXAFS have been measured in the total ion yield mode, using a two stages microchannel plates detector. The experimental conditions are detailed in ref. 4.

3 RESULTS AND DISCUSSION

The sensitivity of the NEXAFS spectroscopy to the chemical state of HCl is illustrated by the Figure 1, which compares the Cl2p NEXAFS spectra of a solid HCl film (50 ML thick deposited at 20 K) with that of the Cl⁻ anion, obtained by absorption of 1L of HCl at 120 K on crystalline water ice film.[6]

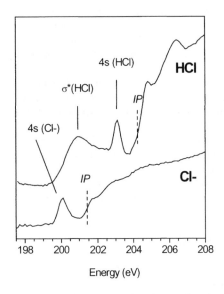

Figure 1 *The Cl2p NEXAFS spectra of the molecular HCl and of the Cl⁻ anion.*

The broad peak at 201 eV results from the $2p_{3/2}$, $2p_{1/2}$ excitations to the σ^*(H-Cl) valence orbital (200.83 eV and 202.15 eV, respectively), and is therefore a fingerprint of the H-Cl valence bond. Then follows the L_34s (203.1 eV) Rydberg state. The ionisation threshold is set at the inflexion point of the continuum (204.8 eV), which is almost superimposed to the L_34p state (204.8 eV) of the L_3np serie. The spectrum of Cl⁻ is strongly different. We observe a narrow peak at 200.1 eV, followed at 201.4 eV by the onset of the ionisation potential (IP), shifted of -3.4 eV with respect to HCl. The direct evidence of the ionization of HCl is the disappearance of the σ^*(H-Cl) state. We assign the remaining peak at 200.1 eV to the lowest excitation of the Cl⁻ ion, $2p^63s^23p^6 \rightarrow 2p^53s^23p^64s$. Its narrowness, similar to that of the 4s Rydberg state in HCl, evidences the atomic character of this line. The downwards shift of the ionization potential compared to HCl is a result of the additional negative charge localized at the Cl atom, increasing the repulsive character of the ion at the ionization threshold and thus decreasing the necessary energy for ionization.

The corresponding PSD-NEXAFS (not presented) have a spectral shape quite similar.[5] We note here that the PSD at the Cl2p-edge is more efficient than the conventional NEXAFS for the detection of HCl. This is due to the fact that PSD detects the H^+ fragments resulting from the H-Cl bond breaking after the photon absorption. When exciting a Cl⁻ ion, no fragment can be emitted since there is no bond-breaking possible, except through indirect fragmentations of the surrounding molecules by, e.g., the Auger electrons of the excited Cl⁻. We estimate the detection limit of molecular HCl on ice to be around 0.005 ML in PSD and 0.05 L for the normal NEXAFS.

Figure 2 shows the Cl2p NEXAFS spectra for increasing HCl exposures on ice at 90 K. The electronic structures are clearly those of the Cl⁻ anion (see figure 1). The PSD spectra also show the characteristic feature of Cl⁻, but not that of molecular HCl, which shows that *HCl is fully ionized on ice at 90 K,* over the whole exposure range.

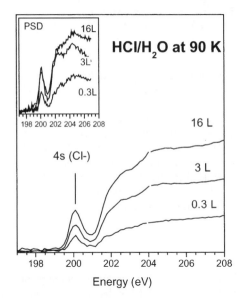

Figure 2 *Cl2p NEXAFS spectra at 90K for increasing HCl exposure on crystalline water ice films. The inset presents the signal recorded in PSD.*

Figure 3 shows the Cl2p NEXAFS spectra for increasing dosing of HCl on ice at 20 K. At this temperature, an HCl multilayer can condense. Thus, as expected, when the exposure raises, the σ*(H-Cl) of molecular HCl becomes clearly visible. For the low exposures however, the electronic features of the Cl⁻ anion prevail, *indicating that HCl is ionised on ice at 20 K.*

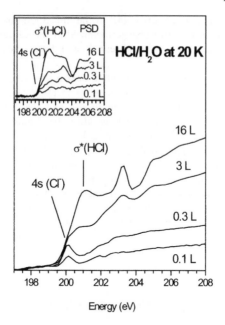

Figure 3 *Cl2p NEXAFS spectra at 20K for increasing HCl exposure on crystalline water ice films. The inset presents the signal recorded in PSD.*

On the inset of the figure 3 are shown the corresponding PSD recording. At 0.1 L and 0.3 L, we observe a mixture of the HCl and the Cl⁻ spectra. The HCl part is not seen on the normal NEXAFS (figure 3) due to its lower sensitivity to HCl. These PSD spectra show that some HCl molecules adsorb on the ice at 20K without ionisation. This molecular contribution begins from the lowest dosing (0.1 L), meaning that a small amount of HCl is molecular, probably in a physisorbed state and/or molecularly adsorbed through single and/or double hydrogen bonds with the surface.[7]

4 CONCLUSION

We have shown unambiguously that HCl ionizes on ice at temperatures as low as 20 K. At low coverages, we have observed some HCl molecules coexisting with the Cl⁻ anion at 20 K. At 90 K, the chlorine species are exclusively in the ionised state.

References

1   M.J. Molina, T.L. Tso, L.T. Molina, F.C. Wang, *Science* 1987, **238**, 1253.
2   S. Solomon, R.R. Garcia, F.S. Rowland, D.J. Wuebbles, *Nature* 1986, **321**, 755.
3   P. Parent, C. Laffon, *J. Phys Chem B* 2005, **109**, 1547.
4   S.C. Park, H.J. Kang, *J. Phys Chem B* 2005, **109**, 5124.
5   F. Bournel, C. Mangeney, M. Tronc, C. Laffon, Ph. Parent, *Surf. Sci.* 2003, **528**, 224.

6   F. Bournel, C. Mangeney, M. Tronc, C. Laffon, Ph. Parent, *Phys. Rev. B.* 2002, **65**, 201404(R).

7   V. Buch, J. Sadlej, N. Uras-Aytemiz, J.P. Devlin, *J. Phys. Chem. A* 2002, **106**, 9374.

8   J.P. Devlin, D.B. Gulluru, V. Buch, *J. Phys Chem B* 2005 **109** 3392

9   M. Kondo, H. Kawanowa, Y. Gotoh , R. Souda, *J. Chem. Phys* 2004, **121**, 8589

10 S. Haq, J. Harnett, A. Hodgson, *J. Phys. Chem. B* 2002, **106**, 3950.

11 K. Bolton, *J. Mol. Struct. (Theochem)* 2003, **632**, 145.

12 K. Bolton, J.B.C. Pettersson, *J. Am. Chem. Soc.* 2001, **123**, 7360.

13 H. Kang, T.H. Shin, S.P. Park, I.K. Kim, S.J. Han *J. Am. Chem. Soc.* 2000, **122**, 9842.

14 S.F. Bahnam, J.R. Sodeau, A.B. Horn, M.R.S. McCoustra, M.A. Chester, *J. Vac. Sci. Technol. A* 1996, **14**, 1620.

15 V.Sadtchenko, C.F. Giese, W.R Gentry, *J. Phys. Chem B* 2000, **104**, 9421.

16 J. Harnett, S. Haq, A. Hodgson, *Surf. Sci.* 2003, **478**, 532.

EFFECTS OF LARGE GUEST SPECIES ON THERMODYNAMIC PROPERTIES OF STRUCTURE-H HYDRATES

T. Makino, T. Sugahara, and K. Ohgaki

Division of Chemical Engineering, Department of Materials Engineering Science, Graduate School of Engineering Science, Osaka University, Japan

1 INTRODUCTION

Structure-H (s-H) hydrate was found in 1987[1] and the single-crystal diffraction data was obtained in 1997[2]. The unit cell of s-H consists of three D-cages (pentagonal dodecahedron, 5^{12}), two D'-cages (dodecahedron, $4^3 5^6 6^3$), and one E-cage (icosahedron, $5^{12} 6^8$). Small (help gas) and large guest species (LGS) are essential to generate the s-H hydrate. The help gas aids in the s-H hydrate formation of LGS that cannot be enclathrated by itself. Some small guest species (ex. CH_4[3-5], Xe[6-8]) have been reported as the help gas of s-H hydrate. The previous study[9] reported that the help gas occupies the D- and D'-cages, while the LGS can occupy only the E-cage.

The s-H hydrates have potential to become a better medium for natural-gas (mixed-gas of CH_4 and impurities) storage and transportation[10]. The s-H hydrates helped by CH_4 can be generated under lower pressure condition than the pure CH_4 hydrate. That is, we can handle the natural-gas hydrate (NGH) under more moderate condition by generating the s-H hydrate. The pressure reduction from pure help-gas hydrate depends largely on the kind of LGS. Some literatures[7,11,12] reported that both the molecular size and the molecular shape of LGS have much effect on the equilibrium pressure of s-H hydrate. A lot of phase equilibrium data for the s-H hydrate systems are required to develop the effective storage and transportation system.

In the present study, CH_4 was adopted as help gas from a viewpoint to develop the gas-hydrate technology mentioned above. Ten C_6-C_8 hydrocarbons, that have various sizes and shapes, were added to CH_4+H_2O system for phase equilibrium measurement. The objective is to study the relations between the equilibrium conditions and the molecular sizes of LGS. The phase equilibrium relations for CH_4+dimethylcyclohexane (DMCH) stereo-isomers, CH_4+methylcyclohexane (MCH), CH_4+cyclooctane (c-Octane), and CH_4+methylcyclopentane (MCP) s-H hydrate systems were measured under four-phase equilibrium condition up to 10 MPa. The four phases were gas, water, liquid LGS, and s-H hydrate phases. In addition, the single-crystals of s-H hydrates were analyzed by Raman spectroscopy in order to obtain Raman spectra about the s-H hydrates. The CH_4+1,1-DMCH and CD_4+1,1-DMCH s-H hydrate systems were analyzed under four-phase equilibrium conditions up to 30 MPa.

2 METHOD AND RESULTS

2.1 Experimental Apparatus for Phase Equilibrium Measurement

Experimental apparatus used in the present study was the same as the one used in the previous work[13]. The details are not mentioned here. Equilibrium temperatures were measured within a reproducibility of 0.02 K using a thermistor probe (Takara D-641) calibrated by a Pt resistance thermometer (25 Ω). The probe was inserted into a hole in the cell wall to measure equilibrium temperatures. Equilibrium pressures were measured by the pressure gauge (Valcom VPRT), which was calibrated by RUSKA quartz Bourdon tube gauge, with an uncertainty of 0.02 MPa.

The CH_4 s-H hydrate was generated under four-phase coexisting condition by the same procedure as the previous study[13]. In order to determine the four-phase equilibrium pressure precisely, the s-H hydrate was formed or dissociated by the pressure control reported in the previous study[13]. When the pressure change became within 0.01 MPa, the system was regarded as the phase equilibrium. Usually, it took about two or three days to establish the equilibrium in the present study. In the high temperature range, it took more than four days to complete the equilibrium.

2.2 Experimental Apparatus for Raman Spectroscopic Analysis

Experimental apparatus, used in the present study, was the same as the one reported previously[14]. The details are not also mentioned here. Equilibrium pressure and temperature were measured by the same procedure mentioned in the section **2.1**. The reproducibility and uncertainty were also the same.

The single crystal of s-H hydrate was prepared with the coexistence of gas, water, and liquid LGS phases by the same procedure as the previous report[15]. We spent more than two weeks to establish the four-phase equilibrium. The single crystal was analyzed by *in situ* Raman spectroscopy by use of a laser Raman microprobe spectrometer with a multi-channel CCD detector. Ar ion laser (wave length: 514.45 nm, generating power: 100 mW), condensed to 2 μm in a spot diameter, was radiated from an object lens onto the samples through the upper sapphire window. Spectral resolution in the present study was 1 cm^{-1}. Raman peaks were calibrated with Raman peaks of Ne in air. The CCD detector was maintained at 140 K by liquid N_2 for heat-noise reduction. The integration time was varied depending on Raman-peak intensities. Raman spectra were fitted by Voigt curve equation.

2.3 Stability Boundaries of the Structure-H Hydrate Helped by CH_4

Figure 1 shows the stability boundaries of s-H hydrates helped by CH_4. The solid circle, open triangle, open circle, open diamond, open reverse-triangle, and solid reverse-triangle stand for the phase equilibria for the CH_4+1,1-DMCH[16], CH_4+MCH[17], CH_4+*cis*-1,2-DMCH[17], CH_4+MCP[18], CH_4+c-Octane[18], and CH_4+*cis*-1,4-DMCH[19] s-H hydrate systems, respectively. The three-phase equilibrium data of pure CH_4 hydrate are also plotted with solid square[13]. The equilibrium pressures of s-H hydrates decrease from that of pure CH_4 hydrate. The slope of stability boundary for each s-H hydrate system (the plot of ln p vs. T is almost linear) is somewhat steeper than that of pure CH_4 hydrate. As reported in the literatures[8,20], these equilibrium curves would cross at high temperature where the s-H hydrate is dissociated and the pure CH_4 s-I hydrate is reconstructed.

The molecular sizes were estimated from the chemical structure, optimized by

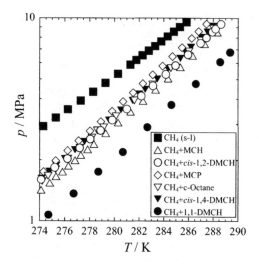

Figure 1 *Stability boundaries of the s-H hydrates helped by CH₄. ■: pure CH₄ hydrate (s-I)[17]; △: CH₄+MCH hydrate (s-H)[17]; ○: CH₄+cis-1,2-DMCH hydrate (s-H)[17]; ◇: CH₄+MCP hydrate (s-H)[18]; ▽: CH₄+c-Octane hydrate (s-H)[18]; ▼: CH₄+cis-1,4-DMCH (s-H)[19]; ●: CH₄+1,1-DMCH (s-H)[16].*

Gaussian 03 (Density Functional Theory, B3LYP method, 6-31(G) basis set), with treating the LGS as spheroids. The optimized structures were calculated with making conditions of one molecule at 280 K and atmospheric pressure. The molecular sizes of LGS and the cavity size of E-cage[2] were summarized in Table 1 with the equilibrium pressures at 280 K. d_{Major} and d_{Minor} stand for the major and minor axes. The major and minor axes of E-cage cavity correspond with c- and a-axes of E-cage cavity, respectively.

1,1-DMCH has the nearest size to the E-cage cavity in the LGS of present study. The s-H hydrate of CH₄+1,1-DMCH has the lower equilibrium-pressure than any other present s-H hydrate systems. The hydrate structure generally becomes more stable when the guest species has more suitable size and shape for the cage cavity. The present result follows this tendency.

The molecular size of MCH is almost same as 1,1-DMCH. However, there is about 1 MPa deviation between the equilibrium pressures of MCH and 1,1-DMCH s-H hydrates at 280 K. The difference of these LGS is only the existence of methyl group. The existence of substituent is implied to be important for the hydrate-structure stability, even if the molecular sizes are similar with each other.

1,1-DMCH, MCH, c-Octane, and cis-1,2-DMCH have the larger minor-axes than the E-cage cavity. The resent study[22] reported that the E-cage is mainly expanded in the direction of a-axis, not much c-axis, to make more space. Those LGS would be encaged distorting the E-cage in the direction of a-axis. In the case of cis-1,4-DMCH, both major and minor axes are larger than those of E-cage cavity. The E-cage of cis-1,4-DMCH s-H hydrate would be distorted much more than those of above four LGS s-H hydrates.

trans-1,2-DMCH has similar size as cis-1,2-DMCH and c-Octane. The molecular size of trans-1,3-DMCH is also similar with that of cis-1,4-DMCH. However, trans-1,2-DMCH, 1,3-DMCH stereo-isomers, and trans-1,4-DMCH in the seven types of DMCH

Table 1 *Summary of the equilibrium pressures and the sizes of LGS and E-cage cavity.*

LGS	p / MPa (280 K)	d_{Major} / nm	d_{Minor} / nm
1,1-DMCH	2.08	0.846	0.706
MCH	2.96	0.845	0.705
c-Octane	3.24	0.815	0.798
cis-1,2-DMCH	3.24	0.844	0.760
cis-1,4-DMCH	3.33	0.891	0.725
MCP	3.58	0.786	0.649
trans-1,2-DMCH	no s-H formation*	0.850	0.802
cis-1,3-DMCH	no s-H formation*	0.931	0.741
trans-1,3-DMCH	no s-H formation*	0.870	0.739
trans-1,4-DMCH	no s-H formation*	0.959	0.721
E-cage**		0.858**	0.684**

* *"no s-H formation" means that the CH₄ s-H hydrate was not generated. The equilibrium pressure is the same as that of pure CH₄ hydrate (5.22 MPa).*
** *The cavity sizes of E-cage were obtained from the X-ray diffraction study[2]. Major (d_{Major}) and Minor (d_{Minor}) axes of E-cage correspond to c- and a-axes, respectively.*

stereo-isomers can not generate the CH₄ s-H hydrate by the same procedures as the other s-H hydrate systems. The equilibrium pressures of hydrate systems added the four LGS (*trans*-1,2-DMCH, *cis*-1,3-DMCH, *trans*-1,3-DMCH and *trans*-1,4-DMCH) are the same as that of pure CH₄ s-I hydrate (5.22 MPa at 280 K). These results speculate that the position of substitutional group is also important for the hydrate-structure stability.

2.4 Raman Spectroscopic Study on Single Crystals of Structure-H hydrate

The larger pressure reduction was observed by adding 1,1-DMCH than any other LGS of present study. 1,1-DMCH was used as LGS for Raman spectroscopy to make the s-H formation easier in the high-pressure optical cell. Figure 2 exhibits the Raman spectra of 1,1-DMCH under four-phase equilibrium condition at 298.28 K and 25.5 MPa for the CH₄+1,1-DMCH hydrate system. The spectra result from the intramolecular C_6 stretching vibration of 1,1-DMCH. Figure 2 (a) is the Raman spectrum of liquid LGS phase. A sharp single-peak was observed at 705 cm⁻¹ and the Raman shift agrees well with the literature[23]. The Raman peak of s-H hydrate phase is shown in Figure 2 (b). Raman shift shows, so-called, blue-shift and it is 2 cm⁻¹ higher than that of liquid LGS phase. The similar Raman peak was detected at 708 cm⁻¹ in the CD₄+1,1-DMCH hydrate system. Both Raman spectra of CH₄+1,1-DMCH and CD₄+1,1-DMCH s-H hydrates do not show significant changes caused by pressurization in the equilibrium pressure range up to 30 MPa. The single peak of 751 cm⁻¹ is derived from the sapphire window of optical cell.

CD₄ was used to avoid the overlap of Raman peaks and to clarify the Raman spectra derived from only 1,1-DMCH, because the Raman peaks of CH₄ and 1,1-DMCH overlap around 2900 cm⁻¹. Figures 3 (a) and (b) are the Raman spectra of 1,1-DMCH observed in the CD₄+1,1-DMCH hydrate system under four-phase equilibrium condition at 294.06 K and 14.2 MPa. The Raman spectrum of liquid LGS phase is shown in Figure 3 (a), that of s-H hydrate phase is Figure 3 (b). Figure 3 (a) also agrees well with the reference[23]. The Raman spectrum of 1,1-DMCH changed apparently by enclathration.

Figure 4 (a) shows the Raman spectra of 1,1-DMCH and CH₄ in the CH₄+1,1-DMCH hydrate system under four-phase equilibrium condition at 298.28 K and 25.5 MPa. All

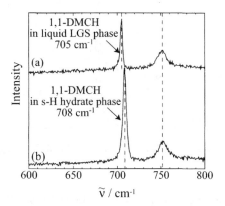

Figure 2   *Raman spectra of intramolecular C_6 stretching vibration of 1,1-DMCH under four-phase equilibrium condition at 298.28 K and 25.5 MPa for the CH_4+1,1-DMCH hydrate system. (a): liquid LGS phase, (b) hydrate phase. The single peak of 751 cm^{-1} is derived from the sapphire window of optical cell.*

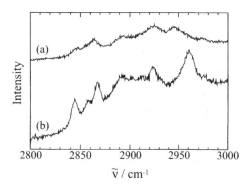

Figure 3   *Raman spectra of 1,1-DMCH for the CD_4+1,1-DMCH hydrate system under four-phase equilibrium condition at 294.06 K and 14.2 MPa. (a): liquid LGS phase, (b): hydrate phase.*

peaks result from guest species enclathrated in the s-H hydrate. The intensive peak around 2910 cm^{-1} originates in encaged CH_4 molecule. The contribution of 1,1-DMCH (Figure 3 (b)) was removed from this spectrum in order to expose the Raman peak of enclathrated CH_4 (the difference caused by help gases is assumed to be negligible in the present study, because the Raman peaks of C_6 stretching vibration and C-H stretching vibration (except for the peaks overlapped by the Raman peaks of CH_4) are almost the same in the CH_4 and CD_4 s-H hydrate systems). The difference spectrum is shown in Figure 4 (b). The single peak has a tail in the lower wave-number range and its shape is obviously different from that of s-I hydrate[24]. This peak would be the convolution of two peaks, because CH_4 is encaged in two kinds of small cages[9]. The recent study[25] reported that the Raman peaks of CH_4 in the D- and D'-cages were detected at 2911 cm^{-1} and 2901 cm^{-1} in the CH_4+2,2-

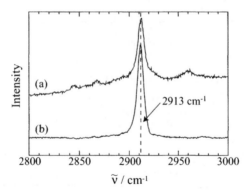

Figure 4   *Raman spectra of CH₄ and 1,1-DMCH for the CH₄+1,1-DMCH hydrate system under four-phase equilibrium condition at 298.28 K and 25.5 MPa. (a): hydrate phase, (b): difference spectrum (after removal of contribution of 1,1-DMCH).*

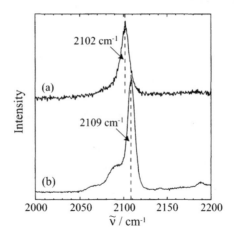

Figure 5   *Raman spectra of CD₄ for the CD₄+1,1-DMCH hydrate system under four-phase equilibrium condition at 294.06 K and 14.2 MPa. (a): hydrate phase, (b): gas phase.*

dimethylbutane s-H hydrate at 150 K. Uchida et al.[25] claimed that the Raman shift might depend largely on the temperature difference during the experiments. Some literatures[12,22,26] and present study suggest that the hydrate cage would be distorted by enclathration of large molecules. The difference of LGS is implied to affect the environment of encaged help-gas.

Incidentally, a broad single-peak of CD₄ was detected at 2102 cm⁻¹ in the CD₄+1,1-DMCH hydrate system as shown in Figure 5 (a). While, in the gaseous CD₄ phase, some peaks were detected around the intensive peak as shown in Figure 5 (b). According to the literatures[27,28], the intensive peak is derived from C-D symmetric stretching vibration mode and the other peaks correspond to the Fermi resonance peaks. The single peak of s-H hydrate phase is suggested to consist of several peaks derived from the C-D stretching vibration and Fermi resonance. The Raman peak of C-D stretching vibration is unsuitable for the quantitative analysis.

3 CONCLUSION

The pressure-temperature relations for the CH_4 s-H hydrates were measured under four-phase equilibrium condition. The s-H hydrate formation was observed in the CH_4+1,1-DMCH, CH_4+MCH, CH_4+c-Octane, CH_4+*cis*-1,2-DMCH, CH_4+*cis*-1,4-DMCH and CH_4+MCP systems. The equilibrium pressure of each s-H system increases in that order at isothermal condition and it is lower than that of pure CH_4 s-I hydrate. The 1,1-DMCH molecule has the best molecular-size to fit the E-cage cavity and constructs the CH_4 s-H hydrate at the lowest pressure in the all LGS of present study.

The single crystals of CH_4+1,1-DMCH and CD_4+1,1-DMCH s-H hydrates were analyzed by means of Raman spectroscopy. The C_6 stretching vibration of enclathrated 1,1-DMCH molecule shows higher Raman shift than that of 1,1-DMCH molecule in liquid LGS phase. The Raman peak of CH_4 enclathrated in the s-H hydrate is broad single-peak that has the long tail in the lower wave-number range.

Nomenclature

p: equilibrium pressure [Pa]
T: equilibrium temperature [K]
d: molecular size or cage-cavity size [nm]

Acknowledgements

The authors are grateful to the Division of Chemical Engineering, Graduate School of Engineering Science, Osaka University for the scientific support by "Gas-Hydrate Analyzing System (GHAS)." One of the authors (T. M.) expresses his special thanks for the center of excellence (21COE) program "Creation of Integrated EcoChemistry of Osaka University".

References

1   J. A. Ripmeester, J. S. Tse, C. I. Ratcliffe and B. M. Powell, *Nature*, 1987, **325**, 135.
2   K. A. Udachin, C. I. Ratcliffe, G. D. Enright and J. A. Ripmeester, *Supramol. Chem.* 1997, **8**, 173.
3   U. Hütz and P. Englezos, *Fluid Phase Equilib.*, 1996, **117(1-2)**, 178.
4   R. Ohmura, S. Matsuda, S. Takeya, T. Ebinuma and H. Narita, *Int. J. Thermophys.*, 2005, **26(5)**, 1515.
5   J.-W. Lee, H. Lu, I. L. Moudrakovski, C. I. Ratcliffe and J. A. Ripmeester, *Angew. Chem., Int. Ed.*, 2006, **45(15)**, 2456.
6   A. P. Mehta and E. D. Sloan, *AIChE J.*, 1996, **42(7)**, 2036.
7   J. A. Ripmeester and C. I. Ratcliffe, *J. Phys. Chem.*, 1990, **94(25)**, 8773.
8   N. Shimada, K. Sugahara, T. Sugahara and K. Ohgaki, *Fluid Phase Equilib.*, 2003, **205**, 17.
9   K. A. Udachin, C. I. Ratcliffe and J. A. Ripmeester, *J. Supramol. Chem.*, 2003, **2(4-5)**, 405.

10   A. A. Khokhar, J. S. Gudmundsson and E. D. Sloan, *Fluid Phase Equilib.*, 1998, **150-151**, 383.
11   A. P. Mehta, T. Y. Makogon, R. C. Burruss, R. F. Wendlandt and E. D. Sloan, *Fluid Phase Equilib.*, 1996, **121(1-2)**, 141.
12   J-W. Lee, H. Lu, I. L. Moudrakovski, C. I. Ratcliffe and J. A. Ripmeester, *Angew. Chem. Int'l Ed'n*, 2006, **45**, 2456.
13   T. Nakamura, T. Makino, T. Sugahara and K. Ohgaki, *Chem. Eng. Sci.*, 2003, **58**, 269.
14   T. Sugahara, T. Makino and K. Ohgaki, *Fluid Phase Equilib.*, 2003, **206**, 117.
15   T. Makino, M. Tongu, T. Sugahara and K. Ohgaki, *Fluid Phase Equilib.*, 2005, 233, 129.
16   T. Hara, S. Hashimoto, T. Sugahara, K. Ohgaki, *Chem. Eng. Sci.*, 2005, **60**, 3117.
17   T. Nakamura, T. Makino, T. Sugahara and K. Ohgaki, *Chem. Eng. Sci.*, 2004, **59**, 163.
18   T. Makino, T. Nakamura, T. Sugahara and K. Ohgaki, *Fluid Phase Equilib.*, 2004, **218**, 235.
19   T. Nakamura, T. Sugahara and K. Ohgaki, *J. Chem. Eng. Data*, 2004, **49 (1)**, 99.
20   T. Y. Makogon, A. P. Mehta and E. D. Sloan Jr., *J. Chem. Eng. Data*, 1996, **41(2)**, 315.
21   Gaussian 03, Revision C.02, M. J. Frisch, G. W. Trucks, H. B. Schlegel, G. E. Scuseria, M. A. Robb, J. R. Cheeseman, J. A. Montgomery, Jr., T. Vreven, K. N. Kudin, J. C. Burant, J. M. Millam, S. S. Iyengar, J. Tomasi, V. Barone, B. Mennucci, M. Cossi, G. Scalmani, N. Rega, G. A. Petersson, H. Nakatsuji, M. Hada, M. Ehara, K. Toyota, R. Fukuda, J. Hasegawa, M. Ishida, T. Nakajima, Y. Honda, O. Kitao, H. Nakai, M. Klene, X. Li, J. E. Knox, H. P. Hratchian, J. B. Cross, C. Adamo, J. Jaramillo, R. Gomperts, R. E. Stratmann, O. Yazyev, A. J. Austin, R. Cammi, C. Pomelli, J. W. Ochterski, P. Y. Ayala, K. Morokuma, G. A. Voth, P. Salvador, J. J. Dannenberg, V. G. Zakrzewski, S. Dapprich, A. D. Daniels, M. C. Strain, O. Farkas, D. K. Malick, A. D. Rabuck, K. Raghavachari, J. B. Foresman, J. V. Ortiz, Q. Cui, A. G. Baboul, S. Clifford, J. Cioslowski, B. B. Stefanov, G. Liu, A. Liashenko, P. Piskorz, I. Komaromi, R. L. Martin, D. J. Fox, T. Keith, M. A. Al-Laham, C. Y. Peng, A. Nanayakkara, M. Challacombe, P. M. W. Gill, B. Johnson, W. Chen, M. W. Wong, C. Gonzalez and J. A. Pople, Gaussian, Inc., Wallingford CT, 2004.
22   S. Takeya, A. Hori, T. Uchida and R. Ohmura, *J. Phys. Chem. B*, 2006, **110(26)**, 12943.
23   J. Lecomte, L. Piaux and O. Miller, *Bull. Soc. Chim. Belg.*, 1936, **45**, 123.
24   S. Nakano, M. Moritoki and K. Ohgaki, *J. Chem. Eng. Data*, 1999, **44**, 254.
25   T. Uchida, R. Ohmura, I. Y. Ikeda, J. Nagao, S. Takeya and A. Hori, *J. Phys. Chem. B*, 2006, **110**, 4583.
26   T. Nakamura, T. Makino, T. Sugahara and K. Ohgaki, *Chem. Eng. Sci.*, 2004, **59**, 163.
27   D. A. Long and E. L. Thomas, *Trans. Faraday Soc.*, 1963, **59**, 1026.
28   D. Bermejo, R. Escribano and J. M. Orza, *J. Mol. Spectro.*, 1977, **65(3)**, 345.

PREDICTION OF THE CELLULAR MICROSTRUCTURE OF SEA ICE BY MORPHOLOGICAL STABILITY THEORY

S. Maus

Geophysical Institute, University Bergen, Allegaten 70, 5007 Bergen, Norway

1 INTRODUCTION

It has long been known that seawater, under most natural conditions, freezes with a cellular interface, consisting of elongated vertical plates[1,2,3]. These are normal to the optical axis which parallels the freezing interface[2]. This lamellar structure involves the entrapment of concentrated brine between the plates and makes sea ice saline. The first quantitative descriptions of the *plate spacing* of sea ice yielded values in the 0.2 to 0.5 mm range for thin ice[4,5,6], ≈ 1 mm at the bottom of thick Arctic sea ice[7,8], and up to ≈ 1.5 mm for an exceptionally thick ice floe[9]. Laboratory experiments with freezing NaCl solutions, to establish the plate spacing's dependence on the growth velocity, indicated a considerable complexity[10]. The plate spacings in the laboratory study were a factor of 2 to 3 smaller than in the field. This discrepancy remained when more systematic field studies were performed[11]. Also the concentration dependence of the plate spacing reported so far is not conclusive[12,10].

Early attempts to relate observations of plate spacing versus growth velocity to some early semi-empirical scaling laws were unsuccessful[12,10]. Meanwhile, the theoretical key to the problem and nowadays classical morphological stability theory (MST) from Mullins and Sekerka[13], became well established[14,15]. Its predictions of the *onset* of cellular instabilities on a *planar* interface were confirmed in freezing experiments of saline solutions in thin growth cells[16,17,18]. An approach to model the *stable* plate spacing of sea ice by MST resulted in predictions an order of magnitude too small[19]. This result is similar to other systems[20,14,21,22].

The coarsening of the initial interface perturbations into a wider stable dendrite spacing has been modelled with some success[23,24]. Scaling laws for cell and dendrite spacings have also been suggested on the basis of analytical approximations[25,20] and numerical studies[26,27]. The approaches have led to reasonable predictions of dendrite spacings in different systems[20,23,22,28]. The present paper first discusses plate spacing observations of sea ice in connection with the latter theories. Then a novel simplistic approach on the grounds of MST is proposed and compared to observations and previously suggested scalings. A second new aspect demonstrated here is the incorporation of a simple parametrisation of solutal convection into MST. It yields a reasonable quantitative modification of the plate spacing during downward growth of sea ice.

2 MORPHOLOGICAL STABILITY

2.1 The Mullins-Sekerka analysis

The classical linear stability theory for a planar interface was formulated in 1964 by Mullins and Sekerka[13]. The theory predicts, under what growth conditions a binary alloy solidifying unidirectionally at constant velocity may become morphologically unstable. Its basic result is a dispersion relation for those perturbation wave lengths that are able to grow, rendering a planar interface unstable. Two approximations of the theory are of practical relevance for the present work. In the thermal steady state, which is approached at large ratios of thermal to solutal diffusivity, and for concentrations close to the *onset of instability*, the characteristic equation of the problem is[29,14,15]

$$\mathcal{G} = 1 + \frac{\mathcal{A}}{4k} - \frac{3\mathcal{A}^{1/2}r}{2} - \frac{\mathcal{A}(1 - 2k)r^2}{4k}, \tag{1}$$

where the variable r is defined as $r^4 = 1 + (2D\omega/V)^2$ and determined by the one real root of

$$r^3 + (2k - 1)r - \frac{2k}{\mathcal{A}^{1/2}} = 0 \tag{2}$$

greater then zero. The non-dimensional numbers which determine the behaviour of equation (1) are an absolute stability parameter

$$\mathcal{A} = \frac{k\Gamma V^2}{mG_cD^2} = \frac{k^2\Gamma V}{m(1 - k)C_\infty D}, \tag{3}$$

and a constitutional supercooling parameter

$$\mathcal{G} = \frac{G_{eff}}{mG_c} = \frac{k}{mC_\infty(k - 1)}\left(\frac{L_vD}{K_i + K_b}\right)\left(1 + \frac{2G_bK_b}{L_vV}\right). \tag{4}$$

In these numbers G_c is the interfacial concentration gradient and under diffusive equilibrium given by

$$mG_c = C_\infty\frac{1 - k}{k}\frac{V}{D} \tag{5}$$

Further, K_b, K_i, G_b and G_i denote thermal conductivities and temperature gradients in the liquid and solid, and D the solute diffusivity in the liquid. L_v, $m = \partial T/\partial C_\infty$ and k are volumetric latent heat of fusion, liquidus slope and interfacial distribution coefficient. C_∞ is the concentration far from the interface and $\Gamma = T_m\gamma_{sl}/L_v$ the Gibbs-Thompson parameter based on the solid liquid interfacial energy γ_{sl} and melting temperature T_m. V is the solidification velocity and $\omega = 2\pi/\lambda$ the wave number of a perturbation.

An approximation to the problem was discussed by Coriell and coworkers[14]. For $\mathcal{A}^{1/6} \ll 1$, which is fulfilled for natural freezing conditions, equation (1) may be written in the approximate form

$$\mathcal{G} = 1 - \left(\frac{27k\mathcal{A}}{4}\right)^{1/3}, \tag{6}$$

while the critical wave length at the *onset of instability* is now given by

$$\lambda_{mi} = 2\pi \frac{D}{V} \left(\frac{2\mathcal{A}}{k^2} \right)^{1/3} = 2\pi 2^{1/3} \left(\frac{D}{V} \right)^{2/3} \left(\frac{\Gamma}{mC_\infty(k-1)} \right)^{1/3}. \tag{7}$$

The second limit of practical interest is the short wavelength limit $\omega D/V \gg 1$ at concentrations far from the onset of stability, when $\mathcal{G} \ll 1$. It is given by[14]

$$\lambda_\Gamma = 2\pi \frac{D}{V} \left(\frac{\mathcal{A}}{k(1-\mathcal{G})} \right)^{1/2}, \tag{8}$$

which in turn yields the most rapidly growing wavelength $\lambda_{max} \approx 3^{1/2}\lambda_\Gamma$.

While the theory has originally been derived for equilibrium freezing conditions[29,14,15], it has later been shown that, in case of the thermal steady state of a large ratio of thermal to solutal diffusivity, it can also be applied to the transient case, provided that C_∞/k in equation (4) is replaced by the instantaneous time-dependent interface concentration[30,31].

2.2 Confirmations and restrictions

At natural growth velocities the stabilising influence of surface tension on the interface is negligible[32] and the stability condition (6) practically becomes $\mathcal{G} < 1$. This has been reasonably confirmed in thin growth cell freezing experiments with an externally imposed temperature gradient[16,17,18]. Wavelengths of the initial perturbations were reported in agreement with the wavelength of most rapidly growing perturbations[18]. However, as mentioned in the introduction, this predictability only holds for the onset of instability. In most noted thin cell studies the distance of initial perturbations changed rapidly by overgrowth to a much coarser cell spacing. The actual plate spacing of sea ice and NaCl ice is, for similar growth velocities, an order of magnitude larger than the wave length at the onset of instability. This cell coarsening has been observed for other systems[20,24,22,33]. Besides an approach to model its transient evolution[23,24], scaling laws for the stable dendrite spacing have been suggested, which gave reasonable predictions for the succinonitrile-acetone binary[20,23]. The approaches will be considered next in connection with observations of aqueous NaCl solutions.

3 PRIMARY SPACING OF CELLULAR AND DENDRITIC ARRAYS

3.1 Scaling laws

Analytical approaches to predict the dendritic spacing via a geometric matching condition[25,20] lead to an equation which in the present notation takes the form

$$a_0 = B \left(\frac{D_s}{V} \right)^{3/4} \left(\frac{\Gamma}{mC_{tip}(k-1)} \right)^{1/4} \left(\frac{1-\mathcal{G}'}{\mathcal{G}'} \right)^{1/2}. \tag{9}$$

All variables are as defined above, with the exception that $\mathcal{G}' = \mathcal{G}C_\infty/(kC_{tip})$ depends on the (unknown) ratio of the concentration C_{tip} at the tip of cells to the far-field bulk value. Here $C_{tip} = C_\infty$ is assumed. For the prefactor, which is related to geometrical constraints at the cell tips, $B = 6.51$ is a reasonable estimate[20,22]. From geometrical considerations this analytic matching approach is not expected to vary by more than ≈ 30 %. It is, however, simplistic and may be problematic under changing growth conditions[22,27].

More recently a similar scaling law has been suggested on the basis of numerical model simulations of *dendritic* array growth[26,27]. In the present notation it takes the form

$$a_0 = 5 \left(\frac{D_s}{V} \right)^{2(1+b/2)/3} \left(\frac{\Gamma}{mC_\infty(k-1)} \right)^{(1-b)/3} (1-\mathcal{G})^{1/2} \left(\frac{k}{\mathcal{G}} \right)^{2(1-b)/3}, \tag{10}$$

where

$$b = 0.3 + 1.9 \, (\mathcal{G}\mathcal{A})^{0.18} . \tag{11}$$

Equation (10) refers to minimum spacings. The numerical model distinguishes between dendritic and cellular shapes. For *cellular* arrays the scaling

$$a_0 = 8.18 k^{-0.335} \left(\frac{D_s}{V} \right)^{0.59} \left(\frac{\Gamma}{mC_\infty(k-1)} \right)^{0.41} . \tag{12}$$

was proposed[26,27].

3.2 Comparison to observations

For a meaningful comparison of the above plate spacing predictions with observations one must ascertain that diffusion is the dominating solute transport mechanism. Inserting the molecular properties of aqueous NaCl into equation (4) gives a critical $C_\infty \approx 0.0013$ ‰[32]. The fact that lake water, with a two orders of magnitude larger $C_\infty \approx 0.1$ ‰, does *not* freeze with a cellular interface thus indicates that natural ice growth is dominated by convective solute transport. However, downward freezing experiments at relatively high freezing velocities[10] may be compared to the theoretical predictions. In the experiments to be discussed the solute entrapment was close to $k \approx 1$ for V above $\approx 5 \times 10^{-4}$ cm/s[34,35]. The following interpretation focuses first on the diffusive regime above this threshold velocity.

In Figure 1 observed plate spacings[10] are compared with the theoretical predictions from equations (9) through (12). Predictions are shown as dashed, dash-dotted and dotted curves, observations as circles. The solid curves are discussed in a later section. At the lower salinity $C_\infty \approx 1$ ‰ the agreement with the *dendritic* model predictions is remarkable (Figure 1a). The principle agreement with the observations at 30 ‰ is less good Figure 1b. The dendritic scalings from Hunt and Trivedi are almost a factor of 2 too large. The observations at $C_\infty \approx 30$ ‰ indicate a shift in the slope of the $a_0 - V$-relation near $V \approx 5 \times 10^{-4}$ cm/s. This slope change is likely associated with convective effects, to be discussed below. It thus appears unlikely that the lack in agreement at high growth velocities can be associated with convection. It is further found that, for both concentrations, the observed exponent b in a power law $a_0 \sim V^{-b}$ is less than predicted by the dendritic scalings.

The discrepancy between observations and predictions is further illustrated in Figure 2, emphasising the concentration dependence at two growth velocities $V_n = 10^{-3}$

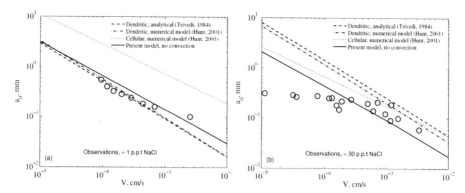

Figure 1 *Plate spacing during downward freezing of (a) 1 ‰ NaCl and (b) 30 ‰ NaCl aqueous solutions[10] compared to analytical and numerical predictions. The solid curve is the wave length λ_{mi} from the present approach.*

and $V_n = 10^{-4}$ cm s^{-1}, to which the observations have been interpolated. At the higher velocity no clear concentration dependence of a_0 is found over the concentration range 1 to 100 ‰. The increase of the spacing with concentration, predicted by the dendritic models is not seen in the observations. The over-prediction reaches a factor 3 to 4 at the highest concentration. The dependence of *cellular* spacings on the solute concentration is opposite to *dendritic* arrays, which is neither observed.

The right hand Figure 2b shows the comparison at a 10 times lower growth velocity $V_n = 10^{-4}$ cm s^{-1}. Due to the likely presence of convection a quantitative agreement can no longer be expected. However, the overall appearance of the data is similar as for the diffusive regime during rapid growth. In the slow growth regime more observations are available and show less scatter. The relation between a_0 and C_∞ is complex and not predicted.

3.3 Cellular versus dendritic shapes

The main aspect that leads to the different parametric forms for cellular and dendritic spacings is the shape, dendrites being parabolic and cells having a flat tip with small curvature[22,26,27]. As cells push much of the solute in front of them, the cellar array solute distribution depends strongly on the distribution coefficient k. From first principles, cellular arrays may be thought to be limited by $a_0 \sim r_{tip}$, where r_{tip} is the tip radius which decreases with C_∞ [20,36,22]. Dendritic arrays involve a further degree of freedom due to inter-dendritic interactions that may lead to an opposite concentration dependence. The details of these interactions and transitions between the classes are not yet fully understood[23,22,26,37].

No single model can explain the much weaker observed concentration dependence of the plate spacing of aqueous NaCl solutions in Figure 2. At low concentrations the predicted cellular spacing from equation (12) is almost an order of magnitude too large (dotted line, Figure 1a). At high concentrations the dendritic models overpredict the actual spacings considerably.

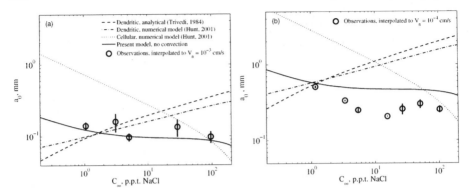

Figure 2 *Plate spacing during downward freezing of NaCl solutions at different salin-*
ities[10], normalised to (a) 10^{-3} and (b) 10^{-4} cm s^{-1} by power law fits for
the regimes $V > 0.4 \times 10^{-3}$ and $0.5 < V < 2 \times 10^{-4}$, with standard devi-
ations given by the bars. Further shown are the analytical and numerical
predictions from previous work and the wavelength λ_{mi} from the present
approach (solid curve).

The observations in Figure 2b might be interpreted as a cellular growth mode
at low concentrations, with a dendritic scaling at larger C_∞. This appears plausible
when considering larger solid fractions in the former and lower solid fractions in the
latter case. Such a simplified interpretation must be viewed with caution. The high
anisotropies in growth kinetics and interfacial energy of ice might modify the results,
even if the latter influence was found to be small in the numerical model[26]. The influence of convection at low V in Figure 2b is another aspect not considered so far.
Progress can be made by the following novel approach to plate spacing selection.

4 A NEW APPROACH TO MARGINAL ARRAY STABILITY

4.1 Basic Solution

The following approach to predict the plate spacing from morphological stability theory is based on two ideas. First, the interface criterion proposed by Bower, Brody and
Flemings[38] is assumed. On the basis of observations, these authors once suggested the
condition

$$-m(C_{int} - C_\infty) = \frac{DG_{eff}}{V}, \qquad (13)$$

for the equilibrium growth of a cellular interface with an effective macroscopic interfacial temperature gradient G_{eff}. Observations of interface concentrations are still
sparser, yet equation (13) has more recently been confirmed by observations identifying C_{int} with the concentration C_{tip} at the tip of cells[37]. Also for cellular freezing NaCl
solutions some observations in agreement with equation (13) exist[39]. In the notation
of morphological stability theory the condition corresponds to $\mathcal{G} = 1$, if the equilibrium
interface concentration C_∞/k in equation (4) is replaced by C_{int}.

The second step differs from previous approaches underlying equations (9) through (12), which all considered the geometry at the tip of cells. The present approach ignores these geometrical constraints of the selection process. Instead, the marginal stability of the *cellular* interface is treated by the same set of equations as the *planar* problem. The Mullins-Sekerka theory from section (2) is applied *macroscopically* with two principal modifications. First, an effective thermal conductivity K_i' near the interface must be defined, with the solid-liquid mixed phase conductivity replacing the pure ice value. As the dendrites or cells are thickening laterally near the interface, this cannot be done by introducing a parallel conduction model, as frequently used in continuum models. The problem involves the solution of the thermal diffusion field near the interface. An approximate analysis[40] suggests that the problem can be treated in a macroscopic sense, modeling the growth by combining the latent heat of fusion of pure ice with an enlarged $K_i' = rK_i'$, where r is ≈ 1.8 for $C_\infty = 35$ ‰ NaCl. The factor r is relatively insensible to the solute concentration in the range 10 and 100 ‰. It reflects the enhancement of solid ice growth by heat flow in the liquid and is limited by lateral conduction. Ignoring this factor does, however, not affect the main features of the following solution. For seawater it merely results in ≈ 30 % larger predictions of a_0.

The second modification is at hand, when realising the fundamental difference between a planar and a macroscopic cellular interface: k' is no longer a material property but becomes a free parameter. The condition of cellular marginal stability is therefore found by solving equation (6) for k'. This maximum possible k' for the interface corresponds to the minimum $(C_{int} - C_\infty)$, corrected by a capillary term implied by $\mathcal{A}$. The wavelength λ_{mi} corresponding to the maximum k' is then readily obtained from equation (7). For a given set of V, C_∞ and the liquid temperature gradient G_b, this maximum k' implies the largest possible wave length. It may be interpreted as the cell spacing at marginal array stability of the macroscopic interface.

4.2 Comparison with observations

The predictions from the proposed marginal stability approach are shown as solid curves in Figures 1 and 2, in comparison with observations and the above discussed models. The solid curves indicate, considering the relation between plate spacing and growth velocity in Figure 1, a reasonable agreement with observations. For both salinities convective effects are expected at growth velocities below $V_n \approx 5 \times 10^{-4}$ cm s^{-1}. In the diffusive high velocity regime, to which the strict applicability of all theoretical scalings is limited, the present predictions are closest to the observed plate spacings.

The performance of the present approach becomes more obvious in Figure 2a, showing the concentration dependence of a_0. It yields only a weak salinity dependence of the plate spacing and is the only model which is consistent with the observations. The predicted concentration dependence cannot be validated by the scattered data, yet is supported by some limited plate spacing observations in brackish water[40].

The quantitative agreement between predictions and observations is, as seen in Figure 2b, worse at lower growth velocities. Due to the increasing influence of haline convection this is not unexpected. The overall data appearance however, including the slight increase in a_0 with concentration, is again reasonably captured by the present simple model approach. The observations indicate an unpredicted minimum in the plate spacing near a concentration of ≈ 10 ‰. This minimum can be explained to a certain degree by the differential influence of solutal convection[40]. The next section demonstrates how solutal convection can be incorporated into the present approach.

5 NATURAL SEA ICE GROWTH

5.1 Solutal Convection

The incorporation of solutal convection into morphological stability theory is here illustrated on the basis of a very simple model. It is assumed that the interfacial solute gradient G_c in the liquid is enhanced by a Nusselt number Nu given by

$$Nu = c_{Nu}Ra^{1/3}, \tag{14}$$

with the solutal Rayleigh Number

$$Ra = \frac{\beta(C_{int} - C_\infty)gH_*^3}{\nu D}. \tag{15}$$

Ra is based on the concentration difference $(C_{int} - C_\infty)$ that corresponds to the maximum k'. β, ν and g are the haline contraction coefficient, kinematic diffusivity ν and gravity acceleration, and H_* is identified with D/V. Observations of the proportionality factor c_{Nu} fall between 0.15 and 0.19 in the high Schmidt number limit[41], and a theoretical limit of 0.152 has been proposed by Chan[42] for a rigid interface. Here $c_{nu} = 0.15$ is used. One further needs to account for the fact that the solute gradient in the porous ice-brine mixture is not modified by convection in the liquid, yet stays at the liquidus temperature. This is done by assuming a lumped concentration gradient G'_c as the average on both sides of the macroscopic interface. It results in

$$G'_c = G_c\frac{(Nu + 1)}{2}. \tag{16}$$

for the convectively modified concentration gradient G'_c.

In the simple model equation (14) the onset of convection corresponds to the growth velocity at which Nu exceeds unity. At larger growth velocities G_c is not modified by convection and $Nu = 1$. Because the convection cannot transport more solute than is available at the interface, there is an upper limit for the Nusselt number at low V. Solute conservation implies the bound

$$Nu < \frac{1 - k'_{eff}}{1 - k'}. \tag{17}$$

where k'_{eff} is the salt fraction retained in the ice. The natural upper bound for Nu is, of course, given by $k_{eff} \approx 0$. Another tentative limit is $(1 - k'_{eff}) \approx 0.5$ and may be proposed on the basis of the assumption, that the downward convective solute flux at the interface, $Nu(1 - k')$, cannot exceed the local upward solute distribution.

The further approach to obtain the convectively modified wavelength λ_{mi} is as follows. The principle modification to the scalings from Section (1) is the rescaling of the diffusion length D/V in equation (7) by the factor (Gc/Gc'), and the replacement of k by k'. The solution is obtained by iteration of equations (7) and (6) which converges rapidly to a pair of Gc' and k' in dependence on V.

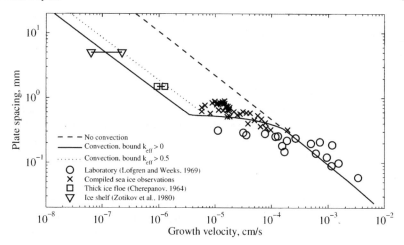

Figure 3 *Predicted plate spacings without (dashed curve) and includ-*
ing the convection model with two upper limits in the Nusselt
number (solid and dotted lines). Crosses refer to observations
compiled from sea ice field studies[40] and circles to laboratory
data[10] at comparable NaCl concentrations from 25 to 31 ‰.
The low velocity points refer to thick arctic sea ice (square)
and the bottom of an Antarctic ice shelf (triangles).

5.2 Observations

In Figure 3 the convectively modified wavelength λ_{mi} is compared to observations
of the sea ice plate spacing a_0 that have been compiled from a number of field studies[40]. These are shown as crosses to contrast them with the so far discussed laboratory
data[10] at comparable concentrations from 25 to 31 ‰ NaCl.

The onset of convection, and thus the deviation of the plate spacing from the strict
diffusive case (dashed line), is, for $C_\infty \approx 35$ ‰, predicted near $V \approx 1.6 \times 10^{-4}$ cm/s.
The resulting flattening in the relationship between a_0 and V is consistent with the
field observations. The laboratory data fall, as mentioned earlier, clearly below the
plate spacings of natural sea ice. This cannot be explained by solutal convection. An
analysis of the growth conditions indicates the presence of strong thermal convection
in the laboratory experiments[40], implying that at growth velocities below 10^{-4} cm/s
the laboratory conditions do not resemble natural sea ice growth.

A rigorous test of the theory requires observations at low growth velocities. Although sparse, some data are available. The squares are observations from the bottom
of a 10 meter thick ice flow[9]. The corresponding growth velocity range has been estimated by assuming an average annual ice surface temperature of -15 to -19 °C[43].
The to date largest a_0 has been observed at the bottom of an Antarctic ice shelf[44].
The corresponding growth velocity is most likely in the range 2 to 7 cm/year[44,45].

Although there is clearly a need for systematic studies at low growth velocities, the agreement of the predictions with this limited set of observations is very promising.

5.3 Discussion

Albeit simplified, the convection model proposed here is consistent with the intro-duced macroscopic interface approach. It implies a reasonable parametrisation of the basic effect of solutal convection: the steepening of the solutal gradient at the cellular interface. The corresponding predicted decrease in the plate spacing is in agreement with the available field observations in Figure 3.

The Nusselt number model, equations (14) through (16), implies a linear increase of Nu with V, once the growth velocity falls below a critical value. In reality a damped increase in Nu might be expected, because strengthening convection decreases the liq-uid fraction on the ice side of the macroscopic interface. Accounting for the latter improves the agreement with observations slightly[40], but the principle result in Figure 3 is not changed. The concentration gradient at the interface may also depend on inter-nal convection[46] and morphological evolution in a high porosity bottom 'skeletal layer' of a few centimeters in thickness[47,35], and even on the internal redistribution of brine via larger channels[47]. Such mechanisms may all be viewed as boundary conditions of a difficult problem to be solved: the interaction of morphology, diffusion and convection in a thin interfacial regime. A detailed resolution of the latter would also require the account of local dendritic selection conditions at the tip of the cells[36,22] and involve a considerable increase in complexity. However, it appears that the present simplified macroscopic approach contains the essential physics that controls the variability in the plate spacing of naturally growing sea ice.

The two available observations at low growth velocities are particularly interesting. They are consistent with the proposed return to the purely diffusive $a_0 - V$-slope, when the Nusselt number reaches its limit implied by k'_{eff} in equation (17). The interfacial k'_{eff} suggested here may be thought to evolve as the interaction of morphological sta-bility, diffusion and convection. Of relevance in this context might be the threshold k of a planar interface below which deep cell grooves are not expected to evolve. It depends on the ratio of solid and liquid thermal conductivities and should be near 0.1 for the ice-water system[48], not inconsistent with observations[34]. Also the marine ice shelf observations at the lowest growth velocity indicate a limiting $k'_{eff} \approx 0.1$ some centimeters from the interface[44]. Figure 3 indicates that the expected dependence of a_0 on the threshold k'_{eff} is weak, but might be resolved experimentally. To make progress more observations and interface model studies are required.

6. SUMMARY

Observations of the plate spacing of saline ice and sea ice have been compared to theo-retically predicted scaling laws. It was found that both the velocity and concentration dependence were not captured by the theories. Such discrepancies are not unique for saline solutions but found for other systems as well[49,22,26,50]. The disagreement has sometimes been explained in terms of history-dependent spacing adjustment[26,51]. However, in any case the numerical predictions proposed as a lower bound on the spacings[26,51]. For freezing of aqueous NaCl solutions they yield too large plate spacings.

A new approach to the problem has been formulated on the basis of a *macroscopic* linear stability analysis. It does not invoke a geometric condition at the tip of cells, on which all earlier theoretical treatments were based. Remarkably, it predicts the plate spacing much closer than any other model. The result indicates that the plate spacing selection of growing sea ice should be viewed as a marginal equilibrium condition of an array of growing plates, rather than the nonlinear evolution process of an initial perturbation, as suggested in previous work[19]. The present view is consistent with the empirical fact, that sea ice almost never grows as a coarsening array from a destabilising planar interface, yet that its lamellar structure evolves from a thin primary layer of randomly oriented disk crystals[35].

It has been illustrated, how the present model can be extended in a simple manner to include the effect of natural solutal convection on the plate spacing. The proposed approach is simple and only involves a rescaling of the solutal gradient lengthscale D/V due to convection. It demonstrates a complex relationship between plate spacing and freezing rate. For most of the natural growth regime the latter does not show a constant slope, explaining why previous empirical correlation approaches have been unsatisfactory. Observations during natural freezing of seawater confirm the plate spacing predictions over five orders of magnitude in the freezing rate. It thus appears that a rather simplified treatment of a macroscopic interface accounts for the main mechanisms of the plate spacing selection during natural sea ice growth. The present results offer new perspectives in modeling and interpretation of microstructural observations of natural sea ice and marine ice shelves.

References

1 D. Walker, *Proc. Royal Soc. London*, 1859, **9**, 609.

2 A. Hamberg, *Svenska Vetenskapsakadamien, Bihang til Handlinger*, 1895, **21**, 1.

3 E. v. Drygalski, *Grönlands Eis und sein Vorland*, vol. 1 of *Grönland-Expedition der Gesellschaft für Erdkunde zu Berlin 1891-1893*, Berlin, Kühl, 1897, 555 pp.

4 T. Fukutomi, M. Saito and Y. Kudo, *Low Temp.Sci.*, 1952, **9**, 113.

5 D. L. Anderson and W. F. Weeks, *Trans. Amer. Geophys. Union*, 1958, **39**, 632.

6 W. F. Weeks and W. L. Hamilton, *The Amer. Mineralogist*, 1962, **47**.

7 A. Assur, in *Arctic Sea Ice*, vol. 598, Proc. Conf.on Arctic Sea Ice, Natl. Acad. Sci., pp. 106–138.

8 W. Schwarzacher, *J. Geophys. Res.*, 1959, **64**, 2357.

9 N. V. Cherepanov, *Trudy Arkt. Antarkt. N.-I. Institut Leningrad*, 1964, **267**, 13.

10 G. Lofgren and W. F. Weeks, *J. Glaciol.*, 1969, **8**, 153.

11 M. Nakawo and N. K. Sinha, *Atmos.-Ocean*, 1984, **22**, 193.

12 P. K. Rohatgi and C. M. Adams, *J. Glaciol.*, 1967, **6**, 663.

13 W. W. Mullins and R. F. Sekerka, *J. Appl. Phys.*, 1964, **35**, 444.

14 S. R. Coriell, G. B. MacFadden and R. F. Sekerka, *Ann. Rev. Mater. Sci.*, 1985, **15**, 119.

15 S. R. Coriell and G. B. McFadden, *Handbook of Crystal Growth*, vol. I, Elsevier, pp. 785–857.

16 C. Körber and M. W. Scheiwe, *J. Cryst. Growth*, 1983, **61**, 307.

17 B. Rubinsky and M. Ikeda, *Cryobiology*, 1985, **22**, 55.

18 C. Koerber, *Quart. Rev. Biophys.*, 1988, **21**, 229.
19 J. S. Wettlaufer, *Europhysics Letters*, 1992, **19**, 337.
20 R. Trivedi and K. Somboonsuk, *Mater. Sci. Engineering*, 1984, **65**, 65.
21 S. DeCheveigne, C. Guthmann and P. Kurowski, *J. Cryst. Growth*, 1988, **92**, 616.
22 B. Billia and R. Trivedi, *Handbook of Crystal Growth*, vol. I, Elsevier, pp. 901–1071.
23 J. A. Warren and J. S. Langer, *Phys. Rev. A*, 1990, **42**, 3518.
24 J. A. Warren and J. S. Langer, *Phys. Rev. E*, 1993, **47**, 2702.
25 J. D. Hunt, in *Solidification and Casting of Metals*, Metals Society London, Sheffield, pp. 3–9.
26 J. D. Hunt and S. Z. Lu, *Met. Mater. Trans. A*, 1996, **27 A**, 611.
27 J. D. Hunt, *Sci. Techn. Adv. Mater.*, 2001, **2**, 147.
28 G. Ding, W. Huang and Y. Zhou, *J. Cryst. Growth*, 1997, **177**, 281.
29 R. F. Sekerka, *J. Appl. Phys.*, 1965, **36**, 264.
30 R. F. Sekerka, in H. S. Peiser, ed., *Crystal Growth*, Pergamon Press, Boston, 20-24 June 1966, pp. 691–702.
31 S. R. Coriell, R. F. Boisvert, G. B. MacFadden, L. N. Brush and J. J. Favier, *J. Cryst. Growth*, 1994, **140**, 139.
32 S. Maus, in W. F. Kuhs, ed., *11th Int. Conf. on Physics and Chemistry of the Ice*, Royal Society of Chemistry, this volume.
33 L. X. Liu and J. S. Kirkaldy, *Acta Metall. Mater.*, 1995, **43**, 2891.
34 W. F. Weeks and G. Lofgren, in *Physics of Snow and Ice*, vol. 1, Hokkaido University, Sapporo, Japan, pp. 579–597.
35 W. F. Weeks and S. F. Ackley, in *The Geophysics of Sea Ice*, vol. 146 of *NATO ASI Series*, Plenum Press, pp. 9–164, ed. by N. Untersteiner.
36 M. E. Glicksman and S. P. Marsh, *Handbook of Crystal Growth*, vol. I, Elsevier, pp. 1075–1122.
37 A. Pocheau and M. Georgelin, *J. Cryst. Growth*, 2003, **250**, 100.
38 T. F. Bower, H. D. Brody and M. C. Flemings, *Trans. Metall. Soc. AIME*, 1966, **236**, 624.
39 J. P. Terwilliger and S. F. Dizio, *Chem. Eng. Science*, 1970, **25**, 1331.
40 S. Maus, Phd thesis, Geophysical Institute, University of Bergen, 5007 Bergen, Norway, 2006.
41 J. R. Selman and C. W. Tobias, *Adv. Chem. Engineer.*, 1978, **10**, 211.
42 S. K. Chan, *Studies Appl. Math.*, 1971, **1**, 13.
43 Y. P. Doronin and D. E. Kheisin, *Morskoi Led (Sea Ice)*, Gidrometeoizdat, Leningrad, 1975, english translation 1977 by Amerind Publishing, New Delhi, 318 pp.
44 I. A. Zotikov, V. S. Zagorodnov and J. V. Raikovsky, *Science*, 1980, **207**, 1463.
45 K. Assmann, H. H. Helmer and A. Beckmann, *Antarct. Sci.*, 2003, **15**, 3.
46 M. G. Worster and J. S. Wettlaufer, *J. Phys. Chem. B*, 1997, **101**, 6132.
47 M. Wakatsuchi, *Contri. Inst. Low Temp. Sci.*, 1983, **A33**, 29.
48 G. J. Merchant and S. H. Davis, *Phys. Rev. B*, 1989, **40**, 1140.
49 R. Trivedi and K. Somboonsuk, *Acta Metall.*, 1985, **33**, 1061.
50 D. Bouchard and J. S. Kirkaldy, *Metall. Mater. Trans. B*, 1997, **28 B**, 651.
51 W. D. Huang, X. G. Geng and Y. H. Zhou, *J. Cryst. Growth*, 1993, **134**, 105.

THE PLANAR-CELLULAR TRANSITION DURING FREEZING OF NATURAL WATERS

S. Maus

Geophysical Institute, University Bergen, Allegaten 70, 5007 Bergen, Norway

1 INTRODUCTION

Whereas lake ice thickens vertically with a planar freezing interface, sea ice growth is cellular[1,2,3,4,5]. A theoretical treatment of the planar-cellular transition under natural ice growth conditions is lacking and the present work is an approach to close this gap.

Cellular freezing of natural waters implies a much larger interstitial solute incorporation in form of liquid brine than planar freezing. The planar-cellular transition may therefore be identified as a change to a milky ice appearance or by a sharp increase in the effective solute distribution coefficient k_{eff}[6,7,8,9]. Systematic observations of the salinity, at which the transition occurs in natural brackish water, are lacking. It is likely that it takes place at concentrations below 3.6 to 4.1 ‰, for which the cellular plate substructure was regularly present in Baltic sea ice[10]. According to reference[11] a transitional salinity of 2 ‰ has been reported by Russian scientists, yet the mentioned studies have remained unpublished. In laboratory experiments transitional salinities between 1 and 50 ‰ have been observed, depending on the strength of convective solute transport due to stirring[7,12,9,8,13,14,15,16]. During freezing of natural waters the main mechanism of solute transport is free convection, and a theoretical treatment needs to consider its dependence on the growth velocity and water salinity.

It is noteworthy that a similar transition has been observed for the crystal orientation. The generally horizontal c-axis in sea ice[2,17] contrasts the occurrence of both orientations in lake ice[18,19,20]. Some limited studies of brackish water ice[21,22] indicate, that a primarily horizontal c-axis orientation is most likely established in a salinity range between 0.5 and 1.3 ‰. According to laboratory experiments the transition depends also on the cooling rate, and a lower transition salinity of ≈ 0.1 to 0.2 ‰ has been reported during very rapid freezing[23]. Although limited, these observations indicate that transitions in the crystal orientation parallel the onset of cellular growth.

Several authors[18,19,9,17] have conceptually related the transition from planar to cellular freezing of ice to the condition of *constitutional supercooling*[24,25]. Here the theoretical framework of morphological stability theory for a planar interface[26] is combined with simple models of diffusion and free convection to derive the transition salinity in dependence on growth conditions.

2 Interface Stability and Constitutional Supercooling

The concept of *constitutional supercooling*[24] is the basic explanation of cellular freezing of alloys and saline solutions. Tiller and coworkers[25] once postulated an appropriate quantitative condition

$$mG_c = -mC_\infty \frac{1-k}{k} \frac{V}{D} > G_b \tag{1}$$

valid for equilibrium freezing. In the steady state the liquidus temperature gradient G_c depends on freezing velocity V, solute diffusivity D, liquidus slope $m = \partial T/\partial C$ and the interfacial solute distribution coefficient k. If G_c exceeds the temperature gradient G_b in the liquid, one has constitutional supercooling near the interface and cellular perturbations may grow.

A more stringent condition was derived by Mullins and Sekerka[26] in their now classical linear stability analysis. It reads

$$C_\infty > \frac{k}{m(k-1)} \left(\frac{L_v D}{K_i + K_b} \right) \left(1 + \frac{2G_b K_l}{L_v V} \right) \frac{1}{\mathcal{S}_\Gamma}, \tag{2}$$

and accounts also for conduction in the solid, latent heat release and the dynamical evolution of infinitesimal perturbations. K_b, K_i, G_b and G_i denote thermal conductivities and temperature gradients in the liquid and solid and L_v is the volumetric latent heat of fusion. $\mathcal{S}_\Gamma$ is a stability function which represents the stabilising effect of surface tension[27]. According to the theory the planar interface may become unstable at a given growth velocity which depends on G_b in the liquid. The corresponding range of wave lengths is shown in Figure 1a for the NaCl-water system at typical lake water concentrations and two values of the G_b. For a liquid temperature gradient of 0.1 K/cm the interface will be stable at velocities below the solid triangle. During the freezing of natural waters the influence of the liquid temperature gradient G_b is often negligible. Inserting $L_v = 305.6$ kJ/kg, $K_i = 2.14$ W/(mK), $K_b = 0.560$ W/(mK), $m = -0.0602$ K/‰ and $D = 7.03 \times 10^{-6}$ cm^2/s as molecular properties at dilute concentrations[28] into equation (2), and noting that $k \approx 10^{-3}$ is small, results in

$$C_\infty \frac{\mathcal{S}_\Gamma}{k} > 1.32 \,°/_{\!\infty} \tag{3}$$

as the condition for instability. The remaining variable $\mathcal{S}_\Gamma$ has been computed for dilute NaCl solutions following reference[27]. It is shown in Figure 1b for plausible values of k and in dependence on the growth velocity. It shows that, as natural growth velocities seldom exceed 2×10^{-4} cm/s, the influence of $\mathcal{S}_\Gamma$ can safely be neglected. In the present application therefore $\mathcal{S}_\Gamma \approx 1$ is justified.

Equation (3) applies strictly to the diffusive equilibrium with $C_{int} = C_\infty/k$ at the interface. However, for large ratios of thermal to solutal diffusivity, it has been shown to hold also in the transient stage[29,30], if C_∞/k is replaced by the time-dependent interfacial concentration $C_{int}(t)$. Hence

$$C_{int}(t) > 1.32 \,°/_{\!\infty} \tag{4}$$

becomes the simplified condition for the planar-cellular transition in natural waters. Equation (4) applies under absence of large temperature gradients and convection.

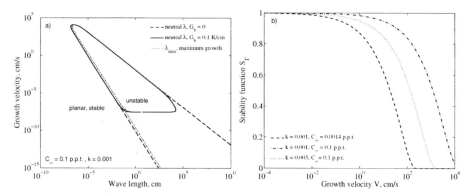

Figure 1 *Interfacial stability for freezing of typical natural lake water: a) Wavelength-velocity diagram for two liquid temperature gradients G_b and b) Stability function S_Γ from Sekerka[27] for dilute NaCl solutions for a range of plausible values of k.*

3 SOLUTE REDISTRIBUTION

To predict the evolution of C_{int} with time a model for the diffusive boundary layer evolution ins needed. In case of a constant solute flux[31] the latter evolves due to

$$C_{int}(t) = C_\infty \left(1 + \frac{2\rho_i}{\sqrt{\pi}\rho_b} V (1-k) \left(\frac{t}{D}\right)^{1/2}\right), \tag{5}$$

where ρ_i and ρ_b are the densities of pure ice and the solution. This form is inaccurate for natural growth conditions, as due to increasing C_{int}/C_∞ the salt flux at the interface is not constant. A frequently used approximate solution to the problem[25]

$$C_{int}(t) = C_\infty \left(1 + \frac{1-k}{k} (1 - exp(-k\tau))\right) \tag{6}$$

with the normalised timescale τ

$$\tau = \left(\frac{\rho_i V}{\rho_b}\right)^2 \frac{t}{D} \tag{7}$$

is valid at large times. The exact solution has been given by Smith et al.[32] as

$$C_{int}(t) = \frac{C_\infty}{2k} \left(1 + erf \left(\frac{\tau}{4}\right)^{1/2} + p\,exp\left((k-1)\,k\tau\right) erfc \left(p \left(\frac{\tau}{4}\right)^{1/2}\right)\right) \tag{8}$$

where

$$p = (2k - 1). \tag{9}$$

The evolution in the relative interfacial concentration obtained with these diffusion models is shown in Figure 2 for short and long times on a linear and logarithmic scale,

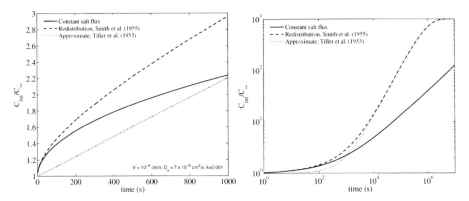

Figure 2   *Relative increase in the interfacial concentrations with time due to different diffusion models for the growth conditions indicated in the figures. Left: short times; right: long times on a log-scale. The dashed line is the exact solution from reference[32].*

respectively. It is seen that the constant salt flux condition is only reasonable for short times, when the approximation (6) from reference[25] cannot be used. At larger time the opposite is the case.

Figure 2 shows that lake water, with $C_\infty \approx 0.1$ ‰, requires, even during rapid growth, $\approx 10^4$ seconds to reach the critical concentration $C_{int} \approx 1.3$ ‰. One may also deduce from equation (3), that a rather dilute solution with $C_\infty \approx 0.0013$ ‰ might become unstable if diffusive equilibrium is reached. According to Figure 2 this takes, even during rapid freezing, ≈ 20 days and 2 meters of ice growth. In a closed container this will, of course, be different, as the released solute increases C_∞.

4 CONVECTION

The next step is to introduce the limiting effect of convection into the problem. A solution based on a nonlinear turbulent boundary layer model has been given by Foster[33]. It states that the critical time t_c at the onset of convection is given as

$$t_c = b \left(\frac{\nu \rho_b}{g \beta V (1-k) C_{int} \rho_i} \right)^{1/2}, \tag{10}$$

where β, ν and g are the haline contraction coefficient, kinematic diffusivity ν and gravity acceleration. In the underlying heuristic model of intermittent convection the interface concentration increases by pure diffusion until t_c is reached. Then a plume of dense fluid breaks off and the process starts again. Foster determined $b = 14$ in equation (10) for a free boundary. In the following a pre-factor of $b = 21$ is used as a more reliable estimate for a rigid boundary[34,28] and the difference will be illustrated in the final section.

Based on the assumption of intermittent boundary layer growth according to equation (8), equation (10) is the desired limiting condition for the interface concentration,

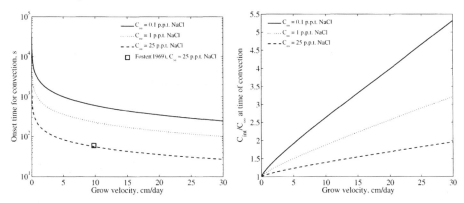

Figure 3 *Critical time (left figure) and relative interfacial concentration (right figure) at the onset of convection obtain by combining equation (10) for a rigid boundary with the diffusion model (8) for different solution salinities. The square is a laboratory observation reported by Foster[35] at 25 ‰.*

in dependence of C_∞ and V. As it is valid for a constant salt flux and C_{int}, equations (10) and (8) have been solved iteratively, and by inserting an effective interface concentration $C'_{int} = (C_{int} + C_\infty)/2$ into equation (10). The computations for the critical time and interface concentration are shown for different far-field concentrations in Figure 3. An observation reported by Foster[35] for the convective onset during the initial freezing of a solution at seawater concentration is shown as a square. It is in reasonable agreement with the predictions of the critical time.

5 RESULTS AND DISCUSSION

Equations (8) and (10) are readily applied to compute, at a given growth velocity, the critical far-field concentration C_∞ at which the interface concentration C_{int} exceeds the cellular instability value of 1.32 ‰. The prediction is shown in Figure 4 for the pre-factors of 21 and 14 in equation (10), resembling rigid and free boundary conditions. It is compared to a number of laboratory studies where the transition has been documented during downward freezing[7,8,36]. The least agreement with the observations from reference[7] might be related to rigorous thermal convection in these experiments[28,37]. For freezing of aqueous KOH[8] one would, due to an almost 2 times larger solute diffusivity, expect a larger concentration. However, the salinity reported in the latter study was the initial C_∞, and the point in Figure 4 might underestimate the intrinsic transition value by 50 %. In the tank experiments from reference[36], which are likely not affected by strong thermal convection, the growth velocity was constant and the changing concentration in the tank was observed continuously. The agreement of the predictions with these data, which are thought to be most reliable, is excellent. It is recalled, that the observed critical time in Figure 3 also was quite close to the prediction for a rigid boundary. The overall agreement indicates that the effect of convection can be reasonably modeled with the proposed simple boundary

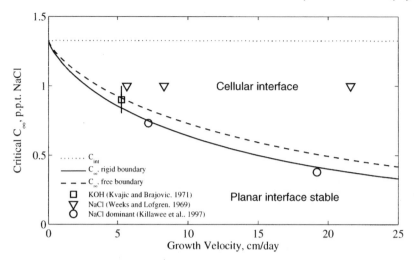

Figure 4   *Planar-cellular transition during downward freezing of saline so-*
lutions in several laboratory studies[7,8,36] compared to the present
predictions of critical interface (dotted line) and bulk far-field
concentrations (dashed and solid curves). The circles are the
most reliable observations in terms of control of the growth con-
ditions.

layer approach, on the basis of intermittency.

Natural ice growth velocities are normally limited by $\approx$ 10 cm/day. This implies, according to Figure 4, a planar-cellular transition in the concentration regime
$0.7 < C_\infty < 1.3$ ‰. The mentioned transition salinity from vertical to horizontal c-
axis orientation in brackish water ice is remarkably similar[21,22], indicating an intimate
connection between cellular growth and preferred crystal orientation. Such a connection can likely be understood in terms of the large anisotropy in the free growth rate
of ice crystals at low supercooling[38,39]. It is expected that, when dynamical cellular
growth exists, this growth advantage parallel to the basal plane involves particularly
rapid geometric selection. Simultaneous observations of the crystal orientation and
planar-cellular transitions are not available yet. These might provide a path to a better understanding of crystal orientation in lake ice sheets, which is still a matter of
debate[18,19,20,40,11].

6 SUMMARY

The present model approach has combined three equations to predict the onset of
cellular growth during freezing of natural waters: (i) constitutional supercooling from
morphological stability theory, (ii) an exact diffusive solute redistribution and (iii) an
intermittent turbulent solutal convection model. The main results are:

- The low growth velocity concentration threshold for cellular instabilities from morphological stability theory is ≈ 1.3 ‰ NaCl.

- The predicted growth velocity dependence of the critical far-field salinity C_∞ for cellular ice growth is in agreement with observations.

- A simple concept of intermittent diffusive boundary layer growth appears to be a reasonable approach to predict maximum interface concentrations during free convection.

- The change in crystal orientation from primarily vertical to horizontal and the planar-cellular transition appear to be closely related.

The quantitative results apply to solutions with NaCl being the dominating salt. They are therefore relevant for the freezing of most natural brackish waters.

References

1   D. Walker, *Proc. Royal Soc. London*, 1859, **9**, 609.
2   A. Hamberg, *Svenska Vetenskapsakadamien, Bihang til Handlinger*, 1895, **21**, 1.
3   E. v. Drygalski, *Grönlands Eis und sein Vorland*, vol. 1 of *Grönland-Expedition der Gesellschaft für Erdkunde zu Berlin 1891-1893*, Berlin, Kühl, 1897, 555 pp.
4   T. Fukutomi, M. Saito and Y. Kudo, *Low Temp.Sci.*, 1952, **9**, 113.
5   W. Schwarzacher, *J. Geophys. Res.*, 1959, **64**, 2357.
6   G. Quincke, *Annalen der Physik, Leipzig*, 1905, **18(11)**, 1.
7   W. F. Weeks and G. Lofgren, in *Physics of Snow and Ice*, vol. 1, Hokkaido University, Sapporo, Japan, pp. 579–597.
8   G. Kvajic and V. Brajovic, *J. Cryst. Growth*, 1971, **11**, 73.
9   K. Wintermantel and W. Kast, *Chemie-Ing.-Technik*, 1973, **45**, 728.
10 A. J. Gow, W. F. Weeks, P. Kosloff and S. Carsey, CRREL REPORT 92-13, Cold Regions Research and Engineering Laboratory, Hanover, 1992.
11 W. F. Weeks and J. S. Wettlaufer, in *The Johannes Weertman Symposium*, The Minerals, Metals and Materials Society, pp. 337–350.
12 A. N. Kirgintsev and B. M. Shavinskii, *Russian Chemical Bulletin*, 1969, **10**, 1528.
13 B. Ozum and D. J. Kirwan, *AICHE Symp. Ser.*, 1976, **72**, 1.
14 G. W. Gross, P. M. Wong and K. Humes, *J. Chem. Phys.*, 1977, **67**, 5264.
15 J. H. Cragin, in *52nd Eastern Snow Conference*, Toronto, Ontario, Canada 1995, pp. 259–262.
16 Y. Matsuoka, M. Kakumoto, K. N. Ishihara and P. H. Shingu, *J. Japan Inst. Metals*, 2001, **65**, 791.
17 W. F. Weeks and S. F. Ackley, in *The Geophysics of Sea Ice*, vol. 146 of *NATO ASI Series*, Plenum Press, pp. 9–164, ed. by N. Untersteiner.
18 P. A. Shumskii, *Osnovy strukturnogo ledovedeniya (Principles of structural glaciology)*, Dover Publications, Inc., 1955, 1964 translated from Russian, 497 pp.
19 C. A. Knight, *J. Glaciol.*, 1962, **4**, 319.
20 B. Michel and R. O. Ramseier, *Can. Geotechnical J.*, 1971, **8**, 36.
21 E. Palosuo, in *Int. Assoc. Sci. Hydrol.*, vol. 54, Gentbrugge, Belgium, pp. 9–14.

22 T. Kawamura, M. A. Granskog, J. Ehn, T. Martma, A. Lindfors, N. Ishikawa,
    K. Shirasawa, M. Leppäranta and R. Vaikmäe, in *Ice in the Environment: Pro-*
    ceedings of the 16th IAHR International Symposium on Ice, Int. Association of
    Hydraulic Engineering and Research, Dunedin, New Zealand, pp. 165–171.
23 E. Palosuo and M. Sippola, in W. D. Kingery, ed., *Ice and Snow*, M.I.T. Press,
    pp. 232–236.
24 J. W. Rutter and B. Chalmers, *Can. J. of Phys.*, 1953, **31**, 15.
25 W. A. Tiller, K. A. Jackson, J. W. Rutter and B. Chalmers, *Acta Metall.*, 1953,
    1, 428.
26 W. W. Mullins and R. F. Sekerka, *J. Appl. Phys.*, 1964, **35**, 444.
27 R. F. Sekerka, *J. Appl. Phys.*, 1965, **36**, 264.
28 S. Maus, Phd thesis, Geophysical Institute, University of Bergen, 5007 Bergen,
    Norway, 2006.
29 R. F. Sekerka, in H. S. Peiser, ed., *Crystal Growth*, Pergamon Press, Boston, 20-24
    June 1966, pp. 691–702.
30 S. R. Coriell, R. F. Boisvert, G. B. MacFadden, L. N. Brush and J. J. Favier, *J.*
    Cryst. Growth, 1994, **140**, 139.
31 H. S. Carslaw and J. C. Jaeger, *Conduction of Heat in Solids*, Clarendon Press,
    Oxford, 2nd edn., 1959, 510 pp.
32 V. G. Smith, W. A. Tiller and J. W. Rutter, *Can. J. Phys.*, 1955, **33**, 723.
33 T. D. Foster, *J. Phys. Oceanogr.*, 1972, **2**, 462.
34 T. D. Foster, *Phys. Fluids*, 1968, **11**, 1257.
35 T. D. Foster, *J. Geophys. Res.*, 1969, **74**, 6967.
36 J. A. Killawee, I. J. Fairchild, J.-L. Tison, L. Janssens and R. Lorrain, *Geochimica*
    and Cosmochim. Acta, 1998, **62**, 3637.
37 S. Maus, in W. F. Kuhs, ed., *11th Int. Conf. on Physics and Chemistry of the*
    Ice, Royal Society of Chemistry, this volume.
38 W. B. Hillig, in R. H. Doremus, B. W. Roberts and D. Turnbull, eds., *Growth*
    and Perfection of Crystals, Wiley and Sons, New York, pp. 350–360.
39 A. S. Michaels, P. L. T. Brian and P. R. Sperry, *J. Appl. Phys.*, 1966, **37**, 4649.
40 P. Hobbs, *Ice Physics*, Clarendon Press, Oxford, 1974, 837 pp.

CRYSTAL GROWTH OF ICE-I/HYDRATE EUTECTIC BINARY SOLUTIONS

C. McCarthy [1,4], K.D. Rieck[2], S.H. Kirby[1], W.B. Durham[3], L.A. Stern[1], and R.F. Cooper[4]

[1] U.S. Geological Survey, 345 Middlefield Rd., Menlo Park, CA 94025
[2] State University of New York at Albany, Albany, NY 12222
[3] U.C. Lawrence Livermore National Laboratory, PO Box 808, Livermore, CA 94450
[4] Brown University, Department of Geological Sciences Providence, RI 02912

1 INTRODUCTION

Reflectance spectra of Europa obtained by Galileo's Near-Infrared Mapping Spectrometer suggest that in addition to water ice, there exists a non-ice, highly-hydrated material found on the surface[1,2]. Although composition of the hydrated material is unknown, laboratory experiments combined with thermochemical knowledge gleaned from carbonaceous chondrites have narrowed the list of possible candidates[3,4]. How the presence of such hydrate(s) would affect the tectonic processes on the surface of Europa is unclear. Although much is known about the rheology of pure ice, very little is known about the effect such a non-ice material would have on the physical properties of ice. Some mechanical interaction amongst phases can be expected; that interaction is dependent to first order on the spatial distribution of phases (i.e. the microstructure). As part of our ongoing investigation of chemical systems likely volumetrically important on Europa, we conducted laboratory experiments to examine the solidification behavior and resulting microstructure/texture of several two-component systems: H_2O-Na_2SO_4, H_2O-$MgSO_4$, and H_2O-$NaCl$.

2 EXPERIMENTAL PROCEDURE

The samples in this study were prepared by a eutectic solidification reaction in which cooperative, side-by-side growth of two solid phases occurs in an undercooled, homogenous liquid solution. This liquid solution was created at room temperature from reagent grade crystals and triply distilled water. Test tubes containing liquid solution corresponding to the most ice-rich stable eutectics (Fig. 1) were immersed in a low-temperature bath at a constant undercooling ($\Delta T \sim 1$-$5°C$). Periodically test tubes containing solidifying samples were briefly (< 20 s duration) removed from the bath, photographed and returned to the bath. Thermocouples positioned within samples recorded the release of latent heat of crystallization, allowing us to monitor the onset of crystallization and quantify growth rate. Once crystallization was complete, the samples were removed from test tubes and cut into 3-5 mm thick slabs with a band saw, all while remaining in a freezer at ~ 253K. Specimens were then transported in contact with liquid nitrogen (77K) to the microscopy lab. Microstructural analysis was performed using images obtained from Cryogenic Scanning Electron Microscopy/Secondary Electron Imaging (CSEM/SEI). Phases were identified using cryogenic x-ray diffraction (XRD) and energy dispersive x-ray spectrometry (EDS) analysis. Compositional gradations in the solidified samples were measured by systematically taking small sections of the samples, melting the sections, and using a digital refractometer to analyze the melt. Results were compared to published refractive index (RI)/composition models, in the case of H_2O-NaCl, or to our own calibration standard, in the case of system H_2O-$MgSO_4$. No such compositional analysis was conducted on system H_2O-Na_2SO_4.

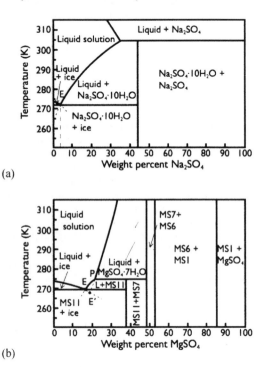

(a)

(b)

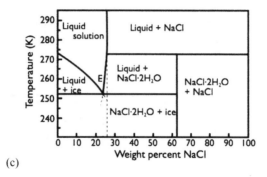

(c)

Figure 1 *Equilibrium phase diagrams for systems (a) H₂O-Na₂SO₄, (b) H₂O-MgSO₄, and (c) H₂O-NaCl at 1 atm. E is the eutectic composition of each system; peritectic and metastable eutectic compositions are denoted by P and E' respectively. Hydrate phases are identified by the level of hydration, i.e. the n in the form MgSO₄·nH₂O, and an abbreviation of the sulfate so that MS7 is MgSO₄·7H₂O, etc. (Diagrams a, b, and c adapted from Kargel[4], Peterson and Wang[5], and Roedder[6], respectively)*

3 RESULTS AND DISCUSSION

3.1 Macroscopic growth

The onset of crystallization within the test tubes occurred wherever a nucleation site was available, typically on the inner wall or on the thermocouple rod. Large, feathery crystals were often observed growing into the sample. To control growth and circumvent the thermodynamic barrier to nucleation, we in some instances used a small bit of seed material with the same bulk composition as the solution. The density of the eutectic solid in all instances is less than that of the liquid solution. Thus, when the seed was deployed, it floated on the surface and crystal growth continued downward with a nearly planar solid-liquid interface.

While photographing samples throughout the solidification process, we observed that in some samples, in addition to a buoyant solid phase, there was also crystallization of a dense phase that sank to the bottom of the tube. The only phases denser than the liquid in the systems studied are primary hydrates. With a composition equal to that of the eutectic, one would not expect crystallization of such a phase. In Fig. 2, an equilibrium phase diagram with metastable extensions is presented to demonstrate what is likely occurring. When the initial eutectic composition (C_E) liquid is slowly cooled, buoyant eutectic solid should form once the eutectic isotherm is reached. However, the ice phase (represented here by α, thus A≡H₂O), which is non-faceting and may have a kinetic advantage over the hydrate phases, may grow dendrites of primary ice ahead of the eutectic. With this ice preferentially removed from the solution, the composition of the remaining liquid travels along the metastable extension of the ice (α) liquidus toward a more salt-rich composition, e.g., $C_1{}^7$. This liquid is hypereutectic: hydrate will crystallize, moving the liquid composition back toward the eutectic. The process can demonstrate some cycling in liquid composition, which damps out with increased crystallization (inset).

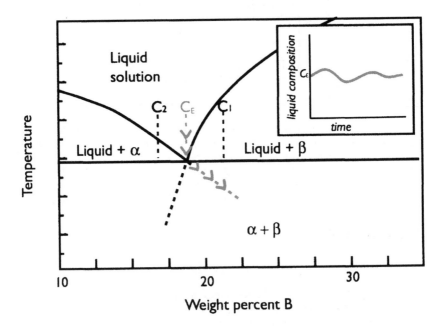

Figure 2 *Portion of a hypothetical A-B binary phase diagram showing a simple eutectic solidification reaction. Compositional variations can occur within the liquid during solidification as a result of relative kinetics of growth of solids α and β. In this case, α grows more easily. When dendrites of phase α grow, the remaining liquid solution becomes more B-rich (C_1). β forms in the resulting hypereutectic liquid, causing the liquid composition to make small oscillations about the initial composition with time (inset). Eventually eutectic solid fills in the spaces around all primary phases.*

Despite the above observation, we discovered only a slight gradation in composition from top to bottom in the sample via refractive index studies. The eutectic microstructures we present below are observed throughout their respective specimens with only minor variations in scale and primary-phase abundance.

3.2 Microstructures of the as-grown eutectics

Crystallization from solution results in the characteristic eutectic microstructures shown in the following figures. The observed microstructures consist of "colonies" of fine, ordered two-phase intergrowths, often interspersed with larger grains of a primary phase. We found that the morphology of the intergrowth within a colony is unique to each ice/hydrate system. The morphologies found in these ice/hydrate systems bear striking resemblance to those found in metallic and ceramic alloys[8,9]. In the interest of controlling microstructure for commercial use, such alloys have been studied extensively. Croker et al.[10] developed a classification scheme for eutectic microstructures and found that they were influenced by the entropy change associated with the crystallization of each phase, the relative volume fraction of phases, and the degree of undercooling. We[11] have extended this analysis to several ice-hydrate systems and so apply the Croker et al.[10] classification labels used below.

3.1.1 H_2O-Na_2SO_4. Figure 3 is a CSEM/SEI showing the "broken lamellar" microstructure of the H_2O-Na_2SO_4 eutectic. The phases present at this composition (4.0 wt%Na_2SO_4) are $Na_2SO_4 \cdot 10H_2O$ (mirabilite) and ice-I. EDS showed that the parallel, blade-like grains standing in relief correspond to the hydrate phase and the matrix is ice-I.

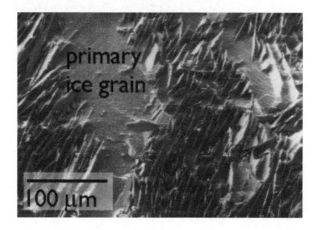

Figure 3   *CSEM/SEI of "broken lamellar" eutectic microstructure for system H_2O-Na_2SO_4. In between large grains of primary ice are eutectic colonies comprised of lamellar mirabilite grains in an ice-I matrix. The bulk composition of this sample is 4.0 wt%Na_2SO_4, which corresponds to that of the eutectic.*

3.1.2 H_2O-$MgSO_4$. The microstructure for the H_2O-$MgSO_4$ system consists of two solid phases interwoven into an intricate fabric. Figure 4 is representative of this system's eutectic microstructure, in which regions with "regular" lamellae border on areas of what Croker et al.[10] refer to as "complex regular" morphology. EDS analysis indicates the dark, recessed phase in the image is ice-I and the light phase is hydrate. Whether the hydrate present is the stable phase, $MgSO_4 \cdot 11H_2O$ (undecahydrate) or the metastable $MgSO_4 \cdot 7H_2O$ (epsomite) is unclear.

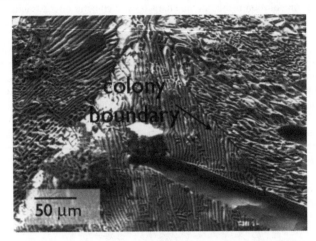

Figure 4   *CSEM/SEI of eutectic microstructure for system H_2O-$MgSO_4$. The lighter phase that stands in relief is hydrate; the recessed, mottled phase in the intergrowth as well as in the large void is water ice. Composition of this sample is 17.3 wt%$MgSO_4$ which corresponds to the stable ice-I/undecahydrate eutectic.*

3.1.3 H_2O-NaCl. The eutectic microstructure for system H_2O-NaCl is shown in Fig. 5. It consists of a mixture of maze-like, "complex-regular" morphology similar to that of H_2O-$MgSO_4$, and "irregular" morphology evinced by angular branching and terminations. XRD analysis for the eutectic sample confirms that the phases present are ice-I and $NaCl \cdot 2H_2O$ (hydrohalite), the latter stands in relief in the CSEM/SEI image with ice-I in the recesses appearing dark and mottled.

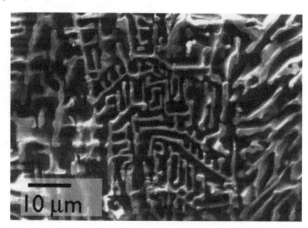

Figure 5 *Eutectic microstructure of the system H₂O-NaCl in which can be found both "complex regular" structure (sides) and the angular, branching structure associated with "irregular" morphology (center). The light phase in the CSEM/SEI is hydrohalite and the dark, recessed phase is ice-I.*

3.3 Mechanical response

Our ultimate interest in the systems studied is how the two-phase material will respond mechanically when subjected to loads such as those associated with Jovian tidal forces on Europa. In polyphase materials, the physical properties depend on the volume fraction of the phases and the interface between the constituents. A mismatch in lattice spacing between the two phases, as is the case with our ice/hydrate systems, results in an incoherent interface. Incoherent interfaces are effective barriers to dislocation motion and in many engineering materials, the high volume of phase boundaries of the eutectic microstructure results in the material being characterized by high stiffness and high damping (attenuation). Indeed, initial mechanical tests on samples corresponding to the eutectic (and metastable eutectic) in system $H_2O-MgSO_4$ have shown that the two-phase eutectic aggregates are distinctly more resistant to flow (i.e., stronger) than pure Ice-I[12]. Moreover, our practical experience in making sample chips by fracturing indicates that the eutectic intergrowths have far higher fracture toughness than that of ice. Preliminary direct measurements of fracture toughness (K_{Ic}) of the NaCl-H₂O eutectic intergrowths by NASA Summer Intern James Pape in our USGS lab indicate that indeed K_{Ic} is 2-4 times that of ordinary ice (Unpublished results, USGS Planetary Ice Lab, Summer 2006). Our continuing research will examine the mechanical energy dissipation of the ice-hydrate aggregates when subjected to cyclic loading. Information gleaned from such work will help constrain models for crustal thickness and surface dynamics on Europa.

4   CONCLUSION

The solidification behavior of several two-component ice-hydrate systems was investigated and the resulting eutectic microstructures examined through CSEM/SEI imaging. Phase identification and composition were characterized with a combination of CXRD, EDS and RI techniques. Continuous lamellae, discontinuous lamellae, and branched, irregular microstructures were observed. The microstructures can be understood in terms of thermodynamics and growth morphology, e.g. faceting, of the constituent phases, the relative volume fractions and degree of undercooling during solidification. Longitudinal composition variations in the samples were observed, though these were modest in magnitude.

Acknowledgements

Support for this research was provided by NASA grant entitled "Experimental Investigation of the Rheologies of Planetary Ices". The authors thank J. Pinkston (USGS, Menlo Park) for his technical support.

References

1   McCord, T.B., Hansen, G.B., Fanale, F.P., Carlson, R.W., Matson, D.L., Johnson, J.V., Smythe, W.D., Crowley, J.K., Martin, P.D., Ocampo, A., Hibbitts, C.A., Granahan, J.C. and the NIMS Team (1998) *Science*, 280, 1242-1244.
2   McCord, T.B., Hansen, G.B., Matson, D.L., Johnson, T.V., Crowley, J.K., Fanale, F.P., Carlson, R.W., Smythe, W.D., Martin, P.D., Hibbitts, C.A., Granahan, J.C. and Ocampo, A. (1999) *J. Geophys. Res*, 104:E5, 11827-11851.
3   Dalton, J.B., Prieto-Ballesteros, O., Kargel, J.S., Jamieson, C.S., Jolivet, J., and Quinn, R. (2005) *Icarus*, 177, 472-490.
4   Kargel, J.S. (1991) *Icarus*, 94, 368-390.
5   Peterson, R.C. and Wang, R. (2006) *Geology*, in press.
6   Roedder, E. (1984) in *Fluid Inclusions: Reviews in Mineralogy, Vol. 12.* Mineralogical Society of America, p. 232.
7   Tiller, W.A. (1991) in *The Science of Crystallization: Macroscopic Phenomena and Defect Generation*, Cambridge Univ. Press, Ch. 4, pp 155-160.
8   Hunt, J.D. and Jackson, K.A. (1966) *Trans. A.I.M.E.*, 236, 843-852.
9   Kerr, H.W. and Winegard, W.C. (1967) *Int. conf. on Crystal Growth, suppl. to The Physics and Chemistry of Solids*, 179-182.
10   Croker, M.N., Fidler, R.S., and Smith, R.W. (1973) *Proc. R. Soc.*, 335:1600, 15-37.
11   McCarthy, C., Cooper, R.F., Kirby, S.H., Rieck, K.D., and Stern, L.A. (submitted to *American Mineralogist*, August 2006)
12   McCarthy, C., Cooper, R.F., Kirby, S.H., and Durham, W.B. (2006) *LPSC XXXVII*, 2467.

X-RAY TOMOGRAPHIC CHARACTERIZATION OF IMPURITIES IN POLYCRYSTALLINE ICE

M. M. Miedaner[1], T. Huthwelker[2], F. Enzmann[1], M. Kersten[1], M. Stampanoni[3], M. Ammann[2]

[1]Environmental Geochemistry, Institute of Geosciences Johannes Gutenberg University, Becher-Weg 21, D-55099 Mainz, Germany
[2]Radio and Environmental Chemistry, Paul Scherrer Institute, CH-5232 Villigen, Switzerland
[3]Swiss Light Source, Paul Scherrer Institute, CH-5232 Villigen, Switzerland

1 INTRODUCTION

Natural ice may grow by vapor deposition, as it is the case for snowflakes. Additionally, ice also grows from the liquid phase. Examples are droplets in tropospheric clouds, which often consist of dilute solutions of acids and salts (e.g. sulfur dioxide, ammonium sulfate, typically at a 10^{-4} M concentration) and also organic acids[1]. Furthermore, ice also forms in brines of higher concentration, such as Sea water (3.5 wt.% NaCl)[2], or in stratospheric aerosols, which consist of highly concentrated solutions (several 10 wt.%) of various inorganic acids[3-6]. When ice forms in such aqueous environments, most impurities are expelled from the crystalline matrix. Hence, multiphase systems are created, which consist of ice, solid salts, gaseous inclusions like air bubbles and highly concentrated inorganic as well as organic solutions.

Ice and its surface host various chemical reactions in nature. Examples include the stratospheric and marine ozone chemistry[2-4]. Such reactions may be located on the ice surface itself, or in the inclusions, which provide a strong ionic environment[7,8]. Hence, to asses the chemical activity of such multiphase systems, it is of interest to study its morphology. Moreover, as micro phases may manifest as a three dimensional network in the ice, it has been predicted theoretically that such structures can play a crucial role for transport processes in ice, even in very clean system, such as ice cores[9].

However, little is known about the morphology of such multiphase-systems. Morphological studies on impurities in ice are rare, and most of them are based on 2D- rather than 3D-information. Optical[10,11] and electron microscopy[12-14], as well as electron micro probe[15] or Raman spectroscopy[16,17] were used to localize impurities in triple junctions, air bubbles, vapor-deposited ice films, and in slices cut from ice cores. However, 2D-scans cannot provide information on higher-order geometrical attributes such as the connectivity of the impurities in veins. X-ray micro-tomography (XMT) overcomes this imaging problem by providing non-destructive cross-sectional as well as three-dimensional object representations from X-ray attenuation mapping. XMT is increasingly used in glaciology to characterize heterogeneous structural features of intact ice samples at a high spatial resolution. Schneebeli and Sokratov[18] performed XMT using a bench-top device with a resolution of about 15-30 µm located in a cold room to elucidate the morphology of snow. Also a cryo-stage for synchrotron XMT was developed by Brzoska and Flin[19-21] which provided a 5 μm resolution. However, a voxel of 1 μm^3 is required to detect inclusions in ice via micro-tomography[22].

This contribution presents images recorded with a novel setup for synchrotron-based XMT. We investigate multiphase systems consisting of typical aerosol constituents, such as mineral dusts and organic components such as octanol, both pure and mixed with a dilute NaBr solution. To gain inclusions of a size suitable to be detected at the yet available spatial resolution (1 μm^3) we had to increase the concentration of impurities significantly above the level found in most natural systems. Therefore, a direct comparison of our results to atmospheric hydrometeors is not possible, and is not the aim of this paper. We rather demonstrate that the presence of organics and inorganic impurities in a solution from which ice freezes has a profound effect on the sample morphology.

2 MATERIALS AND METHODS

2.1 Particle Sampling and Preparation

For all solutions high-purity MilliQ-water (resistivity 0.00055 μS) was used. The sodium bromide, quartz, bentonite, and octanol were of analytical grade (Sigma Aldrich). Emulsions of octanol with pure MilliQ-water or octanol with a 1 wt.% NaBr solution were prepared by filling a vial to one half with the organic solvent and to the other half with the aqueous solution. Suspensions were made from quartz and bentonite (clay) particles (size range 200 nm to 20 μm, 1 g mixed with 100 ml water) to simulate mineral dust components. The "mixed-mode-aerosol" suspensions were generated by adding the minerals or the octanol to the NaBr solution. All these formulations were shaken at 400 rpm for 24 hours prior to freezing in a vial with some head space. Thereby twofold equilibration was reached: The solutions were saturated with respect to both air and octanol.

After this step, fine droplets of these mixtures (0.2 μl) were placed on a hydrophobic paraffin substrate using a 10 μl syringe. Then this substrate was immediately exposed to 240 K in a refrigerator. After 24 hours in this environment, the frozen droplets were gently removed from the paraffin surface using a spatula that had been kept in the refrigerator all the time, and sealed airtight in plastic vials (Zinsser Analytics). All samples were prepared fewer than four days prior to the experiment and stored at 240 K. This temperature was a compromise in order to avoid a strong impact of shape metamorphism on the sample morphology[24,25]. For practical matters, a preliminary comparison of a one-day-old with a four-days-old sample stored at these conditions, showed no changes in the shape of the sphere or distribution and morphology of the inclusions greater than the standard deviation of the geometrical characteristics of the structures measured in this study. Clearly, we cannot completely rule out the impact of shape metamorphism processes, which occur at a temperature of about 240 K. However, in this study we demonstrate the typical internal morphology of artificial samples frozen from a mixed solution. Therefore, metamorphism processes during sample storage are of minor significance for this study.

2.2 Cryotomography

A detailed description of the setup and data pretreatment was given previously[22]. In brief, the samples were transported on dry ice to the XTM end station MS-TOMO of the Swiss Light Source (SLS) at Paul-Scherrer-Institute at Villigen, Switzerland, where all experiments were conducted. Together with sample holders and the tools required for sample handling the samples were stored at all the times in a well isolated box, filled with dry ice. For every scan one ice particle (sample) was placed in a custom-made polyamide cup adjustable to the standard sample mount at this beam line. This cup provided a maximum sample volume of 50 mm³. Ice particles were placed inside it, and the remaining space was filled with freshly molten cycloheptane (melting point approx. 260 K). The presence of the organic solvent mechanically fixes and thermally insulates the sample. The temperature was measured using a PT-100 resistor, which was placed at the bottom of the cup just a few tenth of a millimeter below the ice sample.

The experimental temperature was adjusted using a cold gas flow from a cryojet (Oxford Instruments) directed onto the head of the sample holder. The cup is surrounded by a double walled Kapton-foil cage. Dry nitrogen was flushed between the inner and the outer walls in order to avoid icing in the beam path while recording the tomogram.

For the transport from the dry ice box to the beam line, the sample holder was wrapped in a plastic bag filled with crushed dry ice. The Kapton-foil cage could be slid up onto the cryojet nozzle, which allowed easy access to the sample mount. The polyamide cup was placed into the sample mount, while being in the dry ice bag. Once fixed onto the goniometer stage, the dry ice bag was removed, and the Kapton-cage was slid down to ensure continuous cooling of the sample during final adjustments of the cup.

Then 500 projections were recorded during a 180 degree rotation of the sample holder. In order to correct for beam fluctuations, reference images were taken at every 100th angular step. The projections were used to reconstruct the 3D-image of the investigated object using an appropriate reconstruction algorithm[26,27]. The camera allows a a spatial resolution of 0.7 µm, but the data were immediately hardware-binned, which reduced the resolution to 1.4 µm, but improved the signal-to-noise ratio. The reconstructed slices were cropped to the desired region-of-interest (ROI), and, if necessary, filtered using a 2D edge-preserving smoothing filter (cf. detailed discussion by Miedaner et al.[20]). The final image analysis was performed using the Amira software package (Mercury Systems).

2.3 Data analysis

The relative X-ray mass absorption of each component was coded as grey scale between 0 (strongest absorption in black) and 255 (lowest absorption in white) in the horizontal reconstructed slices. The segmentation was based on threshold values, whereby 0–10 was set for NaBr, 40–120 for ice, and 250–255 for air. Obvious reconstruction artefacts as well as segmented air bubbles inside the cycloheptane matrix were removed manually[22]. For data analysis, the number of pores, their individual volumes (V_i) and surface areas (A_i) were quantified. Connecting voids were identified using the numerical "burning algorithm" method[28]. Additionally, we defined the deviation from the ideal spherical geometry as asphericity[29] Ψ_i:

$$\Psi_i = \frac{\pi^{\frac{1}{3}} (6V_i)^{\frac{2}{3}}}{A_i} \tag{1}$$

Furthermore, the volume-to-surface ratio was determined for each pore as it is common to characterize porous materials[30].

3 RESULTS

3.1 Octanol/H₂O Emulsion

Octanol, a common proxy for liquid organic urban aerosols[31], could be clearly identified by the contrast change at the ice-octanol phase boundaries (cf. Figure 1). The very low contrast between air and octanol inclusions did not allow distinguishing between these two substances. This organic solvent occupied a total volume of $4.1 \cdot 10^{-7} \text{m}^3$. The average

Figure 1   *Octanol inclusions in ice.*

surface area was found to be $2.5 \cdot 10^{-12} \text{m}^2$. The average asphericity reached a value of 0.59, while the volume-to-surface ratio was 2.7 µm.

3.2 Octanol/H$_2$O/NaBr Emulsion

If a 1 wt.-% of NaBr was mixed with octanol prior to freezing, the shape of the resulting inclusions in the frozen sample changed significantly. The convex walls of the ice-octanol boundaries as observed in case of the pure solute, turned into concave ones (cf. Figure 2). The total surface area and total volume also increased significantly compared to solute samples without NaBr. The octanol/brine inclusions formed a very narrow and curly network throughout the whole sample, occupying an average surface area of $8.1 \cdot 10^{-9} \text{m}^2$ and an average volume of $2.2 \cdot 10^{-14} \text{m}^3$. The average asphericity increased to 0.63 compared to the pure octanol/H$_2$O system, while the volume-to-surface ratio decreased to an average value of 1.6 µm.

Figure 2   *NaBr/octanol inclusions in ice. Note the significant difference in shape compared to the inclusions of pure octanol depicted in Figure 1.*

3.3 Quartz Suspensions

Inclusions of quartz particles were found homogeneously distributed over the whole ice sample (cf. Figure 3). The average volume per inclusion was $4.4 \cdot 10^{-18} m^3$, while the average surface area reached a value of $5.7 \cdot 10^{-9} m^2$. The volume-to-surface ratio was found to be $0.77 \mu m$ with a mean asphericity of 0.36.

Figure 3   *Quartz particles in ice frozen from a quartz/MilliQ-water suspension.*

3.4 Quartz/NaBr Suspension

Freezing quartz particles suspended in a 1 wt.-% NaBr solution at a temperature of 240 K leads to small, homogeneously distributed inclusions in the ice (cf. Figure 4). An average inclusion volume of $6.0 \cdot 10^{-16} m^3$ with a surface area of $5.1 \cdot 10^{-10} m^2$ was found. The mean asphericity of these inclusions was 0.68, indicating that they have a rather spherical geometry. The volume-to-surface ratio also reached a low average value of $1.14 \mu m$. Since their asphericity and volume-to-surface ratio are significantly different from the inclusions in pure H_2O, a change in shape due to the addition of salt was observed.

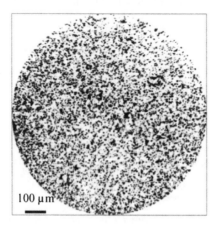

Figure 4   *Quartz inclusions in ice frozen from a NaBr-doped suspension.*

3.5. Bentonite

Figure 5 depicts the ice-trapped clay inclusions. The average impurity void filled a volume of $4.4 \cdot 10^{-13} \text{m}^3$, with an average surface area of $1.3 \cdot 10^{-13} \text{m}^2$. The mean volume-to-surface ratio was 0.36 µm. The clay particles were found to be agglomerated to elongated structures of a rather low asphericity of 0.25, much less than the above described quartz inclusions.

Figure 5   *Bentonite inclusions in ice frozen from a bentonite/MilliQ-water suspension.*

3.6 Bentonite/NaBr Suspension

Bentonite inclusions showed a completely different behavior upon freezing in a NaBr solution. Long, plate-like sheets almost perpendicular to each other were found (cf. Figure 6). An average surface area of $6.0 \cdot 10^{-9} \text{m}^2$, and an average volume of $6.7 \cdot 10^{-15} \text{m}^3$ were determined. The asphericity reached a mean value of 0.35, with a volume-to-surface ratio of 0.79 µm.

Similar to the quartz samples, asphericity and volume-to-surface ratio of the trapped inclusions changed upon addition of the salt - but more drastically in the case of bentonite.

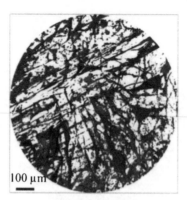

Figure 6   *Bentonite inclusions in ice frozen form a NaBr-doped suspension. Note the significant difference in shape compared to the bentonite particles in pure H_2O (cf. Figure 5).*

4 DISCUSSION

Quartz, bentonite, and octanol, as well as mixtures of these components with a 1 wt.-% NaBr solution, were frozen. The specific surface area, shape, and other geometric parameters were significantly different among the various samples studied. These differences depend on the chemistry of the impurity, as adding a salt to the solution or changing the mineral type had profound effects. In the case of freezing pure octanol/water emulsions, spherical inclusions of the octanol were found, which are caused by the strong hydrophobicity of octanol[31]. However, once an emulsion of a NaBr solution with octanol was made, the surface tension of the salt solution is higher compared to that of pure water[31,32], which obviously influences the shape and distribution of the octanol inclusions. The latter form a highly concave and well-connected network, if NaBr is present, but are convex if the salt is absent. Furthermore, air bubbles could not be identified, which suggests, that all the air could be dissolved in octanol. In addition, all NaBr-precipitates were found mixed into the octanol inclusions. The connectivity of this void network would allow for transport and further chemical reactions inside ice particles, and therefore may alter the chemical behaviour of such particles with respect to the surrounding gas phase as well as during contact with other solution droplets or changes in temperature.

Bentonite and quartz particles formed inclusions homogeneously distributed all over the sample, but of different shape. Both minerals showed a tendency towards agglomeration when they were mixed into a dilute salt solution. This is probably related to the difference in surface tension, which occurs due to the addition of the salt[32]. In particular, in the case of bentonite, one may speculate that also colloidal charging effects forced the inclusions to a more or less perpendicular orientation to each other inside the ice grain matrix. Additionally, the formation of pockets filled by highly concentrated brines occurs during freezing at low growth rates and significantly influences the morphology and distribution of the trapped inclusions[33], too.

5 SUMMARY AND CONCLUSIONS

Cryo-XMT was performed on various types of samples. For these experiments, a new setup was developed. It consists of a polyamide-cup, a cryojet, and a double-walled Kapton-foil-cage. By embedding the sample in cycloheptane inside the polyamide cup, the ice particles are not only fixed mechanically but also favorably isolated against temperature fluctuations. Experiments can be performed at temperatures as low as 230 K with an accuracy of ± 1 K and a stability of ± 3 K. To avoid icing in the path of the beam, all important parts are permanently flushed with dry gaseous nitrogen[22]. A spatial resolution of 1.4 μm was achieved in 1.75 hours per scan.

The systems studied in this work are not found in nature, where impurity concentrations are often lower. However, also in nature ice is frozen from aqueous solutions, leading to complex multiphase systems. The results presented here suggest there is an interdepence of all available impurities on the micro-morpholgy in the multiphase system. Such effects may have a profound impact on chemistry and transport of impurities, as such processes may occur just in the complex, three dimensional inclusions in the ice. It is beyond the scope of this paper to address these issues, but we believe that future applications of micro-tomography with much higher spatial resolution[34] will be well suited to further explore theses issues.

Acknowledgements

Financial support for this research was provided by DFG via the SFB 641 (TROPEIS). A part of this work has been funded by the European integrated project SCOUT O3. This research project has been supported by the European Commission under the 6th Framework Program through the Key Action: Strengthening the European Research Area, Research Infrastructures n°: RII-CT-2004-506008. We are grateful to the PSI-SLS machine and beam line group whose outstanding efforts have made these experiments possible, in particular the practical support provided by the technicians M. Birrer, M. Lange, and D. Meister, and helpful discussions during the experiments by Dr. A. Groso.

References

1   J. Seinfeld and S. Pandis, *Atmospheric Chemistry and Physics*, 1997, Wiley Interscience.
2   G.M. Marion, *Geochim. Cosmochim. Acta*, 2002, **66**, 2499.
3   S. Solomon, *Nature*, 1990, **347**, 347.
4   P.J. Crutzen and F. Arnold, *Nature*, 1986, **324**, 651.
5   Th. Peter, *An. Phys. Chem.*, 1997, **48**, 785.
6   R.S. Gao, P.J. Popp, D.W. Fahey, T.P. Marcy, R.L.Herman, E.M. Weinstock, D.G. Baumgartner, T.J. Garrett, K.H. Rosenlof, T.L. Thompson, P.T. Bui, B.A. Ridley, S.C. Wofsky, O.B. Toon, M.A. Tolbert, B. Kärcher, Th. Peter, P.K. Hudson, A.J.Weinhiemer and A.J. Heymsfield, *Science*, 2004, **303**, 516.
7   K.S. Carslaw, T. Peter and S.L. Clegg, *Rev. Geophys.*, 1997, **35 (2)**, 125.
8   S.L. Clegg and J.H. Seinfeld, *J. Phys. Chem. A*, 2006, **110 (17)**: 5692.
9   A.W. Rempel and J.S. Wettlaufer, *J. Geophys. Res.*, 2002, **107**, B12, 2330.
10   A. Carte, *Proc. Phys. Soc.*, 1961, **77**, 757.
11   L. Keyser and M. Leu, *Microsc. Res. Tech.*, 1993, **25**, 434.
12   P. Barnes, R. Mulvaney, E. Wolff and K. Robinson, *J. Microsc.*, 2002, **205**, 118.
13   P. Barnes, E. Wolff and D. Mallard, *Mic. Res. Tech.*, 2003, **62**, 62.
14   D. Cullen and I. Baker, *Mic. Res. Tec.*, 2001, **55**, 198.
15   R. Mulvany, E. Wolff and K. Oates, *Nature*, 1988, **331**, 247.
16   H. Fukazawa, K. Sugiyama and S. Mae, *J. Phys. Chem.*, 1997, **101**, 6184.
17   H. Ohono, M. Igarashib and T. Hondoha, Earth, *Planet. Sci. Lett.*, 2005, **232 (3)**, 171.
18   M. Schneebeli and A. Sokratov, *Hydr. Proc.*, 2004, **18**, 3655.
19   J.B. Brzoska, C. Coleou, B. Lesaffre, S. Borel, O. Brissaud, W. Ludwig, E. Boller and J. Baruchel, *ESRF Newsletter*, 1999, 22.
20   Flin, F., J.B. Brzoska, B. Lesaffre, C. Coleou and R.A. Pieritz, *J. Phys. D.: Appl. Phys.*, 2003, **36**, A49.
21   Flin, F., J.B. Brzoska, B. Lesaffre, C. Coleou and R.A. Pieritz, *Ann. Glaciol.*, 2004, **38**, 39.
22   M. Miedaner, T. Huthwelker, F. Enzmann, M. Kersten, M. Stampanoni and M. Amman, submitted
24   A. Cabanes, L. Legagneux and F. Domine, *Env. Sci. Tech.*, 2003, **37**, 4, 661.
25   S.C. Colbeck, *J. Appl. Phys.*, 1983, **54**, 2677.
26   G. Herman, *Image reconstruction from projections*, 1980, Academic Press.
27   J. Radon, Ber. Verb. Saechs. Akad. Wiss., Leipzig, *Math. Phys. Kl.*, 1917, **69**, 262.
28   D. Stauffer and A. Aharony, *Introduction to Percolation Theory*, Taylor & Francis, London, 1992.
29   H. Wadell, *J. Geol.*, 1933, **41**, 310
30   S. Brunauer, R. Mikhail and E. Bodor, *J. Coll. Interf. Sci.*, 1967, **24**, 451.
31   Y. Rudich, *Chem. Rev.*, 2003, **103**, 5097.

32   M. Boström, D.R.M. Williams and B.W. Ninham, *Langmuir*, 2001, **17**, 4475.
33   M.G. Worster and J.S. Wettlaufer, *J. Phys. Chem.* B, 1997, **101**, 6132.
34   D. Attwood, *Nature*, 2006, **442**, 642.

EFFECTS OF ADDITIVES AND COOLING RATES ON CRYOPRESERVATION PROCESS OF RAT CORTICAL CELLS

J. Motomura[1], T. Uchida[1]*, M. Nagayama[1], K. Gohara[1], T. Taira[2], K. Shimizu[2], M. Sakai[2]

[1]Division of Applied Physics, Graduate School of Engineering, Hokkaido University, Sapporo 060-8628, Japan
[2]Primary Cell Co., Ltd., Ishikari 061-3242, Japan
*corresponding author

1 INTRODUCTION

Dissociated cells from animals are widely used in in vitro models to enable the investigation of principal biological mechanisms. The cells are incubated with culture medium immediately after collection. Subsequently, they are maintained under suitable conditions by subculturing and exchanging fresh culture medium. Another suitable technique that can be used to obtain cells for experimental studies is the cryopreservation of dissociated cells. Successful cryopreservation and recovery of primary cell cultures would be valuable for extended and prolonged studies.[1] Cryopreserved neurons would be in demand by industries; for example, high-quality neurons are useful for the screening of new drugs.

There are some important factors that need to be considered when cells are to be frozen. These include the type of cryoprotectant used to inhibit the formation of intra- and extracellular ice and the cooling rate of cells.

Cryoprotectants are added to the culture medium in order to protect cells. Glycerine, the first cryoprotectant to be used, was discovered in the 1940s,[2] and it was used for the cryopreservation of bovine sperm. More recently, it has been discovered that other substances such as dimethyl sulfoxide (DMSO),[3] saccharides,[4,5] and proteins also function as cryoprotectants.[6] However, the mechanism of the cryopreservation process in each case remains unclear. In the present study, we compared DMSO and trehalose, which are cryoprotectants of different molecular sizes that may differ in their ability to pass through the cell membrane.

The cooling rate of cells is also an important factor. In slow freezing, the amount of ice that is generated can be controlled by the dehydration of the cell due to the presence of extracellular ice, which results in a higher solute concentration. It is more difficult to freeze cells treated in this manner because the freezing point is depressed at higher solute concentrations. However, under more rapid freezing conditions, ice is generated intracellularly because there is insufficient time for dehydration. Intracellular ice might disrupt cell membranes. Therefore, for cryoprotecting a particular cell with a particular cryoprotectant, there will be a suitable cooling rate.

The cryopreservation technique has been established for some cells.[7] As a result, these cryopreserved cells have been used in many studies. However, in the case of neurons, the appropriate cryopreservation process has not yet been established because of their complex structure, low mitotic activity, and uncertainties regarding the cryoprotection mechanism. Thus, primary culture cells are used for studying neurons.

Neurons show a spontaneous wave of electrical excitation known as an action potential or nerve impulse, which is characterized by pulses that have an amplitude of several dozen microvolts and a width of a few milliseconds. The impulse carries a message without attenuation from one end of the neuron to the other end at speeds as high as 100 m/s or more. This property is important for evaluating the cryopreservation of these cells. Previously, the nerve impulse has been measured using a single electrode. However, this is not useful for evaluating the effectiveness of cryopreservation of a large number of neurons. Recently, a useful device that can be used to measure many nerve impulses has been developed. The multiple electrode device (MED) system simultaneously measures the nerve impulse of recovered neurons.[8–10] Using this MED system, we can observe the normal behavior of these cells.

We studied the effect of cryoprotectants and cooling rates on the cryopreservation of rat cortical neurons. We then evaluated the recovering neurons by observing them by phase-contrast microscopy and by measuring their action potentials with the MED system.

2 MATERIALS AND METHODS

Primary rat cortical cells were prepared from neonatal Sprague-Dawley (SD) rats within 3 days of birth (Primary Cell Co., Ltd.). The dissociated cells contained neurons, glial cells, and fibroblasts. These cells were suspended in the culture medium at 1×10^6 cells/ml solution. The materials that were to be tested as cryoprotectants were then added to the medium. The culture medium consisted of Dulbecco's modified Eagle's medium (DMEM) and 10 vol% fetal bovine serum (FBS). Control dissociated neurons that had not been treated with the cryopreservation were cultured so that they could be compared with the cryopreserved neurons.

Three types of cryoprotectants at various concentrations were selected: DMSO (5–20 wt%), trehalose (3–17 wt%), and a mixture of trehalose (10 wt%) and methanol (8 wt%). A mixture of methanol and trehalose has been reported to be effective for the cryopreservation of salmon sperm.[11] Therefore, this mixture was also used to check its effectiveness in the cryopreservation of neurons.

To study the effect of the cooling rate, we varied the cooling rate of cell suspensions from slow (10 K/min) to rapid (10^2 K/min). The slow cooling-rate experiments were performed by placing the cryotube containing the sample into a freezer (at 193 K). The rapid cooling-rate experiments were performed by placing this tube inside liquid nitrogen (at 77 K). The storage temperature was 193 K for the slow cooling-rate experiments and 77 K for the rapid cooling-rate ones. At temperatures below 193 K, the growth of ice becomes slow. Therefore, we thought that the results will not change over this short storage period.

After storing the samples at each temperature for 1 week, they were rapidly thawed in a warm bath (at 310 K). They were then cultured on poly-L-lysine-coated 24-well plates or the MED probe, which is a plate used to measure nerve impulses, in an incubator (humidified atmosphere including 5% CO_2 at 310 K) after replacing the cryoprotectant with the culture medium. The recovering neurons were first examined by phase-contrast microscopy and compared with control samples. The presence of neurons was also checked by immunofluorescence staining of microtubule-associated protein 2 (MAP2), which is a neuron-specific marker. Apart from this, the number of neurons present was counted by microscopic observation.

To verify the successful recovery of cells, we measured the nerve impulse using the MED 64 system. Neurons were cultured on the MED probe on which 64 electrodes are located. Thus, the nerve impulse can be measured more easily in this system than in a single electrode system.

3   RESULTS AND DISCUSSION

3.1   Control cultured neurons

Initially, we cultured the control neurons without cryopreservation. Neurons cultured on a poly-L-lysine-coated 24-well plate were examined by microscopy and immunofluorescence staining using MAP2 (Figures 1,2). Arrows indicate living neurons.

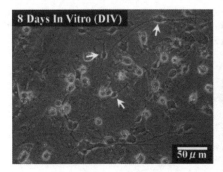 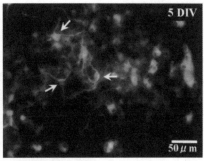

Figure 1 *Control cultured neurons that were not cryopreserved. (phase contrast image).*

Figure 2 *Control cultured neurons stained with MAP2.*

Figure 3 shows the nerve impulse measured with the MED 64 system. (amplitude: 28 μV, width of pulse: 4 ms) In the present study, this was defined as the general nerve impulse of neurons. The MED 64 system is useful for evaluating neuronal function.

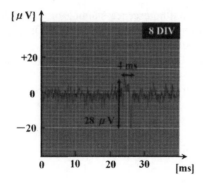

Figure 3 *Nerve impulse of control neurons measured with the MED 64 system.*

In order to check the effectiveness of the cryoprotectants, neurons were cryopreserved with regular culture medium (without cryoprotectants). Recovering cells were not observed in microscopic observations when cryopreservation was carried out using the same experimental protocol (Figure 4). In case of cryopreservation in the absence of cryoprotectants, the ice around the cells disrupts the cell membrane.

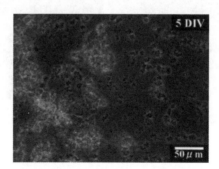

Figure 4 *Recovering cells (after cryopreservation in the absence of cryoprotectants).*

3.2 Cryopreservation with DMSO

Cryopreservation with 5–10 wt% DMSO showed that recovering neurons were present; this was determined by microscopic observations and immunofluorescence staining using MAP2 (Figures 5,6). White arrows show living neurons, and black arrows indicate dead neurons.

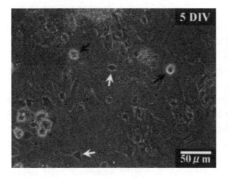 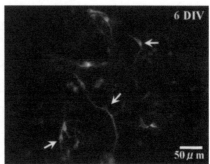

Figure 5 *Recovering neurons (after cryopreservation with 10 wt% DMSO) observed by phase contrast microscopy.*

Figure 6 *Cells stained with MAP2, a neuron-specific marker (after cryopreservation with 10 wt% DMSO).*

The number of neurons recovered after treatment with 10 wt% DMSO was reduced to approximately 60% of that obtained after treatment with 5 wt% DMSO. Cells were not present after cryopreservation with 20 wt% DMSO. These results indicate that neurons were successfully recovered after cryopreservation with low DMSO concentrations. Cryopreservation with 20 wt% DMSO appears to have a toxic effect on the cells.

The morphology of the cells was similar to that of the control neurons at DMSO concentrations of 5 wt% and 10 wt%. Compared to the control cells (as shown in Figure 1), it was observed that in the cryopreserved samples, glial cells abruptly regenerated a few days after thawing. It is suggested that these glial cells help neurons to grow after the latter recover from the cryopreservation conditions.

Furthermore, cryopreservation with 5 wt% DMSO preserves the nerve impulses of the recovering neurons as determined with the MED 64 system (Figure 7).

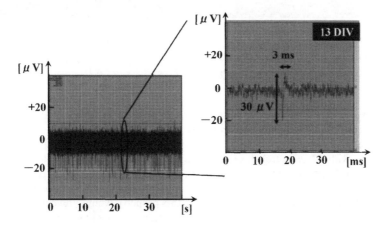

Figure 7 *Nerve impulses of the recovering neurons (with 5 wt% DMSO) as determined with the MED 64 system.*

As shown in the extended signal in Figure 3, the function of the neurons was assayed based on the presence of nerve impulses; these nerve impulses were equivalent to those of neurons that had not been cryopreserved (amplitude: 30 μV, width of pulse: 3 ms). These results indicated that the DMSO-cryopreserved neurons (at DMSO concentrations of 5–10 wt%) can be recovered and cultured without any damage to their morphology or electrical activities.

When the cooling rate was varied, microscopic observations showed that the neurons recovered at slow cooling rates (10 K/min). However, at high cooling rates (10^2 K/min), the recovery of neurons was not observed. This result may be explained on the basis of the freezing process of the cell. The cryopreservation process of DMSO may be explained as follows: DMSO treatment may result in depression of the intracellular freezing point when the extracellular water begins to freeze.[11] After the cells become dehydrated, the intracellular fluid may be vitrified when the temperature approaches the glass-transition point. Thus, no intracellular ice crystals are formed, and DMSO protects the cells.[12]

3.3 Cryopreservation with trehalose

After cryopreservation with 17 wt% trehalose, neurons were assessed by microscopic observations and immunofluorescence staining using MAP2 (Figures 8,9).

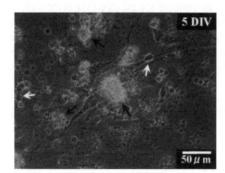 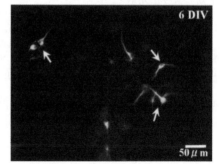

Figure 8 *Recovering neurons (after cryopreservation with 17 wt% trehalose) examined by phase contrast microscopy.*

Figure 9 *Cells stained with MAP2, a neuron-specific marker (after cryopreservation with 17 wt% trehalose).*

Recovering neurons were not observed in the cryopreservation experiments with 3–10 wt% trehalose. Therefore, we consider that trehalose may only function as a cryoprotectant at concentrations of 17 wt% and possibly higher. In the cryopreservation experiments with 17 wt% trehalose, the recovery ratio of neurons was lower than that obtained with 5 wt% DMSO. The nerve impulses of the recovering neurons were not measured in this case because the recovery ratio of neurons was low. It has been reported that trehalose inhibits the growth of ice,[13] and this property would prevent the disruption of the neuronal membrane.[14] To determine if trehalose is a suitable cryoprotectant for neurons, further studies are necessary.

In comparison with trehalose alone, a mixture of trehalose and methanol has been reported to be useful for the cryopreservation of salmon sperm.[15] However, using a mixture of 10 wt% trehalose and 8 wt% methanol, no significant improvement was observed in the cell morphology (Figure 10). In future studies, in order to determine the suitability of the mixture, we have to find additives that heighten the effect of trehalose.

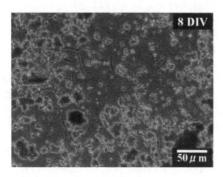

Figure 10 *Recovering neurons (after cryopreservation with 10 wt% trehalose and 8 wt% methanol) examined by phase contrast microscopy.*

4 CONCLUSIONS

We studied the cryopreservation of rat cortical cells using two effective cryoprotectants—DMSO and trehalose—at two different cooling rates. Microscopic observations and immunofluorescence staining with MAP2 revealed that the neurons had recovered after cryopreservation with 5–10 wt% DMSO. The recovering neurons showed nerve impulses that could be studied with the MED 64 system; these impulses were similar to those of neurons that had not been treated with cryopreservation. The neurons did not recover at high DMSO concentrations (20 wt%) because of the toxic effect of DMSO. For the cryopreservation of neurons, cooling in 5 wt% DMSO at 10 K/min was more suitable than at 10^2 K/min. Therefore, it appears that the appropriate conditions for the cryopreservation of cortical neurons are a DMSO concentration of approximately 5 wt% and a cooling rate of 10 K/min.

Some neurons recovered after cryopreservation in 17 wt% trehalose at a cooling rate of 10 K/min as determined by microscopic observations and immunofluorescence staining using MAP2; however, neurons did not recover at lower concentrations of trehalose. These results indicated that trehalose functions as a cryoprotectant only at high concentrations.

The difference in the recovery of cortical neurons in the presence of these additives might be a result of the cryopreservation process that is explained as follows: DMSO is taken up into the cell and allows vitrification of the intracellular solution, leading to the stabilization of the cell membrane. However, in the case of trehalose, the molecules may be too large to permeate into the cell; they may only contact the membrane extracellularly and protect the cell from freezing damage. To clarify the details of this cryoprotection mechanism, we need to study the process of cell freezing.

References

1 F. Otto, P. Gortz, W. Fleischer, M. Siebler, *J. Neurosci. Methods*, 2003, **128**, 173.
2 C. Polge, A.U. Smith, A.S. Darkes, *Nature*, 1949, **164**, 666.
3 T. Negishi, Y. Ishii, S. Kawamura, Y. Kuroda, Y. Yoshikawa, *Neurosci. Lett.*, 2002, **298**, 21.
4 T.E. Honadel, G.J. Killian, *Cryobiology*, 1988, **25**, 331.
5 C.F. Wu, H.C. Tsung, W.J. Zhang, et al., *Reprod. Biomed. Online*, 2005, **11(6)**, 733.
6 J.F. Carpenter, T.N. Hansen, *Proc. Natl. Acad. Sci.*, 1992, **89(19)**, 8953.
7 L.L. Pu, X. Cui, B.F. Fink, D. Gao, H.C. Vasconez, *Plast. Reconstr. Surg.*, 2006, **117(6)**, 1845.
8 S.M. Potter, T.B. DeMarse, *J. Neurosci. Methods*, 2001, **110(1-2)**, 17.
9 S. Honma, W. Nakamura, T. Shirakawa, K. Honma, *Proceedings of the 10th Sapporo Symposium on Biological Rhythm (Sept. 2003 Sapporo)*, 2005, 91.
10 H. Oka, K. Shimono, R. Ogawa, H. Sugihara, M. Taketani, *J. Neurosci. Methods*, 1999, **93(1)**, 61.
11 S. Leekumjorn, A.K. Sum, *Biochim. Biophys. Acta.*, 2006, **8;4C**, 79141.
12 L.-R. Tao, T.-C. Hua, *Ann. N.Y. Acad. Sci.*, 2002, **972**, 151.
13 T. Sei, T. Gonda, Y. Arima, *J. Crystal Growth*, 2002, **240**, 218.
14 A.S. Rudolph, J.H. Crowe, *Cryobiology*, 1985, **22(4)**, 367.
15 N. Koide, *Proceedings of the 8th Trehalose Symposium (Nov. 2004 Tokyo)*, 2005, 49 (In Japanese).

LABORATORY STUDIES OF THE FORMATION OF CUBIC ICE IN AQUEOUS DROPLETS

B.J. Murray* and A.K. Bertram

Department of Chemistry, University of British Columbia, 2036 Main Mall, Vancouver, British Columbia, V6T 1Z1, Canada
*Now at: School of Chemistry, University of Leeds, Woodhouse Lane, Leeds, LS2 9JT, UK. B.J.Murray@Leeds.ac.uk

1 INTRODUCTION

Ice clouds that occur in the upper troposphere have been the focus of intense research in the past few decades.[1, 2] These clouds play an important role in the Earth's climate by scattering and absorbing radiation, although the sign and magnitude of their net climatic impact remains uncertain.[3] They influence the distribution of water vapour, an important greenhouse gas, and they also influence the amount of water that enters the stratosphere, which may have important consequences for polar ozone chemistry.[4] Cirrus ice particles also influence the chemistry of the atmosphere by providing a surface on which heterogeneous chemical reactions can occur.[5] Despite the potential impact of ice clouds on the planet's atmosphere and climate, many important details regarding their formation mechanisms and microphysics remain poorly understood.[6, 7]

The homogeneous freezing of droplets is thought to be an important mechanism in ice cloud formation in the Earth's atmosphere. Recently, it was shown that the metastable cubic crystalline phase of ice can form under conditions relevant for the atmosphere when aqueous droplets freeze.[8, 9] In these experiments Murray and co-workers[8, 9] froze aqueous $(NH_4)_2SO_4$, $(NH_4)_3H(SO_4)_2SO_4$, NaCl and HNO_3 as well as pure water droplets by cooling them down at a modest ramp rate ($10 \, K \, min^{-1}$) and allowing them to freeze homogeneously. They found that the droplets froze dominantly to cubic ice in the aqueous solution droplets below ~195 K – a temperature relevant for the tropical tropopause region. They also found that significant quantities of cubic ice can form in pure water droplets.

The results of Murray and co-workers were in disagreement with what was often assumed in the literature, which was that only the stable hexagonal phase of ice was important in the Earth's troposphere and stratosphere.[10, 11] This assumption came about in part because cubic ice had never been generated in the laboratory under conditions that were relevant to the Earth's atmosphere. Cubic ice had been made in the laboratory using extremely fast cooling rates,[12-16] by annealing the amorphous phase,[17, 18] and by freezing of water in nano-porous silica.[19, 20] Others have employed theoretical arguments to suggest that crystallisation of water droplets begins with nuclei having a cubic structure since an octahedral cubic germ will have a lower free energy than a hexagonal ice germ, hence cubic ice may be kinetically favoured.[14, 21-23] Also, it was found in molecular-dynamics simulations that ice Ic crystallizes in electric fields.[24] However, it was not clear from any of these studies if cubic ice would form in the atmosphere.

Despite the statements made in important text books to the effect that cubic ice is not important in the atmosphere,[10, 11] there are several authors who have proposed cubic ice may form in the atmosphere. Whalley[25, 26] and Riikonen[27] suggest that rare haloes at 28° to the sun may be due to the presence of crystals of cubic ice with an octahedral habit. Mayer *et al.*[13] argued that cubic ice may form in the Earth's atmosphere based on experiments in which cubic ice was produced by the very rapid cooling of water droplets on surfaces held at 190 K. As we suggest above, the relevance to the atmosphere of experiments employing very fast cooling rates is not clear. Murphy also speculated that cubic ice may form in the atmosphere.[28] Hallett *et al.*[29] suggest that layers of cubic ice within ice particles that are predominantly hexagonal may give rise to crystals of three-fold symmetry and extreme habits which are found in the colder parts of the Earth's atmosphere. It has also been suggested that particles of cubic ice can form in the very cold (<150 K) polar summer mesosphere,[30-33] but the current paper is mainly concerned with ice that forms under conditions relevant for the Earth's troposphere and stratosphere.

In this manuscript we reanalyse the data which we have previously reported[8, 9] in order to determine the fraction cubic ice/hexagonal ice in pure water and solution droplets that froze homogeneously. This was not previously reported. We also show that very small (>2 μm) pure water droplets can freeze dominantly to cubic ice when they freeze at ~238 K and are cooled at a modest rate. In our previous paper we showed that a significant amount of cubic ice can form in pure water droplets, but in this previous work we always found that this was mixed with at least 15 % hexagonal ice. The formation of cubic ice in pure water droplets which froze homogeneously at ~235 K, with no detectable hexagonal ice, is a significant finding which has not been reported previously. In addition we review our measurements of the cubic to hexagonal phase change.

2 EXPERIMENTAL

In these experiments, water and aqueous droplets containing $(NH_4)_2SO_4$, HNO_3, $NaCl$ or $(NH_4)_3H(SO_4)_2$ were suspended in an oil matrix, by emulsification, and X-ray diffraction was used to determine the phase of ice that formed when the droplets froze homogeneously. The emulsified samples were prepared by mixing an aqueous solution of known composition with an oil phase (10 wt% lanoline in a hydrocarbon oil) and agitated for about 10 minutes.[34] The agitation process resulted in an emulsion of aqueous droplets with size distributions that approximated to log-normal (see Figure 1). Droplet emulsions with volume median diameters (d_{vm}) of between 5 and 17 μm and with geometric standard deviations of 1.3 - 1.8 could be generated by varying the intensity and duration of agitation. The size of the droplets was determined from digital images taken with an optical microscope equipped with a 50x objective. An example of a typical emulsion sample is illustrated in Figure 1. These emulsion samples were then placed in a temperature controlled cell consisting of an aluminium base and coved with a thin film of Teflon (~30 μm thick) ready to be cooled within the X-ray diffractometer.

The X-ray diffraction measurements were performed with an X-ray diffractometer (Bruker D8 Discover) in a standard Bragg-Brentano reflection configuration. In a typical experiment the temperature of the cell containing the emulsified sample was decreased from ambient to 173 K at a rate of 10 K min[-1]. In order to measure the temperature at which the supercooled liquid droplets homogenously froze to ice, the diffraction angle of a strong reflection associated with both ice Ic and Ih was continually monitored as the cell was cooled. Around ten seconds were required to scan across this peak and during this time the temperature of the cell changed by ~1.5 K. Freezing temperatures were in good agreement with previous measurements.[35] Once at 173 K, the diffraction pattern of the sample between $2\theta = 19 - 50°$ was recorded; this range covered all of the strong ice I_c and ice I_h reflections.[17] We also measured the melting point of ice and any secondary phases that formed as the sample was warmed back up in temperature. These temperatures were in very good agreement with the temperatures predicted by the thermodynamic model of Clegg *et al.*[36] The good agreement of both freezing and melting temperatures indicates that the droplets are not significantly influenced by the presence of the oil emulsion.

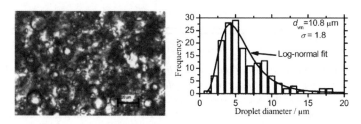

Figure 1 *An image of pure water droplets suspended in an oil emulsion taken with an optical microscope; their size distribution.*

3 RESULTS AND DISCUSSION

3.1 The Crystallisation of pure H₂O droplets

3.1.1 Diffraction Patterns of Frozen Pure Water Droplets. A selection of diffraction patterns of pure water droplets is illustrated in Figure 2. Pattern **a** of Figure 2 is of water droplets that froze at a mean temperature of 234 K and were then warmed to 263 K and annealed for a period of five minutes, before being cooled back to 173 K, the standard temperature at which most diffraction data was measured. This pattern is in excellent agreement with those of hexagonal ice in the literature.[8, 17, 37] Pattern **b** is that of the same sample as in **a**, but before it was annealed. These droplets were cooled from room temperature at a rate of 10 K min⁻¹ to 173 K. This pattern is clearly different from that of hexagonal ice. Inspection of patterns **a** and **b** reveals that all of the peaks exclusive to hexagonal ice are reduced in intensity relative to the peaks common to both cubic and hexagonal ice. This indicates that these pure water droplets froze to a mixture of cubic and hexagonal ice.

The third pattern of Figure 2, is that of very small ($d_{vm} > 2$ μm) droplets that were cooled at a rate of 24 K min⁻¹ from room temperature to 173 K. These droplets froze at a mean freezing temperature of 233 K. Intriguingly, many of the hexagonal peaks are absent, indicating that the droplets did not freeze to any bulk hexagonal ice. This pattern is similar to patterns of cubic ice which are recorded in the literature.[12, 13, 17, 38] The peak close to the position of the 100 peak of hexagonal ice (at ~23°) in pattern **c** and also the region between 22 and 27°, which is significantly raised above the background, are features that are consistent with the presence of hexagonal-like stacking faults in a dominantly cubic structure.[12, 38] Thus, we find that cubic ice with stacking faults in the absence of bulk hexagonal ice can form when small water droplets ($d_{vm} < 2$ μm) homogeneously freeze at modest cooling rates. This is an important result, since previously, bulk water droplets have only been found to freeze to cubic ice, in the absence of bulk hexagonal ice when they are cooled very rapidly at low temperatures.[12-14, 39]

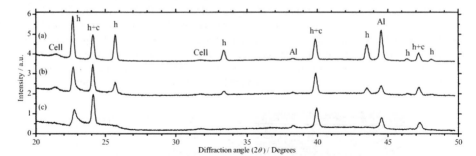

Figure 2   *Diffraction patterns of pure water droplets which were measured at 173 K. Pattern **a** is that of pure water droplets (d_{vm} = 10.2 μm) which froze between 237.5 and 230.4 K and were subsequently annealed at 263 K for five minutes to yield pure hexagonal ice. Peaks exclusive to hexagonal ice are labelled h, those common to cubic and hexagonal are labelled h+c and those due to the cell construction materials and aluminium base are labelled accordingly. Pattern **b** is of the same emulsion, but before it was annealed; i.e. it was cooled from room temperature to 173 K at a rate of 10 K min^{-1}, where this diffraction pattern was measured. Pattern **c** is that of pure water droplets with d_{vm} < 2 μm which were cooled from room temperature to 173 K at a rate of 24 K min^{-1}. These droplets froze at a mean temperature of 233 K.*

3.1.2   Fitting the Diffraction Patterns of Frozen Pure Water Droplets. In Figure 3, we have compared experimental diffraction patterns of frozen water droplets with composite patterns generated by taking a linear combination of the patterns of pure ice I_h (**a**) and ice I_c with stacking faults (**b**). This analysis yields the proportion of cubic and hexagonal ice that forms in aqueous droplets. The best fits were determined by minimizing the sum of the squared differences between the experimental and composite patterns. In fitting the experimental patterns in this way we assume that water droplets freeze to ice that is a mixture of pure hexagonal and stacking faulty cubic. The good fit to the experimental patterns would seem to justify this assumption. From this fitting procedure alone we cannot determine if cubic and hexagonal ice occur within the same droplets, or if a single droplet freezes to a single phase.

When estimating the proportion of stacking faulty ice I_c and ice I_h it was also assumed that the X-ray quantitation constants (integrated intensity per unit mass of ice) for the common peaks are the same for hexagonal ice and cubic ice with stacking faults. Measurements in which we monitor the intensity of the common peaks as cubic ice was annealed to hexagonal ice show that the X-ray quantitation constants are the same to within 5 %. Note that the quantitation constants of the common peaks for pure cubic ice (samples free of stacking faults) will be different from the quantitation constants for hexagonal ice, based on calculations of the diffraction patterns using the POWDER CELL programme[40] and literature data.[17, 37] Evidently, stacking faults appear to influence the quantitation constants of cubic ice. Assuming a 10 % uncertainty in quantitation constants, the proportions of stacking faulty ice I_c and ice I_h only change by at most 5 %.

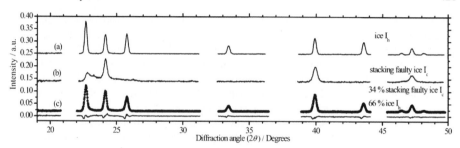

Figure 3   *A composite fit to an experimental diffraction pattern is shown in panel c together with the residual between the fitted and real patterns. The composite pattern is a linear combination of 34 % of the cubic pattern (b) and 66 % of the hexagonal pattern (a). Regions containing significant Bragg peaks from the cell construction materials have been removed from the fit. The broad amorphous background has been subtracted from the patterns.*

3.1.3 The Formation of Cubic Ice in Pure Water Droplets as a Function of Droplet Size. We have investigated the formation of cubic ice in pure water droplets as a function of droplet size. These droplets were cooled at a rate of 10 K min^{-1} to 173 K where a diffraction pattern was measured between 2θ = 20 to 50°. The diffraction patterns are presented elsewhere.[8] We have used the least squared fitting procedure to determine the amount of cubic ice in each frozen emulsion sample and the results are reported in Figure 4. The proportions of cubic ice reported by us previously[8] are a little different to those reported here, because our fitting routine has improved (we now allow the normalisation parameter associated with the composite pattern to vary during the least squared fitting routine, which improves the quality of the fits).

Figure 4a clearly shows that the fraction of cubic ice varies strongly with droplet volume median diameter, with smaller droplets freezing to more cubic ice than larger droplets. In fact, droplets with d_{vm} = 6 μm freeze mainly to cubic ice, whereas when droplets of d_{vm} = 17 μm freeze, they form mainly hexagonal ice.

In order to find an explanation for the dependence of the proportion cubic ice on droplet size we first need to consider our current understanding of ice crystallisation in liquid water. The classical model of crystallisation is that of a process which occurs through two steps, the first being nucleation and the second being growth of the resulting crystal. When a crystal forms in a liquid there will be a surface tension (or surface free energy) between the liquid and the solid, and the addition of extra molecules to this embryo will only become energetically favourable once the embryo reaches some critical size and the bulk energy term becomes dominant.[10] The exact nature of this nucleus is an ongoing topic of research; it may not be a well-defined particle as invoked in the classical model, but rather a less well-ordered structure. However, it is clear that there will be an activation barrier to the formation of this critical cluster, whatever form it takes. Hence crystallisation is often a kinetically controlled process, and as in many chemical reactions, kinetics are likely to play a more important role than thermodynamics in determining the product that initially forms. This may be true for ice: even though ice I_h is the thermodynamically stable phase, a number of theoretical studies[23, 24, 39, 41] and experimental studies[8, 9, 12-14, 21, 22, 42] suggest that ice I_c is the kinetic product and hence the product which first nucleates and grows. A critical germ of ice I_c in an octahedral habit has been shown to have a lower free energy than that of a critical germ of ice I_h; hence the rate at which cubic ice nucleates should be faster than the rate at which hexagonal ice nucleates.[14, 22] This is supported by the empirical Ostwald's rule of stages, which states that the most stable state does not necessarily form when an unstable solution crystallises.[43]

It seems clear that when a droplet initially crystallises it will tend to crystallise to cubic rather than hexagonal ice for the kinetic reasons described above. This alone does not explain the size dependence seen in Figure 4a, but consideration of the processes that occur during crystallisation may provide an explanation. When a water droplet freezes heat is produced since crystallization is an exothermic process. If the heat is not dissipated to the droplet's environment more rapidly than it is produced during crystallization, the temperature of the droplet will increase during freezing, which can allow ice I_c regions in the ice droplet to anneal to ice I_h (it will be shown later in this paper that the ice I_c to ice I_h transition occurs more readily at higher temperatures). Based on equations given in Pruppacher and Klett,[10] and temperatures and thermal properties consistent with our experiments, a droplet of 10 μm in diameter will dissipate heat to the oil matrix at a rate similar to the rate of heat production within the supercooled droplet due to crystallization. As a result, 10 μm particles may not warm up sufficiently to anneal all ice I_c to ice I_h, which is consistent with our experimental results. Smaller droplets will have a greater surface area to volume ratio than larger droplets and will therefore dissipate heat more efficiently. Hence, one would expect the amount that the particle warms up, and therefore the amount of ice I_c, to depend strongly on droplet size, which is in agreement with our observations. In fact, if we make very small droplets and cool them more rapidly, they freeze exclusively to cubic ice (see Figure 2).

Our results are consistent with ice I_c nucleating in all (or most) the droplets, and the final fraction of ice I_c being governed by increase in droplet temperature during crystal growth. From this we infer that ice I_c, rather than ice I_h, is likely to be the crystalline phase that nucleates when water droplets freeze homogeneously at approximately 235 K.

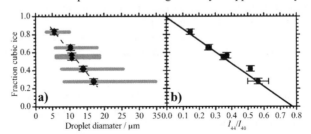

Figure 4   *Panel **a** illustrates the variation of fraction cubic ice as a function of droplet size. The horizontal bars indicate the size range in which 68 % of the droplet volume resides. Panel **b** illustrates the relationship between the fraction of cubic ice and the ratio of the intensities of the hexagonal peak at 44° and the common peak at 40°. The solid line represents the relationship between I_{44}/I_{40} and the amount of cubic ice assuming that the fraction cubic ice = 1-fraction hexagonal = $1-((I_{44}/I_{40})/0.79)$.*

In Figure 4b we have plotted the fraction of cubic ice as a function of the intensity ratio I_{44}/I_{40}, where I_{44} is the intensity of an exclusive hexagonal peak at $2\theta \approx 43.5°$ and I_{40} is a peak common to cubic and hexagonal ice at $2\theta \approx 40.1°$. Previously, we used this intensity ratio to provide a convenient qualitative measure of the amount of ice I_h in the frozen droplets, where $I_{44}/I_{40} = 0.79 \pm 0.3$ indicates pure ice I_h and a value of zero indicates stacking faulty ice I_c.[8, 9] The solid line in Figure 4b is the predicted dependence of fraction cubic as a function of I_{44}/I_{40} assuming that the amount of cubic ice varies linearly with the intensity ratio between 100 % cubic and 100 % hexagonal ice. The close agreement of the fitted data points in Figure 4b with this line clearly shows that this assumption is valid in the case of ice in frozen aqueous droplets. Hence fraction cubic ice = 1-fraction hexagonal = 1-$((I_{44}/I_{40})/0.79)$. We can now use this relationship to re-plot data from our previous publications[8, 9] in terms of fraction cubic.

   3.1.4 The Transformation of Cubic ice to Hexagonal Ice. In this set of experiments the transformation of the cubic fraction of the frozen pure water droplets was investigated as a function of annealing time at several different temperatures between 228 and 263 K. Emulsified droplets of $d_{vm} \approx 10$ μm were cooled at a rate of 10 K min^{-1} to 223 K, and then the temperature was rapidly ramped to the isothermal transformation temperature (T_{trans}). While the frozen droplets were held at $T_{trans} \pm 0.2$ K, the diffraction pattern between $2\theta =$ 39.3° to 44.3° was monitored. This covers the exclusive ice I_h reflection at $2\theta \approx 43.5°$ and the reflection common to both ice I_c and ice I_h at $2\theta \approx 40°$.

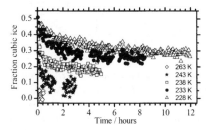

Figure 5   *The fraction of cubic ice as a function of time at several annealing temperatures. See text for details.*

   Using the intensity ratio I_{44}/I_{40} and the relationship defined above (Figure 4) we determined the fraction of cubic ice as a function of time and annealing temperature. This is plotted in Figure 5. The initial proportion of cubic ice was about 50 %, in agreement with Figure 4a for droplets of $d_{vm} \approx 10$ μm, and in all cases the fraction of cubic ice decreases with time. This transformation to stable hexagonal ice clearly proceeds more rapidly at higher temperatures. The results in Figure 5 show that at 228 K a proportion of the ice I_c is very stable, persisting for well over 10 hours. In fact, even after nearly 5 hours at 238 K the ice had not fully relaxed to ice I_h. In contrast, ice I_c at 263 K is rapidly converted to ice I_h, and at 243 K, almost all the ice I_c is converted to ice I_h after ~2 hours.

   To test if X-ray exposure was influencing the ice films during the annealing experiments, we repeated the measurements at 228 K, but this time the shutter to the X-ray source was only opened for enough time to establish the ratio I_{44}/I_{40} once every few hours. The results of this test (not shown) are in agreement within the scatter in the data points in Figure 5 (open triangles) indicating that exposure to X-rays did not significantly affect the ice crystal structure.

   For an in-depth discussion of how the results from our investigation of the lifetime of cubic ice compare to previous measurements reported in the literature the reader is directed to Murray and Bertram.[9]

3.2   The Formation of Ice in Aqueous Solution Droplets

In this set of experiments we investigated the phase of ice that formed when aqueous solution droplets of $(NH_4)_3H(SO_4)_2$, HNO_3, $(NH_4)_2SO_4$, and NaCl froze homogeneously.

The droplets in this study typically had volume median diameters of around 10 μm. In a typical experiment droplets of the desired concentration and composition were cooled from ambient temperatures at a rate of 10 K min⁻¹ to 173 K where a diffraction pattern between 19 and 50° was measured. The intensity ratio I_{44}/I_{40} as a function of droplet freezing temperature, determined from the diffraction patterns measured in these experiments, has been reported previously.[8] In Figure 6 we report the fraction of cubic ice as a function of droplet freezing temperature for each of the solute systems, determined using the relationship described in Figure 4.

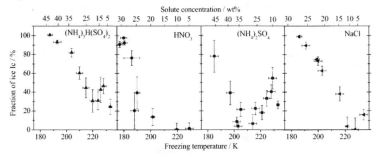

Figure 6   *The fraction cubic ice as a function of freezing temperature that forms in*
(NH₄)₃H(SO₄)₂/H₂O, HNO₃/H₂O, (NH₄)₂SO₄/H₂O, and NaCl/H₂O solution
droplets. In these experiments emulsified droplets of the desired composition
were cooled from room temperature to 173 K at a rate of 10 K min⁻¹ where the
diffraction pattern was measured between 19 and 50°. The intensity ratio was
then determined for each pattern and the relationship described in Figure 4b
was used to find the fraction of cubic ice which is reported in this figure. The
x-error bars represent the temperature range over which we observed
homogeneous freezing, in a single experiment, in addition to the uncertainty in
the absolute temperature.

From an atmospheric scientist's perspective, one of the most interesting conclusions from the data presented in Figure 6 is that at temperatures below ~195 K cubic ice is the dominant product. This may be particularly relevant for the tropical tropopause region of the Earth's atmosphere, where ice clouds commonly form at these low temperatures. Also of interest, it is clear from Figure 6 that the fraction of cubic ice exhibits a strong solute type as well as a strong solute concentration dependence. For example, $(NH_4)_3H(SO_4)_2$ solutions always freeze to form a significant amount of ice I_c, while HNO_3 droplets can freeze to form ~100% ice I_h. These dependencies are not unprecedented: previous studies revealed complex crystallisation behaviour when impurities were added to a system,[43, 44] although the mechanisms for this behaviour are not well understood. It has been suggested that impurities can affect the phases that crystallise in a system by influencing either the phase that nucleates or the phase that grows.[43] Further research is underway in our laboratory in order to improve our understanding these dependencies.

4 SUMMARY AND CONCLUSIONS

In this experimental study we investigated the phase of ice that crystallises in pure water droplets and aqueous solution droplets. We have shown that cubic ice can form in pure bulk water droplets when they are cooled at a modest rate. In fact, small (<2 μm) droplets froze dominantly to cubic ice when cooled at 24 K min⁻¹. We also show that cubic ice which forms in water droplets has complex cubic to hexagonal transition kinetics. The results from this study suggest that cubic ice always nucleates when droplets homogeneously freeze and that it only converts to the stable hexagonal phase if the conditions allow it to (i.e. if the droplets warm significantly during crystallisation to allow the solid-solid transition to become rapid).

We have also demonstrated that cubic ice can form in solution droplets. In fact, below 195 K cubic ice is the dominant product. Above this freezing temperature the amount of cubic ice that forms in a droplet depends on the solute type and solute concentration.

Our results have a number of potentially important atmospheric implications. Firstly, they offer a significant insight into the crystallisation process of droplets of atmospheric relevance and suggest that cubic ice initially forms in droplets and then converts to the stable hexagonal phase. One important consequence of the results presented here is that when considering homogeneous ice nucleation theory[10] in relation to cloud formation, one should consider the pertinent parameters for cubic rather than hexagonal ice. This reinforces the conclusions made by Bartell and co-workers who suggested that cubic ice nucleates in emulsified pure water droplets based on their measurements of freezing large water clusters.[14, 39] Secondly, we have shown that under certain circumstances cubic ice may be the dominant product when atmospheric solution droplets freeze. Since cubic ice will necessarily have a larger vapour pressure than hexagonal ice,[28, 45, 46] a cloud made of cubic ice particles will have a larger vapour pressure. Water vapour is a significant greenhouse gas, hence understanding water vapour distributions around clouds is critically important in understanding the radiative properties of the atmosphere. Ice clouds in the tropical tropopause region also act as a cold trap for air entering the very dry stratosphere. A cloud made of cubic ice may allow more water vapour to enter the stratosphere. It is of interest to note that the mixing ratio of stratospheric water vapour in models is much lower than the mixing ratios that satellite measurements find.[4] Additionally, in a numerical modelling study, Murphy[28] suggested that a cubic to hexagonal ice particle mass transfer mechanism in cirrus clouds may have a significant impact on ice cloud microphysics and lifetime.

References

1 D.K. Lynch, in *Cirrus*, ed. D.K. Lynch, K. Sassen, D.C. Starr and G. Stephens, Oxford University Press, Oxford, 2002, p. 1.

2 K. Sassen, in *Cirrus*, ed. D.K. Lynch, K. Sassen, D.C. Starr and G. Stephens, Oxford University Press, Oxford, 2002, p. 11.

3 J.E. Penner, M. Andreae, H. Annegarn, L. Barrie, J. Feichter, D. Hegg, A. Jayareman, R. Leaitch, D.M. Murphy, J. Nganga and G. Pitari, in *Climate Change 2001: The Scientific Basis. Contributions of Working Group to the Third Assessment Report of the Intergovernmental Panel on Climate Change*, ed. B. Nyenzi and J. Prespero, Cambridge University Press, New York, 2001, p. 291.

4 E. Jensen and L. Pfister, *Geophysical Research Letters*, 2005, **32**, doi:10.1029/2004GL021125.

5 J.P.D. Abbatt, *Chemical Reviews*, 2003, **103**, 4783.

6 P.J. DeMott, in *Cirrus*, ed. D.K. Lynch, K. Sassen, D.C. Starr and G. Stephens, Oxford University Press, Oxford, 2002, p. 102.

7 S. Martin, *Chem. Rev.*, 2000, **100**, 3403.

8 B.J. Murray, D.A. Knopf and A.K. Bertram, *Nature*, 2005, **434**, 202.

9   B.J. Murray and A.K. Bertram, *Physical Chemistry Chemical Physics*, 2006, **8**, 186.
10   H.R. Pruppacher and J.D. Klett, *Microphysics of Clouds and Precipitation*, Kluwer, Dordrecht, 1997.
11   K.C. Young, *Microphysical Processes in Clouds*, Oxford University Press, New York, 1993.
12   I. Kohl, E. Mayer and A. Hallbrucker, *Phys. Chem. Chem. Phys.*, 2000, **2**, 1579.
13   E. Mayer and A. Hallbrucker, *Nature*, 1987, **325**, 601.
14   J.F. Huang and L.S. Bartell, *J. Phys. Chem.*, 1995, **99**, 3924.
15   J. Lepault, D. Bigot, D. Studer and I. Erk, *Journal of Microscopy-Oxford*, 1997, **187**, 158.
16   L.G. Dowell, S.W. Moline and A.P. Rinfret, *Biochim. Biophys. Acta*, 1962, **59**, 158.
17   L.G. Dowell and A.P. Rinfret, *Nature*, 1960, **188**, 1144.
18   P. Jenniskens and D.F. Blake, *Science*, 1994, **265**, 753.
19   D.C. Steytler, J.C. Dore and C.J. Wright, *J. Phys. Chem.*, 1983, **87**, 2458.
20   J. Dore, B. Webber, M. Hartl, P. Behrens and T. Hansen, *Physica A*, 2002, **314**, 501.
21   Y. Furukawa, *J. Met. Soc. Japan*, 1982, **60**, 535.
22   T. Takahashi and T. Kobayashi, *J. Crystal Growth*, 1983, **64**, 593.
23   H. Kiefte, M.J. Clouter and E. Whalley, *J. Chem. Phys.*, 1984, **81**, 1419.
24   I.M. Svishchev and P.G. Kusalik, *Phys. Rev. Lett.*, 1994, **73**, 975.
25   E. Whalley, *Science*, 1981, **211**, 389.
26   E. Whalley, *J. Phys. Chem.*, 1983, **87**, 4174.
27   M. Riikonen, M. Sillanpää, V. Leena, D. Sulluvan, J. Moilanen and I. Luukkonen, *Applied Optics*, 2000, **39**, 6080.
28   D.M. Murphy, *Geophys. Res. Lett.*, 2003, **30**.
29   J. Hallett, W.P. Arnott, M.P. Bailey and J.T. Hallett, in *Cirrus*, ed. D.K. Lynch, K. Sassen, D.C. Starr and G. Stephens, Oxford University Press, Oxford, 2002, p. 102.
30   B.J. Murray and J.M.C. Plane, *Phys. Chem. Chem. Phys.*, 2005, **7**, 3970.
31   B.J. Murray and J.M.C. Plane, *Phys. Chem. Chem. Phys.*, 2003, **5**, 4129.
32   J.M.C. Plane, B.J. Murray, X.Z. Chu and C.S. Gardner, *Science*, 2004, **304**, 426.
33   G.E. Thomas, *Reviews of Geophysics*, 1991, **29**, 553.
34   H.Y.A. Chang, T. Koop, L.T. Molina and M.J. Molina, *J. Phys. Chem. A*, 1999, **103**, 2673.
35   T. Koop, B.P. Luo, A. Tsias and T. Peter, *Nature*, 2000, **406**, 611.
36   S.L. Clegg, P. Brimblecombe and A.S. Wexler, *Journal of Physical Chemistry A*, 1998, **102**, 2137.
37   A. Goto, T. Hondoh and S. Mae, *J. Chem. Phys.*, 1990, **93**, 1412.
38   W.F. Kuhs, D.V. Bliss and J.L. Finney, *J. Phys. Colloq., C1*, 1987, **48**, 631.
39   L.S. Bartell and Y.G. Chushak, in *Water in Confined Geometries*, ed. V. Buch and J.P. Devlin, Springer-Verlag, Berlin, 2003, p. 399.
40   W. Kraus and G. Nolze, *J. Appl. Cryst.*, 1996, **29**, 301.
41   T. Takahashi, *J. Crystal Growth*, 1982, **59**, 441.
42   H. Uyeda and K. Kikuchi, *J. Met. Soc. Japan*, 1976, **54**, 267.
43   J.W. Mullin, *Crystallization*, Elsevier Butterworth-Heinemann, Oxford, 2001.
44   N. Eidelman, R. Azoury and S. Sarig, *J. Cryst. Growth*, 1986, **74**, 1.
45   A. Kouchi, *Nature*, 1987, **330**, 550.
46   J.E. Shilling, M.A. Tolbert, O.B. Toon, E.J. Jensen, B.J. Murray and A.K. Bertram, *Geophys. Res. Lett.*, 2006, **33**, doi:10.1029/2006gl026671.

HYDRATE PHASE TRANSFORMATIONS IMPOSED BY GAS EXCHANGE

M. M. Murshed and W. F. Kuhs

GZG, Abt. Kristallographie, Universität Göttingen, Goldschmidtstrasse 1, 37077 Göttingen, Germany

1 INTRODUCTION

Hydrates of hydrocarbons are stable in a wide range of oceanic and permafrost environments.[1-3] Because of the predominance of methane, generally produced by microbial breakdown of sedimentary organic matters, methane represents more than 99% of the filling in natural gas hydrates.[1,2] If methane hydrate forms above the ocean sea floor within the stability range (e.g. at the contact of upraising gas bubbles with sea water), it will float upwards and dissociate, since crystalline methane hydrate is less dense than water. However, if it forms within sediments in the ocean floor, it will remain bound there.[2] Within the sediment-water column the lowest temperature is usually observed at the sea floor and increases downward through the sediments along the geothermal gradient towards the hot centre of the earth. Therefore gas hydrate can persist in a zone, the so called Gas Hydrate Stability Zone (GHSZ) that extends from the sea floor down to a certain depth. The base of the GHSZ varies mostly by gas chemistry and the local geothermal gradient along with other minor factors such as pore fluid composition (salinity), pore size, and sediment mineralogy. For instance, in the Gulf of Mexico, at a pressure equivalent to 2.5 km water depth, the base of the GHSZ occur at about 21°C for pure methane, but at 23°C for a mixture of about 93% methane, 4% ethane and 1% propane.[2] At the same pressure (2.5 km water depth) for a mixture of about 62% methane, 9% ethane, 23% propane, along with some higher hydrocarbons, the stability temperature increases up to 28°C. It would be interesting to know how at the base of the GHSZ free gases interact with the upwards located hydrates and which types of the hydrate structure are formed either from microbial (methane), thermogenic gases (ethane or higher hydrocarbon) or their mixtures. Because pure methane favours structure type I (sI), while at a certain composition[4] of methane and ethane forms structure type II (sII) and higher hydrocarbons (iso-butane) along with methane forms a hexagonal structure (sH).[5,6] Transformation processes driven by exposing existing gas hydrates to a different gas environment thus may take place in nature but little is known about their kinetics. These processes are interesting also from a more fundamental aspect of hydrate growth and regrowth of which we still have limited knowledge. Thus, to gain more insight we have performed exchange reactions between methane (ethane) hydrate and ethane (methane) gas at temperatures close to 0°C.

Recent studies[7-9] on gas replacement reactions between hydrates and gas phases suggest a possible CO_2 sequestration[10] in the ocean bed by means of a concomitant methane extraction from the hydrates thus preserving the mechanical stability of the host sediments. Yoon et al.[7] reported an in-situ Raman spectroscopy study on the transformation of methane hydrate into carbon dioxide hydrate. Using MAS NMR technique Lee et al.[8] observed the exchange between CO_2 and CH_4 in sI hydrates. A neutron diffraction study by Halpern et al.[9] provided information on the exchange between guest molecules in the hydrate and free molecules in the surrounding phase. In comparison with thermodynamic and even kinetic studies, the work on guest replacement reactions in gas hydrates is still in its infancy. Raman spectroscopy has been widely used for determining the structural identity, cage occupancies, and formation kinetics of gas hydrates; it certainly also can deliver information on guest replacement processes in gas hydrates.[4,11-14] The position and relative intensities of Raman modes both in the C-H and C-C stretch frequency ranges will provide insight into the nature of pure and gas exchanged hydrate phases formed during the transformation in terms of structure type and relative filling of small and large cages. It should be said, at this point, that it is not trivial to establish the thermodynamically stable phase by experiments of short duration as metastable phases can be formed which may persist for rather long times.[15,16] Thus one should be attentive in replacement reactions that metastable phases may form and complicate the picture obtained.

2 EXPERIMENTAL

Methane and ethane hydrates were prepared, starting from spherical ice grains[15] of a size of approximately 50-80 μm, in a high-pressure cell[17] with methane and ethane gases at 270 K and 30 bars for 1 day. The reaction after one day is incomplete but sufficient to provide a full coverage of the ice grains with gas hydrate.[15] For a replacement of methane by ethane, the so prepared methane hydrates were inserted into a high-pressure cell under liquid nitrogen condition, sealed instantly and then exposed to ethane gas. Inversely, ethane hydrates were exposed to a methane atmosphere. Both replacement reactions were carried out at conditions (270 K and 30 bars) similar to the ones at which the pure methane and ethane hydrates were prepared at. Samples were recovered after a preset time of reaction, Raman spectra were taken on parts of the sample, and the remaining sample was then put back into the pressure cell for further reaction covering a period of up to 14 days. Only the Raman data set after 0.25 days of replacement reaction was taken on a new sample, prepared in an identical way to the sample of the consecutive runs. The sealed gas volume was large compared to the gas volume due to hydrate decomposition, the enclathrated component in the free gas is thus expected to be very dilute. It is important to note that the sealed system is refreshed by the subsequent uploading and filled with the pure replacement gas each time. Each of the recovered samples was investigated by Raman spectroscopy carried out on a microfocus DILOR multichannel spectrometer using a Nikon Optiphot microscope. The excitation line at 514.532 nm was produced by a Coherent Innova 70 Ar^+ laser. The scattered radiation was analyzed using a single monochromator with an entrance slit of 0.125 mm, a spectral resolution of 7.8 cm^{-1} and a grating of 1800 grooves/mm. The beam was focused to a spot size of about 10 sq. μm using an SLWD Nikon x20 objective lens. The working temperature was set at 113 K to avoid any hydrate decomposition.

3 RESULTS

The Raman spectra taken at different stages of the methane-ethane replacement reactions are shown in figure 1 and those of ethane-methane replacement reactions in figure 2. The totally symmetric C-H stretching (v_1)[18] bands of methane in the small cavities (5^{12}) are observed at 2911.8 cm^{-1} and 2910.5 cm^{-1} for sI and sII hydrates, respectively. The frequencies at 2901 cm^{-1} and 2900 cm^{-1} are assigned to the v_1 of methane in the large cavities of sI $(5^{12}6^2)$ and sII $(5^{12}6^4)$ hydrates, respectively. The frequencies of the resonance doublet bands[18] at 2883 cm^{-1} and 2939 cm^{-1} refer to ethane encaged in the large cages $(5^{12}6^4)$ of sII hydrates.[4,11-14] In contrast, the resonance doublets at 2885.7 cm^{-1} and 2941.9 cm^{-1} are associated with ethane in the $5^{12}6^2$ cavities of sI hydrate.[4,11-14] The C-C stretching (v_3)[18] bands of ethane in the sI hydrate is observed at 1000.4 cm^{-1} during methane-ethane replacements. In contrast, frequencies at 990.8 cm^{-1} and 1003 cm^{-1} are assigned to v_3 of sII and sI hydrates, respectively, during ethane-methane replacements.

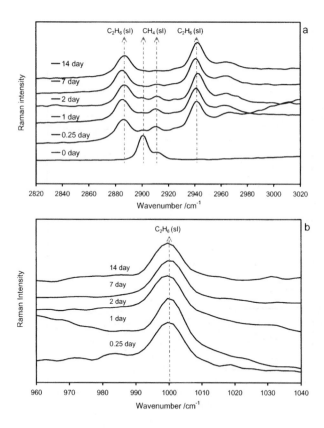

Figure 1 *(a) The totally symmetric C-H stretching (v₁) bands of CH₄ and the resonance doublets for C₂H₆ enclathrated in different cavities of the corresponding hydrate structures, (b) the C-C stretching (v₃) bands of C₂H₆ observed during methane sI-to-ethane sI transformation.*

4 DISCUSSION

The frequencies of v_1 and the resonance doublets of ethane show a lower wavenumber than the corresponding reported values.[4,11-14] Sharma et al.[19] and Uchida et al.[14] argued that the shift towards lower values are likely to be due to the measurement at lower temperature. Tulk et al.[11] and, Schicks and Ripmeester[16] reported identical values of v_1 at 2904 cm^{-1} and 2916 cm^{-1} for methane in the $5^{12}6^2$ and 5^{12} cages, respectively, of sI hydrate. The former authors[11] did not observe any temperature dependent frequency shift while measuring between 10 K and 170 K. On the other hand, Schicks and Ripmeester[16] measured at 273 K and observed no shift of those frequencies even when the initial methane sII hydrate transformed into methane sI one. The resonance doublets of ethane in different hydrates differ as much as 5 cm^{-1} lower in wavenumber than that of Subramanian et al.[4] who distinguished two structure types (sI and sII) based on those frequency values (measured at 274.2 K) in the methane/ethane mixed hydrate systems. Incongruities of the v_1 frequency

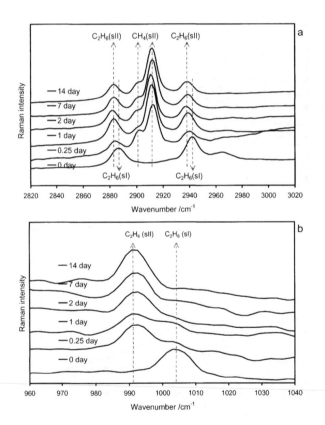

Figure 2   *(a) The totally symmetric C-H stretching (v_1) bands of CH$_4$ and the resonance doublets for C$_2$H$_6$ enclathrated in different cavities of the corresponding hydrate structures, (b) the C-C stretching (v_3) bands of C$_2$H$_6$ observed during ethane substitution from ethane sI hydrates by methane.*

values are also observed in some natural samples measured at a similar temperature. For instance, the natural methane sI hydrates, collected from Congo-Angola basin, show the v_1 values at 2903 cm^{-1} and 2914 cm^{-1}, while the samples from Norwegian margin show the corresponding values at 2900 cm^{-1} and 2911 cm^{-1}.[20] However, it is not clear yet whether the measuring temperature or the formation temperature is the factor for such discrepancies. Measuring at a similar temperature (113 K) we observe identical v_1 values in both methane-ethane and ethane-methane replacement reactions. But the frequency of v_3 for sI differs about 2.5 cm^{-1} from methane-ethane replacement to ethane-methane one. Meanwhile, we cannot conclude that the types of replacements are correlated with this shift of frequencies (C-C regions) of ethane only.

As expected, methane forms sI hydrates, which is clear from the shifts of the v_1 bands. After 6 hours ethane feeding to the methane sI hydrate system the appearance of ethane sI hydrates is identified from the ethane resonance doublets; an observed frequency of 1000.4 cm^{-1} for the v_3 band (Fig. 1b) provides additional support. At this stage the relative intensity ratio of the v_1 bands of methane to the ethane resonance doublets indicates the predominance of ethane sI hydrate. Yoon et al.[7] observed the simultaneous decrease of methane signals, both in the small and large cages, on introducing CO_2, while we see an asymmetric change (the methane signal from small cages is more intense than that of large cages) during the CH_4 sI-to-C_2H_6 sI transformation. This peculiarity is likely to suggest, prima facie, that ethane sI hydrate formation commences through a complete regrowth process while methane hydrate decomposes, as it is not conceivable that methane escapes from the large cages without a destruction of the crystal structure. It further suggests that during methane decomposition a localized high concentration of methane exists at the decomposition interface allowing methane molecules to be encaged in the newly formed ethane-rich hydrate; due to the size difference this will happen predominantly in the small cages. Finally, sometime between 7-14 days, the methane content diminishes as a consequence of the dilution of the gas phase during repeated flashings.

In contrast to the CH_4 sI-to-C_2H_6 sI transformation, we could not observe the counterpart C_2H_6 sI-to-CH_4 sI transformation within the investigated reaction period of 14 days. The pure ethane sI hydrate is clearly identified from the frequency of the resonance doublets at 2885.7 cm^{-1} and 2941.9 cm^{-1} as well as v_3 at 1003 cm^{-1}. After 6 hours of methane admission to the ethane sI hydrate more than one hydrate phase appears, which could be identified based on the Raman spectra both in the C-H and C-C stretch regions. The frequencies of v_1 at 2900 cm^{-1} and 2910.5 cm^{-1} clearly identify methane entrapped in the $5^{12}6^4$ and 5^{12} cages, respectively, of a methane sII hydrate. The relative intensity of those two bands usually is related to the corresponding number of occupied small and large cages of methane sII hydrate. The resonance doublets of ethane at 2883 cm^{-1} and 2939 cm^{-1} refer to ethane in the $5^{12}6^4$ cages of ethane sII hydrate. In this context, the frequency of v_3 band at 990.8 cm^{-1} additionally supports the ethane sII hydrate formation. These observations are consistent with methane sII and ethane sII hydrates forming simultaneously (with some possible admixing of a minority gas component); in-house X-ray diffraction experiments on two samples recovered after 6h and 14d show only the presence of a type II structure supporting this assignment (and excluding the presence of any significant amount of a type I structure). The initial ethane sI hydrate is not completely transformed and remains as long as 2 days (Fig. 2b), which could not be resolved in the C-H stretching regions but is clearly seen in the C-C regions. The transformed state shows no substantial change, even after 14 days in terms of Raman shift or intensity ratio of methane occupancy in the large and small cages (LC/SC).

From a comparative standpoint, it is observed that the ethane sI hydrate formation is faster than the expected methane sI hydrates from the corresponding initial phases. It is well established that enclathration of relatively large molecules tends to strongly stabilize the hydrate structure and thereby lower the equilibrium pressure considerably.[21,22] Therefore, the excess pressure (fugacity) with respect to the equilibrium pressure (fugacity) may be considered in a first approximation as the main thermodynamic driving force for phase changes. Nevertheless, some other salient physical parameters related to the reaction kinetics like the activation energy, subtle local endothermic dissociation or exothermic formation effects, the hydrate microstructure, diffusion limitations and the slightly changing mixed gas composition should also be considered.

5 CONCLUSION

Like in previous reports[4,23] both pure methane and ethane themselves forms sI hydrate. It is worthwhile to mention that in each pure hydrate system the gas feeding was chosen for 24 h to produce only a layer of hydrates on the spherical ice grains to facilitate the subsequent exchange process along with/without the structure transition, which would be distinctly slower[7,24] if the exchange occurs in the bulk system. On the other hand, it should be noted that using only surface layers of gas hydrate exchange reactions at longer time scales may be accompanied by a reaction of untransformed ice into hydrate. A separation of these processes is not easily done using Raman scattering, but could routinely be followed by in situ diffraction techniques.[24] Methane sI hydrates have been exposed to ethane gas, and vice versa, at similar p-T conditions (270 K and 30 bar) both the methane and ethane hydrates were prepared at. In this paper we describe how Raman spectroscopy complemented by X-ray diffraction on recovered samples offers insights into different pure and gas exchanged hydrate phases. Methane sI hydrate transforms into ethane sI hydrates by a destruction/reformation process of the crystalline structure without changing the structure type; it takes as much as 14 days to replace methane completely by ethane. In contrast, when ethane sI hydrate is exposed to free methane gas, at similar p-T conditions, the parent sI structure transforms into methane sII + ethane sII hydrates, which persist even after 14 days with no significant change. Further investigations, in particular in situ neutron diffraction and Raman spectroscopy with better resolution, are in preparation to unravel further details of these exchange processes.

Acknowledgements

We thank Mr. Eberhard Hensel (Göttingen) for his help during the sample preparation. We also thank an anonymous referee for valuable suggestions to improve the manuscript. This is publication no. GEOTECH - 242 of the R&D-Programme GEOTECHNOLOGIEN funded by the German Ministry of Education and Research (BMBF) and German Research Foundation (DFG), Grant (Förderkennzeichen) G0605B.

References

1   K.A. Kvenvolden, *Org. Geochem.*, 1995, **23**, 997.
2   M.D. Max, A.H. Johnson, W.P. Dillon, *Economic geology of natural gas hydrate. Coastal systems and continental margins*, Springer, 2006, p. 105.
3   E.D. Sloan, *Clathrate hydrates of natural gases.* Marcel Dekker, New York, 1998.
4   S. Subramanian, R.A. Kini, S.F. Dec, E.D. Sloan, *Chem. Eng. Sci.*, 2000, **55**, 1981.
5   J.A. Ripmeester, J.S. Tse, C.I. Ratcliffe, B.M. Powell, *Nature*, 1987, **325**, 135.
6   R. Sassen, I.R. MacDodald, *Org. Geochem.*, 1994, **22**, 1029.

7   J.H. Yoon, T. Kawamura, Y. Yamamoto, T. Komai, *J. Phys. Chem. A.*, 2004, **108**, 5057.

8   H. Lee, Y. Seo, Y.T. Seo, I.L. Moudrakovski, J.A. Ripmeester, *Angew. Chem. Int. Ed.*, 2003, **42**, 5048.

9   Y. Halpern, V. Thieu, R.W. Henning, X. Wang, A.J. Schultz, *J. Am. Chem. Soc.*, 2001, **123**, 12826.

10 P.G. Brewer, C. Friederich, E.T. Peltzer, F.M. Orr, *Science*, 1999, **284**, 943.

11 C.A. Tulk, J.A. Ripmeester, D.D. Klug, *Ann. NY. Acad. Sci.*, 2000, **912**, 859.

12 A.K. Sum, R.C. Burruss, E.D. Sloan, *J. Phys. Chem. B.*, 1997, **101**, 7371.

13 S. Subramanian, E.D. Sloan, *J. Phys. Chem. B.*, 2002, **106**, 4348.

14 T. Uchida, S. Takeya, Y. Kamata, I.Y. Ikeda, J. Nagao, T. Ebinuma, H. Narita, O. Zatsepina, B.A. Buffett, *J. Phys. Chem. B.*, 2002, **106**, 12426.

15 D.K. Staykova, W.F. Kuhs, A.N. Salamatin, T. Hansen, *J. Phys. Chem. B.*, 2003, **107**, 10299.

16 J.M. Schicks, J.A. Ripmeester, *Angew. Chem. Int. Ed.*, 2004, **43**, 3310.

17 W.F. Kuhs, E. Hensel, H. Bartels, *J. Phys.: Condens. Matter*, 2005, **17**, 3009.

18 G. Herzberg, *Infrared and Raman Spectra*, 5th Ed. D. Von Nostrand Company Inc. New York, 1951, p. 271.

19 S.K. Sharma, A.K. Mirsa, P.G. Lucey, G.J. Exarhos, C.F. Windisch Jr., *Lunar and Planetary Science*, 2004, **35**, Abstr. 1929.

20 B. Chazallon, C. Fosca, J-L, Charlou, J-P. Donval, C. Bourry, E. Sauter, D. Levaché, *Proc. 5th Int. Conf. on Gas Hydrates*, Trondheim 2005, ref. 2042.

21 E.D. Sloan, F. Fleyfel, *Fluid Phase Equilibria*, 1992, **76**, 123.

22 W.M. Deaton, E.M. Frost, *Gas hydrates and their relation to the operation of natural gas pipe lines*. Bur. Mines Monogr. (U.S.), 1946, **8**, p. 17.

23 M.T. Kirchner, R. Boese, W.E. Billups, L.R. Norman, *J. Am. Chem. Soc.*, 2004, **126**, 9407.

24 W.F. Kuhs, D.K. Staykova, A.N. Salamatin, 2006, *J. Phys. Chem. B*, **110**, 13283.

MECHANISM OF CAGE FORMATION DURING GROWTH OF CH$_4$ AND Xe CLATHRATE HYDRATES: A MOLECULAR DYNAMICS STUDY

H. Nada[1]

[1]National Institute of Advanced Industrial Science and Technology (AIST), Tsukuba 305-8569, Japan

1 INTRODUCTION

Gas hydrates are crystalline forms of water that contain many gas molecule inclusions.[1] Each gas hydrate has a unique molecular-scale structure in which a gas molecule is trapped by a "cage" structure consisting of H$_2$O molecules. Until now, the thermodynamically stability and the structural properties of gas hydrates have been investigated for a lot of gas species by experimental studies.[1] However, the growth mechanism of gas hydrates at the molecular scale, especially the mechanism of cage formation, is still poorly understood. This is because the molecular-scale growth mechanism of gas hydrates is complicated and difficult to elucidate experimentally.

Computer simulations, such as molecular dynamics (MD) simulations, are helpful tools for investigating the growth mechanism of gas hydrates at the molecular scale. So far, MD simulations of the growth of a CH$_4$ hydrate from a concentrated aqueous CH$_4$ solution were carried out at a temperature much lower than 0 °C.[2, 3] However, in real systems, gas hydrates are grown from a two-phase coexistence of liquid water and a gas at temperatures above 0 °C, and for most gas species, the thermodynamically stable concentration of gas molecules in liquid water is much lower than that in a gas hydrate. Therefore, simulations for understanding of the growth mechanism of gas hydrates in real systems should involve dilute aqueous gas solutions at temperatures above 0 °C.

Recently, we carried out an MD simulation of the growth of a CH$_4$ hydrate from a dilute aqueous CH$_4$ solution at a temperature higher than 0 °C.[4] In this study, we also carried out an MD simulation of the growth of a Xe hydrate from a dilute aqueous Xe solution at a temperature higher than 0 °C. In this paper, we discuss the mechanism of cage formation during the growth of both CH$_4$ and Xe hydrates, which were observed in those MD simulations.

2 SIMULATION METHOD

The intermolecular interaction between pairs of H$_2$O molecules was estimated using a six-site model,[5] which was developed for simulations of ice and water near 0 °C. Recently, using the six-site model, the growth of an ice crystal from water near 0 °C was successfully observed in several MD simulations.[6–8] This success clearly demonstrates the suitability of the six-site model for simulating the crystallization of liquid water in real systems.

Therefore, we believe that the model is also suitable for MD simulations of the growth of gas hydrates.

The interactions between pairs of gas molecules were estimated using the LJ potential ($=4\varepsilon\{(\sigma/r)^{12}-(\sigma/r)^{6}\}$, where r is the site-site distance). Two sets of LJ parameters for gas molecules (i.e., ε and σ) were chosen. One was $\varepsilon/k_B = 149$ K (k_B is the Boltzmann constant) and $\sigma = 0.378$ nm, and the other was $\varepsilon/k_B = 231$ K and $\sigma = 0.405$ nm. The former corresponded to the parameter set for a spherically-approximated CH_4 molecule,[9] and the latter for a Xe molecule.[10] The interactions between H_2O and gas molecules were also estimated using the LJ potential, and the ε and σ values for the H_2O–gas LJ potential were determined using the Lorentz–Berthelot rules. Long-range interactions were smoothly truncated at an intermolecular distance of 1 nm using a switching function.[11]

The simulation system contains three phases: a hydrate phase and its decomposed phases consisting of liquid water and gas (Figure 1(a)). The hydrate consists of 736 H_2O and 128 gas molecules. All H_2O molecules in the hydrate were arranged at the lattice sites of the type-I structure,[12] and a gas molecule was placed in all the cages included in the hydrate. Each of the liquid water phases consisted of 368 H_2O molecules, and each of the gas phases consisted of 64 gas molecules. In this study, we prepared two systems. One was the system for the growth of a CH_4 hydrate, and the other for the growth of a Xe hydrate (hereafter, the CH_4 system and the Xe system, respectively). The MD simulation was carried out for each of the two systems independently.

Computation was carried out using a leap-frog algorithm proposed by Fincham[13] with a time step of 1 fs. The total run was 20 ns for the CH_4 system and 14 ns for the Xe system. Temperature and pressure were kept constant at 298 K and 100 MPa, respectively, using a method proposed by Berendsen *et al.*[14] Details of the simulation method is given in Ref. 4.

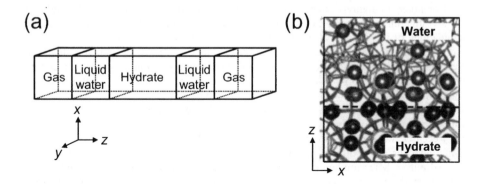

Figure 1 *(a) An illustration of the simulation system used. Periodic boundary conditions were imposed in all three directions. (b) Snapshots of gas molecules and the HB-networks of H_2O molecules at 14.0 ns for the CH_4 system. The dark gray and light gray spheres, respectively, show the gas molecules that were included in the hydrate and the gas at the beginning of the simulation. The dashed line shows the position of the hydrate interface at the beginning of the simulation.*

The pressure of 100 MPa in this simulation is much higher than typical experimental pressures (on the order of 1 or 10 MPa). The reason for the choice of such a high pressure in this simulation was that the frequency of dissolution of gas molecules into liquid water was required to be high enough to observe the growth of each hydrate in a MD simulation. The density of water at 100 MPa (1.05 ± 0.01 g/cm^3 in the six-site model and 1.038 g/cm^3 in experiment) is higher by 4–5 % than the water density at 0.1 MPa. This means that structural properties of water change with increasing pressure. In this study, we assumed that the mechanism of cage formation at 100 MPa is essentially the same as the mechanism at lower pressures. However, more extensive studies about the effect of pressure on the mechanism of cage formation are needed to confirm this assumption.

3 RESULTS AND DISCUSSIONS

The simulations started from a condition in which no gas molecules were dissolved into the liquid water. During the simulations, gas molecules in the gas were gradually dissolved one by one into the liquid water, and the growth of the hydrate from the dilute solution was successfully observed on the interface between the hydrate and the liquid water (hereafter, the interface is referred to as the hydrate interface) for both systems. Then, we analyzed the mechanism of cage formation on the hydrate interface for both systems. In Figure 1(b), snapshots of gas molecules and the hydrogen-bonded networks (HB-networks) near the hydrate interface at the end of the simulation for the CH$_4$ system are shown.

3.1 Ordering of gas molecules

First of all, we will discuss the ordering of gas molecules on the hydrate interface briefly. Figure 2(a) shows the trajectory of a gas molecule that migrated to the hydrate interface. The HB-networks of H$_2$O molecules near the hydrate interface are also shown. It can be seen that, for both systems, on reaching the hydrate interface, the gas molecule was readily arranged at a cage site (large cage site). Note, however, that, when the gas molecule was arranged at the cage site, the H$_2$O molecules surrounding it had not yet formed a cage. Similarly, for both systems, most of the gas molecules that migrated to the hydrate interface were arranged at cage sites prior to cage formation by the surrounding H$_2$O molecules.

The reason for the arrangement of gas molecules at cage sites prior to cage formation by the surrounding H$_2$O molecules can be determined by analyzing the potential energy, U, acting between a gas molecule that is placed on the hydrate interface and the underlying hydrate. Figure 2(b) shows the distribution of the U over all positions of a gas molecule on an x–y plane 0.3 nm distant from the hydrate interface. Four potential minima can be seen (black regions). The positions at which these potential minima appear corresponded to cage sites. This explains why the gas molecules were readily arranged at cage sites, even though the surrounding H$_2$O molecules had not yet formed cages.

In addition, after the beginning of the simulation for the Xe system, a Xe molecule that was arranged at a small cage site under the hydrate interface jumped out from the site to the liquid water and migrated to a large cage site on the hydrate interface (see the Xe molecule that is arranged at a cage site on the hydrate interface in Figure 2(a)). This result is consistent with the fact that some of the small cages included in the thermodynamically stable Xe hydrates are empty.[1]

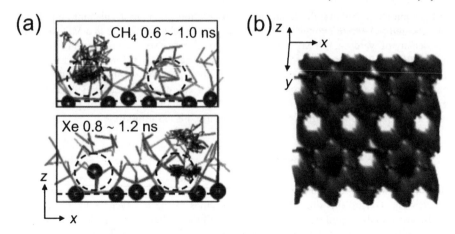

Figure 2   *(a) Trajectory of a gas molecule that ended up at a cage site (large cage site) on the hydrate interface for the CH₄ and Xe systems. The HB-networks of H₂O molecules near the hydrate interface (at 1.0 ns for the CH₄ system and at 1.2 ns for the Xe system) are also shown. The dotted circles show cage sites on the hydrate interfaces. (b) Distribution of U acting between a gas molecule on the hydrate interface and the underlying hydrate over all positions of the gas molecule on the x–y plane 0.3 nm distant from the initial position of the hydrate interface. The distribution of U was calculated using the LJ parameter set for a CH₄ molecule.*

3.2   Cage formation by H_2O molecules

Next, we will discuss the mechanism of cage formation by H_2O molecules on the hydrate interface. Figure 3 shows the time sequence of the HB-networks of H_2O molecules in the first hydration shell around a gas molecule, whose trajectory is illustrated in Figure 2(a), after the gas molecule was arranged at a cage site, for both systems. The ordering of H_2O molecules around the gas molecule can be seen for both systems. However, for both systems, the ordering of H_2O molecules occurred only in the lower part of the first hydration shell (the regions enclosed by solid lines), but did not occur in the upper part of it (the regions enclosed by dotted lines).

Figure 4 shows the time sequence of the H_2O–H_2O and H_2O–gas potential energies, U_{ww} and U_{wg}, in the upper and the lower parts of the first hydration shell around each of the CH₄ and Xe molecules, whose trajectories are illustrated in Figure 2(a). It can be seen that, for both systems, the U_{ww} and U_{wg} in the lower part of the first hydration shell decreased after the arrangement of the gas molecule at a cage site, whereas those in the upper part did not significantly change with time. This also supports that the ordering of H_2O molecules occurred only in the lower part of the first hydration shell. The results given in Figures 3 and 4 strongly suggest that the arrangement of a gas molecule at a cage site is not sufficient for the formation of a cage by the surrounding H_2O molecules.

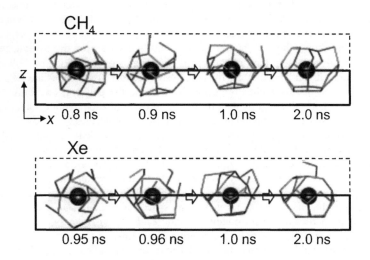

Figure 3 *The time sequence of the HB-networks of H_2O molecules in the first hydration shell around a gas molecule, after the molecule was arranged at a cage site on the hydrate interface for the CH_4 and the Xe systems.*

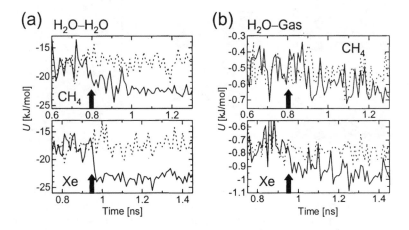

Figure 4 *The time sequence of the potential energy, U, in the upper (dotted lines) and the lower (solid lines) parts of the first hydration shell around a gas molecule for the CH_4 and Xe systems, (a) for the H_2O–H_2O interaction and (b) for the H_2O–gas interaction. Each of the arrows shows the time at which the gas molecule was arranged at a cage site on the hydrate interface.*

In order to search for the regions in which stable polygons that constitute cages of the hydrate were formed by H_2O molecules, we analyzed hydrogen-bonds (HBs) of H_2O molecules near the hydrate interface using the time-averaged coordinates of H_2O molecules over a period of 0.1 ns for each system. The purpose of using the time-averaged coordinates was to search for "solid-like" HBs, i.e., the HBs that were not broken for a long time, like the HBs in the hydrate.

Figure 5(a) shows the distribution of the solid-like HBs near the hydrate interface, which were analyzed using the time averaged coordinates over a period of 13.9–14.0 ns, for each system. It can be seen that, for both systems, most of the solid-like HBs exist in spaces surrounded or sandwiched by gas molecules and the polygons are formed only in those spaces. This result suggests that, during the growth of each hydrate, cage formation occurred only in spaces surrounded or sandwiched by gas molecules.

The occurrence of cage formation only in spaces surrounded or sandwiched by gas molecules is explained as follows. Since both CH_4 and Xe molecules are hydrophobic, H_2O molecules do not form bonds with those gas molecules. Therefore, the H_2O molecules in the spaces form HB-networks by themselves to fit the sizes of the spaces. Moreover, since the spaces are narrow, the movement of each H_2O molecule in the spaces is restricted. Therefore, only the HB-networks formed in the spaces are readily stabilized owing to the hydrophobic properties of the surrounding gas molecules. Consequently, we conclude that the formation of cages is essentially a clustering of confined H_2O molecules in molecular-scale spaces restricted by hydrophobic walls. The mechanism of cage formation, which the present results suggested, is schematically illustrated in Figure 5(b).

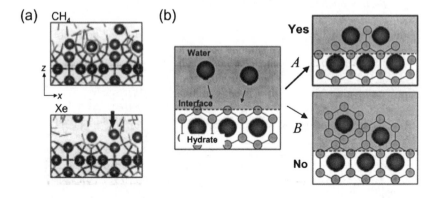

Figure 5   *(a) The distribution of the solid-like HBs near the hydrate interface, which were obtained using the time-averaged coordinates of molecules over a period of 13.9–14.0 ns, for each system. (b) Schematic illustrations of two scenarios of the mechanism of cage formation; scenario A in which only confined H_2O molecules in restricted spaces by gas molecules form stable polygons, and scenario B in which the growth of the clathrate involves the trapping of each gas molecule by cages formed first by H_2O molecules. The present results suggested scenario A.*

Finally, we will discuss briefly differences in the quickness of the ordering of H_2O molecules between the CH_4 and the Xe systems. We observed that the ordering of H_2O molecules around large cage sites occurred more quickly for the Xe system than for the CH_4 system. The quick ordering of H_2O molecules around a large cage site for the Xe system is seen in Figure 4(b), which shows that the U_{ww} in the lower part of the first hydration shell for the Xe system decreased drastically after the arrangement of a Xe molecule at a large cage site, whereas the U_{ww} for the CH_4 system decreased gradually after the arrangement of a CH_4 molecule at a large cage site.

Moreover, we also observed that the ordering of H_2O molecules around small cage sites occurred more slowly for the Xe system than for the CH_4 system. The slow ordering of H_2O molecules around a small cage site for the Xe system is seen in Figure 5(a), which shows that stable polygons were not yet formed around a Xe molecule that was arranged at a small cage site (see the Xe molecule marked by an arrow). For the CH_4 system, the quickness of the ordering of H_2O molecules around large cage sites and that around small cage sites were almost the same.

It is known that a Xe molecule is energetically more stably included in a large cage than in a small cage, whereas the stability of inclusion of a CH_4 molecule in a large cage is almost the same as that in a small cage.[1] It seems that the quickness of cage formation around a gas molecule is strongly related to the stability of cage occupation by the molecule. Unfortunately, the growth observed in the present simulations was not enough to elucidate the relationship between the quickness of cage formation by a gas molecule and the energetic stability of cage occupation by the molecule. More extensive studies about cage formation during the growth of CH_4, Xe and other gas hydrates are needed to elucidate it.

4 CONCLUSIONS

We investigated the mechanism of cage formation during the growth of CH_4 and Xe hydrates by means of MD simulations. The results obtained are summarized as follows.
(1) For both CH_4 and Xe hydrates, cage formation occurred by the following process: gas molecules were arranged at cage sites on the hydrate interface first, and the ordering of H_2O molecules near the gas molecules occurred subsequently.
(2) For both CH_4 and Xe hydrates, cage formation occurred only in spaces surrounded or sandwiched by gas molecules. This result strongly suggested that cage formation is essentially a clustering of H_2O molecules in narrow spaces restricted by hydrophobic walls.
(3) The ordering of H_2O molecules around large cage sites occurred more quickly for the Xe system than for the CH_4 system. Moreover, the ordering of H_2O molecules around small cage sites occurred more slowly for the Xe system than for the CH_4 system. For the CH_4 hydrate, the quickness of the ordering of H_2O molecules around large cage sites and that around small cage site were almost the same. The result implied that the quickness of cage formation by a gas molecule is strongly related to the energetic stability of cage occupation by the molecule.

Acknowledgements

This work was financially supported by a Grant-in-Aid for Scientific Research (C) (No. 17540385) from the Japan Society for the Promotion of Science. This work was done as a Joint Research Program of the Institute of Low Temperature Science, Hokkaido University.

Some of the computations in this work were done using the facilities of the Supercomputer Center, Institute of Solid State Physics, University of Tokyo.

References

1   E.D. Sloan, *Clathrate Hydrates of Natural Gases, Marcel Dekker*, New York, 1998.
2   L.A. Báez, P. Clancy, *Annals of the New York Academy of Science*, 1994, **715**, 177.
3   N.J. English, M.D. MacElroy, *J. Chem. Phys.*, 2004, **120**, 10247.
4   H. Nada, *J. Phys. Chem.*, 2006, **B110**, 16526.
5   H. Nada and J.P. van der Eerden, *J. Chem. Phys.*, 2003, **118**, 7401.
6   H. Nada, J.P. van der Eerden and Y. Furukawa, *J. Cryst. Growth*, 2004, **266**, 297.
7   H. Nada and Y. Furukawa, *J. Cryst. Growth*, 2005, **283**, 242.
8   M.A. Carginano, P.B. Shepson and I. Szleifer, *Mol. Phys.*, 2005, **103**, 2957.
9   L.R. Platt and D.J. Chandler, *J. Chem. Phys.*, 1980, **73**, 3434.
10 H. Tanaka, *Fluid Phase Equilibria*, 1996, **116**, 326.
11 M.J. Vlot, J. Huinink and J.P. van der Eerden, *J. Chem. Phys.*, 1999, **110**, 55.
12 R.K. McMullan and G.A. Jeffrey, *J. Chem. Phys.*, 1965, **42**, 2725.
13 D. Fincham, *Mol. Sim.*, 1992, **8**, 165.
14 H.J.C. Berendsen, J.P.M. Postma, W.F. van Gunsteren, A. DiNola and J.R. Haak, *J. Chem. Phys.* 1984, **81**, 3684.

GROWTH KINETICS ON INTERFACE BETWEEN $\{20\bar{2}1\}$ PLANE OF ICE AND WATER INVESTIGATED BY A MOLECULAR DYNAMICS SIMULATION

H. Nada[1] and Y. Furukawa[2]

[1]National Institute of Advanced Industrial Science and Technology (AIST), Tsukuba 305-8569, Japan
[2]Institute of Low Temperature Science, Hokkaido University, Sapporo 060-0819, Japan

1 INTRODUCTION

The anisotropy in the growth kinetics of ice crystals from water at the molecular level is an important subject that should be investigated in detail, because it is essential for understanding the pattern formation of snow and ice crystals,[1] the freezing of water in biological systems[2] and the formation of acid snow.[3] Until now, several molecular dynamics (MD) simulations have been carried out to elucidate the anisotropy in the growth kinetics of ice crystals from water.[4-7] The anisotropy in the growth kinetics among $\{0001\}$, $\{10\bar{1}0\}$ and $\{11\bar{2}0\}$ planes of an ice crystal, which was obtained in those MD simulations, qualitatively explained the anisotropic growth velocity among those planes of ice crystals grown from water in real systems.

However, for many crystallographic orientations of an ice crystal, the growth kinetics still remains unclear. It should be noted that a macroscopic shape of an ice crystal grown from water is dominated by the anisotropic growth velocity among all crystallographic orientations of the crystal. Therefore, knowledge of the growth kinetics on other planes that are different from $\{0001\}$, $\{10\bar{1}0\}$ and $\{11\bar{2}0\}$ planes will promote better understanding of the relationship between the anisotropy in the growth kinetics and macroscopic shapes of ice crystals grown from water in real systems.

In this study, we focused on the growth kinetics on a $\{20\bar{2}1\}$ plane of an ice crystal (see Figure 1). The shapes of ice crystals grown from pure water are a hexagonal dendrite or circular disk that has flat $\{0001\}$ planes.[1] However, when ice crystals grow from water including type I antifreeze proteins (AFPs), which are known as proteins to inhibit ice growth,[2] the shape of the ice crystals is a hexagonal bipyramid that has flat $\{20\bar{2}1\}$ planes (Figure 1).[2,8,9] Researchers have speculated that the appearance of the flat $\{20\bar{2}1\}$ planes in the presence of AFPs originates from stronger binding of AFPs to the $\{20\bar{2}1\}$ planes than to other crystallographic planes.[8,9] At present, the most practical way to confirm this speculation is to investigate the anisotropy in the growth kinetics among $\{20\bar{2}1\}$ and other planes both in the presence and absence of AFPs by means of simulations, and to compare the results of the anisotropic growth kinetics between in the presence and absence of AFPs. Thus, knowledge of the growth kinetics on a $\{20\bar{2}1\}$ plane of an ice crystal in the absence of AFPs is also important for understanding of the shapes of ice crystals grown from water including AFPs.

In this study, we carried out an MD simulation to investigate the growth kinetics on the interface between a $\{20\bar{2}1\}$ plane of an ice crystal and pure water (hereafter the $\{20\bar{2}1\}$ interface). The growth of several molecular layers was clearly observed on the $\{20\bar{2}1\}$ interface, like the growth on the interface for $\{0001\}$, $\{10\bar{1}0\}$ and $\{11\bar{2}0\}$ planes (the $\{0001\}$, the $\{10\bar{1}0\}$ and the $\{11\bar{2}0\}$ interfaces), which was observed in a previous simulation study.[7] In this paper, the growth velocity on the $\{20\bar{2}1\}$ interface, the structure of the $\{20\bar{2}1\}$ interface during growth and the growth kinetics on the $\{20\bar{2}1\}$ interface, which were analyzed using the simulation data, will be presented.

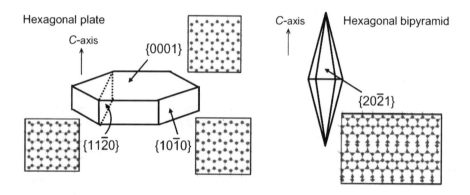

Figure 1 *The $\{0001\}$, the $\{10\bar{1}0\}$, the $\{11\bar{2}0\}$ and the $\{20\bar{2}1\}$ planes of ice I_h. The arrangements of oxygen atoms in those planes are also shown.*

2 SIMULATION METHOD

Figure 2(a) shows the simulation system used, which is a two-phase system containing ice and liquid water. Periodic boundary conditions were imposed in all three directions of the system. At the beginning of the simulation, all H_2O molecules in the ice were arranged at the lattice sites of ice I_h. The ice was oriented in the system so that the $\langle 20\bar{2}1 \rangle$ direction of the ice corresponded to the z direction.

The intermolecular interaction between pairs of H_2O molecules was estimated using a six-site model,[10] which was developed for simulations of ice and water near 0 °C. Recently, using the six-site model, the growth of ice on the $\{0001\}$, the $\{10\bar{1}0\}$ and the $\{11\bar{2}0\}$ interfaces near 0 °C was successfully observed in several MD simulations.[6,7,11] Therefore, we believe that the model is also suitable for the present MD simulation of the growth of ice on the $\{20\bar{2}1\}$ interface.

The molecular geometry of the six-site model is given in Figure 2(a). In the six-site model, there are six interaction sites; an oxygen atom (O), two hydrogen atoms (H), two lone-pair sites (L), and a site M located on the bisector of $\angle HOH$ (=108°). $\angle LOL$ is 111°. OH, OM and OL distances are 0.098, 0.023 and 0.08892 nm, respectively. Point charges of $0.477e$, $-0.044e$ and $-0.866e$ are placed on H, L and M, respectively.

The intermolecular interaction in the six-site model is represented by the sum of the Coulomb potentials acting among the point charges, plus the sum of the Lennard-Jones (LJ) potentials $(=4\varepsilon\{(\sigma/r)^{12}-(\sigma/r)^{6}\}$, where r is the site-site distance) acting on O and H. The values of ε / k_B and σ for the O–O LJ potential are 85.9766 K and 0.3115 nm, respectively, and the values for the H–H LJ potential are 13.8817 K and 0.0673 nm, respectively. The values for the O–H LJ potential, 34.5471 K and 0.1894 nm, are determined using the Lorentz-Berthelot rules.

Following previous studies,[6,7] the intermolecular interaction was modified in the intermolecular distance between 0.95 and 1 nm using a switching function.[12] In this study, we assumed that use of such a switching function did not influence significantly the growth kinetics on the $\{20\bar{2}1\}$ interface and the structure of the $\{20\bar{2}1\}$ interface. More extensive studies about the effect of the long-range interaction on the growth kinetics and the interface structure are required to confirm this assumption.

Computation was carried out using a leap-frog algorithm proposed by Fincham[13] with a time step of 1 fs. Temperature and pressure were maintained at 268 K and 1 atm, respectively, using a method proposed by Berendsen *et al.*[14] As shown in Figure 2(b), the growth of several molecular layers was clearly observed on the $\{20\bar{2}1\}$ interface within the total run of 4 ns.

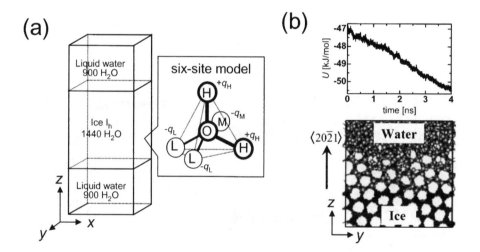

Figure 2 *(a) Illustration of the simulation system used and the molecular geometry of a H₂O molecule in the six-site model. (b) The potential energy, U, of the system as a function of time (upper panel) and snapshots of H₂O molecules near the interface at 2.0 ns. The black and gray spheres, respectively, show the H₂O molecules that were included in the ice and the water at the beginning of the simulation.*

3 RESULTS AND DISCUSSIONS

3.1 Growth velocity

The growth velocity on the $\{20\bar{2}1\}$ interface, which we estimated from the slope of the potential energy, U, as a function of time (the upper panel of Figure 2(b)),[7] was 49 ± 9 cm/s. This value was nearly equal to the growth velocities on the $\{10\bar{1}0\}$ and the $\{11\bar{2}0\}$ interfaces (47 ± 14 cm/s and 40 ± 18 cm/s, respectively), and was much larger than the growth velocity on the $\{0001\}$ interface (27 ± 12 cm/s).[7] The larger growth velocities on the $\{10\bar{1}0\}$, the $\{11\bar{2}0\}$ and the $\{20\bar{2}1\}$ interfaces than on the $\{0001\}$ interface are qualitatively consistent with the experimental result, which showed that macroscopic shapes of ice crystals grown from water were hexagonal dendrites or circular disks that have flat $\{0001\}$ planes.[1]

3.2 Interface structure

We analyzed the distribution of "solid-like" H_2O molecules (sl-H_2O molecules) near the interface to investigate the interface structure. The sl-H_2O molecules were defined as the molecules that have four nearest neighbors and are connected by a hydrogen bond (HB) with each of the nearest neighbors.[7] In this study, in order to search for the H_2O molecules that were in a stable solid-like state for a long time, the sl-H_2O molecules were analyzed using the time-averaged coordinates of H_2O molecules over a period of 0.1 ns.

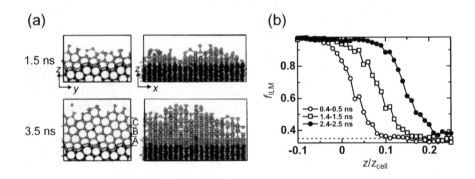

Figure 3 *(a) The snapshots of the sl-H_2O molecules near the interface at 1.5 and 2.5 ns. The dashed lines in the left-hand panels show the geometry of the interface at the beginning of the simulation. (b) The fractions of the sl-H_2O molecules, f_{sl}, along the z direction at three different periods. z_{cell} is the dimension of the system in the z component. f_{sl} for ice is unity and f_{sl} for liquid water is 0.35, which was obtained from a separate MD simulation of bulk liquid water at 268 K and 1 atm (the dotted lines). The origin of z/zc_{ell} was set to the position of the interface at the beginning of the simulation, and the regions in which $z/zc_{ell} > 0$ and $z/zc_{ell} < 0$, respectively, show the ice and the water at the beginning of the simulation.*

Figure 3(a) shows snapshots of the sl-H_2O molecules on the interface at 1.5 and 3.5 ns, which, respectively, were analyzed using time-averaged coordinates over periods of 1.4–1.5 ns and 3.4–3.5 ns. The black and gray spheres, respectively, show the molecules that were included in the ice and the water at the beginning of the simulation. It can clearly be seen that the interface structure during growth was geometrically rough. The diffuse structure of the interface, which can be seen in Figure 3(b), also supports that the interface had a geometrically rough structure. Thus, we conclude that the $\{20\bar{2}1\}$ interface during growth has a geometrically rough structure, like the $\{10\bar{1}0\}$ and the $\{11\bar{2}0\}$ interfaces.

3.3 Growth kinetics

In order to elucidate the growth kinetics on the $\{20\bar{2}1\}$ interface, the water region near the interface was partitioned into three sliced layers, which are named as A, B and C from the one nearest to the interface, as shown in Figure 3(a). The thickness of each layer was approximately 0.38 nm. Then, we analyzed the time-sequence of the hydrogen-bonded networks (HB-networks) in each layer.

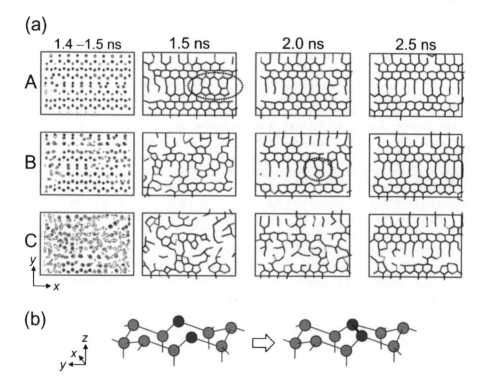

Figure 4 *(a) The HB-networks in layers A, B and C at 1.5, 2.0 and 2.5 ns. Existence probabilities of molecules in each layer during a period of 1.4–1.5 ns are also shown. The probabilities were represented by contour lines in the x–y plane. Dense contour lines indicate high existence probabilities. (b) Schematic illustrations of the 10H_2O patch (left-hand illustration) and the 5-7H_2O patch (right-hand illustration).*

Figure 4 shows the HB-networks in layers A, B and C at 1.5, 2.5 and 3.5 ns. Existence probabilities of molecules in those layers at a period of 1.4–1.5 ns are also shown. It can be seen that the HB-networks in each layer gradually rearranged from a disordered structure to the structure of the $\{20\bar{2}1\}$ plane as time passed. Moreover, since the HB-networks in liquid water spread three-dimensionally in those layers, the rearrangement of the HB-networks occurred in those three layers simultaneously. Thus, we conclude that the growth on the $\{20\bar{2}1\}$ interface occurred by the kinetics of the rearrangement of three dimensional HB-networks, which is essentially the same as the growth kinetics on the $\{10\bar{1}0\}$ and the $\{11\bar{2}0\}$ interfaces.[7]

However, the rearrangement of the HB-networks on the $\{20\bar{2}1\}$ interface was slightly more complicated than the rearrangement on the $\{10\bar{1}0\}$ and the $\{11\bar{2}0\}$ interfaces. As can be seen in Figure 2, when all H_2O molecules are arranged at the ideal lattice sites of the $\{20\bar{2}1\}$ plane, the HB-network that is projected onto the x–y plane contains two kinds of hexagonal patches. One is the hexagonal patch consisting of six H_2O molecules, and the other is the "elongated" hexagonal patch consisting of 10 H_2O molecules (hereafter, $6H_2O$ patch and $10H_2O$ patch, respectively). It should be noted that during growth on the $\{20\bar{2}1\}$ interface, some of the $10H_2O$ patches were formed by a two-step process: first, a pair of pentagon and heptagon (5-7H_2O patch) was formed (see the dotted circles in Figure 4(a)), and second, the 5-7H_2O patch changed to the $10H_2O$ patch.

Figure 4(b) shows schematic illustrations of the molecular arrangement in the $10H_2O$ and the 5-7H_2O patches. In the $10H_2O$ patch, there are two H_2O molecules that make only two HBs with the nearest neighbors (the dark gray molecules in Figure 4(b)). In the 5-7H_2O patch, however, those molecules are located at slightly shifted positions from the ideal lattice sites and are connected to each other by a HB. Thus, all molecules in the 5-7H_2O patch make three or four HBs with the nearest neighbors. This suggests that the formation of the 5-7H_2O patch on the $\{20\bar{2}1\}$ plane is energetically favorable compared to the formation of the $10H_2O$ patch. Of course, in the bulk ice, the structure of the $10H_2O$ patch, in which all molecules are arranged at the ideal lattice sites, is energetically more stable than the structure of the 5-7H_2O patch. This explains why the 5-7H_2O patch was formed first on the interface and it changed to the $10H_2O$ patch as ice grew on the interface.

The formation of the 5-7H_2O patches on the $\{20\bar{2}1\}$ interface implies that the energetically most stable structure of the $\{20\bar{2}1\}$ plane, which corresponds to the potential energy minimum, is partly modified from the structure in which all H_2O molecules are arranged at the ideal lattice sites of the $\{20\bar{2}1\}$ plane. Strictly, four different structures of the $\{20\bar{2}1\}$ plane can be generated by different cut. However, like the structure of the $\{20\bar{2}1\}$ plane that was chosen in this study, all four structures contains H_2O molecules that make two HBs only, which may result in the formation of a partly modified structure as well. More extensive studies are needed to determine the energetically most stable structure of the $\{20\bar{2}1\}$ plane.

4 CONCLUSIONS

An MD simulation of the interface between a $\{20\bar{2}1\}$ plane and water at 268 K and 1 atm was carried out. The growth of several molecular layers was clearly observed on the interface. The simulation results clearly indicated that the interface had a geometrically rough structure. The simulation results also indicated that the growth on the $\{20\bar{2}1\}$ interface occurred by the kinetics of the rearrangement of three-dimensional HB-networks, like the growth kinetics on the $\{10\bar{1}0\}$ and the $\{11\bar{2}0\}$ interfaces.[7] Moreover, during the growth on the $\{20\bar{2}1\}$ interface, a partly modified structure of the $\{20\bar{2}1\}$ plane in which some of the H_2O molecules were not arranged at the ideal lattice sites was formed first, and after that, it changed to the structure in which all H_2O molecules were arranged at the ideal lattice sites of the $\{20\bar{2}1\}$ plane.

The anisotropy in the growth velocity among the $\{20\bar{2}1\}$, the $\{0001\}$, the $\{10\bar{1}0\}$ and the $\{11\bar{2}0\}$ interfaces, which was obtained in the present and the previous simulations,[7] was consistent with that obtained in experimental studies.[1] Thus, the present simulation provided new results that promote better understanding of the relationship between the anisotropy in the molecular-scale growth kinetics and the macroscopic shapes of ice crystals grown from water. Moreover, the growth kinetics on the $\{20\bar{2}1\}$ interface, which was obtained in this simulation study, should also contribute to understanding of the mechanism of ice growth inhibition by AFPs, which are considered to inhibit ice growth on the $\{20\bar{2}1\}$ interface more strongly than other interfaces.

Finally, we would like to comment on simulations for investigating the mechanism of ice growth inhibition by AFPs. So far, several computer simulations have been carried out to investigate the binding energy of a AFP to the surface of a $\{20\bar{2}1\}$ plane in which all H_2O molecules are fixed at the ideal lattice sites.[15-19] However, our simulation results indicated possibility that the energetically most stable structure of the surface of the $\{20\bar{2}1\}$ plane was partly modified from the structure in which all H_2O molecules were arranged at the ideal lattice sites of the $\{20\bar{2}1\}$ plane. Therefore, simulations for investigating the mechanism of ice growth inhibition by AFPs should involve the surface (or interface) of the $\{20\bar{2}1\}$ plane in which all H_2O molecules are free to move. We intend to carry out simulations for investigating the anisotropy in the growth kinetics among several crystallographic planes in which all H_2O molecules are free to move, in the presence of AFPs or other organic molecules.

Acknowledgements

This work was financially supported by a Grant-in-Aid for Scientific Research (C) (No. 17540385) from the Japan Society for the Promotion of Science. This work was done as a Joint Research Program of the Institute of Low Temperature Science, Hokkaido University. The computation in this work was done using the facilities of the Supercomputer Center, Institute of Solid State Physics, University of Tokyo.

References

1   Y. Furukawa and W. Shimada, *J. Cryst. Growth*, 1993, **128**, 234.
2   L.A. Graham, Y.-C. Liou, V.K. Walker and P.L. Davies, *Nature*, 1997, **388**, 727.
3   K. Diehl, S.K. Mitra and H.R. Pruppacher, *Atmos. Environ.*, 1995, **29**, 975.
4   H. Nada and Y. Furukawa, *J. Cryst. Growth*, 1996, **169**, 587.

5   H. Nada and Y. Furukawa, *J. Phys. Chem.*, 1997, **B101**, 6163.
6   H. Nada, J.P. van der Eerden and Y. Furukawa, *J. Cryst. Growth*, 2004, **266**, 297.
7   H. Nada and Y. Furukawa, *J. Cryst. Growth*, 2005, **283**, 242.
8   Y. Yeh and R.E. Finney, *Chem. Rev.*, 1996, **96**, 601.
9   C.S. Strom, X.Y. Liu and Z. Jia, *J. Biol. Chem.*, 2004, **279**, 32407.
10   H. Nada and J.P. van der Eerden, *J. Chem. Phys.*, 2003, **118**, 7401.
11   M.A. Carginano, P.B. Shepson and I. Szleifer, *Mol. Phys.*, 2005, **103**, 2957.
12   M.J. Vlot, J. Huinink and J.P. van der Eerden, *J. Chem. Phys.*, 1999, **110**, 55.
13   D. Fincham, *Mol. Sim.*, 1992, **8**, 165.
14   H.J.C. Berendsen, J.P.M. Postma, W.F. van Gunsteren, A. DiNola and J.R. Haak, *J. Chem. Phys.* 1984, **81**, 3684.
15   M. Lal, A.H. Clark, A.Lips, J.N. Ruddock and D.N.J. White, *Faraday Discuss.*, 1993, **95**, 299.
16   J.D. Madura, A. Wierzbicki, J.P. Harrington, R.H. Maughon, J.A. Raymond and C.S. Sikes, *J. Am. Chem. Soc*, 1994, **116**, 417.
17   A. Cheng and K.M. Merz, Jr., *Biophys. J.*, 1997, **73**, 2851.
18   P. Dalal and F.D. Sönnichsen, *J. Chem. Inf. Comput. Sci.*, 2000, **40**, 1276.
19   A. Jorov, B.S. Zhorov and D.S.C. Yang, *Protein Sci.*, 2004, **13**, 1524.

CHOOSING AN APPROPRIATE WATER MODEL FOR USE IN BIOMOLECULAR SIMULATIONS

D. R. Nutt and J. C. Smith

Computational Molecular Biophysics, Interdisciplinary Centre for Scientific Computing. University of Heidelberg, Im Neuenheimer Feld 368, 69210 Heidelberg, Germany.

1 INTRODUCTION

Even from the start of their development, computers were being applied to the study of chemical systems.[1-7] Early computational models were necessarily crude. For example, rare-gas atoms were first modelled as hard spheres[1] and then later with models including attractive as well as repulsive components, such as the widely-used Lennard-Jones potential.[2] The first simulations of molecular systems were performed on a diatomic molecular liquid in the late 1960s,[3,4] closely followed by the first simulations of liquid water.[5,6]

At around the same time, the computational study of biological systems was beginning with the development of the Consistent Force-Field (CFF) of Lifson and co-workers.[7] With this force-field, the potential energy of any molecule is given by the sum of a number of simple terms describing bond stretching, bond-angle bending, bond twisting, van der Waals interactions and electrostatics:

$$U = \sum_{\text{bonds}} k_b (b - b_0) + \sum_{\text{angles}} k_\theta (\theta - \theta_0) + \sum_{\text{torsions}} k_\phi [1 - \cos(n\phi + \delta)]$$
$$+ \sum_{\text{non-bonded pairs}} \varepsilon \left[\left(\frac{r_0}{r} \right)^{12} - 2 \left(\frac{r_0}{r} \right)^6 \right] + \sum_{\text{partial charges}} \frac{q_i q_j}{r_{ij}} \tag{1}$$

This force-field was then used to carry out energy minimisations on myoglobin and lysozyme,[8] using the structures which had recently become available.[9,10] Since then, the CFF force-field underwent changes and developments (although the form of the original force-field remained virtually unchanged), eventually leading to the now widely used CHARMM[11] and AMBER[12] force-fields, to name but two. Given the functional form of a force-field, it is the subsequent parameterisation which ensures that meaningful results can be obtained. As will be described in greater detail below, the choice of water model to be used in simulations plays a crucial role in the process of force-field parameterisation. Before turning to this aspect, it is first instructive to consider the parameterisation of water models themselves.

1.1   Models for Water

Over the past thirty years, an overwhelming array of different water models have been developed and used in molecular simulations. To date, however, no single water model has been found to give an accurate description of water under all conditions. The recent review by Guillot[13] provides an appraisal of existing water models and suggests some aspects that should be considered in future work.

In the development and parameterisation of water models, two separate approaches can be identified. The first supposes that if the details of the interactions (electrostatic, repulsive and dispersion interactions, for example) between two atoms or molecules can be correctly described, for example by reproducing high-level *ab initio* calculations, the resulting potential will also be suitable for describing larger clusters as well all possible phases. An example of such a potential is provided by the family of anisotropic site–site potentials (ASP) for water developed by Stone and co-workers.[14–16] These so-called *ab initio* interaction potentials, however, suffer from being computationally intensive, leading to limitations in their fields of application (at least with current computational resources). They have nevertheless yielded insight into the structure and properties of small water clusters[17] and the adsorption of water on surfaces.[18–20]

Alternatively, the potential can be parameterised in order to reproduce bulk properties for a chosen phase. Such an approach gave rise to the family of Transferable Intermolecular Potentials (TIP) developed by Jorgensen and co-workers; TIP3P,[21] TIP4P,[21] and TIP5P.[22] These models describe bulk liquid water with an increasing degree of accuracy. However, they are not capable of correctly describing small clusters or other phases since they were not designed or parameterised for this purpose.

Because of this, the computational investigation of other phases, such as water at high pressure or in one of the many ice phases, has required the prior construction of an empirical water model that can describe the features of the chosen phase. This has led to the development of water models such as the MKW (modified Kuwajima–Warshel) model designed to obtain a stable ice XI structure[23] and the GCPM (Gaussian Charge Polarisable Model) designed to describe water over the entire fluid range, including extreme temperatures and pressures.[24]

1.2   Water as a Solvent in Biomolecular Systems

Early biomolecular simulations were carried out either in vacuum or in an environment of fixed dielectric constant in order to reduce the computational expense. In most modern simulations, water is explicitly included in order to describe the system as completely as possible. In some cases, such as very large protein systems, for example myosin,[25] it remains necessary to use one of a range of so-called implicit solvent models such as ACE.[26]

In simulations involving explicit water, it is crucial that a balance exists between water–protein and water–water interactions in order to describe correctly the water–protein interface. This of course depends on the choice of water model to be used with the protein force-field. The developers of most modern force-fields for protein simulations have chosen a specific water model which has then used as the foundation of the force-field parameterisation. In the following we consider the parameterisation of the CHARMM force-field, although other force-fields have used similar approaches.

In both the CHARMM19[27,28] and CHARMM22[29] force-fields, the basis was chosen to be a modified version of the TIP3P water model,[21] since this model provides a satisfactory description of first-shell hydration and the energetics of liquid water while remaining computationally inexpensive. The original model was modified by the addition of van der Waals interaction sites to the H atoms in order to avoid singularities in the integral equation calculations.[27,28] The effects of these additional sites on the properties of the modified water model with respect to the original TIP3P model have been investigated and been found to be small.[30]

The parameters of the peptide backbone were then chosen in such a way that they reproduced the binding energies and structures for a small model system (water interacting with N-methyl acetamide, NMA) calculated using low level quantum mechanical calculations.

1.3 Limitations of the Current Force Field

The TIP3P model (in either the original or modified form) is not a perfect water model. In particular, TIP3P water is found to display too little structuring, with the second peak in the oxygen–oxygen radial distribution function, g_{OO}, missing completely. The isothermal compressibility is too low and the coefficient of the thermal expansion too high. It must also be remembered that the model was parameterised for 1 atm and 25 °C and must be used with caution away from these conditions. This point is reinforced by noting that the freezing temperature of the TIP3P model has recently been calculated to be 146 K.[31]

The TIP3P model is designed for bulk water simulations. Because of this, many-body interactions and polarisation effects are described in an empirical way by exaggerating the dipole moment of each water molecule; the dipole moment of a TIP3P water is 2.35 D, compared to 1.85 D from experiment. Because of this, TIP3P is unable to describe the water dimer minimum energy structure correctly; the binding energy is too high and the O–O distance too short. This suggests that the TIP3P model is not ideal for investigating the details of protein–water interactions, for example, in the case of buried water molecules or water molecules at the protein surface.

In modern simulations, various Ewald summation methods are often used in order to correctly describe the long-range electrostatic interactions. The TIPnP potentials were originally parameterised using truncated Coulomb interactions and using these models with Ewald summation results in changes in both the thermodynamic and kinetic properties.[32]

The approach used in the parameterisation procedure means that once a force-field has been developed, it can only be used together with the selected water model. In principle, use of the CHARMM22 parameter set with water models other than TIP3P may lead to inconsistencies because the water–protein and protein–protein intermolecular parameterisation may not be well balanced.[29] However, whether this is actually the case has not yet been systematically investigated.

It is therefore of significant interest to investigate the behaviour of the CHARMM22 force-field with water models other than the modified TIP3P model. Once this behaviour is understood, it will be possible to choose the most appropriate water model for the simulation being carried out; for example the TIP4P/Ew model when Ewald summation is being used,[32] the TIP5P model if the simulations are being carried out close to the water density maximum at 4 °C,[22] TIP4P/Ice if simulations are to be carried out with ice–water coexistence[33] or even the ASP-W2K model if an accurate description of a small number of surface or buried water molecules are of interest.[16]

In this paper we fix ourselves a more modest aim — to investigate the potential effects of using an unbalanced force-field, a situation which may occur if the CHARMM22 force-field were to be used together with water models other than TIP3P. To this end, we present results of calculations on the water dimer, a benchmark system used in the development of all water models.

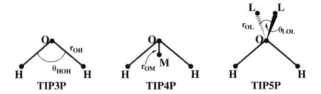

Figure 1   *Structural features of the TIP3P, TIP4P and TIP5P water models*

2 COMPUTATIONAL METHODS

Model calculations were performed using the CHARMM program,[11] version c31b2 together with the CHARMM22 force-field.[29] Geometry optimisations were performed using the Conjugate Gradient algorithm with a tolerance of 2×10^{-5}. To remain consistent with the original parameterisation of the water models, the water molecules were treated as rigid bodies.

2.1 The TIP Family of Water Models

The family of TIPnP water potentials (n = 3, 4 and 5) have the general form[21,22]

$$E_{ab} = \sum_{ij} \frac{q_i q_j e^2}{r_{ij}} + 4\varepsilon \left[\left(\frac{\sigma_O}{r_{OO}} \right)^{12} - \left(\frac{\sigma_O}{r_{OO}} \right)^6 \right] \tag{2}$$

where i and j are the charged sites on molecules a and b separated by a distance r_{ij}, ε_O and σ_O are the van der Waals parameters between the two oxygen sites, which are separated by r_{OO}. The O–H bond lengths are fixed at 0.9572 Å and the H–O–H angle is 105.42°, corresponding to the experimental gas phase values. In the TIP3P model, charges are placed at the O and H atom sites, with a single van der Waals site on O. In the TIP4P model, there is no longer a charge on the O site, instead it is placed at a position corresponding to the molecular centre of mass (M), 0.15 Å from oxygen along the bisector of the H–O–H angle. In the TIP5P model, charges are placed at sites corresponding to lone pair positions (L), 0.7 Å from oxygen. The three models are shown schematically in Figure 1.

In the following, we use the name mTIP3P to differentiate between the original TIP3P model and the CHARMM-modified TIP3P model which has additional van der Waals interaction sites on the H atom. The monomer geometry and parameters are summarised in Table 1. Because of the different definition of the van der Waals interaction energy in CHARMM compared to Equation (2), the σ parameters differ from those presented in the original papers by a factor of $2^{1/6}$.

The TIP family of water potentials represents a useful test-set for investigating the compatibility of alternative water models with the CHARMM22 force-field, since they possess many features common to other water potentials, for example, an interaction site at the centre of mass or at positions corresponding to lone pairs.

Geometry optimisations were performed for linear water dimers, where the O–H...O hydrogen bond is constrained to be linear (see Figure 2 for a definition of the relevant parameters). The water dimers were first constructed using the same water model for donor

Table 1 *Monomer geometry and parameters for the TIPnP water models*

	TIP3P	mTIP3P	TIP4P	TIP5P
q_H / e	0.417	0.417	0.520	0.241
q_O / e	-0.834	-0.834		
q_M / e			-1.04	
q_L / e				-0.241
$\sigma_{OO} / Å$	3.5364	3.5364	3.5399	3.5021
$\varepsilon_{OO} / kcal/mol$	0.1521	0.1521	0.1550	0.16
$\sigma_{HH} / Å$		0.4490		
$\varepsilon_{HH} / kcal/mol$		0.0460		
$r_{OH} / Å$	0.9572	0.9572	0.9572	0.9572
$\theta_{HOH} / °$	104.52	104.52	104.52	104.52
$r_{OM} / Å$			0.15	
$r_{OL} / Å$				0.70
$\theta_{LOL} / °$				109.47

and acceptor molecules (homodimers, in order to illustrate a balanced potential) and then with donor and acceptor molecules described with different water models (heterodimers, a potentially unbalanced description).

3 RESULTS

In Table 2 we present the binding energies and structures for the linear water dimers obtained with the different water models and the different water model combinations.

For the homodimers, it can be noted that the O–O distance is up to 0.3 Å shorter than is found in experiment or in *ab initio* calculations. Likewise, the binding energy is significantly overestimated. These observations are due to the enhanced dipole moment built in to these water molecules to correct for the neglect of polarisation. It can be seen that the addition of van der Waals sites on hydrogen in the mTIP3P model leads to a slight increase in the O–O distance with respect to the original TIP3P definition.

It is interesting to note the significant asymmetry in the results for the heterodimers, in particular with dimers involving a TIP5P molecule. This behaviour can be accounted to the lone pair sites of the TIP5P model. The TIP3P and TIP4P models are quite similar in design, with the main difference being the displacement of the negative charge site along the bisector of the H–O–H angle towards the centre of mass. This small change results in a small asymmetry in the dimer structures and energies when the models are combined. The TIP5P model, however, possesses lone pair sites which create tetrahedrality around the oxygen site. (This is in fact the main reason that the TIP5P model gives a better structural

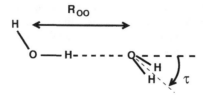

Figure 2 *Structure of the linear water dimer*

Table 2   *Geometry and binding energy for optimised linear water dimers*

Donor	Acceptor	R_{OO} / Å	$\tau/\,^\circ$	ΔE / kcal/mol	E_{vdw}
Homo-dimers					
TIP3P	TIP3P	2.75	27.3	-6.50	1.74
mTIP3P	mTIP3P	2.77	27.4	-6.55	1.50
TIP4P	TIP4P	2.75	46.2	-6.23	1.80
TIP5P	TIP5P	2.68	51.4	-6.78	2.37
Hetero-dimers					
mTIP3P	TIP4P	2.79	50.3	-5.88	1.32
TIP4P	mTIP3P	2.72	21.0	-7.05	2.10
mTIP3P	TIP5P	2.63	51.7	-9.06	3.50
TIP5P	mTIP3P	2.80	30.3	-5.27	1.18
TIP4P	TIP5P	2.53	51.4	-10.60	6.03
TIP5P	TIP4P	2.83	48.8	-4.74	1.00
HF/6-31G*		2.98	56.2	-5.64	
Experiment[a]		2.98 ± 0.02	57 ± 10	-5.4 ± 0.5	

[a] Reference 34 for $(D_2O)_2$

description of bulk water and ice, since it explicitly incorporates the tetrahedrality into the model.) However, the negative charges at the lone pair sites means that the interaction energy with the (no-longer appropriate) charge at the oxygen site of the TIP3P or TIP4P model is overestimated when the sites are adjacent (TIP5P acceptor molecule) or underestimated when the sites are further apart (TIP5P donor molecule). This is a clear example of an "*unbalanced*" potential.

4 CONCLUSIONS

By taking the example of the application of a series of water models to the water dimer, we have demonstrated that the description of a system can be unbalanced if the models are not parameterised for use with each other. The degree of imbalance depends on the difference between the models, a quantity which is difficult to quantify at the moment, although the calculations presented here provide a starting point for such an analysis. This has repercussions on the field of biomolecular simulations, since these results indicate that use of the CHARMM22 force-field with water models other than mTIP3P may lead to imbalance between the water–water and water–protein interactions, as suggested in the CHARMM22 paper.[29] In this context, it is interesting to note that the OPLS force-field[35] claims to be compatible with the TIP3P, TIP4P and SPC water models. To our knowledge, no biomolecular force-field developed to date claims to be compatible with the TIP5P or other more complex models. In cases such as the interaction of biomolecules with ice, either in organisms, in the atmosphere, or in interstellar space, it is clear that a more accurate model than TIP3P will be required to ensure a satisfactory description of the ice phase. To assess whether use of a water model such as TIP5P with the CHARMM22 force-field introduces a significant compatibility problem will require a detailed and careful study of proteins solvated with different water models.

Acknowledgements

We thank the Swiss National Science Foundation for funding through the award of an

Advanced Research Fellowship.

References

1   N. Metropolis, A. Rosenbluth, M.N. Rosenbluth, A. Teller and E. Teller, *J. Chem. Phys.*, 1953, **21**, 1087.
2   W.W. Wood and F.R. Parker, *J. Chem. Phys.*, 1957, **27**, 720.
3   G.D. Harp and B.J. Berne, *J. Chem. Phys.*, 1968, **49**, 1249.
4   B.J. Berne and G.D. Harp, *Adv. Chem. Phys.*, 1970, **17**, 63.
5   J.A. Barker and R.O. Watts, *Chem. Phys. Lett.*, 1969, **3**, 144.
6   A. Rahman and F.H. Stillinger, *J. Chem. Phys.*, 1971, **55**, 3336.
7   M. Bixon and S. Lifson, *Tetrahedron*, 1967, **23**, 769.
8   M. Levitt and S. Lifson, *J. Mol. Biol.*, 1969, **46**, 269.
9   J.C. Kendrew, R.E. Dickerson, B.E. Strandberg, R.G. Hart, D.R. Davies, D.R. Phillips, D.C. Phillips and V.C. Shore, *Nature*, 1960, **185**, 422.
10 C.C.F. Blake, G.A. Mair, A.C.T. North, D.C. Phillips and V.R. Sarma, *Proc. Roy. Soc. Lond. B*, 1967, **167**, 365.
11 B.R. Brooks, R.E. Bruccoleri, B.D. Olafson, D.J. States, S. Swaminathan and M. Karplus, *J. Comp. Chem.*, 1983, **4**, 187.
12 S.J. Weiner, P.A. Kollman, D.A. Case, U.C. Singh, C. Ghio, G. Alagona, S. Profeta and P. Weiner, *J. Am. Chem. Soc.*, 1984, **106**, 765.
13 B. Guillot, *J. Mol. Liq.*, 2002, **101**, 219.
14 C. Millot and A.J. Stone, *Mol. Phys.*, 1992, **77**, 439.
15 C. Millot, J.-C. Soetens, M.T.C. Martins Costa, M.P. Hodges and A.J. Stone, *J. Phys. Chem. A*, 1998, **102**, 754.
16 D.R. Nutt and A.J. Stone, *J. Chem. Phys.*, 2002, **117**, 800.
17 M.P. Hodges, A.J. Stone and S.S. Xantheas, *J. Chem. Phys. A*, 1997, **101**, 9163.
18 O. Engkvist and A.J. Stone, *J. Chem. Phys.*, 1999, **110**, 12089.
19 V. Sadtchenko, G.E. Ewing, D.R. Nutt and A.J. Stone, *Langmuir*, 2002, **18**, 4632.
20 D.R. Nutt and A.J. Stone, *Langmuir*, 2004, **20**, 8715.
21 W.L. Jorgensen, J. Chandrasekhar, J.D. Madura, R.W. Impey and M.L. Klein, *J. Chem. Phys.*, 1983, **79**, 926.
22 M.W. Mahoney and W.L. Jorgensen, *J. Chem. Phys.*, 2000, **112**, 8910.
23 V. Buch, P. Sandler and J. Sadlej, *J. Phys. Chem. B*, 1998, **102**, 8641.
24 P. Paricaud, M. Predota, A.A. Chialvo and P.T. Cummings, *J. Chem. Phys.*, 2005, **122**, 244511.
25 S. Koppole, J.C. Smith and S. Fischer, *J. Mol. Biol.*, 2006, **361**, 604.
26 M. Schaefer and M. Karplus, *J. Phys. Chem.*, 1996, **100**, 1578.
27 W. E. Reiher, III, PhD Thesis, Harvard University, 1985.
28 E. Neria, S. Fischer and M. Karplus, *J. Chem. Phys.*, 1996, **105**, 1902.
29 A.D. MacKerell, Jr., D. Bashford, M. Bellott, R.L. Dunbrack, Jr., J.D. Evanseck, M.J. Field, S. Fischer, J. Gao, H. Guo, S. Ha, D. Joseph-McCarthy, L. Kuchnir, K. Kuczera, F.T.K. Lau, C. Mattos, S. Michnick, T. Ngo, D.T. Nguyen, B. Prodhom, W.E. Reiher, III, B. Roux, M. Schlenkrich, J.C. Smith, R. Stote, J. Straub, M. Watanabe, J. Wiorkiewicz-Kuczera, D. Yin and M. Karplus, *J. Phys. Chem. B*, 1998, **102**, 3586.
30 P. Mark and L. Nilsson, *J. Phys. Chem. A*, 2001, **105**, 9954.
31 C. Vega, E. Sanz and J.L.F Abascal, *J. Chem. Phys.*, 2005, **122**, 114507.
32 H.W. Horn, W.C. Swope, J.W. Pitera, J.D. Madura, T.J. Dick, G.L. Hura and T. Head-Gordon, *J. Chem. Phys.*, 2004, **120**, 9665.

33   J.L.F. Abascal, E. Sanz, R. García Fernández and C. Vega, *J. Chem. Phys.*, 2005, **122**, 234511.
34   R.S. Fellers, C. Leforestier, L.B. Braly, M.G. Brown and R.J. Saykally, *Science*, 1999, **284**, 945.
35   W.L. Jorgensen and J. Tirado-Rives, *J. Am. Chem. Soc.*, 1988, **110**, 1657.

MICRO-RAMAN STUDY OF AIR CLATHRATE HYDRATES IN POLAR ICE FROM DOME FUJI, ANTARCTICA

H. Ohno[*] and T. Hondoh

Institute of Low Temperature Science, Hokkaido University, N19 W8, Sapporo 060-0819, Japan
[*]Present address: Center for Hydrate Research, Department of Chemical Engineering, Colorado School of Mines, Golden, Colorado 80401, USA

1 INTRODUCTION

Bubbles of atmospheric air are trapped in polar ice when snow changes into ice in the upper parts of polar ice sheets. These bubbles are progressively compressed at higher depths and are gradually transformed into air clathrate hydrates below a depth where the overburden pressure exceeds the dissociation pressure of clathrate hydrate.[1]

Air bubbles and clathrate hydrates in polar ice cores have attracted considerable interest because they provide the most direct record of past atmospheric gas compositions (e.g., Raynaud et al.[2]). However, the processes of air clathration in polar ice sheets should be taken into account when considering gas analyses. It is known that extreme fractionation of gases in air inclusions occurs when air bubbles change into clathrate hydrates by the diffusive mass transfer of air molecules between bubbles and hydrates.[3, 4] To estimate the effects of hydrate formation on the distribution of atmospheric gases in deep ice, it is essential to understand the structure and physical properties of natural air hydrates.

The crystal structure of natural air clathrate hydrates has been determined by X-ray diffraction studies.[5-7] Air hydrate forms clathrate structure II. A unit cell of structure II consists of 136 water molecules that form eight large cages (16-hedra) and sixteen small cages (12-hedra). These cages are filled with guest gas molecules. The two main constituents of air—N_2 and O_2, which account for more than 99% of the atmospheric composition—can enter both cavities. Diffraction studies also provide information about cage fillings. The structure refinement of natural air hydrates has revealed that approximately 80−90% of water cages are occupied by guest molecules.[5-7] However, it is difficult to estimate the cage occupancies of N_2 and O_2 molecules separately from diffraction data because the scattering factors of the two types of molecules are close.

Raman spectroscopy is a valuable tool for investigating the composition and distribution of guest molecules in clathrate hydrate. The relative guest composition and cage occupancy can be determined from the intensities of Raman bands. Further, the absolute cage fillings can be estimated from the obtained relative cage occupancy by using a statistical thermodynamic theory.[8, 9] By employing this technique, many researchers have studied the N_2/O_2 gas composition ratios of natural hydrates and obtained important results. It was found that the ratio differs from sample to sample,[10-14] and subsequent studies have shown that this fact is due to selective diffusive flaxes of N_2 and O_2 from bubbles to hydrates.[3, 4] However, no study has obtained information about the cage fillings of air hydrates due to a small difference in the vibrational frequencies between guest molecules in different cavities. Champagnon *et al.*[15] have found that the peaks of Raman signals from N_2 and O_2 molecules of natural hydrates shift slightly when the direction of polarization of incident light changes; they have attributed these findings to guest molecules stretching in large and small cages. Chazallon *et al.*[16] have reported that a Raman band of O_2 molecules in synthetic air hydrate is asymmetrical; the asymmetrical band shape of the O_2 stretching was tentatively attributed to a different guest molecule environment in both types of cages.

We performed high-resolution micro-Raman analyses of air clathrate hydrates in deep ice cores in order to detect signals from O_2 and N_2 guest molecules in large and small water cages separately and investigate the cage fillings of guest molecules within the crystals. The ice cores used in this study were recovered at the Dome Fuji station in East Antarctica.[17] At this site, the transformation from air bubbles to clathrate hydrates begins at approximately 500 m and completes at approximately 1200 m.[18, 19] The recovered ice cores were stored at 223 K; this temperature was sufficiently low to prevent the decomposition of air hydrate.

2 METHOD

2.1 Sample Preparation

We studied air hydrate crystals in the Dome Fuji ice obtained at depths of 780, 1570, and 2450 m (Figure 1). In a cold room at −20 °C, the ice cores containing the hydrate crystals were cut into ice cubes with a volume of $5 \times 5 \times 5$ mm^3 by using a band saw; the surfaces of these cubes were then planed with a microtome.

Figure 1   *Micrograph of air clathrate hydrate in the Dome Fuji ice core obtained at a depth of 1570 m. The scale bar is 100 μm.*

2.2 Experimental Setup

The samples were measured at 243 K and in the range of 15–37 K using two cooling apparatuses. For high temperature measurement, the ice cube was fixed in a cold chamber on the x-y translation stage of a microscope. This chamber also contained an objective lens used for micro-Raman analysis, which enabled direct measurements of the samples without an optical window. During the measurement, the chamber was maintained at 243 ± 0.5 K by regulating the flow rate of cold N_2 gas. The low temperature analysis was performed by using a cryostat equipped with a sample positioning system whose operation is based on the Joule-Thomson effect of He gas (Daikin, UV202CL). A gas compressor of the cryostat was switched off because its operation resulted in vibrations of the sample, thereby disturbing the measurement. Therefore, the temperature of the sample increased from 15 K to 37 K during the measurement.

To improve the spectral resolution, a long-wavelength light-sauce (NEC, GLG-5800) was used. Laser light with a wavelength of 632.8 nm and power of 30 mW was focused to a diameter of approximately 2 μm on the hydrate crystal by using the long-working-distance objective lens with a focal length of 13 mm (Mitutoyo, M Plan Apo 50×). Backscattered Raman spectra were obtained in additive mode using a triple monochromator (Jobin-Yvon, T64000) equipped with 1800 gr./mm gratings (Jobin-Yvon, Plane Holographic Package Aberration Correction Grating), a CCD detector (Jobin-Yvon, Spectraview-2D), and a micro-analysis system (Jobin-Yvon, RSM-500). The Raman frequencies of O_2 and N_2 were calibrated using neon emission lines with wavelengths of 703.24 nm (1582.50 cm^{-1}) and 743.89 nm (2359.53 cm^{-1}), respectively. The measurement errors of the Raman frequencies were better than ±0.1 cm^{-1}. The spectral resolution estimated from the nominal liner dispersion of the spectrometer and entrance slit width (40 μm) was 0.12 cm^{-1} at 750 nm.

3 RESULTS

3.1 Estimation of N_2/O_2 Ratio

When the laser was focused on the hydrate crystal, Raman signals were observed at approximately 1548 and 2324 cm^{-1}. Since the results of the three specimens were almost the same, the spectra of one sample (1570 m) are shown in Figure 2. Based on previous Raman studies on air hydrates,[10, 16] it was considered that the signals at approximately 1548 and 2324 cm^{-1} originated from the O_2 and N_2 stretching modes of the hydrates, respectively. In comparison with the Raman frequencies of ambient air at room temperature, the O_2 and N_2 peak energies decreased by approximately 7 and 6 cm^{-1}, respectively. The N_2/O_2 composition ratios of the hydrates were estimated by using the following formula:

$$Y_{N_2} / Y_{O_2} = \beta I_{N_2} / I_{O_2} \qquad (1)$$

where Y_{N2} and Y_{O2} are the mole fractions and I_{N2} and I_{O2} are the scattering intensities of N_2 and O_2 in air hydrate, respectively. β is a constant that is related to both the difference in Raman-scattering cross sections between the two gases and the wavelength-dependent response of the spectrometer. In this experiment, since the ratio of the Raman intensities of N_2 and O_2 of the ambient atmospheric air with $Y_{N2}/Y_{O2} = 3.7$ was 2.6, the value of β is 1.4. Assuming that the ratio of the scattering cross section of N_2 to that of O_2 does not significantly change by clathration, the composition ratios were estimated as 1.8 at 780 m, 2.0 at 1570 m, and 3.0 at 2450 m. These values were significantly less than the atmospheric ratio of 3.7. Gas fractionation due to hydrate formation[3, 4] was probably responsible for this difference.

3.2 Feature of Spectra

At 243 K, the spectrum of O_2 molecules had a strong band at 1547.7 cm^{-1} and a weak band at approximately 1549.7 cm^{-1}. The strong band was asymmetrical, as reported by Chazallon *et al.*[16]. The spectrum of N_2 molecules had a bimodal shape and its peak positions were at 2323.3 and 2324.0 cm^{-1}. In the temperature range of 15–37 K, the observed Raman bands became narrower. Consequently, the main band of O_2 was resolved into two lines at 1547.6 and 1548.5 cm^{-1}. On the other hand, the weak band was not observed in the low temperature experiment. The two peaks of N_2 stretching modes were observed at 2323.4 and 2323.8 cm^{-1} at lower temperatures.

3.3 Deconvolution of Spectra

The deconvolution of the spectra was achieved by using a peak fitting program (Galactic Industries, GRAMS/AI). The Raman bands were fitted with mixed Gaussian-Lorentzian functions. Results of the deconvolution are shown in Figure 2 and Table 1. For the analyses, we assumed that the main band of O_2 molecules consists of two lines at 243 K. Results show that the peak area of O_2 band 2 in the temperature range of 15-37 K was several percent larger than that at 243 K in all samples, implying this peak included the missing third component. On the other hand, since there was no systematic change in the peak area of N_2 bands with temperature, difference between s values of the high and low temperature measurements might be due to errors of analysis.

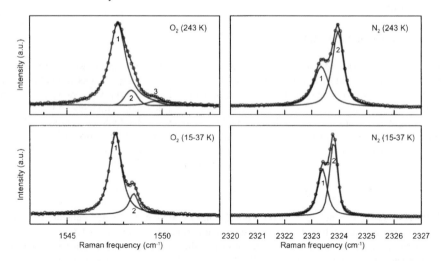

Figure 2   *Raman spectra of the air hydrate crystal in Dome Fuji ice obtained at 1570 m. Lines represent results of peak-fitting analysis. Raman bands are numbered for reference.*

Table 1 *Peak gravity center, v (cm⁻¹), peak area, s (%), peak height, h (a.u.), peak width, w (cm⁻¹), and Lorentz %, l (%), obtained by spectra deconvolution*

sample	T (K)	band	O₂ v	s	h	w	l	band	N₂ v	s	h	w	l
780 m	243	1	1547.71	84.0	1.00	1.11	92.5	1	2323.38	48.8	0.61	0.69	88.1
		2	1548.38	13.9	0.22	0.86	84.0	2	2323.98	51.2	1.00	0.46	70.5
		3	1549.71	3.1	0.05	0.70	34.4						
	15-37	1	1547.57	83.0	1.00	0.87	86.2	1	2323.39	49.8	0.71	0.47	80.6
		2	1548.48	17.0	0.28	0.65	81.2	2	2323.84	50.2	1.00	0.37	45.8
		3	–	–	–	–	–						
1570 m	243	1	1547.68	88.2	1.00	1.04	94.6	1	2323.34	42.5	0.53	0.64	81.6
		2	1548.39	8.4	0.18	0.74	0.0	2	2323.95	57.5	1.00	0.48	67.0
		3	1549.58	3.4	0.05	0.84	75.2						
	15-37	1	1547.56	83.1	1.00	0.78	88.4	1	2323.41	45.7	0.63	0.41	77.6
		2	1548.51	16.9	0.26	0.58	95.7	2	2323.81	54.3	1.00	0.34	47.7
		3	–	–	–	–	–						
2450 m	243	1	1547.71	87.3	1.00	1.07	90.8	1	2323.34	47.4	0.58	0.65	87.7
		2	1548.43	11.2	0.19	0.75	76.0	2	2323.93	52.6	1.00	0.44	63.9
		3	1549.68	1.5	0.03	0.80	0.0						
	15-37	1	1547.54	84.7	1.00	0.81	88.9	1	2323.41	46.4	0.66	0.41	84.3
		2	1548.48	15.3	0.24	0.66	64.0	2	2323.82	53.6	1.00	0.36	46.1
		3	–	–	–	–	–						

4 DISCUSSION

4.1 Origin of Band Splitting

Since guest molecules are isolated by water cages, the host-guest interaction is considered the main reason for the change in the vibrational energy of these molecules by clathration. In structure II hydrate, gas molecules are enclathrated in two types of cavities with different sizes and shapes. The Raman bands of guest molecules in different host cages must be different due to the difference in the interaction with the surrounding water. Therefore, it appears reasonable to assume that the splitting of the two main bands of O_2 and the bands of N_2 originate in guest distribution between the large and small cavities.

The observation of the third band of O_2 is in contrast to the expectation. A weak component comparable to this band has been found in pure O_2 hydrate measured at 200 bar and about 115 K.[20] This finding may be explained by assuming O_2 molecules that doubly occupy the large cages. Although it is generally considered that a large cage cannot contain more than one O_2 molecule because the ratio of the van der Waals diameter of O_2 to the cavity diameter of the cage is approximately 0.63, increasing evidence shows that a certain percentage of the large cavities are doubly occupied by guest molecules at high pressures.[21, 22] Another possible explanation of the weak band is that multiple guest sites in a cavity cause the band splitting. Diffraction studies of oxygen[23] and air hydrate[6] have shown that guest molecules are confined at the centre part of cavities in small cages while those are wandering within large cages, indicating that there are multiple guest sites in a large cavity. Since a large cage does not have a completely spherical shape, there is possibility that some non-equivalent sites for O_2 molecules exist in the cavity.

4.2 Estimation of Cage Occupancy

In order to assign the Raman bands and determine the absolute occupancies of O_2 and N_2 molecules in the small and large cavities, we use the statistical thermodynamic expression derived by van der Waals and Platteeuw.[24] Let us consider an equilibrium state of the ice-hydrate-gas system. Then, the difference between the chemical potential of water molecules in ice, μ_w *(i)*, and that in a hypothetical empty lattice of structure II hydrate, μ_w *(h°)*, is given by

$$\mu_w(i) - \mu_w(h^o) = -RT[\ln(1 - \theta_l) + 2\ln(1 - \theta_s)]/17 \tag{2}$$

where θ_l and θ_s are the fractional occupancies of the large and small cavities, respectively. For air hydrate, the relative cage occupancy, θ_l/θ_s, can be related to the intensities of the Raman bands according to the following equation, under the assumption that the scattering cross sections of N_2 and O_2 molecules are independent of the type of the surrounding water cage:

$$\theta_l / \theta_s = (\theta_{l.O_2} + \theta_{l.N_2})/(\theta_{s.O_2} + \theta_{s.N_2}) = 2(I_{l.O_2} + \beta I_{l.N_2})/(I_{s.O_2} + \beta I_{s.N_2}) \tag{3}$$

where $\theta_{l.O2}$ and $\theta_{l.N2}$ are the large cage occupancies and $\theta_{s.O2}$ and $\theta_{s.N2}$ are the small cage occupancies of O_2 and N_2 molecules, respectively. $I_{l.O2}$ and $I_{l.N2}$ are the Raman intensities of O_2 and N_2 in the large cavities, while $I_{s.O2}$ and $I_{s.N2}$ are those in the small cavities, respectively. By using Equations 1 and 2, the μ_w *(i)* – μ_w *(h°)* value of 937 J/mol,[25] and the results of the deconvolution of the spectra measured in the range of 15–37 K, we calculate the absolute occupancies for four possible cases of the band assignment listed in Table 2. For the calculation, we assume that the ice-hydrate-gas system is in equilibrium when hydrates are formed in the ice sheet and the cage fillings of guest molecules in the hydrates are preserved after hydrate formation. The value of *T* used for the calculation is 227 K, which is the average ice temperature of the bubble to hydrate transition zone at Dome Fuji.[26]

Table 2

	O_2 band 1	O_2 band 2	N_2 band 1	N_2 band 2
Four possible cases of band assignment				
assignment 1	large	small	large	small
assignment 2	large	small	small	large
assignment 3	small	large	large	small
assignment 4	small	large	small	large

4.3 Band Assignment

Table 3 shows the calculations. The calculated values are significantly less than results of previous diffraction studies with the exception of those calculated for assignment 3. θ values obtained by analyzing the X-ray diffraction data of natural air hydrates in polar ice range from 0.77 to 0.91 and average 0.84.[5-7] According to the neutron powder diffraction study of synthetic O_2, N_2, and air hydrates,[22] the fractional occupancies of the large and small cages of air hydrate are 1.06 and 0.84, respectively, at the pressure close to dissociation and temperature of 273 K. Further, assignment 3 is consistent with Raman analyses of pure O_2 and N_2 hydrates. Chazallon[20] has reported that an O_2 vibron peak of O_2 hydrate has a clear shoulder on the high frequency side, whereas a N_2 vibron peak of N_2 hydrate has a weak shoulder on the low frequency side. Considering the fact that there are two times as many small cages as large ones in these hydrates (clathrate structure II), these results indicate that in oxygen hydrate Raman frequency of guest molecules in small cavities is lower than that in large cavities, and in nitrogen hydrate the reverse is true. From these results, we infer that assignment 3 is correct. In order to confirm this assignment, high-resolution Raman analyses of pure components are underway.

Table 3   *Cage occupancy calculated for four possible cases of band assignment θ is the total cage occupancy. θ_l and θ_s represent the fractional occupancies of large and small cavities, respectively. $\theta_{l,O2}$ and $\theta_{l,N2}$ denote the fractional occupancies of O_2 and N_2 molecules in large cavities, while $\theta_{s,O2}$ and $\theta_{s,N2}$ denote those in small cavities, respectively.*

sample	assignment	$\theta_{l,O2}$	$\theta_{l,N2}$	$\theta_{s,O2}$	$\theta_{s,N2}$	θ_l	θ_s	θ
780 m	1	0.48	0.52	0.05	0.26	1.00	0.31	0.54
	2	0.49	0.51	0.05	0.26	1.00	0.31	0.54
	3	**0.16**	**0.83**	**0.39**	**0.41**	**0.99**	**0.81**	**0.87**
	4	0.16	0.83	0.39	0.42	0.99	0.81	0.87
1570 m	1	0.48	0.52	0.05	0.31	1.00	0.36	0.57
	2	0.44	0.56	0.04	0.24	1.00	0.28	0.52
	3	**0.16**	**0.83**	**0.38**	**0.49**	**0.99**	**0.88**	**0.91**
	4	0.14	0.86	0.34	0.36	1.00	0.70	0.80
2450 m	1	0.38	0.62	0.03	0.36	1.00	0.39	0.59
	2	0.35	0.65	0.03	0.28	1.00	0.31	0.54
	3	**0.10**	**0.90**	**0.28**	**0.52**	**0.99**	**0.79**	**0.86**
	4	0.09	0.91	0.24	0.39	1.00	0.64	0.76

4.4 Distribution of Guest Molecules within Hydrate

Our results clearly show a nonuniform distribution of guest molecules in terms of cage filling. The $\theta_{l,N2}/\theta_{l,O2}$ values of 5.2 at 780 m, 5.2 at 1570 m, and 9.0 at 2450 m are significantly larger than the $\theta_{s,N2}/\theta_{s,O2}$ values of 1.1 at 780 m, 1.3 at 1570 m, and 1.9 at 2450 m. In other words, the N_2/O_2 guest composition ratios in the large cavities are considerably larger than those in the small cavities. These results can be explained in terms of the difference between the Langmuir constants of O_2 and N_2 molecules. According to the solid solution model developed by van der Waals and Platteeuw,[24] the fractional occupancy of a molecule M in a cage of type i, $\theta_{i,M}$, is given by

$$\theta_{i,M} = C_{i,M}\, p_M \left/ \left(1 + \sum_{J} C_{i,J}\, p_J \right) \right. \tag{4}$$

where $C_{i,M}$ represents the Langmuir constant of M in the cage of type i and p_M is the fugacity of M. Although the C values of O_2 and N_2 molecules reported differ significantly between previous studies, in large cavities, the Langmuir constant of O_2 tends to be smaller than that of N_2, while in small cavities, it tend to be larger.[22, 27] Therefore, the value of $\theta_{l,N2}/\theta_{l,O2}$ is larger than that of $\theta_{s,N2}/\theta_{s,O2}$. Furthermore, our result is consistent with the calculated stabilities of guest molecules in air hydrates investigated by a molecular orbital method.[28] They have calculated the cohesive energies of water molecules comprising air hydrates with different gas compositions and locations and found that these energies decrease significantly when small cages are occupied by O_2 in stead of N_2. They have also report that these energies are almost independent of guest species in large cavities. Based on these results, they predict that O_2 molecules of air hydrates are preferentially located in small cages.

5 CONCLUSION

High-resolution micro-Raman analyses of O_2 and N_2 molecules of air clathrate hydrates in Dome Fuji ice reveal multiplet features of the vibrational spectra of guest molecules.

By using the statistical thermodynamic model, we have assigned the observed Raman bands and estimated the absolute cage occupancies. The results show that the N_2/O_2 guest composition ratios in the large cages are significantly higher than those in the small cages. The difference between the N_2/O_2 guest composition ratios of the large and small cages is attributed to the difference between the Langmuir constants of O_2 and N_2 molecules. Moreover, our finding is consistent with the preferential occupation of small cavities by O_2 predicted by a molecular orbital method.

Acknowledgements

We thank T. Uchida and A. Hori for their helpful advice. We are also grateful to all participants in the Deep Ice Coring Project at Dome Fuji for their field work, ice sampling, and logistic support. The present research was supported by a Grant-in-Aid for Creative Scientific Research from the Ministry of Education, Culture, Sports, Science and Technology.

References

1   H. Shoji and C.C. Langway Jr., *Nature*, 1982, **298**, 548.
2   D. Raynaud, J. Jouzel, J.M. Barnola, J. Chappellaz, R.J. Delmas and C. Lorius, *Science*, 1993, **259**, 926.
3   T. Ikeda, H. Fukazawa, S. Mae, L. Pepin, P. Duval, B. Champagnon, V.Y. Lipenkov and T. Hondoh, *Geophys. Res. Lett.*, 1999, **26(1)**, 91, doi:10.1029/1998GL900220.
4   T. Ikeda-Fukazawa, T. Hondoh, T. Fukumura, H. Fukazawa and S. Mae, *J. Geophys. Res.*, 2001, **106(D16)**, 17799, doi:10.1029/2000JD000104.
5   H. Anzai, M.S. Thesis, Hokkaido Univ., 1989.
6   T. Hondoh, H. Anzai, A. Goto, S. Mae, A. Higashi and C.C. Langway Jr., *J. Inclusion Phenom. Mol. Recognit. Chem.*, 1990, **8**, 17.
7   S. Takeya, H. Nagaya, T. Matsumoto, T. Hondoh and V.Y. Lipenkov, *J. Phys. Chem. B*, 2000, **104**, 668.
8   A.K. Sum, R.C. Burruss and E.D. Sloan Jr., *J. Phys. Chem. B*, 1997, **101**, 7371.
9   T. Uchida, T. Hirano, T. Ebinuma, H. Narita, K. Gohara, S. Mae and R. Matsumoto, *AIChE J.*, 1999, **45**, 2641.
10 J. Nakahara, Y. Shigesato, A. Higashi, T. Hondoh and C.C. Langway Jr., *Phylos. Mag. B*, 1988, **57**, 421.

11 F. Pauer, J. Kipfstuhl and W.F. Kuhs, *Geophys. Res. Lett.*, 1995, **22(8)**, 969, doi:10.1029/95GL00705.

12 F. Pauer, J. Kipfstuhl and W.F. Kuhs, *Geophys. Res. Lett.*, 1996, **23(2)**, 177, doi:10.1029/95GL03660.

13 F. Pauer, J. Kipfstuhl and W.F. Kuhs, *J. Geophys. Res.*, 1997, **102(C12)**, 26519, doi:10.1029/97JC02352.

14 T. Ikeda, A.N. Salamatin, V.Y. Lipenkov, S. Mae and T. Hondoh, *Ann. Glaciol.*, 2000, **31**, 252.

15 B. Champagnon, G. Panczer, B. Chazallon, L. Arnaud, P. Duval and V. Lipenkov, *J. Raman Spectrosc.*, 1997, **28**, 711.

16 B. Chazallon, B. Champagnon, G. Panczer, F. Pauer, A. Klapproth and W.F. Kuhs, *Eur. J. Mineral.*, 1998, **10**, 1125.

17 Dome-F Deep Coring Group, *Ann. Glaciol.*, 1998, **27**, 333.

18 H. Narita, N. Azuma, T. Hondoh, M. Fujii, M. Kawaguchi, S. Mae, H. Shoji, T. Kameda and O. Watanabe, *Ann. Glaciol.*, 1999, **29**, 207.

19 H. Ohno, V.Y. Lipenkov and T. Hondoh, *Geophys. Res. Lett.*, 2004, **31**, doi:10.1029/2004GL021151.

20 B. Chazallon, *Ph.D. thesis, Georg-August-Universitaet Goettingen*, 1999, in French, http://goopc4.sub.uni-goettingen.de:8080/DB=1/SET=18/TTL=1/SRCH?IKT=1004&TRM=chazallon.

21 E.P. van Klaveren, J.P.J. Michels, J.A. Schouten, D.D. Klug and J.S. Tse, *J. Chem. Phys.*, 2001, **115**, 10500.

22 B. Chazallon and W.F. Kuhs, *J. Chem. Phys.*, 2002, **117**, 308.

23 J.S. Tse, Y.P. Handa, C.I. Ratcliffe and B.M. Powell, *J. Inclusion Phenom.*, 1986, **4**, 235.

24 J.H. van der Waals and J.C. Platteeuw, *Adv. Chem. Phys.*, 1959, **2**, 1.

25 P.B. Dharmawardhana, W.R. Parrish and E.D. Sloan, *Ind. Eng. Chem. Fundam.*, 1980, **19**, 411.

26 T. Hondoh, H. Shoji, O. Watanabe, A.N. Salamation and V.Y. Lipenkov, *Ann. Glaciol.*, 2002, **35**, 384.

27 W.R. Parrish and J.M. Prausnitz, *Ind. Eng. Chem. Process Des. Dev.*, 1972, **11**, 26.

28 A. Hori, S. Horikawa and T. Hondoh, *Proceedings of the 5th International Conference on Natural Gas hydrate*, 2005, 1710.

HIGH PRESSURE NMR OF HYDROGEN-FILLED ICES BY DIAMOND ANVIL CELL

T. Okuchi,[1,3] M. Takigawa,[2] H.K. Mao,[3] R.J. Hemley[3] and T. Yagi[2]

[1] Institute for Advanced Research, Nagoya University, Nagoya 464-8601, Japan
[2] Institute for Solid State Physics, University of Tokyo, Kashiwa, Chiba 277-8581, Japan.
[3] Geophysical Laboratory, Carnegie Institution of Washington, Washington, DC 20015, USA

1 INTRODUCTION

The properties of hydrogen hydrates at high pressures increasingly attract the attention of researchers in solid state physics, chemistry, planetary science, as well as energy engineering science.[1-15] At pressures between 0.2 and 0.4 GPa, a structure II clathrate hydrate crystallizes from a H_2-H_2O mixture upon moderate cooling.[5,6] It has a high H_2:H_2O molar ratio of 1:2 and therefore can be anticipated as a novel clean storage material of hydrogen fuel.[7] At pressures higher than 0.4 GPa, two hydrogen-filled ice phases crystallize from the same mixture. They show even more interesting behaviours, such as a wider H_2:H_2O ratio ranging from 1:6 to 1:1, a some 20 °C higher melting temperature than that of pure ice, and a twice larger compressibility than that of pure ice.[2]

The difference between clathrate hydrates and filled ices lies in their host H_2O framework structure.[16] The H_2 clathrate framework is built up of an assembly of large polyhedral cages. These cages can be simultaneously filled by up to four H_2, stabilizing the loose framework against pressure.[6,8,9] On the other hand, the framework of H_2-filled ices is ice itself; the frameworks of ice II or Ic polymorphs take up H_2 into their intrinsic cavities.[1] The cavities are just as large as one H_2 molecule. Thus, the two kinds of hydrogen hydrates show clear structural contrast, from which we expect different transport properties of the guest molecules. Knowledge of the guest transport in these hydrates has long time been looked for, not only for basic research[8] but also for designing an efficient storage material[11]. This is because it is the principal factor to control the growth and decomposition kinetics of hydrate crystals. However, no results have been reported so far, due to the absence of *in situ* analytical methods at pressures where growth or decomposition occurs. Here we report on our preliminary results on the anticipated property, using a novel high-resolution diamond anvil cell NMR (DAC-NMR) method which has become available recently.

2 MATERIALS AND METHODS

Static pressure of several GPa (10^4 bars) was generated using a specially designed nonmagnetic diamond anvil cell (DAC) within high static magnetic field of 7 Tesla.[17] The DAC was designed to generate a force enough large to compress samples of relatively large volume, which helps to increase NMR signal-to-noise ratio. A diamagnetic gasket with susceptibility matched to the sample played an essential roles to keep the sample field homogeneous to sub-ppm level.[18] We have newly designed an extremely sensitive radio-frequency (rf) probe for solid state NMR analysis. Along the most efficient way to improve the sensitivity of NMR,[19] we tried to reduce the volume of rf coil while keeping the sample at its center. We fabricated a coil of four turns with 1.2 x 2.5 mm^2 in a rectangular cross section and 1.2 mm in length and made of a 75 μm-ϕ silver wire. Note that the coil dimension is much smaller than two opposite diamond anvils with 5 mm in total thickness and 4 mm in diameter. In order to install such a small coil, we laser-drilled each anvil to excavate four straight holes with 150 μm in diameter and 2.5 mm in length, along the direction parallel to the anvil's 1.0 mm-ϕ culet surface. The drilling was made using a commercial tool (COMBI Laser System, Bettonville, Wijnegem, Belgium). This state-of-the-art technique gives an order of magnitude higher signal-to-noise ratio than our previous record.[20] Using this probe we can analyze solid-state samples which give strongly dipolar-coupled broadband spectra. Such samples require a signal-to-noise ratio higher than those giving only sharp resonances which we have so far measured successfully.[21]

A 400 μm-ϕ sample chamber was drilled into slightly pre-indented gasket of 250 μm thickness. This enlarged sample dimension further improved the sensitivity of NMR.[17] We prepared the hydrate sample as follows. We partially filled the sample chamber with desired volume of H_2O along with an air bubble. Piston-cylinder (P-S) pair of the DAC was sealed to avoid H_2O evaporation. The P-S pair was then placed into a gas loading vessel into which H_2 gas at 1500 bar was filled. To exchange the air bubble by H_2, the P-S pair was quickly opened and sealed again within a few seconds. We did not loose any H_2O during this process, which assured that the desired mixing ratio of $H_2:H_2O \sim 1:6$ was obtained. We finally put the rf coil wire through the diamond holes for the NMR analysis. Pressure was determined using the ruby scale before and after each data acquisition. The reproducibility was better than 1 kbar, and the accuracy of the pressure scale was 1.5 kbar up to the highest pressure.

3 RESULTS AND DISCUSSION

The H_2-H_2O two-liquid sample was compressed at room temperature, while optically observing the crystal growth through the diamond window. At 1.9 GPa, the whole H_2O part was suddenly frozen into an inhomogeneous mixture of ices, presumably a mixture of ice VI and H_2-filled ice II. The freezing pressure of H_2O coexisting with H_2 fluid was twice larger than that of pure water. This large difference may be because of large amount of dissolved H_2 into water at the relevant pressure, which is very interesting phenomenon and will be studied by NMR in future.

After the freezing of H_2O fluid part, the filled-ice II swiftly grew from the H_2O-H_2 boundary by consuming the frozen H_2O. The growth rate was apparently determined by H_2 diffusion through the synthesized filled-ice II layer to the unreacted frozen H_2O region. The growth of filled ice was observed by eye and completed within several hours, indicating that the H_2 diffusion in the filled ice is much faster than in other solids. We took NMR spectra after the completion of the growth.

Figure 1 shows the observed NMR spectra together with some relevant sample photographs. A polycrystalline H_2-filled ice II ($H_2:H_2O$ = 1:6) sample with grain size of several microns was finally synthesized, which is an ideal sample for powder NMR analysis (Figure 1a). No preferred orientation was observed. By further increase of the pressure, the filled-ice II decomposed into the mixture of H_2-filled ice Ic ($H_2:H_2O = 1:1$) and pure ice VII polycrystals, after taking about one hour for the reaction. The grain size of the mixture becomes larger than the filled-ice II, but it was still much smaller than the sample dimension, keeping the mixture suitable for powder NMR (Figure 1b). We had expected that static NMR spectra of filled ices would exhibit extreme bandwidth of several tens of kilohertz, because of strong H-H dipolar interaction. Therefore, it was amazing to find that both NMR spectra in Figure 1 have a sharp resonance at the center, indicating an existence of a mobile proton species.

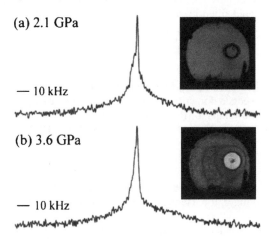

(a) 2.1 GPa

— 10 kHz

(b) 3.6 GPa

— 10 kHz

Figure 1 *Powder NMR spectra and optical photographs of the H_2-H_2O mixture at high pressures and room temperature. The spectra were obtained with a 300 MHz and 300 W solid-state spectrometer. A $\pi/2$-$\pi/2$ solid echo sequence was applied. Typical $\pi/2$ pulse length was 1.5 μs. (a) The H_2-filled ice II phase. The narrow resonance at near center of the spectrum came from the H_2 guests, and the broad resonance came from the ice II framework. The narrow resonance gave a much longer T_2^* than the acquisition time of the spectrum, forcing the peak profile to be smaller and wider than its intrinsic value. The small pore in the photo is H_2 fluid that did not react, which is about a tenth of total H_2 in the sample chamber. Note that the narrow resonance should mainly come from the H_2 guest in the filled ice and not from the H_2 fluid, because the former is one order of magnitude more abundant than the latter, and the former gave smaller relaxation time than the latter. 2698 transients with 1 sec repetition time were averaged. (b) A mixture of H_2-filled ice Ic and pure ice VII phases. The narrow resonance came from the H_2 guests, and the broad resonance came from the H_2O framework of filled-ice Ic and pure ice VII. There was no remaining H_2 in the sample chamber. On further compression from (a), at 3.2 GPa the bright transparent part of polycrystalline filled-ice Ic was formed by the reaction between the fluid H_2 and its surrounding filled-ice II. The remaining filled-ice II then decomposed into the darker part consisting of a mixture of filled-ice Ic and pure ice VII. 8000 transients with 2 sec repetition time were averaged.*

The criterion for such motional narrowing is $\Delta \cdot \tau \ll 1$, where Δ is the static linewidth of the relevant species and τ is its motional correlation time.[22] We reasonably substitute the

$\varDelta$ of filled ices by the ^{1}H linewidth of pure ice polymorphs at 17±0.5 gauss or 36±1 kHz.[23] A reasonable estimation of the correlation time of H_2O molecules in ice polymorphs is the Debye relaxation time τ_D, which is the order of 10^{-4} sec at around room temperature.[24] This value does not satisfy $\varDelta \cdot \tau_D \ll 1$ so that framework H_2O cannot give the sharp peak. Thus guest H_2 is the only possible candidate to give the peak, demonstrating that its correlation time is much smaller than $1/\varDelta$, or its diffusion is quite fast. The remaining broad lines in both spectra of Figure 1 are automatically assigned to the framework H_2O. Their widths are comparable to that of pure ice polymorphs, demonstrating that not only ordered filled-ice II but also disordered filled-ice Ic frameworks are completely static in the filled ices even at room temperature close to the melting.

Acknowledgment

We thank N. Miyajima for helping hydrogen gas sampling into DAC. F. Fujara at TU Darmstadt is acknolwedged for critical comments and manuscript reading. T.O. thanks JSPS Postdoctoral Fellowships for Research Abroad. This work was supported by Ministry of Education in Japan, and DOE (CDAC), NSF, and NASA in USA.

References

1   W.L. Vos, L.W. Finger, R.J. Hemley, and H.K. Mao, *Phys. Rev. Lett.*, 1993, **71**, 3150.
2   W.L. Vos, L.W. Finger, R.J. Hemley, and H.K. Mao, *Chem. Phys. Lett.*, 1996, **257**, 524.
3   Y.A. Dyadin, E.Y. Aladko, K.A. Udachin, and M. Tkacz, *Pol. J. Chem.*, 1994, **68**, 343.
4   Y.A. Dyadin, E.Y. Aladko, A.Y. Manakov, F.V. Zhurko, T.V. Mikina, V.Y. Komarov, and E.V. Grachev, *J. Struct. Chem.* 1999, **40**, 790.
5   Y.A. Dyadin, E.G. Larionov, A.Y. Manakov, F.V. Zhurko, E.Y. Aladko, T.V. Mikina, and V.Y. Komarov, *Mendeleev Commun.*, 1999, 209.
6   W.L. Mao, H.K. Mao, A.F. Goncharov, V.V. Struzhkin, Q. Guo, J. Hu, J. Shu, R.J. Hemley, M. Somayazulu, and Y. Zhao, *Science*, 2002, **297**, 2247.
7   W.L. Mao and H.K. Mao, *Proc. Nat. Acad. Sci.*, 2004, **101**, 708.
8   S. Patchkovskii and J.S. Tse, *Proc. Nat. Acad. Sci.*, 2003, **100**, 14645.
9   K.A. Lokshin, Y. Zhao, D. He, W.L. Mao, H.K. Mao, R.J. Hemley, M.V. Lobanov, and M. Greenblatt, *Phys. Rev. Lett.*, 2004, **93**, 125503.
10 L.J. Florusse, C.J. Peters, J. Schoonman, K.C. Hester, C.A. Koh, S.F. Dec, K.N. Marsh, and E.D. Sloan, *Science*, 2004, **306**, 469.
11 H. Lee, J.W. Lee, D.Y. Kim, J. Park, Y.T. Seo, H. Zeng, I.L. Moudrakovski, C.I. Ratcliffe, and J.A. Ripmeester, *Nature*, 2005, **434**, 743.
12 D.Y. Kim and H. Lee, *J. Am. Chem. Soc.*, 2005, **127**, 9996.
13 S. Alavi, J.A. Ripmeester, and D.D. Klug, *J. Chem. Phys.*, 2005, **123**, 051107.
14 S. Alavi, J.A. Ripmeester, and D.D. Klug, *J. Chem. Phys.*, 2005, **123**, 024507.
15 O.I. Barkalov, S.N. Klyamkin, V.S. Efimchenko, and V.E. Antonov, *JETP Letters*, 2005, **82**, 413.
16 J.S. Loveday, R.J. Nelmes, M. Guthrie, D.D. Klug, and J.S. Tse, *Phys. Rev. Lett.*, 2001, **19**, 215501.
17 T. Okuchi, *Phys. Earth Planet. Inter.*, 2004, **143-144**, 611.

18 T. Okuchi, R.J. Hemley, and H.K. Mao, in *Advances in High-Pressure Technology for Geophysical Applications*, ed. J. Chen, Y. Wang, T. Duffy, G. Shen and L. Dobrzhinetskaya, Elsevier, Amsterdam, 2005, p. 503.

19 D.I. Hoult and R.E. Richards, *J. Magn. Reson.*, 1976, **24**, 71.

20 T. Okuchi, R.J. Hemley, and H.K. Mao, *Rev. Sci. Instrum.*, 2005, **76**, 026111.

21 T. Okuchi, G.D. Cody, H.K. Mao, and R.J. Hemley, *J. Chem. Phys.*, 2005, **122**, 244509.

22 A. Abragam, *The Principles of Nuclear Magnetism*, Clarendon, Oxford, 1961.

23 S.W. Rabideau and E.D. Finch, *J. Chem. Phys.*, 1968, **49**, 4660.

24 V.F. Petrenko and R.W. Whitworth, *Physics of Ice*, Oxford University Press, New York, 1999.

ON THE USE OF THE KIHARA POTENTIAL FOR HYDRATE EQUILIBRIUM CALCULATIONS

N.I. Papadimitriou[1,2], I.N. Tsimpanogiannis[3], A.G. Yiotis[2], T.A. Steriotis[2] and A.K. Stubos[2]

[1]National Technical University of Athens, School of Chemical Engineering, Zografos 15780, Greece
[2]National Center for Scientific Research "Demokritos", Aghia Paraskevi 15310, Greece
[3]Earth & Environmental Sciences Division (EES-6), Los Alamos National Laboratory, Los Alamos, NM 87545,USA

1 INTRODUCTION

Clathrate hydrates (known also as gas hydrates) belong to a large class of crystalline, non-stoichiometric, inclusion-compound materials that are stable within a certain range of pressure and temperature. The host solid framework structure is made up of water molecules, connected through hydrogen bonds that form cavities (cages)[1]. The cavities can be stabilized by the inclusion of small molecules such as CH_4, C_3H_8, CO_2, N_2, Ar, etc. Over 100 different molecules are known to form hydrates.

While hydrate clathrates were discovered almost two hundred years ago[1] (Sir Humphrey Davy, 1810), they were considered, for a long time, a curiosity with no practical application. However, Hammerscmidt[2] during 1934 identified gas hydrates of compounds like methane, ethane, propane and isobutane as the real cause of natural-gas pipelines blocking at low temperatures. Instantly, gas hydrates attracted significant industrial attention, since flow assurance became an important issue for the oil and gas industry. Initial interest was focused on aspects of thermodynamic inhibition of hydrate formation by adding some additional chemical component in the pipeline. Flow assurance is currently an issue with significant economic impact, especially as we move towards producing gas/oil from deep oceanic fields.

In addition to flow assurance gas hydrates are involved in a variety of industrial applications[3,4]. Hydrates are considered as a possible means of transportation of stranded gas, for temporary storage of gas at power plants to be used during peak hours, for water desalination, for sequestration of CO_2 during oceanic disposal, for future energy production from on-shore/off-shore methane hydrate deposits, as a possible cause of global warming due to sudden release of methane from oceanic/permafrost deposits, for H_2 storage as a part of a future hydrogen economy[5,6], for gas separation[7], and as indicators for paleoclimates due to their presence in ice cores. As the scientific community continues to demonstrate that hydrates are promising materials, hydrates will continue to receive increasing attention, as also shown by the increasing number of scientific publications.

Important thermodynamic/transport properties can be either measured experimentally (a process that is associated, in general, with significant cost), or predicted (more desirable since it offers more flexibility). Therefore, there is a need for reliable, theoretically based models that can not only correlate existing experimental data, but, moreover, can be used in a purely predictive/extrapolative mode.

2 BACKGROUND

Van der Waals and Platteeuw[8] (VDWP) were the first to present a theory that describes the thermodynamic equilibrium of hydrates based on principles from Statistical Mechanics. Several modifications[9-12] of the original VDWP model have been utilized in order to expand the limits of its applicability and improve its accuracy. A brief presentation of the equations that describe the modification of the VDWP used in this work is presented below.

The VDWP model and its variants[8-12] are based on the basic equation for phase equilibrium, which is the equality of the chemical potentials of every component in any phase. Specifically for hydrates:

$$\mu_w^H = \mu_w^B \tag{1}$$

where μ_w^H is the chemical potential of water in the hydrate phase and μ_w^B is the chemical potential of water in the bulk water phase (ice or liquid water).

If μ_w^E is the chemical potential of a hypothetical empty hydrate, each of the chemical potentials of water in Eq. (1), can be calculated relatively to μ_w^E. For, the hydrate phase, Van der Waals and Platteeuw have derived the following expression by calculating the great canonical partition function of the enclathrated gas, based on the principles of Statistical Mechanics:

$$\Delta\mu_w^H = -RT\sum_i v_i \, ln\left(1-\sum_j \theta_{ij}\right) \tag{2}$$

where $\Delta\mu_w^H$ is the chemical potential difference between the hydrate and the empty hydrate ($\Delta\mu_w^H = \mu_w^E - \mu_w^H$), R is the gas constant (8.3145 J/mol·K), T is the temperature, v_i is the number of cavities of type i per water molecule and θ_{ij} is the partial occupancy of the cavity of type i by the component j.

The partial occupancies (θ_{ij}) can be calculated through a Langmuir-type equation:

$$\theta_{ij} = \frac{C_{ij} f_j}{1+\sum_j C_{ij} f_j} \tag{3}$$

where f_j is the fugacity of component j in the gaseous phase at equilibrium and C_{ij} is the Langmuir constant for the occupancy of type i cavity by component j.

The Langmuir constants indicate the strength of interaction between each type of cavity and each gas component. For their calculation, the knowledge of the potential energy within the cavity is necessary. The Kihara spherical core potential function is assumed to describe the potential energy inside the cavities accurately enough:

$$\Gamma(r) = 4\varepsilon\left[\left(\frac{\sigma}{r-2\alpha}\right)^{12} - \left(\frac{\sigma}{r-2\alpha}\right)^{6}\right] \qquad r > 2\alpha \tag{4}$$

where $\Gamma(r)$ is the potential energy at distance r from the centre of the core due to water-gas interactions, ε is the energy parameter (depth of the intermolecular potential well), σ is the distance between molecular cores at which $\Gamma(r)=0$ ($\sigma+2\alpha$ is the collision diameter) and α is the core radius.

Adding up the pair-potential energies for the entire cavity we obtain:

$$\omega(r) = 2z\varepsilon\left[\frac{\sigma^{12}}{R^{11}r}\left(\delta^{10} + \frac{\alpha}{R}\delta^{11}\right) - \frac{\sigma^{6}}{R^{5}r}\left(\delta^{4} + \frac{\alpha}{R}\delta^{5}\right)\right] \qquad 0 < r < R-\alpha \tag{5}$$

$$\delta^{N} = \frac{1}{N}\left[\left(1 - \frac{r}{R} - \frac{\alpha}{R}\right)^{-N} - \left(1 + \frac{r}{R} - \frac{\alpha}{R}\right)^{-N}\right] \qquad 0 < r < R-\alpha \tag{6}$$

where $\omega(r)$ is the potential energy at distance r from the centre of the cavity, z is the number of water molecules per cavity (coordination number) and N is the exponent of δ in Eq. (5) (having values: 4,5,10,11).
Finally, the Langmuir constant is given by:

$$C_{ij} = \frac{4\pi}{kT}\int_{0}^{R-\alpha} exp\left(-\frac{\omega(r)}{kT}\right)r^{2}dr \tag{7}$$

where $\pi=3.14159$ and $k=1.38062\cdot10^{-23}$ J/K (Boltzman's constant).
The chemical potential of the bulk phase is a function of temperature, pressure and composition:

$$\frac{\Delta\mu_{w}^{B}}{RT} = \frac{\Delta\mu_{w}^{0}}{RT} - \int_{T_{0}}^{T}\frac{\Delta h_{w}(T)}{RT^{2}}dT + \int_{P_{0}}^{P}\frac{\Delta v_{w}}{RT}dP - ln(a_{w}) \tag{8}$$

where $\Delta\mu_{w}^{0}$ is the chemical potential difference between the bulk water phase and the empty hydrate at reference conditions P_{0} and T_{0} (usually $T_{0} = 273.15$ K and $P_{0} = 0$ Pa), Δh_{w}^{0} is the enthalpy difference between the bulk water phase and the empty hydrate, Δv_{w} is the molar volume difference between the bulk water phase and the empty hydrate, a_{w} is the activity of water in the bulk water phase ($a_{w} = 1.0$ for ice) and P is the pressure.

For the application of the model the values of several parameters need to be determined. These parameters can be divided in three groups:

(i) Structural parameters such as the coordination number, z, and the radius, R, of each cavity.
(ii) Thermodynamic properties such as the difference in the chemical potential, $\Delta\mu^{0}$, the enthalpy, Δh^{0}, and the volume, Δv, of water between the bulk phase (liquid or ice) and the hydrate phase.

(iii) Interaction parameters (parameters of the potential function used to describe the interactions between the solid water molecules and the enclosed gas molecules) such as the energy parameter, ε/k, the distance parameter, σ, and the core sphere radius, a, (in the case when the Kihara spherical core potential function is used).

The structural parameters have been experimentally determined and there is a reasonable consensus on their values. However, a large number of significantly deviating values for the thermodynamic properties and the interaction parameters can be found in the literature[8-16] since various researchers used different data sets to obtain a fitting for those parameters. Considering the values reported by Sloan[1] as reference (for example ε/k=154.54K and σ=3.1650Å for methane, and ε/k=174.40K and σ=3.2614Å for propane) the deviation of other values even exceeds 15%. By fitting any of these parameters to a specific set of hydrate experimental data, a significant improvement in the efficiency of the model, to correlate the particular data, can be achieved. However, it also results in an unavoidable reduction of the models ability to describe successfully other experimental data without recalculation of the fitted parameters. Avlonitis[15] presented a detailed discussion of how experimental data can be utilized to obtain the fitted parameters.

The main objective of this work is to estimate the effect of uncertainty of each of the fitted parameters to the prediction of the equilibrium pressure given by the used VDWP model. For this purpose a sensitivity analysis for each one of the parameters of interest was carried out. In addition, a literature survey was conducted to obtain the reported values for the parameters together with the corresponding approach used for the calculation of those values[8-16].

In principal, the Kihara parameters should be unique for a specific gas and independent of the experimental data used to obtain the parameters. If such a statement holds true, once the parameters are evaluated using some experimental data (e.g. viscosity), then, they can be used (with some degree of confidence) to predict other types of data (e.g. hydrates). However, the Kihara parameters obtained from fitting hydrate equilibrium data show significant deviations from those parameters obtained from second virial coefficient, viscosity or other thermodynamic equilibrium experimental data (Tee et al.[16] obtained the following parameters using second virial coefficient data: ε/k=227.13 K and σ=3.565Å for methane, and ε/k=501.89 K and σ=4.611Å for propane).

3 RESULTS AND DISCUSSION

In order to examine the effect of the Kihara parameters on the predicted hydrate equilibrium pressures, a sensitivity analysis was carried out (see also Cao et al.[17]). In this study we report results for methane (sI hydrate former) and propane (sII hydrate former). The Kihara parameter values, as well as the thermodynamic property values, reported by Sloan were taken as the base-reference case and hydrate equilibrium pressures were calculated by perturbing the reference values in the range $\pm(1\%-10\%)$. On the other hand, the reported thermodynamic parameters $\Delta\mu^0$ and Δh^0 have a wider range, but as it is going to be discussed later, have a less significant effect on the predictions.

Figures 1 and 2 show the results of the parametric studies for methane and propane, respectively. From the results of this study we observe that: **(i)** there is a strong dependence of the predicted hydrate equilibrium pressure on the energy parameter, ε/k, and the distance parameter, σ, **(ii)** there is a less significant dependence on the reference chemical potential difference, $\Delta\mu^0$, **(iii)** there is a minor dependence on the core radius, a, and the reference enthalpy difference, Δh^0, and **(iv)** the quality of predictions is not satisfactory for higher temperatures/pressures (e.g. for values higher than 305 K and 100 Mpa for the case of methane).

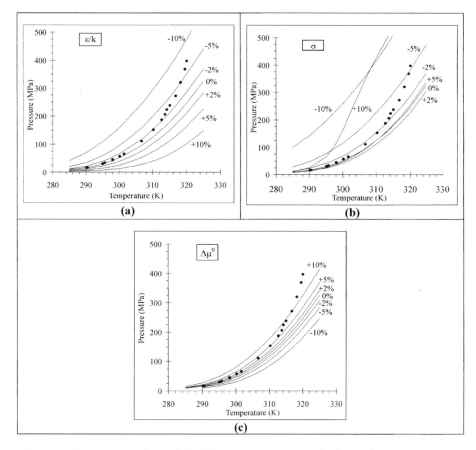

Figure 1 *Sensitivity analysis of the Kihara parameters and the thermodynamic properties for methane. Effect on the predicted equilibrium pressure of: (a) energy parameter, ε/k, (b) distance parameter, σ, and (c) reference chemical potential difference, Δμ⁰.*

In particular, an increase of ε/k by +1% results in doubling the average error (over the entire temperature range examined) in the predicted hydrate equilibrium pressures. For a given temperature, increasing the value of ε/k results in decreasing the predicted value of the hydrate formation pressure. This is rather expected, since the energy parameter is a measure of the strength of the interactions between the enclathrated molecule and the lattice. Lower values of ε/k involve additional difficulties in the gas filling of the cavities, and consequently higher pressures are required to stabilize the hydrate lattice.

In contrast to ε/k, a monotonic variation of σ does not lead to a similar monotonic variation of the predicted hydrate equilibrium pressures. In particular, for the case of methane, a slight increase in σ results in an (not very significant) increase of the equilibrium pressures. However, this tendency seems to invert for further increase of σ and a +10% variation of σ leads in an overestimation of the equilibrium pressures of about 130% in average. The case of propane is more complex and more indicative of the need for very accurate values of the Kihara parameters. While values of σ that are slightly lower or

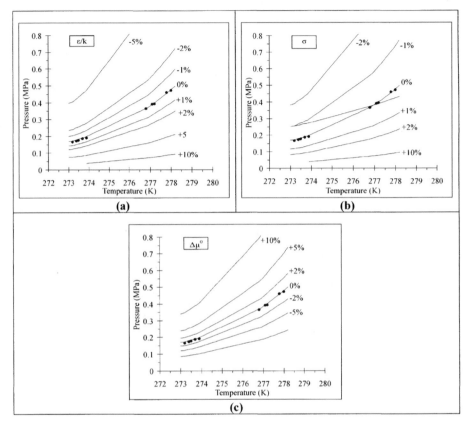

Figure 2 *Sensitivity analysis of the Kihara parameters and the thermodynamic properties for propane. Effect on the predicted equilibrium pressure of: (a) energy parameter, ε/k, (b) distance parameter, σ, and (c) reference chemical potential difference, $\Delta\mu^0$.*

higher than the reference value from Sloan give erroneous results, a value decreased by 10% surprisingly gives better results. A more careful investigation shows that higher values of σ predict that propane occupies only the large cavities of the sII hydrate. A decrease of σ by 10% is significant enough to enable propane to enter the small cavities as well, which is an unphysical prediction. This effect is also demonstrated in Figure 3 where the Langmuir constants are presented as a function of the Kihara parameters ε/k and σ. Due to their exponential dependence on σ, Langmuir constants are very sensitive to even minor variations of σ. For example, while the reference value $\sigma = 3.3093$Å yields, for the small cavity of sII hydrate, an extremely low Langmuir constant of $2.07 \cdot 10^{-9}$MPa^{-1}, a value decreased by 10% ($\sigma = 2.9784$Å) gives a Langmuir constant as higher as ten orders of magnitude ($19\mathrm{MPa}^{-1}$). On the other hand, ε/k doesn't seem to have such an effect on the occupancy of cavities.

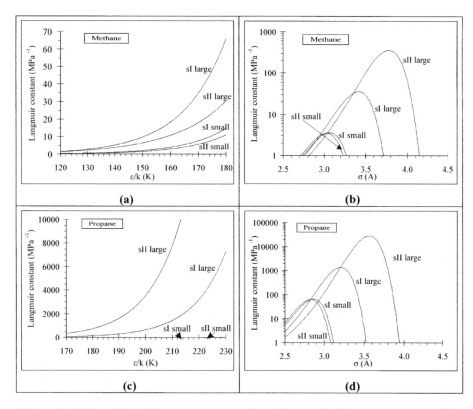

Figure 3 *Effect of the Kihara parameters ε/k and σ, on the Langmuir constants for methane: **(a)** and **(b)**, and for propane: **(c)** and **(d)**.*

4 CONCLUSIONS

A sensitivity analysis was performed to examine the effect of the thermodynamic properties and the Kihara parameters on the hydrate equilibrium calculations. It was demonstrated that the Kihara parameters (ε/k and σ) had a more significant effect on hydrate equilibrium predictions than the thermodynamic properties ($\Delta\mu^0$ and Δh^0) for the cases of methane and propane that were examined in this work. It was observed that parameters obtained from one set of experiments could not always be used in correlating successfully other hydrate experimental data sets. This problem was more pronounced in cases that the fitted parameters were to be used for other properties such as virial coefficients or viscosities. Finally, issues such as satisfactory predictions at very high pressures and multiple cage occupancy need to be considered.

Acknowledgements

Partial funding by the European Commission DG Research (contract SES6-2006-518271/ NESSHY) is gratefully acknowledged by the authors. The work of INT was partially supported by LDRD-DR 20030059DR and LDRD-DR 20040042DR projects, funded by Los Alamos National Laboratory the contribution of which is gratefully acknowledged.

References

1   E.D. Sloan, *Clathrate Hydrates of Natural Gases*, 2nd edn., Marcel Dekker, New York, 1998.
2   E.G. Hammerschmidt, *Ind. Eng. Chem.*, 1934, **26,** 851.
3   E.D. Sloan, *Nature*, 2003, **426**, 353.
4   I. Chatti, A. Delahaye, L. Fournaison, J.P. Petitet, *Energy Conv. Man.,* 2005, **46**, 1333.
5   W.L. Mao, H.K. Mao, A.F. Goncharov, V.V. Struzhkin, Q. Guo, J. Hu, J. Shu, R.J. Hemley, M. Somayazulu, Y. Zhao, *Science*, 2002, **297**,2247.
6   L.J. Florusse, C.J. Peters, J. Schoonman, K.C. Hester, C.A. Koh, S.F. Dec, K.N. Marsh, E.D. Sloan*, Science*, 2004, **306**, 469.
7   S.P. Kang, H. Lee, *Environ. Sci. Technol.*, 2000, **34**, 4397.
8   J.H. Van der Waals and J.C. Platteeuw, *Adv. Chem. Phys.*, 1959, **2, 1**.
9   W.R. Parrish and J.M. Prausnitz, *Ind. Eng. Chem. Proc. Des. Dev.*, 1972, **11,** 26.
10 V.T. John, K.D. Papadopoulos, G.D. Holder, *AIChE J.*, 1985, **31**, 252.
11 G.D. Holder, G.Gorbin, K.D. Papadopoulos, *Ind. Eng. Chem. Fund.*, 1982, **19,** 282.
12 J.B. Klauda and S.I. Sandler, *Ind. Eng. Chem. Res.*, 2000, **39,** 3377.
13 G.D. Holder and J.H. Hand, *AIChE J.*, 1982, **28,** 440.
14 V.T. John, G.D. Holder, AIChE J., 1985, **31**, 252.
15 D. Avlonitis, *Chem. Eng. Sci.*, 1994, **49,** 1161.
16 L.S. Tee, S. Gotoh and W.E. Stuart, *Ind. Eng. Chem. Fund.*, 1966, **5,** 363.
17 Z. Cao, J.W. Tester and B.L. Trout, *J. Phys. Chem. B*, 2002, **106,** 7681.

A NEW STRUCTURE OF AMORPHOUS ICE PROMOTED BY RADIATIONS.

Ph. Parent[1], S. Lacombe[2], F. Bournel[1] and C. Laffon[1]

[1]Laboratoire de Chimie-Physique, Matière et Rayonnement, Université Pierre et Marie Curie- Paris 6 et CNRS, UMR 7614, 11 rue Pierre et Marie Curie, 75231 Paris Cedex 05, France.
[2]Laboratoire des Collisions Atomiques et Moléculaires, UMR 8625, Université Paris Sud 11 et CNRS, 91405 Orsay Cedex, France.

1 INTRODUCTION

There is a poor knowledge of the initial structure of cold ice deposited under vacuum, and, with greater reason, after exposure to radiations. Such cold ice is a dominant component in several astrophysical environments (comets, planetary satellites, and interstellar grains) and its properties are important for the understanding of processes such as molecule formation in the interstellar space and comet outgassing.

Ice films condensed from the water vapour on a cold substrate (T<30 K) has been characterized as a high-density amorphous form of ice, which could be a denser variant of the low-density phase obtained by deposition above 30 K. [1,2] Condensation from the background pressure also leads to ice films that are highly porous at a nanoscale.[3,4] This porosity is lost by warming or by direct deposition of water at T>90 K. Warming ice at 150 K induces the crystallization, whatever the initial structure is.

It is well known that irradiation alters the structure of ice.[5] It is amorphized at T < 80 K by protons, ions,[6-9] photons[9,10] and electrons.[11-13] The structures of the irradiated ices have only been determined in the case of intense electron irradiation, where the high-density amorph was detected.[12,13] Recent molecular dynamic simulations have also shown a densification of the ASW ice when irradiated with 35 eV water molecules,[14] but these simulations also questioned the existence of the high-density phase as the initial structure of ice films deposited at low temperature.

Oxygen K-edge NEXAFS spectroscopy gives valuable informations on the local order of ice.[15-17,4] To gain knowledge on the structure of cold condensed ice films, we have undertaken the NEXAFS study of a porous amorphous ice film before and after a XUV (3-900 eV) irradiation, followed by an annealing to 150 K.

2 EXPERIMENTAL

The experiments were performed at LURE (SuperACO, Orsay) on the SA22 bending magnet beamline equipped with a plane grating spherical mirror monochromator (resolution of approximately 100 meV at the O K-edge, 530 eV). The monochromator was calibrated at the oxygen K-edge by settling the $4a_1$ state of water at 535 eV.[4] The irradiation and the NEXAFS measurements (total electron yield) have been done on the

SACEMOR experiment (base pressure 1×10^{-10} mbar) connected to the beamline. The ice films were deposited on a clean Pt(111) single crystal mounted on a rotatable sample holder cooled by a temperature-controlled liquid helium cryostat. A platinum resistance welded to the substrate measures the temperature; temperature control and ramps (1 K.min^{-1}) were achieved by a feedback system driving the resistive heating of the helium bath. After the ultra pure H_2O was degassed by several freeze-pump-thaw cycles, 120 ML thick ice films were prepared by dosing at 20 K, from the background pressure, 120 Langmuirs (1L = 10^{-6} Torr.s; sticking coefficient =1) of H_2O vapor at a rate of 0.3 L.s^{-1} using a diffuser outlet located at the back of the substrate. In these deposition conditions, the ice film is amorphous and porous. The synchrotron beam was used both for irradiating and for the NEXAFS measurements (with settings avoiding a further irradiation). Preliminary experiments using monochromatic irradiation of a crystalline ice film were performed above the K-edge (550 eV), showing that a cumulated energy dose of 1.65 eV.molecule^{-1} is necessary to reach the saturation of the radiation effects (flux 2.5 10^9 photon.cm^{-2}.s^{-1}; energy flux 1.37 10^9 KeV.cm^{-2}.s^{-1}). We have then used the unmonochromatized "white" beam delivered at the zero-order of the monochromator (bandwidth 3 eV- 900 eV), which gives results similar to the monochromatic exposures, except that the films were processed faster.

3 RESULTS AND DISCUSSION

Figure 1 presents the NEXAFS spectra of the amorphous ice film before and after irradiation at 20 K, followed by an annealing to 150 K (during which no irradiations are made).

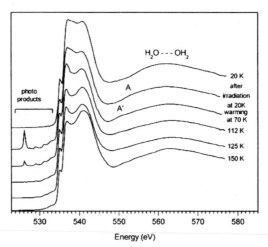

Figure 1   *NEXAFS data of an amorphous ice film deposited at 20 K, irradiated at 20 K and then gradually annealed from 20 K to 150 K. Selected temperature are shown (70 K, 112 K, 125 K, and 150 K).*

The structures below 535 eV correspond to radicals and molecules (OH, O, HO_2, O_2, H_2O_2) produced by the water photolysis, which are not discussed here. On the spectrum recorded just before the irradiation, we observe in the continuum region (E> 545 eV) a single broad resonance labelled H_2O---OH_2. It results from the scattering of the photoelectron with the surrounding oxygen atoms of the first coordination shell.[4] This resonance allows determining the distance between the excited H_2O and its nearest-neighbours,[4,18] its energy being proportional to $1/R^2$, where R is the distance between the O-O atoms in the first shell. On the spectrum recorded after the irradiation, a shoulder "A" appears on the low energy side of H_2O---OH_2. When annealing at 50 K, a second shoulder A' appears around 550 eV. None of these resonances are observed when warming a non-irradiated p-ASW film.[4] At 150 K, faint wiggles superimpose on the H_2O---OH_2 resonance. They correspond to a "multiple scattering" effect within the surrounding molecules, which can only be observed in highly ordered materials. Thus, it shows that the film crystallises. These structures are much weaker than in a crystalline ice film,[4] indicating that some disorder remains in the film.

Figure 2 zooms in on the scattering resonances domain (545 eV- 575 eV) of the NEXAFS spectra presented on the figure 1 (as deposited at 20K, irradiated at 20 K and warmed at 50 K).

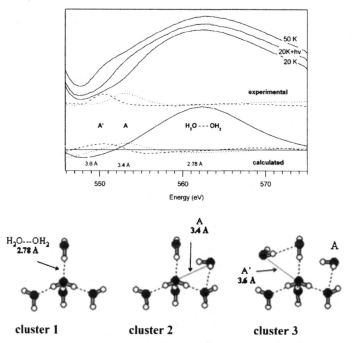

Figure 2 *The scattering resonance domain of the NEXAFS data. The following experimental spectra are shown: as-deposited (20 K), irradiated at 20 K (20 + hν), then warmed at 50 K. The dotted and dashed experimental curves correspond to a subtraction of the experimental data (A: dotted line; A': dashed line) (see text). The calculations of the resonances $H_2O...HO_2$, A and A' are achieved using different clusters which are drawn below the graph (see text).*

The shoulders A and A' are well visible on the raw spectra, but A is better isolated by the subtraction of the data recorded before and after irradiation. A' is also better revealed by the subtraction of the spectrum recorded after irradiation at 20 K with that measured after warming at 50 K. A and A' correspond to two additional scattering resonances resulting from two new scattering centres located around the excited molecule. We have simulated these resonances using self-consistent field full-multiple scattering (MS) calculations with the FEFF-8 code[19] on molecular clusters made of the first coordination shell (to compute the H_2O---OH_2 resonance) plus one or two extra water molecules (to compute the resonances A and A'). The calculated spectra are presented below the experimental data, and the clusters are drawn below the graph. The main H_2O---OH_2 resonance is well simulated with the cluster 1, made of the first coordination shell of ice, i.e. a central water molecule tetrahedrally coordinated with four water molecules, with a mean O-O distance (d_{O-O}) of 2.78 Å. Resonance A appears on the cluster 2, where a single molecule at 3.4 Å has been added to the first coordination shell. Similarly, resonance A' appears at the side of A by adding a further molecule at 3.6 Å (cluster 3). The cluster 2 is thus typical of the local order resulting from the irradiation at 20 K, and the cluster 3 is typical of the effect of the subsequent annealing to 50 K. A closer analysis of the intensities (not shown) indicates that A does not change in intensity between 20 K and 50 K, therefore the site A' does no result of the conversion of A into A'.

We first discuss the structure of the amorphous film before the irradiation, which is expected to be the high-density form, as said in the introduction. In the electron diffraction study of Jenniskens et al.,[2,20] the radial distribution function (RDF) of the high-density phase exhibits an additional contribution at 3.7 Å between the first and the second coordination shells. In the X-ray diffraction study of Narten et al.[1] such kind of contribution was seen, at a distance significantly shorter (3.3 Å). Narten et al proposed that this feature could be due to interstitial molecules located at this distance, or be characteristic of a randomized hydrogen bond network like that of ice II or ice III.[1] Jenniskens et al. proposed that this ice is a collapsed form of the low-density phase I_al. This collapse gives rise to additional oxygen-oxygen distances corresponding to the filling of the structural cavities found between 3 and 4 Å in the low-density ice.[20]

Narten et al. also deduced from their data that one fifth of the water molecules are located at this additional distance. In a RDF picture, this corresponds to 4 nearest-neighbours at 2.76 Å and 1 second neighbour at 3.3 Å. This matches well the atomic surrounding depicted by the cluster corresponding to the structure of ice *after* the irradiation (cluster 2). The local order *before* the irradiation is better described by the 4-coordinated tetrahedron found in the normal amorphous low-density ice and in the crystalline ice (cluster 1). Thus we conclude that the structure of the ice film before the irradiation is not that of the high-density phase but that of the normal low-density phase. In addition, since the irradiated ice has a local order similar to what expected in the high-density phase, we also conclude that the photolysis at 20 K has induced the phase transition from the low-density to the high-density amorph.

An additional site at 3.6 Å appears when warming the irradiated ice at 50 K (cluster 3). This corresponds to a further densification of the lattice of the high-density phase. This structure therefore owns a "very-high" density. Note that it is different from the pressure-induced very-high density ice (VHDA), which RDF presents only a single O-O distance at 3.37 Å.[21] The variation of the local density between the low-density, high density and very-high-density phases can be estimated by assuming that the density of the low-density phase is 0.94 g.cm[-3], and that this corresponds to five water molecules in a sphere of radius equal to the distance to the second oxygen neighbours. The transition to the high-density form results from the addition of one molecule in this sphere, increasing proportionally the initial density of a factor 6/5, which gives a density of 1.13 g.cm[-3] The transition to the very-high density corresponds to the addition of a further molecule, which increases the initial density of a factor 7/5, giving a density of 1.32 g.cm[-3].

4 CONCLUSION

The NEXAFS spectroscopy provides a clear fingerprint of the high-density amorphous ice. This allows showing that before irradiation, the ice film has not the structure of the high-density ice but that of the low-density ice. However, the transition to the high-density form is observed after the irradiation at 20 K: this phase is therefore peculiar to irradiated ice. When warming the irradiated ice at 50 K, we obtain a new form of ice, owning a very-high-density (estimated at 1.32 $g.cm^{-3}$). This phase is specific to warmed irradiated cold amorphous ice.

References

1   A. H. Narten, C. G. Venkatesh, S. A. Rice, *J. Chem. Phys.*, 1977, **64**, 1106.

2   P. Jenniskens, D. F. Blake, *Science*, 1994, **265**, 753.

3   G. A. Kimmel, K. P. Stevenson, Z. Dohnalek, R. S. Smith, B. D. Kay, *J. Chem. Phys.*, 2001, **114**, 5284.

4   Ph. Parent, C. Laffon, C. Mangeney, F. Bournel, M. Tronc, *J. Chem. Phys.*, 2002, **117**, 10842.

5   R. A. Baragiola, *Planet. Space. Sci.* 2003, **51**, 953.

6   G. A. Baratta, G. Leto, F. Spinella, G. Strazzulla, G. Foti, *Astron. Astrophys.*, 1991, **252**, 421.

7   G. Strazzulla G. A. Baratta, G. Leto, G. Strazzulla, G. Foti, *Europhys. Lett.*, 1992, **18**, 517.

8   M. H. Moore, R. L. Hudson, *Astrophys. J.*, 1992, **401**, 353.

9   G. Leto, G. A. Baratta, *Astron. Astrophys.*, 2003, **397**, 7.

10   A. Kouchi, T. Kuroda, *Nature*, 1990, **344**, 134.

11   J. Lepault, R. Freeman, J. Dubochet, *J. Microsc.*, 1993, **132**, RP3.

12   H.-G. Heide, *Ultramicroscopy*, 1984, **14**, 271.

13   H.-G. Heide, E. Zeitler, *Ultramicroscopy*, 1985, **16**, 151.

14   B. Guillot, Y. Guissani, *J. Chem. Phys.*, 2004, **120**, 4366.

15   R. A. Rosenberg, P. R. LaRoe, V. Rehn, J. Stöhr, R. Jeager, C. C. Parks, *Phys. Rev. B*, 1983, **28**, 3026.

16   J. S. Tse, K. H. Tan, J. M. Chen, *Chem. Phys. Lett.*, 1990, **174**, 603.

17   Y. Zubavichus, Y. Yang, M. Zharnikov, O. Fuchs, T. Schmidt, C. Heske, E. Umbach, G. Tzvetkov, F. P. Netzer, M. Grunze, *Chem. Phys. Chem.*, 2004, **5**, 509.

18   Ph. Wernet, D. Testemale, J. –L. Hazemann, R. Argoud, P. Glatzel, L. G. M. Pettersson, A. Nilsson, U. Bergmann, *J. Chem. Phys.*, 2005, **123**, 154503.

19   A. L. Ankudinov, B. Ravel, J. J. Rehr, S. D. Conradson, *Phys. Rev. B*, 1998, **58**, 7565.

20   P. Jenniskens, D. F. Blake, M. A. Wilson, A. Pohorille, *Astrophys. J.*, 1995, **455**, 3398.

21   J. L. Finney, A. Hallbrucker, I. Kohl, A. K. Soper, and D.T. Bowron, *Phys. Rev. Lett.*, 2002, **88**, 225503.

THE RADICAL CHEMISTRY IN IRRADIATED ICE

Ph. Parent[1], S. Lacombe[2], F. Bournel[1] and C. Laffon[1]

[1]Laboratoire de Chimie-Physique, Matière et Rayonnement, Université Pierre et Marie Curie- Paris 6 et CNRS, UMR 7614, 11 rue Pierre et Marie Curie, 75231 Paris Cedex 05, France.
[2]Laboratoire des Collisions Atomiques et Moléculaires, UMR 8625, Université Paris Sud 11 et CNRS, 91405 Orsay Cedex, France.

1 INTRODUCTION

The radiation damages in condensed water are of considerable importance in the fields of biology, medicine, atmospheric chemistry or astrochemistry. The radiations can create molecules and radicals (as H, H_2, O, H_2O_2, HO_2, and O_2) that can strongly react with the irradiated medium and deeply influence its chemistry. Such reactions are extensively studied, but the details of their mechanisms remain conjectural.[1] Previous studies have shown that the NEXAFS spectroscopy at the oxygen K-edge brings informations on the bulk of ice,[2-4] its surface reactivity[5-7] and its photoreactivity as well.[8] We have used this technique to characterize the evolution of the chemical composition of a vapour-deposited crystalline ice film irradiated at 20 K with XUV photons (3-900 eV), and how it evolves when warming the film at 150 K.

2 EXPERIMENTAL

The experiments were performed at LURE (SuperACO, Orsay) on the SA22 bending magnet beamline equipped with a plane grating spherical mirror monochromator (resolution of approximately 100 meV at the oxygen K-edge (530 eV)). The monochromator was calibrated by settling the $4a_1$ state of water at 535 eV.[4] The irradiation and the NEXAFS measurements (recorded in total electron yield) have been done on the SACEMOR experiment (base pressure $1x10^{-10}$ mbar) connected to the beamline. The ice films were deposited on a clean Pt(111) single crystal mounted on a rotatable sample holder cooled by a temperature-controlled liquid helium cryostat. A platinum resistance sensor welded to the substrate measures the temperature. Temperature control and ramps (1 K.min^{-1}) were achieved by a feedback system driving the resistive heating of the helium bath. After the ultra pure H_2O was degassed by several freeze-pump-thaw cycles, 120 ML thick ice films were prepared by dosing at 150 K, from the background pressure, 120 Langmuirs (1L = 10^{-6} Torr.s; sticking coefficient =1) of H_2O vapor at a rate of 0.3 L.s^{-1} using a diffuser outlet located at the back of the substrate. Further UPS experiments performed on these films show that they are well uniform (no signal from the Pt(111) is detected). The ice film deposited is non-porous[9] and crystalline; we do not know if it is the cubic or the hexagonal form.[4]

The synchrotron beam was used both for irradiating and for the NEXAFS measurements (with settings avoiding a further irradiation). Preliminary experiments using monochromatic irradiation of a crystalline ice film were performed above the K-edge (550 eV), showing that a cumulated energy dose of 1.65 eV.molecule^{-1} is necessary to reach the saturation of the radiation effects (flux 2.5 10^9 photon.cm^{-2}.s^{-1}; energy flux 1.37 10^9 KeV.cm^{-2}.s^{-1}). We have then used the unmonochromatized "white" beam delivered at the zero-order of the monochromator (bandwidth 3 eV-900 eV), which gives results similar to the monochromatic exposures, except that the films were processed faster.

3 RESULTS AND DISCUSSION

Figure 1 present the NEXAFS spectrum of the irradiated crystalline ice film. The products observed are OH, O, HO$_2$, O$_2$ and H$_2$O$_2$. The peak at 526 eV is the excitation to the 1π state of the OH radical.[10] The one at 527.4 eV is the excitation to the 2p orbital of the atomic oxygen O(^{3}P).[10] The transition at 528.9 eV is the π^* orbital of the HO$_2$ radical.[8] The peak at 531 eV is the π^* orbital of O$_2$.[11] Last, the broad transition at 533 eV is the σ^* (O-O) orbital of the oxygen peroxide H$_2$O$_2$.[12] The 4a$_1$ state of water is at 535 eV, overlapping the right tail of the H$_2$O$_2$ σ^* state.

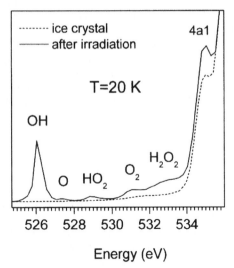

Figure 1 *O K-edge of a crystalline ice film deposited at 150 K and cooled at 20 K, before irradiation (dashed line) and after irradiation (full line). The 4a$_1$ state of water is also seen at 535 eV.*

Figure 2 shows the evolution of these products when warming the film at 125 K (during which no irradiations are made). The integrated intensities of these peaks are plotted on the right panel. These intensities have been obtained after the subtraction of the background (we used the experimental spectrum taken before the irradiation) and a fit of each peak with a Gaussian (the intensities are the area of these Gaussians). For clarity, the intensities of H$_2$O$_2$ and OH have been divided by 3.

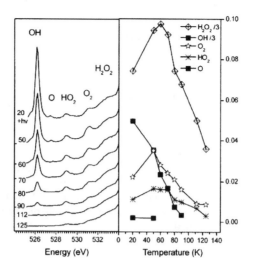

Figure 2 *Evolution with the temperature of OH, O, HO₂, O₂ and H₂O₂. Left panel: pre-edge regions of the NEXAFS of the crystalline film irradiated at 20 K and annealed up to 125 K. Right panel: evolution of the peak intensities with the temperature. The lines guide the eyes. The intensities of H₂O₂ and OH are divided by 3.*

The concentrations of HO₂, O₂ and H₂O₂ reach a maximum around 50 K and then decrease; OH monotonously decreases from 20 K. Note that the concentration of O is almost negligible. We discuss now the possible reaction channels for OH and O, then H₂O₂, O₂ and HO₂. The discussion on the product concentrations at 20 K is based on the thermal and non-thermal reactions that can occur in irradiated ice.[1] At T > 20 K, during the annealing, ice is not irradiated and only thermal reactions are retained.

OH and O: the starting point of the photoreactions is the water dissociation, leading essentially to OH and H. Trapped OH can also dissociate to produce O and H. When the temperature rises, these species can be involved in thermal reactions or diffuse and desorb. Since there is no photochemistry during the warming, the concentration of OH and O can only decrease, as observed.

H₂O₂: the initial production of H₂O₂ is most likely dominates the recombination of two OH:[1]

$$OH + OH \rightarrow H_2O_2 \qquad (1)$$

H₂O₂ can be also produced by the interaction of the water molecules and the secondary electrons through the reaction $H_2O + e^- \rightarrow O^-(^1D) + H_2$ and $O^-(^1D) + H_2O \rightarrow H_2O_2$.[13] When warming the film, the mobility of OH is enhanced by the temperature, favouring the radical recombination and increasing the H₂O₂ concentration, as observed between 20 K and 60 K. Above 60 K, the H₂O₂ concentration drops: the OH concentration becomes too weak for the H₂O₂ synthesis.

O₂: the initial production of O₂ can be related to atomic O by the set of reactions:[1]

$$O + O \rightarrow O_2 \tag{2}$$
$$O + OH \rightarrow O_2 + H \tag{3}$$
$$O + HO_2 \rightarrow O_2 + OH \tag{4}$$
$$O + H_2O_2 \rightarrow O_2 + H_2O \tag{5}$$

O_2 could be also produced by the decomposition of HO_2 after excitations by secondary electrons:[14]

$$HO_2^* \rightarrow O_2 + H \tag{6}$$

The increase in the O_2 concentration with the temperature (10 K - 50 K) cannot be explained by reactions with the atomic oxygen, since its concentration is negligible. O_2 is therefore produced by other reactions involving H_2O_2, HO_2 and OH. From H_2O_2, the exothermic production of O_2 is not possible without O. Only one possible reaction remains: it is the recombination of OH with HO_2:[1]

$$HO_2 + OH \rightarrow O_2 + H_2O \tag{7}$$

HO_2: the initial production of HO_2 can be due to the following the reactions:

$$O + OH \rightarrow HO_2 \tag{8}$$
$$H + O_2 \rightarrow HO_2 \tag{9}$$
$$H_2O_2 + OH \rightarrow H_2O + HO_2 \tag{10}$$

As for O_2, the increase in the HO_2 concentration between 20 K and 50 K cannot be due to the reaction of O with OH since the O concentration is too weak. Reaction of O_2 with H is possible, but the H radicals vanished in this temperature range.[15] HO_2 can only be produced by reaction between OH and H_2O_2.

When the remaining reactions (1), (10) and (7) are put together, we see that OH fuels the following step reaction:

$$OH + OH \rightarrow H_2O_2$$
$$H_2O_2 + OH \rightarrow H_2O + HO_2$$
$$HO_2 + OH \rightarrow O_2 + H_2O$$

Where O_2 is a sub-product of HO_2 and HO_2 is a sub-product of H_2O_2. This process ends around 50 K when the OH concentration is too weak for allowing the initial OH recombinations.

4 CONCLUSION

The NEXAFS spectroscopy at the O K-edge allows the identification of the radicals and molecules created by photolysis, except H and H_2, revealing the presence of OH, O, HO_2, O_2 and H_2O_2 in irradiated ice. Their temperature evolution shows that the creation of HO_2, O_2 and H_2O_2 results from a simple step reaction:

$$OH + OH \rightarrow H_2O_2$$
$$H_2O_2 + OH \rightarrow H_2O + HO_2$$
$$HO_2 + OH \rightarrow O_2 + H_2O$$

References

1   R. E. Johnson, T. I. Quickenden, *J. Geophys. Res.*,1997, **102**, 985.
2   R. A. Rosenberg, P. R. LaRoe, V. Rehn, J. Stöhr, R. Jeager, C. C. Parks, *Phys. Rev. B*, 1983, **28**, 3026.
3   J. S. Tse, K. H. Tan, J. M. Chen, *Chem. Phys. Lett.*, 1990, **174**, 603.
4   Ph. Parent, C. Laffon, C. Mangeney, F. Bournel, M. Tronc, *J. Chem. Phys.*, 2002, **117**, 10842.
5   F. Bournel, C. Mangeney, M. Tronc, C. Laffon, Ph. Parent, *Phys. Rev. B*, 2002, **65**, 201404R.
6   F. Bournel, C. Mangeney, M. Tronc, C. Laffon, Ph. Parent, *Surf. Sci.*, 2003, **528**, 224.
7   Ph. Parent, C. Laffon, *J. Phys. Chem. B*, 2005, **109**, 1547.
8   S. Lacombe, F. Bournel, C. Laffon, Ph. Parent, *Angew. Chem. Int. Ed.*, 2006, **45**, 4559.
9   G. A. Kimmel, K. P. Stevenson, Z. Dohnalek, R. Scott Smith and B. D. Kay, J. Chem. Phys. 2001, **114**, 5284
10 S. Stranges, R. Richter, M. Alagia, *J. Chem. Phys.*,2002, **116**, 3676.
11 Y. Ma, C.T. Chen, G. Meigs, K. Randall and F. Sette, *Phys. Rev. A*, 1991, **44**, 1848.
12 E. Rühl, A.P. Hitchcock, *Chem .Phys.*, 1991, **154**, 323.
13 X. Pan, A. D. Bass, J.-P. Jay-Gerin, L. Sanche, *Icarus*, 2004, **172**, 521.
14 T. M. Orlando, M. T. Sieger, *Surf. Sci.*, 2003, **528**, 1.
15 J. M. Flournoy, L. H. Baum, S. Siegel, *J. Chem. Phys.*, 1962, **36**, 2226.

SPEEDSKATE ICE FRICTION: REVIEW AND NUMERICAL MODEL - FAST 1.0

A. Penny[1], E. Lozowski[1], T. Forest[2], C. Fong[1], S. Maw[3], P. Montgomery[4], N. Sinha[5]

[1]Department of Earth and Atmospheric Sciences, University of Alberta
[2]Department of Mechanical Engineering, University of Alberta
[3]Faculty of Kinesiology, University of Calgary
[4]Mathematics Program, University of Northern British Columbia
[5]Institute for Aerospace Research, National Research Council Canada

1 INTRODUCTION AND REVIEW

The subject of ice friction has been beset with controversy and misunderstanding, in part because of the many orders of magnitude variation in the governing variables (e.g. ice temperature, ice texture, slider speed, slider geometry, normal force). Even a simple question such as, "what effect does ice temperature have on friction?" does not have a simple answer. It is therefore doubtful that a single theory or model could encompass the panoply of physical processes that give rise to ice friction and the diversity of its applications. Similar comments have been made with respect to snow friction.[1] The two classic monographs on ice physics[2,3], devote ten and three pages, respectively to the subject of ice friction, suggesting that our knowledge is very limited, and has actually diminished with time!

A valuable resource is Colbeck's bibliography[4], which lists 192 references on ice friction, dating from 1900 to 1991. Nevertheless, Colbeck has overlooked some significant work such as Fitz et al.,[5] and he lists no publications for the past fifteen years. Rosenberg traces published scientific interest in ice slipperiness and ice adhesion as far back as W. and J.J. Thomson in the 1850's and Michael Faraday in 1859,[6] and there are trails leading back to René Descartes in the 1630's.[7,8]

Modern consensus suggests three principal mechanisms may be involved in ice friction at skating speeds. These are surface premelting, pressure melting, and frictional melting. Colbeck discounts pressure melting as the principal mechanism in skating,[9] while, until now, there has been some controversy as to the significance of premelting vs frictional melting. An adhesion mechanism has been successfully invoked at much lower speeds.[10,11] Applications range from winter sports on ice to ice engineering, with substantial rewards potentially accruing to anyone able to modify ice friction for specific purposes.[12]

Our application is speedskating and our objective is to find ways to reduce the kinetic friction between ice and the blade of a speedskate, without heating the blade, a practice prohibited under rule 276 of the International Skating Union's Special Regulations.[13] The ISU regulations notwithstanding, several patents have been issued to inventors of heated hockey skate blades.[14] Even a small reduction in ice friction could have a big impact in competition, because ice friction can account for up to 25 % of the total resistance encountered by a competitive speedskater.[15] The first quantitative formulation of a frictional meltwater lubrication theory for an ice skate blade was published by Evans et al.,[16] who considered that heat conduction into the ice controls the lubrication process. They, in turn, relied on the earlier work of Bowden and Hughes[17] and Bowden[18], who first convincingly proposed a frictional meltwater lubrication explanation for the low sliding friction on ice and snow. Since Evans et al., there have been few if any advances in ice friction modelling for skate blades. On the experimental side, de Koning et al. have produced the only published measurements of skate blade friction at close to competitive speedskating velocities (~8 ms^{-1}).[15] Kobayashi has published data on ice friction using a loaded skate-borne sled at much lower speeds (<~ 1 ms^{-1}).[19] And Colbeck et al. have published measurements of hockey skate blade temperature rise near the blade-ice interface (at low skating speeds), that suggest frictional meltwater lubrication.[20]

Over the past two decades, there have also been theoretical and experimental advances related to ice friction in other contexts. Some of these have the potential to be imported into a theory of speedskate ice friction.[21,22,23,24,25] Stiffler recognized the significance of the coupling between the lubrication layer hydrodynamics (Couette and squeeze film flow) and its thermodynamics (viscous dissipation, heat conduction and melting).[26,27] Earlier, Furushima had devised a skate blade lubrication theory based on squeeze flow and melting, while ignoring heat conduction.[28] The application of thermohydrodynamics to ice friction has been invoked more recently by Fowler and Bejan,[29] Colbeck,[9] and Summers and Montgomery.[30] There have also been some attempts to apply thermohydrodynamics to curling on ice,[31,32,33,34] and to bobsled ice friction.[35]

This paper attempts to rectify the lack of a published numerical model of speedskate ice friction. We have formulated a mathematical model of the kinetic ice friction resistance encountered by a speedskate blade,[36] and we use it here, in the form of a numerical model to predict the kinetic ice friction coefficient. We call the numerical model **F**riction **A**lgorithm using **S**kate **T**hermohydrodynamics or FAST 1.0. We will validate the model using the published kinetic friction coefficient measurements of de Koning et al.[15] We then speculate about using the model to develop and test new ideas on how to improve speedskating performance by modifying the ice, the blade or the speedskating technique. The recent innovation of the klapskate[37] and the resulting improvements in competition times, suggest that we may not yet have reached the technical or biomechanical limits of speedskating performance.[38,39]

The physical processes involved in speedskate ice friction take place on time scales ranging from microseconds (the momentum diffusion time scale in the lubricating liquid) to minutes (the duration of a race). A very important speedskating time scale is the stroke frequency, which governs the contact time between the blade and the ice. The contact time per stroke varies from about 0.4 to 1.3 s, depending on skater and distance. The biomechanics of skating require that significant changes in a number of important variables occur over this time scale (skate blade angle, normal force etc).[40] Nevertheless, in order to keep the present model simple, we consider only constant velocity motion with the blade vertical, as if the skater were standing upright without stroking, but with his entire weight and some thrust (giving a total normal force of 1.28 times weight), applied to a single blade. The duration of the contact time between the blade and the ice is specified to be comparable to that in actual skating.

We begin by considering the blade-ice contact area. We then formulate the ploughing and shear stress forces, which together comprise the ice friction. Next, we posit the lubrication equation and solve it using finite-differences. Finally, we validate the model results against experimental data and show some additional model predictions.

2 FAST 1.0

2.1 Contact Area

The size and nature of the contact between the slider and the ice is possibly the biggest uncertainty in ice friction. For a stationary skate blade, the *apparent contact area* is the cross-section of the impression left behind in the ice when the blade is removed. The *actual contact area* for a moving blade depends on the size and number of asperities on the surface of both the ice and the blade, and the thickness of any lubricating layer that may separate those asperities. In speedskating, the actual contact area may be close to the apparent contact area.[20] Typically, competition speedskates are polished to give a smooth finish with surface roughness values of 0.1 µm or less. We measured similar values, for the Maple Comet Steel skates used in our experiments, with a PocketSurf III instrument made by Federal. According to our measurements, newly re-surfaced ice has a surface roughness of about one micrometer, but it is possible that ice surface asperities are crushed or melted by the leading edge of the blade. Under our standard speedskating conditions, the thickness of the lubricating melt layer is near 0.5 µm. Hence, the significance of asperity contact for speedskating is uncertain. We sidestep the contact area issue by using an effective ice resistance pressure (hardness), by measuring the apparent contact area of a static skate blade, as a function of ice temperature.

The static force balance for a skater standing on one skate is:

$$l_s w \overline{p}_s = mg \qquad (1)$$

where l_s is the length of the static contact zone, w the width of the contact zone (which we take to be the width of the blade), m the skater's mass, g the local acceleration of gravity, and $\overline{p}_s$ is the static mean resistive stress of the ice. If $\overline{p}_s$ is known, then l_s can be solved for and conversely. We undertook experiments on a cold room ice sheet, at the University of Alberta's Clare Drake Arena, and at the Calgary Olympic Oval to determine $\overline{p}_s$ from measurements of l_s over a range of ice temperatures. The linear regression function was $\overline{p}_s = 3.76 - 0.181T$, where pressure is in MPa and T in Celsius degrees. We use this regression in the model in order to infer l_s from Equation 1. It is likely that the dynamic contact length, l_d, is a fraction of the static contact length, depending on the extent and speed of recovery of the indentation strain, and differences between the static and dynamic resistive stress of the ice. The ratio of the two contact lengths, f_l, is a parameter in the model, and we use 0.5 as a standard value, representing the mid-point of the conceivable range of f_l.

We assume that a speedskate blade has a constant radius of curvature, r, over the contact length. Values of r vary from about 20 to 30 m in long track speedskating. Simple geometry then leads to an estimate of d, the maximum static penetration depth of the blade:

$$d = l_s^2 / 8r \qquad (2)$$

Knowing l_s, w, and d gives us the complete geometry of the apparent contact zone as illustrated in Figure 1.

2.2 Ploughing Force

We assume that a moving blade creates a channel of depth d as it moves across the ice. If $f_l = 0.5$ and there is no strain recovery, the dynamic and static values of d are equal. Part of the resistance encountered by the blade is the force, F_p, required to plough through the ice at the front of the channel. While the actual process could be quite complex involving ice deformation, crushing and fragment clearing, we express it simply in terms of a resisting pressure, p_x:

$$F_p = p_x w d \qquad (3)$$

Estimating p_x is even more difficult than estimating $\bar{p}_s$, and measuring it independently of the total frictional resistance is impossible. Nevertheless, because the contact area is smaller for ploughing and the strain rate is larger, we expect that p_x could significantly exceed $\bar{p}_s$. With this in mind, we use $p_x = f_p \bar{p}_d$ with f_p a parameter in the model which we take to be 1.0.

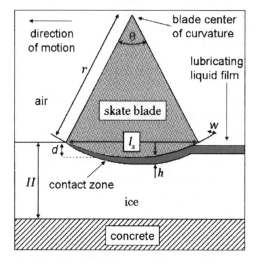

Figure 1   *Skate blade lubrication schematic*

2.3 Shear Stress Force

We begin by positing a thermohydrodynamical equation for the lubricating layer, based on Penny et al.[36] The primary hydrodynamic modelling assumption is the existence of a lubricating melt layer of thickness, h, below the skate blade. We create a time-dependent numerical model by solving the lubrication equation using finite-differences.

Based on the frictional melting layer assumption, we take this part of the skate blade drag to result from the power required to overcome the longitudinal viscous stress exerted on the blade by the lubricating fluid, considered to have a Couette flow. That is:

$$F_s = \mu \frac{V}{h} lw \tag{4}$$

Equation 5 for h, the depth of the lubricating fluid, is based on melting due to viscous dissipation (term 1 on rhs), offset by energy losses due to heat conduction into the ice, both fast (term 3) and slow (term 4), and loss of lubricating fluid through a lateral squeeze flow (term 2) from under the blade. We have ignored heat transfer with the skate blade here, but it is considered in Penny et al.[36] The lubrication equation is:

$$\frac{\partial h}{\partial x} = \frac{\mu V}{h \rho_w l_f} - \frac{mgh^3}{\mu l(w+d)^3 V} - \frac{1}{\rho_w l_f} \sqrt{\frac{k_i \rho_i c_i}{\pi V x}} \Delta T_i - \frac{1}{\rho_w l_f} \frac{k_i}{V H} \Delta T_b \tag{5}$$

The nomenclature is as follows: h lubrication layer thickness; x longitudinal distance along contact zone; μ dynamic viscosity of water; V skating velocity; ρ_w density of water; l_f specific latent heat of freezing at the melting temperature; m skater mass; g acceleration of gravity; l dynamic contact length; w blade width; d skate blade cut depth; k_i thermal conductivity of ice; ρ_i ice density; c_i specific heat capacity of ice; ΔT_i difference between pressure melting temperature and initial ice surface temperature; ΔT_b difference between pressure melting temperature and basal ice temperature; H ice thickness.

Equation 5 is difficult to integrate numerically because of singularities in the first and third terms rhs. One can avoid these singularities by transforming the independent variable to h^2 and the coordinate to $x^{0.5}$. The model code is written in FORTRAN and solves a finite-difference version of Equation 5. Standard values are: blade width 1.1 mm; skater mass 75 kg; rocker radius 25 m; skating velocity 10 ms^{-1}; ice thickness 25 mm; ice surface temperature -9° C; ice basal temperature -18° C; blade contact time 0.75 s; thrust factor 1.28.

3 MODEL RESULTS

Figure 2 shows the thickness of the lubricating layer, h, as a function of distance along the dynamic contact zone. The upstream boundary value of h is the depth of the quasi-liquid layer based on the ice surface temperature.[41] The quasi-liquid layer serves as the lubricant over the first 0.1% of the contact zone. The upper curve represents the frictional melting term (first term on the rhs of Eq (5)). The effects of adding squeeze flow and heat conduction into the ice are shown by the two intermediate curves. The lowest curve combines all factors, and shows that at a fixed reference point on the ice, the depth of the lubricating layer quickly approaches an asymptotic limit close to 0.5 µm under our standard speedskating conditions.

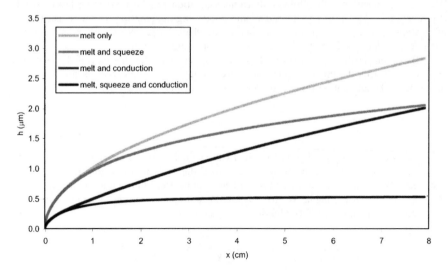

Figure 2 *Lubrication layer thickness vs longitudinal distance along contact zone*

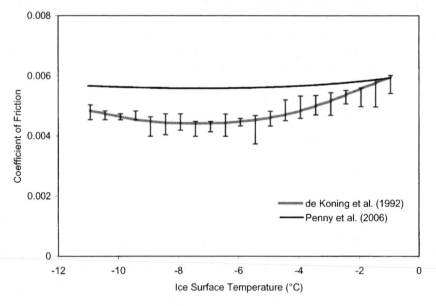

Figure 3 *Modelled and measured coefficient of friction vs ice surface temperature*

Figure 3 compares de Koning's measured friction coefficient[15] with model results, at a nominal skating velocity of 8 ms[-1]. Since we do not have de Koning's raw data, the error bars indicate broadly the range of data values in their graph. Even though the model's computed values are higher than de Koning's measurements, the proximity is encouraging. The model predicts a minimum in the friction coefficient near -7.2° C, while de Koning's measurements show a minimum near –7.5° C. The measured strong upturn in friction for warm ice is also exhibited in the model, but at somewhat higher skating speeds (see Figure 4).

Comparing model results with de Koning's friction coefficient as a function of skate velocity shows a similar tendency of increasing friction coefficient with velocity. However, the model gives a weaker dependence on skating velocity. The stronger measured velocity dependence may be a biomechanical effect arising from the increased thrust at higher velocity.

Figure 4 shows the modeled effect of skating velocity and ice surface temperature on the friction coefficient. Velocities as high as 40 ms[-1] have been considered because of the potential for applying a modified version of our model to skeleton and bobsled runners. Warm ice exhibits a square root dependence on skating velocity, consistent with ignoring heat conduction into the ice and squeeze flow.[36] With cold ice, the heat conduction term becomes more important, and a low coefficient of friction can only be attained with increased skating speed, although the friction coefficient begins to rise again when the speed becomes sufficiently high. For skating velocities close to competitive skating speeds, there is a saddle in the surface and a minimum ice friction coefficient occurs at an intermediate ice temperature, as seen in Figure 3. Because the trend of friction coefficient with ice temperature reverses from low to high skating speeds, the model results suggest that one cannot readily apply or extrapolate the results of low speed friction measurements to much higher speeds.

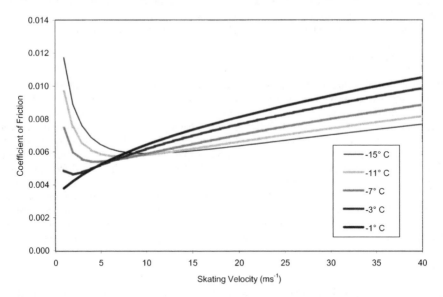

Figure 4 *Effect of ice temperature and skating velocity on coefficient of kinetic ice friction*

4 DISCUSSION AND CONCLUSIONS

FAST 1.0 is the first published numerical model of speedskate ice friction. Using a thermohydrodynamic theory for skate blade lubrication, it is able to estimate the kinetic ice friction coefficient for speedskates over a wide range of skating conditions. It may be possible, in future, to use FAST to investigate the influence of numerous factors on speedskate ice friction, including blade factors such as blade width, rocker radius and thermal conductivity, ice factors such as temperature, hardness and thermal conductivity, and skating factors such as skating speed, skater mass and stroke frequency. With future improvements, the model should also be capable of investigating the effects of various biomechanical factors.

Based on our current experience with the model, we formulate the following principal conclusions:

1. Despite significant simplification of the speedskate ice friction problem, FAST 1.0 is able to predict kinetic ice friction coefficients that are consistent with the published measurements of de Koning et al.,[15] with an error of about 20% or less.

2. Under conditions typical of competitive speedskating, frictional melting, squeeze flow and heat conduction into the ice all play an important role in determining skate blade lubrication. Pressure-induced freezing point depression and the quasi-liquid layer are accounted for in the model, but they play only a minor role in determining the kinetic ice friction coefficient.

3. Because of the complex nature of the temperature and velocity dependence of the kinetic ice friction coefficient, it is important to measure ice friction under conditions that are as close as possible to those where the data will be applied. It is inadvisable to apply or extrapolate measurements and inferences made at low velocities to skating at much higher velocities.

4. Future improvements to FAST will include accounting for more realistic skating biomechanics and application to skeleton and bobsled ice friction. FAST could be a useful vehicle for developing and testing new ideas to improve performance by reducing ice friction.

Acknowledgements

We thank the following for their help with this research. Space does not permit us to elaborate their individual contributions: Johan Bennink, David Chesterman, Sam Colbeck, Rameses D'Souza, Bernie Faulkner, Bob Frederking, Bob Gagnon, Lorne Gold, Mark Greenwald, Ben Jar, Kameron Kiland, Charles Knight, Flo Macapagal, Lasse Makkonen, Mark Messer, Christine Nam, Victor Petrenko, Gerhard Reuter, Jan Rudzinski, Wladyslaw Rudzinski, Charles Ryerson, Martin Sharp, Gary Timco and Monty Wood. We are also grateful to the Clare Drake Arena at the University of Alberta and the Calgary Olympic Oval for allowing us to perform experiments in their facilities. Many thanks are due to the Institute for Aerospace Research, National Research Council of Canada, who hosted the second author during a recent sabbatical leave where the seeds for this research were planted. Last but not least, we are grateful to the Natural Sciences and Engineering Research Council of Canada for a Discovery Grant and an Undergraduate Summer Research Assistantship which funded the research.

References

1   S.C. Colbeck, *A Review of the Processes that Control Snow Friction*, U.S. Army CRREL Report 92-2, 1992, 37.

2   P.V. Hobbs, *Ice Physics,* Clarendon Press, Oxford, 1974, 837.
3   V.F. Petrenko and R.W. Whitworth, *Physics of Ice*, Oxford University Press, 1999, 373.
4   S.C. Colbeck, *Bibliography on Snow and Ice Friction*, US Army Corps of Engineers Cold Regions Research and Engineering Laboratory, Special Report 93-6, 1993, 18.
5   C.D. Fitz, W.H. Ito, H.T. Mantis, G.W. McElrath, R.G. Mokadam, H.E. Staph, W.C. Stolov, K.P. Synestvedt and R.C. Jordan, *Friction on Snow and Ice,* Institute of Technology, Mechanical Engineering Department, University of Minnesota, Minneapolis, 1955.
6   R. Rosenberg, *Why is ice slippery?*, Physics Today, December 2005, 50.
7   J.S. Wettlaufer and J.G. Dash, *Melting below zero*, Scientific American, February 2000, 50.
8   J.G. Dash, A.W. Rempel and J.S. Wettlaufer, *The physics of premelted ice and its geophysical consequences*, Reviews of Modern Physics (in press).
9   S.C. Colbeck, *Pressure melting and ice skating*, American Journal of Physics, 1995, **63**, 888.
10 N. Maeno and M. Arakawa, *Adhesion shear theory of ice friction at low sliding velocities, combined with ice sintering*, Journal of Applied Physics, 2004, **95**, 134.
11 K. Tusima, *Friction of a steel ball on a single crystal of ice*, J. Glaciology, 1977, **19**, 225.
12 V.F. Petrenko, *Ice adhesion and friction: fundamental aspects and applications*, Proceedings 11th International Conference on the Physics and Chemistry of Ice, Bremerhaven, July 2006 (this volume).
13 *Special Regulations Speed Skating and Short Track Speed Skating 2004*, ISU, 2004, 60.
14 J. Furzer and T. Weber, *Heating arrangement for ice skate blades*, U.S. Patent 6988735, Published January 24, 2006.
15 J. de Koning, G. de Groot and G.J. van Ingen Schenau, *Ice friction during speed skating*, J. Biomechanics, 1992, **25**, 565.
16 D.C.B. Evans, J.F. Nye and K.J. Cheeseman, *The kinetic friction of ice*, Proc. Roy. Soc. Lond., 1976, **A347**, 493.
17 F.P. Bowden and T.P. Hughes, *The mechanism of sliding on ice and snow*, Proc. Roy. Soc. Lond., 1939, **A172**, 280.
18 F.P. Bowden, *Friction on snow and ice*, Proc. Roy. Soc. Lond., 1953, **A217**, 462.
19 T. Kobayashi, *Studies of the properties of ice in speed skating rinks*, ASHRAE J., 1973, **73**, 51.
20 S.C. Colbeck, L. Najarian and H.B. Smith, *Sliding temperatures of ice skates*, American Journal of Physics, 1997, **65**, 488.
21 P. Oksanen and J. Keinonen, *The mechanism of friction on ice*, Wear, 1982, **78**, 315.
22 P. Oksanen, *Friction and adhesion of ice*, Publication 10, Technical Research Centre of Finland, Espoo, 1983, 36.
23 L. Makkonen, *Application of a new friction theory to ice and snow*, Annals of Glaciology, 1994, **19**, 155.
24 L. Makkonen, *Surface melting of ice*, Journal of Physical Chemistry B, 1997, **101**, 6196.
25 L. Makkonen, *A theoretical approach to rubber friction on ice*, In: ARTTU Final Report, Helsinki University of Technology, Laboratory of Automotive Engineering, J. Vainikka and H. Pirjola, Eds, 2003, 12.
26 K. Stiffler, *Friction and wear with a fully melting surface*, Journal of Tribology, 1984, **106**, 416.
27 K. Stiffler, *Melting friction and pin-on-disc devices*. Journal of Tribology, 1986, **108**, 105.

28 T. Furushima, *Study of the frictional resistance of skates*, Seppyo, 1972, **34**, 9. (in Japanese with English summary)

29 A.J. Fowler and A. Bejan, *Contact melting during sliding on ice*, Int. J. Heat Mass Transfer, 1993, **36**, 1171.

30 A.E. Summers and P.J. Montgomery, *Investigation of the motion of a fixed rotating solid body over ice with a thin fluid interface*, Advances in Fluid Mechanics V, WIT Press, 2004, 247.

31 M.R. Shegelski, R. Niebergall and M.A. Watton, *The motion of a curling rock*, Canadian Journal of Physics, 1996, **74**, 663.

32 R.A. Penner, *The physics of sliding cylinders and curling rocks*, American Journal of Physics, 2001, **69**, 332.

33 M. Denny, *Curling rock dynamics, towards a realistic model*, Canadian Journal of Physics, 2002, **80**, 1005.

34 B.A. Marmo, and J.R. Blackford, *Ice friction in the sport of curling*, Proceedings 5th International Sports Engineering Conference, Davis, California, 2004.

35 K. Itagaki and N.P. Huber, *Measurement of dynamic friction of ice*, In: Physics and Chemistry of Ice, N. Maeno and T. Hondoh, Eds., Hokkaido University Press, 1992, 212.

36 A. Penny, E.P. Lozowski, T. Forest and P. Montgomery, *A mathematical model of speed skate lubrication and ice friction*, 2006 (in preparation).

37 J. de Koning, H. Houdijk, G. de Groot and M.F. Bobbert, *From biomechanical theory to application in top sports: the Klapskate story*, J. Biomechanics, 2000, **33**, 1225.

38 G.H. Kuper and E. Sterken, *Endurance in speed skating*, Research Report 02C01, University of Groningen, Research Institute SOM, 2002, 17.

39 C. Versluis, *Innovations on thin ice*, Technovations, 2005, **25**, 1183.

40 T.L. Allinger and A.J. van den Bogert, *Skating technique for the straights, based on the optimization of a simulation model*, Medicine and Science in Sports and Exercise, 1997, **29**, 279.

41 J. G. Dash, *Surface melting*, Contemporary Physics, 1989, **30**, 99.

FIRST PRINCIPLES COMPUTATIONAL STUDY OF HYDROGEN BONDS IN ICE I_h

Patricia L. Moore Plummer

Department of Chemistry and Department of Physics and Astronomy, University of Missouri, Columbia, MO 65211 USA Email: plummerp@missouri.edu

1 INTRODUCTION

The hydrogen bond has been a curiosity and a subject of much speculation, discussion, and study for the better part of a century. Perhaps Linus Pauling was the first to recognize its great importance in a multitude of phenomena. He attempted to clarify existing observations and used his unique skills to provide theoretical insights into the roles played by the hydrogen bond[1], particularly in ice. It is the hydrogen bond that is responsible for most, if not all, of the unique properties of water ice. This bond is strong enough to cause water ice to exhibit more stable solid structures than any other material but weak enough to be easily broken at ambient temperatures or pressures. The presence, in each molecule, of two electron lone pair and two hydrogens give water an almost infinite ability to associate and bond with other molecules as well as itself. The facile nature of the hydrogen bond provides ample opportunity for violation of the Bernal-Fowler[2] ice rules and the presence of orientational or Bjerrum[3] defects in all macroscopic samples of ice. This same facile nature allows for considerable cooperative motion of the molecules resulting in the formation, migration and annihilation of both oriential and ionic defects at a much lower energy cost than would be predicted from the strengths of the OH and OH—O bonds.

Both the strength and the flexibility of the hydrogen bond introduce substantial problems for theorists. To date most models for water/ice system have considered the hydrogen bond to be purely electrostatic, sometimes with polarization from the surrounding medium, and the water molecule as a rigid body. Unfortunately this oversimplification has lead to an ever-increasing number of empirical interaction potentials, nearly as many as there are properties to be studied. These potentials have been successful in providing considerable insights into a variety of macroscopic properties but are not able to describe or account for effects that depend on the local environment of the molecules or the lattice when a defect is present. As the sensitivity of experimental methods increases, probing the molecular environment and the hydrogen bond necessitates a corresponding increase in the sophistication of the theoretical treatment of hydrogen bonding.

Even though the capabilities of computer hardware and software have been expanding nearly exponentially, we are still not able to carry out first principle calculations on near-bulk samples of ice. Thus we continue to be dependent on models, either for the ice itself or for the method of calculation. This report continues our use of water clusters[*] to model[4] the interactions in ice. Here we are able to use clusters containing 12 to 26 water molecules to simulate the local environment of the water molecule in ice. In each system, the focus of study is the environment of the molecules participating in four hydrogen bonds. In the clusters that have only 3 or 4 bonded molecules, the structure is optimized without constraints. For the clusters where 2-bonded molecules are present, some of their parameters are constrained in order to maintain an overall "ice-like" structure. In this report we examine the qualitative changes which occur as the environment of a water molecule in the cluster is altered, either by the presence of an ion or by a change in its hydrogen-bonding pattern. The quantitative details are available from the author.

2 METHOD

2.1   Description of the Models.

The smallest cluster employed in this study, containing 12 molecules (Figure 1), is an extension of our studies[4] on the lowest energy structure for 8 molecules, a cube having D_{2d} symmetry. Sample calculations were also made for tetrameric clusters having 16 (Figure 2) and 20 molecules (3 and 4 fused cubes respectively). The second model structure has twenty water molecules organized as a double layer of three of fused five-member rings (Figure 3). Limited calculations employed a 20-moecule ice-like cluster consisting of two layers of three hexagonal bi-layer rings (Figure 4).

Figure 1   *12 molecule Tetrameric Cluster*   **Figure 2**   *16 molecule Tetrameric Cluster*
          (Oxygens in lower plane are obscured by those in the upper plane.)

For studying the effect on structure and electron density of the presence of an ion in the model clusters, a proton is either added or removed from a surface water molecule, thus keeping the clusters isoelectronic. The structure of initial neutral cluster is a minimum on the potential energy surface. The ion-containing cluster is then annealed[**] until the structure is again at an energy minimum. The proton was introduced in a variety of ways. One method creates a H_3O^+ at one of the 3-bonded water sites and the model cluster containing the ion is then annealed until a minimum energy structure is attained. The second method places the proton at various locations near or internal to the cluster and a constant energy trajectory followed. When the potential energy is at a minimum, the cluster structure at that point is then subjected to annealing by removing kinetic energy.

Figure 3 *20 Molecule Pentameric Cluster*   **Figure 4** *20 Molecule Ice Cluster*

2.2   Calculational Details.

The electronic structure calculations were done using the Gaussian 03 quantum mechanical packages.[5] Analyses of the calculations were aided by graphical display of the results. The software packages GausView 03[6] and Molden[7] were used to examine the structural and vibrational data for the stable clusters and to produce the figures.

The density functional methods have been shown to provide an economical method to reliably include electron correlation in electron structure calculations. The hybrid density functionals have been increasingly employed for this purpose. The choice of B3LYP[8] as the density functional method to be employed in this study is justified by increasing evidence of its reliability to study hydrogen-bonded systems.[9] Most DFT approaches have been criticized for not taking proper account of dispersion effects, which can be important in weak hydrogen bonds. The use of the hybrid B3LYP together with a flexible basis appears to substantially reduce or remove this as a concern[10], especially for interactions as strong as those in the water ices. The description of the H-bond interaction is improved by the propensity of the hybrid functionals to contract the density toward the bond and valence regions.

The initial structures were fully optimized at the B3LYP level using the standard split-valence polarized 6-31G(d) basis[11]. Stationary points were evaluated by examining the associated Hessian matrix. The geometries thus identified as stable were used as the initial geometry for optimization, or in the case of the larger clusters, single point energy calculations, with larger basis sets and/or advanced methods.

The choice of basis sets used to describe weakly bound clusters is very important. While adding polarization functions to standard double zeta bases is generally sufficient to provide a reasonable description of 'moderate' to 'strong' hydrogen bonds[12] found in defect free water/ice clusters, additional flexibility in the basis set is required for description of clusters having a negative charge. For a proper description of anions and to allow for polarization and dispersion stabilization, diffuse function are often needed on the hydrogens as well as the heavy atoms.[13] The triple split-valence basis with polarized and diffuse functions, 6-311++G(d,p), satisfies these criteria/needs and has proved reliable in previous studies.[13]

The use of basis sets containing diffuse functions with density functional methods requires added care in the computations. Thus in the B3LYP calculations the evaluation grid was expanded both in terms of additional radial shells and added grid points per shell. Also the SCF convergence criterion was tightened.

The weak binding together with the potential large amplitude motions in these clusters due to many low frequency intermolecular vibrational modes makes the calculation and inclusion of corrections for zero point vibrational energy (ZPVE) important. Therefore ZPVE are calculated for all systems.

The correction for superposition error (BSSE) in the cluster energies for stationary states is calculated using the counterpoise procedure (CP) first described by Boys and. Bernardi[14] and implemented in Gaussion 03 using the procedure of Dannenberg.[15]

3 RESULTS AND DISCUSSION

The hydrogen bond in water ices is very flexible, exhibiting a high degree of bending while maintaining most of its interaction energy. [The potential energy surface is relatively broad and flat around the position of the 'shared' hydrogen.] This flexibility permits the lattice to accommodate the presence of defects, either neutral or charged, without substantial disruption of the oxygen lattice. The hydrogen bond flexibility is due in large measure to the transfer of electrons from O non-bonding (lone pair) orbitals of the acceptor molecule to virtual bond orbital of OH of the donor molecule. Additional transfer from the O lone pair to the OH bond within a given molecule facilitates structural deformation by the increase or decrease of the internal water HOH angle. These observations and those which follow are independent of either the number of molecules in the cluster or the details of the ring shape—tetramer or pentamer. This suggests that the local environment due to the nearest neighbor orientation dominates the subsequent stabilization or migration of an inserted ion defect.

For the size of clusters examined in this study, the lowest energy structures are associated with the smallest dipole moment for a given cluster size. For these model clusters which contain a mixture of 3-bonded (surface-like) and 4-bonded (bulk-like) molecules, the lowest electron density and greatest shift of density from an oxygen lone pair orbital to OH bonding orbitals is observed for double donor (dd) molecules occupying 3-bonded sites (surface-like). The oxygens with highest negative charge are in double donor, double acceptor (dd,aa) molecules occupying bulk-like sites.

3.1 Positive Ion Clusters

For the clusters containing a positive ion, regardless of the method of introduction, the protons prefer to move to 1) reduce the overall dipole of the cluster and 2) to localize the ion on a 4-bonded molecule—from a 'surface' to a 'volume' site. The hydrogens of the H_3O^+ form H-bonds with the adjacent waters while the oxygen's propensity to accept a hydrogen is reduced, leading to an elongated and highly bent h-bond that resembles a distorted L-defect. The positive charge is "shared" between the three H-bonded near neighbors which are now ~0.1e more positive than in the neutral cluster. The migration of the charge from a surface molecule to a volume site relies on the cooperative motions of many of the cluster molecules. A typical migration is illustrated in Figure 6. Soon after a proton is added to a corner O the H-bond from donor neighbors weakened and its donor OH bond lengthens as the proton moves toward the acceptor O occupying a 4-bonded site. Then the acceptor's donated H shifts and an $H_2O-H^+-OH_2$ cluster appears briefly. Then the 'shared' H^+ moves to its acceptor O as its OH bond lengthens. Next the migration is completed with the transfer of the proton to O4. Finally, the donor H-bond to O4 weakens as the O--O distance increases and the H moves slightly away from the O—O line forming pseudo L-defect. This motion is greater in tetrameric cluster systems where the donor molecule has only 3 bonds but decreases with increasing cluster size. Also in the 12-molecule structure, the hydrogen bonds across the body diagonal of the cube from the ion are weakened due to the distortion resulting from the O—O distances around the ion shortening by about 10 percent.

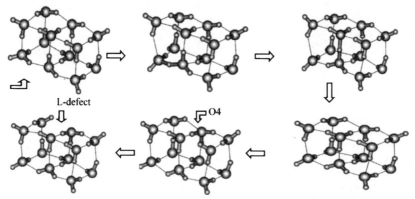

Figure 5 *Sequence of proton moves in annealing of $[H_2O]_{12}H^+$, clockwise from upper left.*

3.2 Negative Ion Clusters.

Removal of a proton which is not participating in a hydrogen bond (from a 'surface' molecule) results, on annealing, in rearrangement of hydrogen bonds similar to that seen for the positive ion. Again the ion moves to a 4-bonded site, this migration being facilitated by cooperative motions of molecules throughout the cluster. The local environment of the negative ion has it as the acceptor of 3 H-bonds while not donating its lone hydrogen. The excess negative charge is shared between these 4 oxygens. Electron density is transferred from the OH bond orbitals of the donors to the anti-bonding orbital of the OH acceptor. Some electron density from the oxygen core of the OH acceptor is back donated to the anti-bonding OH bond of the donors. The oxygen of the acceptor OH retains ~0.15e more than the donors. There is little or no exchange of electrons with the non-donor neighbour, so here; again, the interaction with the fourth neighbour is weakened. Not unexpectively, the librational frequencies for this group is blue shifted by 100-200 cm^{-1} while the associated stretching frequencies are red shifted by ~300 cm^{-1}. The initial and final structures are shown in Figures 6 and 7 for the tetrameric and pentameric clusters.

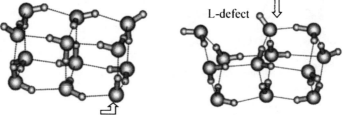

Figure 6 *Initial and final structure for the $[H_2O]_{11}OH^-$ cluster (ion position indicated by arrow)*

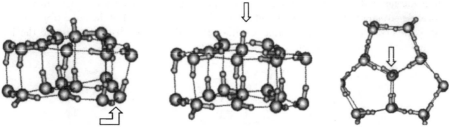

Figure 7   *Initial and 2 views of final structure for the [H₂O]₁₉OH⁻ cluster (ion, indicated by arrow)*

An ion pair was introduced into the tetrameric 12-molecule and the pentameric 20-molecule cluster. All placements of the ion pair in the 12-molecule cluster results in recombination upon annealing. Recombination also occurs in the 20-molecule cluster when the two ions are in the same or adjacent rings. However, a stable zwitter ion results when the ions are initially placed one in the upper level and the other in the lower level with a ring of molecules between. The presence of the ion pair creates very little disruption to the structure or the charge density of the cluster in contrast to that produced by a single ion.

Both the structural and the electron density alteration due to the presence of the ion are evident in the electron density contours. The contours for the neutral and ion containing 12-molecule cluster are shown in Figure 8.

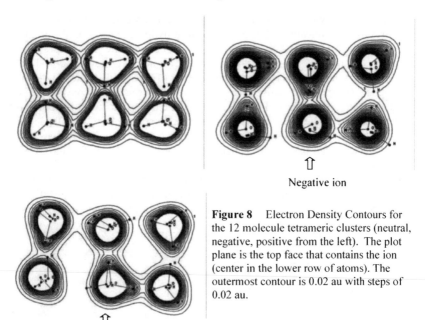

Negative ion

Positive ion

Figure 8   Electron Density Contours for the 12 molecule tetrameric clusters (neutral, negative, positive from the left). The plot plane is the top face that contains the ion (center in the lower row of atoms). The outermost contour is 0.02 au with steps of 0.02 au.

Similar changes in the electron density are found for the 16-molecule cluster but as with the structure, the disruptive effects of the presence of an ion are less.

For the pentameric clusters, the changes due to the ion are even subtler. The major structural change upon ion introduction and subsequent annealing is a decrease in planarity of the rings containing the ion. For the negative ion, the OH⁻group is elevated relative to the rest of the cluster (see Figure 7).

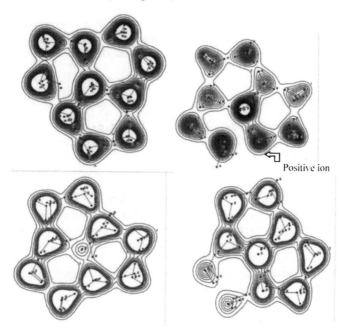

Figure 9   *Electron density contours for pentameric clusters. Plot plane is upper face and contains the ion (upper plots: neutral and positive ion, lower plots zwitter and negative ion). Negative ion is located in the center, positive ion in lower plane at 11 o'clock for zwitter ion and upper plane at 5 o'clock for the positive ion.*

4 SUMMARY AND CONCLUSIONS

Examination of the local environment of dd,aa molecules in the model clusters provide a more realistic understanding of the hydrogen bond in water ices. We find that there is rehybridization of the water OH bond with charge donation to the low –lying unoccupied bonding orbitals from the acceptor molecules.

When an ion is introduced, considerable re-orientation of the hydrogen-bonding pattern is possible without significant structure disruption. Proton migration is really proton transfer, facilitated by the cooperative motions of nearby molecules. These cooperative motions including molecular librations, hydrogen bond distortion, and HOH angle changes are essential (and cannot be described by a rigid body empirical potential model). Proton transfer is favored along paths where the water molecules have the same h-bonding environment (donor—acceptor—donor...) and can be slowed or even halted when the pattern is interrupted. Stable ion pairs can be the result. The presence of such a favored path results in ion migration from a 'surface' to a 'bulk' site.

The presence of an ion is more disruptive the smaller the cluster, especially when the majority of molecules occupy 'surface' sites. In the larger clusters, small changes in the local environment of many molecules facilitate both the migration and the accommodation of the ion.

These qualitative observations are not dependent on basis set size, inclusion of ZPVE nor corrections for BSSE. Additionally, neither Coulombic correlation nor basis sets enlarged above double zeta with polarization change the appearance of the electron charge density. Thus B3LYP with 6-31g(d) can be used to describe even larger, more realistic ice systems.

Acknowledgements

I wish to thank Otho Plummer for many helpful discussions and for graphics support. I also wish to acknowledge the Univ. of Missouri Research Network for computer support.

Notes

* It is also possible to use DFT and periodic boundary conditions to study many ice properties, but when ion defects are present, the cell must contain many more particle to prevent PBC's from altering results. In addition, an accurate description of the 'local' environment is of primary importance. With 'supercells' both the size of basis sets and the type of functional employed are limited.

** A classical molecular dynamics trajectory is run using the quantum mechanical potential surface. Annealing periodically zeroes the kinetic energy.

References

1 L. Pauling, *Nature of the Chemical Bond*, 1965, Cornell Univ. Press, Ithaca, NY
2 J. D. Bernal, R. H. Fowler, *J. Chem. Phys.,* 1933, **1**, 515.
3 N. Bjerrum, K. Danske, *Vid.Selsk-Mat-Fys.Medd*, 1951, **27**, 3; *Science*, 1952, **115**, 385.
4 P. L. M. Plummer, *J. Phys. Chem. B*, 1997, **101**,6247; *Adv. In Sci. & Tech.,*
   Computational Modeling and Simulation of Materials, P. Vincerzini and
   A. Degliesposti, Eds., 1999, **18**, 235; *Canadian J. of Phys.*, 2003, **81**, 325.
5 *Gaussian 03, Revision B.04*, 2003, *Revision D .01*, 2005, Gaussian, Inc., Pittsburgh PA.
6 *GaussView 03*, Gaussian, Inc., copyright 2000-2003 SemiChem, Inc.
7 G. Schatfenaar, J. H. Noordik, *J. Comput.-Aided Mol. Des.* 2000, **14**, 123.
8 A. D. Becke, *J. Chem. Phys.* 1988, **88**, 1053. C. Lee, W. Yang, R. G. Parr, *Phys. Rev. B.,* 1988, **37**, 785.; A. D Becke, *J. Chem. Phys.,* 1993, **98**, 5648.
9 X. Wu, M. C. Vargas, S. Nayak, V. Lotrich, G. Scoles, G., *J. Chem.Phys.,* 2001, **115**, 8748; A. D. Rabuck, G. E. Scusria, *Theor. Chem. Acta,* 2000,**104**, 439.
10 O. Galvez, P. C. Gomez, L. F. Pacios, *J. Chem. Phys,* 2003, **118**, 4878.
11 R. Ditchfiled, W. J. Hehre, J. A. Pople, *J. Chem. Phys.,* 1971, **54**, 724; W. J. Hehre, R. Ditchfield, J. A. Pople, *J. Chem. Phys.,* 1972, **56**, 2257.
12 S. Chattopadhyay, P. L. M. Plummer, *Chem. Phys.,* 1994,**182**, 39.
13 J. S. Lee, S.Y. Park, *J. Chem. Phys.,* 2000,**112**, 230.
14 S. F. Boys, F. Bernardi, *Mol. Phys.,* 1970, **19**, 144.
15 S. Simon, M. Duran, J. Dannenberg, *J. Chem. Phys*, 1996, **105**, 1024.

FREEZING OF WATER ON α-Al$_2$O$_3$ SURFACES

H.H. Richardson,[1] Z.N. Hickman,[1] A.C. Thomas,[1] K.A. Dendramis, [1,2] G.E. Thayer[3] and G.E. Ewing[4]

[1] Department of Chemistry and Biochemistry, Ohio University, Athens, Ohio 45701
[2] current address: Department of Chemistry, University of Washington, Seattle, WA 98195
[3] Department of Biomolecular Materials and Interfaces, Sandia National Labs, Albuquerque, NM 87185
[4] Department of Chemistry, Indiana University, Bloomington, IN 47405

1 INTRODUCTION

Water is not easy to freeze. Purified droplets of 10 μm diameter, a typical size found in clouds, can be cooled to nearly -40 °C. [1,2] and depending upon the conditions, can form cubic ice instead of the more thermodynamically stable hexagonal form.[3] A likely super cooling record is for water clusters of 5 nm diameter that do not freeze until -70 °C.[4] Addition of judiciously selected foreign substances can raise the freezing point of water considerably. Today suspensions of bacterial fragments of *Pseudomonas Syringae* are mixed in with water sprayed over ski slopes to favor production of artificial snow at temperatures only slightly below 0 °C.[5] The choice of AgI as an ice nucleation (IN) agent by Bernard Vonnegut[6] was made because its crystalline lattice constant was a near match to that of the basal face of ice. And AgI in the form of a smoke is still used for cloud seeding operations.

The mechanism of IN, laid out in classical terms, is elegantly described by Fletcher[1] and a survey of effective (and ineffective) agents that catalyze the freezing of water is reviewed by Pruppacher and Klett.[2] More recently, simulations have explored freezing at the molecular level.[7-10]

Despite the vast literature on ice freezing, no uniform picture of the IN process has been put forward. However factors that might favor an effective IN agent have been suggested.[2] Among them are 1) A near match of the IN agent lattice constant with one of the ice lattice constants, 2) The IN agent surface should be hydrophilic, and 3) The IN agent surface should have defects.

To explore these factors we have chosen α-Al$_2$O$_3$ (sapphire) as a model substrate for our IN studies. High purity crystals are readily available. The α-Al$_2$O$_3$ (0001) face has an hexagonal lattice whose constant is within a few percent of that of the basal face of ice.[11] The (0001) face can be annealed into a smooth form and characterized by atomic force microscopy (AFM). The face can be chemically treated to make it hydrophilic. Finally α-Al$_2$O$_3$ can be produced in forms rich in defects.

2 EXPERIMENTAL

2.1   The α-Al₂O₃ surface

Single crystals of α-Al$_2$O$_3$ were purchased from Coating and Crystal Technology
(http://www.coatingandcrystal.com/) with (0001) oriented and polished surfaces (The
specifications given were: < 0.5° orientation accuracy; < 0.5 nm rms roughness). The
surfaces were characterized by us using atomic force microscopy (AFM) and were found
to have large step densities with terrace widths of approximately 50 nm with non-linear
step edges (Figure 1a). To make the surfaces smooth, we annealed the crystals in a
sapphire tube furnace to 1450°C for 24 hours under ambient atmospheric conditions.[12]
After the crystals cooled, they were immediately immersed in 1M HNO$_3$ for 5 minutes to
uniformly hydroxylate the surface.[11] Finally, the surfaces were etched in Aluminum
Etchant 'A'[13] for 20 minutes at an elevated temperature of 35°C. The treated surfaces were
found to be smooth with straight edged steps of height of 1.3 nm and terrace widths of
approximately 500 nm (Figure 1b). The step height (shown by markers in section analysis)
corresponds to the unit crystal 'c' dimension of 1.2988 nm in models with perfect oxygen
termination of α-Al$_2$O$_3$[11]. The particles on the surface are believed to be nanometer size
particulates of Al$_2$O$_3$. The white spots in the image were removed by sonicating the α-
Al$_2$O$_3$ substrate in ultra-pure chloroform before its introduction into the ice nucleation
apparatus.

The study of narrow size distribution α-Al$_2$O$_3$ particles was accomplished by spin
coating their suspension on an α-Al$_2$O$_3$ (0001) face first. Particles of nominal diameter 30
nm and 100 nm (purchased from NanoProducts Corporation,
http://www.nanoproducts.com) were prepared as suspensions in chloroform. After the
nanoparticles were spread, the α-Al$_2$O$_3$ substrate was ready for the nucleation experiments.

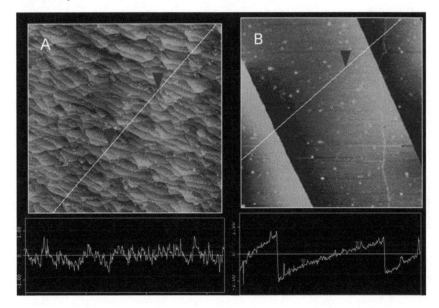

Figure 1 *A. AFM images (tapping mode) of a sapphire surface before it was annealed in the tube furnace. The image is 5.0μm × 5.0 μm, with units on the vertical cross sections in nm. B. AFM image of sapphire after it was annealed and etched in Aluminum Etchant 'A'. The image is 1.0μm × 1.0 μm, units on the vertical cross section are in nm.*

2.2   Methods and procedures

The nucleation of water and ice was monitored with an optical microscope whose objective was inserted into an opening in a purge box, as shown in Figure 2. The α-Al$_2$O$_3$ substrate was held by a sample holder against a peltier cooler with temperature monitoring by two thermocouples.

At the start of an experiment the substrate was cooled under a dry nitrogen purge. When the target temperature was reached the nitrogen flow was reduced and water from the ambient air seeped into the purge box. The relative humidity and temperature inside the purge box (measured with an Omega thermo-hydrometer) at the onset of droplet formation was 7% at 70 °C. Ice nucleation was timed from the first observation of water droplets to the time that crystalline particles formed.

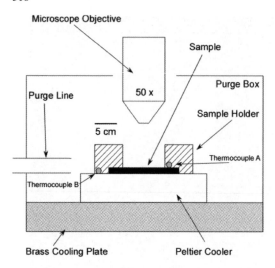

Figure 2   *Ice nucleation apparatus.*

3 RESULTS AND DISCUSSION

3.1   Water nucleation on α-Al₂O₃ (0001)

The α-Al₂O₃ (0001) surface at -20 °C after the formation of water droplets as imaged by optical microscopy is shown in panel A of Figure 3. The outlines of the droplets, typically 10 to 20 μm in diameter, are roughly circular. The interference fringes from the white light illumination enable the height contours to be constructed. This is shown for a representative droplet in panel B of figure 3. The profile allowed the contact angle of 4° to be determined. Larger droplets have a contact angle between 5° and 6°. This measurement demonstrates that the α-Al₂O₃ (0001) surface that we have prepared is indeed hydrophilic.

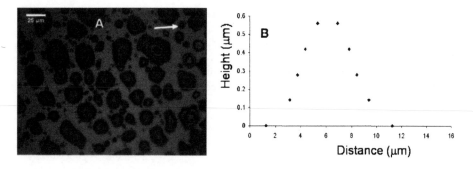

Figure 3   *Water droplets on α-Al₂O₃ (0001) at -20 °C. A. Microscopic images of water droplets. B. Topographical scan of a droplet (arrow in panel A) from its interference fringes.*

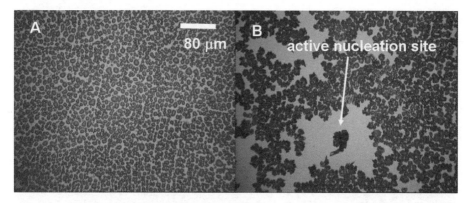

Figure 4   *Ice nucleation on α-Al₂O₃ (0001) at -20 °C. The scale is the same for A and B.*

3.2   Ice nucleation α-Al₂O₃ (0001)

An optical image of water droplets on α-Al₂O₃ (0001) at -20 °C is shown in figure 4A. The droplets, as in figure 3, are roughly 10-20 μm in diameter and separated from each other. They are evenly distributed except for several narrow slightly curved regions running almost vertically where their formation seems to have been inhibited. These regions may be associated with scoring introduced in the polishing process.

After 500 seconds the droplets spontaneously freeze into crystallites as shown in figure 4B. The crystallites, like the droplets, are roughly uniform in diameter but somewhat larger. Unlike the droplets, which were all isolated, the crystallites are connected to neighbors. The single exception to this pattern is the large crystallite labeled "active nucleation site". An explanation for the links among the rest of the crystallites is found in the next figure.

Figure 5 shows the propagation of the ice formation on a finer length scale. Moreover a defect-free region of the smooth α-Al₂O₃ (0001) surface has been chosen for interrogation so that ice does not nucleate at -27 °C until about 10 minutes.

The ice crystallites begin to form in figure 5A. Transformation of a droplet is first evidenced by the disappearance of its smooth boundary being replaced with ragged edges with extensions (pods) growing outward. Next the interference fringes vanish at the freezing of the entire droplet.

The ice growth is fueled by the higher vapor pressure of the droplets. The nucleation process is propagated with a pod. An ice crystallite touches a neighboring droplet, instantly freezing it, the ice nucleation front is indicated by the arrows in figures 5B and 5C. Figure 5D shows the final state of the freezing process with all the crystallites connected.

The significance of the observations in figure 5 is that all the ice crystallites were nucleated, not by α-Al₂O₃ (0001), but by ice itself. Only the occasional active site on Al₂O₃ can nucleate ice.

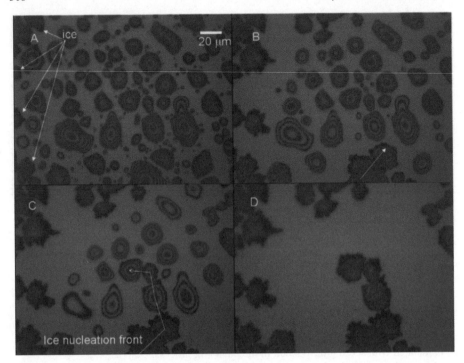

Figure 5   *Ice nucleation on smooth α-Al₂O₃ (0001) at -27 °C. A. water droplets and ice
crystal front. B-C. Ice nucleation of adjacent water droplets by propagation
of ice pods along surface. D. Aftermath of nucleation showing ice crystals on
surface.*

3.3   Ice nucleation from active sites

Figure 6 shows the nucleation behavior at one active nucleation site. Panel A shows the
onset of droplet formation. The active site traps liquid water and converts the water
droplet into ice at a much higher temperature than the surrounding surface. These active
sites will nucleate ice at temperatures 20K higher than the surrounding surface. The
surface temperature in figure 6 is held at -10 °C. Three dimensional dendritic growth of
ice is observed the instant water vapor is introduced. If a layer of water covered the entire
surface then nucleation of this layer would be initiated by the active site. Probing the
nucleation behavior of the surface with this type of experimental procedure would uncover
the nucleation behavior of the small number of defect sites (active nucleation sites). One
might be misled by the nucleation ability of these active sites and conclude that the flat
surface has a high propensity to nucleate ice. The reality is that the flat surface is
extremely passive to nucleating ice but that the active sites with unknown structure are
responsible for the nucleation of ice.

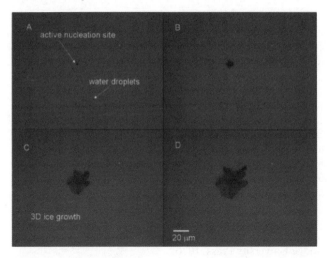

Figure 6 *Ice nucleation on an active site at -10 °C. A. Initial formation of water droplets on the surface with ice nucleation at the active site. B-D. Three dimensional growth of ice crystal.*

3.4 Nucleation by Al₂O₃ nanoparticles.

In these experiments we explored the role of α-Al_2O_3 nanoparticles on the nucleation process. Small particles of α-Al_2O_3 have multiple exposed surface planes that might be good promoters for ice nucleation. After coating the α-Al_2O_3 (0001) surface with 30 nm or 100 nm particles the substrate was cooled to -23 °C and water vapor was introduced. Neither nanoparticle sizes nucleated ice in less than 5 minutes. Although drops of water did form. After 5 minutes ice did form, but this makes Al_2O_3 nanoparticles less effective than the defects on the smooth α-Al_2O_3 (0001) surface.

We have therefore not identified the nature of the active sites on α-Al_2O_3 (0001).

3.5 Thin film water and nucleation on α-Al_2O_3 (0001).

Adsorbed water layers on insulator surfaces have been established since the early work of Langmuir.[14] These layers, often called thin film water, have thicknesses of the order of nanometers.[15] For α-Al_2O_3 the film thickness at room temperature and approaching the equilibrium water vapor pressure, is about 2 nm.[16] The thickness of the film at the low temperatures of our experiments has not been measured. Its role in water droplet and ice nucleation needs to be explored.

4 CONCLUSION

The smooth, hydroxylated, Al_2O_3(0001) surface, despite its favorable lattice match with the basal face of ice and its hydrophilic character, is ineffective at ice nucleating. Nanoparticles of Al_2O_3 are likewise ineffective IN agents. Ice did form on unidentified active sites of $Al_2O_3(0001)$.

In our experiments, ice crystallites appeared, not from IN by $Al_2O_3(0001)$, but by ice nucleating itself. Micrometer size ice crystals send out pods along the surface to touch and nucleate nearby water droplets. The mechanism for the pod growth has not been elucidated.

Acknowledgements

G.E.E. thanks the National Science Foundation for their support. G.E.T. thanks the Center for Integrated NanoTechnologies (CINT), a Department of Energy/Office of Science Nanoscale Science Research Center (NSRC) national user facility, for funding. H.H.R. thanks the NanoBiotechnology Initiative at Ohio University for funding.

References

1   N. H. Fletcher, *'The Chemical Physics of Ice'*, ed. A. Herzenberg, M. M. Woolfson, and J. M. Ziman, Cambridge University Press, 1970.
2   H. R. Pruppacher and J. D. Klett, *'Microphysics of Clouds and Precipitation'*, Kluwer Academic, 1997.
3   E. Mayer and A. Hallbrucker, *Nature*, 1987, **325**, 601.
4   L. S. Bartell and J. F. Huang, *Journal of Physical Chemistry*, 1994, **98**, 7455.
5   www.snowmax.com
6   B. Vonnegut, *J. Appl. Phys.*, 1947, **18**, 593.
7   M. Matsumoto, S. Saito, and I. Ohmine, *Nature*, 2002, **416**, 409.
8   D. R. Nutt and A. J. Stone, *Langmuir*, 2004, **20**, 8715.
9   H. Witek and V. Buch, *Journal of Chemical Physics*, 1999, **110**, 3168.
10 S. Okawa, A. Saito, and T. Matsui, *International Journal Of Refrigeration-Revue Internationale Du Froid*, 2006, **29**, 134.
11 P. J. Eng, T. P. Trainor, G. E. Brown, G. A. Waychunas, M. Newville, S. R. Sutton, and M. L. Rivers, *Science*, 2000, **288**, 1029.
12 C. F. Walters, K. F. McCarty, E. A. Soares, and M. A. Van Hove, *Surface Science*, 2000, **464**, L732.
13 Aluminum Etchant 'A' is an acid cocktail consisting of phosphoric, acetic, nitric, and water in the following ratios 16:1:1:2. Reference L.B. Goetting, T. Deng, and G.M. Whitesides, *Langmuir*, 1999, **15**, 1182.
14 I. Langmuir, *J. Am. Chem. Soc.,* 1918, **40**, 1361.
15 G. E. Ewing, *Journal Of Physical Chemistry B*, 2004, **108**, 15953.
16 H. A. Al-Abadleh and V. H. Grassian, *Langmuir*, 2003, **19**, 341.

NEW HYDROGEN ORDERED PHASES OF ICE

C.G. Salzmann,[1,2] P.G. Radaelli,[3,4] A. Hallbrucker,[1] E. Mayer[1] and J.L. Finney[4]

[1] Institute of General, Inorganic and Theoretical Chemistry, University of Innsbruck, Innrain 52a, A-6020 Innsbruck, Austria
[2] Inorganic Chemistry Laboratory, University of Oxford, South Parks Road, Oxford OX1 3QR, UK
[3] ISIS Facility, Rutherford Appleton Laboratory, Council for the Central Laboratory of the Research Councils (CCLRC), Chilton, Didcot OX11 0QX, UK
[4] Department of Physics and Astronomy, University College London, Gower Street, London WC1E 6BT, UK

1 INTRODUCTION

A full knowledge of the structures of, and transformations between, the many crystalline phases of ice (Figure 1) is important not only in improving our knowledge of the hydrogen bond itself, but also in increasing our understanding of the water molecule in critical chemical, biological, and geological processes.[1, 2] Understanding the behaviour of water at low temperatures is also of potential importance with respect to the state and dynamics of water in biological systems at low temperatures,[3] as well as elsewhere in the solar system.[1, 2] High-pressure low temperature phases of ice in particular are thought to be present on the icy satellites of the outer planets as the result of either meteoroid impacts on their surfaces[4] or the hydrostatic pressure in the ice shells.[5]

In the crystalline phases of ice, individual water molecules are tetrahedrally bonded to their neighbours *via* hydrogen bonding according to the Bernal-Fowler rules.[6] A consequence of this connectivity is that each water molecule may adopt six different orientations. In 1935, Linus Pauling showed that the connectivity of the lattice reduces but not altogether suppresses these degrees of freedom, resulting in residual disorder and hence non-zero entropy at zero temperature.[7] Thermodynamically, the hydrogen disordered phases of ice should, on cooling, transform *via* hydrogen ordering into new, thermodynamically more stable phases, with zero entropy at 0 K.[1] The basic obstacle in achieving these phase transitions is the slowing down of the reorientation of water molecules with decreasing temperature and a consequent freezing-in of disorder. Addition of dopants can accelerate reorientation in some ice phases. Previous attempts to unlock the geometrical frustration in hydrogen bonded ice structures have focused on doping with potassium hydroxide (KOH). They had success in partially increasing the hydrogen ordering in hexagonal ice (ice Ih) and transforming it into hydrogen ordered ice XI.[1, 8-10] Acid dopants, however, had little effect on hydrogen ordering in ice Ih.[10]

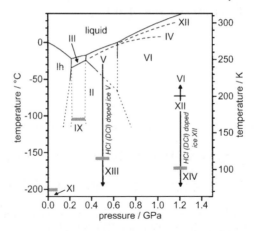

Figure 1   *The phase diagram of ice, including liquidus lines of metastable ices IV and XII (long dashed lines) and extrapolated equilibrium lines at low temperatures (short dashed lines). Hydrogen ordered ices XIII and XIV were prepared by isobaric cooling of HCl (DCl) doped H₂O (D₂O) ice V or XII to 77 K at 0.5 GPa or 1.2 GPa, respectively (indicated by arrows). Cooling of ice XII was started at 190 K, which is 10 K below the temperature where transition to stable ice VI would occur. The temperatures of the hydrogen ordering phase transitions are indicated by grey bars. Transition temperatures for ice V → ice XIII and ice XII → ice XIV are read off from Figure 3 (the pressure dependence of the ordering temperatures is assumed to be small).*

We recently reported that hydrogen disordered ices V and XII transform on acid doping with HCl (DCl) into new hydrogen ordered phases, whose structures were determined by powder neutron diffraction, whereas doping with KOH had little effect.[11-13] These new ice structures were labelled ice XIII for ordered ice V, and ice XIV for ordered ice XII. The space group symmetry of hydrogen disordered ice V is $A2/a$; the monoclinic unit cell contains 28 water molecules,[14-16] and partial ordering of the water molecules is allowed by this space group. Reduction of the space group symmetry occurs on phase transition from ice V to ice XIII, from $A2/a$ to $P2_1/a$.[11]

Ice XII is a metastable phase of ice which occurs mainly in the stability domain of ice VI, but also in those of ices V and II.[17-23] Its structure was first described in 1998 by Lobban et al. who crystallized the tetragonal phase from the liquid at 260 K at a pressure of 0.55 GPa.[17] Later we reported that pure ice XII can be formed in a controlled and reproducible manner by isobaric heating of HDA in the pressure range 0.7 to 1.4 GPa using appropriate heating rates.[22-26] Neutron diffraction showed that the structure of ice XII is hydrogen disordered.[17, 19, 20] In fact, the space group symmetry of ice XII ($I\bar{4}2d$) does not allow any hydrogen ordering and the occupancies of the hydrogen positions are restricted to 0.5.[27] On cooling ice XII, the disordered structure is frozen-in.[19-21] However, doping of ice XII with HCl (DCl) induces a phase transition to an orthorhombic hydrogen ordered phase on cooling under pressure.[11, 13] The new phase of ice has been named ice XIV and has $P2_12_12_1$ space group symmetry. This phase can accommodate varying degrees of hydrogen order and disorder.

Here we report the preparation and structures of the two new hydrogen ordered phases of ice named ice XIII and ice XIV. We further show by relative changes of the lattice parameters that on thermal cycling *in vacuo* between 80 and 125 K, ordered ice XIII transforms in a reversible manner into disordered ice V, and ordered ice XIV into disordered ice XII.

2 METHODS

Doped ice V samples were prepared from solutions containing 0.01 M DCl in D_2O. The solutions were frozen in a piston-cylinder apparatus precooled to 77 K, and the frozen doped ice heated isobarically at 0.5 GPa to 250 K. Thereafter, the samples were cooled from 250 K at 0.8 K min^{-1} to 77 K, decompressed, and recovered under liquid nitrogen (cf. Figure 1). The transition to ice II on cooling at 0.5 GPa was never observed, which is in agreement with ref. 16.

DCl doped D_2O ice XII samples were prepared by following a procedure described in ref. 25. First, DCl doped HDA was prepared by compression of D_2O ice Ih doped with 0.01 M DCl at 77 K.[28] HDA was then heated isobarically starting from 77 K at 1.2 GPa at ~11 K min^{-1} to crystallize pure doped D_2O ice XII.[25] After cooling from 180 K to 77 K at 0.8 K min^{-1}, the samples were decompressed, and recovered in liquid nitrogen (cf. Figure 1).

The recovered samples were characterised in a cryostat (~50 mbar helium, herein refereed to as *in vacuo*) by powder neutron diffraction on the GEM instrument at ISIS.

3 RESULTS AND DISCUSSION

For ice V, space group symmetry reduction and thus formation of a new phase (ice XIII) upon cooling of the DCl doped ice V sample under pressure was indicated by the appearance of (31-2) (cf. curve (1), inset in Figure 2(left)) and (110) reflections in the neutron powder pattern of the recovered sample.

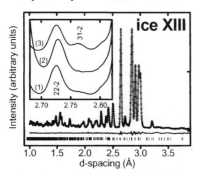 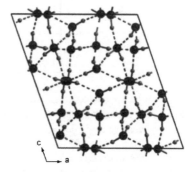

Figure 2 *(left) Observed, calculated, and difference powder neutron diffraction profiles for the hydrogen ordered ice XIII phase observed at 80 K and in vacuo after cooling from 125 K to 80 K at 0.2 K min^{-1}. Tick marks indicate the positions of the Bragg peaks. Inset: magnification of the region where new reflections were observed: (1) after preparation at 80 K, (2) after heating to 125 K, and (3) after slowly cooling from 125 K to 80 K. (right) Unit cell projection of the refined structure: covalent and hydrogen bonds are indicated by solid and dashed lines, respectively.*

The highest-symmetry space group that meets the observed reflection conditions is monoclinic $P2_1/a$ which is a subgroup of the ice V space group ($A2/a$). ($P2_1/a$ was previously proposed by Kamb and La Placa for ordered ice V in an abstract,[29] but their data were never published).

Figure 3(a) shows the changes of the lattice parameters of the DCl doped ice V sample as a function of temperature. Upon heating ordered ice XIII from 80 to 125 K, stepwise changes of the lattice parameters could be observed between ~110 K and ~120 K which went along with the disappearance of the (31-2) and (110) reflections (cf. curve (2), inset in Figure 2(left)). This indicates the phase transition from hydrogen order to disorder. The reversibility of this transition to ordered ice XIII could be demonstrated by cooling back to 80 K at ~0.2 K min^{-1}. In fact, slow cooling *in vacuo* even enhanced the ordering process, as demonstrated by larger changes of the lattice parameters and the increased intensity of the (31-2) reflection (compare curve (3) and (1), inset in Figure 2(left)).

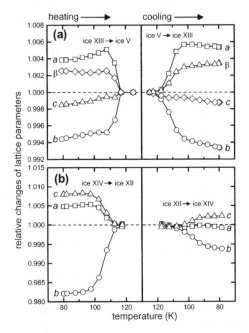

Figure 3   *Relative changes of lattice parameters from powder neutron diffraction during heating and cooling (~0.2 K min^{-1}) in vacuo: (a) heating of an ice XIII sample from 80 to 125 K, and subsequent cooling to 80 K. The onset temperatures of the order/disorder transition are at ~108 K on heating and at ~117 K on cooling. (b) heating of an ice XIV sample from 80 to 118 K, and subsequent cooling to 80 K. The onset temperatures of the order/disorder transition are at ~98 K on heating and at ~107 K on cooling. Lattice parameters are normalised by dividing with the lattice parameters of the hydrogen disordered phases. Solid lines are guides to the eye.*

The structural parameters of ordered ice V (ice XIII) were refined by using the General Structure Analysis System (GSAS) program,[30] by using the values for disordered ice V as a starting point. The refinement of the ordered structure after slow cooling *in vacuo* (cf. Figure 2(left)) revealed the formation of highly ordered molecules as the refined occupation probabilities of the deuterium positions converged within the errors to either 0 or 1 (cf. Table 1 in ref. 11). The unit cell of this hydrogen ordered cell is shown in Figure 2(right). It comprises 28 water molecules with 7 crystallographically distinct oxygen atoms (cf. Table 1 in ref. 11). This makes the $P2_1/a$ structure of ice XIII even more complicated than that of ice V, and thus the most complicated structure among the known crystalline phases of ice.

The ice XIII $\leftrightarrow$ ice V phase transition *in vacuo* was also followed by using Raman spectroscopy.[12] On heating ice XIII from 80 to 120 K, pronounced spectral changes occurred and the spectrum recorded after heating to 120 K is that of hydrogen disordered ice V (refs 31, 32, and Figure 3, curves 6, in our ref. 12). Subsequently, the sample was slowly cooled from 120 K to 80 K at 1 K min^{-1}. The spectrum recorded at 80 K was again that of ordered ice XIII which spectroscopically demonstrated the reversibility of the ice XIII $\leftrightarrow$ ice V phase transition, a result that is consistent with the changes in the lattice parameters shown in Figure 3(a).

The ordering effect in DCl doped D_2O ice XII is visualized most clearly by the splitting of the tetragonal (310) reflection to (130) and (310) orthorhombic reflections (cf. curve (1), inset in Figure 4(left)). The highest symmetry orthorhombic space group that meets the observed reflection conditions is $P2_12_12_1$ which is a subgroup of $I\overline{4}2d$. The unit cell of ordered ice XIV, which contains 12 water molecules with three crystallographic distinct oxygen atoms (cf. Table 2 in ref. 11), is shown in Figure 4(right). Strain may prevent the sample from becoming completely ordered in a similar manner to the transition of ice Ih to ice XI.[33]

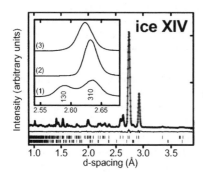 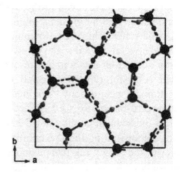

Figure 4 *(left) Observed, calculated, and difference powder neutron diffraction profiles for the hydrogen ordered ice XIV phase observed at 80 K and in vacuo after the preparation. Tick marks indicate the positions of the Bragg peaks, the lower tick marks representing a small ice IV impurity. Inset: magnification of the region where peak splitting was observed for the ordered ice XIV phase: (1) measurement at 80 K after preparation, (2) after heating to 118 K, and (3) after cooling from 118 K to 80 K at ~0.2 K min^{-1}. (right) Unit cell projection of the refined structure: covalent and hydrogen bonds are indicated by solid and dashed lines, respectively.*

Heating of the ordered ice XIV sample from 80 to 118 K *in vacuo* showed, similar to ordered ice XIII, stepwise changes of the lattice parameters between ~95 K and ~115 K (cf. Figure 3(b)). This was accompanied by the disappearance of the orthorhombic splitting which indicates the phase transition from ice XIV to disordered tetragonal ice XII (cf. curve (2), inset in Figure 4(left)). Reversibility of the phase transition was found on cooling. However, the changes of lattice parameters, which correspond to the degree of re-ordering, are less pronounced if compared with the ice V ↔ ice XIII phase transition. For lattice parameter *a*, no significant changes could be observed, and lattice parameters *b* and *c* were found to change back by only 35% and 29%, respectively, of the original stepwise changes observed on heating (cf. Figure 3(b)). The orthorhombic splitting of the 310 reflection could also not be found anymore. Instead only broadening of the 310 reflection could be seen (compare curve (3) and (2), inset in Figure 4(left)). Subsequent cooling under pressure was, however, capable of achieving full reversibility (not shown). Cooling under pressure therefore produces a more ordered ice XIV than cooling *in vacuo*. We speculate that a strain effect may prevent the sample from becoming highly ordered.[11] The existence of strain was deduced from asymmetric peak shapes of the 130 and 310 reflections found in a higher resolution powder neutron diffraction measurement (not shown). The line shapes we observed are strongly reminiscent of those encountered in a variety of materials undergoing martensitic phase transitions. We therefore speculate that regions of incomplete ordering are stabilised by a combination of overall tensile strain and anisotropic strain, due to the different shape anisotropy of the two phases.

Very recently, Tribello et al. identified two hydrogen ordered configurations (ice XIV' and ice XIV'') very close in energy to the ground state by using DFT calculations.[34] They showed furthermore that the difference in energy between these metastable phases and the ground state decreases even further with decreasing pressure. These results suggest that the structure after cooling *in vacuo* might contain an increased fraction of one or more of these metastable states. This scenario could also explain the differences in hydrogen ordering of ice XII between 1.2 GPa and *in vacuo*. A more detailed analysis and discussion of the neutron diffraction data will be presented elsewhere.

The reversibility of the ice XII ↔ ice XIV phase transition was also investigated by using Raman spectroscopy.[13] On heating ice XIV from 80 K to 110 K, the Raman spectra developed to those of disordered ice XII (refs 26, 35, and spectra at 110 K in Figure 2 of ref. 13). Subsequent cooling at 0.5 K min^{-1} led to the weak reappearance of ice XIV lattice modes demonstrating the reversibility of the hydrogen disorder ↔ order phase transition. However, the only weak reappearance of the ice XIV lattice modes shows, consistent with what we have shown here for the changes in lattice parameters (Figure 3(b)), that hydrogen ordering of ice XII is difficult at low pressures.

The remarkable effect of HCl (DCl) as dopant in accelerating the reorientation of water molecules and hydrogen ordering at low temperatures needs to be discussed. At high temperatures, water ices can explore their configurational manifold thanks to the presence of mobile point defects that locally lift the geometrical frustration constraints.[1, 11, 12, 16, 36] The two types of thermally-induced point defects uniquely found in ices are (a) rotational defects in which either two (D defect) or no (L defect) hydrogen atoms are found between neighbouring oxygen atoms, and (b) ionic defects (H_3O^+ and OH$^-$) (cf. refs 1, 11 for discussion). With decreasing temperature, reorientational ordering of water molecules is hampered by the decreasing number and mobility of these point defects. Except for disordered ices III and VII, intrinsic point defects are not sufficient to permit phase transitions from the hydrogen disordered phases to their hydrogen ordered phases. Extrinsic point defects can be introduced by doping ices with impurities, such as in the partial conversion of hydrogen disordered ice Ih to hydrogen ordered ice XI facilitated by KOH doping.[1, 8-10] KOH doping is expected to generate L and OH⁻ defects, whereas HCl doping is thought to produce L and H_3O^+ defects.[1] Both powder neutron diffraction and Raman spectroscopy show that only HCl (DCl) doping induces the phase transitions of disordered ices V and XII to ordered ices XIII and XIV,[11-13] whereas KOH doping had only a minor effect on hydrogen ordering in ices V and XII (cf. Figure 2 in ref. 11, and Figure 3, curve 5, in ref. 12). Thus, the remarkable effect of HCl doping in comparison to that of doping with KOH seems to be caused by H_3O^+ defects.

In calorimetric studies of KOH doped ice V, endothermic peaks observed by Handa et al.[37] on annealing had been attributed previously to an order-disorder phase transition. Subsequent studies of undoped or KOH doped ice V could not confirm a phase transition.[11, 12, 38, 39] Thus, the previous evidence for partial hydrogen ordering in undoped ice V,[16, 27, 37, 39] or in KOH doped ice V,[37, 38] must have been obtained within the ice V phase, which allows considerable hydrogen ordering.[14-16]

4 CONCLUSIONS

Two new hydrogen ordered phases of ice were prepared by cooling the hydrogen disordered ices V and XII under pressure. Previous attempts to unlock the geometrical frustration in hydrogen-bonded ice structures have focused on doping with KOH and have had success in partially increasing the hydrogen ordering in ice Ih and its transition to ice XI. By doping ices V and XII with HCl (DCl), we have prepared hydrogen ordered ice XIII and ice XIV, and we analysed their structures by powder neutron diffraction and Raman spectroscopy. The formation of ice XIII from HCl doped ice V is the first successful example of fully ordering a frozen-in disordered ice phase by using a dopant. The phase transitions between the disordered phases ice V and ice XII, and their corresponding phases ice XIII and ice XIV, are reversible *in vacuo*. At low pressures, these phases are all metastable. Therefore these are, to our knowledge, the first reported examples of reversible phase transitions between two *metastable* ice polymorphs *in vacuo*.

The remarkable effect of HCl (DCl) doping on the reorientation dynamics of hydrogen disordered ice V and ice XII came as a surprise, and it demonstrates that we know very little about the types of extrinsic defects and also the mechanism important for accelerating reorientation in the various disordered phases of ice. These results suggest a way forward in trying to clarify the mechanisms involved. The use of HCl (DCl) doping represents a major step forward in releasing the geometrical frustration that locks in disorder in hydrogen bonded systems, and opens up the possibility of finally completing the phase diagram of ice. In this context, very recent measurements on DCl doped ice VI also show hydrogen ordering, and the refinement of the consequent ice XV is currently under way.

References

1   V. F. Petrenko and R. W. Whitworth, 'Physics of Ice', Oxford University Press, 1999.
2   E. D. Yershov, 'General Geocryology', ed. P. J. Williams, Cambridge University Press, 1998.
3   R. M. Daniel, R. V. Dunn, J. L. Finney, and J. C. Smith, *Annu. Rev. Biophys. Biomol. Structure*, 2003, **32**, 69.
4   S. T. Stewart and T. J. Ahrens, *J.Geophys. Res.*, 2005, **110(E3)**, E03005/1.
5   F. Sohl, T. Spohn, D. Breuer, and K. Nagel, *Icarus*, 2002, **157**, 104.
6   J. D. Bernal and R. H. Fowler, *J. Chem. Phys.*, 1933, **1**, 515.

7   L. Pauling, *J. Am. Chem. Soc.*, 1935, **57**, 2680.
8   Y. Tajima, T. Matsuo, and H. Suga, *Nature*, 1982, **299**, 810.
9   Y. Tajima, T. Matsuo, and H. Suga, *J. Phys. Chem. Solids*, 1984, **45**, 1135.
10 H. Suga, *Pure & Appl. Chem.*, 1983, **55**, 427.
11 C. G. Salzmann, P. G. Radaelli, A. Hallbrucker, E. Mayer, and J. L. Finney, *Science*, 2006, 1758.
12 C. G. Salzmann, A. Hallbrucker, J. L. Finney, and E. Mayer, *Phys. Chem. Chem. Phys.*, 2006, **8**, 3088.
13 C. G. Salzmann, A. Hallbrucker, J. L. Finney, and E. Mayer, *Chem. Phys. Lett.*, 2006, doi:10.1016/j.cplett.2006.08.079.
14 W. C. Hamilton, B. Kamb, S. J. La Placa, and A. Prakash, ed. B. B. N. Riehl and H. Engelhardt, New York, 1969.
15 W. F. Kuhs, C. Lobban, and J. L. Finney, *Rev. High Pressure Sci. Technol.*, 1998, **7**, 1141.
16 C. Lobban, J. L. Finney, and W. F. Kuhs, *J. Chem. Phys.*, 2000, **112**, 7169.
17 C. Lobban, J. L. Finney, and W. F. Kuhs, *Nature*, 1998, **391**, 268.
18 I.-M. Chou, J. G. Blank, A. F. Goncharov, H.-K. Mao, and R. J. Hemley, *Science*, 1998, **281**, 809.
19 M. Koza, H. Schober, A. Tölle, F. Fujara, and T. Hansen, *Nature*, 1999, **397**, 660.
20 M. M. Koza, H. Schober, T. Hansen, A. Tölle, and F. Fujara, *Phys. Rev. Lett.*, 2000, **84**, 4112.
21 C. Salzmann, I. Kohl, T. Loerting, E. Mayer, and A. Hallbrucker, *Journal of Physical Chemistry B*, 2002, **106**, 1.
22 T. Loerting, I. Kohl, C. Salzmann, E. Mayer, and A. Hallbrucker, *J. Chem. Phys.*, 2002, **116**, 3171.
23 I. Kohl, T. Loerting, C. Salzmann, E. Mayer, and A. Hallbrucker, in 'HIGH-DENSITY AMORPHOUS ICE AND ITS PHASE TRANSITION TO ICE XII', ed. V. V. B. e. al., Amsterdam, 2002.
24 C. G. Salzmann, I. Kohl, T. Loerting, E. Mayer, and A. Hallbrucker, *Can. J. Phys.*, 2003, **81**, 25.
25 C. G. Salzmann, E. Mayer, and A. Hallbrucker, *Physical Chemistry Chemical Physics*, 2004, **6**, 5156.
26 C. G. Salzmann, E. Mayer, and A. Hallbrucker, *Phys. Chem. Chem. Phys.*, 2004, **6**, 1269.
27 C. G. Salzmann, I. Kohl, T. Loerting, E. Mayer, and A. Hallbrucker, *Phys. Chem. Chem. Phys.*, 2003, **5**, 3507.
28 O. Mishima, L. D. Calvert, and E. Whalley, *Nature*, 1984, **310**, 393.
29 B. Kamb and S. J. La Placa, *Trans. Am. Geophys. Union*, 1974, **56**, V39 1202.
30 A. C. Larsen and R. B. Von Dreele, 'General Structure Analysis System', LAUR 86-748, Los Alamos National Laboratory Report, 2000.
31 M. J. Taylor and E. Whalley, *J. Chem. Phys.*, 1964, **40**, 1660.
32 B. Minceva-Sukarova, W. F. Sherman, and G. R. Wilkinson, *J. Phys. C: Solid State Phys.*, 1984, **17**, 5833.
33 C. M. B. Line and R. W. Whitworth, *J. Chem. Phys.*, 1996, **104**, 10008.
34 G. A. Tribello, B. Slater, and C. G. Salzmann, *J. Am. Chem. Soc.*, 2006, DOI: 10.1021/ja0630902.
35 C. Salzmann, I. Kohl, T. Loerting, E. Mayer, and A. Hallbrucker, *J. Phys. Chem. B*, 2002, **106**, 1.
36 G. P. Johari, *Chem. Phys.*, 2000, **258**, 277.
37 Y. P. Handa, D. D. Klug, and E. Whalley, *J. Phys. Colloq.*, 1987, **48**, 435.
38 B. Minceva-Sukarova, G. Slark, and W. F. Sherman, *J. Mol. Struct.*, 1988, **175**, 289.
39 B. Minceva-Sukarova, G. E. Slark, and W. F. Sherman, *J. Mol. Struct.*, 1986, **143**, 87.

MICROSCOPIC OBSERVATION AND IN-SITU RAMAN STUDIES ON SOME SINGLE-CRYSTALLINE GAS HYDRATES UNDER HIGH PRESSURE

S. Sasaki, T. Kume, and H. Shimizu

Department of Materials Science and Technology, Gifu University, 1-1 Yanagido, Gifu 501-1193, Japan

1 INTRODUCTION

Gas hydrate is a special class of inclusion compounds in which the hydrophobic guest molecules are trapped in hydrogen-bonded water cages. There are three typical structures of sI, sII, and sH for gas hydrates[1]: cubic sI; two small 5^{12} and six medium $5^{12}6^2$ cages, cubic sII; sixteen small 5^{12} and eight large $5^{12}6^4$ cages, and hexagonal sH; three small 5^{12}, two small $4^3 5^6 6^3$, and one extra large $5^{12}6^8$ cages, where for example $5^{12}6^2$ cage is formed by 12 pentagons and 2 hexagons.

Recent high-pressure X-ray and neutron diffraction studies clarified the pressure-induced phase transformations of argon hydrate,[2,3] krypton hydrate,[4] nitrogen hydrate,[5] and methane hydrate[5-12] at room temperature. These results indicate that the initial sI or sII eventually transforms to the orthorhombic dihydrate (sO) called "filled ice" structure through a hexagonal phase with increasing pressure. On the other hand, the direct observations and in-situ high-pressure Raman studies have revealed the change of the occupation number of host water cages by guest atoms or molecules.[13-19] As it is hard to find out about such a change without structural transformation by means of X-ray and neutron diffraction methods, these results suggest that the microscopic observation and in-situ high-pressure Raman experiment are very effective in finding the change of cage occupancy.

In the cases of nitrogen hydrate and methane hydrate, Raman scattering measurements for intra-molecular vibrations of guest molecules are useful for exploring the change of cage occupancy, because their frequencies are sensitive to the environment of guest molecules. In the NH-I (sII) phase of nitrogen hydrate[15] and the MH-II (sH) phase of methane hydrate,[16-18] the abrupt changes of cage occupancy without structural change were found at 0.50 GPa and 1.3 GPa, respectively, by the Raman measurements for the intra-molecular vibrations of the guest molecules. Accompanying these phase behaviours, fine patches appeared in the hydrate crystals, because the increase of the occupation number of host water cages by guest molecules, namely, the decrease of the hydration number should produce the released water as fine patches in the hydrate crystals. For krypton hydrate,[14] the change of cage occupancy in the KH-III (hexagonal) phase was also observed at 1.0 GPa by the occurrence of fine patches similar to the MH-II phase of methane hydrate.[17] In this paper, we present the advantage of the microscopic observation and the high-pressure

Raman studies on gas hydrates, showing the recent results of krypton hydrate, nitrogen hydrate, and methane hydrate.

2 EXPERIMENT

The direct observations through a microscope and in-situ high-pressure Raman experiments of krypton hydrate, nitrogen hydrate, and methane hydrate have been performed using a diamond anvil cell (DAC). To prepare the sample we loaded guest gas and distilled water into a small sample chamber (0.3 mm in diameter, 0.3 mm in depth) of the DAC together with a tiny ruby chip for pressure calibration. After the loading into the DAC, we made a three-phase equilibrium among water, fluid guest, and gas hydrate at 296 K, then we grew a single crystal of gas hydrate to avoid the unstable and indistinct equilibrium condition of the experiments for the direct observations and in-situ high-pressure Raman scattering measurements. Applied pressures were measured by using the ruby fluorescence method, and its uncertainty was within 0.02 GPa. High-pressure Raman scattering measurements of gas hydrates were carried out in a back-scattering geometry at 296 K. Each compressed gas hydrate sample was basically measured after seasoning of several hours. The sample was excited by a 514.5 nm line of an argon ion laser and the spectra were obtained by a spectrometer (JASCO, NR1800) equipped with a triple polychromater and a liquid-nitrogen-cooled charge-coupled device (CCD) detector. Observations and Raman spectra for each gas hydrate were highly reproducible over several runs of experiments except for NH-I phase of nitrogen hydrate.

3 RESULTS AND DISCUSION

3.1 Krypton Hydrate

As for krypton hydrate, Desgreniers et al.[4] studied the pressure-induced phase transformations of krypton hydrate at room temperature by using X-ray diffraction measurements. They found that the initial cubic sII of krypton hydrate (KH-I) successively transformed to the cubic sI (KH-II), the hexagonal structure (KH-III), and the sO (KH-IV) at 0.3 GPa, 0.6 GPa, and 1.8 GPa, respectively. They also found that the sO phase decomposed at pressures above 3.8 GPa.

Our results of the Raman spectra with corresponding photomicrographs and the pressure dependence of Raman frequency shifts for the lattice mode of krypton hydrate are shown in Figure 1. The visual observation of gas hydrates through a microscope provides us the direct and clear information on phase transformations or the change of the occupation number of host water cages by guest atoms or molecules. The shape of a single crystal of KH-I (sII) is an octahedron, which is suggestive of a single crystalline cubic sII clathrate hydrate. With increasing pressure, the visual changes such as fine patches or grains occurred at 0.45 GPa, 0.75 GPa, 1.0 GPa, and 1.8 GPa. Above 1.8 GPa we could not find any phase change of krypton hydrate at least up to 5.2 GPa in contrast to the X-ray results. Corresponding to these visual changes at 0.45 GPa, 0.75 GPa, and 1.8 GPa, except for 1.0 GPa, the Raman spectra and the pressure dependence of Raman frequency shifts for lattice vibrations in Figure 1 obviously change. These results support the phase transformations determined by X-ray diffraction study.[4] On the other hand the occurrence of fine patches at 1.0 GPa induces no Raman spectral change and no structural transformation. The most plausible origin of this visual change at 1.0 GPa is the increase of the occupation number of host water cages by guest krypton atoms. In order to increase the occupation number while remaining the hexagonal structure of KH-III, the KH-III crystal should partially decomposed and the separated krypton atoms are constrainedly inserted into host water cages inducing multi-occupancy. At the same time the released water remains liquid as fine droplets (patches) in the KH-III crystal. As the above speculation well agrees with the actual experimental results, the increase of the occupation number of host water cage by krypton atoms in KH-III must bring about the visual change at 1.0 GPa. Considering the X-ray diffraction study,[4] we temporarily named the phase between 1.0 GPa and 1.8 GPa KH-III'.

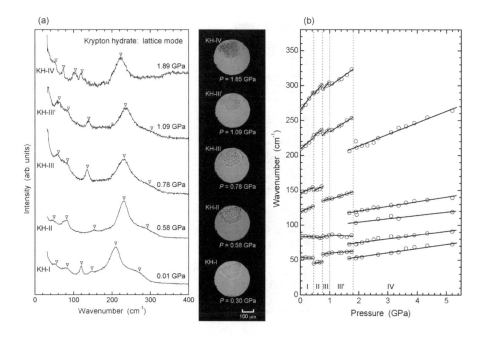

Figure 1 *(a) Raman spectra with corresponding photomicrographs and (b) pressure dependence of Raman frequency shifts for the lattice mode of krypton hydrate at 296 K. Open triangles indicate the peaks from Kr hydrate. Crystals in photomicrographs are krypton hydrate and surroundings are water at 0.30 GPa, 0.58 GPa, 0.78 GPa, and 1.09 GPa and ice-VI at 1.85 GPa.*

3.2 Nitrogen Hydrate

The Raman spectra with corresponding photomicrographs and the pressure dependence of Raman frequency shifts for the N-N stretching mode of nitrogen hydrate are shown in Figure 2. The initial crystal at 0.24 GPa, named NH-I, shows a transparent octahedron-like shape resulting from the cubic sII. Loveday et al.[5] reported that the sII nitrogen hydrate initially transformed to the sH at 0.84 GPa and subsequently transformed to the tetragonal structure (sT) found by Kurnosov et al.[3] and to the sO at around 1.3 GPa and 1.6 GPa, respectively. However, we could confirm only two phase transformations at 0.85 GPa and

1.45 GPa by the visual changes and Raman spectra (see Figure 2). Therefore, there was no transformation to the sT phase in our experimental conditions. According to our results, the observed high-pressure hydrate phases were named NH-II (sH; 0.85 GPa < P < 1.45 GPa) and NH-III (sO; P > 1.45 GPa), respectively. In addition to these phase transformations, we found that the single-crystalline NH-I turned reddish brown at about 0.50 GPa, retaining the outward form, as shown in the photomicrograph at 0.51 GPa of Figure 2. Considering the neutron diffraction study found no structural phase transformation at about 0.50 GPa, we temporarily name this state between 0.50 GPa and 0.85 GPa NH-I'.

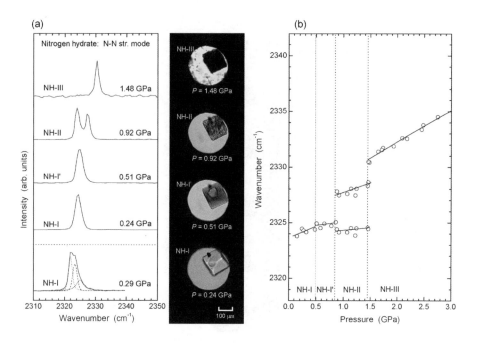

Figure 2   *(a) Raman spectra with corresponding photomicrographs and (b) pressure dependence of Raman frequency shifts for the N-N stretching vibration mode of nitrogen hydrate at 296 K. Crystals in photomicrographs are nitrogen hydrate and surroundings are water at 0.24 GPa, 0.51 GPa, and 0.92 GPa and ice-VI at 1.48 GPa. The asymmetric Raman spectrum at 0.29 GPa, which is often observed depending on the crystal growth conditions, is analyzed into three peaks with frequencies of 2322.0 cm^{-1}, 2323.5 cm^{-1}, and 2325.1 cm^{-1} indicated by broken curves. The low, middle, and high frequency peaks indicated by broken curves originate from the guest nitrogen molecules in the single occupied $5^{12}6^4$ cages, the double occupied $5^{12}6^4$ cages, and the single occupied 5^{12} cages, respectively.[20,21]*

We often obtained the asymmetric Raman spectra clearly composed of three peaks below 0.50 GPa depending on the conditions of crystal growth, such as the Raman spectrum at 0.29 GPa in Figure 2 (a). (Raman frequency shifts of this asymmetric Raman band are removed from Figure 2 (b) as a special case.) This spectrum was analyzed into three peaks with frequencies of 2322.0 cm^{-1}, 2323.5 cm^{-1}, and 2325.1 cm^{-1} indicated by broken curves using the least-squares fitting method. Klaveren et al. calculated the Raman spectra of the nitrogen molecules in the sII nitrogen hydrate with lattice constant 17 Å at 273 K on the assumption that the $5^{12}6^4$ cages are doubly occupied in part by guest nitrogen molecules.[20,21] They showed three characteristic peaks with frequencies of 2330 cm^{-1}, 2332.5 cm^{-1}, and 2335 cm^{-1} in response to the single occupied $5^{12}6^4$ cages, the double occupied $5^{12}6^4$ cages, and the single occupied 5^{12} cages, respectively. Although their values of frequencies are obviously different from our experimental results because of the different conditions between their calculation and our experiments, the qualitative comparison between them is valuable for deducing the origin of Raman signals. On the basis of their results, we can temporarily assign the low, middle, and high frequency peaks to the guest nitrogen vibrations in the single occupied $5^{12}6^4$ cages, the double occupied $5^{12}6^4$ cages, and the single occupied 5^{12} cages, respectively. The strong peak around 2322.0 cm^{-1} corresponding to the single occupied $5^{12}6^4$ cages indicates that the number of the doubly occupied $5^{12}6^4$ cages is less than that of the single occupied $5^{12}6^4$ cages in NH-I (sII). On the other hand, most of $5^{12}6^4$ cages are doubly occupied in NH-I' (sII) and both the $5^{12}6^4$ cages and the 5^{12} cages accommodate the same number of the guest nitrogen molecules in a unit cell of NH-I' (sII). Consequently, the slightly broad single peak in NH-I' (sII) is mainly composed of two peaks with the same intensities and small frequency separation of about 1.5 cm^{-1}. Any spectrum obtained below 0.50 GPa eventually becomes a slightly broad single peak at pressures above 0.50 GPa. Thus, it is thought that the ratio of the number of double occupied $5^{12}6^4$ cages becomes constant above 0.5 GPa. Moreover, since the above change of the occupation number, namely, the increase of the ratio of the number of double occupied $5^{12}6^4$ cages at 0.50 GPa must induces the phase separation of NH-I in part, the released water remains liquid as fine droplets (patches) in NH-I' crystal. This speculation is consistent with the actual experimental results of the occurrence of fine patches at 0.50 GPa and with the pressure dependent high resolution neutron diffraction study on sII nitrogen hydrate which reveals that the $5^{12}6^4$ cages are partially occupied by two nitrogen molecules.[22,23]

In the NH-II (sH) phase, the peak of the N-N stretching vibration obviously splits into two peaks having nearly equal intensities as seen in the Raman spectrum at 0.92 GPa of Figure 2 (a). The sH is composed of three 5^{12} cages, two $4^35^66^3$ cages, and one $5^{12}6^8$ cage per unit cell. Since 5^{12} and $4^35^66^3$ cages have almost the same radii (3.91 Å for 5^{12} cage, 4.06 Å for $4^35^66^3$ cage) and encage at most one guest molecule in each cage, two N-N vibration peaks corresponding to five nitrogen molecules in three 5^{12} and two $4^35^66^3$ cages and to five molecules in one $5^{12}6^8$ cage are expected to appear with nearly equal intensities. These estimations seem to be consistent with the present Raman spectra appearing in NH-II phase. Consequently, the NH-II phase undoubtedly has the sH with five nitrogen molecules in $5^{12}6^8$ cages.

3.3 Methane Hydrate

In recent high-pressure X-ray diffraction,[9-12] neutron diffraction,[5-8] and Raman[16-18] studies on methane hydrate, pressure-induced phase transformations of methane hydrate show that the initial sI (MH-sI) successively transforms to the sH (MH-II) at about 0.9 GPa and to the sO (MH-III) at about 2.0 GPa.

Figure 3 shows the Raman spectra with corresponding photomicrographs and the pressure dependence of Raman frequency shifts for C-H symmetric stretching vibration at 296 K. The photomicrographs at 0.12 GPa and 0.94 GPa show single crystals of MH-sI and MH-II, respectively. The single crystal of MH-II was grown at about 1.0 GPa and 325

K and then was cooled down to 296 K. This single crystal shows a hexagonal pillar which strongly supports the MH-II is the hexagonal structure as suggested by the neutron studies.[5-8] With increasing pressure, the obvious visual changes appeared at 0.90 GPa, 1.30 GPa, and 1.90 GPa. The Raman spectra and the pressure dependence of Raman frequency shifts for the C-H symmetric stretching vibration in Figure 3 also change in response to these visual changes. In MH-sI phase, two characteristic Raman peaks with an intensity ratio of about 3:1 at 0.08 GPa were observed. It is considered that these peaks correspond to the C-H stretching vibrations of methane molecules encaged in six medium $5^{12}6^2$ and two small 5^{12} cages per unit cell, respectively.

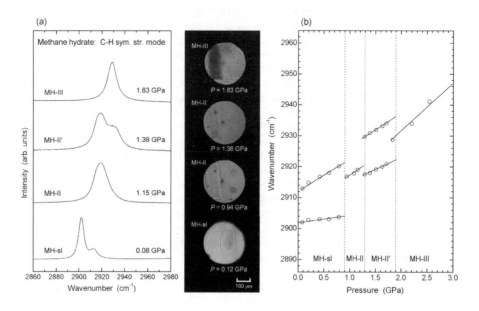

Figure 3   *(a) Raman spectra with corresponding photomicrographs and (b) pressure dependence of Raman frequency shifts for the C-H symmetric stretching vibration mode of methane hydrate at 296 K. Crystals in photomicrographs are methane hydrate and surroundings are water at 0.12 GPa, 0.94 GPa, and 1.36 GPa and ice-VI at 1.83 GPa.*

As for the phase behaviour at 1.30 GPa, although a large number of brown patches appeared in the crystal at about 1.30 GPa as shown in the photomicrograph at 1.36 GPa of Figure 3, the X-ray diffraction[9-12] and neutron diffraction[5-8] studies showed no structural transformation around 1.30 GPa. Therefore we temporarily name this state between 1.30 GPa and 1.90 GPa MH-II'. As discussed in the sections 3.1 and 3.2, the occurrence of fine patches without structural change means the change of cage occupancy. The broad Raman peak in MH-II and the separated Raman peaks in MH-II' in Figure 3 clearly indicate that the cage occupancies are different between MH-II and MH-II', because their frequencies are sensitive to the environment of guest molecules. As a matter of course, this change of cage occupancy at 1.30 GPa induces the decrease of the hydration number, and consequently produces the fine patches by released water in MH-II' crystal. Moreover, the clearly different intensities of two peaks in MH-II' (MH-II) indicate that $5^{12}6^8$ cage in sH includes less guest methane molecules than five, because five guest molecules in $5^{12}6^8$ cage should bring about two peaks with equal intensity, as speculated about the NH-II phase with sH in section 3.2.

4 CONCLUSION

Comparisons of phase transformation pressures between the X-ray diffraction (XRD) and neutron diffraction (ND) studies and the present results by microscopic observations and in-situ Raman scattering measurements for krypton hydrate, nitrogen hydrate, and methane hydrate are summarized in Figure 4. At a glance, one can see the advantage of the microscopic observations and the high-pressure Raman studies on gas hydrates, in terms of the findings of the changes of cage occupancy for KH-I at 1.0 GPa, NH-I at 0.50 GPa, and MH-II at 1.30 GPa. Therefore, it turns out that the direct observations and in-situ high-pressure Raman studies are effective and essential to finding the change of cage occupancy in gas hydrate phases.

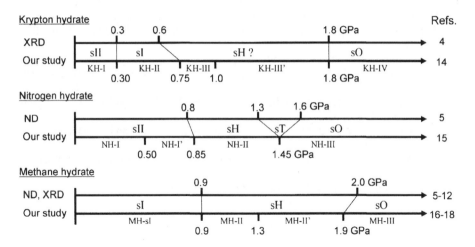

Figure 4   *Comparisons of phase transformation pressures between the X-ray diffraction (XRD) and neutron diffraction (ND) studies and the present results by microscopic observations and in-situ Raman scattering measurements for krypton hydrate, nitrogen hydrate, and methane hydrate.*

References

1   E.D. Sloan, Clathrate Hydrates of Natural Gases, 2nd ed., Marcel Dekker, New York, 1998.
2   A.Y. Manakov, V.I. Voronin, A.V. Kurnosov, A.E. Teplykh, E.G. Larionov, and Y.A. Dyadin, Doklady Phys. Chem., 2001, **378**, 148.
3   A.V. Kurnosov, A.Y. Manakov, V.Y. Komarov, V.I. Voronin, A.E. Teplykh, and Y.A. Dyadin, Doklady Phys. Chem., 2001, **381**, 303.

4   S. Desgreniers, R. Flacau, D.D. Klug, J.S. Tse, CeSMEC (March 24-28, 2003, Florida).
5   J.S. Loveday, R.J. Nelmes, D.D. Klug, J.S. Tse, and S. Desgreniers, Can. J. Phys., 2003, **81**, 539.
6   J.S. Loveday, R.J. Nelmes, M. Guthrie, S.A. Belmonte, D.R. Allan, D.D. Klug, J.S. Tse, and Y.P. Handa, Nature, 2001, **410**, 661.
7   J.S. Loveday, R.J. Nelmes, M. Guthrie, D.D. Klug, and J.S. Tse, Phys. Rev. Lett., 2001, **87**, 215501.
8   J.S. Loveday, R.J. Nelmes, and M. Guthrie, Chem. Phys. Lett., 2001, **350**, 459.
9   H. Hirai, M. Hasegawa, T. Yagi, Y. Yamamoto, K. Nagashima, M. Sakashita, K. Aoki, and T. Kikegawa, Chem. Phys. Lett., 2000, **325**, 490.
10 H. Hirai, Y. Uchihara, T. Kawamura, Y. Yamamoto, and T. Yagi, Proc. Jpn. Acad. B, 2002, **78**, 39.
11 H. Hirai, T. Tanaka, T. Kawamura, Y. Yamamoto, and T. Yagi, Phys. Rev. B, 2003, **68**, 172102.
12 H. Hirai, T. Tanaka, T. Kawamura, Y. Yamamoto, and T. Yagi, J. Phys. Chem. Solids, 2004, **65**, 1555.
13 H. Shimizu, S. Hori, T. Kume, and S. Sasaki, Chem. Phys. Lett., 2003, **368**, 132.
14 S. Sasaki, S. Hori, T. Kume and H. Shimizu, J. Phys. Chem. B, 2006, **110**, 9838.
15 S. Sasaki, S. Hori, T. Kume and H. Shimizu, J. Chem. Phys., 2003, **118**, 7892.
16 H. Shimizu, T. Kumazaki, T. Kume, and S. Sasaki, J. Phys. Chem. B, 2002, **106**, 30.
17 T. Kumazaki, Y. Kito, S. Sasaki, T. Kume and H. Shimizu, Chem. Phys. Lett., 2004, **388**, 18.
18 H. Shimizu, Can. J. Phys., 2003, **81**, 127.
19 H. Shimizu, N. Tada, R. Ikawa, T. Kume, and S. Sasaki, J. Phys. Chem. B, 2005, **109**, 22285.
20 E.P. van Klaveren, J.P.J. Michels, J.A. Schouten, D.D. Klug, and J.S. Tse, J. Chem. Phys., 2001, **115**, 10500.
21 E.P. van Klaveren, J.P.J. Michels, J.A. Schouten, D.D. Klug, and J.S. Tse, J. Chem. Phys., 2002, **117**, 6637.
22 W.F. Kuhs, B. Chazallon, P.G. Radaelli, and F. Pauer, J. Incl. Phenom. Mol. Recognit. Chem., 1997, **29**, 65.
23 B. Chazallon and W.F. Kuhs, J. Chem. Phys., 2002, **117**, 308.

CLATHRATE HYDRATE FORMATION AND GROWTH: EXPERIMENTAL OBSERVATIONS VERSUS PREDICTED BEHAVIOUR

J.M. Schicks[1], M. Luzi[1], J. Erzinger[1], E. Spangenberg[2]

[1]GeoForschungsZentrum Potsdam, Section 4.2, Telegrafenberg, 14473 Potsdam, Germany
[2]GeoForschungsZentrum Potsdam, Section 5.2, Telegrafenberg, 14473 Potsdam, Germany

1 INTRODUCTION

The hydrate formation process can be distinguished into two different scenarios: the clathrate hydrate formation in presence of a free vapour phase and the clathrate hydrate formation process in absence of a free gas phase. Regarding the formation of hydrates in presence of a free gas phase it is known that the initiation process usually occurs at the vapour-liquid interface. However, different models describing the nucleation process exist with two different approaches considering the agglomeration of clusters either in the liquid or vapour phase[1,2]. The model presented by Rodger[3] and Kvamme[4] presumes the transport of the gas molecule to the vapour-liquid interface. Kvamme[4] indicates this step as transport of gas molecules through a stagnant boundary. The gas molecules adsorb at the aqueous surface and water molecules form first partial, then complete cages around the adsorbed species. Those clusters join and grow on the vapour side of the surface until a critical size is achieved which can be caused by the addition of water and gas molecules to existing cavities and/or the joining of cavities along the interface. For this nucleation-at-the-interface-hypothesis the dissolution of the gas molecule in the liquid water phase is not necessary.

A different model presented by Christiansen and Sloan[5] is based on the fact that water molecules form labile water clusters around dissolved gas molecules. The number of water molecules in each water cluster shell depends on the size of the dissolved gas molecules, e.g. 20 for methane and 24 for ethane or 28 for propane. The clusters of the dissolved species combine to form unit cells. The formation rate of a particular hydrate structure depends on the availability of labile clusters with required coordination numbers. With a mixture of methane and propane dissolved in the liquid water phase, hydrates should form more rapidly than if either methane or propane alone are dissolved in the water phase[5]. This cluster nucleation hypothesis is based on the assumption that the guest molecule has to be dissolved in the liquid phase before getting encased into a hydrate lattice.

The clathrate hydrate growth model presented by Englezos and Bishnoi[6,7] is based on crystallization and mass transfer theories. It describes the growth of the hydrate as a three step process. The first step is the transport of the gas molecule into the liquid phase. The second step is the diffusion of the gas molecule through a stagnant liquid diffusion layer which surrounds the hydrate particle. The last step is the incorporation of the gas molecule into the structured water framework of the hydrate particle in the "reaction" layer. Due to the fact that a concentration of the gas molecules into the stagnant liquid layer is not allowed, the diffusion rate of the gas molecule through the stagnant liquid layer and the incorporation rate of the gas molecule into the hydrate structure are equal at steady state. Following Englezos and Bishnoi this rate with which each gas species is being transported to the liquid layer and forms hydrates is proportional to the diffusivity and the fugacity of the gas species in the bulk of the liquid phase. The fugacity of the gas in the bulk of the liquid water phase is comparable to the solubility of the gas in the water phase. Hence, there should be an observable influence of the water solubility of the gas species on its incorporation rate. Similar to the model of Christiansen and Sloan[5] this model implies the dissolution of the gas molecule into the liquid water phase before it will be encased into the hydrate structure. This leads to the conclusion that both, the diffusivity and water solubility of the gas species might have some impact on its inclusion rate into the hydrate.

In order to proof the applicability of the mentioned approaches and to study the incorporation of different gases in the hydrate lattices depending on their properties (solubility, dimension, etc.) we performed investigations on gas hydrates which have been synthesized from gas mixtures as a free gas phase and water. The gas mixtures contain besides methane the isomers of butane (n-butane; iso-butane) and pentane (iso-pentane, 2,2-dimethylpropane), respectively. The exact compositions of the gas mixtures are given in table 1. The experiments and results are described in detail in the diploma thesis of M. Luzi[8].

In case of the absence of a free vapour phase the formation of gas hydrates is limited to the dissolved gas molecules. In this case the variable water solubility of the different components in a gas mixture is supposed to have some influence on the incorporation and the hydrate formation rate. In addition to the assumptions of Christiansen and Sloan[5] mentioned above it is questionable which other factors, such as the presence of sediments in function of a seed crystal, might have an impact on the hydrate formation rate. For this purpose it is absolutely necessary to know how and where the hydrate forms in the sediments. Petrophysical models consider hydrate either as a part of the pore fluid or as part of the solid rock framework. When it is considered as a part of the rock framework it is either assumed as grain within the grain supported framework or as grain cement. Spangenberg[9] modelled the electrical resistivity of hydrate-bearing sediments assuming that hydrate forms as a non-cementing material in the pore space. The results of this modelling correspond very well with the experimental results on glass bead sediments with hydrate formation from a methane saturated water phase without a free gas phase[10]. Dvorkin et al.[11] applied these different models to calculate the sonic wave velocities for the hydrate-bearing formations for the Northwest Eileen State Well No. 2 and compared the results to the measured sonic log. They found that the measured velocities could not be related to the hydrate content with a hydrate cementation model. On the other hand Priest et al.[12] measured seismic velocities on artificial hydrate-bearing sand samples and interpreted the observed velocity increase with hydrate content as a hydrate cementation effect. But in this case the hydrate was formed in a water wet grain skeleton with free methane gas in the pore space so that hydrate formation was forced to the water wetting the quartz grains.

In order to get a better idea of the process of hydrate formation and growth in the pore space of water saturated sediment we realised an experimental set-up to study these processes under the microscope.

2 METHODS AND RESULTS

2.1 Experimental Set-up and Procedure

The main component of the experimental set-up was an optical pressure cell, which can be used in a temperature range between 246 K and 353 K and pressure range between 0.1 MPa and 10.0 MPa. The temperature of the sample cell was controlled by a thermostat and the temperature determined with a precision of ±0.1 K. A pressure controller regulated the pressure with a precision of ±2%. The system pressure was measured with a P3MB from Hottinger Baldwin Messtechnik with a precision of ±0.01% rel. The cooling system of the cell minimized temperature gradients within the sample. One important feature of the experimental set-up was that a gas flow was used. The optical sample cell is described in detail elsewhere[13,14]. For the synthesis of larger amounts of gas hydrates a second pressure cell without an optical window (Figure 1) was integrated in the experimental set-up in parallel and therefore used in the same pressure and temperature range. Similar to the optical pressure cell the gas hydrates are synthesized under gas flow conditions. After 2500 μl of deionized, degassed water was added to the sample cell, the cell was carefully sealed and flushed with the experimental gas mixture before pressurisation. To avoid a temperature gradient in the sample only the bottom of the cell was covered with the mentioned amount of water and the cell was dipped into the cooling liquid. The system was cooled down as rapidly as possible until hydrates were formed. The formation of hydrates could be observed in the optical cell. Raman spectroscopic investigations proved the existence of gas hydrates. After the system was warmed up at constant pressure with the intension of melting most of the hydrate, the temperature was lowered 0.5 K below the decomposition line in order to let the hydrate grow. This use of the larger pressure cell permits the stirring of the sample with a magnetic flea. The experiments have been performed with and without stirring. When the gas hydrate synthesis was completed, the cell was removed from the cooling liquid and quenched in liquid nitrogen. At a temperature of ≤ 173 K the pressure in the cell was decreased to atmospheric pressure. When the temperature of the cell was around 110 K the cell was evacuated to assure that the free gas phase was removed from the system. After that the temperature of the cell was slowly increased until the hydrate dissociated. The released gas was collected and analysed by gas chromatography.

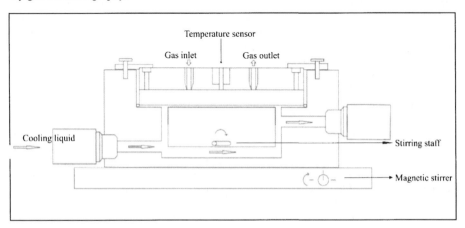

Figure 1 *Sketch of the pressure cell*

The following gas mixtures have been investigated in these studies regarding the formation of clathrate hydrates in presence of a free gas phase.

Table 1 *Composition of gas mixtures used in these studies in mole per cent*

mixture	methane	n-butane	iso-butane	iso-pentane	neo-pentane
1	94 %	3 %	3 %	-	-
2	95 %	2 %	3 %	-	-
3	98 %	-	-	1 %	1 %

For the experiments in absence of a free vapour phase (influences of sediments) the experimental set-up consisted of a gas storage vessel, a gas-water reservoir, a HPLC pump, and a chilled sample cell. Two acrylic glass windows permitted the visual observation of the formation process under the microscope. The cell was equipped with a PT100 temperature sensor and the temperature in the cell was regulated by the PID controller of the cooling system. To simulate the processes in porous media the cell volume can be filled with a layer of granular material. Methane charged water was circulated trough the cell from the gas-water reservoir and pumped back into the gas-water reservoir where it was recharged with methane. The gas-water reservoir was a pressure vessel which was filled up to ¾ with distilled water and pressurised with methane from the gas storage vessel. The system must be filled carefully with water at the beginning of the experiment to assure the complete removal of air from the sample cell. After the system had been filled with water the water in the gas-water reservoir was pressurised to about 14 MPa with methane from the gas storage vessel. The water circulated trough the system at room temperature for a day to come close to gas saturation. When the pressure in the system remained constant the temperature in the sample cell was decreased to about 276 K. The methane charged water from the gas-water reservoir cooled down into the stability field of hydrate when it entered the sample cell and formed methane hydrate.

2.2 Results

In this study we performed experiments to investigate the incorporation rate of gas molecules in hydrates and the formation rate of clathrate hydrates from a liquid water phase in absence and in presence of a free gas phase. In case of a present free gas phase we observed the hydrate formation process in an optical cell. We also analysed the gas composition of the gas which was encased in the hydrate after the decomposition of the hydrates (Figure 2, Table 3). Due to the fact that the aim of this study was to investigate the influence of molecular properties of the guest molecules such as dimension ($\rightarrow$ diffusivity) and water solubility ($\rightarrow$ concentration/fugacity) we focused our observation on the content of isomers which do have the same molecular weight but different molecular dimensions (see Table 2) and solubilities.

The compositions of the gas in the hydrate phase depend on the conditions during the formation process. The isomer compositions of the gas in the hydrate phase of those hydrates which have been formed with stirring are completely different from that which grew without stirring. The ratios of the isomers in the hydrate phase of the stirred system are pretty similar to the ratios of the isomers in the water phase. In contrast the compositions of the gas in the hydrate phase of the not-stirred system show similar trends to the composition of the isomers in the free gas phase. These observations are in particular applicable for the system containing different amounts of iso-butane/n-butane. It is worth to note that the relative amount of the larger isomer n-butane in the hydrate phase of the stirred system is higher than in the gas phase and the liquid water phase.

Regarding the not-stirred system containing neo-pentane and iso-pentane a clear preference for the incorporation of the smaller neo-pentane molecule into the hydrate structure is observable. However, when the system was stirred, the neo-pentane/iso-pentane ratio reverses and the relative content of the larger iso-pentane molecules in the hydrate phase is similar to that in the liquid water phase.

Table 2 *Molecular dimensions of the different guest molecules*

guest molecule	diameter of the guest molecule [Å]
methane	4.36^{15}
iso-butane	6.5^{15}
n-butane	7.1 (gauche-conformation)[16] 6.7 (cis-conformation)[16]
neo-pentane	7.08^{17}
iso-pentane	7.98^{15}

Table 3 *Relative contents of isomers in the gas, water and hydrate phase*

gas mixture	isomer ratio of the gas phase	isomer ratio in the water phase	hydrate composition (not-stirred system)	hydrate composition (stirred system)
gas mixture I 3% iso-butane 3% n-butane 94% methane	iso-C_4H_{10} : n-C_4H_{10} 50% : 50%	iso-C_4H_{10} : n-C_4H_{10} 44% : 56%	iso-C_4H_{10} : n-C_4H_{10} 1. 57.6% : 42.4% 2. 58.9% : 41.1% 3. 53.2% : 46.8% Ø 56.6% : 43.4%	iso-C_4H_{10} : n-C_4H_{10} 1. 31.6% : 68.4% 2. 39.6% : 60.4% 3. 35.6% : 64.4% Ø 35,6% : 64,4%
gas mixture II 3% iso-butane 2% n-butane 95% methane	iso-C_4H_{10} : n-C_4H_{10} 60% : 40%	iso-C_4H_{10} : n-C_4H_{10} 54% : 46%	iso-C_4H_{10} : n-C_4H_{10} 1. 65.3% : 34.7% 2. 61.5% : 38.5% 3. 61.7% : 38.3% Ø 62.8% : 37.2%	iso-C_4H_{10} : n-C_4H_{10} 1. 47.5% : 52.5% 2. 48.0% : 52.0% 3. 54.0% : 46.0% Ø 49.8% : 50.2%
gas mixture III 1% neo-pentane 1% iso-pentane 98% methane	neo-C_5H_{12} : iso-C_5H_{12} 50% : 50%	neo-C_5H_{12} : iso-C_5H_{12} 39% : 61%	neo-C_5H_{12} : iso-C_5H_{12} 1. 78.5% : 21.5% 2. 72.2% : 27.8% 3. 67.3% : 32.7% Ø 72.7% : 27.3%	neo-C_5H_{12} : iso-C_5H_{12} 1. 47.7% : 52.3% 2. 43.1% : 56.9% 3. 46.4% : 53.6% 4. 39.2% : 60.8% Ø 44.1% : 55.9%

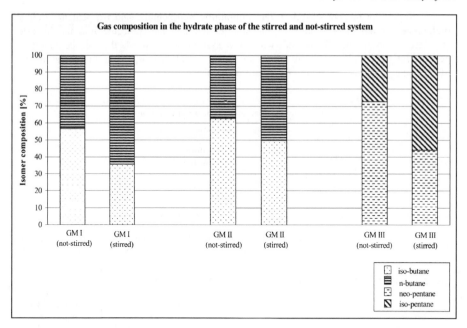

Figure 2 *Gas compositions (isomer) of the hydrate phase of the stirred and not-stirred system*

For the experiments regarding the formation of clathrate hydrates in absence of a free gas phase we used glass beads and beach sand to mimic the influence of sediment grains with different types of fluid-solid interfaces.

At a pressure of about 14 MPa we cooled the system down to about 276 K with the HPLC pump running at a flow rate of about 1 µl/min. We always observed the first hydrate crystals at the top of the sample cell which is an argument for seed crystal formation in the free fluid phase where they rise up to the top of the cell driven by density difference (Figure 3). The fact that the hydrate crystals moved with the pressure pulses caused by the pump strokes indicates that the hydrate crystals did not stick at the acrylic window. When we observed the first hydrate crystals we reduced the flow rate since it controls the methane supply and the hydrate formation rate. Both using glass beads and sand, we did not see hydrate crystals sticking at the grain surfaces. For high hydrate contents the hydrate seems to enclose the grains. The main difference between the experiments with glass beads and sand grains in our experiments is related to the fact that in the sand system hydrate formation starts quicker with less under cooling compared to the glass bead system. Although, the hydrate does not grow at the grain surfaces the fluid-solid interface seems to have an influence on the initiation of seed crystal formation. The mechanism behind that observation is not clear and will be the subject of further investigations.

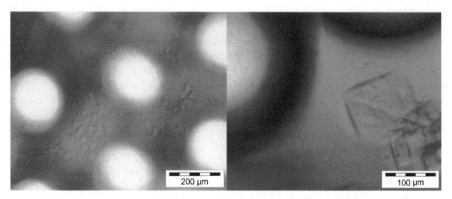

Figure 3 *Hydrate crystal formations in the free fluid phase between glass beads (light balls)*

3 CONCLUSIONS

Regarding the hydrate formation experiments in presence of a free gas phase the results clearly show that diffusivity and water solubility do have an influence on the composition of the gas in the hydrate phase. How strong the influence of those molecular properties is depends on the formation condition. In case of a not-stirred system we assume a hydrate formation at the water-gas interface[1]. For the hydrate growth the gas molecules have to diffuse through the hydrate layer into the water phase or the water molecules have to diffuse through the hydrate layer towards the gas phase. In the first case the diffusivity of the gas becomes a strong impact whereas the water solubility of this gas species becomes a negligible factor. This is reflected in the isomer ratio of the not-stirred systems. As we learned from the model of Englezos and Bishnoi the diffusivity of a gas is inversely proportional to its molar volume. Or, in other words, a small gas molecule can diffuse more easily through the gas phase and/or the hydrate layer and the water phase which was proven by the higher content of the smaller isomer in the hydrate phase. Due to the fact that the isomer ratios in the hydrate phase of the not-stirred system tend towards the isomer ratio of the gas phase it is very likely that the hypothesis of Rodger[3] and Kvamme[4] assuming a hydrate growth towards the gas phase is the more applicable model.

In contrast, stirring of the system avoids the formation of a hydrate layer at the water-gas interface and gas molecules, which have been incorporated into the hydrate phase, can be replaced in the water phase again. In this case the water solubility of the gas is an important factor regarding the incorporation of a gas molecule in the hydrate structure. Since the difference of the molecular dimensions between iso-butane and n-butane which are incorporated in hydrate structures is only between 3% and 8% the difference in diffusivity of these two gases should be significant but not extreme[16]. This effect is interfered by the higher water solubility of the n-butane. Regarding the mixture containing iso-pentane and neo-pentane the influence of the water solubility gets stronger as well. But in this case the difference between the molecule dimensions of neo-pentane and iso-pentane is more significant (11%). Therefore the incorporation rate in this neo-pentane/ iso-pentane system is less dependent on the solubility of the isomer as it is for the iso-butane/n-butane system. For the formation process in a stirred system the hypotheses of Englezos and Bishnoi[6, 7] as well as Christiansen and Sloan[5] are applicable models.

Acknowledgements

We thank Michael Gabriel (GeoForschungsZentrum Potsdam, Section 4.3) for the performance of the gas chromatographic measurements. This is publication no. GEOTECH - 240 of the R&D-Programme GEOTECHNOLOGIEN funded by the German Ministry of Education and Research (BMBF) and German Research Foundation (DFG), Grant 03G0605A.

References

1   B. Kvamme, *IJOPE,* 2002, **12**, 256.
2   M. A. Clarke and P. R. Bishnoi, *Chem. Eng. Sci.,* 2005, **60**, 695.
3   P. M. Rodger, *J. Phys. Chem.,* 1990, **94**, 6080.
4   B. Kvamme, 'A new theory for the kinetics of hydrate formation' in *Proceedings of the 2nd international conference on natural gaz hydrates. Toulouse, June 2-6,* 1996, 139.
5   R. L. Christiansen and E. D. Sloan Jr., 'Mechanisms and Kinetics of Hydrate Formation' in *Annals of the New York Academy of Sciences - International Conference on Natural Gas Hydrates,* eds., E. D. Sloan Jr., J. Happel and M. A. Hnatow, New York Academy of Science, New York, 1994, Vol. 715, 283.
6   P. Englezos, N. Kalogerakis, P. D. Dholabhai and P. R. Bishnoi, *Chem. Eng. Sci.,* 1987, **42**, 2647.
7   P. Englezos, N. Kalogerakis, P. D. Dholabhai and P. R. Bishnoi, *Chem. Eng. Sci.,* 1987, **42**, 2659.
8   M. Luzi, diploma thesis, Universität Potsdam, 2006.
9   E. Spangenberg, *J. Geophys. Res.,* 2001, **106**, 6535.
10   E. Spangenberg, J. Kulenkampff, R. Naumann and J. Erzinger, *Geophys. Res. Lett.,* 2005, **32**, 1.
11   J. Dvorkin, M. B. Helgerud, W. F. Waite, S. H. Kirby and A. Nur, 'Introduction to the Physical Properties and Elasticity Models' in *Natural Gas Hydrate in Oceanic and Permafrost Environments,* ed., M. D. Max, Kluwer Academic Publishers, Dordrecht, 2000, Chapter 20, 245.
12   J. A. Priest, A. I. Best and C. R. I. Clayton, *J. Geophys. Res.,* 2005, **110**, B04102.
13   J. M. Schicks and J. A. Ripmeester, *Angew. Chem. Int. Ed.,* 2004, **43**, 3310.
14   J. M. Schicks, R. Naumann, J. Erzinger, K. C. Hester, C. A. Koh and E. D. Sloan Jr., *J. Phys. Chem. B,* 2006, **110**, 11468.
15   E. D. Sloan Jr., *Clathrate hydrates of natural gases,* 2nd edn., Marcel Dekker, Inc., New York, 1998, 47.
16   D. W. Davidson, S. K. Garg, S. R. Gough, R. E. Hawkins and J. A. Ripmeester, *Can. J. Chem.,* 1977, **55**, 3641.
17   J. A. Ripmeester and C. I. Ratcliffe, *J. Phys. Chem.,* 1990, **94**, 8773.

EFFECT OF SNOW ACCRETION TO THE GPS ANTENNA ON POSITIONING PERFORMANCE

M. Shishido[1], H. Yamamoto[2], S.Iikura[1], T. Endo[1], T. Masunari[3], T. Fujii[1] and K.Okada[4]

[1] Disaster Prevention Technology Division, Railway Technical Research Institute, 2-8-38, Hikari-cho, Kokubunji, Tokyo, Japan
[2] Signaling and Telecommunications Technology Division, Railway Technical Research Institute, 2-8-38, Hikari-cho, Kokubunji, Tokyo, Japan
[3] FURUNO ELECTORIC CO., LTD., 9-52, Ashihara-cho, Nishinomiya, Hyougo, Japan
[4] Civil and Environmental Engineering, Kokushikan Uni., 4-28-1, Setagaya-ku, Tokyo, Japan

1 INTRODUCTION

In recent years, satellite positioning represented by Global Positioning System (GPS) has been in extensive application, and its accuracy and reliability are likely to improve in near future. While some train traffic control systems have already applied GPS as a means of train detection, various technological developments are also under way for the application of GPS.[1] For example, the rotary snow blowers applied the GPS to detect where it locates in the heavy snow.[2] However, if the snow piled up to the GPS antenna, sometimes, the detection is not possible. However, it is unclear what the relationship between the positioning performance of GPS and the depth of piled-up snow to the GPS antennas. This paper reports how dry or wet snow accretion to the GPS antenna decreases the positioning performance.

2 LOCATION AND METHOD

We carried out tests on January 13 and March 4 in 2005 at Shiozawa Snow-Testing Station in Niigata-prefecture, Minami-Uonuma-city located in the northern parts of Japan. In the test, we used the GPS antenna: GSC-001 made by FURUNO ELECTRIC Co. LTD. (Figure 1). It is equipped with a dome-shaped radome. The diameter of its radome is 170mm. To estimate the effect of snow accretion to the GPS antenna on positioning performance, in the test, we set two antennas on the wooden desk. One is covered by snow as "Snow Station" and another was not covered by snow as "Base Station". We installed two antennas with a sufficient distance of 2000mm between them to prevent possible adverse effect to each other. To estimate the effect of snow accretion, we compared two factors. One factor is "Number of Satellites". This is the number of satellites that the GPS antenna could catch. GPS employs more than 24 satellites in about 20000km circular orbits, and they are constantly moving. If the GPS receiver could catch more satellites, the accuracy of the positioning would improve. However, if buildings, terrain, electronic-interference or sometimes even dense foliages could block the signal reception, the accuracy of the GPS positioning performance would be reduced. Of course, GPS typically are not functional indoors, underwater or underground. Therefore, in this test, we count the satellites which have been sufficiently durable to position as Number of Satellites. Another factor is a carrier-to-noise-power-density ratio (C/N_0). In communications, the carrier-to-noise ratio, often written CNR or C/N, is a measure of the received carrier strength relative to the strength of the received noise. High C/N_0 ratios provide better quality of reception, and generally higher positioning accuracy and reliability, than low C/N_0 ratios. Engineers specify the C/N_0 ratio in decibels (dB) between the power in the carrier of the desired signal and the total received noise power. If the incoming carrier strength in microwatts is Pc and the noise level in microwatts, is Pn, then the carrier-to-noise ratio, C/N_0, in decibels is given by the Formula 1.

$$C/N_0 = 10 \times \log10 \,(\, Pc \,/\, Pn \,)$$

Formula 1   *Formura of the carrier-to-noise ratio (C/N_0)*

In the test, the dry compacted snow (The grain size was 0.5mm.) which naturally accumulated on the roof of the Testing Station. Wet snow with defined content of water (c.w.) was generated by adding water to the dry snow artificially. In measuring the content of water, we used a portable calorimeter called Endo-type snow-water content meter.[3]

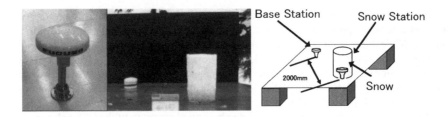

Figure 1   *The picture of the GPS antenna: GSC-001 (FURUNO ELECTRIC Co. LTD.) and the picture and illustration of the Snow Station and Base Station*

Table 1 shows the test procedure. In the test, we piled up the snow in column around the GPS antenna. "Depth of snow" means the height of snow from the top of the GPS radome to the top of the snowfall (Figure 2). Depth of snow was changed from 400mm to 0mm in the dry snow test and from 380mm to 0mm in each wet snow test. The diameter of the snow column is 600mm. The top of the radome is not flat. In Table 1, "0- mm " means that the part of radome was exposed in circle of which diameter was 60mm and "0+ mm" means that the whole head of radome was covered by snow. The angles of direction and elevation from antenna to each satellite are changing from time to time. Similarly, the depth of the snow from the GPS antenna to each satellite has been changing. Therefore, comparing C/N_0, we calculate the depth of the snow to each satellite. Where the satellite situated at a low angle of elevation or relative higher buildings or foliages located adjacent to the GPS antenna, errors in measurement likely to occur due to the effect of a multiplicity of paths (Figure 3).[4,5] The transmitted signal arrivals at the receiver from various directions. Such a phenomenon is called "multipath". It is an unpredictable set of reflections and/or direct waves each with its own degree of attenuation and delay. Figure 4 shows the angle of elevation of the terrain, buildings and foliages around the testing station. The maximum degree of terrain is approximately 10 degrees and the maximum degree of buildings or foliages is approximately 25 degrees. Considering multipath, we only used C/N_0 from the satellites, with an elevation exceeding 30 degrees.

Table 1   *The procedure of the tests*

Dry snow test			Wet snow test								
	Snow Station			Snow Station			Snow Station			Snow Station	
Base Station	Depth of snow	Content of water	Base Station	Depth of snow	Content of water	Base Station	Depth of snow	Content of water	Base Station	Depth of snow	Content of water
	(mm)	(%)		(mm)	(%)		(mm)	(%)		(mm)	(%)
	0			0−※1			0−※1			0−※1	
	40			0+※2			0+※2			0+※2	
	100			2			2			2	
No−snow	190	0		3			3			3	
	280			15			15			15	
	335		No−snow	30	39	No−snow	30	24	No−snow	30	17
	400			45			45			45	
				80			80			80	
				165			165			165	
				250			250			250	
				380			380			380	

※1 Height of snow "0−mm" means that the part of radome has been exposed in circle of which diameter was 60mm.
※2 Height of snow "0+mm" means that the whole head of radome has been exposed.

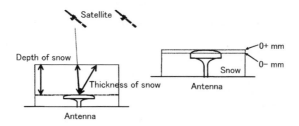

Figure 2   *The illustration of the Snow Station*

3 RESULTS

3.1 Dry snow accretion test

In the dry snow accretion test, we piled up the dry snow at the Snow Station. Depth of snow was changed from 400mm to 0mm. Figure 5 shows the angle of elevation of each satellite. There were nine satellites over the test site. Figure 6 shows the Number of Satellites of the Snow Station and Base Station. The number of Satellites of both sites has been eight while the test. The C/N_0 of the satellites (satellite-3, 19, 27) which had situated more than 30 degrees of elevation was compared between the Snow Station and Base Station (Figure 7). At the Snow Station, the maximum depth of snow has calculated approximately 600mm. C/N_0 of both Snow Station and Base Station has been 50-52 dB during the test, and the value was almost identical regarding each satellite. These results

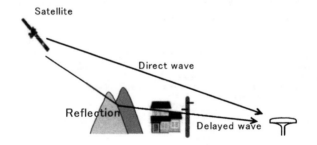

Figure 3   *The illustration of the multipath.*

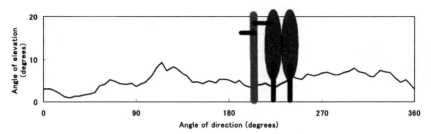

Figure 4   *The angle of elevation of the terrain and structures around the test station*

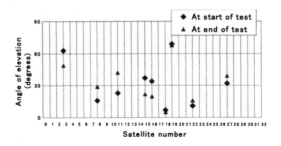

Figure 5   *The angle of elevation of each satellite in the dry snow test*

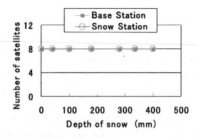

Figure 6   *The Number of Satellites in the dry snow test*

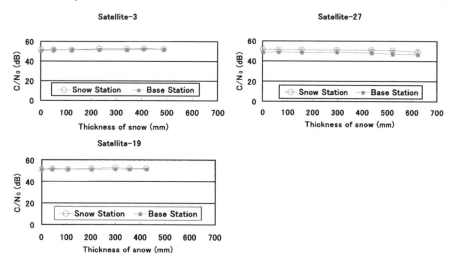

Figure 7   *The C/N₀ of satellite-3, 19, 27 in the dry snow test*

3.2 Wet snow accretion test

In the wet snow accretion test, we piled up the wet snow at the Snow Station. Depth of snow has changed from 380mm to 0mm. Figure 8 shows the distribution of content of water at the Snow Station. We measured this value when the snow column was cut. The content of water was between 38% and 44% (aver. 39%).

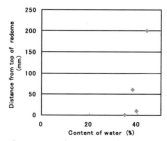

Figure 8   *The distribution of content of water in the snow column*

Figure 9 shows the angle of elevation of each satellite. There were twelve satellites over the test site. Figure 10 shows the Number of Satellites of the Snow Station and Base Station. At the Base Station, the Number of Satellites has been six to eight. At the Snow Station, the Number of Satellites has been zero while the thickness of snow has decreased from 380mm to 80mm, and the Number of Satellite has been increased from three to eight as the thickness of snow decreased from 60mm to 0mm. Between the Snow Station and the Base Station the C/N_0 of the satellites (satellite-11, 20, 28), which situated more than 30 degrees of elevation was compared (Figure 11). At the Base Station, the maximum depth of snow was approximately 580mm. The C/N_0 of Base Station has been between 50dB and 52dB during the test. The C/N_0 of the Base Station was 32dB when the thickness of snow has been changed from 580mm to 80mm. If the C/N_0 was 32dB, the GPS could not position. In addition, the depth has decreased from 80mm to 50mm. the C/N_0 has increased drastically from 32dB to 41dB. If C/N_0 was 41dB means that the GPS could position. In addition, the depth has decreased from 50mm to 0mm. The C/N_0 has increased gradually from 41dB to 52dB. These results mean that the thickness of wet snow affects the accuracy of the GPS positioning performance. However, in the dry snow test, the increase of the depth did not affect the accuracy of the GPS positioning performance. Therefore, we can consider that the water content volume of the snow adversely affects the GPS positioning performance.

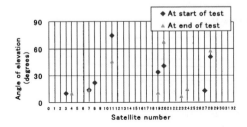

Figure 9   *The angle of elevation of each satellite in the wet snow test (c.w. 39%)*

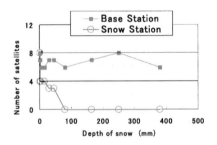

Figure 10   *The Number of Satellites in the wet snow test (c.w. 39%)*

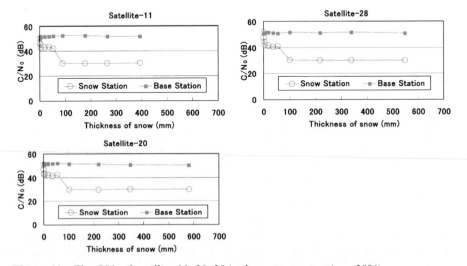

Figure 11   *The C/N_0 of satellite-11, 20, 28 in the wet snow test (c.w. 39%)*

Next, the snow which was regulated content of water as 17% and 24% was accreted to the Snow Station GPS antenna in column, and C/N_0 data was obtained by the same way as c.w. 39% wet snow test. Figure 12 shows the angle of elevation of each satellite in c.w. 17% and c.w. 24% wet snow test. In the c.w.17% wet snow test, four satellites (satellite-2, 4, 7, 13), and in the c.w. 24% wet snow test, five satellites (satellite2, 4, 7, 13, 24) has been situated more than 30 degrees of elevation. In addition, we collect the C/N_0 data of the dry snow test and the c.w. 17%, 24% and 39% wet snow tests when the thickness of snow was less than 50mm. Because, when the thickness was more than 50mm, the C/N_0 would be about 32dB constantly. Figure 13 shows the C/N_0 data of each wet snow test. The C/N_0 of Base Station was obtained at the same time when the C/N_0 of each Snow Station was obtained. The C/N_0 was 52-48dB (aver. 50dB) at the Base Station, 52-47dB (aver. 50dB) at the c.w. 17% Snow Station, 49-44dB (aver 47dB) at the c.w.24% Snow Station, 48-40dB (aver 39dB) at the c.w. 39% Snow Station. This mean larger content of water snow accretion decreased more C/N_0. In addition, if the content of water of accreted snow was less than 24%, until 50mm thickness, the GSC-001 antenna could read the position.

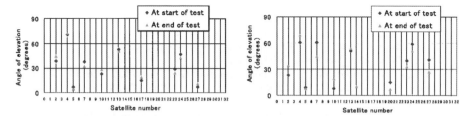

Figure 12 *The angle of elevation of each satellite in the c.w. 17% and 24% wet snow test*

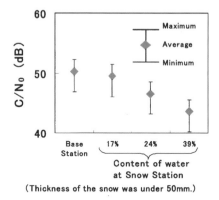

Figure 13 *The C/N_0 data of the dry snow test and the c.w.17%, 24%, 39% wet snow tests*

4 CONCLUSION

In recent years, satellite positioning represented GPS is in extensive use. However, if the snow accreted to the GPS antenna, sometimes, it could not detect. This paper reports how the dry or wet snow accretion to the GPS antenna decreased positioning performance of the GPS.

The accretion of dry snow, of which the maximum thickness was 600mm, did not affect the accuracy of the GPS positioning performance. However, the accretion of wet snow, which was thicker and wetter, decreased more the accuracy of the GPS positioning performance. In this test, when the content of water of accreted snow was more than 24%, and the depth of snow was more than 50mm, the GSC-001 antenna could not read the position.

From these results, we can guess that the water in the snow interrupts the waves from the GPS satellites. However, it is still unclear what the relationship between the type of piled-up snow to the GPS antennas and the positioning performance. When the porosity of the snow is large, the waves may go through the snow easily.

In practical use, if the GPS antenna provided on the roof of trains running through snowy regions, it is preferable that measures taken to prevent the antennae from damage by heavy snowfall.

References

1 H. Yamamoto, *Railway Reseach Review (Japanese)*, 2001, **7**, 45.
2 (article), *JRgazzette (Japanese)*, 2004, **203**, 58
3 K. Kawashima, T. Endo and Y. Takeuchi, *Annals of Glaciology*, 1998, **26**, 103
4 T.Sakai, *Guide of the GPS technology (Japanese)*, 1st Edn., Tokyo Denki University, 2003, p. 93.
5 J.Tsuchiya and H.Tsuji, *Basis of the GPS survey (Japanese)*, Japan Association of Surveyors, 2002, p.44.

GAS HYDRATES IN THE SYSTEM H_2 –CH_4 - H_2O AT PRESSURES OF 4.5 TO 220 MPa AND CONCENTRATIONS OF 0 TO 70 MOL % H_2

S.S. Skiba, E.G. Larionov, A.Yu. Manakov, B.A. Kolesov

Nikolaev Institute of Inorganic Chemistry SB RAS, Novosibirsk, Russian Federation

1 INTRODUCTION

Hydrogen is likely to become the main energy carrier in the future because it is an available, non-toxic and a high-energy fuel. Pure hydrogen gas is required for fuel cells that can be used as basic "reactors" for obtaining electric energy in a hydrogen oxidation process. At the same time, various industrial methods for the production of hydrogen have been proposed, based on the interaction of hydrocarbons (in general methane) that is contained in natural gas, coke-oven gas and the gas obtained by oil treatment with water steam or incomplete oxidation of hydrocarbons. It is natural that the question arises how to isolate hydrogen from such gas mixtures. One method of separation and concentration of multicomponent gas systems is via gas hydrates. By this method components of the mixture can be crystallised as clathrates with associated isolation and regeneration of gases in pure form.[1]

Another important question of hydrogen power engineering is the storage of hydrogen in a form suitable for its subsequent technological use. The task is to develop hydrogen accumulators that can work under technically acceptable P-T conditions and at the same time contain considerable quantities of hydrogen (for instance, according to the estimates of the USA Department of Power Engineering, the minimal amount of hydrogen in accumulators for economically profitable use in motor transport is 6.0 mass-%). Unfortunately, not one of the existing methods (storage in high-pressure vessels, in the liquefied form, adsorbed on various materials, or storage in the form of hydrides) meets the entire set of these requirements.[2-4] In this connection, it seems worthwhile to make attempts aimed at the development of novel methods (materials) to store hydrogen, including a storage in the form of clathrate hydrates.

Clathrate hydrates are crystalline inclusion compounds in which the host framework is formed by water molecules linked through hydrogen bonds. The cavities of the framework are occupied by guest molecules of appropriate size and shape. Hydrates composed of gas molecules as guests and water molecules as the host are called gas hydrates. As a rule, there are only van der Waals interactions between the guest and host subsystems.

As a rule, gas hydrates obtained under the P-T conditions close to those in the near-surface have one of the three following structures: cubic structure I (sI), cubic structure II (sII) or hexagonal structure H (sH). Each of these structures has its own set of polyhedral cavities of different sizes. The better that the size and molecular shape of the guest molecule corresponds to the size and shape of free space in the cavities of host lattice, the higher is the decomposition temperature (melting point) of gas hydrates.

The history of gas hydrate investigation started in 1811 when Davy reported the formation of a crystalline compound of chlorine with water,[5] but investigation of the gas hydrates involving hydrogen has started only recently. At first, a pioneering paper by Namiot et al.[6] stated the existence of solid solutions of hydrogen in Ih ice. Later Vos et al. discovered solid solutions of hydrogen in ice II and Ic.[7, 8] Finally, when analyzing the phase diagram in the pressure range 100 to 360 MPa, the existence of hydrogen hydrate of one in the classical polyhedral structures was assumed.[9, 10] It was established later that this structure is sII.[11] It was also shown[12-14] that small D cavities (5^{12}, in other words, polyhedra with 12 pentagonal faces) in this structure are occupied by only one hydrogen molecule, while large H-cavities ($5^{12}6^4$) contain 2 to 4 hydrogen molecules, depending on temperature. Unfortunately, gas hydrates of pure hydrogen can not be used for hydrogen storage because a pressure above 100 MPa is required for its stability. In addition, the hydrogen content of these hydrates is much 6.0 mass-%.

It is well known that, as a rule, decomposition temperatures of hydrates formed from mixtures of gases that differ in their van der Waals radii are higher than decomposition temperatures of the hydrates of individual components of these mixtures. This is explained by a higher degree of three-dimensional complementarity of the guest molecules and empty space in the host lattice for the hydrates formed from gas mixtures.[15] As the result, the density of three-dimensional arrangement of the hydrate rises, and free the energy decreases; these changes enhance the stability of the hydrate.

Florusse at al.[16] showed that the decomposition temperature of hydrogen hydrate can be increased substantially if large cavities in the sII structure get filled with tetrahydrofuran molecules (THF). Lee at al.[17] discovered that a stable gas hydrate can be obtained if a small part of large cavities is occupied with THF molecules, while other large cavities are filled with hydrogen molecules. The hydrogen content of gas hydrate increases substantially. However, practical application of double hydrates of hydrogen with THF is hindered by the high volatility, toxicity and high cost of tetrahydrofuran.

In this connection, it seemed interesting for us to investigate the system $H_2 - CH_4 - H_2O$ over a wide temperature, pressure and concentration range in order to examine this system for its suitability for hydrogen storage. A mixture of the gas components of this system can serve as highly efficient combined fuel. The purpose of this work is to build a P-T-X phase diagram (P – pressure, T – temperature, X - composition) that allows one to predict the regions of existence of separate components of the system and the thermodynamic parameters of the phase transitions under changeable P-T-X conditions.

2 METHODS AND RESULTS

2.1 Experimental

The decomposition temperatures of hydrates were measured by means of differential thermal analysis (DTA) under the conditions of excess gas in a stainless steel flask that was developed specially for the investigation of hydrate formation with a gaseous guest at high hydrostatic pressure.[18] The hydrate decomposition temperature was measured with a chromel-alumel thermocouple to the accuracy of ±0.3 K. The thermocouple was calibrated with the use of temperature standards. Pressure was measured with a Bourdon-tube pressure gauge. The error of the pressure measurements did not exceed 0.5 %. This procedure was described in more detail previously.[18, 19] The gases used in the investigation contained not less than 99.95 % of the main gas component; they were used without further purification.

The gas – hydrate – water equilibriums were established within about 30 min for any pressure, as determined in separate experiments. The equilibrium was considered to be established if no ice melting peak was observed when the system was heated under the excess of the hydrate-forming gas.

Hydrate samples for Raman scattering and X-ray investigations were prepared as follows. Finely grounded ice (grain size 0.05-0.1 mm.) was loaded into the high-pressure chamber. After that, in order to flush the chamber, we repeatedly pressurized/depressurized it with hydrogen (about 10 MPa) for 3-5 times. Then, hydrogen and methane were consecutively admitted in a proportion of 40% mol. H_2 and 60% mol. CH_4 at the final pressure of 20 MPa. The sample was kept in a freezing chamber at a temperature of -14°C for 2-3 weeks. A pressure drop in the chamber was an evidence of hydrate formation. When we wanted to extract the hydrate for further investigation, we put the high-pressure chamber into liquid nitrogen. A quenched sample was divided into three subsamples. Two of them were used to record X–ray powder diffraction patterns and Raman spectra; one of the subsamples was kept at room temperature to monitor the presence of the hydrate on the basis of gas evolution.

X-ray diffraction studies were performed with quenched samples of hydrates and ice Ih using synchrotron radiation at the 4th beamline of the VEPP-3 storage ring (Budker Institute of Nuclear Physics SB RAS), at a fixed wavelength of 0.3675 Å.[20] A Debye-Scherrer diffraction method was applied. An imaging plate detector MAR3450 (pixel dimension 100 μm) was used to register the diffraction pattern. The distance between the sample and the detector (calibrated against the diffraction pattern of sodium chloride) was 391.0 mm. A fine-ground hydrate sample was placed in an aluminium cell with two foam-coated holes for the primary beam and for the outlet of diffracted radiation.

Raman spectra were recorded with a Triplemate, SPEX spectrometer equipped with a multi-channel detector, LN-1340PB, Princeton Instruments, in a back-scattering geometry. The spectral resolution was 5 cm^{-1}. The 514 nm line of a 50 mW Ar-ion laser was used for spectral excitation. All the spectra were recorded at a temperature near 77 K at atmospheric pressure. The hydrate taken out from the high-pressure chamber was put into a cell that was immersed in liquid nitrogen. This cell had perforation covered with glass. In order to exclude frosting of extraneous compounds on this glass, we maintained it under a flow of helium. The spectra were recorded within two wavenumber intervals: 2700-3000 cm^{-1} (recording C-H vibrations in methane molecule) and 4000-4200 cm^{-1} (recording H-H vibrations in molecular hydrogen).

2.2 Results

The P-T decomposition curve of individual methane hydrate was shown in the graphical form[21] and the corresponding digital experimental data were presented earlier.[22]

The data on the decomposition temperatures of clathrate hydrates of pure hydrogen within the pressure range 90 - 360 MPa were already published in the graphical form.[9, 10] The digital data summarized from these experiments are shown in Table 1.

The experimental data obtained in the present study for $H_2 - CH_4 - H_2O$ system were distinguished by different degrees of reproducibility. The best repeatability of the results was observed for pure methane hydrates and for those experiments in which the hydrogen content of the initial gas mixture was not more than 70 % mol. For higher hydrogen concentration, reproducibility became unacceptably poor, so here we report only the results

Table 1 *Decomposition temperatures for clathrate hydrogen hydrates at high pressure.*

P MPa	90	140	160	184	225	258	290	336	338	364
T °C	-7.0	-3.5	-2.6	-1.6	-0.2	0.6	1.0	1.1	1.2	1.1

for hydrogen concentration in the gas phase near or below 70 % mol. The reduced reproducibility with hydrogen concentrations above this value can be explained by experimental errors accompanying a decrease in the concentration of hydrate-forming gas (methane) and the corresponding increase in the relative amount of admixtures (air) able to get into the system during charging the experimental cell.

For every experimental P-T curve obtained for the three-component system, the data were fitted to the corresponding equations using the least squares method. The coefficients of the polynomial approximation to the decomposition curves for the whole investigated composition range of the three-component system are shown in Table 2.

The experimental data on decomposition temperatures of the three-component system H_2–CH_4-H_2O for 0-70 % (mol.) hydrogen content in the initial gas mixture are shown in Figure 1.

Figure 2 shows possible isobaric profiles of the P-T-X phase diagram; the profiles were constructed using the experimental data on decomposition temperatures of hydrates in the system H_2–CH_4-H_2O. When making these diagrams, we supposed that the methane solubility in the hydrogen hydrate and hydrogen solubility in methane hydrate are negligibly low.

In order to determine the structure of clathrate hydrates formed in this system, we carried out an X-ray powder diffraction investigation. The powder diffraction patterns are shown in Figure 3.

Table 2 *The coefficients of equations T (°C) $= A+B*P+C*P^2+D*P^3+E*lnP$ (P, MPa) that approximate the liquidus lines for the hydrogen - methane – water system at gas pressures up to 220 MPa with the gas in excess. Hydrogen concentration is indicated for the initial gas mixture.*

C_{H2} mol.%	Pressure range (MPa)	A	B	C	D	E
0.0	1.8 - 250	-11.1	-0.179	0.00131	$-2.8*10^{-6}$	10.8
7.5	4.5 - 250	-29.1	-0.004	$4.1*10^{-6}$	$-8.5*10^{-10}$	8.8
10.0	6.4 – 250	-31.2	-0.003	$3.8*10^{-6}$	$-9.0*10^{-10}$	8.9
20.0	5.2 – 250	-28.9	-0.002	$2.8*10^{-6}$	$-6.4*10^{-10}$	8.2
30.0	6.0 – 250	-28.5	0.004	$-1.4*10^{-6}$	$1.8*10^{-10}$	7.5
40.0	6.7 – 250	-40.8	-0.008	$5.8*10^{-6}$	$-1.3*10^{-9}$	10.0
50.0	15.3 – 250	-35.1	0.006	$-3.5*10^{-6}$	$7.5*10^{-10}$	7.4
60.0	19.0 – 250	-23.5	-0.075	0.00057	$-1.2*10^{-6}$	9.3
70.0	42.0 - 250	-15.3	0.027	$-9.7*10^{-6}$	$1.4*10^{-9}$	0.9

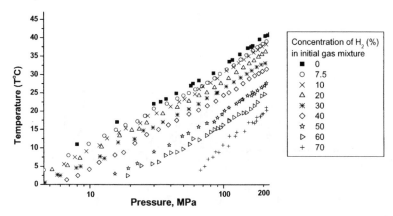

Figure 1   *Decomposition temperatures of the three-component system H_2-CH_4-H_2O for 0-70 % (mol.) hydrogen concentration in the initial gas mixture.*

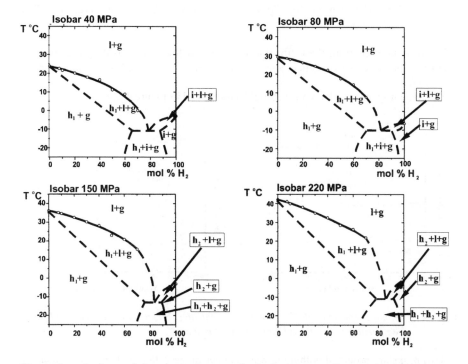

Figure 2   *Isobars of the phase diagram for hydrogen – methane – water at pressures 40, 80, 150 and 220 MPa. o-experimental data. h_1–methane hydrate, h_2–hydrogen hydrate, l–aqueous solution with gas , g–gaseous phase.*

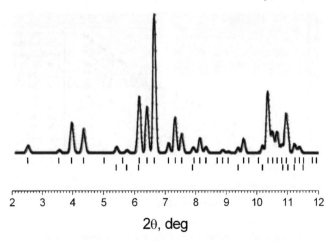

Figure 3   *Synchrotron X–ray powder diffraction pattern of quenched hydrate obtained at a wavelength of 0.3675 Å. Vertical tick-marks correspond to the calculated positions of reflections for the hydrate of cubic structure 1 (upper row) and of ice I_h (lower row). The diffractogram was recorded at a temperature of 77K and atmospheric pressure*

It can be inferred from the diffraction patterns that the investigated hydrate may be assigned to the structure sI (lattice constant 11.86Å at –140°C) with some admixture of unreacted ice.

The Raman spectra of the same material are presented in Figure 4.

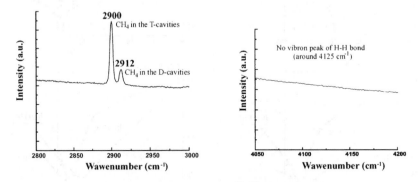

Figure 4   *Raman spectra of the quenched hydrate. Spectra were recorded at a temperature of 77K and atmospheric pressure*

Two peaks are observed within the wavenumber range 2700-3000 cm⁻¹ corresponding to C-H vibrations in methane molecule. The peak at 2900 cm⁻¹ corresponds to the vibrations of methane molecules trapped in large T-cavities of sI structure. The peak at 2912 cm⁻¹ corresponds to those of methane molecules contained in small D-cavities.[23] No bands were observed within the wavenumber range 4050-4200 cm^{-1} where the adsorption band corresponding to H-H vibrations in hydrogen molecule was to be observed.

It was shown previously[24] that methane molecules in the individual methane hydrate formed in the CH_4-H_2O system at increased pressure occupy almost all the large T and small D cavities in sI. It may be concluded from Fig. 2 and 4 that methane molecules in H_2-CH_4-H_2O system also occupy all the cavities in the hydrate leaving no place for hydrogen molecules (at least within the concentration range 0-70% mol. H_2 and at pressures 20 MPa (pressures of X–Ray and Raman – experiments) and higher). This may be due to the fact that H_2 molecules are less complementary to T and D cavities than CH_4 molecules and cannot compete with the latter for positioning in a cavity even in the case when more than one[12-14] hydrogen molecule can be located in a T cavity.

The indirect confirmation of these conclusions can be found in[25], where it was shown that methane – hydrogen mixtures can be effectively separated by means of methane hydrate formation at a pressure less than 5.0 MPa and temperature about 0 °C.

3   CONCLUSIONS

1. The data on decomposition temperatures of gas hydrates in the system $H_2 - CH_4 - H_2O$ at pressure up to 250 MPa were obtained.
2. A general form of the phase diagram for the system $H_2 - CH_4 - H_2O$ is proposed within a broad temperature, pressure and concentration range.
3. X–ray powder diffraction patterns and Raman spectra of the hydrates formed in the system under investigation were obtained.
4. It was shown that the hydrates formed from mixtures with initial hydrogen contents of 0 - 70 mol. % and at pressures 20 MPa and higher have sI structure with cavities likely to be fully occupied by methane molecules; thus hydrogen molecules are not easily included in the cavities of the hydrate. Therefore, the system $H_2 - CH_4 - H_2O$ in the form of gas hydrate cannot be used to store hydrogen under these conditions.

Acknowledgement

Financial support from the Center for Hydrate Research, Department of Chemical Engineering, Colorado School of Mines is gratefully acknowledged.

References

1   S.Sh. Byk, Yu.F. Makagon, V.I. Fomina, *Gas Hydrates*, Chemistry, Moskau, 1980 (rus)
2   L. Schlapbach, A. Zuttel, *Nature*, 2001, **414**(15), 353.
3   A.C. Dillon, M.J. Heben, *Appl. Phys. A*, 2001, **72**, 133.
4   A. Zuttel, Naturwissenschaften, 2004, .**91**, 157.
5   H. Davy, *Phil. Trans. Roy. Soc.* (London), 1811, **101**, №30, Alembic Club Reprints №9, 58.
6   A.Yu. Namiot, E.V. Bukhgalter, *J. Structural Chem.*, 1965, **6**, 911 (rus.).
7   W.L. Vos, L.W. Finger, R.J. Hemley, H.Mao, *Phys. Rev. Lett.*, 1993, **71**, № 19, 3150.
8   W.L. Vos, L.W. Finger, R.J. Hemley, H. Mao, *Chem. Phys. Lett.*, 1996, **257**, 524.
9   Yu.A Dyadin, E.G. Larionov, A.Yu. Manakov, F.V. Zhurko, E.Ya. Aladko, T.V. Mikina and V.Yu. Komarov, *Mendeleev Commun.*, 1999, 209.

10 Yu.A Dyadin, E.G. Larionov, E.Ya. Aladko, A.Yu. Manakov, F.V. Zhurko, T.B. Mikina, V.Yu. Komarov and E.V. Grachev, *J. Structural Chem.*, 1999, **40**, № 5, 974 (rus.).

11 W.L. Mao, H. Mao, A.F. Goncharov, V.V. Struzhkin, Q. Guo, J. Hu, J. Shu, R.J. Hemley, M. Somayazulu and Y. Zhao, *Science*, 2002, **297**, 2247.

12 K.A. Lokshin, Y. Zhao, D. He, W.L. Mao, Ho-Kwang Mao, R.J. Hemley, M.V. Lobanov, and M. Greenblatt, *Phys. Rev. Lett.*, 2004, **93**, №12, 1.

13 K.A Lokshin, Y. Zhao, M.V. Lobanov, M. Greenblatt, *Proceedings of the Fifth International Conference on Gas Hydrates*, **v. 2**, Trondheim, Norway, June 12-16, 2005, 428.

14 K.A.Lokshin, Y. Zhao, W.L. Mao, H. Mao, R.J. Hemley, M.V. Lobanov, and M. Greenblatt, *Proceedings of the Fifth International Conference on Gas Hydrates*, **v. 5**, Trondheim, Norway, June 12-16 2005, 1601.

15 E.G. Larionov, F.V. Zhurko, and Yu.A. Dyadin, *J. Structural Chem.*, 2002, **43**, №6,985.

16 L.J. Florusse, C.J. Peters, J. Schoonman, K.C. Hester, C.A. Koh, S.F. Dec, K.N. Marsh, E.D. Sloan, *Science*, 2004, **306**, 469.

17 H. Lee, J.W. Lee, D.Y. Kim, J. Park, Yu.T. Seo, H. Zeng, I.L. Moudrakovski, C.I. Ratcliffe, and J.A. Ripmeester, *Nature*, 2005, **434**, 743.

18 Yu.A. Dyadin, E.G. Larionov, D.S. Mirinskij, T.V. Mikina, E.Ya. Aladko, and L.I. Starostina, *J. Incl. Phenom.* 1997, **28**, 271.

19 Yu. A. Dyadin, E.G. Larionov, T.V. Mikina and L.I. Starostina, *Mendeleev Comm.*, 1996, 44.

20 A.G. Ogienko, A.V. Kurnosov, A.Y. Manakov, E.G. Larionov, A.I Ancharov, M.A. Sheronov, A.N. Nesterov, *J. Phys. Chem. B, 2006,* **110**, 2840.

21 Yu.A. Dyadin, E.Ya Aladko, E.G. Larionov, *Mendeleev Commun.*, 1997, 34.

22 E.Ya. Aladko, Yu.A. Dyadin, A.Yu. Manakov, F.V. Zhurko and E.G.Larionov, *J. Supramolecular Chem.*, 2002, **2**, 369.

23 A.K. Sum, R.C. Burruss, E.D. Sloan, *J.Phys.Chem. B*, 1997, **101**, 7371.

24 A. Klapproth, E. Goreshnik, D. Staykova, H. Klein, and W. F. Kuhs, *Can. J. Phys.*, 2003, **81**, 503.

25 G.J. Chen, C.Y. Sun, C.F. Ma and T.M. Guo, *Proceedings of the Fourth International Conference on Gas Hydrates*, Yokohama, Japan, May 19-23, 2002, 1016.

CHEMISTRY INDUCED BY IMPLANTATION OF REACTIVE IONS IN WATER ICE

G. Strazzulla[1], G. Leto[1], F. Spinella[1] and O. Gomis[2]

[1]INAF-Osservatorio Astrofisico, via S. Sofia 78, I-95123 Catania, Italy
[2]Depart.de Fisica Aplicada Escuela Politecnica Superior Alcoy, UPV, Spain

1 INTRODUCTION

Solid icy surfaces are observed both in the interstellar medium as mantles on silicatic or carbonaceous grains[1-3] and on many objects in the Solar System.[4] In space, these icy targets are continuously bombarded by energetic ions from solar wind and flares, planetary magnetospheres, stellar winds and galactic cosmic rays. When an energetic ion collides with an icy target produces physico-chemical modifications in the latter. The study of those effects is based on laboratory ion irradiation experiments carried out under physical conditions as close as possible to the astrophysical ones.[5-10]

Reactive ions (e.g., H, C, N, O, S) induce all of the effects of any other ion including the synthesis of molecular species originally not present in the target. In addition these ions have a chance, by implantation in the target, to form new species containing the projectile. An ongoing research program performed at our laboratory has the aim to investigate the implantation of reactive ions in frozen ices by using IR spectroscopy.[11-14] Implantation experiments are particularly relevant for application to objects in the Solar System where ices on planetary surfaces, comets, etc., are much thicker than the penetration depth of the relevant ion populations. In fact there is a possibility to detect, on planetary surfaces, molecular species not necessarily native of that object but formed by implantation of incoming reactive ions. As an example, the presence of SO_2 on Europa and Callisto,[15-17] has been interpreted as possibly formed by magnetospheric sulfur implanted in water ice.[15,18] The suggested presence of hydrated sulfuric acid on Europa has been also correlated with a radiolytic sulfur cycle.[19] Frozen layers of pure CO_2 have been irradiated with 1.5 keV protons and evidenced the formation of carbonic acid (H_2CO_3),[12] later suggested to account for a 3.88 μm NIMS absorption feature on Callisto.[20]

Here we present results obtained by implanting carbon, nitrogen and sulfur ions in water ice at 16 and 77 K. Carbon implantation produces carbon dioxide with a production yield of about 0.5 molecules per impinging ion. Nitrogen implantation does not produce any N-bearing species that is detectable by IR spectroscopy. We find that after sulfur implantation hydrated sulfuric acid is probably formed while we have been able to find only an upper limit to the production yield of SO_2 ($Y \leq 0.025$ SO_2-molecules ion[-1]). All of the ions have also the capability to synthesize hydrogen peroxide. The results are discussed in the light of the relevance they have for the chemistry induced by magnetospheric ions colliding with the surfaces of the icy Jovian moons.

2   METHOD AND RESULTS

2.1 Experimental procedure

Experiments are performed in a vacuum ($P \approx 10^{-7}$ mbar) chamber faced to an IR spectrometer.[9] The spectra here shown have a resolution of 1 cm^{-1} and a sampling of 0.25 cm^{-1}. Water ice films are prepared by vapor deposition on IR-transparent substrata at 15-80 K. Ions are obtained from a 200 kV ion implanter (Danfysik). Ion currents are maintained low enough (100's nano-ampere cm^{-2}) to avoid macroscopic heating of the target. The ion beam sweeps the target to obtain uniform irradiation on an area of about 3 cm^2 i.e. much larger than the area traveled by the IR beam. The substrate plane forms an angle of 45 degrees with the IR beam and the ion beam (these latter being mutually perpendicular) so that before, during and after irradiation, spectra can be taken without tilting the sample.

We deposited water ice films having a thickness of 1-10 μm. The sample thickness is, in all cases, larger than the penetration depth of the incoming ions. Targets have been implanted with C, N or S ions. The list of used ions, energies and target temperatures is given in Table 1. The use of isotopically labeled carbon is due to an unavoidable (small) contamination of carbon dioxide in the target sample.

Reactive incoming ions deposit energy in the target, as unreactive ions also do. Each incoming ion having energies between 15 and 200 keV, breaks 10^3-10^4 bonds and a large number of new molecules can be formed per single incoming ion by recombination of fragments of the irradiated species. In addition only reactive ions have a chance to form species that include the projectile with a maximum yield of one molecule per incoming ion.

One of the effects of ion irradiation is the destruction of molecules of the target. Molecular fragments recombine and may form species originally not present. By measuring the integrated band area (in cm^{-1} and in an optical depth scale) and by knowing the integrated absorbance of that band (cm molecule^{-1}) it is possible to measure the column density (molecules cm^{-2}) of a given species.

In blank experiments we stay at low T (15 K) for hours to verify if any contaminant is deposited: the only contaminant we detected was water that is deposited at a rate of about 1 angstrom min^{-1}. Typically the experiments are conducted within a couple of hours and an extra deposition of about 100 angstroms of water ice is expected.

Table 1 *Ion, energy and temperature used for implantation experiments in icy H_2O films.*

Ion	Energy (keV)	Temperature (K)
N^+	15	16
N^+	30	77
N^+	30	16
$^{13}C^+$	30	77
$^{13}C^+$	30	16
S	200	16
S	200	77

2.2 Results

Figure 1 shows the IR spectra of H_2O ice as deposited (16 K) and after implantation of 30 keV $^{13}C^+$ ions at two different ion fluences (6 and 20 x 10^{15} ions cm^{-2}). The Figure clearly exhibits the formation of H_2O_2 that is formed whatever is the used ion as already well known[21-23] and $^{13}CO_2$ that is formed because of the implantation of $^{13}C^+$ ions.

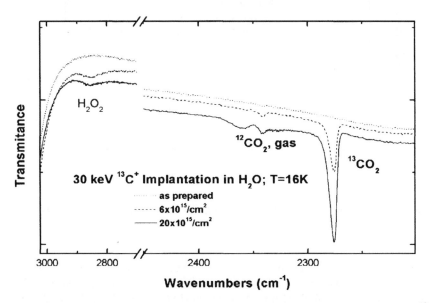

Figure 1 *IR spectra of H_2O ice as deposited (16 K) and after implantation of 30 keV $^{13}C^+$ ions at two different ion fluences (6 and 20 x 10^{15} ions cm^{-2})*

The amount of $^{13}CO_2$ (molecules cm^{-2}) synthesized in experiments performed at 16 and 77 K is obtained by measuring the integrated band area (in cm^{-1} and in an optical depth scale) and by knowing the integrated absorbance of that band (7.6x10^{-17} cm molecule^{-1}).[14,24] The measured production yields (molecules ion^{-1}) are Y= 0.47±0.02 at 16 K and Y= 0.42±0.04 at 77 K i.e. the yields are, within the experimental errors, identical at the two different temperatures. It has been also noticed that $^{13}CO_2$ is promptly formed after implantation of ^{13}C in water ice. ^{13}CO is barely visible only at high ion fluences. Such a result has beeen interpreted as an evidence that the production of carbon monoxide is not due to the direct implantation of carbon but to the destruction of carbon dioxide in a mixture with water. This confirms results obtained by irradiation of H_2O ice with 10 keV C^+ ions that evidenced the formation of $^{12}CO_2$ but not CO.[25] The chemistry of irradiated water starts with the dissociation to H+OH. A plethora of neutral, excited and ionized species are also formed: H, H_2, O, O_2, H_2O, HO_2, H_2O_2.[10,26] These species recombine and e.g. they form hydrogen peroxide. If the implanted ion is reactive it can participate to the chemistry. In the case of C implantation carbon dioxide might be formed by direct addiction of C to O_2 that is one of the products of water radiolysis or by: $HO_2+C \rightarrow H+CO$

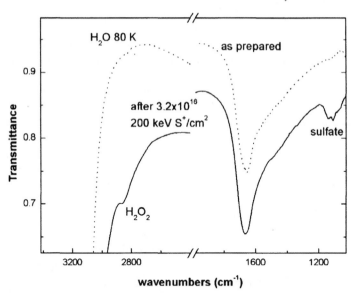

Figure 2 *IR spectra of H₂O ice as deposited (80 K) and after implantation of 200 keV S⁺ ions (3.2 x 10¹⁶ ions cm⁻²)*

In Figure 2 we show the infrared spectra of frozen (80 K) water before and after implantation with 200 keV S⁺ ions (3.2x10¹⁶ ions cm⁻²). It is clear the appearance of a band centered at about 2850 cm⁻¹ (3.51 μm) attributed to H_2O_2 and of a broad band centered at about 1100 cm⁻¹ (9.09 μm) that we attribute to a sulfate, most probably hydrated sulfuric acid (Strazzulla et al 2006, in preparation). We searched for the formation of SO_2, by looking at its most intense bands at about 1340 cm⁻¹ and 1150 cm⁻¹. We have been able to find only an upper limit to the production yield of SO_2 ($Y \leq 0.025$ SO_2-molecules ion⁻¹). The upper limit has been obtained by measuring the integrated band areas (in cm⁻¹ and in an optical depth scale) of the most intense bands at about 1340 cm⁻¹ and 1150 cm⁻¹ and by knowing the integrated absorbance of those bands (in cm molecule⁻¹). The measured areas are very small and hardly distinguishable from the noise.

We have also conducted experiments by implanting 15 keV N⁺ ions in pure H_2O ice and we searched for the formation of species such as N_2O, NO and NO_2 by looking at their features at about 2235, 1878 and 1612 cm⁻¹ respectively.[27] A comparison between IR spectra obtained after implantation of N⁺ ions with that obtained after irradiation of a mixture $H_2O:N_2$=1:1 with 60 keV Ar⁺⁺ demonstrated that N-bearing species are easily formed after irradiation of mixtures already containing nitrogen, but there was no clear evidence of their formation after implantation of nitrogen in water.[14]

It is also important to notice that we have no evidence of the formation of NH and CH after implantation of N and C. An upper limit to the formation of CH bonds is, by number, on the order of a few percent of the implanted carbon.

The present and previous experiments have demonstrated that it is not possible to predict when implantation will produce a given species but by doing the experiments.

Oxygen implantation in frozen methane does not produce the "expected" CO and CO_2.[28] N implantation in frozen mixtures $H_2O:CH_4$ produces a X-OCN band but implantation of O in frozen mixtures $H_2O:N_2$ does not.[29-30]

2.2 Discussion

The discussed results may be relevant to explain the presence of some species observed on the icy moons in the Jovian system. On Callisto, and Ganymede water ice is dominant but features have been observed at 3.4 (2940 cm^{-1}), 3.5 (2857 cm^{-1}), 3.88 (2580 cm^{-1}), 4.05 (2470 cm^{-1}), 4.25 (2350 cm^{-1}) and 4.57 (2190 cm^1) μm.[31-32] The prime candidates for the five bands are: C-H, H_2O_2, S-H, SO_2, CO_2, and CN respectively. It has been also suggested that H_2CO_3 from a carbon radiolysis cycle is a more likely candidate for the band at 3.88 μm as measured in laboratory.[20] On Europa water ice is also the dominant species with traces of a hydrate compound, H_2O_2, SO_2, and CO_2. A still open question is to understand if those species are native from the satellites or are synthesized by exogenic processes such as ion irradiation.

The results we obtained at 77 K, a temperature more appropriate to the jovian moons, are qualitatively and quantitatively the same as those at 16 K. Moreover, after implantation, we warmed the samples and obtained spectra at progressively higher temperatures: Carbon dioxide is still there at 140 K when water ice completed the crystallization process, and it remains trapped until water is lost by sublimation. This gives us confidence that our findings can be extrapolated to environments with a large range of temperatures.[14]

The amount of CO_2 present in a condensed phase on the surfaces of Callisto, and Ganymede has been suggested to be on the order of $3x10^{17}$ molecules cm^{-2}.[31-32] On Europa sulfur has been supposed to be in three forms, polymerized sulfur, SO_2 (on the order of $2x10^{17}$ molecules cm^{-2}) and sulfuric acid hydrate.[19] In the next we calculate the times necessary to produce those amounts of carbon and sulfur dioxide on the Galilean moons.

The fluxes of sulphur (keV-MeV) ions at the surfaces of the galilean moons are in the second column of Table 2.[33] Although the abundance of carbon ions has not been quantified, the number ratio C/S has been considered and found to increase from about 0.2 near by Europa to about unity near Callisto for Voyager measurements in the 1 MeV nucl^{-1} range.[34-35] Here we use C/S= 0.2 at Europa, 0.5 at Ganymede and 1 at Callisto.

Table 2 *Fluxes of sulphur ions,[33] carbon ions (estimated by us), and times necessary to produce $3x10^{17}$ CO_2 molecules cm^{-2} by C-ion implantation and $2x10^{17}SO_2$ molecules cm^{-2} by S-ion implantation.*

Satellite	S-ions cm^{-2} s^{-1} (measured)	C-ions cm^{-2} s^{-1} (estimated)	Time (years) to produce $3x10^{17}$ CO_2 molec cm^{-2}	Time (years) to produce $2x10^{17}$ SO_2 molec cm^{-2}
Europa global	$9x10^6$	$1.8x10^6$	$1.3x10^4$	$2.8x10^4$
Ganymede polar	$3.2x10^6$	$1.6x10^6$	$1.5x10^4$	$7.9x10^4$
Ganymede equator	$5.8x10^4$	$2.9x10^4$	$8.2x10^5$	$4.4x10^6$
Callisto global	$3.6x10^5$	$3.6x10^5$	$6.6x10^4$	$7.0x10^5$

Consequently the estimated flux of C keV-MeV ions is reported in the third column of Table 2 where times necessary to produce $3x10^{17}$ CO_2 molecules cm^{-2} by C-ion implantation and $2x10^{17}SO_2$ molecules cm^{-2} by S-ion implantation are also reported. These latter have been calculated by using the experimentally measured production yields.

Several mechanisms, among which ion irradiation, have been invoked to explain the presence of carbon or sulfur dioxide frozen on the surfaces of the galilean satellites. Ion irradiation has been in fact suggested to be responsible for the formation of O_2 and O_3 observed on Ganymede.[36-37] The dissociation of water molecules, a C-bearing species, has been suggested to induce the formation of bonds with oxygen-bearing fragments leading to the synthesis of carbon dioxide.[31-32] CO_2 production by ion irradiation of H_2O ice on top of carbonaceous materials has been also discussed.[38] We have quantitatively measured the yield of formation of carbon dioxide by implantation of carbon ions. Our conclusion is that although a relevant quantity of CO_2 can be formed, however this is not the dominant formation mechanism provided that the fluxes of C-ions are correct. This mainly because we would expect more CO_2 on the polar region of Ganymede than at the equator but observations indicate that carbon dioxide is clearly more abundant in the equatorial region than in the south polar region.[31-32] Moreover the escape rate of CO_2 from the Callisto's atmosphere has been estimated to be of about $6x10^6$ cm^{-2} s^{-1} that requires a continuous supply from the surface;[39] C-ion implantation followed by desorption of the formed carbon dioxide cannot be enough (see Table 2).

Experiments of implantation of reactive ions are not numerous and consequently their application has been little discussed. A noteworthy exception is the case of SO_2. It has been observed on Europa and Callisto and it has been interpreted as formed by magnetospheric sulfur implanted in water ice.[15,18] We have found only an upper limit to the formation yield of sulfur dioxide after S implantation that does not permit us to be conclusive about the effective relevance of the process at Europa. The finding of the formation of a sulfate, most probably hydrated sulfuric acid, could be of extreme relevance to support the suggested radiolytic sulfur cycle at Europa.

References

1   D.C.B. Whittet and A.G.G.M. Tielens. In: Pendleton, Y., Tielens, A.G.G.M. (Eds.), From Stardust to Planetesimals. In: Astron. Soc. Pac. Conf. Ser., Vol. 122. ASP, San Francisco, 1997 p. 161.

2   Y.J. Pendleton and J.E. Chiar. In: Pendleton, Y., Tielens, A.G.G.M. (Eds.), From Stardust to Planetesimals. In: Astron. Soc. Pac. Conf. Ser., Vol. 122. ASP, San Francisco, 1997, p. 179.

3   S.A. Sandford, L.A. Allamandola and M.P. Bernstein. In: Pendleton, Y., Tielens, A.G.G.M. (Eds.), From Stardust to Planetesimals. In: Astron. Soc. Pac. Conf. Ser., Vol. 122. ASP, San Francisco, 1997, p. 201.

4   B. Schmitt, C. de Bergh and M. Festou, M. (Eds.), 1998. Solar System Ices. Kluwer, Dordrecht, Netherlands.

5   R.E. Johnson. In: Schimitt, B., De Bergh, C., Festou, M. (Eds.), Solar System Ices. Kluwer, Dordrecht, Netherlands, 1998, p. 303.

6   R.A. Baragiola, R.A. Vidal, W. Svendsen, J. Schou, M. Shi, D.A. Bahr and C.L. Atteberry, NIMB 2003, **209**, 294.

7   M.H. Moore, R.L. Hudson, and P.A. Gerakines, Spectrochim. Acta A 2001, **57**, 843.

8   G. Strazzulla, In: Schimitt, B., de Bergh, C., Festou, M. (Eds.), Solar System Ices. Kluwer, Dordrecht, Netherlands, 1998, p. 281.

9   G. Strazzulla, G.A., Baratta and M.E. Palumbo, Spectrochim. Acta A 2001, **57**, 825.

10   R.E. Johnson, R.W. Carlson, J.F. Cooper, C. Paranicas, M.H. Moore and M.C. Wong. In: Bagenal, F., McKinnon, W., Dowling, T. (Eds.), Jupiter: Planet, Satellites, and Magnetosphere. Cambridge Univ. Press, Cambridge, UK, 2004, p. 485.

11   G. Strazzulla, J.R. Brucato, M.E. Palumbo and M.A. Satorre, NIMB, 1996, **116**, 289.

12   J.R. Brucato, M.E. Palumbo and G. Strazzulla. Icarus, 1997, **125**, 135.

13   G. Strazzulla, G.A. Baratta, M.E. Palumbo and M.A. Satorre NIMB 2000, **166-167**, 13

14   G. Strazzulla, G. Leto, O. Gomis, M.A. Satorre, Icarus, 2003, **164**, 163.

15   A.L.Lane, R.M. Nelson and D.L. Matson, Nature 1981, **292**, 38.

16   K.S. Noll, R.E. Johnson, M.A. McGrath and J.J. Caldwell, Geophys. Res. Lett. 1997, **24**, 1139.

17   K.S. Noll, H.A. Weaver and A.M. Gonella, J. Geophys. Res. 1995, **100**, 19057.

18   N.J. Sack, R.E. Johnson, J.W. Boring and R.A. Baragiola, Icarus 1992, **100**, 534.

19   R.W. Carlson, R.E. Johnson and M.S. Anderson, Science 1999, **286**, 97.

20   W. Hage, K.R. Liedl, A. Hallbrucker and E. Mayer, Science 1998, **279**, 1332.

21   M.H. Moore and R.L. Hudson, Icarus 2000, 145, 282.

22   O. Gomis, M.A. Satorre, G. Strazzulla and G. Leto, Planet. Space Sci. 2004, **52**, 371.

23   M.J. Loeffler, U. Raut, R.A. Vidal, R.A. Baragiola and R.W. Carlson, Icarus 2006, **180**, 265

24   G. Strazzulla, G. A. Baratta, G. Compagnini, M. E Palumbo and M. A. Satorre, A&A 1998, **338**, 292

25   J.P. Bibring and F. Rocard, Adv. Space Res. 1984, 4, 103.

26   R.A. Baragiola, R.A. Vidal, W. Svendsen, J. Schou, D. A. Bahr and C. L. Attebeny NIMB, 2003, **209**, 294

27   R.B. Bohn, S.A. Sandford, L.J. Allamandola and D. Cruikshank, Icarus 1994, **111**, 151.

28   M.E. Palumbo, Adv Sp Res 1997, **20**, 1637.

29   L. Ottaviano, M.E. Palumbo, and G. Strazzulla 2000. In What are the prospect for Cosmic Physics in Italy? (S. Aiello and A. Blanco Eds.) p. 149. Conf Proc 68, SIF Bologna, Italy.

30   G. Strazzulla. In From stardust to Planetesimals (Y.J. Pendleton, and A.G.G.M. Tielens, Eds.), p. 423. Astron. Soc. Pac. Conf. Series Book, Vol. 122. S. Francisco. 122, 1997, p. 423.

31   T.B McCord, R. W. Carlson, W. D. Smythe, G. B. Hansen, R. N. Clark, C. A. Hibbitts, F. P. Fanale, J. C. Granahan, M. Segura, D. L. Matson, T. V. Johnson and P. D. Martin, Science 1997, **278**, 271

32   T.B. McCord, G. B. Hansen, R. N. Clark, P. D. Martin, C. A. Hibbitts, F. P. Fanale, J. C. Granahan, M. Segura, D. L. Matson, T. V. Johnson, R. W. Carlson, W. D. Smythe and G. E. Danielson, J. Geophys. Res. 1998, **103**, 8603.

33   J.F. Cooper, R. E. Johnson, B. H. Mauk, H. B. Garrett and N. Gehrels, Icarus 2001, **149**, 133.

34   D. Hamilton, G. Gloeckler, S. M. Krimigis and L. J. Lanzerotti J. Geophys. Res. 1981, **86**, 8301.

35   C.M.S. Cohen, , E.C. Stone and R. S. Selesnick, J. Geophys. Res. 2001, 106, **29**,871.

36   J.R. Spencer, W. Calvin and J. Person, J. Geophys. Res. 1995, **100**, 19049.

37   R.E. Johnson and W. A. Jesser, Ap. J. Lett. 1997, **480**, L79.

38   O. Gomis and G. Strazzulla, Icarus 2005, **177**, 570.

39   R.W. Carlson, Science 1999, **283**, 820

STRUCTURE H HYDRATE KINETICS STUDIED BY NMR SPECTROSCOPY

R. Susilo,[1,2] I. L. Moudrakovski,[1] J. A. Ripmeester,[1] and P. Englezos[2]

[1] Steacie Institute for Molecular Science, National Research Council of Canada, Ottawa, ON, K1N 0R6, Canada
[2] Department of Chemical and Biological Engineering, University of British Columbia, Vancouver, BC, V6T 1Z3, Canada

1 INTRODUCTION

The knowledge on thermodynamics and kinetics of gas hydrates is important to further explore, develop and establish the technology for the safe and economic design of hydrate-based processes. They provide the information on hydrate formation conditions and reaction rates at which the phase transformation occurs. Fortunately thermodynamic studies are well established and hydrate phase equilibrium data and prediction software are readily available. However the study of hydrate kinetics, especially on structure H (sH) hydrate is very limited. sH hydrate has unique properties because two guest molecules of different sizes are required and it has the largest simple cage among all known hydrate structures. Hence, larger molecules like methylcyclohexane can fit into the cavity and smaller molecules may have multiple occupy in the large cage. The addition of a large molecule as a guest substance (LMGS) also stabilizes the crystal and reduces the equilibrium pressure. This is seen as an attractive opportunity and valid potential for gas storage application via gas hydrates[1-3].

Hydrate kinetics studies have been carried out primarily through gas uptake measurements in a semi-batch reactor with mechanical mixing or water spraying through a nozzle. The results indicate that the rate of hydrate formation depends on the magnitude of the driving force and the type of LMGS[4-6]. Both these studies used methane as the small cage guest substance along with LMGS to form the complex sH hydrate. LMGSs are non-aqueous liquids (NAL) under the experimental conditions used, and thus hydrate formation is in a gas-liquid water-non aqueous liquid three phase system. This is challenging because mass transfer plays an important role here and without homogeneous mixing such measurements may not be repeatable and reproducible. It has been shown recently that the conversion to hydrate is quite an inhomogeneous process and that the observation of gradual conversion in bulk samples only arises as a result of averaging over many local environments[7]. Consequently, the kinetics through gas uptake measurements is in reality "average kinetics" over the whole sample. The question is how they are related to kinetics obtained with microscopic techniques like NMR corresponding to the bulk environment. More specifically, how the interaction between each molecule in each phase may explain the kinetics results obtained from bulk measurements. Identification of factors which may influence hydrate kinetics allows better strategy for designing and developing hydrate based technology.

The objective of the present study is to monitor the kinetics of sH hydrate during the formation and decomposition involving methane with neohexane, methylcyclohexane, and tert-butyl methyl ether as the LMGS at microscopic level. This enables a direct comparison with the previously reported macroscopic kinetic data from Lee et al. (2005) as well as from Mori's group.[5,6]. This microscopic approach is also expected to provide additional information that explains the phenomena found during the bulk measurements. Hydrate kinetics study at a molecular level have been reported but were restricted to simpler gas-liquid water or gas-ice systems.[8-10] The presence of non-aqueous liquid phase adds more complication to the measurements which already has limitations such as low sensitivity or long data acquisition times. A long induction period may also be a problem for such complex system because no mixing is possible. In this study, hydrates were grown from ice powder instead of liquid water to significantly reduce the waiting period due to induction time. Interference between signals from each component and/or phase that might be encountered was tackled by monitoring both deuterium (^{2}H) and proton (^{1}H) NMR spectroscopic signatures in isotopically enriched samples. Hence the rate of methane hydrate growth, decomposition and the diffusion in NAL were able to be quantified. The distribution of LMGS between ice particles was also observed by using proton micro-imaging NMR. This work will be the first attempt to monitor sH hydrate growth and decomposition kinetics using NMR spectroscopy.

2 EXPERIMENTAL SECTION

The experiments were performed on a Bruker DSX 400 NMR instrument at the NMR facility located at NRC in Ottawa, Canada. A custom-designed static sample 5 mm NMR probe for handling pressures up to ~ 350 bars was employed. The probe is connected to a gas feed line equipped with a pressure gauge. The temperature during the experiment was controlled using a Bruker BVT 3000 temperature control unit. The ^{1}H and ^{2}H NMR spectra were calibrated with a measured quantity of a mixture of H_2O/D_2O of known concentration at 298 K. The chemical shift of proton and deuterium from water was assigned to be 4.7 ppm. The probe was cooled before loading the ice and LMGS. Hydrates were formed from ~0.15 g of freshly ground ice particles loaded and packed into the NMR cell along with stoichiometric excess LMGS at 253K and ~4.5MPa. The zero time of the measurements was set when the cell was pressurized with 99%-deuterated methane (CD_4).

^{1}H and ^{2}H NMR spectra were recorded every 5 minutes for 20 hours. Spectra were acquired with 16 scans and 3 s delay time. Signals from the gas and liquid phase appeared as sharp Lorentzian lines whereas those from the solid phase appeared as broad Gaussian lines. Unfortunately, the broad ^{1}H signals from ice and LMGS in hydrate phase were not distinguishable as the nuclear dipolar broadening was much larger than the chemical shift range for ^{1}H. However the ^{2}H NMR spectra took the form of broad and sharp peaks, which corresponded to the amount of methane in the gas/dissolved in LMGS and the hydrate phase respectively. The chemical shift positions for both sharp and broad peaks are very close, within the region of -1.5 ± 0.5 ppm. The broad signals are shifted slightly to higher field from the sharp signals. Methane in the gas and dissolved in LMGS phase appear as two sharp peaks but were very close to each other and hence are considered as one sharp peak. The spectra were then analyzed with 'dmfit', a solid-state NMR program provided by Massiot et al. (2002)[11].

LMGS interaction with ice particles was also observed by the ¹H micro-imaging NMR techniques. This experiment was performed on a Bruker Avance 200 Spectrometer. Multi-slice spin-echo pulse sequences with Gaussian selective pulses were employed. In most experiments, three slices of 500-μm thickness with a separation of 2 mm were acquired simultaneously in planes parallel to the axis of the cell. The 128 × 128 acquisition matrix with eight scans was accumulated to obtain good signal-to-noise ratio. The experiments were carried out in a cell capable of handling pressures up to ~350 bar connected to a high-pressure handling system. The experimental arrangement is described in detail elsewhere.[7]

3 RESULTS AND DISCUSSION

A typical time-evolution of the ²H NMR spectrum for the Ice+NH+CD₄ system during hydrate formation at 253K and ~4.5 MPa over a 20 hour period is shown in Figure 1. A sequence of sharp Lorentzian lines appears after the cell is pressurized at time t=0, which corresponds to methane-d4 in the gas phase and dissolved in the NAL phase. Broad lines appear as shoulders on the sharp lines as ice is transformed into the solid hydrate phase. The amplitude of the hydrate signal is much lower than that of the gas/dissolved phase due to the broad spectral contributions. Hence the induction time and the early stages of hydrate growth are not so clearly visible in Figure 1 but hydrate formation does become evident as more hydrate is formed with time, as also shown in the inset. Integrated methane intensities in the gas+dissolved and solid phase during the 20 hours reaction time are normalized to the maximum intensity and shown in Figure 2.

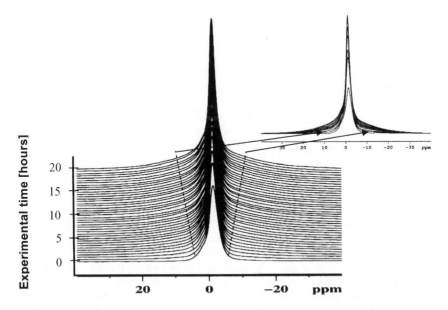

Figure 1 *Deuterium (²H) NMR spectra evolution of Ice+Neohexane+CD₄ at 253K and ~4.5 MPa for 20 hours. Sharp signal corresponds to methane in the gas and dissolved in LMGS. Broad signal correspond to the hydrate phase.*

Methane intensity in the gas+dissolved phase increases and reaches saturation in the first ~3 hours except in the case of MCH. The maximum amount of methane in the dissolved phase is limited by the solubility. However the intensity reported in this study does not necessarily correspond to the solubility unless the exact amount of LMGS and signal contributed from the gas phase are known. The LMGS amount was not measured because it was very small (less than 0.1 g). In the absence of NAL, gas diffusion was also observed which suggests the presence of a gas film. After 3 hours, the intensity stays constant for some time before it drops slowly due to hydrate formation indicated by the pressure drop. Methane diffuses relatively faster in NH and TBME and the slowest in MCH. The intensity from the gas/dissolved phase corresponds to the difference between methane dissolution and hydrate formation rates because both of them occur simultaneously. The diffusion appears to be faster than hydrate growth in most cases, as its intensity continues to grow up to the saturation level with hydrate growth occurring at the same time. Methane intensity in the hydrate phase keeps on increasing during the 20 hour period. Induction times were observed for NH (<60 min) and the MCH system (~2 hrs) but not for TBME and ice only. Hydrate generally grows relatively fast for the first ~5 hours before it slows down due to the formation of a hydrate film on the surface which hinders further contact between the ice, methane, and LMGS. The hydrate growth rate with TBME is the largest followed by NH, and MCH. This trend is exactly what was observed in gas uptake measurements in a well-mixed semi-batch reactor[4] and spray system.[5]

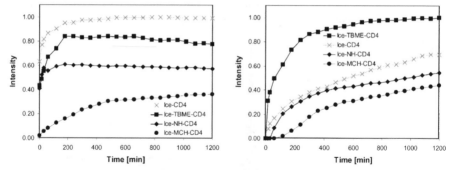

Figure 2   *Normalized intensity evolution of methane in the gas + dissolved in non-aqueous liquid phase (left) and hydrate phase (right) at 253K and ~4.5MPa*

The conversion of ice into hydrate was calculated from the calibrated intensity values with the assumption that the large cage is fully occupied by LMGS and both small and medium cages are 80% occupied by methane. Such occupancy values were reported from ^{13}C solid-state NMR[12] and single crystal X-Ray diffraction[13] study. The occupancy values were not determined because of the very small amount of hydrate formed and difficulties in recovering the hydrate for ^{13}C solid state NMR analysis. It is important to note that the occupancy values vary with operating conditions and hence the calculated conversion is strongly dependent on the methane occupancy. The conversions obtained after 20 hours at 253K were still far below 100%, as shown in Figure 3. The TBME system (~35%) has the highest degree of conversion followed by NH (~19%), ice only (~18%) and MCH (~16%). It is important to note that in sI hydrate, methane can also fill the large cage so that its methane content is higher than in sH hydrate. Hence the actual conversion with the same methane content depends on the hydrate structure, which will be less for sI hydrate. The conversion rate in this study for sI methane hydrate is comparable to that observed in neutron diffraction studies[14,15].

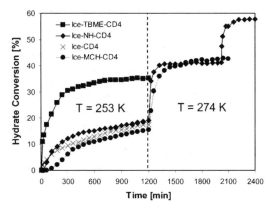

Figure 3 *Ice to hydrate conversion rate, assuming both 5^{12} and $4^3 5^6 6^3$ are 80% occupied with methane, T = 253K for the first 20 hours and T = 274K after 20 hours*

Due to slow conversion rates for both NH and MCH system, the experiments were continued after the 20 hours period but this time with the temperature ramping to a point above the ice point (274 K). As seen from Figure 3, further ice conversion into hydrate was indeed achieved. The additional hydrate growth was instantaneous for both systems, which can be explained as follows. The melting of unreacted ice to liquid water inside the hydrate shell causes a volume reduction and pressure drop inside the hydrate shell. This may decompose the hydrate slightly on the inside of the shell and well crack the hydrate film. Hence, more gas may diffuse into the shell due to the pressure gradient and create new contact with water and LMGS. The hydrate film acts to seed new hydrate which then allows rapid and instantaneous hydrate growth. The exothermic reaction from hydrate formation can also be thermally balanced by the endothermic ice melting. This phenomenon was documented by ^{1}H micro-imaging study of methane reacting with ice particles.[16] Methane solubility in LMGS is also higher at lower temperature. Thus excess methane upon ice melting can go directly into hydrate without gas diffusion limitation.

We note that the continuous ice/water transformation into hydrate inside a hydrate-coated particle requires an expansion in volume because of the different densities of the various phases, which is approximately double the volume reduction from ice melting to water. This expansion may occur gradually or suddenly by rupture of the hydrate shell. The former is typically seen in the first stage of rapid hydrate growth after melting the ice. However the conversion was limited to ~40% and no further significant hydrate growth observed afterward, likely because of pressure build-up inside the ice shell as ice converted to hydrate. Unexpected hydrate growth was observed in the NH system at ~2000 minutes and further conversion was achieved quickly up to ~58%. This phenomenon was probably due to rupture of the hydrate layer which created new access sites for the occluded water. It is noted that similar behaviour was observed in the gas uptake experiments but with MCH.[4]

The decomposition of the hydrate was also monitored for a 3 hour period for the systems with NH and TBME by lowering the pressure to atmospheric. As seen in Figure 4 (left) the hydrate decomposed very quickly during the first hour especially that with TBME as guest. A small amount of residual methane, corresponding to methane dissolved in NAL, was also seen (not shown in the figure). It was also observed that the hydrate with TBME decomposed completely. The system with NH did not dissociate completely in three hours. A very slow decomposition rate was observed after the initial fast period. This is probably due to larger hydrate content of the NH system (~60%) as compared to TBME (~35%) so some hydrate in the core may be isolated by the formation of ice film that preserves it longer. However it is clearly seen that the decomposition rate of the TBME system is faster than that of the NH system. In one experiment with neohexane (NH), hydrate was reformed using ice that had experienced hydrate formation. This reformation experiment followed the previous experiment reported above after decomposing the hydrate. The so called 'memory ice' proved to speed up the kinetics, as seen in Figure 4 (right). Memory effects were also reported on Xenon hydrate reformation experiments[17]. It is unclear whether residual hydrate structure still exists in ice particles. However it is certain is that the system has been saturated with methane so even though small amount of methane is left dissolved in LMGS or even in the hydrate phase but the effect to hydrate growth seems to be significant.

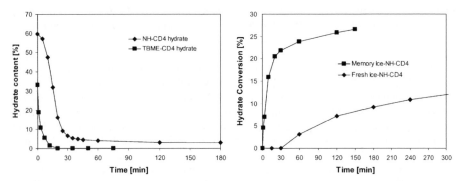

Figure 4   *Hydrate decomposition rate at 253K and 1 atm for NH and MCH system (left) followed by reformation at 253K and ~4.5MPa for the NH system (right)*

^{1}H micro-imaging NMR experiments were employed to image the proton density in the system and observe the interaction between LMGS and ice. Protons from ice are invisible in the standard micro-imaging experiment, hence the liquid LMGS ^{1}H spin density can be observed selectively during hydrate formation and appears as a bright area in the image, whereas ice and the hydrate are invisible (dark). As seen in Figure 5, the distribution of TBME is reasonably homogeneous, in contrast to that of NH and MCH. The image shows that TBME fills the empty spaces between ice particles whereas the interaction/contact of NH and MCH with the ice particles is poor. Some ice particles are still isolated from MCH even after 20 hours without pressurization. NH and MCH preferably stay on the top of the sample or near the wall of the sample tube. It appears that TBME has a strong affinity towards ice that will play an important role in speeding up the formation kinetics, unlike the other NALs that are hydrophobic. The measurements on liquid-liquid equilibria between NAL and water also indicated the strong affinity of TBME for water.[18] Hence TBME has better wetting with ice than NH or MCH system.

The differences in LMGS wetting towards ice may explain further the observed phenomena during the hydrate kinetic measurements. Hydrates formation rate in the TBME system is the fastest because TBME has a better wetting property with ice than NH and MCH. Hence the intensity of methane dissolved in TBME is also the highest. Although the diffusion rate of methane in NH is almost at the same magnitude as in TBME, the hydrate growth is slower due to poor wetting with ice. MCH system has the highest mass transfer resistance, as it does not wet ice particles well and slow methane diffusivity resulting in very slow hydrate kinetics

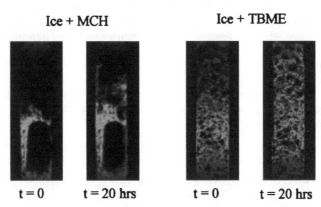

Figure 5   *Distribution of LMGSs and diffusion of LMGS (MCH and TBME) between ice particles at atmospheric condition observed by ^1H micro-imaging NMR*

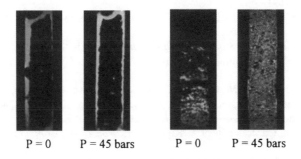

Figure 6   *Proton density image of Ice+MCH with different degree of ice packing: the left images correspond to highly packed ice and the right images correspond to not so highly packed ice*

4 CONCLUSIONS

NMR spectroscopy was successfully employed to monitor the dynamics of sI and sH methane (CD_4) hydrates growth and decomposition. Methane diffusion, hydrate growth and decomposition rates vary among the systems studied. Ice to hydrate conversion is limited by the formation of hydrate films that prevent further mass transport. Temperature ramping above the ice point helps to enhance further hydrate conversion. Hydrate growth rates obtained by NMR were found to agree with the bulk measurements, with TBME exhibiting the fastest kinetics. Good agreement was also found with the reported neutron diffraction study for sI methane hydrate. The imaging study (MRI) showed that the contact of TBME with ice particles was reasonably homogeneous but not so for the NH and MCH systems. Ice packing density also influenced the interaction with LMGS. The majority of hydrate decomposes very quickly at atmospheric condition however small hydrate content still remains for a longer time. The residual of methane in hydrate and dissolved in LMGS phase may explain the faster kinetic during hydrate reformation. Mass and heat transport play important roles in hydrate formation/decomposition besides the intrinsic kinetics.

Acknowledgements

The financial support of National Sciences and Engineering Research Council of Canada (NSERC) is greatly appreciated. RS gratefully acknowledges the financial support from Canada Graduate Scholarship (CGS).

References

1   A.A. Khokhar, J.S. Gudmundsson, E.D. Sloan, *Fluid Phase Equilib.*, 1998, **150**, 383.
2   Y.H. Mori, *J. Chem Ind & Eng China*, 2003, **54**, 1.
3   P. Englezos, J.-D. Lee, *Kor. J. Chem. Eng.*, 2005, **22**, 671.
4   J.D. Lee, R. Susilo, P. Englezos, *Energy & Fuels*, 2005, **19**, 1008.
5   H. Tsuji, R. Ohmura, Y.H. Mori, *Energy & Fuels*, 2004, **18**, 418.
6   H. Tsuji, T. Kobayashi, R. Ohmura, Y.H. Mori, *Energy and Fuels*, 2005, **19**, 869.
7   I.L. Moudrakovski, G.E. McLaurin, C.I. Ratcliffe, J.A. Ripmeester, *J. Phys. Chem. B*, 2004, **108**, 17591.
8   T. Komai, S.-P. Kang, J.-H. Yoon, Y. Yamamoto, T. Kawamura, M. Ohtake, *J. Phys. Chem. B*, 2004, **108**, 8062.
9   R. Kini, S.F. Dec, E.D. Sloan Jr., *J. Phys. Chem. A*, 2004, **108**, 9550.
10   J.-H. Yoon, T. Kawamura, Y. Yamamoto, T. Komai, *J. Phys. Chem. A*, 2004, **108**, 5057.
11   D. Massiot, F. Fayon, M. Capron, I. King, S. Le Calve, B. Alonso, J-O. Durand, B. Bujoli, Z. Gan, G. Hoatson, *Magnetic Resonance in Chemistry*, 2002, **40**, 70.
12   Y.-T. Seo and H. Lee, *Kor. J. Chem. Eng.*, 2003, **20**, 1085.
13   K.A. Udachin, C.I. Ratcliffe, J.A. Ripmeester, *Proc.4th Int. Conf. Gas Hydrates*, 2002, 604.
14   X. Wang, A.J. Schultz, Y. Halpern, *Proc.4th Int. Conf. Gas Hydrates*, 2002, 455.
15   D.K. Staykova, W.F. Kuhs, A.N. Salamantin, T. Hansen, *J. Phys. Chem. B*, 2003, **107**, 10299.
16   I.L. Moudrakovski, C.I. Ratcliffe, G.E. McLaurin, B. Simard, J.A. Ripmeester, *J. Phys. Chem. A*, 1999, **103**, 4969.
17   I.L. Moudrakovski, A.A. Sanchez, C.I. Ratcliffe, J.A. Ripmeester, *J. Phys. Chem. B*, 2001, **105**, 12338.
18   R. Susilo, J.-D. Lee, P. Englezos, *Fluid Phase Equilib.*, 2005, **231**, 20.
19   R. Susilo, J.A. Ripmeester, P. Englezos, *unpublished results*.

DIELECTRIC RELAXATION OF ICE SAMPLES GROWN FROM VAPOR-PHASE OR LIQUID-PHASE WATER

I. Takei

Centre for Development in Education, Hokuriku University, Kanazawa 920-1181, Japan

1 INTRODUCTION

Ice samples generally exhibit a dielectric dispersion with a relaxation time τ of $5\text{-}6\times10^{-5}$ sec at -10 °C and an activation energy of 0.57-0.64 eV for τ adjacent temperatures.[1-6]

The dielectric relaxation process of ice can be understood in terms of proton behavior; namely, the concentration and movement of Bjerrum defects (L- and D-defect) and ionic defects (H_3O^+ and OH^-), which are thermally created in the ice lattice.[6-10] We know that ice samples highly doped with HF or HCl show a dielectric dispersion with a short relaxation time τ and low activation energy of τ.[6,11-13] The decreases in the relaxation time and activation energy are explained by the increasing concentrations of the above-mentioned point defects in the ice samples with increasing levels of dopant.[12]

Snow and polar snow(firn)-ice show similar decreases.[14-20] There is however a problem that some sample kinds of ice, such as polar snow-ice[16-20] and hoarfrost,[21] with lower impurity concentrations, exhibit an unaccounted for decrease in the relaxation time τ. The cause of this decrease may not be related to the impurity content. The change in the dielectric behavior probably results from not only the presence of impurities, but also the inclusion of gas, dislocations, vacancies and grain boundaries/surfaces in the ice samples.

In this paper, we discuss the relaxation process in ice samples grown from the liquid or vapor phase, and attempt to reveal differences in dielectric behavior between ice samples grown by the different processes. There is a possibility that the dielectric properties of ice are affected by the process by which it grew.

2 RELAXATION PROCESS OF ICE SAMPLES

Ice samples have a main dispersion induced by reorientation of the water molecules and proton conduction with movement of the point defects. Here, we discuss values of the relaxation time τ of the main dispersion of ice samples reported in the literature and measured by the present authors. For convenience in experimental measurements, we define two classification of ice sample as bulk ice and ice particle aggregates corresponding to two types of growth, liquid phase growth and vapor phase growth.

Measurements of the dielectric characteristics of samples of ice and ice particle aggregates were conducted with a LCR-meter (HIOKI, 3531). Most measurements were performed at frequencies between 50 Hz and 5 MHz and at temperatures between −15 °C and −0 °C. The basic accuracy of LCR meter is as being less than ±2% in a frequency range of 100 Hz to 1 MHz, ±3.5% below 100 Hz, and ±7% above 1 MHz.

The dielectric relaxation processes of matter can be analyzed with an empirical model of dielectric dispersion, for example, the one described by Havriliak-Negami's equation.[22] We analyzed dielectric data obtained for our samples using a model of complex permittivity $k*$ with two dispersions (the main and the low-frequency dispersion of a space charge effect) and conductivity σ_o (caused by electrode discharge), as follows;

$$k^* = k' - j\,k'' = k_\infty + \frac{\Delta k_1}{\left(1+(j\omega\tau_1)^{\alpha_1}\right)^{\beta_1}} + \frac{\Delta k_2}{\left(1+(j\omega\tau_2)^{\alpha_2}\right)^{\beta_2}} - j\frac{\sigma_o}{\omega\varepsilon_o} \qquad (1)$$

where k' and k'' are the real and imaginary parts of the permittivity, j is the imaginary unit, k_∞ is the high-frequency permittivity, ε_o and ω are the permittivity of a vacuum and angular frequency, Δk_i, τ_i, and α_i and β_i are the dispersion strength, relaxation time, and α and β factors of the dispersion i (=1, 2), respectively. In the case of α_i=1 and β_i=1, the dielectric dispersion is known as the Debye-type, and for $0<\alpha_i<1$ and β_i=1 the Cole-Cole type,[23] and for α_i=1 and $0<\beta_i<1$ the Davidson-Cole type.[24]

2.1 Ice grown from liquid phase (bulk ice)

2.1.1 Pure ice sample. Figure 1(a) shows results of the relaxation time τ by Auty and Cole,[1] whose data for polycrystalline ice have been some of the most quoted reproducible results reported, the results reported by von Hippel *et al.*[3] whose data for single crystal ice comprise the third component in six dielectric dispersions analyzed by computer, and

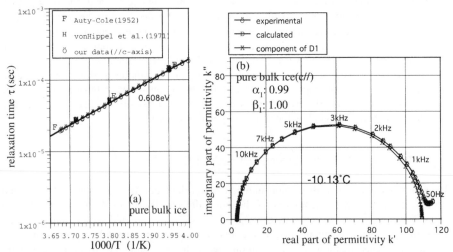

Figure 1 *For pure bulk ice samples, (a) Temperature dependence of the dielectric relaxation time τ and (b) Cole-Cole plots of pure ice crystal (parallel to the c-axis) at -10 °C. The dielectric dispersion is of the Debye type (α=0.99, β=1.00).*

results obtained by the present authors for single crystal. The ice sample measured was obtained from thin plate-ice frozen on a surface of de-ionized and de-gassed water in a Teflon vessel in a laboratory freezer. The sample was 20 mm in diameter and 1 mm in thickness and was attached with a main electrode (1.4 mm diameter) and a guard ring. The dielectric relaxation time τ of the ice sample had an absolute value of 4.8×10^{-5} sec at -10 °C and the activation energy was 0.608 ± 0.001 eV around that temperature. The activation energy of τ of pure ice has been reported to be in the range of 0.60 eV to 0.64 eV by computer fitting.[3,6] Analysis results obtained for our sample at -10 °C described with Cole-Cole plots are shown in Fig. 1(b); (O) denotes experimental data, (+) fitted data calculated with equation (1), and (•) a component of the main dispersion (D1). A dielectric dispersion of the Debye type ($\alpha=0.99$, $\beta=1.00$) is observed.

    2.1.2 Impure ice sample. Dielectric properties of impure ice have been reported by many researchers.[6,9,11-13] The overall effect of increasing the fluoride or chloride concentration is to decrease the absolute value of τ and the activation energy. Camplin *et al.*[9] reported that the relaxation time τ of HF-doped ice decreases with increasing concentration above 4.4×10^{21} HF molecules per m^3 (8.0×10^{-6} mol/l, sample No. 20.11) at -10 °C. The activation energy of τ for highly HF-doped ice is 0.26 eV. This value can be understood from the migration energy of Bjerrum defects in the ice lattice under a defect model[10] for the dielectric properties of ice. Data analysis for these impure ice samples has almost always been carried out using a Debye type dispersion model.

    2.1.3 Clathrate ice samples. Prior to the results reported by Auty and Cole,[1] researchers had faced difficulties obtaining reproducible results for the dielectric properties of ice. The use of well-degassed ice samples has enabled the obtainment of reproducible results. This indicates that the existence of gas molecules in ice affects the dielectric properties. Clathrate ice shows a short relaxation time and low activation energy.[25]

2.2 Ice grown from vapor phase (bulk ice)

    2.2.1 Hoarfrost. One kind of ice grown from the vapor phase is hoarfrost. Measurement of the dielectric properties of bulk samples of hoarfrost has been carried out by Itagaki[26] using small Hg-electrodes. Although the purpose of these measurements was to elucidate the dielectric properties of dislocation-free ice, dielectric relaxation time τ with a small dispersion strength has been reported. The absolute values of τ show small values of $1.1 \times 10^{-5} - 3.0 \times 10^{-5}$ sec at -10 °C. The activation energy of τ has not been given.

    2.2.2 Polar deep ice. Polar deep ice is made of snow under a compressing process. The dielectric properties of polar ice core samples have been reported as having small values of relaxation time τ and activation energy.[16-20] The observation of small τ values for the core ice samples suggests that Bjerrum defects are more numerous in polar ice than in ordinary ice. The impurity concentration of polar ice is not sufficiently high to decrease the τ value; the HCl concentration is about 2×10^{-6} mol/l for Byrd core ice.[19] Since we know that polar deep ice has structures of clathrate gas hydrate, imperfection in the structures and the existence of gas molecules in the ice lattice seem to affect the dielectric properties.[16] It is well known that the dielectric properties of ice samples derived from polar deep ice that has melted and refrozen are similar to those of ordinary ice.[19]

2.3 Ice grown from liquid phase (ice particles aggregate)

    2.3.1 Crushed ice samples. Figure 2 shows the dielectric properties of samples consisting of packed crushed ice particles, which were prepared by crushing single crystal

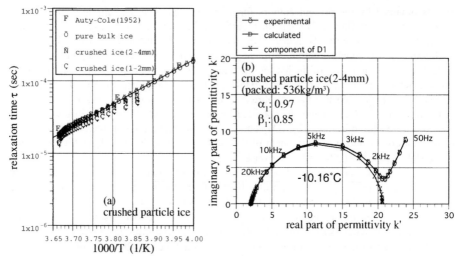

Figure 2   *For packed samples of crushed ice particles, (a) Temperature dependence of the dielectric relaxation time τ and (b) Cole-Cole plots of the sample (after 400 h of annealing at -1 °C, 2-4 mm particle size, 536 kg/m³ packed density) at -10 °C. The dielectric dispersion is of the Davidson-Cole type (α=0.97, β=0.85).*

ice grown from de-ionized and de-gassed water by the Bridgman method. The crushed ice particles had heterogeneous shapes. The packed samples were 90 mm in diameter and 10 mm in thickness and were attached with a main electrode (56 mm diameter) and a guard ring. Because of bond growth between ice particles with time, the samples were annealed at -1 °C for 200-400 h in order to obtain reproducible measurements. The relaxation time τ and the activation energy of τ for the crushed ice samples at -10 °C are similar to those of ordinary ice (Fig. 2(a)). The dielectric dispersion, for the sample with a particle size of 2-4 mm and a packed density 536 kg/m³ after 400 h of annealing at -1 °C, is almost of the Davidson-Cole type (α=0.97, β=0.85, Fig. 2(b)). The dispersion in the first stage of the annealing process only negligibly corresponded to the Davidson-Cole type. The data obtained for the sample was also analyzed using a multi-dispersion model: the main dispersion of the crushed ice samples consisted of two Debye type dispersion components of 2×10^{-6} sec and 1×10^{-5} sec in the first stage of the annealing process at -1 °C.

   2.3.2 Spherical particle (quick-grown) ice.   Figure 3 shows the measured dielectric properties of a sample consisting of packed spherical ice particles, which were prepared from de-ionized and de-gassed water droplets quickly frozen in liquid nitrogen. The ice particles were nicely spherical with diameter of about 3 mm and had an opaque and milky appearance. The values of τ obtained for the as-grown sample of packed spherical ice particles show more scatter and are markedly smaller than those obtained for the sample annealed for 700 h at -1 °C (Fig. 3(a)). The relaxation time τ at -10 °C for the annealed sample has the small value of 1.2×10^{-5} sec, and the activation energy of τ around -10 °C is the value of 0.276 ±0.005 eV, which is lower than that of ordinary ice. The dielectric dispersion, for the sample with a packed density of 524 kg/m³ after 700 h of annealing at -1 °C, is of the Davidson-Cole type (α=1.00, β=0.85, Fig. 3(b)). The dispersion in the first stage of the annealing process was also of the Davidson-Cole type (α=0.96, β=0.78, after 2 h of annealing at -1 °C).

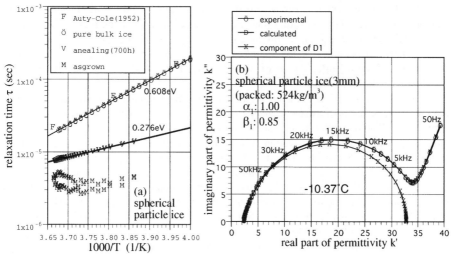

Figure 3 *For a packed sample of spherical ice particles, (a) Temperature dependence of the dielectric relaxation time τ and (b) Cole-Cole plots of the sample (after 700 h of annealing at -1 °C, 3 mm diameter, 524 kg/m³ density) at -10 °C. The dielectric dispersion is of the Davidson-Cole type (α=1.00, β=0.85).*

2.4 Ice grown from vapor phase (ice particle aggregates)

2.4.1 Hoarfrost samples. Figure 4 shows the dielectric properties of samples consisting

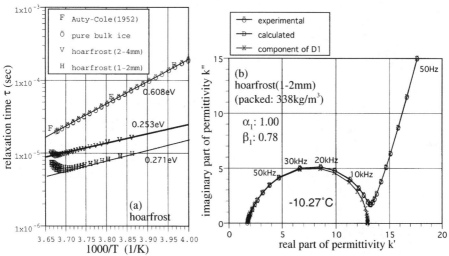

Figure 4 *For packed samples of hoarfrost particles, (a) Temperature dependence of the dielectric relaxation time τ and (b) Cole-Cole plots of the sample (after 240 h of annealing at -1 °C, 1-2 mm particle size, 338 kg/m³ packed density) at -10 °C. The dielectric dispersion is of the Davidson-Cole type (α=1.00, β=0.78).*

of packed hoarfrost particles. The hoarfrost used was naturally grown from water vapor in a laboratory freezer. The relaxation time τ of the packed hoarfrost samples at -10 °C has a small value and the activation energy of τ around -10 °C is about 0.25-0.27 eV (Fig. 4(a)). The dielectric dispersion, for the sample with a particle size of 1-2 mm and a packed density of 338 kg/m^3 after 240 h of annealing at -1 °C, is of the Davidson-Cole type (α=1.00, β=0.78, Fig. 4(b)).

    2.4.2 Snow samples. A short relaxation time (high relaxation frequency) has been reported for the dielectric dispersion of many snow samples including polar shallow ice core.[14,15,17,20] The relaxation time τ of the snow samples of the present study at -10 °C also have small values and the activation energy of τ around -10 °C is rather small at about 0.21-0.22 eV (Fig. 5(a)). The dielectric dispersion, for a new snow sample with a packed density of 338 kg/m^3 after 240 h of annealing at -2 °C, is of the Havriliak-Negami type (α=0.89, β=0.86, Fig. 5(b)). The dielectric properties of snow samples having such types of dispersion, including the Davidson-Cole and the Cole-Cole type, are most probably affected by particle shape, particle size and impurity. A granular snow sample consisting of snow that partly melted and then refroze or was granulized at the melting temperature, exhibits a larger value of τ than other snow samples that have not undergone such a melting process (Fig. 5(a)), although the granular snow sample has a high conductivity.

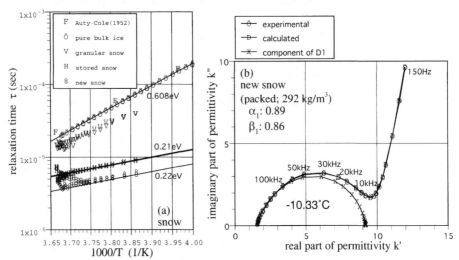

Figure 5   *For snow samples, (a) Temperature dependence of the relaxation time τ and (b) Cole-Cole plots of the sample (new snow, 292 kg/m^3 packed density, after 240 h of annealing at -2 °C) at -10 °C. The dispersion is the Havriliak-Negami type (α=0.89, β=0.89).*

3 DISCUSSION

The results described above indicate two interesting facts: many samples consisting of packed ice particles exhibit a dielectric dispersion of the Davidson-Cole type, and ice samples grown from the vapor phase exhibit a shorter relaxation time τ and a lower activation energy of τ than ordinary ice samples grown from the liquid phase.

Regarding the first fact, bulk ice samples seem to exhibit a dielectric dispersion of the Debye type with a single relaxation time. On the other hand, samples consisting of packed ice particles show a dielectric dispersion of the Havriliak-Negami type ($0<\alpha<1$, $0<\beta<1$), and particularly of the Davidson-Cole type ($\alpha=1$, $0<\beta<1$) in many cases. The relaxation time in the case of dielectric dispersion of the Davidson-Cole type generally has a distribution shifted to the high-frequency side. We can, however, also analyze data obtained for the ice samples consisting of packed ice particles by using a model of two (the high-frequency and the main) dispersions instead of the Davidson-Cole type. For the dielectric dispersion on the high-frequency side, the possibility that it is a dispersion of the Maxwell-Wagner type has been pointed out.[20] Since the crushed ice sample of the present study (made of ice grown from the liquid phase) shows a high-frequency component in addition to the main dispersion even in the first stage of the annealing process, the dispersion at high frequencies is not introduced from dielectric characteristics of ice because such a dispersion is not observed for bulk ice. This dispersion must be mainly introduced from the arrangement structure of ice particles in the sample.

A sample consisting of packed ice particles has numerous cavities and ice particle surfaces inside, and the number of dangling bonds oriented with a hydrogen nucleus at the surface of ice tends to increase. Therefore we can expect that the number of L-defects near the ice particle surfaces also tends to increase and protons near the ice particle surface can easily move. The dispersion of the Davidson-Cole type may be caused by proton behaviors near the ice particle surfaces in a sample consisting of packed ice.

Regarding the second fact, the experimental results described above suggest that ice grown from the vapor phase has a short relaxation time and small activation energy. This means that Bjerrum defects are more numerous in ice grown from the vapor phase than in ordinary ice. One way in which defects can arise is from impurities dissolved in the ice. Some ice samples grown from the vapor phase, such as hoarfrost and polar snow and ice, however, do not contain a sufficient amount of impurities to decrease the relaxation time. We know that samples of polar ice that have been melted and refrozen show dielectric properties similar to those of ordinary ice in the laboratory.[19] Temperate glacier ice[16,18] and granular snow (Fig. 5(a)), which have at some time experienced melting, also show dielectric properties approaching those of ordinary ice. There is an additional cause for a decrease in the relaxation time besides the presence of impurities. For ice grown from the vapor phase, there should be a cause relating to the crystal growth process. The facts mentioned above suggest that the relaxation process of ice is different not only for pure and impure ice but also for ice samples grown from the vapor and liquid phases.

Reproducibility of measurements of the dielectric characteristics of ice samples may be related to factors such as aging, impurities, inclusion of gas, and the presence of crystal imperfections. Inclusion of gas is one of the most important factors preventing realization of reproducible measurements. Using ice samples grown from well-degassed water, Auty and Cole[1] were able to measure ice samples reproducibly.

Ice grown from the vapor phase is expected to have many lattice imperfections such as vacancies and inclusion of gas. This hypothesis can explain previous reports on snow, hoarfrost, and polar ice samples, which were considered to have a low impurity concentration, yet exhibited dielectric relaxation times shorter than those of samples that had been melted and refrozen and samples of ordinary ice. There is a possibility that these imperfections introduce differences in dielectric relaxation process between samples of ice grown from vapor-phase and liquid-phase water. Further study of other evidence is needed to elucidate such differences, for example, via a rigorous investigation of good-quality bulk hoarfrost samples.

4 CONCLUSION

Bulk ice samples show dielectric dispersion of the Debye type. On the other hand, samples consisting of packed ice particles tend to show dielectric dispersion of the Davidson-Cole type.

Dielectric properties differ between ice samples grown from the vapor phase and the liquid phase: the relaxation time and activation energy of ice grown from the vapor phase have lower values than of liquid growth ice. This difference suggests that vapor-phase growth introduces a crystal imperfection (such as vacancies and inclusion of gas) with increasing Bjerrum defects.

References

1   R.P. Auty and R.H. Cole, *J. Chem. Phys.*, 1952, **20**, 1309.
2   O. Wörz and R.H. Cole, *J. Chem. Phys.*, 1969, **51**, 1546.
3   A. von Hippel, D.B. Knoll and W.B. Westphel, *J. Chem. Phys.*, 1971, **54**, 134.
4   R. Ruepp, in *Physics and Chemistry of Ice*, eds., E. Whalley, S.J. Jones and L.W. Gold, Royal Society of Canada, Ottawa, 1973, pp.179-186.
5   R. Taubenberger, in *Physics and Chemistry of Ice*, eds., E. Whalley, S.J. Jones and L.W. Gold, Royal Society of Canada, Ottawa, 1973, pp.187-193.
6   G.C. Camplin and J.W. Glen, in *Physics and Chemistry of Ice*, edited by E. Whalley, S.J. Jones and L.W. Gold, Royal Society of Canada, Ottawa, 1973, pp.256-261.
7   C. Jaccard, *Helv. Phys. Acta*, 1959, **32**, 89.
8   C. Jaccard, *Phys. Kondens. Mater.*, 1964, **3**, 99.
9   I. Takei and N. Maeno, *Journal de Physique*, 1987, **48**, Colloque Cl, 121.
10   V.F. Petrenko and R.W. Whitworth, In *Physics of ice*, Oxford University Press, 1999, ch.4, p.60.
11   A. Steinemann, *Helv. Phys. Acta*, 1957, **30**, 581.
12   G.C. Camplin, J.W. Glen and J. G. Paren, *J. Glaciol.*, 1978, **21**, 123.
13   G.W. Gross, I.C. Hayslip and R.N. Hoy, *J. Glaciol.,*, 1978, **21**, 143.
14   D. Kuroiwa. In *the physics and mechanics of snow as a material*, Cold Regions Research and Engineering Laboratory. (CRREL Monogr. II-B), 1962, p.63.
15   C.M. Keeler, *CRREL Res. Rep.*, 1969, **271**.
16   J.G. Paren, in *Physics and Chemistry of Ice*, eds., E. Whalley, S.J. Jones and L.W. Gold , Royal Society of Canada, Ottawa, 1973, pp.262-267.
17   N. Maeno, *Memories of National Institute of Polar Research*, 1978, Special Issue No.**10**, 77.
18   J.W. Glen and J.G. Paren, *J. Glaciol.*, 1975, **15**, 15.
19   W.J. Fitzgerald and J.G. Paren, *J. Glaciol.*, 1975, **15**, 39.
20   J.G. Paren and J.W. Glen, *J. Glaciol.*, 1978, **21**, 173.
21   I. Takei and N. Maeno, *Ann. Glaciol.*, 2001, **32**, 14.
22   S. Havriliak and S. Negami, *J. Poly. Sci.* Part C, 1966, **14**, 99.
23   K.S. Cole and R.H. Cole, *J. Chem. Phys.*, 1941, **9**, 341.
24   D.W. Davidson and R.H. Cole, *J. Chem. Phys.*, 1951, **19**(12), 1484.
25   S.R. Gough and D.W. Davidson, in *Physics and Chemistry of Ice*, eds., E. Whalley, S.J. Jones and L.W. Gold , Royal Society of Canada, Ottawa, 1973, pp.51-55.
26   K. Itagaki, *J. Glaciol.*, 1978, **21**, 207.

ESR OBSERVATION OF SELF-PRESERVATION EFFECT OF METHANE HYDRATE

K. Takeya[1], K. Nango[1], T. Sugahara[1], A. Tani[2], and K. Ohgaki[1]

[1] Division of Chemical Engineering, Graduate School of Engineering Science, Osaka University, 1-3 Machikaneyama, Toyonaka, Osaka 560-8531, Japan
[2] Department of Earth and Space Science, Graduate School of Science, Osaka University, Osaka University, 1-1 Machikaneyama, Toyonaka, Osaka 560-0043, Japan

1 INTRODUCTION

Gas hydrates have a curious structure formed from many cages constructed by hydrogen-bond framework of water molecules and included gaseous molecules which are called guest species. The structure can be stabilized by inclusion of a second component, guest species such as methane, ethane, carbon dioxide and xenon.[1] There are several types of hydrates, such as structure I, II and H, depending on the guest molecule size.[2, 3, 4] Structure I and II have a cubic unit cell, while structure H has a hexagonal one. If the cages of structure I are fully occupied with compound hydrate former M, chemical formula of the clathrate hydrate is $M\bullet5.75H_2O$.

Methane hydrates have attracted much attention as future energy resources because of the enormous amounts of those deposits. Because methane has fewer carbon atoms than all other fossil fuels and the amount of exhaust CO_2 is relatively small when it burns, hydrates are considered to be cleaner energy resources. Various researchers have reported that they have accumulated extensively in permafrost regions and in sediments beneath the deep ocean floor.[5, 6]

Methane hydrate has an unclear property called the self-preservation or anomalous preservation effect, i.e., slow dissociation of the hydrate under thermodynamically unstable conditions.[7] This slow dissociation is thought to be caused by a layer of ice covering hydrate surfaces that forms from the dissociated hydrate.[8] However, little is known about this preservation mechanism. For example, it is still unclear that this effect is not observed between 193-240 K but above 240 K.[9]

Electron spin resonance is a spectroscopic method of observing resonance absorption of microwave power by unpaired electron spins in a magnetic field. Magnetic moment of electron spin, μ_s occurs in the opposite direction to the spin momentum, S. This is expressed as $\mu_s = - g\beta S$, where g is the g-factor (2.0023 for a free electron) and β is the Bohr magnetron. The Hamiltonian, H, of an electron in a magnetic field, H, is written as $H_s = -\mu_s \cdot H_0 = g\beta S \cdot H_0$. Eigenvalues of the Hamiltonian, H_s are $E_1 = +1/2\ g\ b\ H_z$ and $E_2 = -1/2\ g\ b\ H_z$ where the direction of the quantization is the z-axis. The transition between the two Zeeman levels can be induced by an electromagnetic wave with the frequency of v. The condition of the resonance absorption of the electromagnetic wave is obtained as

$\Delta E = E_1 - E_2 = g\beta H_z = h\nu$, where h is a Plank constant.

In a previous study,[10] we have reported the thermal stability of gamma-ray induced methyl radicals in methane hydrate using an electron spin resonance (ESR) method. Gamma-ray irradiation induces methyl radicals and hydrogen atoms in methane hydrate from reaction $CH_4 + h\nu \rightarrow CH_3^- + H^-$. Their concentrations are approximately p.p.m order to original methane molecules. Although induced hydrogen atoms decay below 120 K, induced methyl radicals are isolated in hydrate cages and the thermal stability of methyl radical follows thermodynamic stability of methane hydrate. The decay of the methyl radical is a dimer reaction from their migration in the hydrate crystal and generates ethane molecules. It proceeds by two different temperature-dependent processes, that is, the respective activation energies of these processes are 20.0 ± 1.6 kJ/mol for the lower temperature region of 210-230 K and 54.8 ± 5.7 kJ/mol for the higher temperature region of 235-260 K.[10] In the latter temperature region, where the self-preservation appears, the activation energy agrees well with the enthalpy change of methane hydrate into liquid water and gaseous methane.[11] These findings suggest that methane hydrate dissociates into liquid (supercooled) water and gaseous methane in the temperature range of 235-260 K. Furthermore, it is reported that propane hydrate decomposes into metastable (supercooled) water and gas.[12] This hypothesis has also been proposed by others about the possible role of supercooled water in methane hydrate dissociation.[13] We think that this supercooled water plays a significant role in hydrate dissociation above 235 K. It is considered that there is another model in self-preservation effect due to this supercooled water.

The purpose of this paper is to discuss a dissociation model of methane hydrate during the self-preservation effect through observation of methyl radicals. The decay of the induced methyl radicals requires another methyl radical beyond the hydrate cages, and thus this decay should connect to the hydrate dissociation or migration of methyl radicals. We report the ESR study of property of gamma-ray induced methyl radicals in methane hydrate before and after annealing at 250 K.

2 METHOD AND RESULTS

2.1 Experimental

2.1.1 Sample Preparation. As the sample preparation in the present study is the same to the previous one,[10] the details are not mentioned here. The polycrystalline sample was synthesized in a high-pressure cell made of stainless steel. Methane hydrates were prepared in a temperature range of 277-279 K and a pressure range of 5-10 MPa. Hydrate particles of about 1 mm in diameter were picked out and kept in plastic vials at 77 K.

2.1.2 Experimental Procedures. The samples in the vials were immersed in liquid nitrogen and irradiated by γ-rays at a dose of 10 kGy using a source of [60]Co. Several pieces (about 5 mg) were picked out and put in the ESR sample tube at 77 K for ESR measurements. Five samples labeled samples A-E were prepared and measured by ESR at 120 K. The samples were annealed at 250 K for 1 hour (A), 3 hours (B, C, D) and 6 hours (E). One end of the tube was then sealed and the second γ-ray irradiation was performed to the samples A-E as the same condition of the first one. The annealed samples labeled samples A'-E' were measured by ESR again at 120 K. The samples D and D' were also measured at 250 K by ESR during annealing processes at 250 K for 3 hours. All annealing experiments performed with nitrogen flow unit within ESR measurements unit at 0.1 MPa.

In addition, another irradiated methane hydrate (sample F) sample was annealed at 253 K for 3 hours (sample F') were measured by Raman spectroscopy. A scheme of experimental procedure is shown in Figure 1.

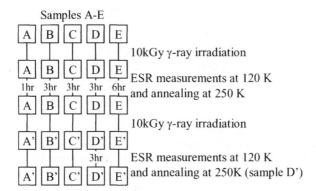

Figure 1 *Scheme of experimental procedure for ESR measurements in the present study*

2.1.3 ESR Measurements. The polycrystalline samples A-E and A'-E' were measured by a commercial ESR spectrometer (JEOL RE1X) with 100 kHz field modulation of 0.1 mT using a nitrogen gas flow unit system (JEOL ES-DVT2). This system was also used for annealing the samples A-E. The signal intensity was determined by peak amplitude of ESR spectrum, since the line shape did not change throughout the measurements.

2.1.4 Raman Spectroscopy Measurements. The methane molecules in hydrate were analyzed by Raman spectroscopy by use of a laser Raman microprobe spectrometer (JASCO NSR-1000). The temperature of hydrate crystal was kept constant at 253 K during Raman measurements and heating the sample F by use of a freezing stage (Linkam LK-FDCSI).

2.2 Results and discussion

Irradiated methane hydrates give the same signal as methyl radical ($g = 2.0029$) before and after annealing. The ratio of induced methyl radical to the original methane molecules is in the p.p.m range. The observed ESR signal intensities of induced methyl radicals in samples A-E and A'-E' are shown in Table 1. The intensities varied among the samples because of different sample masses in the ESR tubes. The amount of methyl radicals is proportional to the amount of remaining methane in the hydrate, because the dose response for the signal intensity of methyl radicals in methane hydrate by gamma-ray irradiation is proportional to amount of the methane molecules.[14] After annealing at 250 K, all samples contained nearly 80 % of the initial gases content.

Table 1 *ESR signal intensity normalized by the standard sample of magnesium oxide with*
Mn²⁺ as impurities of the methyl radicals in the samples A-E and A'-E' and
the ratio.

	ESR intensity (Initial)	ESR intensity after annealing	Methane amount after annealing (%)
Sample A and A'	84.5/Mn	76.9/Mn	91.0
Sample B and B'	207/Mn	159/Mn	76.8
Sample C and C'	4.65/Mn	3.72/Mn	80.0
Sample D and D'	69.2/Mn	57.5/Mn	83.1
Sample E and E'	104/Mn	83.8/Mn	80.6

Figure 2 shows the microwave power dependence of methyl radicals in methane hydrate irradiated before (sample C) and after (sample C') annealing at 250 K. There is no significant difference between the power dependences of these samples. This dependence is connected to the peripheral hydrate structure of the radical. The microwave power dependence connects closely to relaxation and saturation of electron spin. If the microwave power is so high as to pump out the spins in the lower level or the relaxation time is too long for the population to be restored, a decrease in the signal intensity is observed. This relaxation is controlled by relation between radical and surrounding or peripheral structure.

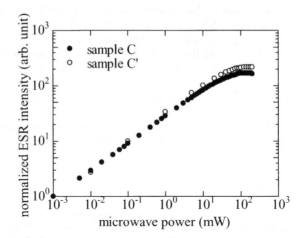

Figure 2   *Microwave power dependence of the methyl radical in γ-ray irradiated*
methane hydrate before and after annealing. The ESR intensities are
normalized at 0.001 mW.

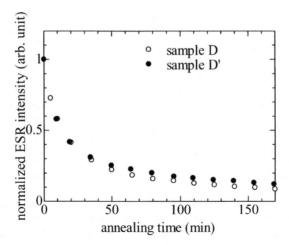

Figure 3 *Isothermal annealing curves at 250 K for the methyl radical in the samples D and D'.*

The results of isothermal annealing experiments at 250 K for the methyl radical in hydrates (D, D') are shown in Figure 3. Each ESR intensity on the vertical axis was normalized by the initial intensity before annealing. The decay of induced methyl radicals in the sample D proceeds, even though 83.1 % of the initial methane molecules remain in the sample after first annealing as discussed above. There are no significant differences in the decay rate between the first and second annealings. The thermal stability of methyl radicals in water ice is less stable than that in hydrate.[15] These results suggest that apparently they keep the same hydrate crystal structure.

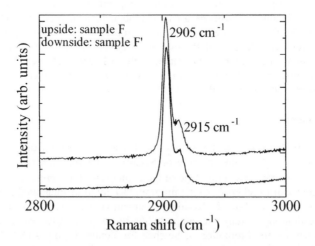

Figure 4 *Raman spectra of methane hydrate in the samples F and F'*

Methane has the active Raman spectrum of the C-H symmetric stretching vibration mode. Typical Raman spectra of the C-H symmetric stretching vibration mode for methane in the methane hydrate before and after annealing (samples F and F') are shown in Figure 4. In the case of methane hydrate, the spectrum splits into two peaks. The split of Raman peak indicates that methane is trapped in both small (5^{12}) and large ($5^{12}6^2$) cages. The larger peak corresponds to the methane in the large cage and the smaller one corresponds to the small cage. The energy difference between the two peaks is caused by the differences in characteristic volume of each cage.[16] The persistence of characteristic Raman spectra of methane in hydrate before and after annealing indicates that the hydrate structure persists after annealing.

These results show that the decay of methyl radicals proceeds at 250 K without the release of methane from methane hydrate crystal. Methane hydrate crystals during self-preservation allow the methyl radical decay. Each methyl radical is isolated in hydrate cages. This decay reaction of methyl radicals needs methyl radical migration in hydrate crystal or diffusion beyond the hydrogen-bond hydrate framework of themselves, because this is a dimerization reaction.[10] What mechanism operates in methane hydrate during the self-preservation? We propose a model that methyl radicals can diffuse beyond hydrogen-bond framework without release of gas from hydrate crystal during the self-preservation effect. In a hydrate crystal, the lattice vibrations in each single crystal become higher amplitude with increasing temperature, so methyl radicals can diffuse beyond each hydrate cage and undergo the dimer reaction, generating the ethane molecules, even though a major fraction of the hydrogen-bonding is preserved in the hydrate crystal structure. Because this event occurs inside of hydrate single crystals without the release of methane, a large fraction of the methane is retained in hydrate during self-preservation.

3 CONCLUSION

Gamma-irradiated methane hydrates both before and after annealing at 250 K give ESR signals characteristic of methyl radical ($g = 2.0029$). The 80 % of the starting amount of methane in hydrate is preserved throughout high temperature (250 K) annealing.

The decay of methyl radicals progresses in the methane hydrate at 250 K without the release of methane from hydrate crystals. The properties of methyl radical, which are microwave power dependence and thermal stability, do not change after annealing. In addition, the Raman spectrum does not change after 3 hours at 253 K. Therefore, methyl radicals can diffuse beyond their original hydrate cages and undergo a dimerization reaction to generate ethane molecules.

Acknowledgement

We thank Drs. H. Sato, C. Yamanaka and M. Katsura for helpful discussions. We are indebted to Dr. T. Ikeda at the Institute of Scientific and Industrial Research of Osaka University for ⁶⁰Co γ-ray irradiation. One of the authors (K. T.) expresses special thanks to 21st century center of excellence (COE) program "Creation of Integrated EcoChemistry". We are also grateful to Prof. T. Kasai (a cooperative research member of COE) for valuable suggestions. This study was supported by the grant from Research Fellowships of the Japan Society for the Promotion of Science for Young Scientists (No. 17-9677). We thank Dr. S. H. Kirby for helpful review.

References

1   E. D. Sloan, Jr. *Clathrate Hydrates of Natural Gases*, second ed. Marcel Dekker Inc., New York, 1998.
2   M. von Stackelberg, *Naturwissenschaften*, 1949, **36**, 327.
3   H. R. Müller and M. von Stackelberg, *Naturwissenschaften*, 1952, **39**, 20.
4   J. A. Ripmeester, J. S. Tse, C. I. Ratcliffe and B. M. Powell, *Nature*, 1987, **325**, 135.
5   M. D. Max, and A. Lowrie, *J. Pet. Geol.*, 1996, **19**, 41.
6   R. D. Stoll and G. M. Bryan, *J. Geophys. Res.*, 1979, **84**, 1629.
7   V. S. Yakushev and V. A. Istomin, *Physics and Chemistry of Ice* (Hokkaido University Press, Sapporo, 1992) 136
8   S. Takeya, W. Shimada, Y. Kamata, T. Ebinuma, T. Uchida, J. Nagao and H. Narita, *J. Phys. Chem. A*, 2001, **105**, 9756.
9   L. A. Stern, S. Circone, S. H. Kirby and W. B. Durham, *J. Phys. Chem. B*, 2001, **105**, 1756
10 K. Takeya, K. Nango, T. Sugahara, K. Ohgaki, and A. Tani, *J. Phys. Chem. B*, 2005, **109**, 21086.
11 Y. P. Handa, *J. Chem. Thermodyn.*, 1986, **18**, 915.
12 V. P. Mel'nikov, A. N. Nesterov, A. M. Reshetnikov, *Doklady Earth Sciences*, 2003, **389A**, No.3, 455.
13 S. Circone, L. A. Stern, S. H. Kirby, *J. Phys. Chem. B*, 2004, **108**, 5747.
14 K. Takeya, A. Tani, T. Yada and M. Ikeya, *Appl. Radiat. Meas.*, 2005, **62**, 371.
15 K. Kanosue, Ph. D. Thesis, University of Osaka, Osaka, 2000.
16 S. Nakano, M. Moritoki, and K. Ohgaki, *J. Chem. Eng. Data*, 1999, **44**, 254.

INVESTIGATION OF THE STRUCTURAL DISORDER IN ICE Ih USING NEUTRON DIFFRACTION AND REVERSE MONTE CARLO MODELLING

L. Temleitner and L. Pusztai

Research Institute for Solid State Physics and Optics, Hungarian Academy of Sciences, Budapest, P.O. Box 49., H-1525, Hungary

1 INTRODUCTION

Although the crystallographic structure of the most common form of ice, phase Ih, has been known for many decades[1], its proton-disordered local structure is still being discussed[2,3]. As a demonstration of this fact, we mention that the geometry of water molecules in ice Ih has only recently been determined accurately, with the help of rather sophisticated quantum mechanical calculations[3].

Apart from Bragg-scattering (which is the main concern of classical crystallography), the experimentally measurable neutron diffraction pattern of ice Ih displays a rather characteristic diffuse signal, as was shown, e.g., in Ref.[4]. In most crystallographic studies, this part of the powder pattern is considered as backgr ound' and is subtracted from the measured data, since it has no effect on the translational symmetry of the crystal. Diffuse scattering, on the other hand, carries information on the local (dis)order which for ice Ih, that was described as a þroton glass' (as opposed to a þroton crystal') [3], may apparently be an important and integral part of the structure.

Total neutron scattering data (including diffuse scattering) from ice Ih have been made use by Floriano et al.[5], who derived parameters concerning only the molecular structure via Fourier transforming the measured structure factor. Later, the approach of Nield et al.[6,7,8] has included Reverse Monte Carlo (RMC) structural modelling[9] of single crystal neutron diffraction data. Their method, which did not consider Bragg-scattering during the calculations, applied a special computer code dubbed as RMCX'. Although RMCX has not been further developed over the past decade, the work of Nield and co-workers showed that studying the diffuse scattering contribution may lead to a deeper understanding of the behaviour of ice Ih.

In this contribution, we report a neutron powder diffraction study of ice Ih at 120 and 200 K. The data, including Bragg and diffuse scattering, were interpreted by the RMCPOW (Reverse Monte Carlo for POWder diffraction data) method[10], using large (containing, e.g., 8x8x8 unit cells, that is, more than 2000 molecules) three dimensional models. RMCPOW is based on the standard RMC algorithm[9] and aims at matching the full powder diffraction signal. Description of the diffuse scattering part is done by moving atoms around while maintaining the long range ordering of the crystal. The resulting particle configurations are analysed in terms of partial radial distribution functions and cosine distribution of bond angles – similarly to a liquid/amorphous structure. RMCPOW is faster and more general than RMCX; in addition, powder diffraction data are a lot easier to take and handle than single crystal data. For these reasons it is foreseen that the combination described in the present contribution (powder diffraction + RMCPOW modelling) may be used more widely in the future.

2 NEUTRON DIFFRACTION EXPERIMENT

Neutron powder diffraction experiments on fine D_2O ice powder have been carried out using the NPD two-axis diffractometer installed at Studsvik NFL (Sweden; the facility was, regrettably, phased out in 2005). A Ge (2 2 0) monochromator provided a neutron beam with a wavelength of 1.47 Å, so that a scattering vector (Q) range of $0.3 - 8$ Å^{-1} could be covered. The sample was filled in a 10 mm diameter vanadium container; the container then was placed within a standard CCR. In order to minimise the effects of preferred orientation of ice crystallites, the whole setup was rotated during the experiment. Measurements on the empty sample holder, as well as of a bulk vanadium rod (as incoherent scatterer) have also been performed, at the temperatures of the ice measurements (120 and 200K). Raw data have been treated the same way as liquids/amorphous diffraction results, i.e., corrections for instrumental background, detector efficiency, absorption and multiple scattering, as well as absolute normalisation (with respect to an incoherent scatterer) have been carried out.

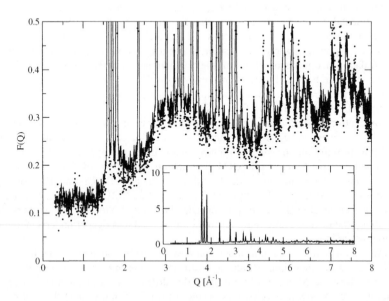

Figure 1 *Large panel: measured powder diffraction pattern of (D_2O) ice Ih at 120 K (symbols) and 200 K (solid line); the focus is on the diffuse scattering part. Insert: powder diffraction pattern of ice Ih at 200 K.*

Figure 1 shows the corrected and normalised total scattering functions, $F(Q)$, of ice Ih at two temperatures, 120 and 200 K. The enhanced level of diffuse scattering at 200 K may be observed.

3 STRUCTURAL MODELLING – RESULTS

The RMCPOW calculations were started from initial configurations that obeyed the 'ice-rules'[11] and possessed no net dipole moments. For generating the starting coordinates, a slightly modified version of the algorithm of Barkema et al.[12] has been applied. By using supercells of 2^3, 4^3, 8^3 and 11x12x7 unit cells of hexagonal ice I it could be established that results to be presented here show no dependence on the system size (apart from statistics). We found the 8^3 unit cell system convenient (large enough but the computational costs were not too high) and therefore, properties of such systems will be discussed below.

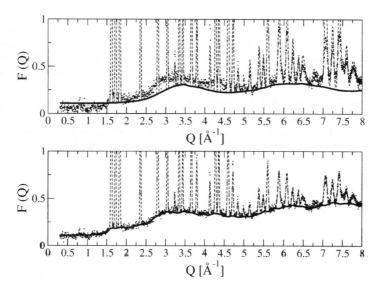

Figure 2 *Upper panel: measured powder diffraction pattern of (D₂O) ice Ih at 120 K (symbols) as compared to that of the initial configuration (dashed line); the diffuse part is also shown (thick solid line). Lower panel: RMCPOW fit to the measured signal (dashed line); the diffuse part is also given (thick solid line).*

Figures 2 and 3 compare measured and simulated total scattering functions (tsf) for ice Ih at 120 and 200 K, respectively. The upper panels of Figures 2 and 3 compare tsf's for the initial stages of the RMCPOW modelling. It is obvious that Bragg scattering, as far as peak positions are concerned, is well described by the starting configurations, although intensities may not always match the measured values. Note, however, that the shape and intensity of the diffuse part had to change enormously between the initial and final states of the calculations.

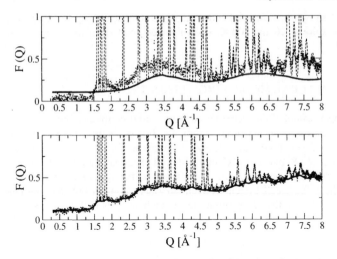

Figure 3 *Upper panel: measured powder diffraction pattern of (D₂O) ice Ih at 200 K (symbols) as compared to that of the initial configuration (dashed line); the diffuse part is also shown (thick solid line). Lower panel: RMCPOW fit to the measured signal (dashed line); the diffuse part is also given (thick solid line).*

We must declare here that the entire study should be considered as a 'pilot project', with numerous details still to be clarified/refined. Still, the above finding seems to indicate that an important aspect of the structure of ice Ih, represented by the measured diffuse scattering, could not be captured sufficiently well by existing models of the crystalline structure.

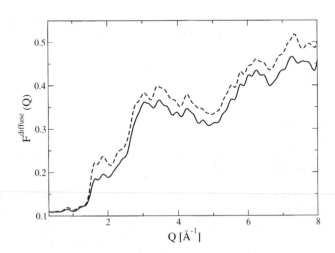

Figure 4 *Diffuse scattering from (D₂O) ice Ih at 120 K (solid line) and 200 K (dashed line), as calculated from RMCPOW.*

In Figure 4, diffuse scattering contributions are compared for the 120 and 200 K states. As it could be expected, the curves show similar features, except around the first maximum, just below 2 Å⁻¹. This – small, but apparent – difference may indicate that the structure might vary as the temperature increases beyond the level that would be expected on the basis of the stronger thermal vibrations. For being able to - at least – speculate about the origin of this variation we must switch from reciprocal to real space.

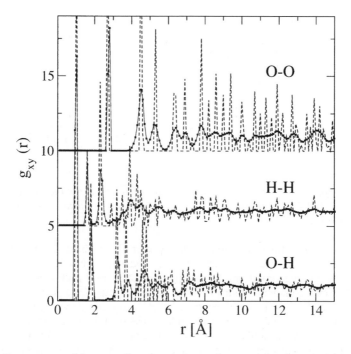

Figure 5 *Comparison of the partial pair correlation functions for the initial (dashed lines) and final (solid lines) stages at 120 K. In both cases, systems of 8³ unit cells were applied.*

Initial and final O-O, O-H and H-H partial radial distribution functions (prdf) at 120 K are compared in Figure 5. The changes between the two stages are obvious, particularly as far as the O-O prdf is concerned: every single narrow initial peak has broadened (although still recognisable in the final stage). Most of these changes may be attributed to the effects of thermal vibrations, which are absent in the starting particle arrangement. The situation is less clear in this respect for the H-H and O-H prdf's which were affected by the 'proton-glass' nature of (even ideal) ice Ih. In any case, it has to be acknowledged that the differences between initial and final prdf's are the real space manifestations of the differences between initial and final diffuse scattering, shown in Figure 2 for the 120 K system. (It may be instructive to notice that the O-O part of Figure 5 can be considered as a demonstration of the difference between ideal and real crystalline behaviour.)

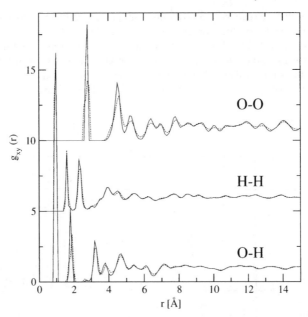

Figure 6 *Comparison of the partial pair correlation functions for the 120 K (solid lines) and 200 K (dashed lines) states. In both cases, systems of 8³ unit cells were applied.*

Figure 6 provides the comparison of the three prdf's obtained for the 120 and 200 K states. The two sets of curves are rather similar, as it can be expected for the material in the given temperature range (far from the melting point). The H-H prdf's are nearly identical; there are, however, small alterations in terms of the O-O and O-H prdf's. For the former, the fist and second maxima seem to be shifted slightly towards higher r values, whereas for the latter, it is the first and third intermolecular peaks (just below 2 and just below 4 $Å^{-1}$, respectively) that behave similarly. The exact value of the shift is hard to tell at this stage, because of the – for this purpose, too – coarse r-spacing that was applied while calculating prdf's. (The given value of 0.1 Å was required for achieving reasonable statistics.) It will be necessary to take a closer look at these small variations (e.g., by further increasing the size of the system and thus allowing for a better r-resolution); what can be stated already at this stage is that the changes are the r-space manifestations of the difference of the diffuse scattering signal found to be present between the investigated two states of ice Ih (see Figure 4). It can also be suggested that the differences are connected to the hydrogen bonding network of ice Ih: hydrogen bonds appear to be somewhat longer at 200 K.

From the particle configurations provided by RMCPOW modelling it is possible to calculate any characteristics of the structure, including ones that go beyond two-particle properties. One of the simplest possibilities is to consider angular distributions. In Figure 7, two of these distributions, connected directly to hydrogen bonding, are presented for the starting configuration and for the final stages at 120 and 200 K. The distribution of the cosines of O...O...O angles show a continuous (and symmetric) broadening as temperature increases; the position of the maximum, however, remains at -1/3, which is the value for the tetrahedral angle. The distribution of the O-H...O angles also indicates distortion of the straight angle that characterises hydrogen bonds in the ideal ice Ih crystal. The difference between the tails at 120 and 200 K seems to be a direct consequence of the slightly different diffuse scattering contributions at the two temperatures (see Figure 4).

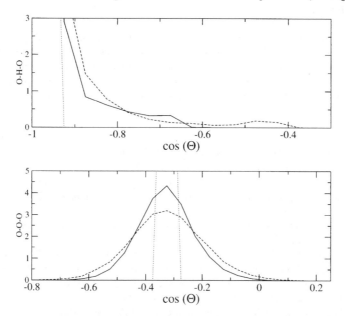

Figure 7 *Cosine distributions of O-H...O (upper panel) and O...O...O (lower panel) angles ("-" stands for intra-, whereas "..." stands for intermolecular connections). Dotted line: initial configuration; solid line: 120 K; dashed line: 200 K.*

4   SUMMARY AND OUTLOOK

It is shown that both the Bragg and the diffuse scattering parts of neutron powder diffraction data on ice Ih can be interpreted simultaneously by constructing large models of the structure that are consistent with the measured total scattering functions within errors. The RMCPOW algorithm[10] proved to be readily applicable for the purpose.

It is found that proton disorder on its own cannot be responsible for the measured level and shape of diffuse scattering. The present results, particularly the O-H and O-O partial radial distribution functions and the distribution of the O-H...O hydrogen bond angles suggest that small changes of the hydrogen bonded network are most probably responsible for the slightly different shape of the diffuse scattering signal at 120 and 200 K.

In this pilot study, ice Ih was considered in a 'safe' temperature range, between 120 and 200 K. For improving the reliability of the results presented here, an obvious extension would be an X-ray diffraction measurement (also of the 'total scattering type') and the subsequent RMCPOW modelling of both the neutron and X-ray data together. Using such a combination, approaching the melting point would be easier to achieve; investigations on this 'high temperature' ice Ih are expected to provide valuable information on the behaviour hydrogen bonded network of water molecules during the process of melting. It is also worth mentioning that similar studies may be carried out on other phases of ice, too, provided that the appropriate total scattering type diffraction data are available.

Acknowledgment

LP is grateful to the staff of Studsvik NFL (Sweden) for their hospitality while staying with them and for their help with the numerous experiments on ice Ih powder. He is also grateful to Dr. V.M. Nield, for her help with the initial ice powder experiment on the SLAD diffractometer in Studsvik. Financial support for this work was provided by the OTKA (Hungarian Basic Research Fund), grant Nos. T048580 and T043542.

References

1   S.W. Peterson and H.A. Levy, *Acta Cryst.*, 1957, **10**, 1105.
2   S.W. Rick and A.D.J. Haymet, *J. Chem. Phys.*, 2003, **118**, 9291.
3   J.-L. Kuo, M.L. Klein and W.F. Kuhs, *J. Chem. Phys.*, 2005, **123**, 134505.
4   A.J. Leadbetter, R.C. Ward, J.W. Clark, P.A. Tucker, T. Matsuo and H. Suga, *J. Chem. Phys.*, 1985, **82**, 424.
5   M.A. Floriano, D.D. Klug, E. Whalley, E.C. Svensson, V.F. Sears and E.D. Hallman, *Nature*, 1987, **329**, 821.
6   V.M. Nield and R.W. Whitworth, *J. Phys.: Condens. Matter*, 1995, **7**, 8259.
7   M.N. Beverley and V.M. Nield, *J. Phys.: Condens. Matter*, 1997, **9**, 5145.
8   M.N. Beverley and V.M. Nield, *J. Phys. Chem. B*, 1997, **101**, 6188.
9   R.L. McGreevy and L. Pusztai, *Molec. Simul.*, 1988, **1**, 359.
10 A. Mellergård and R.L. McGreevy, *Acta Cryst.*, 1999, **A55**, 783.
11 J.D. Bernal and R.H Fowler, *J. Chem. Phys.*, 1933, **1**, 515.
12 G.T. Barkema and M.E.J. Newman, *Phys. Rev. E*, 1998, **57**, 1155.

FIRST–PRINCIPLES CALCULATION OF STRUCTURE AND DYNAMICAL PROPERTIES OF ORIENTATIONAL DEFECTS IN ICE

N. Tsogbadrakh and I. Morrison

Joule Physics Laboratory, Institute for Materials Research, University of Salford, Salford, M5 4WT, UK. E lmail: Tsogbadrakh@num.edu.mn and I.Morrison@salford.ac.uk

1 INTRODUCTION

In ice there are five categories of point defects associated with the movement of atoms or molecules from one site to another: molecular defects, protonic defects, ionic or atomistic impurities, electronic defects and combined defects[1]. Molecular defects involve the displacement of entire water H_2O molecules, leaving a vacancy (region where a molecule is absent from its usual site) or an interstitial (an extra molecule occupying a site in a cavity). Ionic or atomistic impurities involve the substitution of one atom for another. Electronic defects are associated with ionized molecules or trapped electrons. Combined defects entail the occurrence of two or more of the above kinds of defect. Their most important properties are associated with their movement by the thermally activated processes[1]. The defects considered in this study were first proposed by N. Bjerrum are specific to ice–like structures and are referred to as '*protonic point defects*' to distinguish them from more conventional point defects such as vacancies and interstitials. There are four types of protonic point defects in ice: two ionic and two '*Bjerrum or orientational defects*'[2]. In orientational defects the Bernal lFowler's ice rules[3] are broken such that in a Bjerrum D defect there exist two hydrogens between nearest neighbour oxygens, and in a Bjerrum L defect the bonds between nearest neighbour oxygens exist without any hydrogen (Figure 1). These Bjerrum D and L defects are generated by molecular rotations, and can travel to an adjacent site via a thermal activated process.

The protonic point defects diffuse through bulk ice, and their motions follow a zig lzag path along appropriately oriented bonds, reorienting both molecules and bonds along its track. The Bjerrum D and L defects migrate by the molecular rotations of 120°. Such jumps are proposed to provide a mechanism for molecular reorientation in different relaxation phenomena (mechanic, dielectric, nuclear magnetic) and transport properties (self diffusion, tracer diffusion, conductivity). The migration of the Bjerrum defects is considered as the dominant process for mechanical and dielectric relaxation in pure ice. After the initial formation, the pair of the Bjerrum D and L defects can be further separated by successive molecular rotations, eventually creating a pair of independent, or ideal D and L defects.

The structure of this paper is as follows: in section 2 we briefly describe the formation mechanism of the Bjerrum defects used in this study and their possible ideal diffusion pathways, and in section 3 we introduce jump rates and diffusion coefficients in solids. In section 4 we describe the computational methods used in our study. In section 5 we present the results of density functional based computational studies of the structure and dynamics of the Bjerrum defects for tetrahedrally fully coordinated ice Ih, and compare with experimental values. Finally we set out our conclusions in section 6.

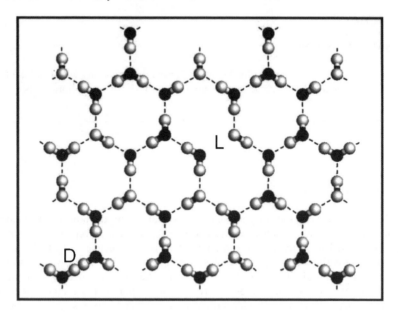

Figure 1   *The schematic illustration of the Bjerrum D and L defects in Ice Ih*

2 GEOMETRY OF BJERRUM DEFECTS IN ICE

2.1   Formation of Ideal Bjerrum D and L defects

In order to generate the ideal Bjerrum D and L defects in ice Ih, we have created a $2 \times 2 \times 2$ supercell from a unit cell with 16 water molecules containing a representative sample of molecular orientations[4]. The supercell consists of 128 water molecules. We first generated a pair of Bjerrum defects at one edge of the supercell by the molecular rotation of a rigid water molecule by 120°. These two defects were then separated away from each other by a geometric transformation corresponding to 8 ideal Bjerrum jumps of the L defect after the formation of the defect pair. The ideal Bjerrum L defect was thus generated at the centre of the supercell, and the ideal Bjerrum D defect remained at the edge of the supercell. A full relaxation of all atomic positions was then performed for this structure, see section 4.

2.2   Possible Ideal Diffusion Pathways

In ice Ih, there are two different geometrical rotational mechanisms for the diffusion of the Bjerrum D and L defects through bulk ice[5]: *i.*) a rotation of the central rigid water molecule about an OIH bond direction, *ii.*) a rotation of the central rigid water molecule about a lone lpair direction (Figure 2).

In the first case one hydrogen bond, associated with the central rotating molecule, joined to one of oxygen atoms of a nearest neighbour water molecule is broken. The other covalent bond of the central water molecule remains fixed, and the hydrogen bond collinear with this covalent bond is not broken. In the second case both the hydrogen bonds of the central water molecule are broken. In both the cases the position of oxygen atom for a central water molecule remains fixed.

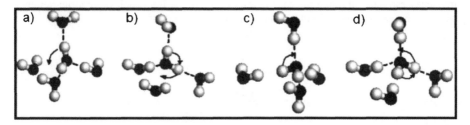

Figure 2 *The two different ideal diffusion pathways of the Bjerrum D and L defects in the tetrahedral arrangement for Ice Ih. Here a) the migration of the L defect about an O–H bond direction, b) the migration of the L defect about a lone–pair direction, c) the migration of the D defect about an O–H bond direction, d) the migration of the D defect about a lone–pair direction.*

2.3 Saddle Points of Diffusion Pathways

We assumed that the activated complex ('*saddle point*') for migration of the Bjerrum defects will be roughly at the midway point of a rigid molecular rotation by 120°, and consider that the molecular rotations are done in thermal equilibrium. Since a molecular rotation of 60° of a rigid water molecule participating in the Bjerrum D and L defects clearly involves straining or breaking at least two of its three original hydrogen bonds, one might expect a sizable contribution to the activation energy of motion. In our model these defect containing tetrahedral structures are described by core clusters containing 5 water molecules within the supercell with a total of 128 water molecules. These clusters include the first nearest neighbour of the ideal individual defects. In order to find saddle points on the possible ideal diffusion pathways, we will perform a controlled search within these core regions of the defect containing clusters, and barrier heights will be established, see section 4.

3 JUMP RATES AND DIFFUSION COEFFICIENTS

For the interpretation of experimental observations on ice the microscopic picture of the diffusion process is established through the evaluation of atomic jump rates. In transition state theory the atomic jump lattempt frequency appears as a ratio of two partition functions of which the numerator involves the potential at the saddle point on top of the potential ridge, through which all jump trajectories in configuration space must cross. The Vineyard theory approximates the relevant potential surfaces harmonically described. Using this transition state theory we can find the jump rate of a particular protons as follows[6]:

$$\Gamma(T) = v^* \exp\left(\frac{-E_m}{k_B T}\right), \tag{1}$$

where k_B is the Boltzmann constant, T is the environmental temperature, E_m is the activation energy of motion for the Bjerrum defects and v^* is the atomic jump attempt frequency estimated through the calculation of vibrational frequencies of configurations at both the energy minima and saddle point, which is obtained as[6]:

$$v^* = \frac{\prod_{j=1}^{3N} v_j}{\prod_{j=1}^{3N-1} v_j'}, \tag{2}$$

where $v = \omega / 2\pi$, and v is expressed in units of sec[11], and the denominator product does not include the imaginary frequency at the saddle point. The atomic jump–attempt frequency is very different from a simple Einstein frequency, or any single frequency exhibited in physical space. As stated by G. H. Vineyard, the atomic jump attempt frequency is the ratio of the product of the $3N$ normal vibrational frequencies v_j of the entire system at the initial point of the transition to the $3N-1$ normal vibrational frequencies v_j' of the system constrained at the saddle point configuration[6].

Finally the diffusion coefficient for the Bjerrum defects is found as[7]:

$$D(T) = \frac{1}{12} d_{O-O}^2 \cdot \Gamma_{tot}(T), \tag{3}$$

where d_{O-O} is the oxygen–oxygen distance of defect containing bonds and $\Gamma_{tot}(T)$ is the total jump rate of the Bjerrum defects, which is defined as:

$$\Gamma_{tot}(T) = 2\Gamma_{OH}(T) + 2\Gamma_{LP}(T), \tag{4}$$

where $\Gamma_{OH}(T)$ and $\Gamma_{LP}(T)$ are the jump rates of a particular proton via the rotations about O–H bond and lone–pair directions respectively.

4 COMPUTATIONAL METHODS

The calculations of structure optimizations and vibrational frequencies were performed within the framework of density functional theory[8, 9] using both the LCAO approach as employed in the SIESTA code[10] and the plane wave approach as employed in the CASTEP code[11]. The valence electrons of atoms in ice are described by the self–consistent solution to the Kohn Sham equations with the PBE GGA exchange correlation functional. In order to integrate over the Brillouin zone, we used 1 K point in both of the codes. Periodic boundary conditions are employed. In the CASTEP code the interaction of valance electrons is described by ultrasoft pseudopotentials. The cut off energy of plane waves was chosen as 350 eV, corresponding to medium precision as defined in user guide. In the

SIESTA code, the interactions of valence electrons with the atomic ionic cores are described by the norm–conserving pseudopotentials with the partial core correction of 0.6 au. on the oxygen atom. We used the optimized–*'double zeta plus polarization (DZP)'* basis sets with medium localization in the SIESTA code. A mesh cut loff energy of 350 Ry, which defines the equivalent plane wave cut–off for the grid, was used. The forces on atomic ions are obtained by the Hellman lFeynman theorem and were used to relax atomic ionic positions to the minimum energy. The atomic forces within the supercell were minimized to within 0.035 eV/Å and 0.05 eV/Å in the SIESTA and CASTEP codes respectively.

In order to find relaxed structures during the saddle point searches in the defect containing clusters, on each rigid molecular rotation we performed a constrained structure optimization in which the position of the rotating hydrogen inside of the cluster was fixed. The positions of all atoms outside of the cluster were also fixed. In this case the number of degree of freedom for our supercell decreases up to 42, which correspond to 14 atoms associated with the 5 water molecules in the defect containing cluster within the relaxed supercell. Following this procedure the force on the fixed hydrogen in the cluster is not completely relaxed. The atomic forces in the cluster should be full relaxed at both the energy minima and saddle point. In order to relax this residual force on the fixed hydrogen in the defect containing cluster, we made a shift of 0.01 Å along the direction of the residual force, and then the structure optimization was done again, this was repeated until the residual force becomes less than 0.1 eV/Å.

In order to find vibrational frequencies within the defect containing tetrahedral structures at both the energy minima and saddle points, we used a finite different method[4] from the calculation of the *ab initio* total energy and forces, and created a zone centre restricted force constant matrix using the VIBRA program in the SIESTA code. In the calculations of restricted force constant the shift is chosen to be 0.02 Å. It was not possible to do this using the CASTEP code due to computational power restrictions. The dimension of the restricted force constant matrix corresponds to the number of degree of freedom in the each cluster. Therefore we have found 45 normal vibrational frequencies of the defect containing clusters at both the energy minima and saddle point. Using these normal vibrational frequencies the atomic jump lattempt frequencies for each possible ideal diffusion pathway were estimated using Eq. 2.

5 RESULTS

5.1 Static Result

In our relaxed defect structure the ideal Bjerrum D and L defects were separated from each other by the sixth nearest neighbour, and the separation distance of the ideal Bjerrum defects was found to be 12.085 Å and 11.542 Å by the SIESTA and CASTEP codes respectively. The Bjerrum D and L defects containing $O_1...O_2$ distances were found to be 2.869 Å and 3.364 Å respectively by the SIESTA code, and 3.607 Å and 3.357 Å respectively by the CASTEP code. These distances are increased dramatically compared to non ldefective ice due to electrostatic repulsion. For the Bjerrum L defect the $O_1...O_2$ distances are the same in both the SIESTA and CASTEP codes. For the Bjerrum D defect distortion was observed on the $O_1...O_2$ bond by the SIESTA code. Therefore the $O_1...O_2$ distance is shorter than the $O_1...O_2$ distance calculated by the CASTEP code. The averaged value of the remaining oxygen loxygen distances in the defect containing tetrahedral arrangements was relatively unchanged by the CASTEP code. The average O lO lO angle of a tetrahedral structures has decreased to be 108.66°II08.95° in the ideal defect containing tetrahedral structures. The average values of both the OIH bond length and the HIOIH angle for water molecules were also unchanged in the case of the CASTEP code. We have observed that the Bjerrum D and L defects in bulk ice Ih contribute to local configurations (bend angle and covalent bond length) of the water molecules within the first and second nearest neighbours of defect containing tetrahedral structures, but the contribution decreases significantly at the third nearest neighbour.

In the study of bulk ice by R. Podeszwa and V. Buch[12], the Bjerrum D and L defects containing oxygenIoxygen distances obtained to be (3.4I4.0) Å and typically 4.0 Å respectively, as compared to the mean minimum oxygenIoxygen distance of 2.785 Å for hydrogen bonds in the empirical potential. These results obtained in the empirical potential are in reasonable agreement with our above results. For the Bjerrum L defect, the mean first and second nearest neighbour distances were 2.785 Å and 4.55 Å respectively[12]. These results obtained in the empirical potential are in excellent agreement with our above results. In quantum chemical study of a cluster with 8 water molecules[13], the L defect containing oxygenIoxygen distance increased to 4.73 Å, but the remaining oxygen–oxygen distances were found to be (2.76I2.96) Å. In our calculation by both the SIESTA and CASTEP codes, the formation energy was found to be 1.208 eV and 1.420 eV respectively. These results were in reasonable agreement with a recent value of (1.268±0.04) eV calculated by the VASP code[14]. The first theoretically value of formation energy for Bjerrum defects in pure ice was (1.11±0.05) eV[15].

5.2   Activation Energies of Motion for the Ideal Bjerrum Defects

We shown the activation energies corresponding to a saddle point on each possible diffusion pathway associated with the L defect in Table 1.

Table 1 *The activation energies of motion for the ideal Bjerrum*
L defect on each diffusion pathway (meV)

	SIESTA	CASTEP
O–H bond direction	204.26	212.53
lone–pair direction	145.69	

M. D. Newton first estimated the activation energies of motion for the Bjerrum L defect on a $(H_2O)_5$ cluster using *ab initio* molecular orbitals[5]. He found it to be 143.10 meV and 166.95 meV via rotations of the central water molecule about OIH bond and loneIpair directions respectively. Recently it has been found to be (0.120±0.02) eV in bulk ice using the VASP code[14]. Although our results for the activation energy of motion are in the range of experimental values, they are slightly higher than other theoretical results.

5.3   Temperature Dependencies of Jump Rates and Diffusion Coefficients

We have estimated the atomic jump–attempt frequencies and jump rates for the L defect in our structure for both diffusion pathways. We have presented the atomic jump–attempt frequencies in Table 2. Using Eqs. (1) and (4) we calculate the total jump rate of the ideal Bjerrum L defect in bulk ice Ih at T>150 K. In our calculation the jump rate for the ideal Bjerrum L defect at 110° C was obtained as:

$$\Gamma_{tot}^{L}(-10^{0}C) = 1.19 \times 10^{11} s^{-1} \qquad (5)$$

From the result of the molecular dynamics simulation by R. Podeszwa and V. Buch[12], the jump rate of the Bjerrum L defect at 230 K is shown in the range $(1.0-5.5)\times10^{11}s^{-1}$. These values agree with the experimental value of $2\times10^{11}s^{-1}$ at $110°$ C[16, 17].

Table 2 *The atomic jump-attempt frequencies for the Bjerrum*
L defects as calculated by the SIESTA code

	O–H bond direction	lone–pair direction
$v^*(THz)$	108.169	28.503

We have presented the temperature dependence of the total jump rate of the ideal Bjerrum L defect in bulk ice Ih in Figure 3a.

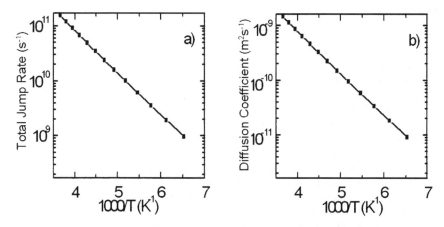

Figure 3 *The temperature dependencies of the total jump rate and diffusion coefficient for the Bjerrum L defect in bulk Ice Ih in the panel a) and b) respectively.*

Finally we have estimated the diffusion coefficient for the ideal Bjerrum L defect, and have found to be:

$$D^l(-10^0 C) = 1.12\times10^{-9} m^2 s^{-1} \tag{6}$$

We have presented the temperature dependence of the diffusion coefficient for the ideal Bjerrum L defect in Figure 3b. This calculated diffusion coefficient for ideal Bjerrum L defect is shown to be higher than the self diffusion coefficient for molecular defects[1, 7].

6 CONCLUSION

We have observed that the ideal Bjerrum D and L defects in bulk ice Ih contribute to local configurations of the water molecules within the first and second nearest neighbours of defect containing tetrahedral structures, but the contribution significantly decreases at the third nearest neighbour. The defect containing oxygen–oxygen distances increase dramatically due to electrostatic repulsion. The formation energy of a pair ideal Bjerrum D and L defects in ice Ih was found to be 1.208 eV and 1.420 eV using the SIESTA and CASTEP codes respectively. In addition we have established the temperature dependencies of the jump rate and diffusion coefficient for the ideal Bjerrum L defect in bulk ice Ih. Further work is ongoing to calculate diffusion coefficients for the D defect.

Acknowledgements

We thank the ORS Awards Scheme of United Kingdom Scholarships for International Research Students and the Salford University for the financial support during this work.

References

1   V.F. Petrenko and R.W. Whitworth, *Physics of Ice,* Reprinted edn, Oxford University Press, 2002.
2   N. Bjerrum, *Science,* 1952, **115,** 385.
3   J.D. Bernal and R.H. Fowler, *J. Chem. Phys.,* 1933, **1,** 515.
4   I. Morrison, J. lC. Li, S. Jenkins, S.S. Xantheas and M.C. Payne, *J. Phys. Chem. B,* 1997, **101,** 6146
5   M.D. Newton, *J. Phys. Chem.,* 1983, **87,** 4288.
6   G.H. Vineyard, *J. Phys. Chem. Solids,* 1957, **3,** 121.
7   L. Onsager and L.K. Runnels, *J. Chem. Phys.,* 1969, **50,** 1089.
8   P. Hohenberg and W. Kohn, *Phys. Rev.,* 1964, **136,** B864.
9   W. Kohn and L.J. Sham, *Phys. Rev.,* 1965, **140,** A1133.
10   J.M. Soler, E. Artacho, J.D. Gale, A. Garcia, J. Junquera, P. Ordejon and D. Sanchez – Portal, *J. Phys.: Condens. Matter,* 2002, **14,** 2745.
11   M.D. Segal, P.J.D. Lindan, M.J. Probert, C.J. Pickard, P.J. Hasnip, S.J. Clark and M.C. Payne, *J. Phys.: Condens. Matter,* 2002, **14,** 2717.
12   R. Podeszwa and V. Buch, *Phys. Rev. Lett.,* 1999, **83,** 4570.
13   P.L.M. Plummer, *9th Cimtec–World Forum on New Materials, Symposium–I. Computational Modeling and Simulation of Materials,* eds., P. Vincenzini and A.Degli Esposti, 1999, pp. 235.
14   M. de Koning, A. Antonelli, A.J.R. da Silva and A. Fazzio, *Phys. Rev. Lett.,* 2006, **96,** 075501.
15   H. Granicher, *Proceedings of the Royal Society of London: Series A, Mathematical and Physical Sciences,* 1958, **247,** pp. 453.
16   H. Granicher, *Phys. Kondens. Materie,* 1963, **1,** 1.
17   N. Maeno, *Physics of Ice,* Hokkaido University Press, Sapporo, Japan, 1980, Translation into Russian by A.I. Leonov, ed., V.F. Petrenko, Nauka, Moscow, Russia, 1988.

EFFECTS OF ADDITIVES ON FORMATION RATES OF CO_2 HYDRATE FILMS

T. Uchida[1], I.Y. Ikeda[2], R. Ohmura[3], S. Tsuda[2]

[1] Division of Applied Physics, Graduate School of Engineering, Hokkaido University, Sapporo 060-8628, Japan
[2] National Institute of Advanced Industrial Science and Technology (AIST), Sapporo 062-8517, Japan
[3] Department of Mechanical Engineering, Keio University, Yokohama 223-8522, Japan

1 INTRODUCTION

Naturally occurring clathrate hydrates are found in marine sediments and in permafrost. Because they contain a large amount of methane, they are thought to have potential as an unconventional energy resource. At the same time, however, clathrate hydrates are a serious problem for the gas and oil industries, because they form easily under suitable conditions at the sites of natural gas production, transportation, and processing. The inhibition and control of hydrates in pipelines adds tremendously to gas production costs.[1]

The usual technology for hydrate inhibition includes heating, pressure reduction and the addition of chemicals. The additives are classified into two types: the thermodynamic inhibitors and the kinetic inhibitors. The thermodynamic inhibitors, such as NaCl, methanol and glycol, shift the equilibrium conditions. These additives have been shown to inhibit the hydrate formation successfully, but typically they are required to work at relatively higher concentrations ranging from 15 to 50wt% of the free water phase.[1] On the other hand, certain types of polymers act as kinetic inhibitors that can delay the start of nucleation and thereby slow the hydrate growth rate at low concentration[2], such that hydrate masses will not grow sufficiently large to cause blockage during the residence time of the liquid phase in the pipeline.

These methods are often economically unsound or environmentally destructive. For example, it has been reported that methanol injection costs the oil and gas industry about $500 million per year.[3] Also, it would be advantageous to discover naturally occurring additives from an environmental perspective. Antifreeze proteins (AFPs) and antifreeze glycoproteins (AFGPs) are considered to be potential inhibitors, since they are thought to inhibit freezing in a non-colligative manner by binding to the surface of ice crystals and delaying ice crystal growth rates.[4,5] Whether AFPs might also inhibit the formation of clathrate hydrates has been questioned by the proposed mechanism of the inhibition of ice growth by AFPs.

A number of recent experimental and simulation studies have shown that two different AFPs are capable of inhibiting the growth of clathrate hydrates. Zeng et al.[6-8] studied the effect of type I and an insect AFPs on the formation of tetrahydrofuran (THF) clathrate hydrate by observing changes in the crystal morphology of THF hydrate and by determining the induction time for nucleation. They found that AFPs were more effective than the kinetic inhibitor polyvinylpyrrolidone (PVP). Zeng et al.[9] performed a Monte

Carlo simulation of ice-crystal growth to study the mechanism of inhibition of AFPs on the surface of the THF hydrate and propane hydrate. They found that most of the octahedron surfaces of the THF hydrate were covered with AFP molecules, which could reduce the growth rate of the THF hydrate only allowing plate growth perpendicular to that surface. It is thus necessary to look for other experimental evidences to clarify the common features of AFPs on the inhibition of the clathrate hydrate formation.

Other potential inhibitors with less environmental impact might be selected from among the cryoprotecting materials naturally produced in animals. One such material is trehalose, which consists of fructose and glucose rings connected by a glycosidic bond, and is known to decrease the melting point of ice below the molar freezing-point depression.[10,11] This material is expected to act as a thermodynamic inhibitor for clathrate hydrates.

In the present study, we investigated the effect of two cryoprotectants, type-III AFP and trehalose, on the crystal growth rate of CO_2 hydrate, a type-I gas hydrate. This is the first investigation of the effect of type-III AFP on the inhibition of clathrate hydrate, which has been confirmed to be a cryoprotectant.[12] It is also the first examination of the effect of trehalose on the crystal growth of clathrate hydrates. On the other hand, type-I clathrate hydrates, such as the CO_2 hydrate used here, have a crystal structure just slightly different from that of type-II clathrate hydrates, such as THF hydrate and propane hydrate. Uchida et al.[13,14] have extensively investigated the growth mechanisms of CO_2-hydrate films formed on the interface between CO_2 phase and water. The use of an experimental technique that has been used previously would be advantageous for the comparison of results with other conditions, such as pure water or NaCl solutions, and to accumulate the useful knowledge on the hydrate inhibition mechanism to search for feasible materials with less environmental impact.

2 MATERIAL AND METHODS

We used type-III AFP (RD3-N1; retrieved from Antarctic eel pout) and trehalose (research-grade; Hayashibara Biochemical Laboratories, Inc.) as the inhibitors. The characteristics of this AFP are described in detail in Miura et al.[12] We added a small amount of AFP (from 0.01 to 1 mg ml^{-1} water) to distilled and deionized water (approximately 18 MΩ in resistivity) for the experiments. Trehalose was added at concentrations ranging from 1 to 50 wt%. CO_2 was delivered from Hokkaido Air Water, Inc., and its purity was approximately 99%.

The experimental procedures were almost the same as in the previous studies.[13,14] In brief, a pressure vessel (approximately 10 cm^3 in volume) equipped with two optical windows was filled with liquid CO_2 at pressures ranging from 4 to 6 MPa at room temperature. Then a small amount of solution (prepared to the experimental concentrations of additives in advance) was added to form a water droplet in the vessel. The temperature was controlled within $\pm$ 0.2 K by the cooling jacket that covered the vessel. After setting the system temperature to the experimental condition, the formation and growth process of CO_2 hydrate film on the water droplet were observed by a microscope and CCD camera system, and recorded using an S-VHS time-lapse video. During the temperature change, the rates of temperature change did not controlled. Several sequences of the hydrate formation were tested to measure the induction period τ and the supercooling $\Delta T = T_f - T_e$, where T_f is the temperature in the vessel at the time the hydrate started to grow and T_e is the equilibrium temperature of CO_2 hydrate at the experimental pressure, and to observe the lateral growth process of CO_2 hydrate film.

3 RESULTS AND DISCUSSIONS

3.1 Effects of type-III AFP on the formation process of the CO_2-hydrate film

Figure 1 shows a typical snapshot of the CO_2-hydrate film formed at the boundary between the AFP-solution droplet and the surrounding CO_2.

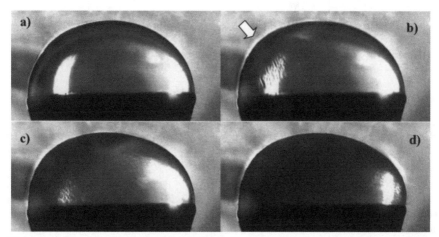

Figure 1 *CO_2-hydrate film on an 0.01 mg ml⁻¹ AFP-solution droplet at P = 5.5 MPa and T = 264.2 K. The droplet is on a stainless-steel stand (approximately 5mm in diameter). a) The initial droplet shape. b) The hydrate film begins to form at the left-top of the droplet (shown by an arrow). (c)-(d): CO_2 hydrate film grows laterally along the droplet surface from left to right. Each image is about 0.4 s apart.*

Figure 2 shows the temperature change of typical experimental run with various concentrations of AFP solution. Arrows in this figure indicate the formation times of CO_2 hydrate film on the solution droplet. The origin of the time axis (t=0) was defined as the time when the temperature reached T_e at the experimental pressure. Because the concentration of AFP was very small, the value of T_e for this experiment was taken as the same as that for pure CO_2 hydrate. From Figure 2, we estimated both τ and ΔT.

Table 1 summarizes the results obtained for CO_2-hydrate film formation in AFP-solution experiments. In each column, the upper value is that obtained in the first run, while the lower number in brackets is the average for all subsequent runs with its variation. For AFP-0.01 mg ml-1 samples, we showed the values in brackets are <average +/- standard deviation> for 8 times repeating experiments. Table 1 shows that τ was estimated as ranging between 60 and 100 min, which is much longer than the periods observed on the pure water droplet.[13] It is concluded that CO_2 hydrate formation was clearly inhibited by the presence of AFP molecules, even though the concentration was as small as 10^{-3} ppm.

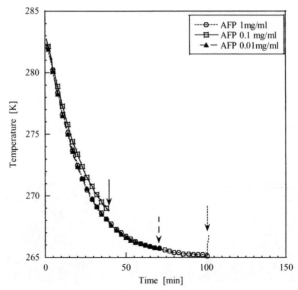

Figure 2 *Temperature profiles of the CO₂-AFP solution system at various AFP concentrations during the induction period. Each arrow indicates a formation point of CO₂ hydrate film on the droplet.*

When we compare τ and ΔT of the first run with those of the subsequent average ones, we find that the memory effect is small, or even be recovered, in the solutions with a small AFP concentration. The recovering of memory effect tended to larger in higher AFP concentrations. This is a very interesting phenomenon, and has been reported to be unique for AFPs,[8] although this effect should be confirmed by statistical analysis in future studies.

Zeng et al.[7,8] reported the formation ratio of THF hydrate and propane hydrate in type-I AFP solution within 24 hours. The supercooling for THF hydrate formation was approximately 4.4 K and that for propane hydrate was 7.4 K. These small supercooling conditions would make the induction period longer than that in the present study. Although a direct comparison between their results and the present ones is difficult, the effectiveness of both types of AFP on the hydrate inhibition seems to be higher than that of other kinetic inhibitors. Since the AFP concentration is low enough not to change the physical properties of the dominant part of water, such as the melting point or the surface tension, the effective mechanism of the growth inhibition would consist of disturbing the nucleation process.

Table 1 *Summary of CO₂-hydrate formation on the droplet of AFP solutions*

AFP concentration	τ [min]	ΔT [K]
1 mg ml⁻¹	101 (50 ± 33)	17.9 (17.0 ± 2.1)
0.1 mg ml⁻¹	63 (31 ± 8)	16.8 (12.5 ± 1.3)
0.01 mg ml⁻¹	71 <73 ± 33>	17.0 <18.2 ± 0.7>
pure water[13]	~ 20	< 10

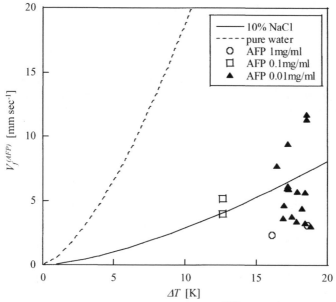

Figure 3 *Lateral growth rate of CO_2 hydrate film $v_f^{(AFP)}$ as a function of supercooling ΔT and for various AFP concentrations. The dotted line and solid line indicate the previous data on pure water and 10 wt% NaCl solution, respectively.*[14]

Using the time sequence of images in Figure 1, we measured the lateral growth rate of CO_2 hydrate by observing the position of the front edge of the hydrate film on the droplet surface. Figure 3 shows the lateral growth rate $v_f^{(AFP)}$ at each AFP concentration as a function of ΔT. In this figure, data from previous studies[13,14] are also shown (dotted and solid lines for pure and 10wt% NaCl solution, respectively). This figure indicates that the lateral growth rates obtained in the present study varied from 2 to 10 mm sec^{-1} as ΔT varied from 12 to 18 K. The lateral growth rates of CO_2-hydrate film on AFP solutions were much smaller than those on pure water, and were of the same order as those on the NaCl solution with approximately 10 wt%.[14] The dependence of $v_f^{(AFP)}$ on ΔT in each concentration is not clear. The inhibition effects of AFP also seem to have weak dependence on AFP concentrations.

Zeng et al.[7] estimated the effectiveness of the growth inhibition of the type-I AFP on the type-II propane hydrate. It is also difficult to compare these data directly with those obtained in the present study. However, type-III AFP can work as an inhibitor at a concentration 10^{-2} times smaller than type-I AFP. It is thus clear that the effectiveness of the growth inhibition of the type-III AFP on the type-I CO_2 hydrate is much larger than that of the classic inhibitors. This allows us to hypothesize that type-III AFP may have various functional sites fitting on the hydrate (and ice) crystal because its shape is spherical. Since the growth inhibition effect of AFPs on clathrate hydrates was observed even though their concentrations were too low to affect the equilibrium condition shift, we consider that the inhibition process is mainly due to the disturbance of growth kinetics. More experimental efforts, especially the direct observations of molecular scale, and theoretical approaches are necessary to reveal the detailed mechanism of the inhibition.

3.2 Effects of trehalose on the formation process of the CO_2-hydrate film

The formation process of CO_2-hydrate film on the trehalose-solution droplet was almost the same as shown in Figure 1: CO_2 hydrate nucleated somewhere on the droplet, and thethin film of CO_2 hydrate laterally grew at the interface between CO_2 and solution. At higher trehalose-concentration, the nucleation was observed to occur near the top of the droplet for several times as observed on the NaCl solutions.[14]

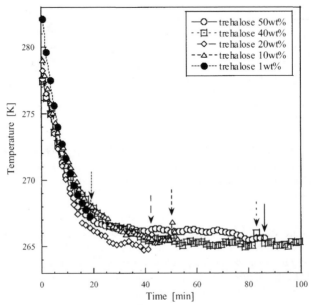

Figure 4   *Temperature profiles of CO_2-trehalose solution system at various trehalose concentrations during induction period. Each arrow indicates the formation point of CO_2 hydrate film on the droplet.*

Figure 4 shows a *T-t* diagram of some experimental runs with different trehalose concentrations. The origin of the time axis and arrows are the same as in Figure 2. This picture indicates that τ varies from approximately 20 minutes to more than 80 minutes at temperatures even at supercooling conditions. This is the same tendency as seen for other hydrate inhibitors, so trehalose is considered to be an effective inhibitor.

Table 2 summarizes the results of CO_2 hydrate film formation obtained in trehalose-solution experiments. For both τ and ΔT columns, the upper value is that obtained in the first run, while the lower number in brackets is the average for all subsequent runs with its variation. For trehalose-50 wt% samples, we showed the values in brackets are <average +/- standard deviation> for 8 times repeating experiments. The equilibrium temperature shift of CO_2 hydrate in each trehalose solution $\Delta T_e = T_e - T_e^0$, where T_e^0 is the dissociation temperature of CO_2 hydrate at the same pressure in pure water system[1], is also roughly estimate by the dissociation process observations of CO_2-hydrate film during the heating process. The increasing rates were more than 1 K min^{-1}, which were not controlled. So the accuracy of the estimation would be approximately +/- 0.9 K. Compared to the effect of NaCl on the equilibrium conditions, ΔT_e measured in the present study is found to be of almost the same order. Therefore, trehalose may function as a thermodynamic inhibitor.

Table 2: *Summary of CO_2-hydrate formation on the droplet of trehalose solutions*

trehalose concentration	τ [min]	ΔT [K]	ΔT_e [K] ($\pm$ 0.9 K)
50 wt%	84 <42 $\pm$ 39>	5.8 <7.7 $\pm$ 3.2>	3.1
40 wt%	83 (139 $\pm$ 149)	12.9 (11.6 $\pm$ 2.6)	2.3
20 wt%	42 (101 $\pm$ 58)	14.0 (13.8 $\pm$ 0.2)	0.5
10 wt%	50 (51 $\pm$ 10)	13.8 (16.2 $\pm$ 2.5)	0.7
1 wt%	19	14.9	0
pure water[13]	~ 20	< 10	-

This table shows that τ becomes remarkably longer at trehalose concentrations higher than 10 wt%, and that ΔT is within an almost similar range at different concentrations. Therefore it is suggested that the inhibition effect of trehalose on the CO_2 hydrate nucleation is larger at a trehalose concentration higher than 20 wt% in the solution.

It is noted that τ in the same concentration scattered vary widely, and the difference of τ between the first run and those of all subsequent runs is included in the variation. The memory effect in the trehalose solutions was, therefore, not clearly observed in the present study. This may indicate that the nucleation process of CO_2 hydrate crystals in the trehalose solution is slightly complicated. Further systematic studies are required to clarify this problem.

Then the growth rate of CO_2 hydrate films on the trehalose-solution droplet $v_f^{(treha)}$ was examined in the same manner as in the previous section. Figure 5 shows $v_f^{(treha)}$ in each trehalose solution as a function of ΔT. Data from previous studies[13,14] are also included in the figure for purpose of comparison (dotted and solid lines for pure and 10 wt% NaCl solution, respectively). It can be seen that the $v_f^{(treha)}$ values were on the order of 10^{-0} mm sec^{-1} as ΔT varied from 4 to 18 K. $v_f^{(treha)}$ was found to be smaller than those on pure water and similar to those of the NaCl solution with similar concentration (approximately 10 wt%).[14] Therefore it was concluded that trehalose is one of the effective inhibitors of clathrate hydrates. At a similar ΔT condition of approximately 13 K, $v_f^{(treha)}$ tends to have been smaller at the higher trehalose-concentration solutions. This concentration dependence may suggest that trehalose works as a kinetic inhibitor. The large distributions of $v_f^{(treha)}$ at the same concentration observed at large ΔT conditions may have resulted from the variety of nucleation conditions. This kind of variation was also found in the experiments of pure water.[13]

However, the value of $v_f^{(treha)}$ at a trehalose concentration of 50 wt% is almost independent of ΔT. This is not the case for thermodynamic inhibitors such as NaCl,[14] but as shown in the present study, it is the case for kinetic inhibitors such as type-III AFP.

The result obtained by the transmission electron microscope observations of the freeze-fracture replica on trehalose solutions[15] suggested that the ice-crystal nucleation is enhanced in the solution with higher trehalose concentration. Based on these experimental results and assuming a similar nucleation process between ice and clathrate hydrates, we can speculate that trehalose, as a kinetic inhibitor, may not inhibit the nucleation but may work as an anti-agglomerant, keeping small hydrate particles dispersed as they form.

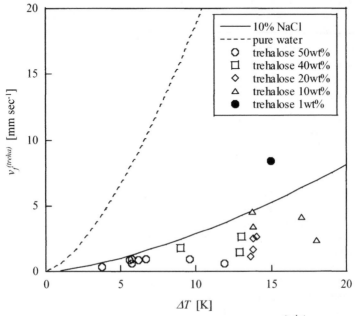

Figure 5   Lateral growth rate of CO_2 hydrate film $v_f^{(treha)}$ as a function of supercooling ΔT and for various trehalose concentrations. Dotted line and solid line indicate the previous data on pure water and 10 wt% NaCl solution, respectively.[14]

Another possible mechanism of trehalose molecules as a kinetic inhibitor is mentioned below. In the growth process of CO_2 hydrate, trehalose may work as the kinetic inhibitor that prevents the rate-determining process of the crystal formation at the reaction site which might have small dependence on ΔT. Trehalose has been observed to prevent ice-crystal growth with the reduction of the free-water providing because trehalose strongly hydrated in the solution.[11,15] This effect is found apparently when the trehalose concentration increases beyond the intrinsic hydration number of trehalose molecules. It is thus reasonable that the kinetic effect of trehalose on the inhibition of the hydrate formation would be resulted from the smaller supplement of free water from the solution of higher trehalose concentrations.

4 CONCLUSIONS

The inhibition effects of type-III AFP and trehalose, two cryoprotecting materials produced in animals, on type-I CO_2 clathrate-hydrates were examined. For comparison with the results of a previous study in which the lateral growth rates of CO_2-hydrate film were dependent on temperature, pressure and NaCl concentration, the solution droplet was observed in a high pressure vessel filled with CO_2. Type-III AFP was found to increase the induction period and to reduce the lateral growth rate of CO_2-hydrate films. It worked well at low concentrations, indicating that AFP works as a kinetic inhibitor. It was also indicated that AFP would weaken the memory effect of CO_2-hydrate formation. Trehalose had similar inhibition effects on both the induction period and the lateral growth rate, but it had little apparent concentration-dependence on them. Since trehalose also causes the equilibrium conditions of the CO_2 hydrate to shift to lower temperatures, it works not only as a thermodynamic inhibitor but also as a kinetic inhibitor, especially as an anti-agglomerant.

Acknowledgements

Part of this work was supported financially by the Industrial Technology Research Grant Program in '01 from New Energy and Industrial Technology Development Organization (NEDO) entitled "Studies on Energy Translation Technology Using Clathrate Compounds" (00B60016d) and by a Grant-in-Aid for Scientific Research from the Japan Society for the Promotion of Science (JSPS) entitled "Activity control of cells with structurizations of water" (#17340125). Trehalose was donated by Dr. H. Chaen and Dr. M. Kubota (Hayashibara Biochemical Labs., Inc.). This work was done in AIST with the support of the Methane Hydrate Research Laboratory.

References

1   E.D. Sloan, Jr., *Clathrate Hydrates of Natural Gases, 2nd ed.*, Marcel Dekker, Inc., New York, 1998.
2   J.P. Lederhous, J.P. Long, A. Sum, R.L. Christiansen, E.D. Sloan, Jr., *Chem. Eng. Sci.*, 1996, **51**, 1221.
3   E.D. Sloan, Jr., *Gas Res. Inst. Topical Rept.* GRI-91/0302, Gas Res. Inst., Chicago, 1992.
4   P.L. Davies, J. Baardsnes, M.J. Kuiper, V.K. Walker, *Philos. Trans. R. Soc. London B*, 2002, **357**, 927.
5   Y. Yeh, R.E. Feeney, *Chem. Rev.*, 1996, **96**, 601.
6   H. Zeng, L.D. Wilson, V.K. Walker, J.A. Ripmeester, *Proc. 4th Int. Conf. Gas Hydrates, Yokohama, May 19-23, 2002*, 2002, 526.
7   H. Zeng, L.D. Wilson, V.K. Walker, J.A. Ripmeester, *Can. J. Phys.*, 2003, **81**, 17.
8   H. Zeng, L.D. Wilson, V.K. Walker, J.A. Ripmeester, *J. Am. Chem. Soc.*, 2006, **128**, 2844.
9   H. Zeng, A. Brown, B. Wathern, J.A. Ripmeester, V.K. Walker, *Proc. 5th Int. Conf. Gas Hydrates, June 12-16, 2005, Trondheim, Norway*, 2005, **1**, 1.
10   T. Sei, T. Gonda, Y. Arima, *J. Crystal Growth*, 2002, **240**, 218.
11   T. Sei, T. Gonda, *Cryobiol. Cryotechnol.*, 2004, **50**, 93.
12   K. Miura, S. Ohgiya, T. Hoshino, N. Nemoto, T. Suetake, A. Miura, L. Spyracopoulos, H. Kondoh, S. Tsuda, *J. Biolog. Chem.*, 2001, **276**, 1304.
13   T. Uchida, T. Ebinuma, J. Kawabata, H. Narita, *J. Crystal Growth*, 1999, **204**, 348.

14   T. Uchida, I.Y. Ikeda, S. Takeya, T. Ebinuma, J. Nagao, H. Narita, *J. Crystal Growth*,
     2002, **237-239**, 383.
15   T. Uchida, M. Nagayama, T. Shibayama, K. Gohara, submitted to *J. Crystal Growth*.

RIPPLE FORMATION MECHANISM ON ICICLES UNDER A THIN SHEAR FLOW

K. Ueno[1] and N. Maeno[2]

[1] Graduate School of Engineering, Nagoya University, Nagoya 464-8603, Japan.
E-mail: ueno@fcs.coe.nagoya-u.ac.jp
[2] Professor Emeritus of Hokkaido University, Hanakawa Minami 7-2-133,
Ishikari 061-3207, Japan. E-mail: maenony@ybb.ne.jp

1 INTRODUCTION

When icicle grows ring-like ripples appear on its surface as seen in **Figure 1** (a). **Figure 1** (b) is a vertical cross section of an icicle with ripples. Many tiny air bubbles are trapped in just upstream region of any protruded part of the icicle, and line up in the upward direction during icicle growth.[1] This observation shows that an initial flat solid-liquid (ice-water) interface not only becomes unstable to form ripples but also moves upwards.

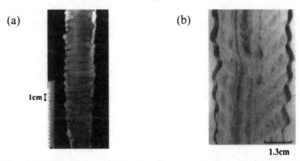

(a)   (b)

1cm

1.3cm

Figure 1   *(a) Ripples on an icicle. (b) Vertical cross section of an icicle with ripples.*

A similar ripple pattern of ice can be experimentally produced within a gutter of span width $l = 3$ (cm) on an inclined plane.[2] This apparatus was set in a cold room of $-8 \pm 1°$ C and water of $0°$ C was continuously supplied from the top of the plane at the rate $Q = 160$ (ml/hr). The average wavelength λ_{mean} of the ripples of ice was measured for various angles θ of the inclined plane and the relationship $\lambda_{mean} \sim 0.86/(\sin\theta)^{0.76}$ (cm) was obtained. The mean thickness h_0 of the liquid film can be expressed as the Nusselt equation $h_0 = [2\nu/(g\sin\theta)]^{1/3}[3Q/(2l)]^{1/3}$,[3] where g is the gravitational acceleration, ν is the kinematic viscosity of the liquid. The value of h_0 can be changed by controlling Q/l or θ. The experimental result shows that λ_{mean} increases with the decrease in the angle. Since h_0 also increases with the decrease in the angle for a fixed value of Q/l, this indicates that h_0 is one of the characteristic length in determining the wavelength of ripples.

Theoretical investigations have never been made on the morphological instability of solidification front in crystal growth from flowing liquid film with one side being a free surface. Since the thickness of the liquid film is very thin, we must take into account the shape of the liquid-air surface when the solid-liquid interface is disturbed. We need to determine heat flux in each phase under the hydrodynamic and thermodynamic boundary conditions imposed at each interface. The first theoretical attempt was made by Ogawa and Furukawa. [4] They determined the wavelength of ripples quantitatively and also predicted that the ripple pattern moves in the downstream direction. However, their theoretically predicted dependence $\lambda_{max} \sim 0.47/(\sin\theta)^{1/3}$, was largely different from the experimental result. Furthermore, the direction of migration of ripples was opposite to the observation. One of the present authors (K.U.) recently proposed a new mechanism of destabilization and/or stabilization and migration of the interface.[5, 6] His result, $\lambda_{max} \sim 0.97/(\sin\theta)^{0.64}$, was fairly in good agreement with the experimental result and ripples were shown to move to the upstream direction as observed in **Figure 1** (b).

The above study is not complete in that the comparision of theoretical and experimental results was made only at a single value of Q/l $(=53.3$ [(ml/hr)/cm]). However, for natural icicles Q/l does not necessarily take a definite value since the values of Q , radii R, and $l = 2\pi R$ vary depending on the surrounding environmental conditions. In the present paper, we present the observation result showing the relation between the wavelength of ripples and the diameter, and the computational result of the wavelength of ripples for various values of Q/l. More detailed comparison of previous models is also made in particular reference to the difference in boundary conditions.

2   THEORETICAL FRAMEWORK

We show here only the theoretical framework developed in the previous paper.[5] Our analysis is restricted in two dimensions of the vertical plane (x, y) of the inclined plane, and for simplicity we assume that the region of the crystal is semi-infinite. The x axis is parallel to the inclined plane and the y axis is normal to it . We choose a reference frame moving in the y direction with a mean growth rate $\overline{V}$ with respect to a fixed laboratory frame of reference. The origin of y axis is the solid-liquid interface. An experimental observation shows that the decrease in the temperature and the increase in the wind speed in the air lead to the increase in $\overline{V}$. The typical values of $\overline{V}$ measured are on the order of 10^{-6} (m/s).[1] Here we assume that $\overline{V}$ takes a constant value and there is no wind in the air.

2.1 Governing equations

The equations for the temperature in the liquid T_l, the solid T_s and the air T_a are given by $\partial T_l/\partial t + u\partial T_l/\partial x + v\partial T_l/\partial y = \kappa_l(\partial^2 T_l/\partial x^2 + \partial^2 T_l/\partial y^2)$, $\partial T_i/\partial t - \overline{V}\partial T/\partial y = \kappa_i(\partial^2 T_i/\partial x^2 + \partial^2 T_i/\partial y^2)$ $(i = s, a)$, where t is time, κ_l, κ_s and κ_a are the thermal diffusivities of the liquid, solid and air, respectively. u and v are the velocity components in the x and y directions and obey the Navier-Stokes equations,

$$\partial u / \partial t + u \partial u / \partial x + v \partial u / \partial y = -1 / \rho_l \partial p / \partial x + v(\partial^2 u / \partial x^2 + \partial^2 u / \partial y^2) + g \sin\theta, \quad (1)$$

$$\partial v / \partial t + u \partial v / \partial x + v \partial v / \partial y = -1 / \rho_l \partial p / \partial y + v(\partial^2 v / \partial x^2 + \partial^2 v / \partial y^2) - g \cos\theta \quad (2)$$

and the equation of continuity $\partial u / \partial x + \partial v / \partial y = 0$, where p is the pressure and ρ_l is the liquid density. Using the stream function ψ, u and v can be derived from $u = u_0 \partial \psi / \partial y$ and $v = -u_0 \partial \psi / \partial x$, where u_0 is the surface fluid velocity.

2.2 Boundary conditions

Both velocity components at the solid-liquid interface $y = \zeta(t,x)$ must satisfy $u|_{y=\zeta} = 0$ and $v|_{y=\zeta} = 0$. The kinematic condition at the liquid-air surface $y = \xi(t,x)$ is $\partial \xi / \partial t + u|_{y=\xi} \partial \xi / \partial x = v|_{y=\xi}$. At the liquid-air surface the shear stress must vanish: $\partial u / \partial y|_{y=\xi} + \partial v / \partial x|_{y=\xi} = 0$ and the normal stress including the stress induced by the surface tension γ of the liquid-air surface must balance the pressure P_0 of the atmosphere: $- p|_{y=\xi} + 2\rho_l v \partial v / \partial y|_{y=\xi} - \gamma \partial^2 \xi / \partial x^2 = -P_0$. We refer to these as the hydrodynamic boundary conditions.

The continuity of the temperature at the solid-liquid interface is $T_l|_{y=\zeta} = T_s|_{y=\zeta}$. The continuity of heat flow across the solid-liquid interface is $L(\overline{V} + \partial \zeta / \partial t) = K_s \partial T_s / \partial y|_{y=\zeta} - K_l \partial T_l / \partial y|_{y=\zeta}$, where L is the latent heat of solidification per unit volume, and K_l and K_s are the thermal conductivities of the liquid and solid, respectively. The continuity of the temperature at the liquid-air surface is $T_l|_{y=\xi} = T_a|_{y=\xi} = T_{la}$, where T_{la} is a temperature at the liquid-air surface. The continuity of heat flow across the liquid-air surface is $- K_l \partial T_l / \partial y|_{y=\xi} = -K_a \partial T_a / \partial y|_{y=\xi}$, where K_a is the thermal conductivity of the air. We refer to these as the thermodynamic boundary conditions.

2.3 Linear stability analysis

In the following analysis, we use the same values of physical properties as those used in the previous paper.[5] As non-dimensional numbers, we define the Reynolds number, $Re \equiv u_0 h_0 / v = [3/(2v)](Q/l)$; the Peclet number, $Pe \equiv u_0 h_0 / \kappa_l = [3/(2\kappa_l)](Q/l)$; the Froude number, $F \equiv u_0 /(gh_0)^{1/2}$ and the Weber number, $W = \gamma /(\rho_l h_0 u_0^2)$. For $Q = 160$ (ml/hr), $l = 3$ (cm) and $\theta = \pi/2$, these non-dimensional numbers take values of $Re \sim 1.23$, $Pe \sim 17.1$, and $W \sim 1.42 \times 10^3$.

We separate ξ, T_l, T_s, T_a, ψ and p into unperturbed steady fields and perturbed ones. As seen in **Figure 1** (a), the ring-like structure encircles the icicle and there is no azimuthal variation on the surface of the icicle. Therefore, it is sufficient to consider only a one dimensional perturbation of the solid-liquid interface in the x direction, i.e., we assume the form $\zeta(t,x) = \zeta_k \exp[\sigma t + ikx]$, where k is the wavenumber, and $\sigma = \sigma_r + i\sigma_i$, σ_r being the amplification rate and $v_p \equiv -\sigma_i / k$ being the phase velocity of the perturbation, and ζ_k is a small amplitude of the interface. We suppose that the amplitudes of the corresponding perturbations of the liquid-air surface, the temperature of the liquid, solid and air, the stream function and pressure are to be of the order of ζ_k.

The following calculations are based on a linear stability analysis taking into account only the first order of ζ_k. Furthermore, we use two approximations. One is the long wavelength approximation: [7] the wavelength of perturbation of the solid-liquid interface is much larger than the mean thickness h_0 of the liquid film, then we can define a small dimensionless wavenumber $\mu = kh_0$. The other is the quasistationary approximation: we can neglect the time dependence of perturbed temperature in each phase and that of perturbed fluid flow in the thin liquid film because these fields respond relatively rapidly to the slow development of the perturbation of solid-liquid interface.

3   RIPPLE FORMATION AND MIGRATION

Under the long wavelength and quasistationary approximations and with the use of the linearized forms of the hydrodynamic and thermodynamic boundary conditions, first, we solve the Orr-Sommerfeld equation for the amplitude of perturbed part of the stream function from the Navier-Stokes equations. Second, we solve the equation for the amplitude of perturbed part of the temperature in the liquid film. The dispersion relation for the fluctuation of the solid-liquid interface is determined by the use of these solutions. From the real and imaginary part of this dispersion relation, we obtain the amplification rate σ_r and the phase velocity $v_p = -\sigma_i / k$ as follows: [6]

$$\sigma_r = \overline{V} / h_0 \left[-3\alpha\mu Pe / 2 + \mu(36 - 3\alpha\mu Pe / 2) + n\mu\{-7\alpha\mu Pe / 10 - \alpha^2 + \mu(36 - 7\alpha\mu Pe / 10)\} \right] / (36 + \alpha^2), (3)$$

$$v_p = -\overline{V} / \mu \left[-\alpha^2\mu Pe / 4 + \mu(6\alpha + 9\mu Pe) + n\mu\{6\alpha - 7\alpha^2\mu Pe / 60 + \mu(6\alpha + 21\mu Pe / 5)\} \right] / (36 + \alpha^2), (4)$$

where $\alpha = \mu Re\cos\theta / F^2 + \mu^3 ReW = 2\cot\theta h_0 k + (2/\sin\theta)a^2 h_0 k^3$ represents the restoring forces due to gravity and surface tension, [7] $a = [\gamma /(\rho_l g)]^{1/2}$ is the capillary length [3] and $n = K_s / K_l = 3.92$. A typical value of the capillary length a of water is about 2.8 (mm). Since the value of W is very large as estimated above, the μ^3 term in α can not be neglected.

Using the solution to the Orr-Sommerfeld equation, we can derive the relation between the amplitude ξ_k of the liquid-air surface and the amplitude ζ_k of the solid-liquid interface: $\xi_k = 6/(6 - i\alpha)\zeta_k$. Since the value of α is small in the lower wavenumber regions, the effect of restoring forces on the liquid-air surface can be neglected, then, the amplitude of the liquid-air surface is almost the same as that of the solid-liquid interface. On the other hand, since the value of α is large in the higher wavenumber regions, the effect of restoring forces on the liquid-air surface becomes effective, then the shape of the liquid-air surface tends to be flat. This indicates that the shape of the liquid-air surface can be changed depending on the wavelength of the solid-liquid interface. It is found that the capillary length a is an important characteristic length to determine the shape of the liquid-air surface.

The amplification rate σ_r of eq. (3) and the phase velocity v_p of eq. (4) versus the wavenumber k, are shown by the solid lines in FIG. 4 and FIG. 5 in ref. 6, respectively. It should be stressed that although the change in $\overline{V}$ due to the environmental conditions affects the magnitude of σ_r, the wavenumber at which σ_r takes a maximum value does not change. From this most unstable mode we define the wavelength λ_{max} of ripples . If we estimate the value of v_p at the most unstable mode, the value is about $-0.6\,\overline{V}$. The negative sign indicates that the solid-liquid interface moves to the upstream direction, which is consistent with the observation in **Figure 1** (b).

The Ogawa-Furukawa model [4] is different from our model in the following points. One is the neglect of the effect of restoring forces α, then, the shape of the liquid-air surface have the same amplitude as that of the solid-liquid interface. The other is the difference in the thermodynamic boundary conditions, i.e., $T_l\,|_{y=\zeta}=T_s\,|_{y=\zeta}=T_m$ and $T_l\,|_{y=\xi}=T_a\,|_{y=\xi}$ are imposed at the solid-liquid interface and the liquid-air surface, respectively, where T_m is the equilibrium melting temperature. As a result, they obtained [4] $\sigma_r=\overline{V}k[(1-239(\mu Pe)^2/10080)/\{(1-239(\mu Pe)^2/10080)^2+(5\mu Pe/12)^2\}]$ and $v_p=\overline{V}[5\mu Pe/12/\{(1-239(\mu Pe)^2/10080)^2+(5\mu Pe/12)^2\}]$, which are shown by the dotted lines in FIG. 4 and FIG. 5 in ref. 6, respectively. Although there exists a wavenumber that σ_r takes a maximum value, the dependence of λ_{max} on $\sin\theta$ was largely deviated from the experimental results as mentioned in the introduction. Furthermore, the sign of v_p is opposite to that of our result. The dispersion relation derived from our model contains two characteristic lengths, h_0 and a. On the other hand, the dispersion relation derived from the Ogawa-Furukawa model contains only h_0. This difference in both models leads to the different dependence of λ_{max} on $\sin\theta$.

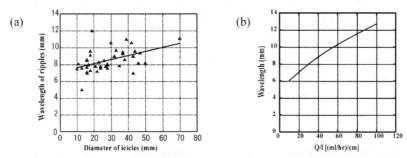

Figure 2 *(a) Observed wavelength of ripples vs diameter for natural icicles. (b) Calculated wavelength of ripples vs* Q/l *[(ml/hr)/cm].*

Figure 2 (a) shows the relation between the wavelength of ripples and the diameter. The data were measured on photographs which were taken in many winters at various sites in Hokkaido in Japan. The growth conditions such as air temperature, water supply rate, wind speed, etc. are all different and the number of data (41) is not enough to deduce any definite conclusions, but a weak trend of increase in the wavelength with the increase in diameter can be noticed. Majorities of wavelength are within 7~11 (mm) and the mean is 8.5 (mm). **Figure 2** (b) shows the wavelength calculated from our σ_r by varying the value of Q/l at $\theta=\pi/2$. Since the typical values of Q over icicles are on the order of tens of ml/hr and icicle radii are usually in the range of 1~10cm, [1] we have varied the value of Q/l over the range of $10\sim100$ [(ml/hr)/cm]. The wavelength increases slowly with the increase in Q/l.

4   MECHANISM OF MIGRATION AND STABILIZATION

In **Figure 3**, the thick solid lines between ice and water and that between water and air represent the solid-liquid interface $\mathrm{Im}[\zeta/h_0]$ and the liquid-air surface $1+\mathrm{Im}[\xi/h_0]$, respectively. Here Im denotes the imaginary part of its argument. The thick dashed lines are distribution of heat flux q_l-q_s at $\mathrm{Im}[\zeta/h_0]$ and that of heat flux q_a at $1+\mathrm{Im}[\xi/h_0]$, respectively, where q_l , q_s and q_a are defined as $q_l \equiv \mathrm{Im}[-K_l\partial T_l'/\partial y\,|_{y=\zeta}]$, $q_s \equiv \mathrm{Im}[-K_s\partial T_s'/\partial y\,|_{y=\zeta}]$ and $q_a \equiv \mathrm{Im}[-K_a\partial T_a'/\partial y\,|_{y=\xi}]$.

We can express $q_l-q_s = L\,|\sigma|\sin[k(x-\mathrm{v}_pt)-\phi]\exp(\sigma_rt)\zeta_k$, where $|\sigma|=\sqrt{\sigma_r^2+\sigma_i^2}$, $\sigma_r=|\sigma|\cos\phi$, $\sigma_i=-|\sigma|\sin\phi$, and ϕ is the phase difference between q_l-q_s and $\mathrm{Im}[\zeta/h_0]=\sin[k(x-\mathrm{v}_pt)]\exp(\sigma_rt)\zeta_k/h_0$. The fine dotted lines represent the isotherm.

First, we consider the configurations in our model as shown in **Figure 3** (a) and (b). In this case, the thick dashed lines are shifted to the upward direction against the thick solid lines, that is, $\phi<0$. Then $\sigma_i>0$ and $\mathrm{v}_p=-\sigma_i/k<0$. The negative sign of v_p is consistent with that of the solid line in FIG. 5 in ref. 6. In the configuration of $-\pi/2<\phi<0$ as shown in **Figure 3** (a), $\sigma_r>0$ and the solid-liquid interface is unstable. In unperturbed state, we assume that $T_s=0°$ C in the ice and the temperature at the water-air surface is T_{la}. Its value is $-0.06°$ C from the heat continuity equation when $\overline{V}=10^{-6}$ (m/s). After the interface is perturbed, the temperature over the water-air surface remains at the constant $T_{la}=-0.06°$ C, while the temperature of the ice-water interface is not necessary to be kept at $T_m=0°$ C. In fact, we can see extremely small deviation of temperature from $T_m=0°$ C in **Figure 3** (a) and (b), and the deviation is much larger than the change in T_m due to the curvature effect of the solid-liquid interface. Consequently, there is heat flow q_s from the solid-liquid interface to the ice in the regions of $T_s>T_m$, while there is heat flow q_s from the ice to the interface in the regions of $T_s<T_m$. This deviation effect of temperature from $T_m=0°$ C is reflected in the second terms in eqs. (3) and (4).

The heat flux q_a is large at any protruded part of liquid-air surface, where the heat transfer by thermal diffusion into the air is very large because the temperature gradient is large. Since the value of α is small at the wavenumber in **Figure 3** (a), the amplitudes of both interfaces are almost the same and phase difference between them can not be seen. Therefore, this seems to result in faster cooling and hence to promote freezing at any protruded part of the solid-liquid interface. However, the actual heat flux q_l-q_s is shifted to the upward direction by ϕ against the solid-liquid interface. This indicates that the part of largest temperature gradient is not necessary at the protruded part of the solid-liquid interface.

The mean growth rate of solidification is given by $\overline{V}=K_l\overline{G_l}/L$, where $\overline{G_l}$ is unperturbed temperature gradient in the liquid film. In the regions of $q_l - q_s > 0$, where the total heat flux is larger than the mean heat flux $K_l \overline{G_l}$, the growth rate of the solid-liquid interface is larger than $\overline{V}$ and then the interface tends to grow much faster. On the other hand, in the regions of $q_l - q_s < 0$, where the total heat flux is smaller than $K_l \overline{G_l}$, the growth rate of the solid-liquid interface is smaller than $\overline{V}$ and then the interface tends to melt. The phase shift of heat flux to the upward direction leads to growth in just the upstream regions of any protruded part and to melting in just the downstream regions of any protruded part at the reference frame of moving with $\overline{V}$. Consequently, the solid-liquid interface not only becomes to be unstable but also moves in the upward direction, which is consistent with the observation in **Figure 1** (b). In the configuration of $-\pi < \phi < -\pi/2$ as shown in **Figure 3** (b), $\sigma_r < 0$ and the solid-liquid interface is stable. In this configuration, $q_l - q_s < 0$ at any protruded part of the solid-liquid interface and then the interface tends to melt, while $q_l - q_s > 0$ at the depression part of it and then the interface tends to grow. Consequently, the interface will become to be flat and ripples disappear. This is a new mechanism of stabilization of the solid-liquid interface.

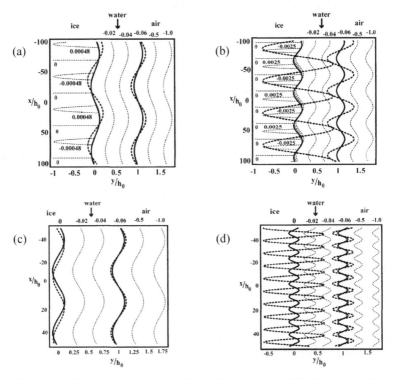

Figure 3 *Schematic illustrations of the solid-liquid interface $Im[\zeta/h_0]$ and the liquid-air surface $1 + Im[\xi/h_0]$ (thick solid lines), and the distributions of net heat flux $q_l - q_s$ at $Im[\zeta/h_0]$ and heat flux q_a at $1 + Im[\xi/h_0]$ (thick dashed lines). The present model: (a) $k = 628\ m^{-1}$, (b) $k = 1256\ m^{-1}$. The Ogawa-Furukawa model: (c) $k = 1256\ m^{-1}$, (d) $k = 6280\ m^{-1}$.*

Finally we consider the configurations in the Ogawa-Furukawa model as shown in **Figure 3** (c) and (d). Since only obscure interpretations were given in the previous paper, [4] we give quantitative interpretations of the destabilization and/or stabilization of the solid-liquid interface. In **Figure 3** (c) and (d), the thick dashed lines are shifted to the downward direction against the thick solid lines, that is, $\phi > 0$. Then $\sigma_i < 0$ and $v_p = -\sigma_i / k > 0$. The positive sign of v_p is not consistent with the observation in **Figure 1** (b). Likewise in **Figure 3** (a) and (b), we put $T_s = 0°C$ and $T_{la} = -0.06°C$ in the unperturbed state. In the Ogawa-Furukawa model, the temperature is not constant over the water-air surface when the surface is perturbed. As a result, the temperature at any protruded part of liquid-air surface is below $-0.06°$ C, while the temperature at any depression part of surface is above $-0.06°$ C, but the change in temperature is very small and we can not see it clearly in **Figure 3** (c) and (d). On the other hand, the temperature of the ice-water interface remains at the constant $T_m = 0°$ C after the interface is perturbed. Therefore, the heat flow q_s is absent. In the configuration of $0 < \phi < \pi/2$ as shown in **Figure 3** (c), $\sigma_r > 0$ and the solid-liquid interface is unstable, while in the configuration of $\pi/2 < \phi < \pi$ as shown in **Figure 3** (d), $\sigma_r < 0$ and the interface is stable.

5   CONCLUSION

The observation of natural icicles showed the weak increase of ripple wavelength with icicle diameter.   The wavelength also increased slowly with increasing water supply rate per width, which was calculated from the present theoretical model that takes account of the effect of the restoring forces due to gravity and surface tension on the liquid-air surface in the icicle growth under a thin shear flow. The two models for the destabilization and/or stabilization of the solid-liquid interface under a thin shear flow were compared and discussed in some detail. It was concluded that different boundary conditions in the two models lead to the completely different final results.

Acknowledgements

This work was supported by a Grant-in-Aid for the 21st Century COE "Frontiers of Computational Science".

References

1   N. Maeno, L. Makkonen, K. Nishimura, K. Kosugi and T. Takahashi,   *J. Glaciol.*, 1994, **40**, 319.
2   S. Matusda, Master's Thesis, Institute of Low Temperature Science, Hokkaido University, 1997.
3   L. D. Landau and E. M. Lifschitz, *Fluid Mechanics*, Pergamon Press, London, 1959.
4   N. Ogawa and Y. Furukawa, *Phys. Rev. E*, 2002, **66**, 041202.
5   K. Ueno, *Phys. Rev. E*, 2003, **68**, 021603.
6   K. Ueno, *Phys. Rev. E*, 2004, **69**, 051604.
7   T. B. Benjamin, *J. Fluid Mech.*, 1957, **2**, 554.

MOLECULAR SIMULATIONS OF WATER FREEZING: BRINE REJECTION AND HOMOGENEOUS NUCLEATION

L. Vrbka and P. Jungwirth

Institute of Organic Chemistry and Biochemistry, Academy of Sciences of the Czech Republic, and Centre for Biomolecules and Complex Molecular Systems, Flemingovo nam. 2, 16610 Prague 6, Czech Republic. E-mail: lubos.vrbka@uochb.cas.cz

1 INTRODUCTION

Water is probably the most important and the most intensely studied substance on Earth. It is the solvent of life and it is also of vital importance in many aspects of our existence, ranging from cloud microphysics to its key role as a solvent in many chemical reactions. The familiar process of water freezing is encountered in many natural and technologically relevant processes. In this contribution, we discuss the applicability of the methods of computational chemistry for the theoretical study of two important phenomena. Namely, we apply the molecular dynamics (MD) simulations to the study of brine rejection from freezing salt solutions and the study of homogeneous nucleation of supercooled water.

Sodium chloride and other common inorganic salts are very poorly soluble in ice, with solubilities in the micromolar range at best.[1,2] Liquid water, however, is capable of dissolving molar amounts of these salts. This property is clearly demonstrated by the fact that the salinity of the sea ice is much smaller than that of seawater. When the aqueous NaCl solution freezes above its eutectic temperature (-21.1 °C)[3] it solidifies as almost pure ice. The salt is rejected into the surrounding unfrozen solution. Salt concentration and temperature gradients due to latent heat release are established across the freezing front as freezing proceeds. This results in the formation of macroscopic instabilities and the initially planar freezing front becomes corrugated.[4-6]

When freezing occurs on the surface of the ocean, the rejected salt increases the density of underlying water masses. This leads to massive circulations in the ocean influencing the global climate.[7,8] Brine rejection is also proposed to play important role in the thundercloud electrification.[9] Supercooled cloud water droplets containing salt particles originating from the soluble cloud condensation nuclei can freeze upon impact. Dissolved salt is then rejected from the freezing water and concentrates on the surface of the graupel. The electrification then occurs via the collisions between neat ice crystals and salt-covered graupels.

Brine rejection is thus an extremely interesting and also important process with wide natural, atmospheric, and technological consequences (with desalination as a prominent example). It has, therefore, been subject to experimental[10] as well as theoretical research.[11] In most cases, water in nature freezes heterogeneously, i.e., in contact with crystallization nuclei such as small pieces of ice or minerals. When such contact is avoided, water can be supercooled to very low temperatures (up to −30 °C) before homogeneous freezing takes place.[12] Droplets of supercooled water are known to be formed in the stratosphere and upper troposphere.[13,14] Homogeneous nucleation is the controlling mechanism for the formation of cirrus and polar stratospheric clouds, affecting directly the radiative balance of the Earth.

There exists an ongoing scientific debate whether homogeneous nucleation starts at the surface or rather in the bulk of the water droplet. The experimental evidence in this case is ambiguous. It has been concluded recently that, based on existing thermodynamic nucleation models, as well as laboratory and atmospheric data, ice nucleation at the surface can be neither confirmed nor disregarded.[15] Recent laboratory experiments indicate that homogeneous ice freezing is a volume-proportional process with surface nucleation being potentially important for small droplets with radii below 20 μm.[16] On the contrary, higher freezing temperatures and faster freezing rates were observed for droplets with an ice-forming nucleus placed close to the surface compared to a bulk location.[17] It has been suggested recently that molecular simulations could provide further information concerning the surface-initiated homogeneous nucleation.[18]

Thanks to advancement of the computer technology computer simulations of processes related to water freezing are becoming feasible. The greatest advantage of the calculations is that they can provide insight to the structure and dynamics of the system at an atomic level, with resolution often inaccessible to experimental techniques.

Crystallization usually involves very long time scales (at least when compared to time scales of routine calculations) and complicated potential energy landscapes. Computer simulations of this process are, therefore, considered to be difficult in general. In a series of papers, Haymet and coworkers investigated the structure and dynamics of the ice/water interface.[19-21] In their approach, the pre-built patches of water and ice were put together to create the interface. The necessity to simulate the highly improbable creation of the crystallization nucleus was thus avoided. Similar setup was used by other groups to assess various properties of the ice/water interface.[22-26]

These simulations can be easily extended to systems containing solutes, namely simple salts. However, most of the reported simulation studies concerned only the behavior of the solute at the stable ice/water interface.[27,28] Only recently results of successful simulations of the brine rejection process involving a moving ice/solution boundary have been reported.[11,29,30]

Water freezing was observed in simulations of systems subjected to an electric field,[31] in confined water,[32] and in (non-dynamic) Monte Carlo calculations.[33] However, there are to the best of our knowledge only two successful MD simulations of water freezing "from scratch", i.e., without any bias introduced by initial conditions (existing crystallization nucleus or external electric potential).[34,35]

A key point in the simulations is the choice of the interaction potential. There exist many different water models optimized for different purposes.[36] Each model shows better or worse agreement with particular water properties. These models are mostly fitted to describe liquid water. Therefore, their use for the simulations of ice can be tricky. Namely, one of the properties that is often described incorrectly is the melting temperature. A comprehensive comparison of the most common water models with respect to the melting temperature of water has been published recently.[37] Values in the wide range of 190 – 270 K were obtained. It is, therefore, always necessary to choose between the computer efficiency of the model and the quality of the description of water properties, although a more complicated model does not always mean better description.

We applied the MD technique to the simulation of brine rejection from freezing salt solutions employing the rigid 3-site simple point charge (SPC/E) water model[38,39] to obtain the microscopic picture of this very important natural process. Further, we used a recently proposed 6-center water potential[40] (abbreviated NE6) for simulation of the water freezing "from scratch" performed in a slab geometry. The ultimate goal was to provide a computational view on the surface-induced homogeneous nucleation.

2 COMPUTATIONAL METHODOLOGY

2.1 Brine rejection

Our calculations can be viewed as an extension of the work of Bryk and Haymet.[27] They studied the behavior of the Na^+ and Cl^- ions at the static ice/water interface on the nanosecond timescale. We were interested in the solidification process itself and in the expulsion of the ions into the remaining liquid. This required using longer simulation times (hundreds of nanoseconds). For the purpose of our research, we used the SPC/E water potential.[38] Potential parameters for sodium and chloride ions were taken as the non-polarizable set from Ref. 41.

For the construction of the simulation cells we employed the approach of Hayward and Haymet.[20] A brief outline of the procedure is given below while a more detailed description can be found in our recent papers.[11,29] Pre-equilibrated cells of liquid water and ice were combined together to create systems with alternating solid/liquid phases. Salt ions were then introduced into the liquid. After the application of the periodic boundary conditions, infinite slabs in the xy-plane of ice next to a liquid salt solution were formed. We then let these cells to evolve at different temperatures and observed the time needed to freeze the remaining liquid part of the sample. We also monitored the positions of the ions.

After the equilibration of the simulation cells, the production runs were performed at the target temperatures with the timestep of 1 fs. Long-range electrostatic interactions were treated using the smooth particle mesh Ewald method (SPME). Calculations were run in the NPT ensemble, with anisotropic pressure coupling and target pressure of 1 atm, using Berendsen barostat. The total length of the simulations ranged from several hundreds of nanoseconds to a microsecond.

2.2 Homogeneous nucleation

We used the newly proposed six-center water potential (NE6) for the simulations.[40] A simulation cell with approximate dimensions 13.5 x 15.5 x 30.0 $Å^3$ containing 192 water molecules was elongated in the z-direction to 100 Å. Replication of the original unit cell twice and three times in the z-direction and elongation to 180 and 270 Å provided the simulation cells with 384 and 576 water molecules, respectively. Subsequent application of periodic boundary conditions yielded infinite slabs in the xy-plane, each possessing two air/water interfaces perpendicular to the z-axis.

The production runs followed the system preparation and adjustment of the target temperature. Newtonian equations of motion were solved with a time step of 1 fs, with the total length of simulations in the range of 100 – 500 ns. The calculations were run in the NVT ensemble, with long-range electrostatics treated again using the SPME method. The melting temperature of the NE6 water model is estimated to be close to 0 °C. The simulations of the smallest cell peformed at different temperatures indicated that supercooling the sample to 250 K provided the fastest freezing rates. Therefore, simulations of the larger systems were performed at this temperature as well.

3 RESULTS AND DISCUSSION

3.1 Brine rejection

We have simulated freezing of a salt solution in contact with a patch of cubic ice Ic. The properties of this ice form (density, heat capacity, etc.) are very similar to those of the most common hexagonal ice Ih. Ice Ic is actually a metastable form of the ice Ih. There exist several important processes where cubic rather than hexagonal ice is formed. This type of ice is present in the upper atmosphere and can play important role in cloud formation.[42-44] It was also reported that freezing in nanopores leads preferentially to cubic ice.[45] Very recently, cubic ice was reported to be the dominant ice crystal modification during the freezing of water droplets (with radius up to 15 nm) and thin water films (up to 10 nm thick).[46] These dimensions are comparable to the initial thickness of the ice patch in our simulations.

We performed series of simulations for a range of temperatures around the melting point of the SPC/E water model (215 K, i.e., significantly below the experimental value).[37] Despite this deficiency, this very simple model provides very reasonable description of other water and ice properties.[21] Below, we introduce a relative temperature scale with the melting point assigned a value of 0°.

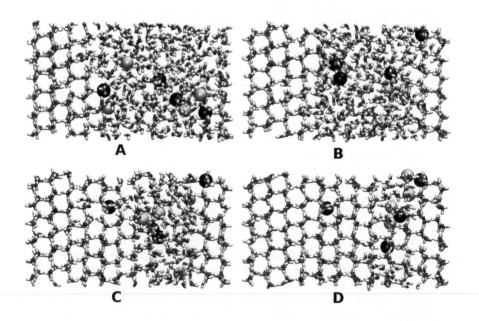

Figure 1   *Snapshots from the MD simulation of the brine rejection with 4 NaCl pairs. Na^+ and Cl^- are given as light and dark spheres, respectively. Snapshots correspond to A) 200 ns, B) 400 ns, C) 600 ns, D) 815 ns simulation time.*

First, it was necessary to establish a robust simulation protocol for reproducible freezing of neat water in contact with a patch of ice. The results can be summarized as follows. In previous studies of the ice/water coexistence rather short simulation times were used (up to 2 ns).[20,21,46] Our simulations show, however, that hundreds of nanoseconds are needed to freeze the nanometer-scale unit cell of SPC/E water and even melting close to 0° takes several to tens of nanoseconds.[11,29] Depending on the size of the system melting was observed in the temperature range of -10 - 0°.

Then simulations of salt solutions in contact with ice were performed. In general, melting temperature decreased and freezing was observed at -15°. Also, the time needed to freeze the sample grew dramatically with increasing salt concentration. These are direct demonstrations of the kinetic and thermodynamic antifreeze effects of the added salt at the molecular level.

Representative snapshots from brine rejection simulations are given on Figure 1. It can be clearly seen that both sodium and chloride are rejected by the freezing front into the remaining liquid solution (save a single chloride which got trapped in the ice lattice). At the end of the simulation, a thin layer of unfrozen concentrated brine solution with glassy character is formed. There is a slow tendency to incorporate more of the remaining water molecules into the ice lattice; however this process exceeds the time scale of present simulations.

A more detailed analysis revealed that the freezing proceeds by the following mechanism. First, there is a fluctuation (decrease) of salt density near the freezing front followed by the buildup of the new ice layer.

3.2 Homogeneous nucleation

We have also simulated homogeneous freezing of slabs of neat water. The main outcome of the simulations is shown in Figure 2. It shows snapshots from the MD simulation of the medium-sized simulation cell. Only water molecules in the central unit cell are displayed. When periodic boundary conditions are employed, an extended slab is formed. In the figure, it is easy to locate the interfacial regions (top and bottom of the picture) and the bulk region located between the two interfaces.

The first crystallization nucleus forms in the subsurface approximately at 45 ns. The second subsurface nucleus occurs 110 ns after the beginning of the simulation. Freezing proceeds independently from both nuclei and the freezing fronts meet at 160 ns, forming a somewhat disordered contact zone. Cubic ice Ic forms predominantly in the simulations and many defects can be identified in the ice lattice. Freezing to cubic ice is consistent with the recent findings that small droplets and thin water films prefer this crystal modification over hexagonal ice Ih.[47]

Similar results were also obtained for other simulations of systems of varying sizes. In all cases the initial crystallization nucleus forms in the subsurface and freezing then proceeds towards the bulk.

In one case we observed concurrent formation of two ice nuclei; one in the subsurface and one in the bulk. The freezing then proceeded from both nuclei. Figures displaying snapshots from these simulations and further analysis can be found elsewhere.[48]

The preferred nucleation in the subsurface can be due to several factors. First, the volume change upon the phase transition (change in the molar volume upon freezing is positive for water) will direct the process to the region of lower density, i.e., towards the interface. However, the water molecules at the very interface are very disordered and undercoordinated and cannot freeze.

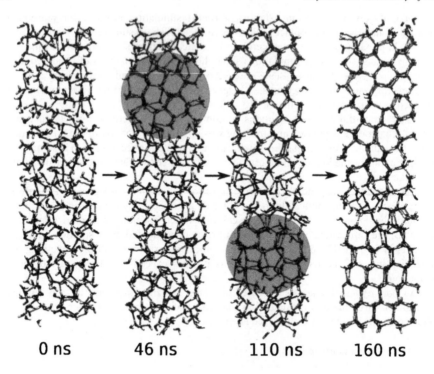

0 ns   46 ns   110 ns   160 ns

Figure 2   *Snapshots from the MD simulation of the slab with 384 water molecules. Both*
crystallization nuclei (shaded regions) form in the subsurface. Water freezes
mostly as cubic ice, with many defects.

There exists an electric field at the interface establishing the surface potential
estimated to be in the order of 100 mV.[49,50] Such potential difference across the sub-
nanometer wide interfacial region is equivalent to very strong electric field. It has been
shown recently that such field can be used to enable ice nucleation.[31]

Molecules at the interface and in the subsurface also have larger mobility than those in
bulk phase. Their reorientation into positions corresponding to ice lattice can be expected
to be faster than in the bulk. This can be viewed as a kinetic advantage for the creation of
the ice nucleus near the interface.

4 CONCLUSION

This contribution shows how molecular simulations can be used to provide better insight
into highly relevant natural and technological processes of ice nucleation and freezing.

We investigated freezing of water and salt solutions by means of molecular dynamics
simulations. We first established a robust simulation protocol for water freezing and than
applied this approach to the study of the brine rejection process. Brine rejection was
observed for a series of systems with varying salt concentration. We showed the anti-freeze

effect of the salt on the molecular level and proposed a mechanism correlating local fluctuations in the salt ion density and the progression of the freezing front.

Computer simulations were also used to show that the crystallization nucleus is more likely to form in the subsurface than in the bulk phase of the water slab. This result can have far reaching atmospheric implications. It has been suggested that formation of an ice nucleus at the interface would be hampered by contamination of the surface by organic surfactants.[12] The effect of the adsorbed material will surely propagate towards the subsurface as well, however it will be smaller than in the topmost layer. Therefore, the anthropogenic emissions should have an effect on the radiative balance of the Earth atmosphere. This effect should, however, be smaller than predicted using the assumption of surface nucleation.

Acknowledgements

We would like to thank the **META**Centrum project for generous allocation of computer time. Support from the Czech Ministry of Education (grant LC512) is gratefully acknowledged. L.V. acknowledges the support from the granting Agency of the Czech Republic (grant 203/05/H001).

References

1   V. F. Petrenko and R.W. Whitworth, *Physics of Ice,* Oxford University Press, Oxford, 1999.
2   G.W. Gross, P. M. Wong, and K. Humes, *J. Chem. Phys.,* 1999, **67**, 5264
3   R. E. Dickerson, *Molecular Thermodynamics,* W. A. Benjamin, Pasadena, 1969.
4   B. Rubinsky, *J. Cryst. Growth*, 1983, **62**, 513.
5   M. G. Worster and J. S. Wettlaufer, *J. Phys. Chem. B*, 1997, **101**, 6132.
6   W.W. Mullins and R. F. Sekerka, *J. Appl. Phys.*, 1964, **35**, 444.
7   K. Aagaard and E. C. Carmack, *J. Geophys. Res.*, 1989, **94**,14485.
8   A.Y. Shcherbina, L. D. Talley, and D. L. Rudnick, *Science*, 2003, **302**, 1952.
9   P. Jungwirth, D. Rosenfeld, and V. Buch, *Atmos. Res.*, 2005, **76**, 190.
10   K. Nagashima and Y. Furukawa, *J. Cryst. Growth*, 2000, **209**, 167.
11   L. Vrbka and P. Jungwirth, *Phys. Rev. Lett.*, 2005, **95**, 148501.
12   A. Tabazadeh, Y. S. Djikaev and H. Reiss, *Proc. Natl. Acad. Sci.*, 2002, **99**, 15873.
13   H. Prupacher and J. D. Klett, Microphysics of Clouds and Precipitation, Kluwer Academic Publishers, Dordrecht, 1998.
14   M. Del Guasta, M. Morandi, L. Stefanutti, S. Balestri, E. Kyrö, M. Rummukainen, V. Rizi, B. Stein, VC. Wedekind, B. Mielke, R. Mathey, V. Mitev and M. Douard, *J. Aerosol Sci.*, 1998, **29**, 357.
15   J. E. Kay, V. Tsemekhman, B. Larson, M. Baker and B. Swanson, *Atmos. Chem. Phys. Discuss.*, 2003, **3**, 3361.
16   D. Duft and T. Leisner, *Atmos. Chem. Phys. Discuss.*, 2004, **4**, 3077.
17   R. A. Shaw, A. J. Durant and Y. Mi, *J. Phys. Chem. B*, 2005, **109**, 9865.
18   S. Sastry, *Nature*, 2005, **438**, 746
19   B. B. Laird and A. D. J. Haymet, *Chem. Rev.*, 1992, **92**, 1819.
20   J. A. Hayward and A. D. J. Haymet, *J. Chem. Phys.*, 2001, **114**, 3713.
21   T. Bryk and A. D. J. Haymet, *J. Chem. Phys.*, 2002, **117**, 10258.
22   L. A. Baez, P. Clancy, J. Chem. Phys 103, 1995, 9744
23   H. Nada, Y, Furukawa, J. Phys. Chem. B 101, 1997, 6163
24   H. Nada, J. P. van der Eerden, Y. Furukawa, J. Cryst. Growth 266, 2004, 297

25 M. A. Carignano, P. B. Shepson, I. Szleifer, Mol. Phys. 103, 2005, 2957
26 T. Ikeda-Fukazawa, K. Kawamura, J. Chem. Phys. 120, 2004, 1395
27 T. Bryk and A. D. J. Haymet, *J. Mol. Liq.*, 2004, **112**, 47
28 E. Smith, T. Bryk and A. D. J. Haymet, *J. Chem. Phys.*, 2005, **123**, 2005, 034706.
29 L. Vrbka and P. Jungwirth, *J. Mol. Liq.*, 2006, in press.
30 M. A. Carignano, E. Baskaran, P. B. Shepson and I. Szleifer, *Ann. Glaciology*, 2006, in press.
31 I. M. Svishchev and P. G. Kusalik, *Phys. Rev. Lett.*, 1994, **73**, 975.
32 K. Koga, H. Tanaka, X. C. Zeng, *Nature*, 2000, **408**, 564.
33 R. Radhakrishnan and B. L. Trout, *J. Am. Chem. Soc.*, 2003, **125**, 7743.
34 M. Matsumoto, S. Saito and I. Ohmine, *Nature*, 2002, **416**, 409.
35 M. Yamada, S. Mossa, H. E. Stanley and F. Sciortino, *Phys. Rev. Lett*, 2002, **88**, 195701.
36 B. Guillot, *J. Mol. Liq.*, 2002, **101**, 219.
37 C. Vega, E. Sanz and J.L.F. Abascal, *J. Chem. Phys,*, 2005, **122**, 114507.
38 H. J. C. Berendsen, J. R. Grigera and T. P. Straatsma, *J. Phys. Chem.*, 1987, **91**, 6269.
39 T. Bryk and A. D. J. Haymet, *Mol. Sim.*, 2004, **30**, 131.
40 H. Nada and J. P. van der Eerden, *J. Chem. Phys.*, 2003, **118**, 7401.
41 D. E. Smith and L. X. Dang, *J. Chem. Phys.*, 1994, **100**, 3757.
42 E. Whalley, *Science*, 1981, **211**, 389.
43 E. Whalley, *J. Phys. Chem.*, 1983, 87, 4174.
44 B. J. Murray, D. A. Knopf and A. K. Bertram, *Nature (London)*, 2005, **434**, 202.
45 K. Morishige and H. Uematsu, *J. Chem. Phys.*, 2005, **122**, 044711.
46 H. Nada and Y. Furukawa, *J. Cryst. Growth*, 1996, 169, 587.
47 G. P. Johari, J. Chem. Phys. 122, 2005, 194504.
48 L. Vrbka, P. Jungwirth, *J. Phys. Chem. B*, 2006, in press.
49 M. A. Wilson, A. Pohorille and L. R. Pratt, *J. Chem. Phys.*, 1988, **88**, 3281.
50 M. Paluch, *Adv. Colloid. Interface Sci.*, 2000, **84**, 27.

IMPLICATIONS FOR AND FINDINGS FROM DEEP ICE CORE DRILLINGS — AN EXAMPLE: THE ULTIMATE TENSILE STRENGTH OF ICE AT HIGH STRAIN RATES

F. Wilhelms[1], S. G. Sheldon[2], I. Hamann[1] and S. Kipfstuhl[1]

[1]Alfred-Wegener-Institute for polar and marine research, Columbusstraße, D-27568 Bremerhaven, Germany, E-mail: fwilhelm@awi-bremerhaven.de
[2]Ice & Climate Group, Department of Geophysics, Niels Bohr Institute, Juliane Maries Vej 30, DK-2100 Copenhagen Ø, Denmark

1 INTRODUCTION

At present, several deep ice core drilling operations have just been finished and new ventures are in the planning phase. In this paper we discuss ice properties related data attained when drilling the Dronning Maud Land (DML) deep ice core[1] in the framework of the European Project for Ice Coring in Antarctica (EPICA). The drill site is located[2] in East Antarctica at 75° S, close to the Greenwich meridian. The pull force rating of the winch to break the core at the end of a drill run is one of the key design parameter for ice coring winches. Amongst drilling personnel it is common knowledge that core breaks are getting harder in warm ice. However this has never been quantified, but is consistent with the experience based on safety tests for ice-screws. The engineers of the German Alpine Club's security-council quote the warm ice (just below the melting point) to be tougher and attribute this to its "more plastic behaviour"[3]. When designing heating for railway switches, engineers want to estimate under which conditions the frozen blade can be ripped from the track. The literature on tensile strength of ice suggests a decrease with increasing temperature for low strain rates[4,5] (in the order of 10^{-6} ... 10^{-5} s^{-1}), however the latest review of all compiled data suggests that for low strain rates the tensile strength of ice is independent from temperature[6], but for high strain rates this might be completely different as crack healing acts or cracks become blunt[7]. The core breaks presented here represent much higher strain rates and should be well in a crack-nucleation regime[8]. On the other hand we frequently observed refrozen water on the drill head, when drilling at temperatures above −10 °C. Liquid water at the cutters could either act as lubricant or even heal micro-cracks that were initiated by the cutting process. Both processes would lead to a reduction of the number of initial micro-cracks and thus strengthen the ice core. The remarkably smooth ice core surfaces we observed towards the bedrock at higher temperatures also suggest the presence of liquid water during the cutting process.

2 ICE CORE DRILLING PROCEDURE

2.1 Experimental setup

Deep ice cores are drilled by cutting a ring into the glacier. The inner remaining cylinder moves into a tube, the so called core barrel. After a certain depth increment, depending on the used drill system between about 0.8 m to more than 4 m, the cylindrical ice core is broken off from the bore-hole's bottom and hoisted to the surface. The drill system's layout is described elsewhere[9,10]. Figure 1 gives an overview about the drill system. During the breaking and hoisting process the core is held by the three core catchers (Figure 1a) and the force is applied by pulling with a winch (Figure 1b). Even though the core catchers are designed to introduce fracture into the core, they quite often abrade the core's surface and the abrasion wedges the core into the core barrel. Thus, the pulling force is in most core breaks applied to the core's entire cross section. The applied pulling force is measured with a strain gauge, calibrated in kilogramme (force), which is situated in the drill towers top pulley. When breaking the core, one spools in the cable with a speed of a few cm/s. The maximum tension is recorded automatically and is noted as "core break", as well as the final plumbing depth, for each respective drill run.

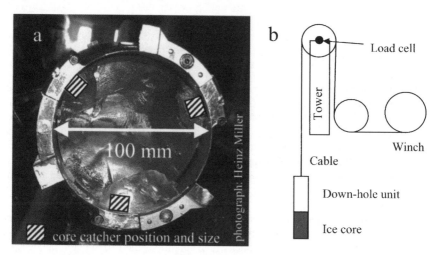

Figure 1   *The experimental setup; a: Drill head with cutters and hatched core catcher area; b: The core catchers hold the core when pulling with the winch and recording the break force with the load cell.*

2.2 Core-break data treatment

From the recorded force data the cable and the drill's weight in the drilling liquid are subtracted and the data are converted into true force readings. The confinement pressure does not contribute to the net force, as the drill is permeable to fluid and thus the fluid can be easily displaced around the ice core. For each core break the cable depth plumbing is also recorded, so that one has a core break versus depth data set. In total there are about 1500 core breaks, so that on average there are about 40 core breaks in a depth interval of 50 m, which is used to average the data for further investigation. The core break errors are the quadratic sum of 6 % systematic error from load cell calibration and the standard error of the mean of each depth interval.

2.3 Temperature and pressure logging of the hole

The bore-hole was logged several times with a logging tool, thus measuring temperature, pressure and the geometry of the bore-hole. The tool is described elsewhere[11], relevant for the discussion here is the temperature measurement with an absolute precision of better than 50 mK. The pressure in the hole deviates by less than 500 kPa from the ice pressure over the whole range and can be calculated roughly from the given drilling depth by multiplying with the average drill liquid/ice density of 0.92 Mgm^{-3} and the acceleration of gravity of 9.8 ms^{-2}. It thus increases linear from about 0.1 MPa (atmospheric pressure) at the top of the hole to about 25 MPa at the bottom of the hole in 2774 m depth.

3 THE TEMPERATURE DEPENDENCE OF ICE'S ULTIMATE TENSILE STRESS

3.1 Ultimate tensile stress calculated from break strength

Figure 2 presents the dependence of break strength versus temperature for raw (grey dots) and averaged (error bars) data. As the ultimate tensile stress, or also called fracture stress, is the average force per area to break the sample, the fracture stress is strictly proportional to the break strength with the factor of inverse cores cross section $1/(\pi \, (0.049 \text{ m})^2)$. The right axis of the Figure is thus the fracture stress to break off the core. The top axis of the Figure presents the coincident depth and pressure for the respective ordinate temperature.

3.2 Comparison with published data and consistency check

As we present an increase of tensile fracture strength at high strain rates for the first time, we can not compare them directly to previously published data. But we can compare certain aspects of our data with the literature and thus check for consistency.

3.2.1 Tensile fracture stress reading at a certain temperature. For polycrystalline ice with a grain size of 10 mm one finds a fracture stress of about 800 kPa at –10 °C reported in the literature[8] for high strain rates. This finding is consistent with the observed fracture stress in this study of about 700 kPa at 263 K.

3.2.2 Variation with crystal size. The dependence of fracture stress on crystal size is well established in the literature[8], thus one could argue that the observed temperature dependence is truly a change in crystal size? This kind of artefact can be excluded, as the fracture stress decreases with increasing grain size[8]. In deep ice cores in general and, also observed in the core from DML here, the grain size increases with depth and as temperature also increases with depth, one should expect a decrease of fracture stress with increasing temperature. The contrary is the case and the increase in fracture stress with temperature might even be stronger than observed here.

3.2.3 Variation with crystal orientation fabrics. One could also argue that the observed increase is an artefact due to a change in the fabric's crystal orientation of the ice core. For the DML core discussed here the fabric changes from a random orientation steadily to a girdle type fabric at a depth of about 1000 m and then quite suddenly within an interval of about 20 m around 2040 m to a single maximum type fabric, where it remains further down. Thus, in the interval of dramatic change, the fabric does not change and has changed before without a significant change in fracture strength. Thus the textural and fabric changes that occur in an ice core can be excluded to be the source of our observed increase of fracture stress with depth.

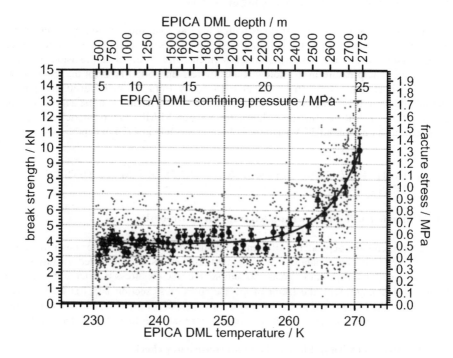

Figure 2   *The fracture stress of polar ice under high strain rates; from the raw breaking strength (dots), averaged values with error bars (6% systematic error and standard error of the mean) have been calculated in 50 m intervals. By division with the cross section one yields the fracture stress of ice. The solid line is a least squares curve fit to the data.*

3.3   An empirical fit in the temperature range above 230 Kelvin

A more succinct description is given by an empirical curve to the averaged data. Fitting an exponential function by minimizing the χ^2, yields a best fit for: (fracture stress)/kPa = 505 + 1.4E–20*exp((absolute temperature)/5.17 K).

4   CONCLUSION

We report an effect quantitatively that seems to be well known in practical engineering, but is to our knowledge not yet described in the scientific literature. Astonishing is that the ordinate found in section 3.3 is the same as the ordinate in the theory to describe (fracture stress)/kPa = 510 + 30 (m/grain size)$^{1/2}$ in terms for crack nucleation[8]. This is consistent with the observation in the cited work, as at high strain rates the fracture stress is crack nucleation controlled. In the cited work Schulson explains the discrepancies to the theoretical value for the grain size dependent factor with a thermally activated mechanism that relaxes the stress concentrated at the grain boundaries. A quasi liquid layer at the grain boundaries could help to relax piled dislocations. Liquid water in the bore hole could reduce the number of introduced cracks at the ice core surface or even heal cracks. At present, there seems to be no theory available to describe the fracture stress adequately. Thus our empirical fit will provide needed information to engineering branches, as e.g. railway engineers who required data on ultimate tensile stress of ice from us to estimate under which conditions frozen railway switches would simply break the ice freezing the blade to the track. These data will serve until systematic tests under laboratory conditions are available. If liquid water plays a significant role in the bore hole, the machining procedure of samples for stress tests should be addressed carefully. Machining at high temperatures to minimise the number of induced cracks at the sample surface should be considered.

Acknowledgements

We thank Paul Duval for his fruitful questions and comments during the review process. This work is a contribution to the European Project for Ice Coring in Antarctica (EPICA), a joint European Science Foundation/European Commission scientific programme, funded by the EU and by national contributions from Belgium, Denmark, France, Germany, Italy, the Netherlands, Norway, Sweden, Switzerland and the United Kingdom. The main logistic support was provided by IPEV and PNRA (at Dome C) and AWI (at Dronning Maud Land). This is EPICA publication no. 163 and AWI publication awi-n16131.

References

1   EPICA community members, *Nature*, 9 November 2006, **444**, 195 - 198, doi:10.1038/ nature05301.
2   C. Drücker, F. Wilhelms, H. Oerter, A. Frenzel, H. Gernandt and H. Miller, *Mem. Natl Inst. Polar Res., Spec. Issue,* 2002, **56**, 302–312.
3   C. Semmel and D. Stopper, *DAV Panorama, ISSN 1437–5923*, 2005, **2**, 91–95.
4   T. R. Butkovich, *SIPRE Res. Rep.,* 1954, **11**, 12. (reproduced in P. V. Hobbs, Clarendron Press, Cambridge, 1974, 333.)
5   E. M. Schulson, P. N. Lim and R. W. Lee, *Philos. Mag. A,* 1984, **49**, 353–363.
6   E. M. Schulson, personal communication, 2006.
7   E. M. Schulson, S. G. Hoxie and W. A. Nixon, *Philos. Mag. A,* 1989, **59**, 303–311.
8   E. M. Schulson, *Journal de Physique, Colloque C1, supplément au n° 3,* **48**, 1987, C1-207–C1-220.
9   S. J. Johnsen, N. S. Gundestrup, S. B. Hansen, J. Schwander and H. Rufli, *Mem. Natl Inst. Polar Res., Spec. Issue,* 1994, **49**, 9–23.
10 L. Augustin and A. Antonelli, *Mem. Natl Inst. Polar Res., Spec. Issue,* 2002, **56**, 226–244.
11 N. S. Gundestrup, H. B. Clausen and L. B. Hansen, *Mem. Natl Inst. Polar Res., Spec. Issue,* 1994, **49**, 224–233.

ISOTHERMAL AMORPHOUS-AMORPHOUS-AMORPHOUS TRANSITIONS IN WATER

K. Winkel[1], W. Schustereder[1,2], I. Kohl[1], C. G. Salzmann[1,4], E. Mayer[1], T. Loerting[1,3]

[1] Institute of General, Inorganic and Theoretical Chemistry, University of Innsbruck, Innrain 52a, A-6020 Innsbruck, Austria
[2] Max-Planck-Institute for Plasma-Physics, EURATOM Association, Boltzmannstr. 2, D-85748 Garching, Germany
[3] Institute of Physical Chemistry, University of Innsbruck, Innrain 52a, A-6020 Innsbruck
[4] Inorganic Chemistry Lab, University of Oxford, South Parks Road, OX1 3QR United Kingdom

1 INTRODUCTION

Polyamorphism is a fundamental phenomenon, which is believed to be linked to the occurrence of two or even more liquids in one-component systems[1]. It is furthermore thought that a possible occurrence of multiple liquids in one-component systems is intimately linked with their anomalous properties, such as negative thermal expansion coefficients or isothermal compressibility maxima[2-4]. Polyamorphism has been observed since its discovery on the example of water[5] in many one-component substances[6-8]. Probably the most extensively investigated "anomalous" liquid is water, which shows "typical" behaviour at high temperatures, whereas its anomalies become stronger on cooling into its supercooled state[9-12]. For instance the thermal expansion coefficient is positive at T>277 K, but negative at T<277 K. The anomalies disappear not only on heating, but also on increasing the pressure to P>200 MPa, where e.g. the dynamic viscosity and the self diffusion coefficient show the typical pressure dependence found for most of the liquids, e.g., organic liquids, van-der-Waals liquids or metal melts. To explain these anomalies "water's second critical point" has been postulated[13]. There are supposedly two liquid states, which can not be discerned at the "second critical point", which are called "low density liquid" (LDL) and "high density liquid" (HDL) and which are separated by a first-order phase transition. This second critical point is postulated to be at a pressure p~17 – 340 MPa and at a temperature T~145 – 230 K[14-18]. Unfortunately, this p-T area is not accessible for direct experiments with liquid water, since most of it is located in the "no man's land", where only crystalline ice is observed.

Therefore, experiments are performed on "immobilized liquids", or in other words on amorphous water (also called vitreous water or glassy water)[19]. Currently, three structurally distinct amorphous states of water are known: low- (LDA)[5,20], high- (HDA)[20,21] and very high- (VHDA) density amorphous ice[22,23]. We emphasize that HDA is not a well defined state but rather comprises a number of substates. It has been suggested to use the nomenclature "uHDA" ("unrelaxed HDA")[24], "eHDA" ("expanded HDA")[24] and/or "rHDA" ("relaxed HDA")[25,26] to account for this. Even though no signs of micro-crystallinity have been found in neutron or X-ray diffraction studies, it is unclear whether

HDA and VHDA are truly immobilized liquids, i.e., if they are truly related to a high-(HDL) and very-high density liquid (VHDL). It has been argued that HDA is in fact a mechanically collapsed crystal unrelated to HDL[27]. Whereas there seems to be a discontinuity in density upon transforming LDA to HDA[5], which supports the liquid-liquid transition hypothesis, the experimental data is ambiguous whether there is also a discontinuity in density upon transforming HDA to VHDA[28]. Hence, it is unclear if there might even be a third or fourth critical point in water's phase diagram related to a liquid-liquid transition in a one component system as proposed in some computer simulations[29-33], whereas others show only two critical points[34-37]. Some hints for multiple liquid-liquid transitions have been provided from an infrared spectroscopy study[38], but clear experimental evidence is missing.

Herein, we expand on the discussion of our recently observed isothermal amorphous-amorphous-amorphous transition sequence[23]. We achieved to compress LDA in an isothermal, dilatometric experiment at 125 K in a stepwise fashion *via* HDA to VHDA. However, we can not distinguish if this stepwise process is a kinetically controlled continuous process or if both steps are true phase transitions (of first or higher order). We want to emphasize that the main focus here is to investigate transitions between different amorphous states *at elevated pressures* rather than the annealing effects observed *at 1 bar*. The vast majority of computational studies shows qualitatively similar features in the metastable phase diagram of amorphous water (cf. e.g. Fig.1 in ref. 39): at elevated pressures the *thermodynamic equilibrium line* between HDA and LDA can be reversibly crossed, whereas by heating at 1 bar the *spinodal* is irreversibly crossed. These two fundamentally different mechanisms need to be scrutinized separately.

2   METHODS

In brief, 300 mg of LDA are prepared in a piston-cylinder apparatus with a bore diameter of 8 mm by compressing ice Ih to HDA at 77 K to 1.5 GPa and by subsequent heating of HDA at 0.020 GPa. This protocol was shown to reliably produce LDA with the maximum of the first diffraction peak at $2\theta=24°$ (d=0.37 nm)[40]. After quenching to 77 K and recovery to 1 bar samples are characterized using powder X-ray diffraction. All diffractograms were recorded at 83 – 88 K in θ-θ geometry employing Cu-Kα rays (λ=0.154 nm) using a Siemens D-5000 diffractometer equipped with a low-temperature camera from Anton Paar. The sample plate was in horizontal position during the whole measurement. Installation of a "Goebel mirror" allowed us to use small amounts of sample without distortion of the Bragg peaks. Neither any crystalline by-products nor any remnants of HDA can be detected using powder X-ray diffraction. LDA is encased in a container of indium, which is a ductile metal also at 77 K, and which prevents shock-wave heating of the ice on compression to pressures up to 2 GPa[41]. LDA is then compressed isothermally at 20 MPa/min at 77 K, 100 K or 125 K, quenched to 77 K at elevated pressures and finally decompressed to ambient pressure at -20 MPa/min. Typically, the temperature varies by ±1 K at 125 K at 100 K, and up to ±2 K during the transition events, whereas it is essentially constant for experiments performed at 77 K.

3   RESULTS

In Fig. 1a the raw data for three compression/decompression experiments up to 1.5 GPa starting with LDA at 125 K are shown. The piston displacement measured after precompression to 0.1 GPa serves as the origin (piston displacement = 0). Despite of the slight temperature oscillations up to ±2 K the reproducibility is excellent - all three curves are almost identical. There is a sharp step in piston displacement of about 1.1 mm in the range 0.4 – 0.5 GPa and a broader step of about 0.6 mm in the range 0.8 – 1.0 GPa. After decompression an overall densification of 1.6 mm is observed in all three cases. The midpoint of the transformation for the first density step attributed to the LDA→HDA transformation varies in the pressure range 0.44 – 0.48 GPa for the three experiments shown in Fig.1a and additional eight experiments done in the same manner with the exception that quenching and decompression was initiated at 0.7 GPa (not shown). This variation is possibly related to small differences in temperature or to other effects beyond our control. The statistical variation of nominal pressures of transformation from LDA to HDA at 125 K is hence ±0.02 GPa. The piston displacement at a pressure of 0.6 GPa varies by no more than ±0.15 mm for eleven experiments, which implies a variation of about 14% in the step height.

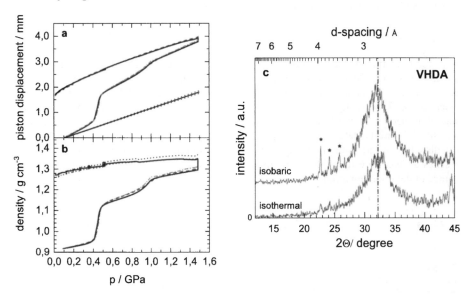

Figure 1   *(a) Three curves of "raw" piston displacement data obtained on compressing LDA at 125 K (top curves) together with the apparatus correction $d^0(p)$ (straight line at bottom). (b) "Processed" density data (see text) (c) Powder X-ray diffractograms of the final recovered states (bottom) in comparison with the powder X-ray diffractogram of VHDA obtained by annealing HDA at 1.1 GPa[22] (top). The dashed line is intended to guide the eye to the location of the maximum of the first broad diffraction peak. Stars indicate positions of reflexes caused by traces of Ih, which had formed by condensation of water vapour during transfer of sample onto the precooled X-ray sample holder (cf. ref. 42). The sample itself is fully amorphous.*

The X-ray diffractogram obtained from the recovered state after one of these experiments is shown in Fig. 1c (bottom). This X-ray diffractogram is directly compared to an X-ray diffractogram obtained from a recovered sample of VHDA as prepared by annealing HDA to ~160 K at 1.1 GPa (top)[22]. Both diffractograms are very similar, with a first broad diffraction maximum at $2\theta=32.3\pm0.2°$ and can both be defined as VHDA (in accordance with the definition that all HDA states annealed under pressures above 0.8 GPa should be called VHDA)[26].

In order to calculate *in situ* densities from the piston displacement data a reference density is required, which is in our case the density $\rho(LDA)=0.92\pm0.02$ g/cm³ at 125 K [43]. Furthermore, it is necessary to subtract the piston displacement obtained in a blind experiment, i.e., the same experiment, but without ice (bottom curve in Fig. 1a). This blind experiment was previously called "indium correction"[20,28]. However, the correction functions $d^0(p)$ recorded with *350 mg of indium and no ice* (solid line) and the correction functions recorded with *no indium and no ice* (dashed line) shown in Fig. 2c indicate that the correction function is dominated by the compressibility of the piston-cylinder apparatus and the frame of the material testing machine, whereas the indium contribution is negligible. The *in situ* density $\rho(p)$ can be calculated using the mass of ice m(ice I_h) and the cross-section of the bore A from the difference in piston displacement $\Delta d(p)$ ("real experiment" d(p) minus "blind experiment" $d^0(p)$) as

$$\rho(p) = \rho(LDA) \times (A^{-1}\times m(ice\ I_h)\times\rho(ice\ I_h)^{-1} - \Delta d(p))^{-1} \times A^{-1}\times m(ice\ I_h)\times\rho(ice\ I_h)^{-1} \quad (1)$$

The results of this correction procedure are shown in Fig. 1b. The density steps can be quantified to be ~20% at 0.45 GPa and ~5% at 0.95 GPa. The second density step decreases to ~1% upon employing higher compression rates of 600 or 6000 MPa/min, whereas the first density step remains essentially unchanged. This might imply that the activation barrier from LDA to HDA is lower than the activation barrier from HDA to VHDA [23]. From the data shown in Fig. 1b, the isothermal compressibility κ_T can easily be calculated by differentiation (cf. Fig. 3, bottom panel in ref. 23). It shows two maxima and reaches at least five times the isothermal compressibilities κ_T of LDA and HDA in the course of the transformation.

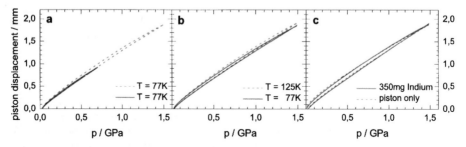

Figure 2   *Correction function $d^0(p)$ obtained using different experimental parameters. (a)Piston-cylinder apparatus lined with 300 mg of indium upon compression to 1.5 GPa at 77 K (dashed line) as compared to compression to 0.7 GPa (solid line). The hysteresis upon decompression is evident. (b) Effect of temperature increase from 77 K (solid line) to 125 K (dashed line). (c) Comparison of compression/decompression curves for piston-cylinder apparatus lined with 300 mg of indium (solid line) and without indium (dashed line). The effect of different compression rates ranging from 6 MPa/min to 6 GPa/min was found to be negligible.*

Figs. 2a-c show various blind experiments $d^0(p)$ obtained in a compression/decompression cycle. Fig.2a compares a blind experiment conducted on compressing up to 0.7 GPa (solid line) and up to 1.5 GPa (dashed line), all other parameters being equal. It is evident that there is a hysteresis of up to 0.1 mm at intermediate pressures and also a slight hysteresis after full decompression. The hysteresis is much more pronounced for the experiment conducted up to 1.5 GPa. It is, therefore, necessary to use two separate correction functions $d^0(p)$ for the compression part and the decompression part, especially for experiments conducted in the range 0-1.5 GPa, or even higher. It is not valid to extrapolate correction functions $d^0(p)$ to higher pressures.

For our previous calculation of the density of VHDA after recovery to 77 K and 1 bar (cf. Fig. 1 in ref. 23) we have assumed a constant bore diameter of 8.0 mm, and have arrived at an unreasonably high value of ~1.31 g/cm^3, rather than the accepted VHDA density of 1.25±0.01 g/cm^3 measured by two independent methods, namely flotation of VHDA in a liquid N_2/Ar mixture and isobaric expansion of VHDA to LDA by heating at 1 bar and quenching to 77 K[22]. Furthermore, we saw spurious effects, such as a slight increase of density on decompressing from 1.5 GPa to 1.0 GPa. For the graph shown in Fig. 1b we use a bore diameter of 7.8 mm, since we want to take into account the fact that a ~100 μm thick indium foil is wrapped around the sample. We do not take into account the contraction of the bore diameter upon cooling from ambient conditions to 77 K, since we do not know the thermal expansivity of our steel cylinder. Using 7.8 mm for the bore diameter, a density of 1.27 g/cm^3 results for VHDA, which is close to the accepted value, and the spurious density increase upon decompression disappears as well. Using a diameter less than 7.8 mm we arrive at a density of 1.25 g/cm^3. We, therefore, conclude that the exact diameter available to the sample is important in quantifying its density. Since we do not know it exactly the density given in Fig. 1b is accurate only to ±0.02 g/cm^3. In addition to the error by the unknown bore diameter, the error in the density of LDA of ±0.02 g/cm^3 also contributes to the total error[43].

We want to point out that the densities reported in Fig.1b are lower than the densities reported by some of us in a different study[26]. For instance at 1.0 GPa the density in Fig. 1b reads as 1.26 g/cm^3, whereas it reads as 1.33 g/cm^3 in Fig.2 of ref. 26. We believe that the latter densities are the correct ones to believe a) because highly accurate densities of crystalline material *at elevated pressures* were used as a reference for the calculation (whereas for the data in Fig.1b the density $\rho(LDA)=0.92\pm0.02$ g/cm^3 *at 1bar* serves as the reference) and b) because these data refer to crystallization temperature, e.g., ~165 K at 1.0 GPa rather than to 125 K. An increase in temperature at elevated pressures was recently shown to cause significant structural relaxation effects in HDA[24]. Whereas at a pressure of ~0.2 GPa this causes the formation of "expanded HDA" of lower density[24], it is very likely that at much higher pressures such as 1.0 GPa similar relaxation effects cause the formation of "compressed HDA" of higher density. That is, on increasing the temperature from 125 K to 165 K at 1.0 GPa structural relaxation may increase the density from 1.26 g/cm^3 to 1.33 g/cm^3. For the reason of this missing equilibration at 125 K we are not able to say if the second density step is related solely to a kinetic process or a combined thermodynamic and kinetic process.

Fig.2b shows that the temperature effect is much smaller than the hysteresis effect shown in Fig. 2a, at least in the range from 77 K to 125 K. The difference amounts to less than 0.1 mm at a pressure of 1.5 GPa and much less than 0.1 mm at a pressure of 0.7 GPa. For comparison quenching of VHDA from 125 K to 77 K at 1.5 GPa causes a change of ~0.20 mm (cf. Fig. 1a) and quenching of HDA from 125 K to 77 K at 0.7 GPa causes a change of ~0.17 mm (cf. Fig. 3a). That is, the thermal expansion of HDA and VHDA exceeds the thermal expansion of the apparatus and indium. Nevertheless, the temperature effect on the apparatus correction is not negligible. As already mentioned above, Fig.2c shows that, at least for compression experiments, indium does not contribute significantly to the correction function. Please note, that this might be different for isobaric heating/cooling experiments.

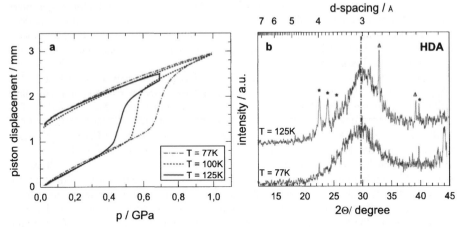

Figure 3   *(a) "Raw" piston displacement data for the 1st density step obtained on compression of LDA at 77 K (dashed line), 100 K (dotted line) and 125 K (solid line). The final density after recovery calculates as 1.15±0.01 g/cm³ in all three cases. (b) Powder X-ray diffractograms of the recovered samples compressed at 125 K (top) and 77 K (bottom). For details cf. the caption to Fig. 1c. Triangles indicate peaks arising from indium contamination.*

In Fig. 3a we show three pressure vs. piston displacement curves obtained on compression of LDA and related to the first density step. The main difference is the compression temperature. Whereas the second density step is not observable at 77 K at pressures up to 1.6 GPa[5], the first density step is observable also at these temperatures. This might again imply that the activation barrier from LDA to HDA is lower than the activation barrier from HDA to VHDA. As also observed by Mishima[44] the pressure required to transform LDA to HDA decreases as the temperature increases, namely from ~0.70 GPa at 77 K to ~0.55 GPa at 100 K and to ~0.45 GPa at 125 K. The height of the step remains the same. All decompression curves of HDA reach a final piston displacement of ~1.4 mm, which implies that the densities of all three samples are roughly equal after recovery. It amounts to 1.15±0.01 g/cm³ after the correction procedure described above. Please note that the dashed curve at 100 K was recorded without quenching to 77 K, i.e., a conversion back to LDA does not take place also at 100 K. In Fig. 3b we show the X-ray diffractograms of the samples compressed at 77 K (bottom) and 125 K (top). On ignoring the sharp features arising from tiny amounts of condensed hexagonal ice and indium both diffractograms are very similar with a first broad maximum at 2θ~30.5±0.3°. However, both are clearly different from the diffractograms shown in Fig. 1c and show the maximum at about the position expected for HDA[20,45-49]. We want to emphasize that the diffractogram shown here (Fig.3b, top) is different from the one shown as diffractogram B before by us[23,28] in the sense that the first broad diffraction maximum is shifted by about 2θ~1°. Also heterogeneity as indicated by small angle scattering[50] is no longer evident. The diffractogram to rely on is the one shown in Fig.3b rather than the earlier one, which probably suffered from slight annealing at 1 bar during sample transfer, which is known to be at the origin of such a shift in the diffraction maximum[20,51,52] and to induce heterogeneity[50]. The X-ray diffractograms of HDA can not be obtained by co-adding X-ray diffractograms of LDA and VHDA in different ratios. This typically produces diffractograms showing a broad double maximum and a minimum in between located at $2\theta \sim 30°$ ($d \sim 0.30$ nm). According to this X-ray diffraction evidence, HDA produced upon isothermal compression of LDA is, therefore, not a mixture of LDA and VHDA up to the length scales probed with our equipment, i.e., ~ 7 Å. On the other hand neutron diffraction shows that all the amorphous states obtained in the course of the HDA→LDA transition at ~ 0.3 GPa and 130 K can be expressed as a linear combination of LDA and HDA[53]. This supports the picture of a first order phase transition between LDA and HDA at 130 K, whereas the transition between LDA and VHDA at 125 K is clearly not a single first-order phase transition, but rather a combined process.

4 CONCLUSIONS

We conclude that we have observed an amorphous-amorphous-amorphous transition from LDA to VHDA *via* HDA. We can not distinguish if these are phase transitions in the sense of a first or higher order transition or if these are continuous annealing processes governed by accelerating/decelerating kinetics. In the case of a-SiO_2, the only other case for which such an amorphous-amorphous-amorphous transition has been observed, the authors infer that it is not a first-order transition[54]. Furthermore, we discuss how the raw pressure vs. piston displacement data is converted to pressure vs. sample density data by subtracting the blind experiment without ice, which was previously called "indium correction". We show that the indium itself, used as a lubricant at 77 K, contributes negligibly to the "apparatus correction". We furthermore point out that calculation of accurate absolute $\rho(p)$ curves requires knowledge of the diameter available to the sample, as well as an "apparatus correction" recorded at exactly the same conditions both for the compression as well as for the decompression step (hysteresis and temperature effects!), which we did not properly account for previously[23]. This explains why we (erroneously) calculated a density of 1.31 g/cm^3 for VHDA at 77 K and 1 bar in Fig. 1 of ref. 23. On doing the correction procedure properly we obtain a value of 1.27 ± 0.02 g/cm^3 from the density data, which is close to the accepted value of 1.25 ± 0.01 g/cm^3.

Acknowledgements

We are thankful to Prof. Erminald Bertel for discussions and grateful to the Austrian Science Fund FWF (project no R37-N11) for financial support. I.K. acknowledges support by the Austrian Academy of Sciences (APART-programme).

References

1   P. G. Debenedetti, *Nature*, **392**, 127 (1998).
2   C. A. Angell, R. D. Bressel, M. Hemmati, et al., *PCCP*, **2**, 1559 (2000).
3   H. Tanaka, *Europhys. Lett.*, **50**, 340 (2000).
4   V. V. Brazhkin and A. G. Lyapin, *J. Phys.: Cond. Matter*, **15**, 6059 (2003).
5   O. Mishima, L. D. Calvert, and E. Whalley, *Nature*, **314**, 76 (1985).
6   S. M. Sharma and S. K. Sikka, Prog. *Mater. Sci.*, **40**, 1 (1996).
7   P. H. Poole, T. Grande, C. A. Angell, et al., *Science*, **275**, 322 (1997).
8   P. F. McMillan, *J. Mater. Chem.*, **14**, 1506 (2004).

9   C. A. Angell, in *Water Compr. Treatise* (1982), Vol. 7, p. 1.
10 P. G. Debenedetti, *J. Phys.: Cond. Matter*, **15**, R1669 (2003).
11 P. G. Debenedetti and H. E. Stanley, *Physics Today*, **56**, 40 (2003).
12 C. A. Angell, *Annu. Rev. Phys. Chem.*, **55**, 555 (2004).
13 P. H. Poole, F. Sciortino, U. Essmann, et al., *Nature*, **360**, 324 (1992).
14 E. G. Ponyatovsky, V. V. Sinitsyn, T. A. Pozdnyakova, *JCP*, **109**, 2413 (1998).
15 H. E. Stanley, M. C. Barbosa, S. Mossa, et al., *Physica A*, **315**, 281 (2002).
16 S. B. Kiselev and J. F. Ely, *J. Chem. Phys.*, **116**, 5657 (2002).
17 R. C. Dougherty, *Chem. Phys.*, **298**, 307 (2004).
18 H. Kanno and K. Miyata, *Chem. Phys. Lett.*, **422**, 507 (2006).
19 P. G. Debenedetti and F. H. Stillinger, *Nature*, **410**, 259 (2001).
20 O. Mishima, L. D. Calvert, and E. Whalley, *Nature*, **310**, 393 (1984).
21 T. Loerting, I. Kohl, W. Schustereder, et al., *Chem. Phys. Chem.*, **7**, 1203 (2006).
22 T. Loerting, C. Salzmann, I. Kohl, et al., *PCCP*, **3**, 5355 (2001).
23 T. Loerting, W. Schustereder, K. Winkel, et al., *Phys. Rev. Lett.*, **96**, 025702 (2006).
24 R. J. Nelmes, J. S. Loveday, T. Straessle, et al., *Nature Physics*, **2**, 414 (2006).
25 M. Guthrie, J. Urquidi, C. A. Tulk, et al., *Phys. Rev. B*, **68**, 184110 (2003).
26 C. G. Salzmann, T. Loerting, S. Klotz, et al., *PCCP*, **8**, 386 (2006).
27 J. S. Tse, *J. Chem. Phys.*, **96**, 5482 (1992).
28 T. Loerting, K. Winkel, C. G. Salzmann, et al., *PCCP*, **8**, 2810 (2006).
29 I. Brovchenko, A. Geiger, and A. Oleinikova, *J. Chem. Phys.*, **118**, 9473 (2003).
30 S. V. Buldyrev and H. E. Stanley, *Physica A*, **330**, 124 (2003).
31 I. Brovchenko, A. Geiger, and A. Oleinikova, *J. Chem. Phys.*, **123**, 044515 (2005).
32 P. Jedlovszky and R. Vallauri, *J. Chem. Phys.*, **122**, 081101 (2005).
33 I. Brovchenko and A. Oleinikova, *J. Chem. Phys.*, **124**, 164505 (2006).
34 N. Giovambattista, H. E. Stanley, and F. Sciortino, *Phys. Rev. E*, **72**, 031510 (2005).
35 N. Giovambattista, H. E. Stanley, and F. Sciortino, *Phys. Rev. Lett.*, **94**, 107803 (2005).
36 R. Martonak, D. Donadio, and M. Parrinello, *Phys. Rev. Lett.*, **92**, 225702 (2004).
37 R. Martonak, D. Donadio, and M. Parrinello, *J. Chem. Phys.*, **122**, 134501 (2005).
38 D. E. Khostariya, A. Zahl, T. D. Dolidze, et al., *Chem. Phys. Chem.*, **5**, 1398 (2004).
39 E. G. Ponyatovsky, *J. Phys.: Cond. Matter*, **15**, 6123 (2003).
40 C. G. Salzmann, I. Kohl, T. Loerting, et al., *PCCP*, **5**, 3507 (2003).
41 I. Kohl, E. Mayer, and A. Hallbrucker, *PCCP*, **3**, 602 (2001).
42 C. G. Salzmann, E. Mayer, and A. Hallbrucker, *PCCP*, **6**, 5156 (2004).
43 V. F. Petrenko and R. W. Whitworth, *Physics of Ice* (Oxford University Press, 1999).
44 O. Mishima, *J. Chem. Phys.*, **100**, 5910 (1994).
45 L. Bosio, G. P. Johari, and J. Teixeira, *Phys. Rev. Lett.*, **56**, 460 (1986).
46 A. Bizid, L. Bosio, A. Defrain, et al., *J. Chem. Phys.*, **87**, 2225 (1987).
47 H. Schober, M. M. Koza, A. Tolle, et al., *Phys. Rev. Lett.*, **85**, 4100 (2000).
48 M. Guthrie, C. A. Tulk, C. J. Benmore, et al., *Chem. Phys. Lett.*, **397**, 335 (2004).
49 J. S. Tse, D. D. Klug, M. Guthrie, et al., *Phys. Rev. B*, **71**, 214107 (2005).
50 H. Schober, M. Koza, A. Tölle, et al., *Physica B*, **241-243**, 897 (1998).
51 C. A. Tulk, C. J. Benmore, J. Urquidi, et al., *Science*, **297**, 1320 (2002).
52 M. M. Koza, H. Schober, H. E. Fischer, et al., *J. Phys. Cond. Matt.*, **15**, 321 (2003).
53 S. Klotz, T. Straessle, R. J. Nelmes, et al., *Phys. Rev. Lett.*, **94**, 025506 (2005).
54 A. G. Lyapin, V. V. Brazhkin, E. L. Gromnitskaya, et al., *NATO Science Series, II: Mathematics, Physics and Chemistry*, **81**, 449 (2002).

MECHANICAL STRENGTH AND FLOW PROPERTIES OF ICE-SILICATE MIXTURE DEPENDING ON THE SILICATE CONTENTS AND THE SILICATE PARTICLE SIZES

M. Yasui[1] and M. Arakawa[1]

[1] Graduate School of Environmental Studies, Nagoya University, Chikusa-ku, Furo-cho, 464-8602, Nagoya, Japan

1 INTRODUCTION

Icy satellites have an icy crust and mantle and are composed of mixtures of ice and silicates.[1] Stripe patterns showing layered structures with different sand contents have been discovered on the polar cap on Mars.[2] There could be a wide range of temperature conditions from the melting point of ice to below 100 K in icy satellites and on Mars. Therefore, it is important to study the rheological properties of ice-silicate mixtures at wide temperatures to elucidate the tectonics of icy satellites and the flow dynamics of ice sheets on Mars.

The rheological properties of ice-solid particle mixtures have been studied by several researchers. Nayar et al. (1971) performed creep tests by using mixtures of ice and amorphous silica particles with a size of about 15nm, in which the volume fraction was below 1%. They found that the creep rates decreased simply as the volume fraction was increased.[3] Hooke et al. (1972) also performed creep tests by using mixtures of ice and fine sands with a size of about 180μm, in which the volume fraction ranged from 0 to 35%. They found that in some cases the creep rates of ice with low sand content were higher than that of pure ice, while the creep rates of ice with higher sand content decreased exponentially as the sand content was increased.[4] In relation to planetological implications, Durham et al. (1992) performed compression tests by using ice-solid particle mixtures at a confining pressure that ranged from 10 to 50MPa and in a wide temperature range from -196 to -50°C. They used mixtures with a volume fraction ranging from 0.1 to 56%, and with solid particles ranging in size from 1 to 150μm. They found that the strength of mixtures with low volume fractions was almost the same as or slightly weaker than the strength of pure ice. In contrast, the strength of mixtures with a volume fraction of 56% was higher than that of pure ice.[5] Mangold et al. (2002) performed triaxial creep tests using mixtures of ice and quartz grains ranging in size from 0.02 to 2mm in which the volume fraction ranged from 52 to 75%. They found that the viscosity of the mixtures was higher than that of pure ice.[6]

The previous studies agree that the mechanical strength of ice and solid particle mixtures with particle contents higher than 50% is greater than that of pure ice. However, with particle contents less than 50%, the results of the cases that have been reported are in disagreement, showing that the strength of such mixtures is both greater and less than the strength of pure ice. This disagreement may result from the fact that solid particles of various sizes and shapes were used in the previous tests, which means that the samples used in each study could have different internal structures. Therefore, we carried out deformation experiments on ice-silicate mixtures systematically in order to study the effects of silicate particles on mechanical behavior. In this study, we focused on examining the silica contents, particle sizes and distribution of solid particles.

2   EXPERIMENTAL METHODS

The sample was prepared by mixing ice grains (0.3-1mm) with spherical glass beads that were either 1μm or 1mm in size. The bead content ranged from 0 to 50wt.%. The ice grains were produced by crushing commercial ice blocks, and they were sieved to sort out the grain sizes.

We prepared two types of samples. The first sample was made from ice grains, glass beads and liquid water. The ice particles were mixed with the glass beads to distribute them evenly. This granular mixture was put into a cylindrical mold, and the pore spaces between the grains were filled with cold water (0°C). The mixture was then frozen in a cold room at −10°C. The sample had a cylindrical shape with a diameter of 30mm and a length of 60mm. We refer to this type of sample as a frozen sample (f.s.). The second type of sample was made from only ice grains and glass beads; liquid water was not used. The evenly mixed ice grains and glass beads were put into a metal cylinder and evacuated to a final pressure of less than 10kPa. The mixture was then compressed by a piston at a pressure of 10MPa for about 10 minutes. The sample was removed from the cylinder and kept in a cold room of −10°C for one day. The sample had a cylindrical shape with a diameter of 20mm and a length of 40mm. We refer to this type of sample as a pressure-sintered sample (p.s.s.).

Table 1   *Experimental conditions*

Series No.	wt.%	beads type[a]	preparation method[b]	strain rate range $[s^{-1}]$	number of samples
S1	10	small	p.s.s.	$2.0 \times 10^{-5} - 8.3 \times 10^{-4}$	4
S2	10	small	f.s.	$8.5 \times 10^{-5} - 2.9 \times 10^{-3}$	3
S3	30	small	f.s.	$2.9 \times 10^{-6} - 2.9 \times 10^{-3}$	16
S4	50	small	p.s.s.	$2.1 \times 10^{-5} - 9.7 \times 10^{-4}$	4
S5	50	small	f.s.	$8.6 \times 10^{-5} - 2.9 \times 10^{-3}$	2
S6	10	large	p.s.s.	$2.1 \times 10^{-5} - 8.2 \times 10^{-4}$	5
S7	10	large	f.s.	$9.1 \times 10^{-5} - 9.4 \times 10^{-5}$	2
S8	30	large	f.s.	$2.8 \times 10^{-6} - 6.8 \times 10^{-4}$	16
S9	50	large	f.s.	8.5×10^{-5}	1

[a] The small bead-type is defined SiO_2 amorphous silica particles with an average diameter of 1μm. The large bead-type is defined as glass beads with an average diameter of 1mm.
[b] f.s. stands for "frozen sample." p.s.s. stands for "pressure-sintered sample."

It could be important for the deformation experiments to describe the sample properties such as the ice grain size, elongation, and preferred orientation. The ice grain size of f.s. was roughly estimated to be almost the same as the initial grain size (0.3 – 1mm), and the ice grain size of p.s.s. was smaller than the initial grain size. It was difficult to prepare the thin section of the sample with silica particles, so that we could not measure the physical properties such as preferred orientations and grain sizes in detail.

Using these two types of samples, we performed uniaxial compression tests under constant strain rates from 2.8×10^{-6} to $2.9 \times 10^{-3} s^{-1}$ in a cold room at $-10°C$. The experimental conditions used in this study are summarized in Table 1. Nine series of the deformation tests were performed so as to study the dependencies on each physical parameter.

3   RESULTS AND DISCUSSION

3.1   Stress-Strain Curves

The stress-strain curves of the frozen sample with different bead sizes at $8.5 \times 10^{-5} s^{-1}$ are shown in Figure 1. At the initial stage of deformation, the stress increases linearly, and then the slope of the curve gradually decreases. The point at which the slope changes is called the yielding point. After that, the stress increases to its maximum level. After the peak, the stress does not become constant and continues to decrease in most cases.

Comparing these two curves, it is remarkable that the interval between the yielding point and the maximum stress point on the 1mm-glass-bead-sample curve is shorter than that on the 1μm-glass-bead-sample curve, and that the 1mm-glass-bead-sample curve shows the stress sharply decreasing beyond the maximum stress. Also, the maximum stress of the 1μm-glass-bead-sample curve is greater than that of the 1mm-glass-bead-sample curve.

In this study, the total strain ranged from 10 to 20%. In the case of the 1mm glass bead sample, many glass beads were squeezed out from the interior and concentrated on the surface. The number of beads squeezed out increased as the silica content and total strain increased. The ratio of beads size to specimen size was too large, so it might cause the squeeze-out effect.

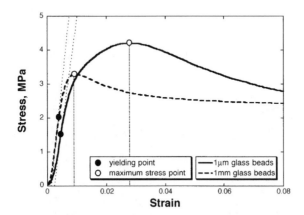

Figure 1   *Stress-strain curves of the 1μm glass bead sample (solid line) and the 1mm glass bead sample (dotted line) at the strain rate of $8.5 \times 10^{-5} s^{-1}$. Both samples have a silica content of 50wt.%.*

3.2   Effect of Particle Sizes and Silica Contents

The maximum stress on the stress-strain curve in each experiment was plotted against the strain rate for the frozen sample in Figure 2. The relationship between the maximum stress (σ) and the strain rate ($\dot{\varepsilon}$) can be expressed in the power law equation,

$$\dot{\varepsilon} = A\sigma^{n}.\tag{1}$$

This equation is almost equivalent to the flow law derived from creep tests.[7] A and n are the material constants, and are shown in Table 2.

The results for the f.s. with different silica sizes are shown in Figure 2. The silica contents of S3 and S8 were both 30wt.%. The difference between the two relationships is not very large, but it is clearly recognizable; the maximum stress of the 1mm glass bead sample is smaller than that of the 1μm glass bead sample. Also, the difference between these two lines becomes small as the strain rate decreases. On the other hand, as shown in Figure 3, in the case of the pressure-sintered sample with a silica content of 10wt.%, the 1mm glass bead sample was stronger than the 1μm glass bead sample, and the difference in strength between them grew large as the strain rate decreased.

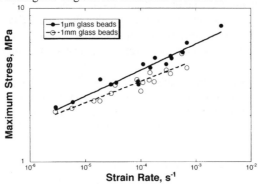

Figure 2   *Maximum stress versus strain rate of frozen samples with a silica content of 30wt.%*

Table 2   *Power index, n, and constant, A, in Eq. (1)*

Series	A $s^{-1}(MPa)^{-n}$	n [a]
S1	1.14×10^{-5}	3.40 (0.18)
S2	8.25×10^{-8}	5.58 (0.61)
S3	5.92×10^{-8}	5.42 (0.29)
S4	9.49×10^{-6}	5.37 (0.28)
S5	2.85×10^{-10}	8.78
S6	9.12×10^{-8}	5.63 (0.28)
S8	4.11×10^{-8}	6.45 (0.40)
pure ice [8]	5.69×10^{-7}	3.31 (0.79)

[a] The numbers in parentheses give the s.d.

The effect of silica contents on the flow law can be clearly recognized in Figure 3. In the case of the pressure-sintered sample with 1μm glass beads, the sample with a silica content of 10wt.% was stronger than the sample with a silica content of 50wt.%. These lines cross at the strain rate of $1.5 \times 10^{-5} s^{-1}$. The difference of the slopes in Figure 3 could be caused by the different contributions of the deformation mechanisms in each case. In Figure 4a, which shows the pressure-sintered sample with a silica content of 10wt.% (S1), the ice grains can be seen to partly deform and become fine. Thin bead layers can be observed among the granular ice grains after the deformation. This suggests that the bulk deformation of the sample could be caused by the deformation of the ice grains themselves. The schematic illustration of the sample structure is shown in Fig.4c. On the other hand, in the pressure-sintered sample with a silica content of 50wt.% (S4), a thick silica layer covered the ice grain and isolated each ice grain completely. Thus, it might be possible that the ice grains did not contact each other or that direct contact among the ice grains could be rare. Therefore, we can speculate that the deformation might be caused by grain boundary sliding; in other words, the sliding could occur in the thick silica layers, which means that the sliding friction activated in the silica layers might control the deformation.

3.3 Effect of Sample Preparation Methods

Next, we consider the effect of the sample preparation methods. The relationships between the maximum stress and the strain rate in samples that were made by two different methods are shown in Figure 3. The main difference in the preparation methods is whether or not cold water was used to make the sample. A and n in Eq. (1) are shown in Table 2. In the case of the 1μm glass bead samples, the frozen samples were twice or three times stronger than the pressure-sintered samples. A comparison of the results for S1 and S2, which had the same silica contents, shows that the p.s.s. of S1 had a value of n of 3.4, which is smaller than the value of n in S2, which is 5.6 (Table 2). The n of S1, at 3.4, is almost the same as the value derived for polycrystalline pure ice. However, the n of S2, at 5.6, is much higher than the value for polycrystalline ice. S1 shows a maximum stress about half of that of S2 at $10^{-4} s^{-1}$, which means that the S1 sample was significantly weaker than the S2 sample. Comparing samples with silica contents of 50 wt.%, the f.s. of S5 is about 3 times stronger than the p.s.s. of S4 and both series show high n values between 5 and 9.

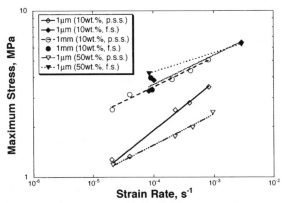

Figure 3 *Maximum stress versus strain rate of the glass bead samples (p.s.s. and f.s.)*

It is very interesting to find that in the case of the 1μm glass bead sample, the p.s.s. is always weaker than the f.s. The boundary layers of the frozen sample might include a matrix of not only glass beads but also ice formed by the freezing of the cold water that was added (Figure 4b,d). In Figure 4b, it is clearly shown that the sample is separated into two areas, a black area containing the glass beads and ice matrix, and a transparent area showing visible ice grains. We noticed that none of the visible ice grains became plastically deformed after the sample was strained up to 20%. Therefore, we speculate that the internal deformation could have been concentrated in the black area consisting of a frozen mixture of silica particles and the ice matrix. However, in the pressure-sintered sample, the boundary layers did not include the ice matrix; they consisted of only glass beads. Thus, the contact area among the ice grains could have been reduced by these layers and could have depressed the cohesion among the ice grains. This reduction of cohesion among the ice grains might have been the cause of the decreased strength of the pressure-sintered sample.

In the case of the 1mm glass bead samples, both the frozen sample and the pressure-sintered sample had the same strength, as shown in the comparison of S6 with S7. It is possible that both the frozen sample and the pressure-sintered sample had the same internal structure whether or not cold water was used.

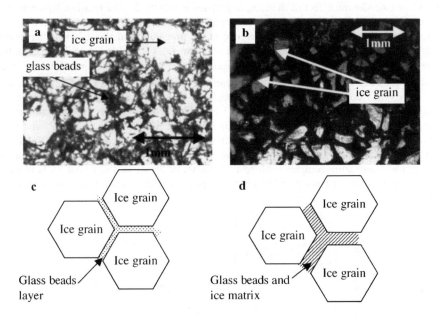

Figure 4   *Thin sections of the 1μm glass bead samples observed by microscope*
a)   Pressure-sintered sample with a silica content of 10wt.%
b)   Frozen sample with a silica content of 30wt.%
c)   Schematic illustration showing the sample structure observed in a)
d)   Schematic illustration showing the sample structure observed in b)

3.4 Relationships between Silica Contents and Flow Parameters

The effect of silica content on the flow law of the frozen sample containing 1μm silica beads is shown in Figure 5. This figure shows the flow laws of S2, S3, S5 and pure ice, which correspond to silica contents of 10, 30, 50, and 0 wt.%, respectively (Table 2). The most important feature observed in the figure is that the slope of each line changes from 1/3.3 to 1/8.8 with the increase of the silica content. Because of the difference in the slope, the lines intersect at a strain rate between 10^{-5} and 10^{-3} s^{-1}, so that the relative positions of the lines are switched around the intersection. This means that the strength of the sample with the highest silica content was the highest, and that the strength decreased as the silica content decreased at a strain rate lower than 10^{-5} s^{-1}. For convenience, we will use the word strength to refer to the maximum strength (σ) as defined in Eq. (1). The dependence of the silica content on the strength is quite complicated between 10^{-5} and 10^{-3} s^{-1}. The differences in strength among the samples were not very large, but the order of the lines changes many times with the strain rate regardless of the narrow range of the differences in strength.

The sample strength at different silica volume fractions is shown in Figure 6 for the purpose of comparison with the previous experimental results. The strength is given in terms of relative strength, which means that the stress of the flow law at each strain rate is normalized by the stress at the strain rate of pure ice derived in each experiment. Below, we plot the relative strength of the frozen sample containing 1μm silica beads obtained from Figure 5 at the strain rates of 10^{-4}, 10^{-5}, and 10^{-6} s^{-1}. This clearly shows that the relative strength at 10^{-4} s^{-1} drops to below 1 with the increase of the silica volume fraction, f. But the dependence of the strength on f drastically changes at strain rates higher than 10^{-5} s^{-1}; that is, the strength increases with f. Another important feature is that the growth rate of the dependence on f increases with the increase of the strain rate.

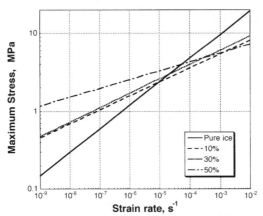

Figure 5 *Comparison of the frozen samples with pure ice*[8]

The experimental results given by Durham et al.(1992) and Hooke et al.(1972) are shown in Figure 6, and each result is fitted using exponential functions in order to interpolate the results. We chose the ultimate stress data for minerals including quartz sand, whose size was ~100μm at a strain rate of 3~4x10^{-4} s^{-1}, a temperature of 223K and a confining pressure of 50 and 100 MPa in Durham et al. Hooke et al. performed creep tests at a low stress range of about 0.6 MPa and a temperature of about –9°C. Therefore, we recalculated their results to show the stress at 10^{-8} s^{-1} according to the flow law obtained at each level of silica content. The relative strength obtained by Durham et al. has a weak positive dependence on f and is consistent with our expected result at 10^{-5} s^{-1}. Our strain rate is about one order of magnitude smaller, but it might be caused by the difference in the particle size and shape. The relative strength obtained by Hooke et al. at the strain rate of 10^{-8} s^{-1} shows greater dependence on f and is in good agreement with our result obtained at 10^{-6} s^{-1}. One of the reasons for the difference in the strain rate of about 2 orders of magnitude might come from the difference in the particle size and shape; they used natural sand of irregular shape and with a size of about 180μm.

We conclude from this comparison with previous studies that the particle size and shape could affect the strength a great deal, but the general features seen in Figure 6 could be maintained in any cases, irrespective of the difference in the shape and size of the particles. The slope in Figure 6 changes with the strain rate between negative and positive, and becomes steeper at the lower strain rate. These complex behaviors of the strength dependencies on the particle contents are caused by the flow law index, n, which increases with increases in the particle contents. Mars polar ice sheets is likely to fall into the low strain rate regime ($<10^{-5}$) where adding silica increases the maximum stress (strengthens the ice).

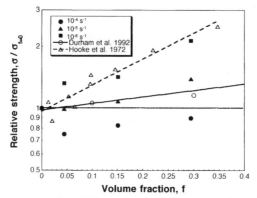

Figure 6   *Relative strength vs Volume fraction of particles. f.s. values at different strain rates are compared with the results of previous studies.*

4   CONCLUSION

We found that silica particles had many effects on the mechanical behavior of ice. The most important effect was that the power law index, n, of the flow law increased with the increase of the silica contents. The relative strength changed with the silica contents and strain rates according to variations in n of the flow law. Thus, we found that the slope of the relative strength on the silica contents varied from negative to positive with the decrease of the strain rate. We also found that the sample preparation method had an important effect on the strength. The pressure-sintered sample including 1μm silica beads was always weaker than the frozen sample under the same physical conditions because the cohesion force among the ice grains was reduced by the silica layers, which prevent direct contact from occurring among the ice grains.

Acknowledgements

We thank Dr. S. Watanabe for his useful discussion and S. Nakatsubo of the Contribution Division of the Institute of Low Temperature Science, Hokkaido University, for his technical help. We also thank K. Hori and K. Murazawa for their help with preparing samples in a cold room. This work was partly supported by a grant-in-aid for scientific research (Grant 17340127) from the Japan Ministry of Education, Science, Sports and Culture and was partly supported by the Grant for Joint Research Program of the Institute of Low Temperature Science, Hokkaido University.

References

1   M. Arakawa and A. Kouchi, *"Impact processes of ice in the solar system" in High - Pressure Shock Compression of Solids V*, Springer-Verlag, New York, 2002, p.199.
2   S.M. Milkovich and J.W. Head III, *J. Geophys. Res.*, 2005, **110**, E01005.
3   H.S. Nayar, F.V. Lenel, and G.S. Ansell, *J. Appl. Phys.*, 1971, **42**, 3786.
4   R.L. Hooke, B.B. Dahlin and M.T. Kauper, *J. Glaciol.*, 1972, **11**, 327.
5   W.B. Durham, S.H. Kirby and L.A. Stern, *J.Geophys. Res.*, 1992, **97**, 20883.
6   N. Mangold, P. Allemand, P. Duval, Y. Geraud, and P. Thomas, *Planetary and Space Science*, 2002, **50**, 385
7   M. Mellor and D.M. Cole, *Cold Regions Science and Technology*, 1982, **5**, 201.
8   M. Arakawa and N. Maeno, *Cold Regions Science and Technology*, 1997, **26**, 215.

ADSORPTION OF ANTIFREEZE PROTEIN AND A COMMERCIAL LOW DOSAGE HYDRATE INHIBITIOR ON HYDROPHILIC AND HYDROPHOBIC SURFACES

H. Zeng[1], V.K. Walker[2] and J.A. Ripmeester[1,2*]

[1]National Research Council of Canada, Ottawa, Ontario K1A 0R6, Canada,
[2]Queen's University, Kingston, Ontario K7L 3N6, Canada
* To whom correspondence should be sent

1   INTRODUCTION

Antifreeze proteins (AFPs) and glycoproteins (AFGPs) have been found in a variety of organisms that are exposed to seasonally low temperatures.[1] These proteins lower the freezing point without significantly affecting the equilibrium melting point, thereby earning the alternative designation, thermal hysteresis (TH) proteins. Generally, it is suggested that the inhibition activities of AF(G)Ps come from their ability to adsorb specifically to some faces of ice crystal embryos.[2,3] Because of their novel TH properties as well as their ability to prevent ice recrystallization, applications of AF(G)Ps for cryopreservation in the health and food industries have been explored.[4] Recently, we have reported that the AFPs can also inhibit the formation of clathrate hydrate,[5,6] an ice-like inclusion compound usually composed of natural gas components and water; this inhibition activity offers another avenue for potential applications. Remarkably, AFPs from fish and insect also showed the ability to eliminate the "memory effect", i.e. the faster reformation of clathrate hydrate after melting.[5,6] Despite these observations, the interaction mechanism(s) of AF(G)Ps on ice or gas hydrate formation is not well understood.

Traditional methods for the study of AFPs and the interaction with ice include ice etching, nanoliter osmometry, ice growth in a capillary, and the inhibition of ice recrystallization. For ice etching,[7] ice hemispheres are grown initially from water then further grown in the presence of AF(G)Ps. The adsorption of the proteins on specific faces of the ice crystal results in etching patterns that reveal adsorption planes. A nanolitre osmometer,[8] or the capillary method[9] is used to determine the TH by visually determining the melting point and the temperature at which a single small crystal grows rapidly. Microscopic observation of ice crystal size after periods at temperatures close to melting is the basis of the ice recrystallization inhibition assay.[10]

All of these methods make use of ice seed crystals in solution and thus provide information on the effects of AF(G)Ps on the growth of the ice crystal, but no information on the role of these proteins in ice nucleation, the initial stage of ice formation. Observing phase transitions in micron-sized emulsions appears promising, but it has been reported that there is no significant inhibition of homogeneous nucleation with AFGP in emulsified droplets.[11] On the other hand, the effect of unavoidable impurities[12] in any bulk sample or on the container wall results in heterogeneous nucleation. The induction time for the nucleation of microliter-sized droplets (0.4 µl) of Type III AFP solutions at different degrees of supercooling has been reported,[12] but the relatively large sizes of the water droplets (i.e. ∅ ~1mm) suggest the effect of AFP on ice growth, but not nucleation, was examined. Thus, the adsorption of AF(G)Ps on the nucleating impurities in the solution may provide insight into the heterogeneous nucleation of ice.

The formation of gas hydrate during the production and transportation of natural gas and oil has been a serious problem for the energy industry.[13] Traditionally, thermodynamic inhibitors such as methanol are used. More recently, low dosage hydrate inhibitors (LDHIs) have been studied.[14] LDHIs can be classified into anti-agglomerants (AAs) and kinetic inhibitors (KIs), which prevent the agglomeration of small gas hydrate particles or retard the growth of tiny gas hydrate crystals, respectively. LDHIs adsorb to clathrate hydrates and inhibit crystal growth.[13,14] Interestingly, AFPs appear to have better inhibition activities on a model tetrahydrofuran (THF) hydrate than a commercial kinetic inhibitor, poly(N-vinylpyrrolidone) (PVP), and also on natural gas hydrates.[5,6] Because these proteins are not easily classified according to the traditional inhibitor classification, we have recently suggested a new scheme.[15] However, in addition to the ability of AFPs from fish and insects to inhibit clathrate hydrate growth, the observation that these proteins eliminate the memory effect suggests that they will also effect the heterogeneous nucleation of clathrate hydrates.

The quartz crystal microbalance (QCM) is an excellent tool for these investigations since the frequency change produced by the adsorption on the surface of a piezoelectric crystal can be used to assess the mass (to a few ng/cm^2) of the adsorbent using the Sauerbrey equation.[16] Since the adsorbed protein layers can have some degree of structural flexibility or viscoelasticity that is undetectable by the determination of the resonance frequency alone, the energy loss, or dissipation factor (D), due to the shear of the adsorbent on the crystal in aqueous solution must also be determined.[17,18] The technique is termed QCM-D and as well as representing an improvement in the study of biomolecular-surface interactions, it presents an opportunity to observe the adsorption of AFP and PVP, on a model nucleator with a hydrophilic surface.

Although AF(G)Ps and LDHIs are distinct, they both inhibit the growth of crystals. Neither AFGP nor PVP are reported to significantly affect ice nucleation,[11] and similarly, we have shown that AFPs and PVP did not affect homogeneous nucleation of THF hydrate.[5] It is not known if these two types of inhibitors can adsorb to other hydrophilic surfaces, however silica is an ubiquitous impurity and common to both these systems. Thus, it is of interest to determine the effects of these inhibitors on heterogeneous nucleation of ice/clathrate hydrate.

2 EXPERIMENTAL SECTION

Purified Type I fish AFP from winter flounder (wfAFP) was kindly provided by A/F Protein Canada Inc. Dr. E. D. Sloan (Colorado School of Mines, USA) kindly provided the PVP (30K) sample. The wfAFP and PVP solutions were prepared with ultrapure water (18.2 mΩ·cm at 298K, produced by Milli-Q® ultrapure water purification systems, Millipore, Billerica, U.S.A.) directly at 25.0 μM. Samples were assessed for their ability to adsorb to SiO_2 or polystyrene (PS).

All measurements were conducted with a quartz crystal microbalance (Q-Sense D300, Q-Sense AB, Gothenburg, Sweden) with 5-MHz AT-cut crystals (Q-Sense AB). The sensor crystal was coated with SiO_2 (QSX-303) or polystyrene (QSX-305) on the gold electrode on one side of the sensor. The sensor crystal was cleaned in a UV-O_3 chamber and then placed in a 250 μl measurement chamber with the coated surface facing the testing liquid and the other side facing air. Before measurements, purified water was equilibrated at 299.0 ± 0.02 K (controlled by a Peltier element) and then introduced into the measurement cell. Once a stable baseline was achieved, 1.5 ml AFP or PVP solution was equilibrated at the same temperature and ~0.5 ml aliquots of the solution was introduced into the measurement chamber at a time, replacing the water in the cell. Frequency shift (f) and dissipation factor (D) measurements were sampled at a rate of ~1Hz with a sensitivity of <0.5 Hz and 1×10^{-7}, respectively.

To observe the desorption of AFP/PVP from the SiO_2/PS surface, after the resonance frequency reached a plateau, 1.5ml MQ-water was equilibrated at the same temperature and ~0.5 ml MQ-water was introduced into the measurement cell to rinse the macromolecule-covered surface. The rinsing process was repeated three times for each sample.

Contact angle measurements were conducted on a FTÅ 100 contact angle meter for SiO_2-coated QCM sensor crystals that were immersed in a solution of 0.25mM wfAFP or PVP for 60 min, then rinsed with MQ-water and dried by compressed air. Three independent measurements were done for each sample.

3 RESULTS AND DISCUSSION

The resonance frequency of the quartz crystal started to decrease when wfAFP or PVP adsorbed onto the SiO_2 surface (Fig. 1). The value of f decreased smoothly and reached a plateau. The final f value indicates the adsorbent mass on the surface and D represents the viscoelastic properties of the adsorbed molecules.[17,18] The wfAFP had low, relatively constant D values and as adsorption progressed, the D values for wfAFP remained almost constant. In contrast, D values for PVP increased throughout the assay period.

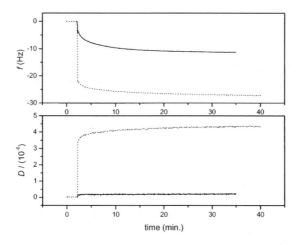

Figure 1 *A representative graph showing the adsorption of wfAFP (25 μM, solid line) and PVP (25 μM, dotted line) on the SiO_2 surface at 299.0 K. Frequency shift (f) and dissipation factor (D) vs. time.*

The ratios, $R = \Delta D/\Delta f$, for the adsorption of wfAFP and PVP to SiO_2 are shown in Fig. 2. A change in the absolute value of R comes from three factors, the adsorbed layer, the trapped liquid and the bulk solution.[17,18] A large absolute value of R, indicates a porous, flexible adlayer with trapped liquid. When wfAFP and PVP formed adlayers on the SiO_2 surface, they showed two different stages of adsorption. Initially, the AFP molecules formed porous, non-rigid adlayers (note the D factor shifts in Fig. 2). Subsequently, as the adsorption of wfAFP progressed, $|R|$ remained constant. This is consistent with the presence of a rearrangement, such that the adlayer becomes a more compact, rigid film with less trapped water after some time. It needs to be pointed out that the gap between –5 to –20 Hz in Fig. 2B is due to the fast adsorption of PVP.

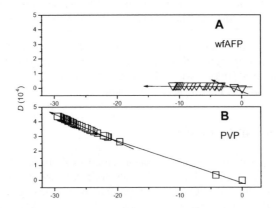

Figure 2 *Adsorption behaviour of inhibitors on silica. (A) wfAFP (▽) and (B) PVP (□) at 25 μM adsorb to the SiO₂ surface, represented as D vs. f. Different adsorption steps are indicated by the arrows.*

After being rinsed with ultra-pure water, the resonance frequency of the wfAFP-covered silica surface decreased by about 2-3%, indicating a small loss of a limited amount of wfAFP on the silica surface. Strikingly, rinsing three times removed almost 25% of the PVP from the silica surface (Fig. 3).

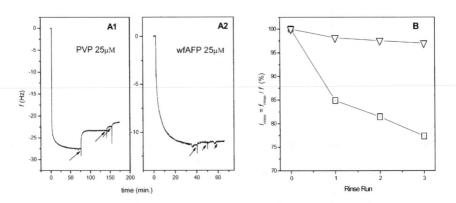

Figure 3 *The effect of rinsing the silica surface. After the adsorption of wfAFP and PVP (both at 25 µM) the silica surface was rinsed (3 x 0.5 ml) with MQ water. The change of frequency with each rinse (marked with arrows) is shown for PVP (A1) and wfAFP (A2). Fig. 3(B) shows the adsorption mass loss percentage (I_{rinse}) of wfAFP (∇) and PVP ($\square$) at each rinse with triangles.*

The adsorption of wfAFP and PVP (25.0 µM) to a hydrophobic polystyrene surface was also observed (Fig. 4). Both of these inhibitors formed a porous, non-rigid film on the hydrophobic surface. It is noticeable that the adsorption of wfAFP onto the hydrophobic surface is completely different from that onto a hydrophyllic surface and the adsorption consists of two steps in which the adlayer film is much looser for the second step.

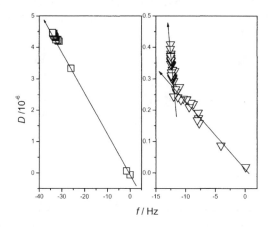

Figure 4 *The adsorption behaviour of inhibitors on a hydrophobic surface. Adsorption of wfAFP (∇) and PVP ($\square$) (25 µM) on the PS surface is represented as D vs. f. Different adsorption steps are indicated by arrows.*

The contact angles for wfAFP- or PVP-covered silica surfaces were also measured (Table 1). After adsorption, the contact angle for the wfAFP-covered silica surface was almost double that of the PVP-covered one, indicating a more hydrophobic surface with AFP adsorption.

Table 1 *Contact angle measurements for wfAFP- or PVP-covered SiO_2 surface.*

Samples on silica surface	Contact angle (degrees)
MQ-Water	5.17 ± 0.32
PVP 0.25mM, rinsed	27.49 ± 1.99
wfAFP 0.25mM, rinsed	46.15 ± 1.41

As is well known, a suitable contaminant or "sympathetic" surface is needed to induce heterogeneous nucleation.[19] This study suggests that AFP changes the characteristics of these surfaces. It appears to adsorb or "poison" nucleation sites, including impurities such as hydrated oxides of Si or Fe, or hydrophilic container walls, and thus reduces the probability of subsequent ice/clathrate hydrate formation. The QCM-D analyses shown for SiO_2 (Fig. 2) indicate that wfAFP forms a compact film on the silica surface, while PVP forms a loose film as indicated by the D values. This dissimilarity likely makes a great difference in the energy barrier for heterogeneous nucleation of ice/clathrate hydrate on the surface. Further, it suggests a mechanism for the different inhibition activities of the proteins and the polymer. As explained in Fig. 5, because the PVP film formed on the SiO_2 surface was looser, more solution was therefore trapped in the adlayer film. In comparison, wfAFP formed a more compact film on the hydrophilic surface, and thus less solution would have access to the nucleating surface, resulting in inhibition of heterogeneous nucleation.

The AFP has hydrophilic and hydrophobic residues arranged as patches on the faces of silica. In contrast, PVP has hydrophilic residue groups randomly exposed to the silica surface. These residues do not have a regular arrangement, and we speculate that the PVP molecules can easily interact with each other as well as aggregate on the SiO_2 surface, forming a loose film (Fig. 5).

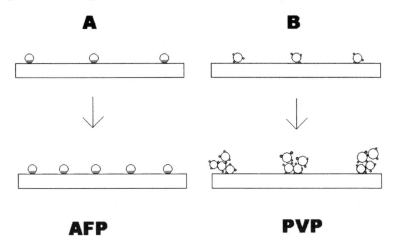

Figure 5 *An explanatory scheme showing the different adsorption status of AFPs/PVP on the silica surface. A and B represent the reorganization processes during the adsorption of AFPs and PVP onto SiO₂ surface. It should be noted that the inhibitors do not need to entirely cover the silica surface, but sufficient numbers of molecules are required to make it unfavorable for the addition of water molecules on the SiO₂ surface. The large circles represent the cross-section of the AFP/PVP molecules in which the grey parts represent the hydrophilic side of AFP and the small grey circles represent the hydrophilic groups of PVP.*

The observation that these solute molecules rearrange on the surface agrees with the report of a rearrangement of AFP molecules on the ice surface.[20] An early stage of adsorption can be pictured as isolated adsorbed molecules, conformationally disordered and randomly distributed on the surface of the substrate. The final stage involves close-packed adsorbents with relatively uniform molecular orientation and conformation.[21] We speculate that initially, not all wfAFP molecules would contact the silica surface with their "ice-contacting" residues on one side of the α-helix, and misorientated molecules would rearrange until there was a good match on the surface lattice, allowing AFPs to adsorb. Finally, AFP likely forms a rigid layer on the surface, which means that the interaction of AFP molecules with each other is also an important factor. Because the nucleating sites on the hydrophilic SiO_2 surface are covered by AFP, and fewer water molecules can diffuse to the surface due to the compact AFP film (see Fig. 5), it becomes much more difficult for heterogeneous nucleation to occur.

The change of surface properties is crucially important for the heterogeneous nucleation of ice/hydrate and the inhibition activities of ice/hydrate inhibitors. Since Type I wfAFP gave a contact angle almost two times that of PVP (Table 1), it changed the surface properties to a greater degree. This is in agreement with our previous observations that AFPs can act as stronger hydrate inhibitors than PVP.[5] A simplified model (Fig. 5) shows stylized cross-sections of AFP and PVP on the SiO_2 surface. When AFP molecules adsorb and form a rigid film on the silica surface, more hydrophobic residues of the protein are exposed to the water. PVP molecules form a looser film on the silica surface with some hydrophilic groups of PVP molecules exposed to water, giving the PVP-covered silica surface a contact angle smaller than that of AFP-covered one. It should be noted also that adsorption to a nucleating site may be quite different from the adsorption to a dynamic ice surface. To explore this, mutations to the ice-binding residues in AFPs may be useful.[22]

One has to compare these QCM-D results carefully with those from commonly used methods for the effects of AFPs on ice growth, since ice and silica surfaces have quite different surface patterns. It is possible that this could make a difference in the adsorption behaviors of molecules on the surfaces. We do suggest that adsorption of AFPs on silica and ice/clathrate hydrate is closely related to the effects of AFPs during the heterogeneous nucleation stage especially, but that there is also relevance to the crystal growth stage. This study focuses on the former process, but with caution, it may also be useful for the understanding of the latter process.

In our previous studies, we suggested that the "memory effect" could occur because of the imprinted surface of the impurities. When a hydrate crystal grows on an impurity surface, the surface geometry of the hydrated or hydroxylated silica may well be altered so that when the hydrate melts the surface is now a better nucleator of hydrate than it was in the fresh supercooled solution.[5] The AFPs can eliminate the memory effect by adsorbing to the altered particle surfaces, which may well be stronger adsorbers of AFPs than during the first nucleation cycle. Based on the observation of the change of adsorbed mass after rinsing (Fig. 3), we propose a related mechanism to explain the difference between these two macromolecules' with respect to the "memory effect" of clathrate hydrates.[5,6] When clathrate hydrates form, hydrophilic impurity particles, such as silica, are virtually unavoidable. If wfAFP or PVP are present in solution, the impurity surface will be partially covered with wfAFP or PVP. As discussed above, AFP molecules form a rigid and compact film while PVP molecules form looser films and, as a result, different inhibition activities are observed under hydrate-forming conditions. After the clathrate hydrate is formed, the impurity surface area not having AFP/PVP molecules will be available for nucleating clathrate hydrate. When the clathrate hydrate is decomposed at modest temperatures, the silica surface is essentially "rinsed" due to the movement of the melted solution. Because only limited amounts of wfAFP appear to be rinsed away from the silica surface, much of the surface is still covered with AFP molecules, and in any case, liberated AFP molecules are free to adsorb to the impurity surface during the melting. Therefore the kinetics of AFP inhibition on clathrate hydrate does not change appreciably during the reformation of hydrate.[5,6] In contrast, since PVP covers the silica surface more loosely, a large area of the impurity surface will be exposed to the solution after the "rinsing". Even if free PVP adsorbs again to the silica surface, this second adsorption is also looser and allows water and hydrate cages access to the hydrophilic surface, with the result that heterogeneous nucleation of clathrate hydrate will occur more readily. This will give rise to the observed "memory effect".[5]

One has to distinguish our studies here from the more popular studies on the adsorption of AFPs or LDHIs on ice or gas hydrate since the former focuses on the effect of AFP/LDHI on heterogeneous nucleation of ice/gas hydrate and the later focuses on the effect on the effects on the ice/hydrate growth. Thus, the studies represented in this study deals with impurity surfaces but not ice/hydrate surface. The authors think that this is an important aspect for the understanding of good ice/hydrate inhibition since the nucleation process is heterogeneous and must involve impurity surfaces.

It is worthwhile to investigate the adsorption of AFPs at a temperature slightly below the equilibrium melting point of ice (i.e. in supercooled AFP solutions). Due to the limitation of the instrument used in this study, this has not been done in our lab. However, adsorption of AFPs and PVP at 291K, the lowest temperature available on this instrument, did not show significant differences (not shown).

In summary, we have shown that QCM-D provides a method to examine the adsorption behaviour of macromolecules on hydrophilic surface and has provided information about the re-arrangement of AFPs at the liquid-solid interface. Significantly, the change of the surface properties by the adsorbed macromolecular inhibitors appears to be crucial for the inhibition of the heterogeneous nucleation of ice/hydrate formation. The results provide insight to the understanding of the differences between these macromolecules and their inhibitionary activities on the heterogeneous nucleation of ice and/or gas hydrate as well as memory effect.

References

1   P.L. Davies, J. Baardsnes, M.J. Kuiper and V.K. Walker, *Phil. Trans. R. Soc. Lond. B. Biol. Sci.* 2002, **357**, 927

2   C.A. Knight, *Nature* 2000, **406**,249

3   S.P. Graether, M.J. Kuiper, S.M. Gagne, V.K. Walker, Z.C. Jia, B.D. Sykes and P.L. Davies, *Nature* 2000, **406**,325

4   R.F. Feeney and Y. Yeh, *Trends Food Sci. Tech.* 1998, **9**,1

5   H. Zeng, L.D. Wilson, V.K. Walker and J.A. Ripmeester, *J. Amer. Chem. Soc.* 2006, **128**, 2844

6   H. Zeng, I. Moudrakovski, V.K. Walker and J.A. Ripmeester, *AIChE J.* 2006, in press

7   C.A. Knight, C.C. Cheng and A.L. DeVries, *Biophys. J.* 1991, **59**,409

8   A. Chakrabartty and C.L. Hew, *Eur. J. Biochem.,* 1991, **202**, 1057

9   A.L. DeVries, *Science* 1971, **172**,1152

10   C.A. Knight, J. Hallet and A.L. DeVries, *Cryobiology* 1988, **25**,55

11   F. Franks, J. Darlington, T. Schenz, S.F. Mathias, L. Slade and H. Levine, *Nature* 1987, **325**,146

12   N. Du, X.Y. Liu and C.L. Hew, *The J. Biol. Chem.* 2003, **278**,36000

13   E. D. Sloan, Jr., in *Clathrate Hydrates of Natural Gases,* 2nd Edn. Marcel Dekker Inc. New York, 1998, ch. 8, p538

14   C.A. Koh, R.E. Westacott, W. Zhang, K. Hirachand, J.L. Creek and A.K. Soper, *Fluid Phase Equil.* 2002, **194**,143

15   H. Zeng, V.K. Walker and J.A. Ripmeester, *Proc. 5th Inter. Conf. Gas Hydrate*, 2005, **4**, 1295

16   G. Sauerbrey, *Z. Phys.* 1959, **155**,206

17 M. Rodahl, F. Hook, A. Krozer, P. Brzezinski and B. Kasemo, *Rev. Sci. Instrum.* 1995, **66**,3924

18 M. Rodahl, F. Hook, C. Fredriksson, C. Keller, K. Krozer, P. Brzezinski, M. Voinova and B. Kasemo, *Faraday Discuss.* 1997, **107**,229

19 J. W. Mullin, in *Crystallization,* 4th Edn., Butterworth-Heinemann. 2001, ch. 5, p.192

20 L. Chapsky and B. Rubinsky, *FEBS Lett.* 1997, **412**, 241

21 D.K. Schwartz, *Annu. Rev. Phys. Chem.* 2001, **52**,107

22 J. Baardsnes, L.H. Kondejewski, R.S. Hodges, H. Chao, C. Kay and P.L. Davies, *FEBS Lett.* 1999, **463**,87

DIFFUSION, INCORPORATION, AND SEGREGATION OF ANTIFREEZE GLYCOPROTEINS AT THE ICE/SOLUTION INTERFACE

S. Zepeda[1], H. Nakaya[1], Y. Uda[1], E. Yokoyama[2], Y. Furukawa[1]

[1]Institute of Low Temperature Science, Hokkaido University, Sapporo 060-0819, Japan,
[2]Computer Center, Gakushuin University, Tokyo 171-8588, Japan

1   INTRODUCTION

Antifreeze proteins (AFPs) and glycoproteins (AFGPs) are found in nature in many cold weather organisms including fish, plants, and insects.[1] These proteins suppress the freezing temperature of the blood serum in fish just enough, about 1 degree, to keep them from freezing in their supercooled environments and inhibit ice recrystallization enough to reduce damage in freeze tolerant organisms, making them essential for survival. To date workers have described at least 5 distinct classes of these proteins, AFP types I-IV and AFGPs, as well as two distinct types of insect AFPs. The Type I and Type II AFPs are alanine and cysteine rich, respectively, while the AFGPs are alanine rich as well as glycosylated.[1] They exist in a variety of structures ranging from $\alpha-$ or $\beta-$helical to globular and yet to some extent they all accomplish the same function. The proteins are thought to bind to the surface of ice and inhibit growth via a local freezing temperature suppression derived from the Gibbs-Thomson effect. Hydrogen bonding surfaces on the protein are found that compliment specific crystallographic planes of the bulk-like terminated ice surface. Although many similar studies have been carried out for nearly all of the antifreezes, no direct evidence for this "ice binding" phenomenon or clear evidence for the exact nature of the mechanism is available and this is necessary to understand how these proteins function. However, the colligative effect for the amount of protein in the fish is far too small to account for the observed freezing suppression while the melting temperature essentially remains unchanged. The proteins are also known to modify the microscopic and gross morphology of the ice crystal leading to a consensus that the mechanism must be a surface phenomenon. Our focused efforts here are to learn about the protein interaction with the ice surface. Here, the temperatures of interest are only a few degrees below the melting point, where ice is known to have an extensive surface melting or quasi-liquid layer (qll) that only augments the difficulty of directly probing an already complex system.[2,3]

   In the present work, we study ice crystal growth in AFGP solutions with phase contrast and fluorescence microscopies in a 1-directional growth apparatus. With fluorescence microscopy we have directly visualized the protein dynamics at the interface of a growing ice crystal. Contrary to previous understandings, the proteins become incorporated into veins and not directly into the crystal matrix.[7] This indicates that the proteins only weakly adsorb to the interface. Under slower growth conditions no veins are present and the protein builds up at the interface of the growing ice crystal and the diffusion of the protein away from the interface can be directly visualized allowing us to calculate the diffusion constant.

2 EXPERIMENTAL

AFGP 6,7,8 samples were prepared from the blood serum of *Trematomas borchgrevinki* as described previously with no further purification.[5] The sample contained mostly AFGP-8 and small amounts of AFGP 6 and AFGP 7 as verified by gel electrophoresis. Proteins were labelled at the N-terminal with flourescein isothiocyanate (FITC). Activity was checked by verifying the ability to halt ice growth in solution during free solution growth from a capillary as described elsewhere.[6] The labelled proteins also showed growth modification identical to the native proteins. The product was then placed in dialysis membranes that were then placed in 18 MΩ deionized water until the solution resistance was in the MΩ range. Imaging was carried out with an inverted microscope equipped with phase contrast accessories and a mercury lamp for fluorescence imaging (IX-71, Olympus Corporation, Tokyo, Japan). Ice crystals were grown in a home built 1-directional growth apparatus. This apparatus is similar to one described by Furukawa et al with a few adaptations for use with the inverted microscope.[7] Briefly, the growth cell consists of two glass cover slides (76mm x 25mm x 1mm) with nylon fishing line (0.05mm) as a spacer. A temperature gradient was established by setting up two copper plates cooled by thermoelectric coolers with a PID temperature controller (Melcor Corporation, Trenton, New Jersey). Temperature on each end was maintained to within 4 mK during the experiments. Ice was nucleated at -15°C and the sample was then repetitively melted and grown until a single crystal remained in the growth cell. This crystal was then melted back until it was released from the glass so as to float to ensure that the c-axis would be perpendicular to the growth cell when growth was finally reinitiated. All samples were grown with the c-axis perpendicular to the growth cell. The growth cell was placed on a Teflon slider and this was pulled towards the cold side to grow the ice crystal. The movement was controlled with a stepping motor at velocities between 0.5 μm/s and 15 μm/s. Experiments for data shown were carried out in a temperature

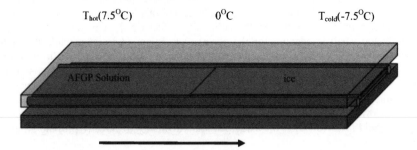

Figure 1   *Schematic representation of the growth cell unit. The top and bottom represent the glass cover slides (76mm x 25mm x 1mm) and the gold tubes represent the nylon fishing line. Silicone is used on the outer space between the cover slides to hold the unit together. The final dimensions of the growth cell are 76mm x 20mm x 0.05 mm*

gradient of 1.5 K/mm. Raw images were analyzed using ImageJ software to obtain the concentration profiles shown in all the figures.[8] First, the background was determined by using a blank sample, pure water. This was subtracted from the protein sample images. The intensity was then normalized to the intensity in the solution and then multiplied by the prepared solution concentration to give the protein distribution throughout the entire image. To calculate the diffusion constant a flat interface was grown at 0.5 μm/s or 1.0 μm/s until protein was clearly accumulated at the interface and an exponential concentration profile was obtained. The slider was then stopped to allow the protein to diffuse away from the interface. Images were taken at least every two minutes for 40 minutes.

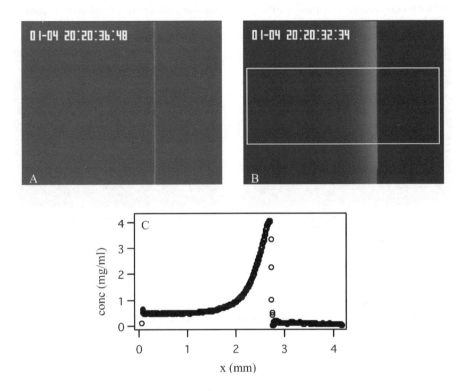

Figure 2 a) *Phase contrast image of the solution ice interface of a 0.1 mg/ml sample grown at 0.5 μm/s. b) Corresponding fluorescent image taken approximately 4 seconds after the phase contrast image. c) Concentration profile of the blue box in b). Each point represents the average intensity along the width of the box and x represents the position along the length of the box. Both images are 4.2 mm across.*

3   RESULTS AND DISCUSSION

In order to directly observe the AFGP interaction with the ice surface, 1-directional growth experiments were carried out. Figure 2a shows a typical structure for the ice interface at growth velocities less than 1 μm/s.   Generally, one obtains a very flat interface and the solid completely fills the space between the glass cover slides. Figure 2b clearly shows the distribution of the protein. The solid appears completely dark and the concentration drops to the background levels. This directly shows that the protein is entirely rejected from the ice crystal during growth.   As the ice crystal continues to grow the protein accumulates at the interface (figure 1B) to concentrations nearly an order of magnitude greater than the original solution concentration effectively sampling a wide range of both solution and surface concentrations. At these elevated concentrations the interface and solid remain the same and and yet the amount of protein in the solid still remains lower than the detectable

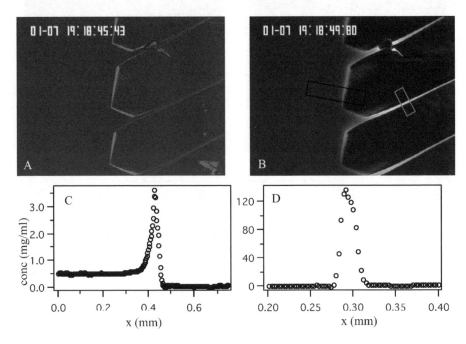

Figure 3   *a) Phase contrast image of the solution ice interface of a 0.1 mg/ml sample grown at 5 μm/s. b) Corresponding fluorescent image taken approximately 4 seconds after the phase contrast image. c) and d) Concentration profiles of the dark and light colored rectangles in b), respectively. Each point in c) and d) represent the average intensity of the width of the rectangles in b. X indicates the position along the length of the box from left to right and from top to bottom of the dark and light colored rectangle, respectively.*

limits. This finding is contrary to previous understandings of the antifreeze mechanism for freezing suppression activity of AFPs: that is strong binding or "irreversible binding" to the surface is a requirement, which in turn would result in protein incorporation into the crystal matrix[9] It seems unlikely that if the proteins were irreversibly bound that the amount incorporated into the ice would fall below the background levels. An irreversible mode of function would allow proteins to more readily escape incorporation into the crystal.

As one speeds up the growth rates instabilities occur at the interface and a transition into a different growth mode occurs. The interface is no longer flat and now forms finger like structures with veins where the protein becomes incorporated into the solid. Although in general it is difficult to obtain 3-d information from a 2-d image, the facets that develop at the interface are prismatic or pyramidal as is generally seen with previous experiments. One possibility for this change in growth morphology can be due to constitutional (compositional) supercooling. As one speeds up the growth rate a larger amount of protein builds up at the interface resulting in a lowering of the local freezing temperature, constitutional supercooling. When the initial change in growth form occurs the crystal growth speeds up compared to the growth of the flat interface growth as the crystal exits the supercooled region. Further growth does not result in further instabilities or any overall change in the growth pattern and the protein is further trapped in the veins. The veins remain liquid for longer periods; effectively it is supercooled to lower temperatures than the ice front when if freezes. Figure 3d shows that the amount of protein in the veins is over two orders of magnitude greater than the solution concentration.

At this higher velocity it appears that the ice crystal may have protein inside. Carefully comparing the phase contrast (figure 3a) and fluorescent images the features that appear in both images readily coincide with each other. Similarly as in the veins between the growth fingers, the protein becomes trapped between the ice and the cover glass Figure 3c shows the concentration distribution for a section of the finger that has no structures seen in the phase contrast image or no space between the cover glass and the ice. The concentration in the solid becomes negligible as little or no protein is detected in the ice crystal itself. Earlier experiments by Knight and DeVries show that AFGPs preferentially adsorb to specific ice planes that become etched during sublimation of the ice crystal. They also showed that the adsorbed proteins then become trapped in the ice crystal along

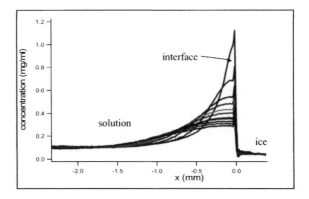

Figure 4 *Diffusion profile of the protein at 2-minute intervals. The first measurement is that with the highest peak at the interface.*

the adsorbed planes. Here the growth was carried out from a cold finger in solution, accessing all the possible crystallographic orientations. One possibility could be that the incorporation occurs in a similar fashion to incorporation of proteins as in our higher velocity experiments.

The diffusion constant was calculated as described in the experimental section. Typical concentration profiles of the diffusion process are shown in Figure 4. The value obtained was 1.0×10^{-10} m^2/s ($\pm0.1\times10^{-10}$ m^2/s). This is good agreement with the value of 0.9×10^{-10} m^2/s obtained by Krishnan et al.[10] Furthermore, this validates our 1-directional growth technique to study the protein kinetics at a growing interface.

4 CONCLUSIONS

"Antifreeze" function in fish was first discovered over 50 years ago by Scholander and co-workers and still today there is much debate about the exact mechanism.[11, 12] Even with the discovery and advancement of many powerful surface techniques the exact nature of the surface of ice remains a difficult question further complicating an accurate description of the protein surface interaction. Here we directly explore the nature of the surface interaction of AFGPs with ice. Our studies indicate that at all growth rates the protein is expelled from the ice crystal matrix as no significant amount of protein is detected. At higher growth rates the AFGP incorporate into the solid ice structure in veins that develop from growth instabilities at the interface while at the slower growth velocities no such growth patterns develop and no such incorporation occurs. This is true over a wide range of both concentrations and growth rates. With the data presented it becomes difficult to envision a Gibbs-Thomson mechanism of function and it is necessary to relax the irreversible binding requirement for the mode of function for this AFGP. Such a mechanism would necessarily lead to the protein becoming incorporated into the ice crystal. The emerging picture for the mechanism is one that includes mobile protein kinetics at the surface. This is more realistic if one considers the highly dynamic nature of the ice interface, especially near the melting point.[13] Further work needs to be carried out to explore the mechanism of the larger AFGPs as well as the other non-glycosylated AFPs.

Acknowledgements

The authors acknowledge Yin Yeh and the late Robert E. Feeney for the AFGP sample and Shuichiro Matsumoto for the AFGP labelling. This research was partially supported by the Ministry of Education, Science, Sports and Culture, Grant-in-Aid for Scientific Research (B), 15340136, 2003, and by the Grant of the Ground Research for Space Utilization promoted by the Japan Space Forum (JSF). S. Zepeda acknowledges support from the Japan Society for the Promotion of Science (JSPS).

References

1   Y. Yeh and R. E. Feeney, *Chem. Rev.*, 1996, **96**, 601-617.
2   H. Bluhm, D. F. Ogletree, C. S. Fadley, Z. Hussain and N. Salmeron, *J. Phys.: Cond. Matt.*, 2002, **14**, L227-L233.
3   Y. Furukawa, M. Yamamoto and T. Kuroda, *J. Cryst. Growth*, 1987, **82**, 665-677.
4   C. A. Knight, E. Driggers and A. L. DeVries, *Biophys. J.*, 1993, **64**, 252-259.
5   D. T. Osuga and R. E. Feeney, *J. Biol. Chem.*, 1978, **253**, 5338-5434.

6   S. Zepeda, Y. Uda, Y. Furukawa, manuscript in preparation
7   Y. Furukawa, N. Inohara and E. Yokoyama, *J. Cryst. Growth*, 2005, **275**, 161-174.
8   M. D. Abramoff, P. J. Magelhaes and S. J. Ram, *Biophot. Int.*, 2004, **11**, 36-42.
9   C. Knight, A. Wierzbicki, R. A. Laursen and W. Zhang, *Cryst. Growth Des.*, 2001, **1**, 429-438.
10 V. V. Krishnan, W. H. Fink, R. E. Feeney and Y. Yeh, *Biophys. Chem.*, 2004, **110**, 223-230.
11 P. F. Scholander, L. Van Dam, J. W. Kanwisher, H. T. Hammmel and M. S. Gordon, *J. Cell. Comp. Physiol.*, 1957, **49**, 5-24.
12 P. F. Scholander, W. Flagg, V. Walters and L. Irving, *Physiol. Zool.*, 1953, **26**, 67-92.
13 H. Nada and Y. Furukawa, *J. Cryst. Growth*, 2005, **283**, 242-256.

Subject Index